IDENTIFICATION OF PHYSICAL SYSTEMS

IDENTIFICATION OF PHYSICAL SYSTEMS

APPLICATIONS TO CONDITION MONITORING, FAULT DIAGNOSIS, SOFT SENSOR AND CONTROLLER DESIGN

Rajamani Doraiswami, Chris Diduch and Maryhelen Stevenson
University of New Brunswick, Canada

WILEY

This edition first published 2014

Registered office
John Wiley & Sons Ltd, The Atrium, Southern Gate, Chichester, West Sussex, PO19 8SQ, United Kingdom

For details of our global editorial offices, for customer services and for information about how to apply for permission to reuse the copyright material in this book please see our website at www.wiley.com.

Library of Congress Cataloging-in-Publication Data

Doraiswami, Rajamani.
Identification of physical systems : applications to condition monitoring, fault diagnosis, softsensor, and controller design / Rajamani Doraiswami, Chris Diduch, Maryhelen Stevenson.
pages cm
Includes bibliographical references and index.
ISBN 978-1-119-99012-3 (cloth)
1. Systems engineering. 2. Systems engineering–Mathematics. I. Diduch, Chris. II. Stevenson, Maryhelen. III. Title.
TA168.D66 2014
620.001′1–dc23

2013049559

A catalogue record for this book is available from the British Library.

ISBN 9781119990123

Set in 9/11pt Times by Aptara Inc., New Delhi, India
Printed and bound in Singapore by Markono Print Media Pte Ltd

1 2014

Contents

Preface

The topic of identification and its applications is interdisciplinary, covering the areas of estimation theory, signal processing, and control with a rich background in probability and stochastic processes and linear algebra. Applications include control system design and analysis, fault detection and isolation, health monitoring, condition-based maintenance, fault diagnosis of a sensor network, and soft sensing. A soft sensor estimates variables of interest from the process output measurements using a maintenance-free software, instead of a hardware, device.

The Kalman filter forms the backbone of the work presented in this book, in view of its key property, namely, the residual is a zero-mean white noise process if and only if there is no mismatch between the model of the system and that employed in the design of the Kalman filter. This property is termed the *residual property* of the Kalman filter. The structure of the identification model is selected to be that of the Kalman filter so as to exploit the residual property. If the error between the system output and its estimate obtained using the identified model is a zero-mean white noise process, then the identified model is the best fit to the system model. Fault detection and isolation is based on analyzing the Kalman filter residual. If the residual property holds, then it is asserted that there is a fault and the residual is then further analyzed to isolate the faulty subsystems. A bank of Kalman filters is employed for fault diagnosis in a sensor network because of the distributed nature of the latter. At the core of a soft sensor is a Kalman filter, which generates the estimates of the unmeasured variable from the output measurements.

Chapters 1–5 provide a background for system identification and its applications including modeling of signals and systems, deterministic and random signals, characterization of signals, and estimation theory, as explained below.

Modeling of Signals and Systems

Chapter 1 describes state-space and linear regression models of systems subject to disturbance and measurement noise. The disturbances may include effects such as the gravity load, electrical power demand, fluctuations in the flow in a fluid system, wind gusts, bias, power frequency signal, dc offset, crew or passenger load in vehicles such as space-crafts, ships, helicopters, and planes, and process faults. Measurement noise is a random signal inherent in all physical components. The most common source of noise is thermal noise, due to the motion of thermally agitated free electrons in a conductor.

A signal is modeled as an output of a linear time-invariant system driven by an impulse (delta) function if it is deterministic, and by a zero-mean white noise process if it is random. An *integrated model* is developed that includes the model of the disturbance and the measurement noise. The integrated model is driven by both the system input and the zero-mean white noise processes that generate the random disturbances and measurement noise. This model sets the stage for developing the model of the Kalman filter for the system.

Temporal and Spectral Characterization of Signals: Correlation and Spectral Density (Coherence)

Characterization of the signals in terms of the correlation and its frequency-domain counterpart, the power spectral density, are treated in Chapter 2. The correlation is a measure of the statistical similarity or dissimilarity between two waveforms, whereas the magnitude-squared coherence spectrum measures the spectral similarity or dissimilarity between two signals. This measure (coherence spectrum) is the frequency-domain counterpart of the correlation coefficient which preserves only the shape of the correlation function. The coherence spectrum is widely used in many areas including medical diagnosis, performance monitoring, and fault diagnosis. Non-parametric identification of a system in terms of its frequency response may be obtained simply by dividing the cross-power spectral density of the system input and its output, by the spectral density of the input at each frequency. The non-parametric identification serves to cross check the result of the parametric identification of the system.

Estimation theory: Estimation theory is a branch of statistics with wide application in science and engineering. It especially forms the backbone of the system identification field. Because of its pivotal role in System identification, estimation theory has been thoroughly addressed in three complete chapters (Chapters 3–5) which provide its foundational knowledge and results.

Estimation of a Deterministic Parameter

The problem of estimating a deterministic parameter from noisy measurements is developed in Chapter 3. The measurement model is assumed to be linear. The output measurement is a linear function of the unknown parameter with additive noise. The probability density function of the measurement is not restricted to be Gaussian. Commonly occurring heavy tailed probability density functions (PDFs) such as the Laplacian, the exponential, and Cauchy PDFs are considered. If the PDF governing the measurement is unknown except for its mean and variance, a worst-case PDF, which is a partly Gaussian and partly exponential, is employed. The worst-case PDF is derived using min-max theory. A thin-tailed PDF, such as the Gaussian one, characterizes "good" measurement data, while those with thick tails characterize bad measurement data. A popular lower bound used in estimation theory to define the efficiency of an estimator, namely the Cramer – Rao lower bound, is derived for the error covariance of the estimator. An estimator is termed efficient if its estimation error covariance is equal to the Cramer–Rao lower bound, and is unbiased if the true value of the parameter is equal to the expected value of its estimate. Approaches to the estimation of deterministic non-random parameters are developed, including maximum likelihood and the least-squares methods.

Estimation of Random Parameter

Chapter 4 deals with random parameter estimation. It is shown that the optimal estimate in the sense of minimum mean-squared error is the conditional mean of the parameter given the past and present measurement. Extension of the random parameter estimation to the optimum mean-squared error estimation of random process, including the states and output of a system, leads to the development of the Kalman filter.

Least-Squares Estimation

Chapter 5 deals with the widely used least-squares method for estimating the unknown deterministic parameter from the measurement signal which is corrupted by colored or white noise. Properties of the least-squares estimate are derived. Expressions for the bias error and covariance of the estimation error are obtained for the case when the number of data samples is both finite and infinitely large. The least-squares and its generalized version of the weighted least-squares method produces an estimator that is unbiased and is the best linear unbiased estimator (BLUE estimator). Most importantly, the optimal estimate is a solution of a set of linear equations which can be efficiently computed using the Singular Value Decomposition (SVD) technique. Moreover, the least-squares estimation has a very useful geometric interpretation. The residual can be shown to be actually orthogonal to the hyper-plane generated by the columns of the data matrix. This is called the orthogonality principle.

Kalman Filter

The Kalman filter is developed in Chapter 6. It is widely used in a plethora of science and engineering applications including tracking, navigation, fault diagnosis, condition-based maintenance, performance (or health or product quality) monitoring, soft sensing, estimation of a signal of a known class buried in noise, speech enhancement, and controller implementation. It also plays a crucial role in system identification as the structure of the identification model is chosen to be the same as that of the Kalman filter. It is an optimal recursive estimator of the states of a dynamical system that is very well suited for systems that can be non-stationary, unlike the Wiener filter which is limited to stationary processes only. Further extensions of the Kalman filter have taken it into the realms of non-Gaussian and nonlinear systems and have spawned a variety of powerful filters, ranging from the well-known extended KF to the most general particle filter (PF). In a KF, the system is modeled in a state-space form driven by a zero-mean white Gaussian noise process. The Kalman filter consists of two sets of equation, a static (or algebraic) and a dynamic equation. The dynamic equation, or state equation, is driven by the input of the system and the residual. The algebraic (or static) equation, also known as the output equation, contains an additive white Gaussian noise which represents the measurement noise. The Kalman filter is designed for the *integrated model* of the system formed of the models of the system, the disturbance and the measurement noise. There are two approaches to deriving the Kalman filter: one relies on the stochastic estimation theory and the other on the deterministic theory. A deterministic approach is adopted herein. The structure of the Kalman filter is determined using the *internal model principle*, which establishes the necessary and sufficient condition for the tracking of the output of a dynamical system. In accordance with this principle, the Kalman filter consists of (i) a copy of the system model driven by the residuals and (ii) a gain term, termed the Kalman gain, used to stabilize the filter. The internal model principle provides a mathematical justification for the robustness of the Kalman filter to noise and disturbances and for the high sensitivity (i.e., a lack of robustness) of the mean of the residuals to model mismatch. This property is judiciously exploited in designing a Kalman filter for applications such as performance monitoring and fault diagnosis. The Kalman filter computes the estimates by fusing the *a posteriori* information provided by the measurement, and the *a priori* information contained in the model which governs the evolution of the measurement. The covariance of the measurement noise, and that of the plant noise, quantify the degree of belief associated with the measurement and model information, respectively. The estimate of the state is obtained as the best compromise between the estimates generated by the model and those obtained from the measurement, depending upon the plant noise and the measurement noise covariances.

System Identification

Chapter 7 considers various methods used for system identificationm including the classical least-squares, the high-order least-squares, the prediction error, and the subspace methods. Given that a very large class of dynamical systems can be modeled by rational transfer functions, the identification of these systems therefore deals with the problem of estimating the unknown parameters which characterize them completely, namely the numerator and denominator coefficients of their transfer functions. The vector formed of these coefficients is termed the feature *vector.* The least-squares identification technique is an extension of the classical least-squares method to the case when the matrix relating the output of the system to the feature vector is not constant; but is a function of the past inputs and the outputs. The basic principle behind a general identification method is as follows: (i) choose a model, for example a transfer function model, based initially on physical laws governing the system, (ii) estimate the feature vector by minimizing the residual, which is the error between the output of the system and its estimate from the assumed model, (iii) verify whether the error between the system output and its estimate, is a zero-mean white noise process. If it is not a white noise process, refine the structure (the order of the numerator, the order of the denominator, and the delay) of the model and repeat the previous steps. It is interesting to note here that by choosing a model (i.e., by fixing its structure), system identification is reduced to a parameter estimation problem. In order to prevent overfitting the data, a well-known criterion for model order selection, such as the Akaike Information Criterion, is used. In order to meet the above requirements and to comply with the internal model principle, the structure of the assumed model is chosen to be the same as that of the Kalman filter of the model in view of its residual property. It is shown that the widely-used high-order least-squares and the prediction error methods are derived from the *residual model* that relates the Kalman filter residual, to the system input and output in terms of the feature vector. The subspace method, however, is derived directly from the Kalman filter model relating the residual, system input and output.

Closed loop Identification

Chapter 8 extends the identification to systems operating in closed loop, In practice, and for a variety of reasons (for e.g., analysis, design and control), it is often necessary to identify a system that must operate in a closed-loop fashion under some type of feedback control. These reasons could also include design of high performance controller, safety issues, the need to stabilize an unstable plant and /or improve its performance while avoiding the cost incurred through downtime if the plant were to be taken offline for test. In these cases, it is therefore necessary to perform closed-loop identification. Applications include aerospace, magnetic levitation, levitated micro-robotics, magnetically-levitated automotive engine valves, magnetic bearings, mechatronics, adaptive control of processes, satellite-launching vehicles or unstable aircraft operating in closed-loop and process control systems. There are three basic approaches to closed-loop identification, namely direct, an indirect, and a two-stage. Using a direct approach, there may be a bias in the identified subsystem models due mainly to the correlation between the input and the noise. Further, the open loop plant may be unstable. The two-stage approach is emphasized. It consists of two stages of identification. In the first stage, the closed-loop transfer functions relating the system input to all the measured outputs are identified. In the second stage, using the estimated outputs from the first stage, the (open loop) subsystem transfer functions are estimated.

Applications of Identification: Chapters 9, 10, 11, 12, and 13 deal with the applications of identification.

Fault Diagnosis

Fault diagnosis is developed in Chapter 9. As the complexity of engineering systems increases, fault diagnosis, condition-based maintenance, and health monitoring become of vital practical importance to ensure key advantages, such as the system's reliability and performance sustainability, reduction of down time, low operational costs, and personnel safety. Fault diagnosis of physical systems is still a challenging problem and continues to be a subject of intense research both in industry and in academia, in view of the stringent and conflicting requirements in practice for a high probability of correct detection and isolation, low false alarm probability, and timely decision on the fault status.

A physical system is an interconnection of subsystems including the actuators, and sensors and plants. Each subsystem is modeled as a transfer function that may represent a physical entity of the system such as a sensor, actuator, controller, or other system component that is subject to faults. Parameters, termed herein as *diagnostic parameters*, are selected so that they are capable of monitoring the health of the subsystems, and may be varied either directly or indirectly (using an emulator) during the off-line identification phase. An emulator is a transfer function block which is connected at the input with a view to inducing faults in a subsystem which may arise as a result of variations in the phase and the magnitude of the transfer function of a subsystem. An emulator may take the form of a gain or a filter to induce gain or phase variations. A fault occurs within a subsystem when one or more of its diagnostic parameters vary. A variation in the diagnostic parameter does not necessarily imply that the subsystem has failed, but it may lead to a potential failure resulting in poor product quality, shut down, or damage to subsystem components. Hence a proactive action such as condition-based preventive maintenance must be taken prior to the occurrence of a fault.

A unified approach to both detection and isolation of a fault is presented based on Kalman filter residual property. The fault detection capability of the Kalman filter residual is extended to the task of fault isolation. It is shown that the residual is a linear function in each of the diagnostic parameters when other parameters are kept constant, that is, it is multi-linear function of the diagnostic parameters. A vector, termed *influence vector*, plays a crucial role in the fault isolation process. The influence vector is made of elements that are partial derivatives of the feature vector with respect to each diagnostic parameter. The nominal fault-free model of the system and the influence vectors are estimated off-line by performing a number of experiments to cover all likely operating scenarios rather than merely identify the system at a given operating point. Emulators, which are transfer blocks, are included at the accessible points in the system such as inputs, the outputs, or both in order to at mimic the likely operating scenarios. This is similar in spirit to the artificial neural network approach where a training set comprising data obtained from a number of representative operating scenarios is presented so as to capture completely the behaviour of the system.

The decision to select between the hypothesis that the system has a fault, and the alternative hypothesis that it does not, is extremely difficult in practice as the statistics of both the noise corrupting the data and the model that generated these data are not known precisely. To effectively discriminate between these two important decisions (or hypotheses), the Bayes decision strategy is employed here as it allows for the inclusion of the information about the cost associated with the decision taken, and the *a priori* probability of the occurrence of a fault. The fault isolation problem is similarly posed as a multiple hypothesis testing problem. The hypotheses include a single fault in a subsystem, simultaneous faults in two subsystems, and so on until simultaneous faults in all subsystems. For single fault, a closed form solution is presented. A fault in a subsystem is asserted if the correlation between the measured residual and one of a number of hypothesized residual estimates is maximum.

Modeling and Identification of Physical Systems

In Chapter 10, modeling and identification of physical systems is given. The theoretical underpinnings of identification and its applications are thoroughly verified by extensive simulations and very well corroborated by the practical implementation laboratory scale systems including: (i) a two-tank process control system, (ii) a magnetically-levitated system, and (iii) a mechatronic control system. In this chapter, a mathematical model of the physical system derived from physical laws is given. The input–output data obtained from experiments performed on theses physical system are used to evaluate the performance of the identification, fault diagnosis, and soft sensor schemes. The closed-loop identification scheme developed in the earlier chapters is employed.

Fault Diagnosis of Physical Systems

Chapter 11, treats model-based fault diagnosis of physical systems. The background on the Kalman filter, closed-loop identification, and fault diagnosis are given in earlier chapters, Case studies in fault diagnosis of a laboratory-scale two-tank process control system and the position control are presented.

Fault Diagnosis of a Sensor Network

In Chapter 12, a model-based fault diagnosis scheme is developed for a sensor network of a cascade, parallel, and feedback combination of subsystems. The objective is to detect and isolate a fault in any of the subsystems and measurement sensors which are subject to disturbances and/or measurement noise. The approach hinges on the use of a bank of Kalman filters (KF) to detect and isolate faults. Each KF is driven by either a pair (i) of consecutive sensor measurements or (ii) of a reference input and a measurement. It is shown that the KF residual is a reliable indicator of a fault in subsystems and sensors located in the path between the pair of the KF's input. A simple and efficient procedure is developed that analyzes each of the associated paths and leads to both the detection and isolation of any fault that occurred in the paths analyzed. The scheme is successfully evaluated on several simulated examples and a physical fluid system exemplified by a benchmarked laboratory-scale two-tank system to detect and isolate faults, including sensor, actuator, and leakage ones. Further, its performance is compared with those of the Artificial Neural Network and fuzzy logic-based model-free schemes.

Soft Sensor

A model-based soft sensor is proposed for estimating unmeasured variable for high performance, fault tolerant, and reliable control systems is considered in Chapter 13. A soft sensor can be broadly defined as a software-based sensor. Soft sensors are invaluable in numerous scientific and industrial applications where hardware sensors are either too costly to maintain and/or too dangerous or impossible to physically access. Soft sensors act as the virtual eyes and ears of operators and engineers looking to draw conclusions from processes that are difficult – or impossible – to measure with a physical sensor. With no moving parts, a soft sensor offers a maintenance-free method for a variety of data acquisition tasks that serve numerous applications, such as fault diagnosis, process control, instrumentation, signal processing, and the like. In fact, they are found to be ideal for use in aerospace, pharmaceutical, process control, mining, oil and gas, and healthcare industries. It is anticipated that a wave of soft sensing will sweep through the measurement world through its increasing use in smart phones nowadays. Soft sensing is already providing the core component of the new and emerging area of smart sensing. The design and use of a soft sensor is illustrated in this chapter in the specific and important area of robust and fault tolerant control. A soft sensor uses a software algorithm that derives its sensing power from the use of an Artificial Neural networks, a Neuro-fuzzy system, Kernel methods (support vector machines), a multi-variate statistical analysis, a Kalman filter, or other model-based or model-free approaches (Angelov and Kordon, 2010). A model-based approach using a Kalman filter for the design of a soft sensor is proposed here. The nominal model of the system is identified by performing a number of experiments to cover various operating regimes using emulators. The proposed scheme is evaluated on a simulated, as well as a laboratory-scale, velocity control system.

Nomenclature

Vector Norm

$$x = [x_1 \quad x_2 \quad x_3 \quad \cdots \quad x_n]^T$$

$$\|x\|_2 = \sqrt{x^T x} = \sqrt{\sum_{i=1}^{n} x_i^2} = \|x\| \qquad \text{2-norm of } x$$

$$\|x\|_1 = \sum_{i=1}^{n} |x_i| \qquad \text{1-norm of } x$$

$$\|x\|_\infty = \max_{i=1,2\ldots n} \{|x_i|\} \qquad \infty\text{-norm of } x$$

Matrix Norm

Let A be a nxm with elements $\{a_{ij}\}$

$\|A\| = \sup_{x,\|x\|=1} \{\|Ax\|\}$ is a 2-norm of a matrix A. It is also called a spectral norm of a matrix A. It is the largest singular value of A or the square root of the largest eigenvalue of the positive semi-definite matrix $A^T A$.

$\|A\|_1 = \sup_{x,\|x\|_1=1} \{\|Ax\|_1\} = \max_j \left\{ \sum_i |a_{ij}| \right\}$ is a 1-norm of a matrix A. It is the largest absolute column sum of the matrix.

$\|A\|_\infty = \sup_{x,\|x\|_\infty=1} \{\|Ax\|_\infty\} = \max_i \left\{ \sum_j |a_{ij}| \right\}$ is a ∞-norm of a matrix A. It is the largest absolute row sum of the matrix.

$\|A\|_F = \sqrt{\sum_{i=1}^{n} \sum_{j=1}^{m} |a_{ij}|^2} = \sqrt{trace\{A^T A\}}$ is a Frobenius-norm of a matrix A. The Fresenius norm is often used in matrix analysis as it is easy to compute; for example, to determine how close the two matrices A and B are. It is not an induced norm. $\sqrt{I} = \sqrt{n}$.

A useful inequality between 1, 2, and ∞ norms is given by

$$\|A\| \leq \sqrt{\|A\|_\infty \|A\|_1}$$

I	identity matrix
SVD	Singular Value Decomposition $A = USV^T$ where U is nxn, V is mxm, and S is nxm matrices, U and V are unitary matrices.
$diag(a_1, a_2, \ldots, a_n)$	nxn diagonal matrix with a_i as its ith diagonal element
det(A)	determinant of A
trace(A)	trace of A

$\boldsymbol{A}^{\dagger}$	pseudo inverse of $\boldsymbol{A}$
$\lambda(\boldsymbol{A})$	eigenvalue of $\boldsymbol{A}$
$\sigma(\boldsymbol{A})$	singular values of $\boldsymbol{A}$
$\sigma_i(\boldsymbol{A})$	ith singular values of $\boldsymbol{A}$
$\sigma_{\min}(\boldsymbol{A})$	smallest singular value of $\boldsymbol{A}$
$\sigma_{\max}(\boldsymbol{A})$	largest singular value of $\boldsymbol{A}$
$\|A\|$	2-norm (spectral norm) of $\boldsymbol{A}$: $\|A\| = \sigma_{\max}(\boldsymbol{A})$
$[a_{ij}]$	matrix with a_{ij} as its ith row and jth column element
$A(i,j)$	ijth element of $\boldsymbol{A}$
$\boldsymbol{A}(:j)$	jth column vector of $\boldsymbol{A}$
$\boldsymbol{A}(i:)$	ith row vector of $\boldsymbol{A}$
$Re\{\alpha\}$	real part of a complex α
$\mathrm{Im}\{\alpha\}$	imaginary part of a complex α
$\boldsymbol{A} \geq \boldsymbol{B}$	$\boldsymbol{A}$-$\boldsymbol{B}$ is positive semi-definite
$\boldsymbol{P}_r$	projection operator of $\boldsymbol{H}$:$\boldsymbol{P}_r = \boldsymbol{H}(\boldsymbol{H}^T\boldsymbol{W}\boldsymbol{H})^{-1}\boldsymbol{H}^T\boldsymbol{W} = \boldsymbol{H}\boldsymbol{H}^{\dagger}$
$\boldsymbol{I} - \boldsymbol{P}_r$	orthogonal complement projector
$\boldsymbol{W}$	weighting matrix
$\boldsymbol{1}$	vector of all ones
$\boldsymbol{1}_1$	matrix of all ones
$span\left(\boldsymbol{h}_1 \quad \boldsymbol{h}_2 \quad \cdots \quad \boldsymbol{h}_M\right) = \left\{y : y = \sum_{i=1}^{M} \alpha_i \boldsymbol{h}_i\right\}$	is the linear space generated by $\left\{\boldsymbol{h}_1 \quad \boldsymbol{h}_2 \quad \cdots \quad \boldsymbol{h}_M\right\}$
$\Re$	field of real numbers
$\Re^n$	Euclidian space of $nx1$ real vectors
$\Re^{nxm}$	Euclidian space of nxm real matrices

Probability and Random Process

PDF	Probability Density Function
i.i.d.	independent identically distributed
$f_y(\boldsymbol{y})$	PDF of $\boldsymbol{y}$
$E[x]$	Expectation of x
$E[x\|y]$	Conditional Expectation: expectation x given y
μ_x	mean value of x: $\mu_x = E[x(k)]$
$\boldsymbol{\Sigma}_v$	covariance of $\boldsymbol{v}$
σ_v	variance of $\boldsymbol{v}$
$\ln f_y(\boldsymbol{y})$	log of $f_y(\boldsymbol{y})$
$\boldsymbol{I}_F$	Fisher information
$\dfrac{\delta f_y(\boldsymbol{y})}{\delta \theta}$	partial derivative
$L(\boldsymbol{\theta}\|y) = f_y(\boldsymbol{\theta}\|\boldsymbol{y})$	likelihood function
$l(\boldsymbol{\theta}\|y) = \ln f_y(\boldsymbol{\theta}\|\boldsymbol{y})$	log-likelihood function

Transforms

$\Im(.)$	Fourier transform of $(\cdot)$
$\Im^{-1}(.)$	inverse Fourier transform of $(\cdot)$
$\mathbb{Z}(\cdot)$	z-transform of $(\cdot)$
z	z-transform variable
$\mathbb{Z}^{-1}(\cdot)$	the inverse z-transform of $(\cdot)$
ω	frequency in radians per second

f frequency in hertz (Hz)
FFT Fast Fourier Transform

Operations on Signals

$r_{xx}(m)$ correlation of $x(k)$ and $y(k)$
$x(k) \circ y(k)$ convolution of $x(k)$ and $y(k)$
$E_{xy}(z)$ energy spectral density of $x(k)$ and $y(k)$
$P_{xy}(z)$ power spectral density of $x(k)$ and $y(k)$
E_x energy signal $x(k)$
P_x power signal $x(k)$
$\arg\left\{\max_x (f(x))\right\}$ the value of x that maximizes $f(x)$
$\arg\left\{\min_x (f(x))\right\}$ the value of x that minimizes $f(x)$

$$sign(x) = \begin{cases} 1 & x > 0 \\ 0 & x = 0 \\ -1 & x < 0 \end{cases}$$

Signals and Systems

$\boldsymbol{y}$ output
$\boldsymbol{x}$ state of a system
$\hat{\boldsymbol{y}}$ estimate of the output $\boldsymbol{y}$
$\boldsymbol{e}$ residual: $\boldsymbol{e} = \boldsymbol{y} - \hat{\boldsymbol{y}}$
r (reference) input
w disturbance affecting the system
v measurement noise at the output
SNR Signal to Noise Ratio
$\delta(m) = \begin{cases} 1 & m = 0 \\ 0 & m \neq 0 \end{cases}$ Kronecker delta function
$G(z)$ transfer function of a system
$(\boldsymbol{A}, \boldsymbol{B}, \boldsymbol{C})$ state-space model
AR Auto Regressive
MA Moving Average
ARMA Auto Regressive and Moving Average
ARMAX Auto Regressive and Moving Average with external input
FIR Finite Impulse Response
PRBS Pseudo Random Binary Signal
SISO Single Input, Single Output
SIMO Single Input, Multiple Output
MIMO Multiple Input, Multiple Output
$\boldsymbol{\psi}(k)$ data vector
$\boldsymbol{H}$ data matrix
$\boldsymbol{\theta}$ feature vector
$\hat{\boldsymbol{\theta}}$ estimate of $\boldsymbol{\theta}$

Estimation

ML Maximum Likelihood
LS Least-Squares

PEM	Prediction Error Method
SM	Subspace Method
HOLS	High-Order Least-Squares
CRLB	Cramér–Rao Lower Bound
MMSE	Minimum Mean-Square Estimation
AIC	Akaike information criterion

Control

PI	Proportional Integral
PID	Proportional, Integral, and Derivative
IM	Internal Model
H_∞	H-infinity
$S(z)$	Sensitivity function
$T(z)$	Complementary sensitivity function
PWM	Pulse Width Modulator

Fault Diagnosis

γ	diagnostic parameter
$\boldsymbol{\Omega}_i$	influence vector of γ_i
$lr(\boldsymbol{e})$	log-likelihood ratio
$t_s(\boldsymbol{e})$	test statistics
η_{th}	threshold value
H_i	ith hypothesis
FD	Fault Diagnosis
FDI	Fault Detection and Isolation
SN	Sensor Network
KF	Kalman Filter
ANN	Artificial Neural Network
FL	Fuzzy Logic
ANFIS	Adaptive Neuro-Fuzzy Inference System

1

Modeling of Signals and Systems

1.1 Introduction

A system output is generated as a result of the input, the disturbances, and the measurement noise driving the plant. The input is a termed as "signal" while the disturbance and the measurement noise are termed as "noise." A signal is the desired waveform while a noise is as an unwanted waveform, and the output is the result of convolution (or filtering) of the signal and the noise by the system. Examples of signal include speech, music, biological signals, and so on, and examples of noise include 60 Hz power frequency waveform, echo, reflection, thermal noise, shot noise, impulse noise, and so on. A signal or noise may be deterministic or stochastic. Signals such as speech, music, and biological signals are stochastic: they are not exactly the same from one realization to the other. There are two approaches to characterize the input–output behavior of a system:

- Non-parametric (classical or FFT-based) approach.
- Parametric (modern or model-based) approach.

In the parametric approach, the plant, the signal, and the noise are described by a discrete-time model. In the non-parametric approach, the plant is characterized by its frequency response, and the signal and the noise are characterized by correlation functions (or equivalently by power spectral densities). Generally, FFT forms the basic algorithm used to obtain the non-parametric model. Both approaches complement each other. The parametric approach provides a detailed microscopic description of the plant, the signal, and the noise. The non-parametric approach is computationally fast but provides only a macroscopic picture.

In general, the signal or the noise may be classified as deterministic and random processes. A class of deterministic processes – including the widely prevalent constants, exponentials, sinusoids, exponentially damped sinusoids, and periodic waveforms – are modeled as an output of a Linear Time Invariant (LTI) system driven by delta function. Essentially the model of a deterministic signal (or noise) is the z-transform of the signal (or noise).

We frequently encounter non-deterministic or random signals, which appear everywhere and are not analytically describable. In many engineering problems one has to analyze or design systems subject to uncertainties resulting from incomplete knowledge of the system, inaccurate models, measurement

Identification of Physical Systems: Applications to Condition Monitoring, Fault Diagnosis, Soft Sensor and Controller Design, First Edition. Rajamani Doraiswami, Chris Diduch and Maryhelen Stevenson.

errors, and uncertain environments. Uncertainty in the behavior of the systems is commonly handled using following approaches:

1. *Deterministic approach:*
   The uncertainty is factored in the analysis and particularly the design by considering the worst case scenario.
2. *Fuzzy-logic approach*:
   The fuzziness of the variable (e.g., small, medium, large values) are handled using the mathematics of fuzzy logic.
3. *Probabilistic approach:*
   The uncertainty is handled by treating the variables as random signals.

In this chapter, we will restrict ourselves to the probabilistic approach. These uncertainties are usually modeled as random signal inputs (noise and disturbances) to the system. The measurements and the disturbances affecting the system are treated as random signals. Commonly, a fictitious random input is introduced to mimic uncertainties in the model of a system. Random signals are characterized in terms of statistical terms that represent average behavior when a large number of experiments is performed. The set of all the outcomes of the experiments is called an *ensemble* of time functions or equivalently a *random process* (or *stochastic process*). A random signal is modeled as an output of a LTI system driven by zero-mean white noise, unlike a deterministic signal which is modeled as an output with delta function input. A class of low-pass, high-pass, and band-pass random signals are modeled by selecting an appropriate LTI system (filter).

An output of a system is mostly affected by disturbances and measurement noise. The disturbance and the measurement noise may be deterministic waveforms or a random process. Deterministic waveforms include constant, sinusoid, or a periodic signals, while the random waveform may be a low-pass, a band-pass, or a high-pass process. An integrated model is obtained by augmenting the model of the plant with those of the disturbance and the measurement noise. The resulting integrated model is expressed in the form of a high-order difference equation model, such as an Auto Regressive (AR), a Moving Average (MA), or a Auto Regressive and Moving Average (ARMA) model. The input is formed of the plant input, and the inputs driving the disturbance and measurement noise model inputs, namely delta functions and/or white noise processes. Similarly, an augmented state-space model of the plant is derived by combining the state-space models of the plant, the disturbances, and the measurement noise.

This integrated model is employed subsequently in the system identification, condition monitoring, and fault detection and isolation. The difference equation model is used for system identification while the state-space model is used for obtaining the Kalman filter or an observer.

A model of a class of signals (rather than the form of a specific member of that class) is developed from the deterministic or the stochastic signal model by setting the driving input to zero. A model of a class of signals such as the reference, the disturbance, and the measurement noise is employed in many applications. Since the response of a system depends upon the reference input to the system, and the disturbances and the measurement noise affecting its output, the desired performance may degrade if the influence of these signals is not factored into the design of the system. For example, in the controller or in the Kalman filter, the steady-state tracking will be ensured if and only if a model of the class of these signals is included and this model is driven by the tracking error. The model of the class of signal which is integrated in the controller, the observer, or the Kalman filter is termed as the *internal model* of the signal. The internal model ensures that given output tracks the reference input in spite of the presence of the disturbances and the measurement noise waveform corrupting the plant output.

Tables, formulae, and background information required are given in the Nomenclature section.

1.2 Classification of Signals

Signals are classified broadly as deterministic or random; bounded or unbounded; energy or power; causal, anti-causal, or non-causal.

1.2.1 Deterministic and Random Signals

Deterministic signals can be modeled exactly by a mathematical expression, rule, or table. Because of this, future values of any deterministic signal can be calculated from past values. For this reason, these signals are relatively easy to analyze as they do not change, and we can make accurate assumptions about their past and future behavior.

Deterministic signals are not always adequate to model real-world situations. Random signals, on the other hand, cannot be characterized by a simple, well-defined mathematical equation. They are modeled in probabilistic terms. Probability and statistics are employed to analyze their behavior. Also, because of their randomness, average values from a collection of signals are usually studied rather than analyzing one individual signal.

Unlike deterministic signals, stochastic signals – or random signals – are not so nice. Random signals cannot be characterized by a simple, well-defined mathematical equation and their future values cannot be predicted. Rather, we must use probability and statistics to analyze their behavior. Also, because of their randomness, average values from a collection of signals obtained from a number of experiments are usually studied rather than analyzing one individual outcome from one experiment.

1.2.2 Bounded and Unbounded Signal

A signal $x(k)$ is said to be *bounded* if

$$|x(k)| < \infty \quad \text{for all } k \tag{1.1}$$

That is, a bounded signal assumes a finite value for all time instants. At no time instant $x(k)$ goes to infinity. A signal, which is not bounded, is called an *unbounded* signal.

1.2.3 Energy and Power Signals

The signals are classified according to different characteristics, namely energy or power signals. The energy of a signal $x(k)$ defined for $k \geq 0$ and denoted E_x is defined as

$$E_x = \lim_{N\to\infty} it \sum_{k=0}^{N-1} |x(k)|^2 \tag{1.2}$$

The magnitude-squared value of $x(k)$ is used so as to include both complex and real-valued signals. The energy E_x of a signal $x(k)$ may be finite or infinite. The average power of a signal $x(k)$, denoted P_x, is defined as

$$P_x = \lim_{N\to\infty} it \frac{1}{N} \sum_{k=0}^{N-1} |x(k)|^2 \tag{1.3}$$

The power P_x may be finite or infinite. From the definitions of energy and power, a signal may be classified as follows:

The Number of Data Samples N is Infinitely Large

1. An energy signal if E_x is non-zero and is finite

$$0 < E_x = \lim_{N\to\infty} it \sum_{k=0}^{N-1} |x(k)|^2 < \infty \tag{1.4}$$

2. A power signal if its power P_x is non-zero and finite

$$0 < P_x = \lim_{N\to\infty} it \frac{1}{N} \sum_{k=0}^{N-1} |x(k)|^2 < \infty \tag{1.5}$$

3. A signal is said to be neither if both the energy and power are not finite.

It can be seen that if the energy E_x is finite, then its power $P_x = 0$. On the other hand if E_x is infinite then P_x may be finite or infinite.

The Number of Data Samples N is Finite

In the case when N is finite, it is usually called an energy signal, although it could also be termed as power signal.

$$E_x = \sum_{k=0}^{N-1} |x(k)|^2 \tag{1.6}$$

1.2.4 Causal, Non-causal, and Anti-causal Signals

Signals are classified depending upon the time interval over which they are defined, that is whether they are defined for all time, only positive time, or only negative time intervals. Let $s(k)$ be some signal:

1. The signal $s(k)$ is said to be *causal* if it is defined only for positive time intervals, and is zero for negative time intervals

$$s(k) = 0, \quad k < 0 \tag{1.7}$$

2. The signal $s(k)$ is said to be *anti-causal* if it is defined only for negative time intervals, and is zero for positive negative time intervals

$$s(k) = 0, \quad k > 0 \tag{1.8}$$

3. The signal $s(k)$ is said to be *non-causal* if it is defined for all time intervals, both positive and negative.

Figure 1.1 shows (a) causal, (b) anti-causal, and (c) non-causal signals at the.

1.2.5 Causal, Non-causal, and Anti-causal Systems

Similarly to signals, the system is classified as causal, non-causal, or anti-causal as follows:

1. A system is said to be *causal* (also termed *non-anticipative* or *physical*) if its output at the present time instant $y(k)$ is a function of input at the present and past time instants $u(k - i)$ for $i \leq 0$, and not a function of inputs at future time instants $u(k + i)$ for $i > 0$. Equivalently the impulse response of a causal system is zero for all time instants $k < 0$.
2. A system is *non-causal* if its output at the present time instant $y(k)$ is a function also of future input values in addition to the past input values. That is, $y(k)$ is a function of input at the present and past time instants $u(k - i)$ for $i \leq 0$, as well as inputs at some future time instants $u(k + i)$ for $i > 0$.

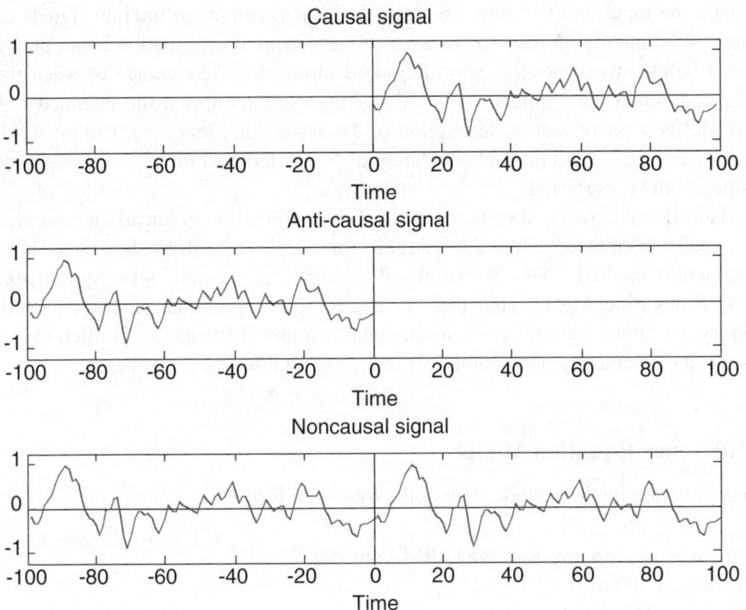

Figure 1.1 Causal, anti-causal and non-causal signals

3. A system is *anti-causal* system if its output $y(k)$ is a function solely of future and present input values $u(k+i)$ for $i \geq 0$. That is the output does not depend on past input values $u(k-i)$ for $i < 0$.

One can also define a system as causal, non-causal, and anti-causal if its impulse response is causal, non-causal, and anti-causal respectively.

1.3 Model of Systems and Signals

The mathematical model is assumed to be linear and time invariant. It may be described in the

1. *Time domain*
2. *Frequency domain*

1.3.1 Time-Domain Model

Generally the mathematical model of a physical system is a set of interconnected differential/difference and algebraic equations. From modeling, analysis, design, and implementation points of view, it is convenient to represent the given system using one of the following:

1. *Difference equation model*: a set of nth order difference equations
2. *State-space model*: a set of n first-order difference equations

Both models are equivalent from the point of view of the input–output relationship. The difference equation model describes the input–output relationship: the model relates the present and the past inputs and the outputs. Hence they are called *input–output* models. The state-space models, on the other hand, include not only the input and the output but also *internal variables* or internal signals called *states*, providing thereby an internal description and hence the structural information about the system. Both state-space and input–output models are widely used in practice. The choice between the two types of models depends upon the application, the available information, and the intended objective. The state-space model is a vector-matrix description of the system and lends itself to powerful modeling, analysis, design, and realization or implementation of controllers and filters. For system identification, an input–output model is preferred.

A state-variable denoted $x(k)$ of a system is defined as a minimal set of linearly independent variables, termed states, such that knowledge of the states at any time k_0 plus the information on the input excitation $u(k)$ for subsequently applied $k \geq k_0$, are sufficient to determine the state of the system $x(k)$ and hence the output $y(k)$ at any time, $k \geq k_0$. Thus the state is a compact representation of the past history of the system; to know the future, only the present state and the input are required. Further, the state variable representation can conveniently handle multiple input and multiple output systems.

1.3.1.1 Difference Equation Model

The difference equation model includes one of the following forms:

1. Auto-Regressive and Moving Average (ARMA) model.
2. Auto-Regressive (AR) model.
3. Moving Average (MA) model.

Auto-Regressive and Moving Average Model:
This is a difference equation model given by

$$\sum_{i=0}^{n_a} a_i y(k-i) = \sum_{i=0}^{n_b} b_i u(k-i) \tag{1.9}$$

where $\{a_i\}$ with the leading coefficient $a_0 = 1$ are the coefficients of the auto-regressive (AR) part of the ARMA model, and $\{b_i\}$ are the MA coefficients of the moving average (MA) part of the ARMA model. The input $u(k)$ and the output $y(k)$ are assumed scalar. The term on the left,

$$y(k) + a_1 y(k-1) + a_2 y(k-2) + a_3 y(k-3) + \cdots + a_{n_a} y(k-n_a), \tag{1.10}$$

is a *convolution* of the output $\{y(k-i)\}$ and the AR coefficients $\{a_i\}$. The term on the right,

$$b_0 u(k) + b_1 u(k-1) + b_2 u(k-2) + b_3 u(k-3) + \cdots + b_{n_b} u(k-n_b), \tag{1.11}$$

is a *convolution* of the input $\{u(k-i)\}$ and the coefficients $\{b_i\}$. If $b_i = 0$ for $i = 0, 1, 2, \ldots, n_d - 1$, then the model has a delay n_d given by

$$y(k) + \sum_{i=1}^{n_a} a_i y(k-i) = \sum_{i=n_d}^{n_b} b_i u(k-i) \tag{1.12}$$

If $b_0 \neq 0$, that is if the delay $n_d = 0$, the system is said to have a direct transmission path from the input $u(k)$ to the output $y(k)$. An input applied at any a time instant will affect the output at the same time instant, whereas if $b_0 = 0$, the system will exhibit inertia: the effect of the input at instant k_0 will affect the output at a later time instant $k > k_0$. The degree of the AR part n_a is termed the order of the difference equation.

The Auto-Regressive Model is given by:

$$y(k) + \sum_{i=1}^{n_a} a_i y(k-i) = b_0 u(k) \tag{1.13}$$

The AR model is a special case of the ARMA model when all the coefficients $\{b_i\}$ except the leading coefficient are zero. If $b_i = 0$ for $i = 1, 2, \ldots, n_d - 1$, then the model has a delay n_d and is given by

$$y(k) + \sum_{i=1}^{n_a} a_i y(k-i) = b_{n_d} u(k - n_d) \tag{1.14}$$

The AR model is employed in signal processing applications, including speech, biological signals, and spectral estimation.

The Moving Average Model is given by:

$$y(k) = \sum_{i=0}^{n_b} b_i u(k-i) \tag{1.15}$$

The MA model is a special case of the ARMA model when all the coefficients $\{a_i\}$ except the leading coefficient a_0 are zero. The MA model is generally employed in adaptive filters and their applications, as the MA model is always stable.

1.3.1.2 State-Space Model

The state variable model is formed of (i) a state equation which is a set of n simultaneous first-order difference equations relating n states $x(k)$ and the input $u(k)$, and (ii) an output equation, which is an algebraic equation that expresses the outputs $y(k)$ as a linear combination of the states $x(k)$ and the current input $u(k)$. It is a vector-matrix equation given by

$$\begin{aligned} \boldsymbol{x}(k+1) &= \boldsymbol{A}\boldsymbol{x}(k) + \boldsymbol{B}u(k) \\ y(k) &= \boldsymbol{C}\boldsymbol{x}(k) + Du(k) \end{aligned} \tag{1.16}$$

where $\boldsymbol{x}(k)$ is $nx1$ the vector of states, $u(k)$ and $y(k)$ are scalars; $\boldsymbol{A}$ is the nxn matrix, $\boldsymbol{B}$ is the $nx1$ vector, and $\boldsymbol{C}$ is the $1xn$ vector; $\boldsymbol{x}(k) = [\, x_1(k) \;\; x_2(k) \;\; x_3(k) \;\; \ldots \;\; x_n(k)\,]^T$. A state-space model is expressed compactly as $\boldsymbol{A}, \boldsymbol{B}, \boldsymbol{C}, D$. In the case when the direct transmission term $D = 0$, that is when the system is strictly proper, a triplet $(\boldsymbol{A}, \boldsymbol{B}, \boldsymbol{C})$ is employed to denote a state-space model. Figure 1.2 shows the state-space model formed of the state and the output equations.

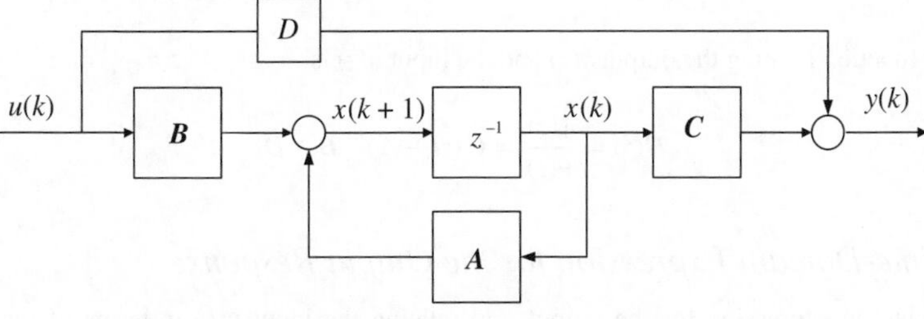

Figure 1.2 State-space model $(\boldsymbol{A}, \boldsymbol{B}, \boldsymbol{C}, D)$

1.3.2 Frequency-Domain Model

Taking the z-transform of ARMA model (1.12) given by

$$y(z) = H(z)u(z) \tag{1.17}$$

$$H(z) = \frac{z^{-n_d}\left(b_{n_d} + b_{n_d+1}z^{-1} + b_{n_d+2}z^{-2} + \cdots + b_{n_b}z^{-n_b+n_d}\right)}{1 + a_1z^{-1} + a_2z^{-2} + a_3z^{-3} + \cdots + a_{n_a}z^{-n_a}} \tag{1.18}$$

where $H(z)$ is the transfer function, and $u(z)$ and $y(z)$ are respectively z-transforms of $u(k)$ and $y(k)$. We frequently use the same notation for time and frequency-domain variables, except that the respective arguments are different. Similarly the z-transforms of AR and MA models (1.14) and (1.15) yield respectively all-pole and all-zero models

$$y(z) = \left(\frac{b_{n_d}z^{-n_d}}{1 + a_1z^{-1} + a_2z^{-2} + a_3z^{-3} + \cdots + a_{n_a}z^{-n_a}}\right)u(z) \tag{1.19}$$

$$y(z) = \left(b_0 + b_1z^{-1} + b_2z^{-2} + b_3z^{-3} + \cdots + b_{n_b}z^{-n_b}\right)u(z) \tag{1.20}$$

The ARMA, AR, and MA model may be classified as follows:

1. the ARMA model has both poles and zeros
2. the AR model has only poles (zeros are all located at the origin)
3. the MA model has only zeros (poles are all located at the origin)

1.4 Equivalence of Input–Output and State-Space Models

1.4.1 State-Space and Transfer Function Model

The state-space model in the time domain may be expressed in the frequency domain by relating only the input and output by expressing them in terms of the states. Taking the z-transform of the of the state and the output Eq. (1.16) we get

$$\boldsymbol{x}(z) = (z\boldsymbol{I} - \boldsymbol{A})^{-1}\boldsymbol{B}z\boldsymbol{x}(0) + (z\boldsymbol{I} - \boldsymbol{A})^{-1}\boldsymbol{B}u(z) \tag{1.21}$$

$$y(z) = \boldsymbol{C}\boldsymbol{x}(z) + Du(z) \tag{1.22}$$

where $\boldsymbol{x}(0)$ is an initial condition. Substituting for $\boldsymbol{x}(z)$ from Eq. (1.21) in Eq. (1.22), the expression for the output $y(z)$ in terms of the input $u(z)$ and the initial condition $\boldsymbol{x}(0)$ becomes;

$$y(z) = \boldsymbol{C}(z\boldsymbol{I} - \boldsymbol{A})^{-1}\boldsymbol{B}z\boldsymbol{x}(0) + [\boldsymbol{C}(z\boldsymbol{I} - \boldsymbol{A})^{-1}\boldsymbol{B} + D]u(z) \tag{1.23}$$

The transfer function relating the output $y(z)$ and the input $u(z)$ is:

$$H(z) = \frac{y(z)}{u(z)} = \boldsymbol{C}(z\boldsymbol{I} - \boldsymbol{A})^{-1}\boldsymbol{B} + D \tag{1.24}$$

1.4.2 Time-Domain Expression for the Output Response

Let us compute an expression for the output $y(k)$ relating the input $u(k)$ in terms of the state-space matrices $\boldsymbol{A}, \boldsymbol{B}, \boldsymbol{C}$, and D. Taking the inverse z-transform of Eq. (1.21), the expression of the state $\boldsymbol{x}(k)$ in terms of the past k inputs $u(k-i) : i = 0, 1, 2, \ldots, k-1$ and the initial condition $\boldsymbol{x}(0)$ becomes:

$$\boldsymbol{x}(k) = A^k \boldsymbol{x}(0) + \sum_{i=1}^{k} \boldsymbol{A}^{k-i} \boldsymbol{B} u(i-1) \tag{1.25}$$

The expression for $y(k)$ relating the initial condition $x(0)$ and the past inputs input $\{u(k-i)\}$ becomes

$$y(k) = \boldsymbol{C}\boldsymbol{A}^k \boldsymbol{x}(0) + \sum_{i=1}^{k} \boldsymbol{C}\boldsymbol{A}^{k-i} \boldsymbol{B} u(i-1) + Du(k) \tag{1.26}$$

1.4.3 State-Space and the Difference Equation Model

For a given difference equation or a transfer function $H(z)$, there are an infinite number of realizations of the state-space model $\boldsymbol{A}, \boldsymbol{B}, \boldsymbol{C}, D$. We will use the observer canonical form of the state-space signal model as it is convenient for the analysis and design of filters such as the Kalman filter.

1.4.4 Observer Canonical Form

Consider the input–output models, namely the ARMA model given by Eq. (1.9). Without loss of generality and for notational convenience assume that the order of the AR and the MA parts of the difference equations are the same: $n_a = n_b = n$. A block diagram representation of the equivalent frequency domain model is shown in Figure 1.3. It is formed of n unit delay elements represented by z^{-1}, the AR and MA coefficients $-a_i$ and b_i respectively are indicated on the arrows. The output of ith delay element is labeled as a state $x_i(k)$. As a consequence the input to the ith delay element is $x_i(k+1)$.

Using Figure 1.3 the dynamical equation of the state-space model $(\boldsymbol{A}, \boldsymbol{B}, \boldsymbol{C}, D)$ becomes

$$\boldsymbol{x}(k+1) = \begin{bmatrix} -a_1 & 1 & 0 & . & 0 & 0 \\ -a_2 & 0 & 1 & . & 0 & 0 \\ -a_3 & 0 & 0 & . & 0 & 0 \\ . & . & . & . & . & . \\ -a_{n-1} & 0 & 0 & . & 0 & 1 \\ -a_n & 0 & 0 & . & 0 & 0 \end{bmatrix} \boldsymbol{x}(k) + \begin{bmatrix} b_1 - b_0 a_1 \\ b_2 - b_0 a_2 \\ b_3 - b_0 a_3 \\ . \\ b_{n-1} - b_0 a_{n-1} \\ b_n - b_0 a_n \end{bmatrix} u(k) \tag{1.27}$$

$$y(k) = [1 \quad 0 \quad 0 \quad 0 \quad 0 \quad 0]\, \boldsymbol{x}(k) + b_0\, u(k)$$

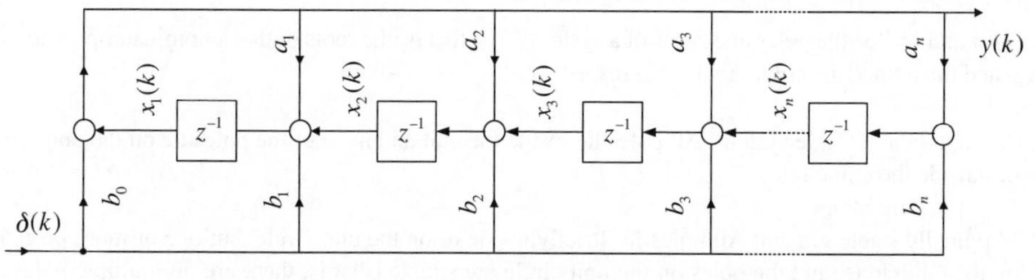

Figure 1.3 Block diagram representation of an ARMA model

If $b_0 = 0$, then the state-space model $(\boldsymbol{A}, \boldsymbol{B}, \boldsymbol{C}, D)$ has the same the same $\boldsymbol{A}$, and $\boldsymbol{C}$, but D and $\boldsymbol{B}$ will be different. Setting $b_0 = 0$ we get

$$\begin{aligned} \boldsymbol{B} &= [\, b_1 \quad b_2 \quad b_3 \quad . \quad b_{n-1} \quad b_n \,]^T \\ D &= 0 \end{aligned} \tag{1.28}$$

Remark When $b_0 = 0$, the state-space model has no direct transmission term, that is $D = 0$. In this case the state-space may simply be obtained from the difference equation model by copying the negative values of the denominator coefficients $\{-a_i\}$ in the first column of the $\boldsymbol{A}$ matrix, and the numerator coefficients $\{b_i\}$ in column vector $\boldsymbol{B}$.

1.4.5 Characterization of the Model

We may characterize the discrete-time model given by the difference Eq. (1.12), the transfer function model expressed in terms of the z-transform variable z^{-1}(1.18), and the state-space model (1.27) similar to the continuous-time model based on the *time delay* n_d, which is analogous to *relative degree*. The model is said to be

1. *Proper* if $n_d \geq 0$.
2. *Strictly proper* if it is proper and $n_d \geq 1$.
3. *Improper* if $n_a < n_b$.

If $b_0 \neq 0$, or equivalently if the delay $n_d = 0$, the system is said to have direct a transmission path from the input $u(k)$ to the output $y(k)$. In this case, the state-space model will have the direct transmission term $D \neq 0$. These definitions are analogous to the definitions strictly proper, proper, and improper fractions in describing the ratios of integers. For example, $1/2$ is a proper fraction, while $3/2$ is an improper fraction. Further, the definitions are consistent with those of the continuous-time models. Unlike the case of the continuous time model, there is no restriction on the relative degree $n_a - n_b$ (in case of the continuous time system the transfer function must be proper, to ensure causality). The discrete-time model given above is always causal for all values of n_a and n_b. Further, there is no loss of generality in assuming that the model is improper $n_b > n_a$ as an appropriate number of zero coefficients may be included in the denominator so that the numerator and denominator orders are equal; that is, additional terms in the denominator $D_s(z)$ with zero coefficients $\{a_i = 0 : n_a < i \leq n_b\}$ are added. Similarly if $n_b > n_a$, there is no loss of generality in assuming that the numerator and denominator orders are equal $n_b = n_a$ as an additional term with zero coefficients $\{b_i = 0 : n_b < i \leq n_a\}$ may be included.

Without loss of generality, and for notational simplicity, we will assume $n_b = n_a$ wherever possible.

1.4.6 Stability of (Discrete-Time) Systems

Let $\{p_i\}$ and $\{z_i\}$ be the poles and zeros of a system $H(z)$, that is, the roots of the denominator polynomial $D(z)$ and the numerator polynomial $N(z)$ respectively:

1. Asymptotically stable system: All poles lie inside the unit and none of the poles lie on the unit circle or outside the unit circle.
   $|p_i| < 1$ for all i.
2. Marginally stable system: All poles lie strictly inside or on the unit circle, but one or more poles lie on the unit circle, and the poles on the unit circle are simple (that is, there are no multiple poles on the unit circle).

1. $|p_i| \leq 1$ for all i.
2. If $|p_i| = |p_j| = 1$ then $i \neq j$.
3. Unstable system: At least one pole lies outside the unit circle.
   At least one pole p_i satisfies $|p_i| > 1$.
4. A polynomial is said to be stable if all its roots are strictly inside the unit circle

Note: In this work a stable system refers to an asymptotic stable system, that is, all the poles are strictly inside the unit circle.

1.4.7 Minimum Phase System

Definition *A transfer function $H(z)$ is said to be minimum phase if its poles and zeros are stable (poles and zeros are both strictly inside a unit circle). That is $|p_i| \leq 1$ and $|z_i| < 1$ for all* i. *Equivalently $H(z)$ is minimum phase if $H(z)$ and its inverse $H^{-1}(z)$ are both stable. One may peek at Figure 1.10 in a later section where the poles and the zeros of minimum phase filters are shown. Minimum phase transfer function has many applications including modeling of signals, identification, and filtering.*

1.4.8 Pole-Zero Locations and the Output Response

1. The pole location determines the form of the output response. In general the closer the pole is to the origin of the unit circle, the faster is the response, and the closer the pole is to the unit circle the more sluggish is the response. The impulse/step response of a first-order discrete time system is oscillatory if the poles are located on the unit circle.
2. Complex poles and their zeros must always occur in complex conjugate pairs.
3. Poles or zeros located at the origin do not affect the frequency response.
4. To ensure stability, the poles must be located strictly inside the unit circle while zeros can be placed anywhere in the z-plane.
5. Zeros on the unit circle may be located to create a null or a valley in the magnitude response. The presence of a zero close to the unit circle will cause the magnitude response to be small at frequencies that correspond to points of the unit circle close to the zero. Attenuation at a specified frequency range is obtained by locating a zero in the vicinity of the corresponding frequencies on the unit circle. A stop band response is obtained by locating zeros on or close to the unit circle.
6. The presence of a pole close to the unit circle will amplify the magnitude response in the vicinity of the corresponding frequencies on the unit circle. Amplification at a specified frequency range is obtained by locating a pole in the vicinity of the corresponding points on the unit circle. Thus a pole has the opposite effect to that of a zero.
7. A narrow transition band is achieved by placing a zero close to the pole along (or near) the same radial line and close to the unit circle. This property is exploited to obtain a sharp notch or a narrow peak for a specified frequency.
8. The AR term is responsible for spectral peaks while the MA term is responsible for spectral valleys.
9. A signal with finite duration has a MA model while a signal with infinite duration has at least one pole: an AR or ARM model.

1.5 Deterministic Signals

A class of deterministic signals, which are the impulse response of LTI systems, include constant signals (dc signals), exponential signals, sinusoids, damped sinusoids, and their combinations. A signal $s(k)$ is generated as an output of a linear time invariant discrete-time system $H(z)$ whose input $u(k)$ is a delta function $\delta(k)$. The input $u(k)$ is

$$u(k) = \delta(k) = \begin{cases} 1 & \text{if } k = 0 \\ 0 & \text{else} \end{cases} \tag{1.29}$$

1.5.1 Transfer Function Model

We will obtain the signal model of $s(k)$ in the frequency domain relating the z-transform of the output $s(z)$ and the z-transform of the input $u(z)$. As the z-transform of the delta function is unity, $\delta(z) = 1$ we get

$$u(z) = 1 \tag{1.30}$$

Hence the signal model becomes

$$s(z) = H(z) \tag{1.31}$$

This shows that the signal model is equal to the z-transform of the signal. Figure 1.4 shows the model of a signal $s(k)$. A delta function input $\delta(k)$ and its Fourier transform $u(\omega)$ are shown on the left at the input side while the output $s(k)$ and its Fourier transform $s(\omega)$ are shown on the right at the output side of the block diagram. The input–output block diagram is sandwiched between the input and output plots.

A signal is generated as an output of a proper rational transfer function, which is a ratio of two polynomials, excited by a delta function input. In other words the z-transform of the signal $s(z)$ is a is a ratio of two polynomials in $\{z^{-i}\}$, namely the numerator polynomial $N_s(z)$ and denominator polynomial $D_s(z)$:

$$s(z) = \frac{N(z)}{D(z)} \tag{1.32}$$

where $D(z) = 1 + \sum_{i=1}^{n_a} a_i z^{-i}$, $N(z) = \sum_{i=0}^{n_b} b_i z^{-i}$; n_a is the model order, which is the order of the denominator polynomial; n_b is the order of the denominator polynomial; $a_i : i = 0, 1, 2, \ldots, n_a$ and $b_i : i = 0, 1, 2, \ldots, n_b$ are respectively the coefficients of $D(z)$ and $N(z)$.

1.5.2 Difference Equation Model

The signal model in the time domain is merely the inverse z-transform of $s(z)$, or equivalently the impulse response of the signal model $H_s(z) = s(z)$. Cross multiplying, and substituting the expressions for $N_s(z)$ and $D_s(z)$, Eq. (1.32) becomes

$$\left(1 + \sum_{i=1}^{n_a} a_i z^{-i}\right) s(z) = \sum_{i=0}^{n_a - 1} b_i z^{-i} \tag{1.33}$$

Taking the inverse z-transform yields a linear difference equation with constant coefficients driven by the delta function

$$s(k) + \sum_{i=0}^{n_a} a_i s(k-i) = \sum_{i=0}^{n_a - 1} b_i \delta(k-i) \tag{1.34}$$

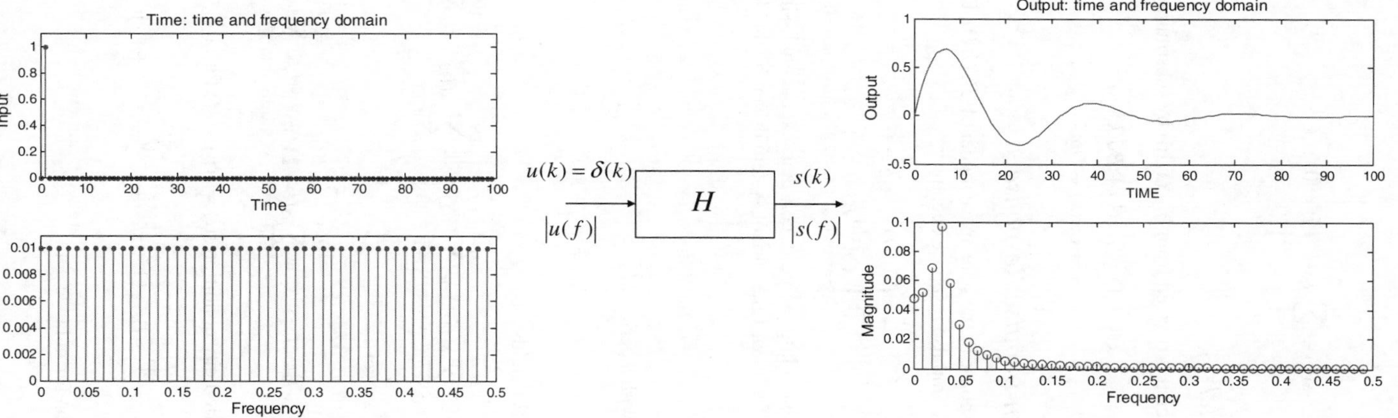

Figure 1.4 Input–output model of a deterministic signal

Since $\delta(k-i)=0$ *for* $i>n_a$, the time domain model reduces to the following homogenous difference equation

$$s(k)+\sum_{i=0}^{n_a}a_i s(k-i)=0\ :\ k>n_a \tag{1.35}$$

1.5.3 State-Space Model

The state space model for the signal is derived from Eq. (1.16) by substituting $u(k)=\delta(k)$

$$\begin{aligned}\boldsymbol{x}(k+1)&=\boldsymbol{A}\boldsymbol{x}(k)+\boldsymbol{B}\delta(k)\\ s(k)&=\boldsymbol{C}\boldsymbol{x}(k)+D\delta(k)\end{aligned} \tag{1.36}$$

1.5.4 Expression for an Impulse Response

The signal $s(k)$ is the impulse response of the state-space model $(\boldsymbol{A},\boldsymbol{B},\boldsymbol{C},D)$. Using Eq. (1.26) and setting the initial condition $\boldsymbol{x}(0)=\boldsymbol{0}$ the impulse response becomes:

$$s(k)=\begin{cases} D & k=0\\ \boldsymbol{C}\boldsymbol{A}^{k-1}\boldsymbol{B} & k>1\end{cases} \tag{1.37}$$

Models of various waveforms, including periodic, constant, exponential, sinusoids, and damped sinusoids, are derived as an output of a linear time variant system driven by the delta function.

1.5.5 Periodic Signal

Consider a periodic signal $s(k)$ with a period M.

$$s(k)=s(k+M) \tag{1.38}$$

Computing the z-transform of $s(k)$ yields

$$s(z)=\sum_{k=0}^{\infty}s(k)z^{-k}=\sum_{i=0}^{M-1}s(i)z^{-i}+\sum_{i=M}^{2M-1}s(i)z^{-i}+\sum_{i=2M}^{3M-1}s(i)z^{-i}+\sum_{i=3M}^{4M-1}s(i)z^{-i}\dots. \tag{1.39}$$

Changing the index summation we get

$$s(z)=\sum_{i=0}^{M-1}s(i)z^{-i}+\sum_{i=0}^{M-1}s(i+M)z^{-i-M}+\sum_{i=0}^{M-1}s(i+2M)z^{-i-2M}+\sum_{i=0}^{M-1}s(i+3M)z^{-i-3M}\dots. \tag{1.40}$$

Simplifying by invoking the periodicity condition $s(i)=s(i+\ell M)$ for all ℓ, we get

$$s(z)=\sum_{i=0}^{M-1}s(i)z^{-i}\left(1+z^{-M}+z^{-2M}+z^{-3M}+\cdots\right) \tag{1.41}$$

Using the power series expansion formula we get

$$s(z)=\frac{\sum\limits_{i=0}^{M-1}s(i)z^{-i}}{1-z^{-M}} \tag{1.42}$$

where the numerator and the denominator coefficients are

$$a_i = \begin{cases} 0 & i = 1, 2, \ldots, M-1 \\ -1 & i = M \end{cases} \quad \text{and} \quad b_i = \begin{cases} s(i) & i = 0, 1, 2, \ldots, M-1 \\ 0 & i = M \end{cases}.$$

The difference equation model becomes

$$s(k) - s(k-M) = b_0\delta(k) + b_1\delta(k-1) \cdots + b_{M-1}\delta(k-M+1) \tag{1.43}$$

A state-space model is obtained from the difference equation model by using the simple method of converting the difference Eq. (1.34) to the corresponding state space Eq. (1.27) with $\boldsymbol{A}, \boldsymbol{B}, \boldsymbol{C}$ being MxM, $Mx1$ and $1xM$ matrices and D being scalar given by

$$\boldsymbol{x}(k+1) = \begin{bmatrix} 0 & 1 & 0 & 0 & 0 \\ 0 & 0 & 1 & 0 & 0 \\ . & . & . & . & . \\ 0 & 0 & 0 & 0 & 1 \\ 1 & 0 & 0 & 0 & 0 \end{bmatrix} \boldsymbol{x}(k) + \begin{bmatrix} b_1 \\ b_2 \\ . \\ b_{M-1} \\ b_0 \end{bmatrix} \delta(k) \tag{1.44}$$

$$s(k) = [\,1 \quad 0 \quad . \quad 0 \quad 0\,]\,\boldsymbol{x}(k) + b_0\delta(k)$$

1.5.6 *Periodic Impulse Train*

Let $s(k)$ be the unit impulse train with a period M given by

$$s(k) = \sum_{\ell=0}^{\infty} \delta(k - \ell M) \tag{1.45}$$

A frequency-domain model is obtained by taking the z-transform of $s(k)$ and is given by

$$s(z) = \sum_{\ell=0}^{\infty} z^{-M\ell} \tag{1.46}$$

Using the power series expansion formula $\sum_{\ell=0}^{\infty} z^{-\ell M} = \frac{1}{1-z^{-M}}$ we get

$$s(z) = \frac{1}{1 - z^{-M}} \tag{1.47}$$

where $a_i = \begin{cases} 0 & i = 1, 2, \ldots, M-1 \\ -1 & i = M \end{cases}$ and $b_i = \begin{cases} 1 & i = 0 \\ 0 & else \end{cases}$.

The difference equation model becomes

$$s(k) - s(k-M) = \delta(k) \tag{1.48}$$

Similarly to the case of periodic signal, a state-space model $(\boldsymbol{A}, \boldsymbol{B}, \boldsymbol{C}, D)$ where $\boldsymbol{A}$, $\boldsymbol{B}$, $\boldsymbol{C}$ are MxM, $Mx1$ and $1xM$ matrices and D is scalar is given by

$$\begin{aligned}
\boldsymbol{x}(k+1) &= \begin{bmatrix} 0 & 1 & 0 & 0 & 0 \\ 0 & 0 & 1 & 0 & 0 \\ . & . & . & . & . \\ 0 & 0 & 0 & 0 & 1 \\ 1 & 0 & 0 & 0 & 0 \end{bmatrix} \boldsymbol{x}(k) + \begin{bmatrix} 0 \\ 0 \\ . \\ 0 \\ 1 \end{bmatrix} \delta(k) \\
s(k) &= [\,1 \quad 0 \quad . \quad 0 \quad 0\,]\,\boldsymbol{x}(k) + \delta(k)
\end{aligned} \tag{1.49}$$

Remark An impulse train is a special case of a periodic waveform with identical $\boldsymbol{A}$ and $\boldsymbol{C}$ but different $\boldsymbol{B}$. An impulse train is a mathematical artifice to model periodic waveforms including the excitation input for a voiced speech waveform and sampled waveform.

1.5.7 A Finite Duration Signal

Let $s(k)$ be some finite duration signal given by

$$s(k) = \begin{cases} b_k & 0 \le k \le M-1 \\ 0 & else \end{cases} \tag{1.50}$$

The finite duration signal is a degenerate case of a periodic signal when the period is infinite. The difference equation model is

$$s(k) = \sum_{i=0}^{M-1} b_i \delta(k-i) \tag{1.51}$$

Computing the z-transform yields

$$s(z) = \frac{\sum\limits_{i=0}^{M-1} b_i z^{-i}}{1} \tag{1.52}$$

This model may be interpreted to be a rational polynomial whose denominator is unity, that is its leading coefficient is unity when the rest of the denominator coefficients are all zeros, $a_i = 0$. The numerator coefficients are $b_i = 0$, $i = 1, 2, \ldots, M$. A state-space model $(\boldsymbol{A}, \boldsymbol{B}, \boldsymbol{C}, D)$ is given by

$$\begin{aligned}
\boldsymbol{x}(k+1) &= \begin{bmatrix} 0 & 1 & 0 & 0 & 0 \\ 0 & 0 & 1 & 0 & 0 \\ . & . & . & . & . \\ 0 & 0 & 0 & 0 & 1 \\ 0 & 0 & 0 & 0 & 0 \end{bmatrix} \boldsymbol{x}(k) + \begin{bmatrix} b_1 \\ b_2 \\ . \\ b_{M-1} \\ 0 \end{bmatrix} \delta(k) \\
s(k) &= [\,1 \quad 0 \quad . \quad 0 \quad 0\,]\,\boldsymbol{x}(k) + b_0 \delta(k)
\end{aligned} \tag{1.53}$$

Comment *The model (1.51) may also be termed a MA model or a Finite Impulse Response (FIR) filter, while the rest of the models, namely the AR and ARMA models, may be termed as Infinite Impulse Response (IIR) filters.*

1.5.8 Model of a Class of All Signals

We have so far considered the model of a given waveform, such as a constant and a sinusoid. A model that generates a class of signals is derived for (i) a class of constants of different amplitudes, (ii) sinusoids of identical frequency but different phases and amplitudes, and (iii) a class of periodic waveforms of the same period such as a square wave, triangular wave, and an impulse train. A model of a class signal is simply an unforced version of the model of a particular member of the class, and is required in many applications including the design of controller, observer, and Kalman filter. A difference equation model of a class of signal is derived by setting the forcing term to zero:

$$s(k) + \sum_{i=0}^{n} a_i s(k-i) = 0 \text{ for all } k \tag{1.54}$$

It is an unforced system and the form of the signal $s(k)$ output depends upon the initial conditions. It models a class of all signals whose z-transforms have the same denominator $D(z)$ but different numerator $N(z)$. A state-space model for a class of signals is derived from Eqs. (1.135) and (1.36) setting the forcing input to zero:

$$\begin{aligned} \boldsymbol{x}(k+1) &= \boldsymbol{A}\boldsymbol{x}(k) \\ s(k) &= \boldsymbol{C}\boldsymbol{x}(k) \end{aligned} \tag{1.55}$$

This model generates a class of signals whose state-space model $(\boldsymbol{A}, \boldsymbol{B}, \boldsymbol{C}, D)$ have the same system matrices $\boldsymbol{A}$ and $\boldsymbol{C}$ but different $\boldsymbol{B}$ and D. Using Eq. (1.26), the output $s(k)$ generated depends upon the initial condition $\boldsymbol{x}(0)$ given by:

$$s(k) = \boldsymbol{C}\boldsymbol{A}^k\boldsymbol{x}(0) \tag{1.56}$$

Table 1.1 gives the difference equation and state-space model of a typical class of all signals including constants, sinusoids, and ramps.

1.5.8.1 Noise Annihilating Operator

In many applications, noise or unwanted signals, such as a constant or ramp type bias, and 60 Hz power frequency signals are annihilated if the class to which the signal belongs is known. Let $\upsilon(k)$ be some unwanted signal whose model is given by Eq. (1.54) or Eq. (1.55). The noise annihilating operation is a simple exploitation of the noise model

$$\upsilon(k) + \sum_{i=1}^{n} a_i \upsilon(k-i) = 0 \text{ for all } k > n \tag{1.57}$$

Thus the operator $D(z) = 1 + \sum_{i=1}^{n} a_i z^{-i} = |z\boldsymbol{I} - \boldsymbol{A}|$ is a noise annihilation filter if $D(z)$ is the denominator polynomial of the noise model, or equivalently $(\boldsymbol{C}, \boldsymbol{A})$ is its state-space model.

Table 1.1 Model of a class of signal

Signal	State-space model	Difference equation model
$y(k) = \alpha$	$x(k+1) = x(k)$ $y(k) = x(k)$	$y(k) - y(k-1) = 0$
$y(k) = \beta\rho^k$	$x(k+1) = \rho x(k)$ $y(k) = x(k)$	$y(k) - \rho y(k-1) = 0$
$y(k) = \beta_0 + \beta_1 k + \beta_2 k^2$	$\mathbf{x}(k+1) = \begin{bmatrix} 2 & 1 \\ -1 & 0 \end{bmatrix} \mathbf{x}(k)$ $y(k) = \begin{bmatrix} 1 & 0 \end{bmatrix} \mathbf{x}(k)$	$y(k) - 2y(k-1) + y(k-2) = 0$
$y(k) = a\sin(\omega k + \varphi)$	$\mathbf{x}(k+1) = \begin{bmatrix} 2\cos\omega & 1 \\ -1 & 0 \end{bmatrix} \mathbf{x}(k)$ $y(k) = \begin{bmatrix} 1 & 0 \end{bmatrix} \mathbf{x}(k)$	$y(k) - 2\cos\omega y(k-1) + y(k-2) = 0$

1.5.9 Examples of Deterministic Signals

A class of signals, such as constant signals (dc signals), exponential signals, sinusoids, damped sinusoids, square waves, and triangular waves, is considered. A frequency domain model relating the z-transform of the input and that of the output, and a time model expressed in the deference equation form are obtained

Example 1.1 Constant signal:

$$s(k) = \begin{cases} 1 & k \geq 0 \\ 0 & else \end{cases} \tag{1.58}$$

The frequency-domain model becomes

$$s(z) = \frac{1}{1 - z^{-1}} \tag{1.59}$$

The time-domain model expressed in the form of a difference equation becomes

$$s(k) = s(k-1) + \delta(k) \tag{1.60}$$

The state-space model (A, B, C, D) becomes

$$\begin{aligned} x(k+1) &= x(k) + \delta(k) \\ s(k) &= x(k) + \delta(k) \end{aligned} \tag{1.61}$$

Example 1.2 Exponential signal:

$$s(k) = \begin{cases} \rho^k & k \geq 0 \\ 0 & else \end{cases} \tag{1.62}$$

Frequency-domain model is

$$s(z) = \frac{1}{1 - \rho z^{-1}} \tag{1.63}$$

Time-domain model becomes

$$s(k) - \rho s(k-1) = \delta(k) \tag{1.64}$$

The difference equation model governing the class of all constant signals is given by

$$s(k) = \rho s(k-1) \tag{1.65}$$

The initial condition $s(0)$ will determine a particular member of the class, namely $s(k) = s(0)\rho^k$. The state-space model for an exponential is given by

$$\begin{aligned} x(k+1) &= \rho x(k) + \rho\delta(k) \\ s(k) &= x(k) + \delta(k) \end{aligned} \tag{1.66}$$

Example 1.3 Sinusoid

$$s(k) = \cos(\omega_0 k + \varphi_0) \tag{1.67}$$

The frequency-domain model is

$$s(z) = \frac{(1 - z^{-1}\cos\omega_0)\cos\varphi_0 - z^{-1}\sin\omega_0\sin\varphi_0}{1 - 2z^{-1}\cos\omega_0 + z^{-2}} \tag{1.68}$$

The time-domain model is

$$s(k) - 2\cos\omega_0 s(k-1) + s(k-2) = \cos\varphi_0\delta(k) - \left(\cos\omega_0\cos\varphi_0 + \sin\omega_0\sin\varphi_0\right)\delta(k-1) \tag{1.69}$$

For the case of $s(k) = \cos\omega_0 k$ the model (1.69) becomes

$$s(k) - 2\cos\omega_0 s(k-1) + s(k-2) = \delta(k) - \cos\omega_0\delta(k-1) \tag{1.70}$$

Hence the frequency-domain model becomes

$$s(z) = \frac{(1 - z^{-1}\cos\omega_0)}{1 - 2z^{-1}\cos\omega_0 + z^{-2}} \tag{1.71}$$

The state-space model using Eq. (1.70) becomes

$$\begin{aligned} \begin{bmatrix} x_1(k+1) \\ x_2(k+1) \end{bmatrix} &= \begin{bmatrix} 2\cos\omega_0 & 1 \\ -1 & 0 \end{bmatrix} \begin{bmatrix} x_1(k) \\ x_2(k) \end{bmatrix} + \begin{bmatrix} \cos\omega_0 \\ -1 \end{bmatrix} \delta(k) \\ s(k) &= [1 \quad 0] \begin{bmatrix} x_1(k) \\ x_2(k) \end{bmatrix} + \delta(k) \end{aligned} \tag{1.72}$$

For the case of $s(k) = \sin \omega_0 k$, the model (1.69) becomes

$$s(k) - 2\cos \omega_0 s(k-1) + s(k-2) = \sin \omega_0 \delta(k-1) \tag{1.73}$$

The frequency-domain model is

$$s(z) = \frac{z^{-1} sin\omega_0}{1 - 2z^{-1}\cos \omega_0 + z^{-2}} \tag{1.74}$$

The class of all sinusoids of frequency ω_0 is a solution of the following homogenous equation

$$s(k) - 2\cos \omega_0 s(k-1) + s(k-2) = 0 \tag{1.75}$$

A state-space model using Eq. (1.73) is given by

$$\begin{aligned} \begin{bmatrix} x_1(k+1) \\ x_2(k+1) \end{bmatrix} &= \begin{bmatrix} 2\cos \omega_0 & 1 \\ -1 & 0 \end{bmatrix} \begin{bmatrix} x_1(k) \\ x_2(k) \end{bmatrix} + \begin{bmatrix} \sin(\omega_0) \\ 0 \end{bmatrix} \delta(k) \\ s(k) &= [1 \quad 0] \begin{bmatrix} x_1(k) \\ x_2(k) \end{bmatrix} \end{aligned} \tag{1.76}$$

Example 1.4 Exponentially weighted sinusoid

$$s(k) = \rho^k \cos(\omega_0 k + \varphi_0) \tag{1.77}$$

The frequency-domain model is

$$s(z) = \frac{(1 - z^{-1}\rho \cos \omega_0)\cos \varphi_0 - z^{-1}\rho \sin \omega_0 \sin \varphi_0}{1 - 2z^{-1}\rho \cos \omega_0 + \rho^2 z^{-2}} \tag{1.78}$$

The time-domain model is

$$s(k) - 2\rho \cos \omega_0 s(k-1) + \rho^2 s(k-2) = \cos \varphi_0 \delta(k) - \left(\rho \cos \omega_0 \cos \varphi_0 + \rho \sin \omega_0 \sin \varphi_0\right) \delta(k-1) \tag{1.79}$$

For the case of $s(k) = \rho^k \cos \omega_0 k$ the model becomes

$$s(k) - 2\rho \cos \omega_0 s(k-1) + \rho^2 s(k-2) = \delta(k) - \rho \cos \omega_0 \delta(k-1) \tag{1.80}$$

The state-space model is

$$\begin{aligned} \begin{bmatrix} x_1(k+1) \\ x_2(k+1) \end{bmatrix} &= \begin{bmatrix} 2\rho \cos \omega_0 & 1 \\ -\rho^2 & 0 \end{bmatrix} \begin{bmatrix} x_1(k) \\ x_2(k) \end{bmatrix} + \begin{bmatrix} \rho \cos \omega_0 \\ -\rho^2 \end{bmatrix} \delta(k) \\ s(k) &= \begin{bmatrix} 1 & 0 \end{bmatrix} \begin{bmatrix} x_1(k) \\ x_2(k) \end{bmatrix} + \delta(k) \end{aligned} \tag{1.81}$$

For the case of $s(k) = \rho^k \sin \omega_0 k$, Eq. (1.78) becomes

$$s(z) = \frac{z^{-1}\rho \sin \omega_0}{1 - 2\rho z^{-1} \cos \omega_0 + \rho^2 z^{-2}} \tag{1.82}$$

The state-space model becomes

$$\begin{bmatrix} x_1(k+1) \\ x_2(k+1) \end{bmatrix} = \begin{bmatrix} 2\rho \cos \omega_0 & 1 \\ -\rho^2 & 0 \end{bmatrix} \begin{bmatrix} x_1(k) \\ x_2(k) \end{bmatrix} + \begin{bmatrix} \rho \sin(\omega_0) \\ 0 \end{bmatrix} \delta(k)$$
$$s(k) = [1 \quad 0] \begin{bmatrix} x_1(k) \\ x_2(k) \end{bmatrix} \tag{1.83}$$

Remarks A larger class of signals may be obtained, including the class of all polynomials, exponentials, the sinusoids, and the weighted sinusoids, by

- Additive combination
- Multiplicative combination (e.g., amplitude modulated signals, narrow-band FM signals)
- Sums of products combination

Example 1.5 Periodic waveform
Square waveform
Consider a model of a square wave with period $M = 4$ with

$$s(i) = \begin{cases} 1 & i = 0, 1 \\ -1 & i = 2, 3 \end{cases} \tag{1.84}$$

The frequency-domain model of a square wave using Eq. (1.42) becomes

$$s(z) = \frac{b_0 + b_1 z^{-1} + b_2 z^{-2} + b_3 z^{-3}}{1 - z^{-4}} \tag{1.85}$$

where $b_i = s(i)$. Substituting for $\{b_i\}$ from Eq. (1.84), the difference equation model becomes

$$s(k) - s(k-4) = \delta(k) + \delta(k-1) - \delta(k-2) - \delta(k-3) \tag{1.86}$$

Using the expression for a periodic waveform Eq. (1.44) we get

$$\boldsymbol{x}(k+1) = \begin{bmatrix} 0 & 1 & 0 & 0 \\ 0 & 0 & 1 & 0 \\ 0 & 0 & 0 & 1 \\ 1 & 0 & 0 & 0 \end{bmatrix} \boldsymbol{x}(k) + \begin{bmatrix} 1 \\ -1 \\ -1 \\ 1 \end{bmatrix} \delta(k)$$
$$s(k) = [1 \quad 0 \quad 0 \quad 0]\boldsymbol{x}(k) + \delta(k) \tag{1.87}$$

Triangular wave

Consider a model of a triangular wave with period $M = 4$ with values $s(0) = 0$, $s(1) = 1$, $s(2) = 0$, $s(3) = -1$. The frequency-domain model of a square wave using Eq. (1.42) becomes

$$s(z) = \frac{b_0 + b_1 z^{-1} + b_2 z^{-2} + b_3 z^{-3}}{1 - z^{-4}} = \frac{z^{-1} - z^{-3}}{1 - z^{-4}} \tag{1.88}$$

The difference equation model becomes

$$s(k) - s(k-4) = \delta(k-1) - \delta(k-3) \tag{1.89}$$

Using the expression for a periodic waveform Eq. (1.44) we get

$$\begin{aligned} \boldsymbol{x}(k+1) &= \begin{bmatrix} 0 & 1 & 0 & 1 \\ 0 & 0 & 1 & 0 \\ 0 & 0 & 0 & 1 \\ 1 & 0 & 0 & 0 \end{bmatrix} \boldsymbol{x}(k) + \begin{bmatrix} 1 \\ 0 \\ -1 \\ 0 \end{bmatrix} \delta(k) \\ y(k) &= [\,1 \quad 0 \quad 0 \quad 0\,]\boldsymbol{x}(k) \end{aligned} \tag{1.90}$$

Figure 1.5 shows typical deterministic signals, namely exponential, sinusoid, damped sinusoid, square wave, and triangular wave. Signals are generated as outputs of a linear system excited by delta functions. The top four figures are the impulse responses while the bottom four figures are the corresponding Fourier transforms.

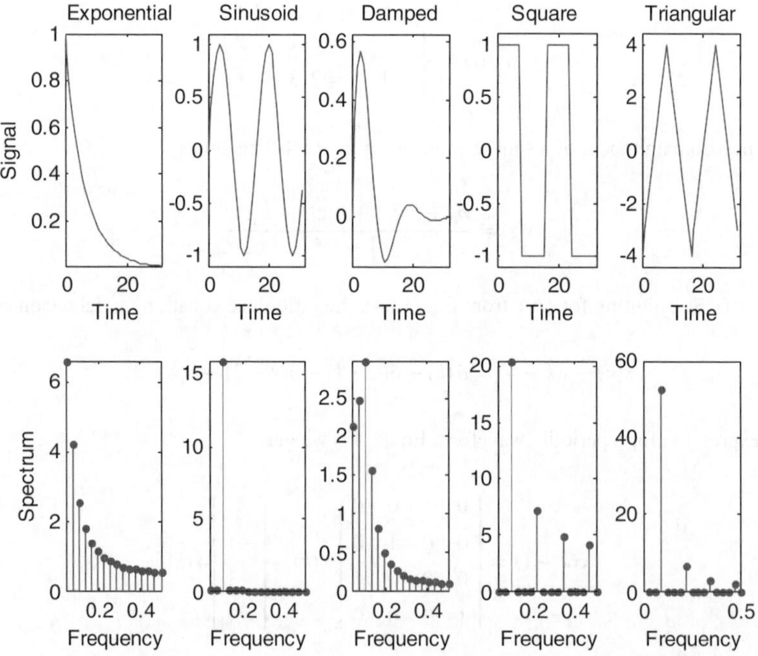

Figure 1.5 Typical examples of deterministic signals and their magnitude spectra

Typical examples of periodic waveform include speech waveform (in particular vowels), and biological waveforms such as phonocardiogram waveform (heart sounds) and Electrocardiogram waveform (ECG). One way to identify vowels is to verify whether their DTFT is a line spectrum. Likewise, from the line spectrum of the heart sound one can determine the heartbeat.

1.6 Introduction to Random Signals

We will restrict the random signal – also called a random process, a random waveform, or a stochastic process – to a discrete-time process. A random signal, denoted $x(k,\xi)$, is a function of two variables, namely time index $k \in \Gamma \subseteq \mathfrak{R}$. and an outcome $\xi \in S$ of an experiment where $\mathfrak{R}$ is a field of real numbers, Γ is a subset of $\mathfrak{R}$ which in the discrete-time case is a set of all integers, and S is the sample space of all outcomes of an experiment plus the null outcome, $x(k,\xi) : Sx\Gamma \to \mathfrak{R}$.

1. $x(k,\xi)$ is a family of functions or an *ensemble* if both outcome ξ and time k vary.
2. $x(k,\xi_i)$ is a time function if $\xi = \xi_i$ is fixed and k varies. It is called a *realization* of the stochastic process.
3. $x(k,\xi)$ is a *random variable* if $k = k_i$ is fixed and ξ varies.
4. $x(k,\xi)$ is a *number* if both $\xi = \xi_i$ and $k = k_i$.

Generally, the dependence on ξ is not emphasized: a stochastic process is denoted by $x(k)$ rather than $x(k,\xi)$. The random signal at a fixed time instant $k = i$, is a random variable and its behavior is determined by the PDF $f_X(x(i), i)$ which is function of both the random variable $x(i)$ and the time index i. As a consequence, the mean, the variance, and the auto-correlation function will be functions of time. The mean $\mu_x(i)$, and the variance $\sigma_x^2(i)$ of $x(k)$ at $k = i$ are

$$\mu_x(i) = E\,[x(i)] \tag{1.91}$$

$$\sigma_x^2(i) = E[(x(i) - \mu_x(i))^2] \tag{1.92}$$

The auto-cross-correlation $r_{xx}(i,j)$ of $x(k)$ evaluated at $k = i$ and $k = j, j > i$

$$r_{xx}(i,j) = E\,[x(i)x(j)] \tag{1.93}$$

1.6.1 Stationary Random Signal

A random signal $x(k)$ is strictly stationary of order p if its joint PDF of any p random variables $\{x(k_i) : i = 1, 2, \ldots, p\}$ where $k_i : i = 1, 2, \ldots, p$ is a subsequence of the time index k remains shift invariant for all $k_i : i = 1, 2, \ldots, p$ and time shift m

$$f_X(x(k_1), x(k_2), \ldots, x(k_p)) = f_X(x(k_1 + m), x(k_2 + m), \ldots, x(k_p + m)) \tag{1.94}$$

The joint statistics of $\{x(k_i) : i = 1, 2, \ldots, p\}$ and $\{x(k_i + m) : i = 1, 2, \ldots, p\}$ are the same for any p subsequence and any time shift:

1. A stochastic process is strict sense stationary of order $p = 1$ if for all k and m

$$f_X\,(x(k)) = f_X\,(x(k + m)) \tag{1.95}$$

The statistics of $x(k)$ will the same for all k and m. For example, the mean and the variance will be time invariant

$$E[x(k)] = E[x(k+m)] = \mu_x \tag{1.96}$$

$$E[(x(k)-\mu_x)^2] = E[(x(k+m)-\mu_x)^2] = \sigma_x^2 \tag{1.97}$$

2. A stochastic process is strict sense stationary of order $p = 2$ if for all time indices k and ℓ, and time shift m

$$f_X(x(k), x(\ell)) = f_X(x(k+m), x(\ell+m)) \tag{1.98}$$

The statistics of $x(k)$ will the same for all k and m. For example, the mean and the variance will be time invariant while the auto-correlation will be a function of the time difference $k - \ell$:

$$E[x(k)] = E[x(\ell)] = \mu_x \tag{1.99}$$

$$E[(x(k)-\mu_x)^2] = E[(x(\ell)-\mu_x)^2] = \sigma_x^2 \tag{1.100}$$

$$r_{xx}(k,\ell) = E[x(k)x(\ell)] = r_{xx}(k-\ell) \tag{1.101}$$

3. If the stochastic process is stationary of the order p, then the mean and the variance will be will be constant, the auto-correlation will be a function of the time difference and higher order moments will be the shift invariant.

1.6.1.1 Wide-Sense Stationary Random Signal

A special case of stationary random signal, termed *wide-sense stationary random signal*, satisfies only stationary properties of the mean, variance, and auto-correlation function. In many practical applications, random signals are assumed to be wide-sense rather strictly stationary to simplify the problem without affecting the acceptable accuracy of the model. The mean and the variance are assumed constant while the correlation is assumed to be a function of the time difference similar to those of strict-sense stationary random signal of order $p = 2$. The mean and the variance are constant and the correlation is a function of the time difference given by

$$E[x(k)] = \mu_x \tag{1.102}$$

$$E[(x(k)-\mu_x)^2] = \sigma_x^2 \tag{1.103}$$

$$E[x(k)x(\ell)] = r_{xx}(k-\ell) \tag{1.104}$$

A stochastic process that is stationary of the order 2 is wide-sense stationary. However a wide-sense stationary is not stationary of the order 2 unless it is a Gaussian process. Further, a wide-sense stationary process is also strict-sense stationary if the PDF of the random signal is a Gaussian process.

1.6.2 Joint PDF and Statistics of Random Signals

We will extend the characterization of a signal to two random signals. Let two random signals $x(k)$ and $y(k)$ be defined at different time instants by p random variables $\{x(k_i) : i = 1, 2, \dots, p\}$ where $k_i : i = 1, 2, \dots, p$ and q random variable $\{y(\ell_i) : i = 1, 2, \dots, q\}$ where $\ell_i : i = 1, 2, \dots, q$. These random variables are completely characterized by the joint PDF $f_{XY}(x(k_1), x(k_2), \ldots, x(k_p), y(\ell_1), y(\ell_2), \ldots, y(\ell_q))$. Random signals $x(k)$ and $y(k)$ are statistically independent if

$$\begin{aligned} & f_{XY}\left(x(k_1), x(k_2), \ldots, x(k_p), y(\ell_1), y(\ell_2), \ldots, y(\ell_q)\right) \\ & = f_X\left(x(k_1), x(k_2), \ldots, x(k_p)\right) f_Y\left(y(\ell_1), y(\ell_2), \ldots, y(\ell_q)\right) \end{aligned} \tag{1.105}$$

1.6.2.1 Strict-Sense Stationary of Order *p*

Two random signals $x(k)$ and $y(k)$ are strictly stationary of order p if their joint PDF of any p random variables $\{x(k_i) : i = 1, 2, \ldots, p\}$ where $k_i : i = 1, 2, \ldots, p$, and any p random variable $\{y(\ell_i) : i = 1, 2, \ldots, p\}$ where $\ell_i : i = 1, 2, \ldots, p$ is a subsequence of the time index ℓ remains shift invariant for all $k_i : i = 1, 2, \ldots, p$, $\ell_i : i = 1, 2, \ldots, p$ time shift m

$$\begin{aligned} & f_{XY}\left(x(k_1+m), x(k_2+m), \ldots, x(k_p+m), y(\ell_1+m), y(\ell_2+m), \ldots, y(\ell_p+m)\right) \\ & = f_{XY}\left(x(k_1), x(k_2), \ldots, x(k_p), y(\ell_1), y(\ell_2), \ldots, y(\ell_p)\right) \end{aligned} \tag{1.106}$$

Strict-sense stationary of orders 1 and 2 are defined similar to Eqs. (1.95) and (1.98). As a consequence the expectation of a product of (i) any function $h_x(x(k))$ of $x(k)$ and (ii) any function $h_y(y(k))$ of $y(k)$ is a product of their expectations

$$E[h_x(x(k))h_y(y(k)] = E[h_x(x(k))]E[h_y(y(k)] \tag{1.107}$$

1.6.2.2 Wide-Sense Stationary Random Signals

The mean μ_x and the variance σ_x^2 of $x(k)$, and the mean μ_y and the variance σ_y^2 of $y(k)$ are all assumed constant, while the cross-correlation is assumed to be a function of the time difference:

$$r_{xy}(k-\ell) = E\left[x(k)y(\ell)\right] \tag{1.108}$$

1.6.2.3 Quasi-Stationary Ergodic Process

In practical applications, a stochastic process $x(k)$ may have a deterministic component:

$$x(k) = x_r(k) + x_d(k) \tag{1.109}$$

where $x_r(k)$ is a purely random process with zero mean and variance σ^2, and $x_d(k)$ is a purely deterministic process. The mean, the variance, and the correlation of $x(k)$ are given by:

$$E\left[x(k)\right] = x_d(k) \tag{1.110}$$

$$E[(x(k) - \mu_x)^2] = \sigma^2 \tag{1.111}$$

$$\begin{aligned} r_{xx}(k,\ell) &= E[x(k)x(\ell)] = E[(x_r(k) + x_d(k))(x_r(\ell) + x_d(\ell))] \\ &= E[x_r(k)x_r(\ell)] + E[x_r(k)x_d(\ell)] + E[x_d(k)x_r(\ell)] + E[x_d(k)x_d(\ell)] \end{aligned} \tag{1.112}$$

Since $x_r(k)$ is a zero-mean and stationary process, and $\ell = k - m$, we get:

$$E[x(k)x(k-m)] = r_{x_r x_r}(m) + x_d(k)x_d(k-m) \tag{1.113}$$

Clearly the stochastic process with the deterministic component is not stationary as the mean is a function of the time and the correlation does not depend upon the time lag $m = k - \ell$, although the variance is constant.

Let us consider an example where the deterministic component is a sinusoid of frequency f, $x_d(k) = \sin(2\pi fk)$, and the random component is a zero-mean stationary stochastic process with variance $\sigma_{x_r}^2$ given by:

$$x(k) = v(k) + \sin(2\pi fk) \tag{1.114}$$

The mean, variance, and the correlation of $x(k)$ are given by:

$$E\left[x(k)\right] = \sin\left(2\pi fk\right) \tag{1.115}$$

$$E\left[\left(x(k) - \mu_x\right)^2\right] = \sigma_{x_r}^2 \tag{1.116}$$

$$r_{xx}\left(k, k-m\right) = r_{x_r x_r}\left(m\right) + \sin\left(2\pi fk\right)\sin\left(2\pi f\left(k-m\right)\right) \tag{1.117}$$

Using the trigonometric formula, $\sin A \sin B = \frac{1}{2}\left[\cos(A - B) - \cos(A + B)\right]$, we get:

$$r_{xx}\left(k, k-m\right) = r_{x_r x_r}\left(m\right) + \frac{1}{2}\cos\left(2\pi fm\right) - \frac{1}{2}\cos\left(2\pi f\left(2k-m\right)\right) \tag{1.118}$$

Remark Except when the deterministic component is a constant, a stochastic process which is the sum of a zero-mean stationary stochastic process and a deterministic component is not stationary using the classical definition of stationarity. In practice this case is prevalent, and hence the definition of stationarity is extended to handle this situation. The following definition is widely employed:

Definition *A random signal $x(k)$ is said to be quasi-stationary [1] if the mean $\mu_x\left(k\right) = E\left[x(k)\right]$ and correlation function $r_{xx}(k, k-m) = E\left[x(k)x(k-m)\right]$ are bounded:*

$$\left|\mu_x(k)\right| < \infty \tag{1.119}$$

$$\left|r_{xx}\left(k, k-m\right)\right| < \infty \tag{1.120}$$

$$\lim_{N\to\infty}\left\{\frac{1}{N}\sum_{k=0}^{N-1} r_{xx}(k, k-m)\right\} \to r_{xx}\left(m\right) \tag{1.121}$$

Let us verify whether the example Eq. (1.114) of a stochastic process with deterministic sinusoidal component is quasi-stationary. Using the expression for the correlation given in Eq. (1.118), taking an average of N *correlations yields:*

$$\frac{1}{N}\sum_{k=0}^{N-1} r_{xx}\left(k, k-m\right) = \frac{1}{N}\sum_{k=0}^{N-1}\left[r_{x_r x_r}\left(m\right) + \frac{1}{2}\cos\left(2\pi fm\right) - \frac{1}{2}\cos\left(2\pi f\left(2k-m\right)\right)\right] \tag{1.122}$$

Simplifying we get:

$$\frac{1}{N}\sum_{k=0}^{N-1} r_{xx}\left(k, k-m\right) = r_{x_r x_r}\left(m\right) + \frac{1}{2}\cos\left(2\pi fm\right) - \frac{1}{2}\frac{1}{N}\sum_{k=0}^{N-1}\cos\left(2\pi f\left(2k-m\right)\right) \tag{1.123}$$

Let us consider the summation term on the right. Let L be the maximum number of periods of the sinusoid contained in the time interval $(0 : N-1)$ *such that* $N - LM \leq M$, $M = 1/f$ *is the period of the sinusoid. The last term becomes:*

$$\frac{1}{N}\sum_{k=0}^{N-1}\cos(2\pi f(2k-m)) = \frac{1}{N}\sum_{k=0}^{LM-1}\cos(2\pi f(2k-m)) + \frac{1}{N}\sum_{k=LM}^{N}\cos(2\pi f(2k-m)) \quad (1.124)$$

On the right, the first term is zero and the second term is a constant. Taking the limit as $N \to \infty$, *expression (1.123) becomes:*

$$\lim_{N\to\infty}\frac{1}{N}\sum_{k=0}^{N-1}r_{xx}(k,k-m) = r_{x_rx_r}(m) + \frac{1}{2}\cos(2\pi fm) - \lim_{N\to\infty}\frac{1}{2}\frac{1}{N}\sum_{k=0}^{N-1}\cos(2\pi f(2k-m)) \quad (1.125)$$

The last term on the right is zero and we get:

$$\lim_{N\to\infty}\frac{1}{N}\sum_{k=0}^{N-1}r_{xx}(k,k-m) = r_{x_rx_r}(m) + \frac{1}{2}\cos(2\pi fm) \quad (1.126)$$

This shows that the stochastic process with a sinusoidal component is quasi-stationary.

Remark If (i) the mean and the correlation of the stochastic process are bounded, (ii) the purely random component is stationary, and (iii) the deterministic is a constant then the stochastic process with a deterministic component is stationary. If the stochastic process is stationary, then it is quasi-stationary, and not *vice versa*. The quasi-stationary finds applications in many areas including system identification.

1.6.2.4 Orthogonal Random Signals

Random signals $x(k)$ and $y(k)$ are orthogonal if the cross-correlation is zero

$$r_{xy}(m) = 0 \text{ for all } m \quad (1.127)$$

1.6.2.5 Uncorrelated Random Signals

Random signals $x(k)$ and $y(k)$ are uncorrelated if

$$C_{xy}(m) = E[(x(k)-\mu_x)(y(k-m)-\mu_y)] = 0 \quad (1.128)$$

where $C_{xy}(m) = E[(x(k)-\mu_x)(y(k-m)-\mu_y)]$ is called cross-covariance of $x(k)$ and $y(k)$. The covariance and correlation are identical if the random signals are zero-mean.

Comment *The definition of the uncorrelated random variable is confusing as it is defined in terms of the covariance instead of correlation, while orthogonality is defined in terms of correlation function.*

1.6.3 Ergodic Process

The statistics of a random signal is obtained by averaging over an ensemble of realizations (a collection of all realizations of random signals) to estimate the statistical parameters such as the mean, variance, and auto-correlation. In practice one does not have an ensemble of realizations and only a segment of a single realization is available. From a practical point of view, characterizing a signal as random is useful only if we can estimate its statistical parameters from a single realization of the data. Fortunately for a class of random processes, one may be able to estimate the statistical parameters from a single realization, and the process whereby the statistical parameters may be estimated from one realization is called an *ergodic process*. In other words a random process is ergodic if the statistics can be determined from a single realization of the process, that is, an ensemble average may be estimated using a time average of a single realization. In many applications, it is sufficient to estimate the mean, the variance, and the correlation from a single realization. Hence the definition of ergodicity is restricted to the statistical parameters, namely the mean and the auto-correlation, and hence the stochastic process is defined as *mean-ergodic* if the mean is ergodic and *correlation-ergodic* if the correlation is ergodic. The property of ergodicity makes sense only for a stationary process. Let the estimates of the mean, variance, and auto-correlation of $x(k)$ be respectively $\hat{\mu}_x$, $\hat{\sigma}_x^2$, $\hat{r}_{xx}(m)$ and the estimate of the cross-correlation of $x(k)$ and $y(k)$ is $\hat{r}_{xy}(m)$. The above estimates $\hat{\mu}_x$, $\hat{\sigma}_x^2$, $\hat{r}_{xx}(m)$ are computed using a single realization of the random signal $x(k)$, while $\hat{r}_{xy}(m)$ is computed from single realizations of $x(k)$ and $y(k)$. The ensemble averages, which are the true statistical averages of the mean, variance, auto-correlation, and cross-correlation, are respectively μ_x, σ_x^2, $r_{xx}(m)$, and $r_{xy}(m)$.

1. $x(k)$ is *ergodic in the mean* if $\hat{\mu}_x$ is asymptotically equal to μ_x

$$\hat{\mu}_x = \underset{N\to\infty}{\lim it}\frac{1}{N}\sum_{k=0}^{N-1} x(k) = E\left[x(k)\right] = \mu_x \tag{1.129}$$

2. $x(k)$ is *ergodic in the variance* if $\hat{\sigma}_x^2$ is asymptotically equal to σ_x^2

$$\hat{\sigma}_x^2 = \underset{N\to\infty}{\lim it}\frac{1}{N}\left(\sum_{k=0}^{N-1} x(k) - \hat{\mu}_x\right)^2 = E\left[\left(x(k) - \mu_x\right)^2\right] = \sigma_x^2 \tag{1.130}$$

3. $x(k)$ is *ergodic in the auto-correlation* if $\hat{r}_{xx}(m)$ is asymptotically equal to $r_{xx}(m)$

$$\hat{r}_{xx}(m) = \underset{N\to\infty}{\lim it}\frac{1}{N}\sum_{k=0}^{N-1} x(k)x(k-m) = E\left[x(k)x(k-m)\right] \tag{1.131}$$

4. $x(k)$ and $y(k)$ are *ergodic in the cross-correlation* if $\hat{r}_{xy}(m)$ is asymptotically equal to $r_{xy}(m)$

$$\hat{r}_{xy}(m) = \underset{N\to\infty}{\lim it}\frac{1}{N}\sum_{k=0}^{N-1} x(k)y(k-m) = E\left[x(k)y(k-m)\right] \tag{1.132}$$

In other words, a random process is ergodic if the time average equals the ensemble averages. When a process is ergodic, all statistical information such as the mean, the variance, and the autocorrelation may be derived from just one realization.

Figure 1.6 shows an ensemble of a stochastic process formed of four realizations. The figure on the left shows the four realizations of $x(k)$ and that on the right of $y(k)$. Values of the random signal, $x(k)$ and $y(k)$ at time instants $k = 50$ and $k = 60$ are indicated on each of the four realizations.

1.7 Model of Random Signals

A stochastic process is modeled similarly to a deterministic waveform. Unlike a delta function, an elementary process, called a white noise process, forms a building block for generating a large class of random signals.

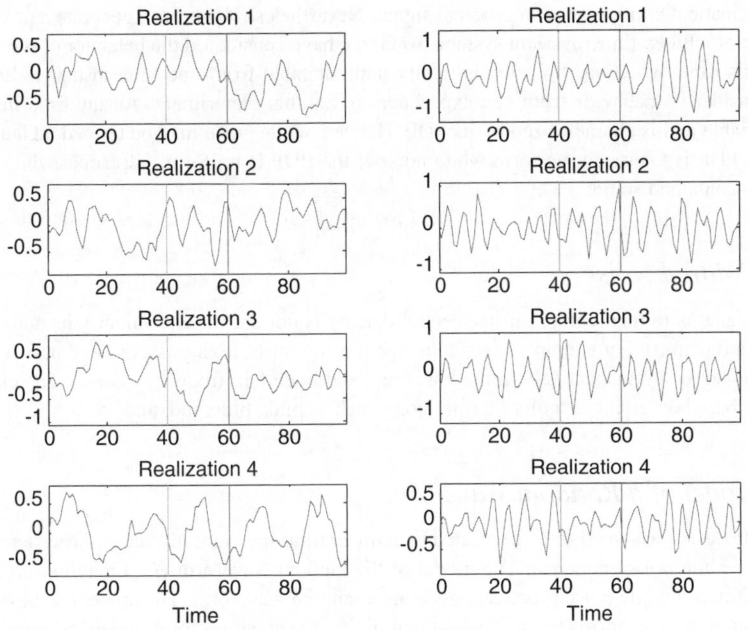

Figure 1.6 Ensembles of two random signals

1.7.1 White Noise Process

This is an elementary signal that serves as a building block to generate a random signal similar to the delta function: the delta function is a building block for a deterministic signal whereas the white noise is a building block for the random signal. White noise is a sequence of independent and identically distributed random (i.i.d.) variables.

1.7.1.1 Properties

1. It is a zero-mean process

$$E[v(k)] = \mu_v = 0 \tag{1.133}$$

2. The correlation function is a delta function

$$r_{vv}(\mathrm{m}) = E\left[v(k)v(k-m)\right] = \sigma_v^2\delta(\mathrm{m}) = \begin{cases} \sigma_v^2 & m = 0 \\ 0 & m \neq 0 \end{cases} \tag{1.134}$$

3. Equivalently its power spectral density (Fourier transform of auto-correlation) is a constant

$$P_{vv}(f) = \Im(r_{vv}(\mathrm{m})) = \sigma_v^2 \text{ for all } f \tag{1.135}$$

This sequence is historically called white because of its analogy to white light, which contains all frequencies with constant power spectrum. The white noise sequence is purely random: knowing, $v(\mathrm{k})$ in no way can be used to predict $v(\mathrm{k}+1)$, that is conditional probability

$$P\{v(k)|v(k-1), v(k-2), \ldots v(2), v(1), v(0)\} = P\{v(k)\} \tag{1.136}$$

It is a mathematical entity and not a physical signal. Nevertheless it serves to generate a physical signal as an output of a linear time invariant system. So far we have considered the behavior of the white noise as regards its time evolution. Its value is totally unpredictable from one time instant to another time instant. Consider its behavior from one experiment to another experiment. At any time instant it is a random variable and is characterized by its PDF. Hence a white noise may be termed a Gaussian white noise if the PDF is Gaussian, uniform white noise if the PDF is uniform, Laplacian white noise if its PDF is Laplacian, and so on.

1.7.2 Colored Noise

A random signal is termed colored if its spectral density is not flat (or equivalently its auto-correlation is not a delta function). For example, a random signal is low-pass, high-pass, or band pass if its spectral bandwidth is dominant respectively in the low, high, and in certain frequency bands. The classification may also be based on giving a "color" terminology such as pink, blue, red, and so on.

1.7.3 Model of a Random Waveform

A random process is assumed to be generated by a linear time-invariant discrete-time system excited by a zero-mean white noise sequence. The model of the random waveform $H(z)$ may be referred to as a *filter* which filters a white noise process to generate a random waveform. The model can be expressed in (i) input–output transfer form (ii) an input–output difference equation form, or (iii) a state-space form. The difference equation model takes the form of an ARMA, AR, or MA model:

1. *Input–output transfer function model*

$$s(z) = H(z)v(z) \tag{1.137}$$

where $H(z) = \dfrac{N(z)}{D(z)} = \dfrac{b_0 + b_1 z^{-1} + b_2 z^{-2} + \cdots + b_{n_b} z^{-n_b}}{1 + a_1 z^{-1} + a_2 z^{-2} + a_3 z^{-3} + \cdots + a_{n_a} z^{-n_a}}$

2. *Difference equation model*
   The difference equation model expressed as an ARMA, AR, and MA model is similar to those given in Eqs. (1.9), (1.13), and (1.15) except that the input driving the model is a zero mean white noise process, $u(k) = v(k)$.
3. *State-space model*
   The state variable model is formed of (i) a state equation which is a set of n simultaneous first-order difference equations relating n states $x(k)$ and the input $v(k)$, and (ii) an output equation which is an algebraic equation that expresses the outputs $s(k)$ as a linear combination of the states $x(k)$ and the current input $v(k)$. It is a vector-matrix equation given by

$$\begin{aligned} \boldsymbol{x}(k+1) &= \boldsymbol{A}\boldsymbol{x}(k) + \boldsymbol{B}v(k) \\ s(k) &= Cx(k) + Dv(k) \end{aligned} \tag{1.138}$$

where $\boldsymbol{x}(k)$ is $nx1$ vector of states, $u(k)$ and $y(k)$ are scalars; $\boldsymbol{A}$ is nxn matrix, $\boldsymbol{B}$ is $nx1$ vector, and $\boldsymbol{C}$ is $1xn$ vector; $\boldsymbol{x}(k) = [\,x_1(k)\;\; x_2(k)\;\; x_3(k)\;\; \ldots\;\; x_n(k)\,]^T$.

1.7.3.1 Ergodicity of a Filtered White Noise Process

A class of random process $x(k) = x_r(k) + x_d(k)$ formed of pure random and deterministic components are an ergodic process if the following conditions hold:

1. $x_r(k)$ is a filtered white noise process, $x_r(z) = H(z)v(z)$
2. $x(k)$ is quasi-stationary.

The proof that $x(k)$ is ergodic is given in [1].

1.7.4 Classification of the Random Waveform

The power spectral density captures the frequency content of a random process. The power spectral density is the Fourier transform of the correlation, and is widely used as it is computationally efficient to implement. A power spectral density indicates how the energy/power is distributed over the frequency. The power spectral analysis finds application in many fields including medical diagnosis, speech recognition, performance monitoring, fault diagnosis, and market analysis to name a few.

It is assumed that the random signal is stationary and ergodic in the mean and correlation. Then the power spectral density given by

$$P_{ss}(f) = \lim_{N\to\infty} it \frac{1}{2N+1} E[|s_N(f)|^2] \tag{1.139}$$

where $s_N(f) = \sum_{k=-N}^{N} s(k)e^{-2\pi fk}$. Using the expression (1.137) we get

$$s_N(f) = H(f)v_N(f) \tag{1.140}$$

Substituting Eq. (1.140) in Eq. (1.139) we get

$$P_{ss}(f) = |H(f)|^2 \lim_{N\to\infty} it \frac{1}{2N+1} E\left[|v_N(f)|^2\right] = |H(f)|^2 P_{vv}(f) \tag{1.141}$$

Using the expression for $P_{vv}(f)$ of a zero-mean white noise process given in Eq. (1.135) we get

$$P_{ss}(f) = |H(f)|^2 \sigma_v^2 \tag{1.142}$$

where $P_{ss}(f)$ is the power spectral density of the random process $s(k)$. In view of the expression for the power spectral density Eq. (1.142), the power spectral density depends upon the magnitude-squared of the frequency transfer function $|H(f)|^2$. In other words, the spectral content of power spectral density of the random process depends upon the magnitude response $|H(f)|^2$ of the filter. For example, a random process is low, high, band-pass, or band-stop if $H(f)$is a low, high, band-pass, or band-stop filter respectively.

The time and the frequency domain behavior of the filter $H(z)$ are given in a later section.

A random signal is generated as an output of a band-pass filter excited by a zero-mean *white noise process* $v(k)$ as shown in Figure 1.7. A white noise input $v(k)$ and its power spectral density $P_{vv}(f)$, which is the magnitude-squared Fourier Transform of $v(k)$, where f is the frequency, is shown on the left at the input side while the output $s(k)$ and its power spectral density $P_{ss}(f)$ are shown on the right at the output side of the block diagram. The input–output block diagram is sandwiched between the input and output plots.

1.7.5 Frequency Response and Pole-Zero Locations

The Fourier transform of a discrete time signal, termed the Discrete-Time Fourier Transform, is the z-transform of the signal evaluated on the unit circle in the z-domain. The relation between the z-transform and the Discrete-Time Fourier Transform (DTFT) variables is $z = e^{j\omega} = e^{j2\pi f}$, where $f = F_c/F_s$

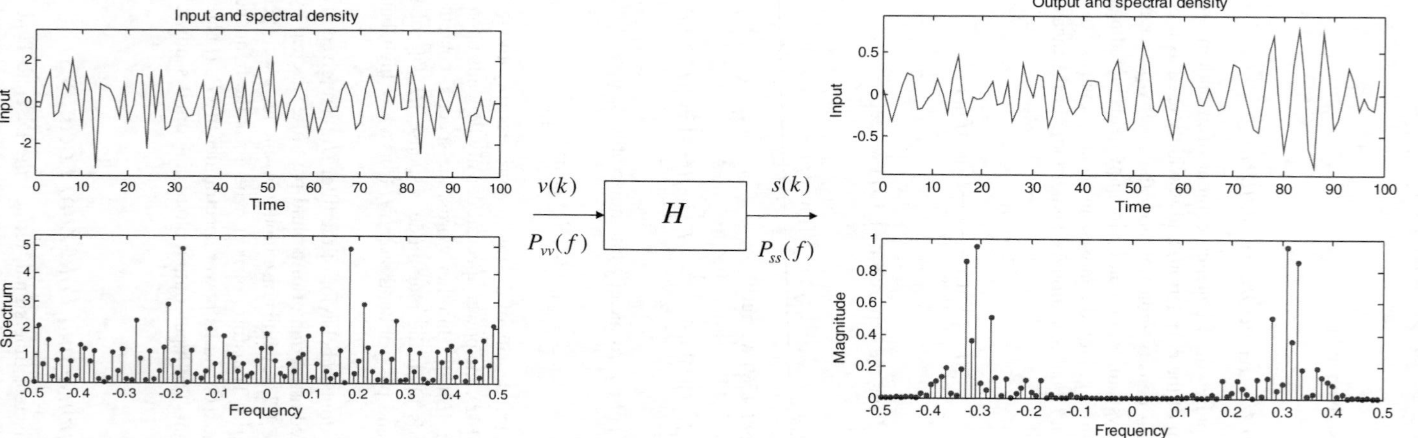

Figure 1.7 Input–output model of a random signal

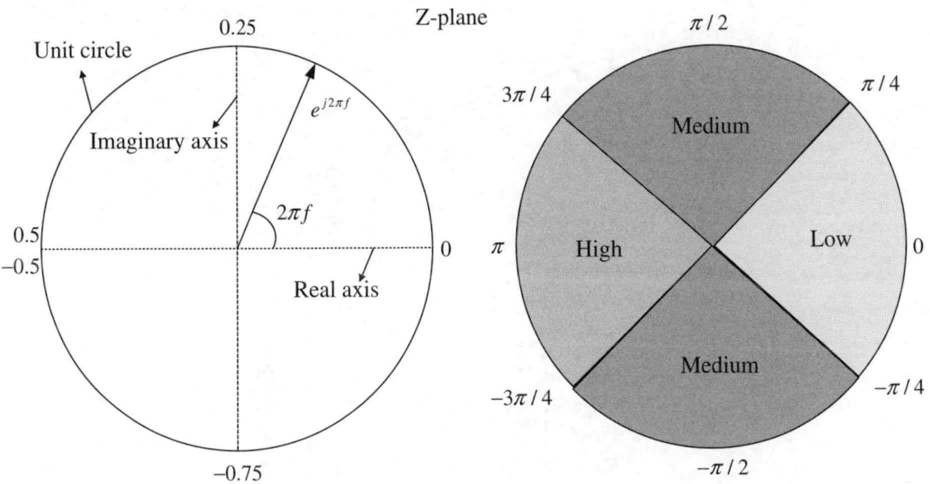

Figure 1.8 DTFT and the z-transform, and low-, mid-, and high-frequency sectors

is a normalized frequency where F_c and F_s are respectively the frequency of the continuous-time signal and sampling rate. The units of f, F_c and F_s are respectively cycles per sample, cycles per second, and samples per second, while ω has the unit of radians per sample. See the subfigure on the left of Figure 1.8. The frequency f is restricted to the interval $-0.5 \leq f \leq 0.5$ whereas the frequency of the continuous time signal F_c covers the entire real line $(-\infty \quad \infty)$. Further DTFT of a discrete-time signal, $s(f)$ is periodic with period unity, $s(f) = s(f + \ell)$ where ℓ is an integer. To obtain the Fourier Transform the relationship between z and ω namely $z = e^{j2\pi f}$ is used. The subfigure on the left in Figure 1.8 shows the unit vector $z = e^{j2\pi f}$ sweeping a unit circle in the z-plane as the frequency variable f varies in the range $-0.5 \leq f \leq 0.5$. The frequency f is the angle of vector $z = e^{j2\pi f}$ measured with respect to the real axis. The frequency regions defined arbitrarily as (i) $-0.125 \leq f < 0.125$ (ii) $0.125 \leq f < 0.375$, and $-0.375 \leq f < -0.125$ (iii) $-0.375 < f \leq 0.375$ characterize a low-frequency, a mid-frequency, and a high-frequency region. Equivalently, the sectors swept by the unit vector $z = e^{j2\pi f}$ with phase angle $\varphi = 2\pi f$ covering $-\pi/4 \leq \varphi < \pi/4$, $\pi/4 \leq \varphi < 3\pi/4$, and $-3\pi/4 \leq \varphi < -\pi/4$, and $-3\pi/4 < \varphi < 3\pi/4$ represent low-, medium-, and high-frequency sectors as shown in Figure 1.8 on the right.

1.7.5.1 Evaluation of the Magnitude Response

The magnitude response of an LTI system at a given frequency may be easily evaluated using vectors from the zeros and poles of the system to a point on the unit circle. The magnitude of the frequency response $|H(f)|$ is:

$$|H(f)| = \frac{\prod_i N_i}{\prod_i D_i} \tag{1.143}$$

where N_i and D_i are respectively the distance from the ith zero and the jth pole to the tip of the unit vector $z = e^{j2\pi f}$ as shown in 1.9 for a low, a high, a band-pass, and a notch filter.

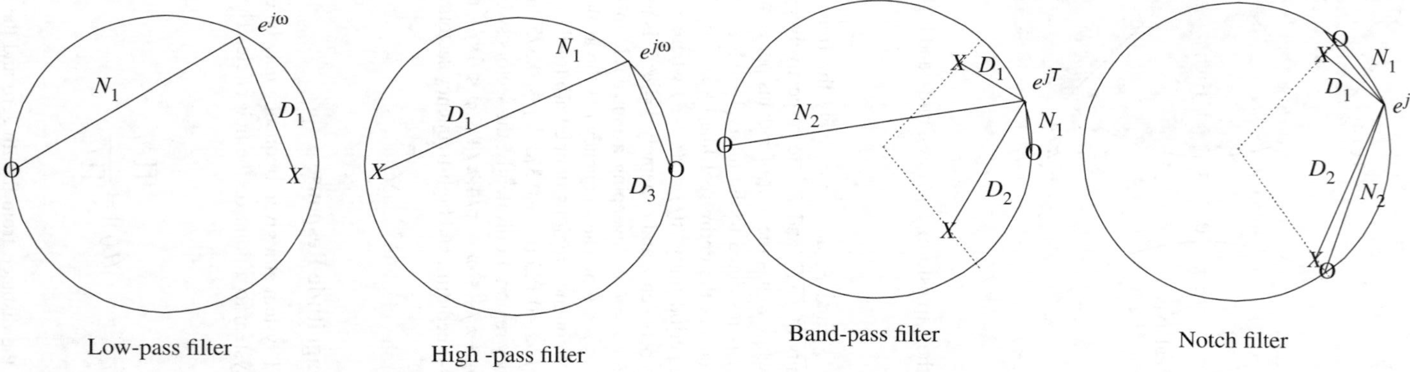

Figure 1.9 Magnitude response evaluation: low-, high-, band-pass, and notch filters

1.7.5.2 Pole-zero Locations and Frequency Response

The frequency response of a system depends upon the location of is poles and zeros. A peak in the frequency response results from complex conjugate poles close to the unit circle, while a valley results from complex conjugate zeros close to the unit circle. Complex conjugate poles and zeros close to each other will create a narrow peak if the poles are close to the unit circle and a notch if the zeros are closer to the unit circle. A peak in the frequency response will amplify spectral components of the input close to the frequency of the peak; and a valley in the frequency response will attenuate spectral components close to the frequency of the valley.

The unit disk in the z-plane may be divided into low-frequency, high-frequency, and mid-frequency regions as indicated in Figure 1.8 by low, high, and mid. If the poles (zeros) are located in low-frequency regions this will enhance (attenuate) the low frequency components of the input.

Assumption: minimum phase model *The LTI discrete-time model of the random signal $H_s(z)$ is assumed to be a minimum phase transfer function with both the poles and the zeros located strictly inside the unit circle, unlike the case of deterministic signal model where the model may have poles on the unit circles and there is no restriction on the location of the zeros. The minimum phase system and its inverse are both stable and causal. The rationale for the assumption of minimum phase stems from the following properties of a minimum phase system.*

In many applications, including system identification, signal detection, pattern classification, and Kalman filer applications, a colored random noise corrupting the data is converted to a zero-mean white noise process by filtering the input and the output by a *whitening filter*. A whitening filter is the inverse of the random signal model. Consider the random signal generated as an output of minimum phase system $H_s(z)$ given by Eq. (1.137). Multiplying the random signal output $s(z)$ by the whitening filter $H_s^{-1}(z)$ we get

$$s_f(z) = v(z) \tag{1.144}$$

where $s_f(z) = H_s^{-1}(z)s(z)$ is the filtered random signal. The filtered random signal $s_f(k)$ is a zero-mean white noise process and is also termed an *innovation process*. In Figure 1.10 the model of the colored random signal is shown on the left while the whitened random signal obtained by the whitening operation is shown on the right.

A similar approach is employed to recover a signal distorted by some LTI system.

In many cases, a magnitude squared transfer function model (or its subsystem transfer function model such as that of the signal) is identified from the power spectral densities of the input and the output data. The phase information in the data is ignored. It can be shown that the transfer function model can be determined uniquely from the identified magnitude squared transfer function if the system is minimum phase.

There is no loss of generality in assuming a system to be minimum phase, as any rational stable transfer function $H(z)$ may be decomposed into a minimum phase transfer function $H_{\min}(z)$ and an all pass transfer functions $H_{all}(z)$:

$$H(z) = H_{\min}(z)\, H_{all}(z) \tag{1.145}$$

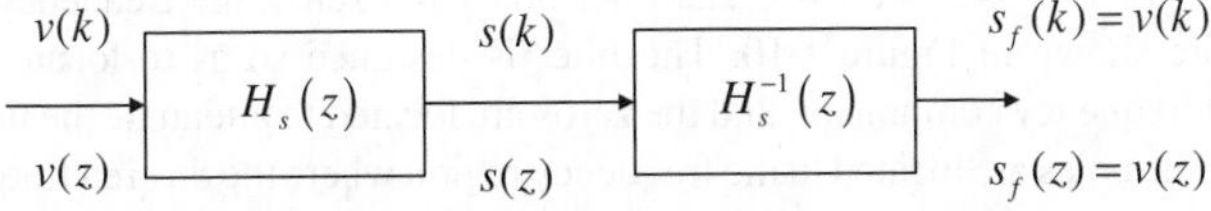

Figure 1.10 Generation of random signal and the whitened random signal

1.7.6 Illustrative Examples of Filters

Example 1.6 Low-pass filter
The low-pass filter has two real poles at 0.9 so as to amplify the low-frequency components of the input and two real zeros at –0.9 to attenuate high-frequency components:

$$H_s(z) = \frac{0.0028\left(1 + 1.8z^{-1} + 0.81z^{-2}\right)}{1 - 1.8z^{-1} + 0.81z^{-2}} \tag{1.146}$$

Example 1.7 High-pass filter
The high-pass filter has two real poles at –0.9 in the high-frequency region to accentuate high-frequency components and two real zeros at 0.9 to attenuate the low-frequency components.

$$H_s(z) = \frac{0.0028\left(1 - 1.8z^{-1} + 0.81z^{-2}\right)}{1 + 1.8z^{-1} + 0.81z^{-2}} \tag{1.147}$$

Example 1.8 Band-pass filter
The band-pass filter has complex conjugate poles $0.6930 \pm j0.6930 = 0.98\angle \pm \pi/4$ which amplify frequency $f_0 = 0.1250$ cycles per sample and the zeros at –0.98 and 0.98 attenuate both low- and high-frequency components of the input:

$$H_s(z) = \frac{\left(1 - \rho z^{-1}\right)\left(1 + \rho z^{-1}\right)}{1 - 2\rho\cos\omega_0 z^{-1} + \rho^2 z^{-2}} = \frac{\left(1 - 0.98z^{-1}\right)\left(1 + 0.98z^{-1}\right)}{1 - 1.3859z^{-1} + 0.9604z^{-2}} \tag{1.148}$$

Example 1.9 Band-stop filter
The band-stop filter has complex conjugate poles $0.6237 \pm j0.6237 = 0.7840\angle \pm \pi/4$ and zeros $0.6930 \pm j0.6930 = 0.9800\angle \pm \pi/4$ symmetrically located on radial lines at angles $\pm\pi/4$. As a consequence it notches out the frequency $f_0 = 0.1250$ cycles per sample.

$$H_s(z) = \left(\frac{1 - 2z^{-1}\rho_n \cos 2\pi f_0 + \rho_n^2 z^{-2}}{1 - 2z^{-1}\rho_d \cos 2\pi f_0 + \rho_d^2 z^{-2}}\right) = \left(\frac{1 - 1.3859z^{-1} + 0.9604z^{-2}}{1 - 1.3166z^{-1} + 0.8668z^{-2}}\right) \tag{1.149}$$

where $\rho_n = 0.98$, $\rho_d = 0.93$.

Figure 1.11 shows (i) the poles and the zeros, (ii) the magnitude of the frequency response, and (iii) the impulse response of low-pass, high-pass, band-pass, and band-stop filters. It clearly shows that frequency response is influence by the pole-zero location.

1.7.7 Illustrative Examples of Random Signals

Typical random signals include a low-pass, high-pass, band-pass, and band-stop random signals which are generated by filtering a white noise by a low-pass, a high-pass, a band-pass, and a band-stop filter respectively, given in Eqs. (1.146)–(1.149). The filter poles and zeros, the frequency response, and the impulse response are shown in Figure 1.10. The filter is designed so as to locate the poles so as to enhance the desired frequency component, and the zeros are located to attenuate the undesired frequency components. Hence the poles are located in the frequency region where the desired frequency components are located and the zeros in the region where the unwanted components are located. See Figures 1.8 and 1.9. For example, the poles of the low-pass (high-pass) filter are located in low-pass (high-pass)

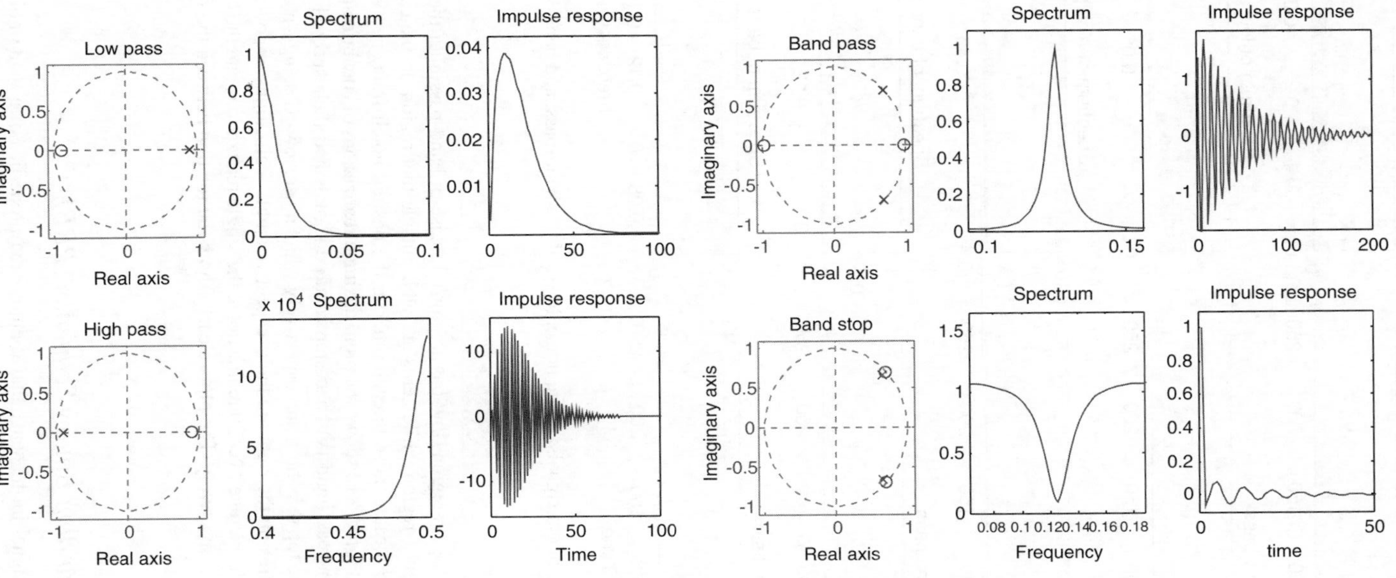

Figure 1.11 Pole-zero location frequency and impulse response

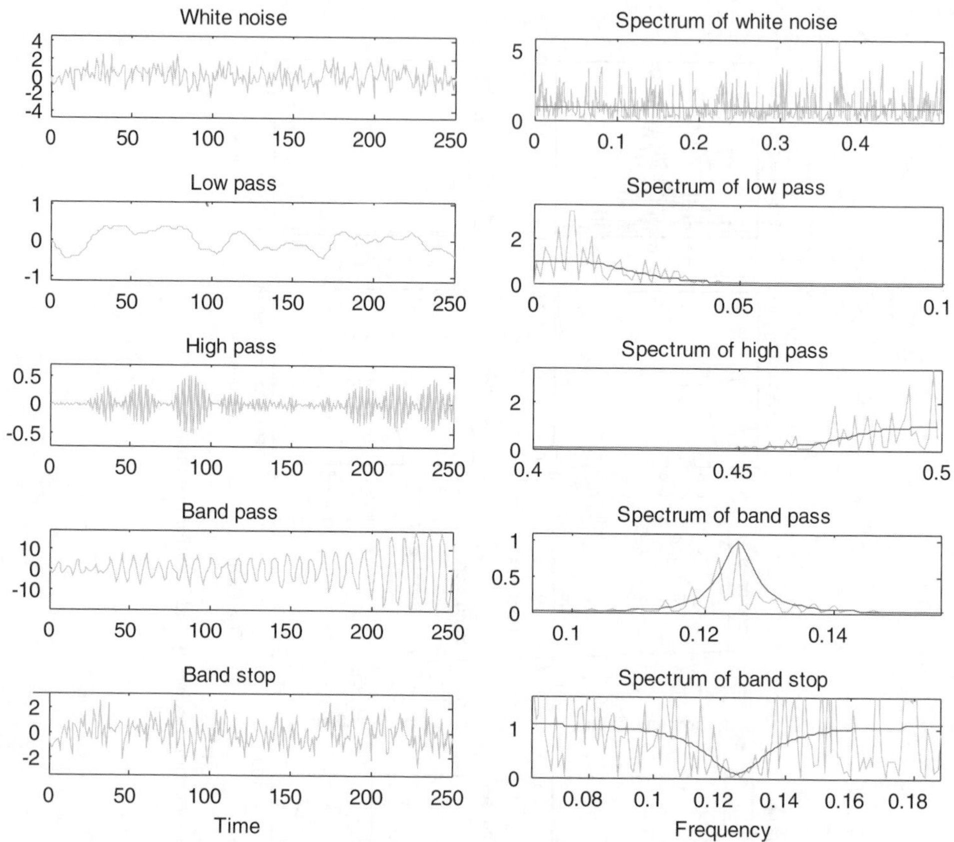

Figure 1.12 Random signals: white noise, low-, high-, band-pass, and band-stop signals

region, zeros in the high-pass region (low-pass region). In case of band-pass filter the poles are located at the band-pass frequency region while zeros are well outside that region. In the case of notch filter, the poles and zeros are located close to each other in the desired notch frequency with zeros located close to the unit circle. Figure 1.12 shows the plots of single realizations of (i) white noise, (ii) low-pass, (iii) high-pass, (iv) band-pass, and (v) band-stop random signals and their spectral components. The plots on the left from the top to bottom are respectively white noise, low-pass, high-pass, band-pass, and band-stop signals. Figures on the right are their respective single realizations of power spectral density as well as their ensemble average over 30 realizations: the single realizations and the ensemble averages are jagged and smooth respectively. The plots clearly show the relation between the filter response and the filter type.

1.7.8 Pseudo Random Binary Sequence (PRBS)

A PRBS is a random signal in the sense that it enjoys the properties of a white noise sequence. It is "pseudo" because it is deterministic and periodic, unlike a random white noise sequences. It is generated by shift registers with feedback. A profound treatment of the state-space model, the correlation, and the

Table 1.2 Binary arithmetic

Addition	Multiplication
$0 + 1 = 1$	$1 \times 1 = 1$
$1 + 0 = 0$	$1 \times 0 = 0$
$1 + 1 = 0$ *with a carry '1'*	$0 \times 0 = 0$

power spectral density of the PRBS is given in [2]. The state-space model of the PRBS signal, which is generated by a p-bit shift register, is given by

$$\boldsymbol{x}(k+1) = \begin{bmatrix} a_1 & a_2 & a_3 & . & a_p \\ 1 & 0 & 0 & 0 & 0 \\ 0 & 1 & 0 & 0 & 0 \\ . & . & . & . & . \\ 0 & 0 & 0 & 1 & 0 \end{bmatrix} \boldsymbol{x}(k)$$
$$y(k) = [0 \quad 0 \quad 0 \quad . \quad 1]\boldsymbol{x}(k) \tag{1.150}$$

where $\boldsymbol{x}(k)$ is a $px1$ binary state vector and the output $y(k)$ is the PRBS . The state-space model generates PRBS $y(k)$ which switches between 0 and 1. It is period $M = 2^p - 1$ and M is called the maximum length of the PRBS.

The binary arithmetic operation is performed to compute the state $x(k)$ and the output $y(k)$. The binary arithmetic and multiplication operations are listed in Table 1.2 below.

A two-state PRBS, $s(k)$, whose values switch between a and $-a$, can be derived from the output $y(k)$ of the state-space model (1.150) as:

$$s(k) = 2a\,(y(k) - 0.5) \tag{1.151}$$

The auto-correlation function $r_{ss}(m)$ and the power spectral density $P_{ss}(f)$ of the odd waveform $s(k)$ over one period N are similar to that of a white noise sequence given by:

$$r_{ss}(m) = \begin{cases} 1 & \textit{if } m = 0 \\ -\frac{1}{M} & \textit{otherwise} \end{cases} \tag{1.152}$$

$$P_{ss}(f) = \frac{1}{M^2}\delta(f) + \frac{M+1}{M^2}\sum_{\ell=1}^{M-1}\delta\left(f - \frac{\ell}{M}\right) \tag{1.153}$$

The PRBS finds applications in many areas, including communications, encryption, and simulation (for example in system identification).

Example A PRBS with values +1 and –1 is generated using the state-space model (1.150) and (1.151). Figure 1.13 shows the diagram of the PRBS generated using a 5-bit shift register with feedback from the

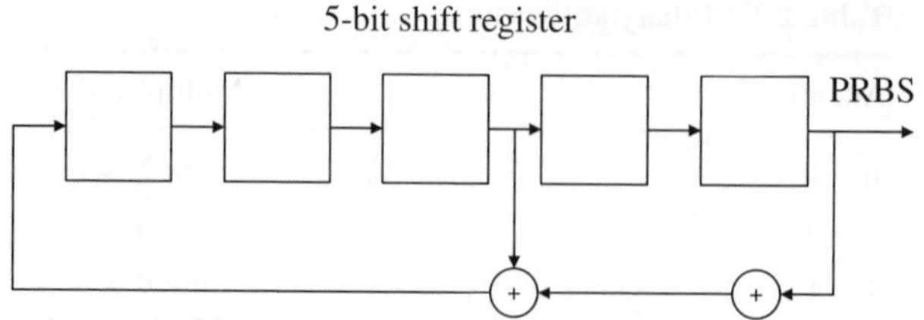

Figure 1.13 PRBS generator by shift register

third and the fifth registers. The state-space model is

$$\boldsymbol{x}(k+1) = \begin{bmatrix} 0 & 0 & 1 & 0 & 1 \\ 1 & 0 & 0 & 0 & 0 \\ 0 & 1 & 0 & 0 & 0 \\ 0 & 0 & 0 & 0 & 0 \\ 0 & 0 & 0 & 1 & 0 \end{bmatrix} \boldsymbol{x}(k) \tag{1.154}$$

$$y(k) = [0 \quad 0 \quad 0 \quad 0 \quad 1]\,\boldsymbol{x}(k)$$

In Figure 1.14, the PRBS waveform $s(k)$, its auto-correlation $r_{ss}(m)$, and the power spectral density $P_{ss}(f)$ are shown in subfigures (a), (b), and (c) respectively.

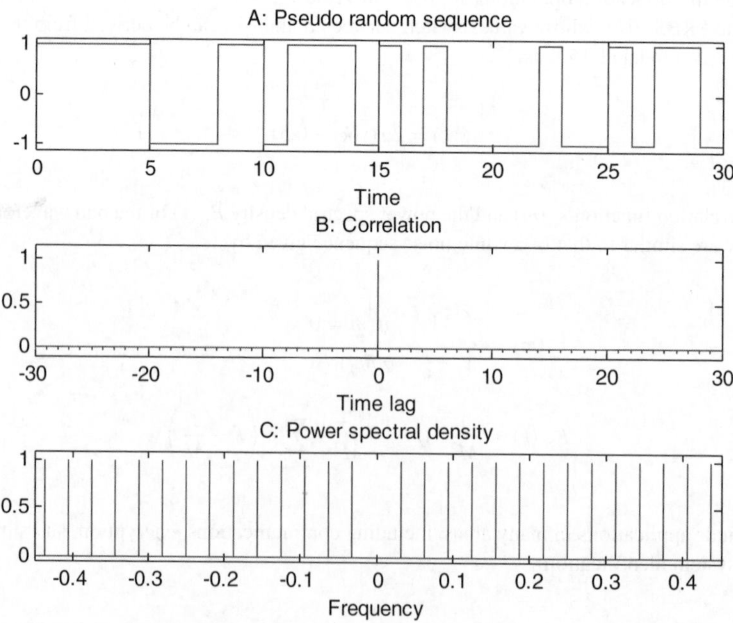

Figure 1.14 PRBS signal, its auto-correlation and power spectral density

1.8 Model of a System with Disturbance and Measurement Noise

The model of a physical plant relating the input and the output plays a crucial role in model-based approaches in applications including controller design, soft sensor, performance monitoring, and fault diagnosis. An integrated model of a system is developed by augmenting the model of the plant with the model of disturbance and the model of the measurement noise.

A model of typical system formed of a plant subject to disturbances and measurement noise is derived by combining the models of the plant, the disturbance, and the measurement noise. The input to the system, the disturbance, and the noise are in general deterministic and/or random signals. The disturbance represents any input that affects the system but cannot be manipulated (unlike the input $r(k)$), such as a gravity load, electrical power demand, fluctuations in the flow in a fluid system, wind gusts, bias, power frequency signal, DC offset, crew or passenger load in vehicles such as spacecrafts, ships, and helicopter and planes, and faults. The output of the system is corrupted by measurement noise. The analysis, design, and monitoring of the behavior of the system requires a model of not only of the plant but also of the disturbance and the noise affecting its input and the output.

1.8.1 Input–Output Model of the System

An input –output of a plant is typically represented by a transfer function of the plant $G_p(z)$ with an input $r(k)$, disturbance $w(k)$, and an output $y(k)$ which is corrupted by an additive measurement noise $v(k)$. The transfer function $G_p(z) = \frac{N_p(z)}{D_p(z)}$ is formed of a cascade connection of two subsystems $G_{p1}(z) = \frac{N_{p1}(z)}{D_{p1}(z)}$ and $G_{p2}(z) = \frac{N_{p2}(z)}{D_{p2}(z)}$ such that $G_p(z) = G_{p1}(z)G_{p2}(z)$. The input $u(k)$ is accessible and can be manipulated, while the disturbance affecting the system $w(k)$, in general, is not accessible or measured. The input–output model of the plant is given by:

$$y(z) = G_p(z)r(z) + G_{p2}(z)w_p(z) + v_p(z) \tag{1.155}$$

where $y(z)$, $r(z)$, $w_p(z)$ and $v_p(z)$ are respectively the z-transforms of the output, $y(k)$, the input $u_p(z)$, the disturbance $w_p(k)$, and the measurement noise $v_p(k)$.

1.8.1.1 Disturbance Model

The disturbance $w(k)$ enters at a node between $G_{p1}(z)$ and $G_{p2}(z)$. The disturbance $w_p(k)$ may be deterministic process $w_d(k)$, a random process $w_r(k)$, or in general their combination:

$$w_p(z) = w_r(z) + w_d(z) \tag{1.156}$$

The random component w_r is modeled as an output of a LTI system $H_w(z) = \frac{N_w(z)}{D_w(z)}$ excited by a zero-mean white noise process $w(k)$:

$$w_r(z) = H_w(z)w(z) \tag{1.157}$$

Since the exact form of the determinist component w_d is generally not known except for the class to which it may belong, it is modeled as an output of an unforced LTI system:

$$D_{dw}(z)w_d(z) = N_{dw}(z) \tag{1.158}$$

where $D_{dw}(z)$ is some known polynomial, and $N_{dw}(z)$ is unknown and depends upon the initial conditions.

1.8.1.2 Measurement Noise Model

The output of the system $y(k)$ is corrupted by a measurement noise v_p. The measurement noise is modeled similar to that of the disturbance. The measurement noise may be deterministic process $v_d(k)$, a random process $v_r(k)$, or their combination given by

$$v_p(z) = v_r(z) + v_d(z) \tag{1.159}$$

The random component v_r is modeled as an output of a $H_v(z) = \frac{N_v(z)}{D_v(z)}$ excited by a zero-mean white noise process $v(k)$

$$v_r(z) = H_v(z)v(z) \tag{1.160}$$

Since the exact form of the determinist component v_d is generally not known except for the class to which it may belong, it is modeled as an output of an unforced LTI system:

$$D_{dv}(z)v_d(z) = N_{dv}(z) \tag{1.161}$$

where $D_{dv}(z)$ is some known polynomial, and $N_{dv}(z)$ is generally unknown and depends upon the initial conditions.

Comment *Generally the components of the disturbance $w_r(k)$ and the measurement noise $v_r(k)$ are respectively a low-pass random process and a high-pass random process. They are obtained by filtering the zero mean white noise process with low- and high-pass filters. The deterministic component of the disturbance $w_d(k)$ and that of the measurement noise $v_d(k)$ belong to the class of signals such as constants, sinusoids, and ramps.*

1.8.1.3 Integrated Input–Output Model

A general case of the disturbance and noise models is considered by assuming that the disturbance and the measurement noise include both deterministic and random process components. The overall block diagram is shown in Figure 1.15.

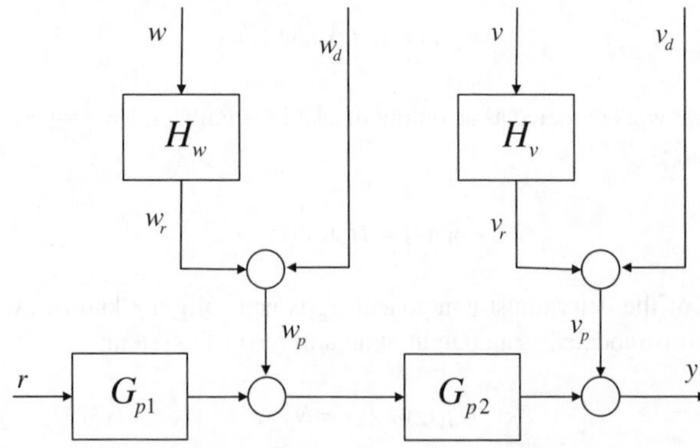

Figure 1.15 System with input, output, disturbance, and measurement noise

The model of the plant (1.155) and, assuming that the disturbance and the measurement noise include both deterministic and random process components, the disturbance model (1.156) and the measurement noise model (1.159), are combined:

$$y(z) = G_p(z)r(z) + G_{p2}(z)w(z) + H_v(z)v(z) + G_{p2}(z)w_d(z) + v_d(z) \tag{1.162}$$

Expressing (i) the rational functions in terms of their numerator and denominator polynomial, (ii) multiplying both sides of the equation by $D_d(z)$ which is the least common multiple of polynomial $D_{dw}(z)$ and $D_{dv}(z)$, and (iii) expressing $D_p(z) = D_{p1}(z)D_{p2}(z)$ we get the following polynomial model

$$D_y(z)y(z) = N_u(z)r(z) + D_{p2}(z)N_{p2}(z) + N_{u_w}(z)\,w(z) + N_{u_v}(z)v(z) + N_d(z) \tag{1.163}$$

where

$$D_y(z) = D_d(z)D_p(z)D_w(z)D_v(z) = 1 + \sum_{i=1}^{n_a} a_i z^{-i};$$
$$N_u(z) = N_p(z)D_d(z)\,D_w(z)D_v(z) = \sum_{i=1}^{n_b} b_i z^{-i};$$
$$N_{u_v}(z) = D_d(z)D_p(z)D_w(z)N_v(z) = \sum_{i=0}^{n_v} b_{vi} z^{-i};$$

$N_{u_w}(z) = D_d(z)D_{p1}(z)N_{p2}(z)D_v(z)N_w(z) = \sum_{i=0}^{n_w} b_{wi}z^{-i}$; $\{a_i\}$, $\{b_i\}$, $\{b_{vi}\}$, $\{a_{wi}\}$ are the coefficients respectively of $D_y(z)$, $N_u(z)$, $N_{u_v}(z)$, and $N_{u_w}(z)$ and $N_d(z)$ is the term associated with the deterministic term $G_{p2}(z)w_d(z) + v_d(z)$ after cross-multiplication.

Comment *The operator $D_d(z)$ is a noise annihilation operator. Multiplying both sides by $D_d(z)$ of Eq. (1.162) annihilates the deterministic components $G_{p2}(z)w_d(z)$ and $v_d(z)$ in finite time instants as shown by the annihilation operation (1.57).*

Assumptions: Minimum phase model

1. *The plant $G_p(z)$ is assumed to be stable. The roots of the denominator polynomial $D_p(z)$ are strictly inside the unit circle.*
2. *The random disturbance and the random measurement noise transfer functions $H_w(z)$ and $H_v(z)$ are assumed to be minimum phase, that is the roots of the numerator and the denominator polynomials $N_w(z)$ and $D_w(z)$, $N_v(z)$, and $D_v(z)$ are strictly inside the unit circle.*
3. *It is assumed that the white noise inputs $w(k)$ and $v(k)$ are uncorrelated.*

1.8.1.4 Difference Equation Model

Taking the inverse z-transform of Eq. (1.163) we get

$$y(k) = -\sum_{i=1}^{n_a} a_i y(k-i) + \sum_{i=1}^{n_b} b_i r(k-i) + v_o(k) \tag{1.164}$$

$$v_o(k) = \sum_{i=0}^{n_w} b_{wi} w(k-i) + \sum_{i=0}^{n_v} b_{vi} v(k-i) \tag{1.165}$$

where the equation error term $v_o(k)$ represents the totality of the effects of the disturbances, and the measurement noise affecting the output $y(k)$: It is a colored noise.

1.8.2 State-Space Model of the System

1.8.2.1 State-Space Model of the Plant

Expressing the plant model (1.162) in state-space form $(\boldsymbol{A}_p, \boldsymbol{B}_p, \boldsymbol{C}_p, D_p)$ we get

$$\begin{aligned} \boldsymbol{x}_p(k+1) &= \boldsymbol{A}_p\boldsymbol{x}_p(k) + \boldsymbol{B}_p u_p(k) + \boldsymbol{E}_p w_p(k) \\ y(k) &= \boldsymbol{C}_p\boldsymbol{x}_p(k) + D_p u_p(k) + v_p(k) \end{aligned} \tag{1.166}$$

1.8.2.2 State-Space Model of the Noise and Disturbance

Expressing the disturbance model (1.157) in state-space form $(\boldsymbol{A}_w, \boldsymbol{B}_w, \boldsymbol{C}_w, D_w)$ we get

$$\begin{aligned} \boldsymbol{x}_w(k+1) &= \boldsymbol{A}_w\boldsymbol{x}_w(k) + B_w w(k) \\ w_r(k) &= \boldsymbol{C}_w\boldsymbol{x}_w(k) + D_w w(k) \end{aligned} \tag{1.167}$$

Expressing the measurement noise model (1.160) in state-space form $(\boldsymbol{A}_v, \boldsymbol{B}_v, \boldsymbol{C}_v, D_v)$ we get

$$\begin{aligned} \boldsymbol{x}_v(k+1) &= \boldsymbol{A}_v\boldsymbol{x}_v(k) + \boldsymbol{B}_v v(k) \\ \boldsymbol{v}_r(k) &= \boldsymbol{C}_v\boldsymbol{x}_v(k) + D_v v(k) \end{aligned} \tag{1.168}$$

The deterministic models of the noise and the disturbance (1.158) and (1.161) is expressed in state-space form $(\boldsymbol{A}_d, \boldsymbol{C}_d)$:

$$\begin{aligned} \boldsymbol{x}_d(k+1) &= \boldsymbol{A}_d\boldsymbol{x}_d(k) \\ \begin{bmatrix} w_d(k) \\ v_d(k) \end{bmatrix} &= \boldsymbol{C}_v\boldsymbol{x}_d(k) \end{aligned} \tag{1.169}$$

where $(\boldsymbol{A}_d, \boldsymbol{C}_d)$ is a minimal realization, $\boldsymbol{C}_d = \begin{bmatrix} \boldsymbol{C}_{wd} \\ \boldsymbol{C}_{vd} \end{bmatrix}$.

1.8.2.3 State-Space Model of the System

Figure 1.16 shows how these three state-space models of Eqs. (1.166), (1.167), (1.168), (1.169), and the algebraic models (1.156) and (1.159) fit together.

Combining the plant model $(\boldsymbol{A}_p, \boldsymbol{B}_p, \boldsymbol{C}_p, D_p)$, the disturbance model $(\boldsymbol{A}_w, \boldsymbol{B}_w, \boldsymbol{C}_w, D_w)$, the measurement noise model $(\boldsymbol{A}_v, \boldsymbol{B}_v, \boldsymbol{C}_v, D_v)$, and $(\boldsymbol{A}_d, \boldsymbol{C}_d)$, we get

$$\begin{aligned} \boldsymbol{x}(k+1) &= \boldsymbol{A}\boldsymbol{x}(k) + \boldsymbol{B}\boldsymbol{u}(k) \\ y(k) &= \boldsymbol{C}\boldsymbol{x}(k) + \boldsymbol{D}\boldsymbol{u}(k) \end{aligned} \tag{1.170}$$

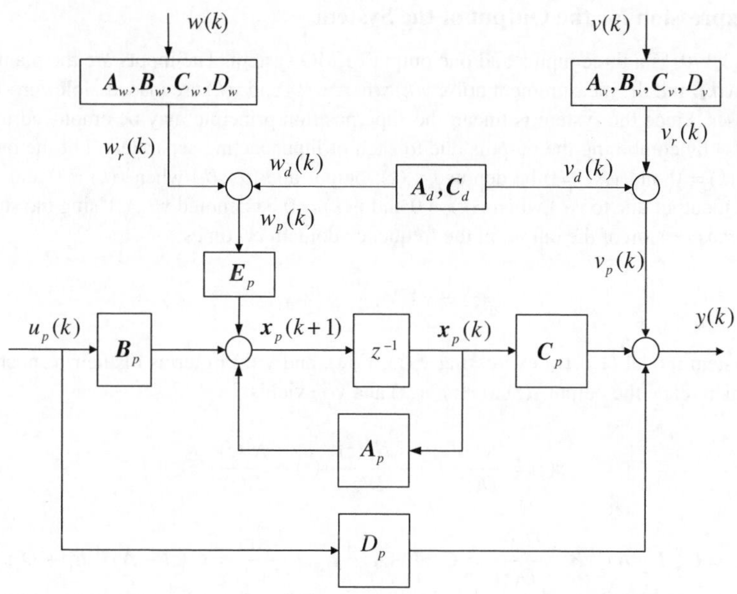

Figure 1.16 State-space model of the plant with noise and disturbance

where

$$\boldsymbol{x}=\begin{bmatrix}\boldsymbol{x}_p\\ \boldsymbol{x}_w\\ \boldsymbol{x}_v\\ x_d\end{bmatrix},\ \boldsymbol{u}=\begin{bmatrix}r\\ w\\ v\end{bmatrix},\ \boldsymbol{A}=\begin{bmatrix}\boldsymbol{A}_p & \boldsymbol{E}_p\boldsymbol{C}_w & 0 & \boldsymbol{E}_p\boldsymbol{C}_{wd}\\ 0 & \boldsymbol{A}_v & 0 & 0\\ 0 & 0 & \boldsymbol{A}_v & 0\\ 0 & 0 & 0 & \boldsymbol{A}_d\end{bmatrix},\ \boldsymbol{B}=\begin{bmatrix}\boldsymbol{B}_p & \boldsymbol{E}_p\boldsymbol{D}_w & \boldsymbol{0}\\ \boldsymbol{0} & \boldsymbol{B}_w & \boldsymbol{0}\\ \boldsymbol{0} & \boldsymbol{0} & \boldsymbol{B}_v\end{bmatrix}$$

$$\boldsymbol{C}=\begin{bmatrix}\boldsymbol{C}_p & 0 & \boldsymbol{C}_v & \boldsymbol{C}_{vd}\end{bmatrix},\boldsymbol{D}=\begin{bmatrix}D_p & 0 & D_v\end{bmatrix}$$

The integrated state space model is shown in Figure 1.17.

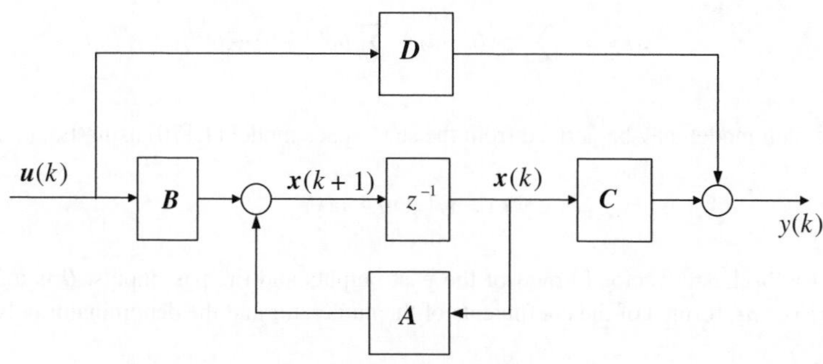

Figure 1.17 Integrated state-space model of the system

1.8.2.4 Expression for the Output of the System

The system (1.170) is a three inputs and one output MIMO system. The inputs are the plant input $r(k)$, disturbance $w(k)$, and the measurement noise $v(k)$ where $w(k)$ and $v(k)$ are inaccessible zero-mean white noise processes. Since the system is linear, the superposition principle may be employed to determine the output $y(k)$ by combining the outputs due to each of inputs acting separately. Let the output due to $r(k)$ when $w(k) = 0$ and $v(k) = 0$ be denoted $y_r(k)$; output due to $w(k)$ when $r(k) = 0$ and $v(k) = 0$ be denoted $y_w(k)$; output due to $v(k)$ when $r(k) = 0$ and $w(k) = 0$ be denoted $y_v(k)$. Using the superposition principle, the expression of the output in the frequency domain becomes:

$$y(z) = y_r(z) + y_w(z) + y_v(z) \tag{1.171}$$

Using the system model (1.170), expressing $y_r(z)$, $y_w(z)$, and $y_v(z)$ in terms of their respective transfer functions which relate the output $y(z)$ to $r(z)$, $w(z)$ and $v(z)$ yields:

$$y(z) = \frac{N(z)}{D(z)}r(z) + \frac{N_w(z)}{D(z)}w(z) + \frac{N_v(z)}{D(z)}v(z) \tag{1.172}$$

where $\dfrac{N(z)}{D(z)} = \boldsymbol{C}(z\boldsymbol{I} - \boldsymbol{A})^{-1}\boldsymbol{B}$; $\dfrac{N_w(z)}{D(z)} = \boldsymbol{C}(z\boldsymbol{I} - \boldsymbol{A})^{-1}\boldsymbol{E}_w$; $\dfrac{N_v(z)}{D(z)} = \boldsymbol{C}(z\boldsymbol{I} - \boldsymbol{A})^{-1}\boldsymbol{E}_v + D_v$; Since the denominator polynomial $D(z) = |z\boldsymbol{I} - \boldsymbol{A}|$ is the same for each of the individual outputs, we may rewrite the expression (1.172) by multiplying both sides by $D(z)$ as:

$$D(z)y(z) = N(z)r(z) + \upsilon(z) \tag{1.173}$$

where $\upsilon(z) = N_w(z)w(z) + N_v(z)v(z)$. Expressing the numerator polynomial $N(z)$ and the denominator polynomial $D(z)$ in the terms of their respective coefficients $\{b_i\}$ and $\{a_i\}$ as $N(z) = \sum_{i=1}^{n} b_i z^{-i}$ and $D(z) = 1 + \sum_{i=1}^{n} a_i z^{-i}$, (1.173) becomes:

$$y(z) = -\sum_{i=1}^{n} a_i z^{-i} y(z) + \sum_{i=1}^{n} b_i z^{-i} r(z) + \upsilon(z) \tag{1.174}$$

1.8.2.5 Linear Regression Model

Taking the inverse z-transform of Eq. (1.174) yields

$$y(k) = -\sum_{i=1}^{n} a_i y(k-i) + \sum_{i=1}^{n} b_i r(k-i) + \upsilon(k) \tag{1.175}$$

A linear regression model may be derived from the state-space model (1.170) using Eq. (1.175) as:

$$y(k) = \boldsymbol{\psi}^T(k)\boldsymbol{\theta} + \boldsymbol{\upsilon}(k) \tag{1.176}$$

where $\boldsymbol{\psi}(k)$ is a $2n$x1 data vector formed of the past outputs and the past inputs, $\boldsymbol{\theta}$ is a $2n$x1 vector, termed *feature vector*, formed of the coefficients of the numerator and the denominator polynomials

$$\psi^T(k) = [-y(k-1) \;\; -y(k-2) \;\; \ldots \;\; -y(k-n) \;\; r(k-1) \;\; r(k-2) \;\; \ldots \;\; r(k-n)] \tag{1.177}$$

$$\theta = [a_1 \;\; a_2 \;\; \ldots \;\; a_n \;\; b_1 \;\; b_2 \;\; \ldots \;\; b_n]^T \tag{1.178}$$

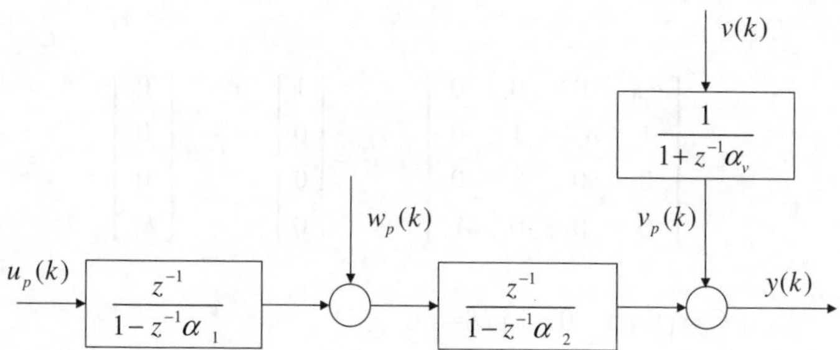

Figure 1.18 System with constant disturbance and low-pass measurement noise

Comment *The analysis, design, and implementation of the system are considerably simplified as one needs to consider only an augmented system model which is driven by plant input and zero mean white noise processes (instead of a plant driven by the plant input, colored random disturbance, and colored random measurement noise). Nice properties of white noise, such as the constant spectral density (as the white noise is an uncorrelated random signal), may be exploited.*

1.8.3 Illustrative Examples in Integrated System Model

Example 1.10 Consider a system with constant deterministic disturbance and a high-pass random measurement noise as shown in Figure 1.18.

State-space models

Plant model

$$\begin{aligned} \boldsymbol{x}_p(k+1) &= \boldsymbol{A}_p\boldsymbol{x}_p(k) + \boldsymbol{B}_p u_p(k) + \boldsymbol{E}_p w_p(k) \\ y(k) &= \boldsymbol{C}_p x_p(k) + v_p(k) \end{aligned} \tag{1.179}$$

where

$$\boldsymbol{A}_p = \begin{bmatrix} \alpha_{p1} & 0 \\ 1 & \alpha_{p2} \end{bmatrix},\ \boldsymbol{B}_p = \begin{bmatrix} 1 \\ 0 \end{bmatrix},\ \boldsymbol{E}_p = \begin{bmatrix} 0 \\ 1 \end{bmatrix},\ \boldsymbol{C}_p = [0 \quad 1];\ \alpha_{p1} = 0.5,\ \alpha_{p2} = 0.9$$

Disturbance model: constant disturbance

$$\begin{aligned} x_w(k+1) &= A_w x_w(k) \\ w_p(k) &= C_w x_w(k) \end{aligned} \tag{1.180}$$

where $A_w = 1,\ C_w = 1$

Measurement noise model: low-pass random noise

$$\begin{aligned} x_v(k+1) &= A_v x_v(k) + B_v v(k) \\ v_p(k) &= C_v x_v(k) + D_v v(k) \end{aligned} \tag{1.181}$$

where $A_v = 0.7$ $B_v = 0.7,\ C_v = 1,\ D_v = 1.$

The state-space model (1.170) becomes

$$\begin{aligned} \boldsymbol{x}(k+1) &= \boldsymbol{A}\boldsymbol{x}(k) + \boldsymbol{B}\boldsymbol{u}_p(k) + \boldsymbol{F}\boldsymbol{v}(k) \\ y(k) &= \boldsymbol{C}\boldsymbol{x}(k) + Dv(k) \end{aligned} \tag{1.182}$$

where

$$A = \begin{bmatrix} \alpha_{p1} & 0 & 0 & 0 \\ 1 & \alpha_{p2} & 1 & 0 \\ 0 & 0 & A_w & 0 \\ 0 & 0 & 0 & A_v \end{bmatrix}, \quad B = \begin{bmatrix} 1 \\ 0 \\ 0 \\ 0 \end{bmatrix}, \quad F = \begin{bmatrix} 0 \\ 0 \\ 0 \\ B_v \end{bmatrix}$$

$$C = [0 \quad 1 \quad 0 \quad 1] \; D = D_v$$

Transfer function model
The plant model is

$$G_p(z) = G_{p1}(z)G_{p2}(z) = \frac{z^{-2}}{\left(1-\alpha_{p2}z^{-1}\right)\left(1-\alpha_{p2}z^{-1}\right)} = \frac{z^{-2}}{1-\left(\alpha_{p1}+\alpha_{p2}\right)z^{-1}+\alpha_{p1}\alpha_{p2}z^{-2}} \tag{1.183}$$

where $G_{p1}(z) = \dfrac{z^{-1}}{1-\alpha_{p1}z^{-1}}$; $G_{p2}(z) = \dfrac{z^{-1}}{1-\alpha_{p2}z^{-1}}$;
The disturbance model

$$w(z) = \frac{z^{-1}}{1-z^{-1}} \tag{1.184}$$

The measurement noise model

$$v(z) = H_v(z)v_v(z) = \frac{v_v(z)}{1+\alpha_v z^{-1}} \tag{1.185}$$

where $\alpha_v = 0.7$
The model of the integrated system given in Eq. (1.162) becomes

$$y(z) = \frac{z^{-2}u_p(z)}{\left(1-\alpha_{p1}z^{-1}\right)\left(1-\alpha_{p2}z^{-1}\right)} + \frac{z^{-2}}{\left(1-\alpha_{p2}z^{-1}\right)\left(1-z^{-1}\right)} + \frac{v_v(z)}{1+\alpha_v z^{-1}} \tag{1.186}$$

Rewriting as two inputs, u, and u_v, and one output y system using Eq. (1.163), we get

$$D_y(z)y(z) = N_u(z)u_p(z) + N_v(z)u_v(z) + N_d(z) \tag{1.187}$$

where $D_d(z) = 1 - z^{-1}$,

$$\begin{aligned} D_y(z) &= \left(1-\alpha_{p1}z^{-1}\right)\left(1-\alpha_{p2}z^{-1}\right)D_d(z)\left(1-\alpha_v z^{-1}\right) \\ &= \sum_{i=0}^{4} a_i z^{-i} = 1 - 1.7z^{-1} + 0.17z^{-2} + 0.84z^{-3} - 0.315z^{-4} \\ N_u(z) &= z^{-2}(1-z^{-1})\left(1-\alpha_v z^{-1}\right) = \sum_{i=1}^{4} b_i z^{-i} = z^{-2} - 0.3z^{-3} - 0.7z^{-4} \\ N_v(z) &= (1-z^{-1})\left(1-\alpha_{p1}z^{-1}\right)\left(1-\alpha_{p2}z^{-1}\right) = \sum_{i=0}^{3} b_{vi} z^{-i} = 1 - 0.7z^{-1} - 2,23z^{-2} + 2.695z^{-3} - 0.765z^{-4} \\ N_d(z) &= z^{-2} + 0.2z^{-3} - 0,35z^{-4} \end{aligned}$$

Difference equation model
The difference equation model (1.164) becomes

$$y(k) = -\sum_{i=1}^{4} a_i y(k-i) + \sum_{i=1}^{4} b_i u_p(k-i) + \sum_{i=1}^{4} b_{wi}\delta(k-i) + \sum_{i=1}^{3} b_{vi} v_v(k-i) \tag{1.188}$$

Since $\delta(k-i) = 0$ for $i \geq 1$ we get

$$y(k) = -\sum_{i=1}^{4} a_i y(k-i) + \sum_{i=1}^{4} b_i u_p(k-i) + \sum_{i=1}^{3} b_{vi} v_v(k-i) = 0, \text{ for } k > 4 \tag{1.189}$$

Figure 1.19 shows the plots of the output $y(k)$ when the input $u_p(z)$ is a unit step. Subfigure (a) at the top shows the output $y(k)$ when the input $u_p(k)$ is a unit step, $u_w(k) = 0$ and $u_v(k) = 0$ representing the ideal case when the disturbance and the measurement noise are both absent; subfigure (b) at top shows the effect of disturbance on the output when the input $u(k) = 0$, $u_w = \delta(k)$ and $u_v = 0$; subfigure (c) at top shows the effect of the measurement noise affecting the output when the input $u_p(k) = 0$, $u_w(k) = 0$, and $u_v(k) = v_v(k)$ a zero-mean white noise process; subfigure (d) at the bottom shows the output when the input $u(k)$ is a unit step, $u_w(k) = 0$ and $u_v(k) = v_v(k)$, that is when there is no disturbance; subfigure (e) at the bottom shows the output when the input $u(k)$ is a unit step, $u_w(k) = \delta(k)$ and $u_v(k) = 0$, that is when there is no measurement noise; subfigure (f) at the bottom shows when the input $u_p(z)$ is a unit step, $u_w(k) = \delta(k)$ and $u_v(k) = v_v(k)$ when there is both measurement noise and disturbance.

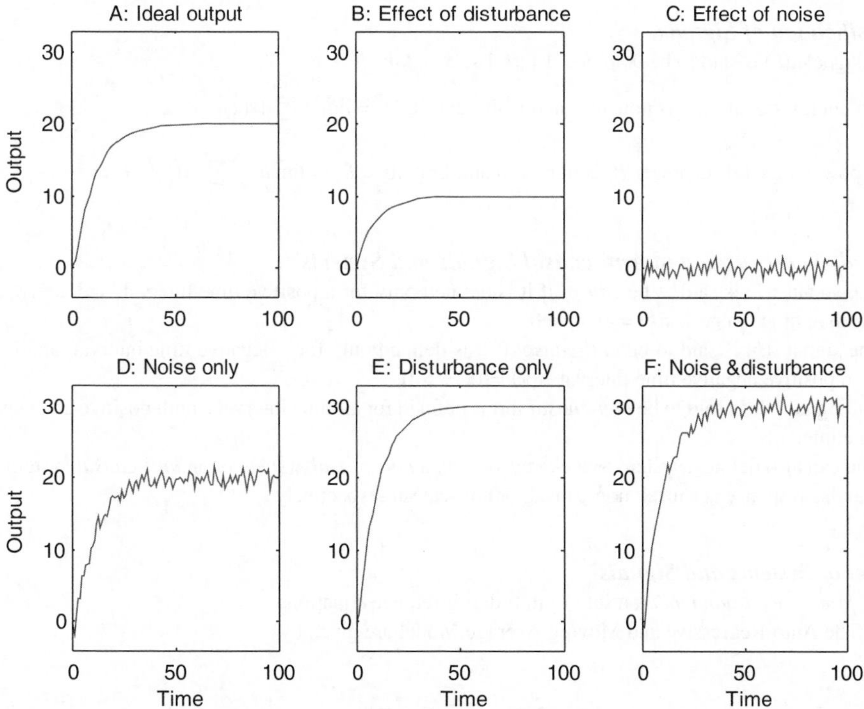

Figure 1.19 Output, disturbance, and measurement noise of the integrated model

Comments *Thanks to the linearity of the models, the superposition principle may be employed to determine the output by combining the outputs due to each of the inputs acting separately. For example, consider subfigures (a) and (b), where the outputs are respectively due only to the plant input* u_p *and the input* v_w. *Subfigure (e) shows the output due to both inputs, which thanks to linearity is the sum of the outputs shown in subfigures (a) and (b). Similarly the outputs shown in subfigures (d) and (f) are respectively the sum of the outputs shown in (a) and (c), and the sum of the outputs in (a), (b), and (c). It must, however, be emphasized that a physical system is not generally linear, and a linearized model may be used to enjoy the vast resources available in the analysis and design of linear system.*

The order of the plant model is 2, while the order of the integrated model is 4, which the sum of the orders of the plant, the disturbance, and the measurement noise models. The high-order difference equation model (1.164), (or (1.188)) is employed to identify the plant model from the input–output data (u_p, y). *The augmented state-space model (1.182) (or (1.182)) is embodied in an observer or Kalman filter.*

In the illustrated example the objective is to compute the plant output given the plant input and the models of the noise and the disturbance. It may be termed a "forward" problem. In practice however, the problem we face is a "reverse or inverse" problem. We are given the plant input and the output corrupted by noise and disturbance, the objective in Eq. for (i) the system identification problem is to estimate the plant model from the input–output data and (ii) for the observer or Kalman filter application, to estimate the "true" output which is the ideal noise and disturbance-free output of the plant. Referring to Figure 1.16, the corrupted output shown in subfigure (f) is given and the objective is to estimate the output shown in subfigure (a).

1.9 Summary

Classification of Signals

1. A signal $x(k)$ is said to be *bounded* if $|x(k)| < \infty$ *for all k*
2. An energy signal if E_x is non-zero and is finite: $0 < E_x = \lim_{N\to\infty} it \sum_{k=0}^{N-1} |x(k)|^2 < \infty$
3. A power signal if its power P_x is non-zero and finite:$0 < P_x = \lim_{N\to\infty} it \frac{1}{N} \sum_{k=0}^{N-1} |x(k)|^2 < \infty$

Causal, Non-causal, and Anti-causal Signals and Systems

1. The signal $s(k)$ is said to be *causal* if it is defined only for a positive time interval, and is zero for a negative time interval: $s(k) = 0,\ k < 0$.
2. The signal $s(k)$ is said to be *anti-causal* if it is defined only for a negative time interval, and is zero for a positive negative time interval: $s(k) = 0,\ k > 0$.
3. The signal $s(k)$ is said to be *non-causal* if it is defined for all time intervals, both positive and negative time intervals.
4. One can also define a system as a *causal system*, a *non-causal system*, or an *anti-causal* system if its impulse response is causal, non-causal, or anti-causal respectively.

Model of Systems and Signals

1. *Difference equation model*: a set of nth order difference equations
   - The Auto-Regressive and Moving Average Model are given by

$$\sum_{i=0}^{n_a} a_i y(k-i) = \sum_{i=0}^{n_b} b_i u(k-i)$$

- *Auto-Regressive (AR) model*

$$\sum_{i=0}^{n_a} a_i y(k-i) = b_0 u(k)$$

- *Moving Average (MA) model*

$$y(k) = \sum_{i=0}^{n_b} b_i u(k-i)$$

2. *State-space model*: a set of n first-order difference equations

$$\begin{aligned} \boldsymbol{x}(k+1) &= \boldsymbol{A}\boldsymbol{x}(k) + \boldsymbol{B}u(k) \\ y(k) &= \boldsymbol{C}\boldsymbol{x}(k) + Du(k) \end{aligned}$$

Observable Canonical Form

$$\begin{bmatrix} x_1(k+1) \\ x_2(k+1) \\ x_3(k+1) \\ . \\ x_{n-1}(k+1) \\ x_n(k+1) \end{bmatrix} = \begin{bmatrix} -a_1 & 1 & 0 & . & 0 & 0 \\ -a_2 & 0 & 1 & . & 0 & 0 \\ -a_3 & 0 & 0 & . & 0 & 0 \\ . & . & . & . & . & . \\ -a_{n-1} & 0 & 0 & . & 0 & 1 \\ -a_n & 0 & 0 & . & 0 & 0 \end{bmatrix} \begin{bmatrix} x_1(k) \\ x_2(k) \\ x_3(k) \\ . \\ x_{n-1}(k) \\ x_n(k) \end{bmatrix} + \begin{bmatrix} b_1 - b_0 a_1 \\ b_2 - b_0 a_2 \\ b_3 - b_0 a_3 \\ . \\ b_{n-1} - b_0 a_{n-1} \\ b_n - b_0 a_n \end{bmatrix} u(k)$$

$$y(k) = [1 \quad 0 \quad 0 \quad . \quad 0 \quad 0]\boldsymbol{x}(k) + b_0 u(k)$$

Frequency-domain Model

$$y(z) = H(z)u(z) \text{ where } H(z) = \frac{z^{-n_d}\left(b_{n_d} + b_{n_d+1}z^{-1} + b_{n_d+2}z^{-2} + \cdots + b_{n_b}z^{-n_b+n_d}\right)}{1 + a_1 z^{-1} + a_2 z^{-2} + a_3 z^{-3} + \cdots + a_{n_a} z^{-n_a}},$$

1. *Proper* if $n_d \geq 0$.
2. *Strictly proper* if it is proper and $n_d \geq 1$.
3. *Improper* if $n_a < n_b$.
4. $b_0 \neq 0$, $n_d = 0$, or $D = 0$:the system is said to have a direct transmission path.
5. *Stable* (also termed *asymptotically stable*) if all the poles lie inside the unit circle and *unstable system* if at least one pole lies outside the unit circle.

Minimum Phase System

A transfer function $H(z)$ is said to be minimum phase if its poles and zeros are stable.

Model of Deterministic and Random Signals

Signal model

1. *Frequency-domain model*

$$s(z) = H(z)u(z)$$

2. *Time-domain model*

$$s(k) + \sum_{i=1}^{n_s} a_i s(k-i) = \sum_{i=0}^{n_s} b_i u(k-i)$$

3. *State-space model*

$$\begin{aligned}\boldsymbol{x}(k+1) &= \boldsymbol{A}\boldsymbol{x}(k) + \boldsymbol{B}u(k)\\ y(k) &= \boldsymbol{C}\boldsymbol{x}(k) + Du(k)\end{aligned}$$

where $u(k) = \begin{cases} \delta(k) & \text{if deterministic} \\ v(k) & \text{if random} \end{cases}$ and $u(z) = \begin{cases} 1 & \text{if deterministic} \\ v(z) & \text{if random} \end{cases}$

Introduction to Random Signals

1. *Strict-sense stationary of order* p

$$\begin{aligned}&f_{XY}\left(x(k_1+m), x(k_2+m), \dots, x(k_p+m), y(\ell_1+m), y(\ell_2+m), \dots, y(\ell_p+m)\right)\\ &\quad = f_{XY}\left(x(k_1), x(k_2), \dots, x(k_p), y(\ell_1), y(\ell_2), \dots, y(\ell_p)\right)\end{aligned}$$

2. *Wide-sense stationary random signal*

$$\begin{aligned} E[x(k)] &= \mu_x \\ E[(x(k)-\mu_x)^2] &= \sigma_x^2 \\ E[x(k)y(\ell)] &= r_{xy}(k-\ell) \end{aligned}$$

Quasi-stationary process

$$\begin{aligned} |E[x(k)]| &< \infty \\ |E[x(k)x(k-m]| &< \infty \\ \hat{r}_{xx}(k, k-m) &= \lim_{N\to\infty}\left\{\frac{1}{N}x(k)x(k-m)\right\} \to r_{xx}(m) \end{aligned}$$

Orthogonal random signals

$$r_{xy}(m) = 0 \ \text{ for all } \ m$$

Uncorrelated random signals

$$C_{xy}(m) = E[(x(k)-\mu_x)(y(k-m)-\mu_y)] = 0$$

Ergodic process :

1. $x(k)$ is *ergodic in the mean* if $\hat{\mu}_x$ is asymptotically equal to μ_x

$$\hat{\mu}_x = \lim_{N\to\infty} it\, \frac{1}{N}\sum_{k=0}^{N-1} x(k) = E[x(k)] = \mu_x$$

2. $x(k)$ is *ergodic in the variance* if $\hat{\sigma}_x^2$ is asymptotically equal to σ_x^2

$$\hat{\sigma}_x^2 = \lim_{N\to\infty} it\, \frac{1}{N}\left(\sum_{k=0}^{N-1} x(k) - \hat{\mu}_x\right)^2 = E[(x(k) - \mu_x)^2] = \sigma_x^2$$

3. $x(k)$ is *ergodic in the auto-correlation* if $\hat{r}_{xx}(m)$ is asymptotically equal to $r_{xx}(m)$

$$\hat{r}_{xx}(m) = \lim_{N\to\infty} it\, \frac{1}{N}\sum_{k=0}^{N-1} x(k)x(k-m) = E\left[x(k)x(k-m)\right]$$

4. $x(k)$ and $y(k)$ are *ergodic:*

$$\hat{r}_{xy}(m) = \lim_{N\to\infty} it\, \frac{1}{N}\sum_{k=0}^{N-1} x(k)y(k-m) = E\left[x(k)y(k-m)\right]$$

White Noise Process

$$E[v(k)] = \mu_v = 0$$

$$r_{vv}(\mathrm{m}) = E\left[v(k)v(k-m)\right] = \sigma_v^2\delta(m) = \begin{cases} \sigma_v^2 & m = 0 \\ 0 & m \neq 0 \end{cases}$$

$$P_{vv}(\omega) = \sigma^2$$

Classification of Random Waveform

$$P_{ss}(f) = |H(f)|^2\,\sigma_v^2$$

Whitening Filter

$$y_f(z) = H^{-1}(z)y(z) = v(z)$$

Model of a Class of All Signals

$$s(k) + \sum_{i=0}^{n} a_i s(k-i) = 0 \text{ for all } k$$

$$\begin{aligned} \boldsymbol{x}(k+1) &= \boldsymbol{A}\boldsymbol{x}(k) \\ s(k) &= \boldsymbol{C}\boldsymbol{x}(k) \\ s(k) &= \boldsymbol{C}\boldsymbol{A}^k\boldsymbol{x}(0) \end{aligned}$$

Noise Annihilation Filter

$$\upsilon(k) + \sum_{i=1}^{n} a_i\upsilon(k-i) = 0 \text{ for all } k > n$$

Pseudo-Random Binary Sequence (PRBS)

$$\boldsymbol{x}(k+1) = \begin{bmatrix} a_1 & a_2 & a_3 & . & a_p \\ 1 & 0 & 0 & 0 & 0 \\ 0 & 1 & 0 & 0 & 0 \\ . & . & . & . & . \\ 0 & 0 & 0 & 1 & 0 \end{bmatrix} \boldsymbol{x}(k)$$

$$y(k) = \begin{bmatrix} 0 & 0 & 0 & . & 1 \end{bmatrix} x(k)$$

Integrated Model of a System

$$\boldsymbol{x}_p(k+1) = \boldsymbol{A}_p x_p(k) + \boldsymbol{B}_p u_p(k) + \boldsymbol{E}_p w(k)$$
$$y(k) = \boldsymbol{C}_p \boldsymbol{x}_p(k) + D_p u_p(k) + v(k)$$

Frequency-domain model

$$y(z) = G_p(z)u_p(z) + G_{p2}(z)H_w(z)u_w(z) + H_v(z)u_v(z)$$

Difference equation Model

$$y(k) = -\sum_{i=1}^{n_a} a_i y(k-i) + \sum_{i=1}^{n_b} b_i u(k-i) + \sum_{i=0}^{n_w} b_{wi} v_w(k-i) + \sum_{i=0}^{n_v} b_{vi} v_v(k-i)$$

State-space model of the system
Combining the plant, the disturbance, and the measurement noise models we get

$$\boldsymbol{x}(k+1) = \boldsymbol{A}\boldsymbol{x}(k) + \boldsymbol{B}\boldsymbol{u}(k)$$
$$y(k) = \boldsymbol{C}\boldsymbol{x}(k) + \boldsymbol{D}\boldsymbol{u}(k)$$

$$\boldsymbol{x} = \begin{bmatrix} \boldsymbol{x}_p \\ \boldsymbol{x}_w \\ \boldsymbol{x}_v \end{bmatrix}, \boldsymbol{u} = \begin{bmatrix} u_p \\ u_w \\ u_v \end{bmatrix}, \boldsymbol{A} = \begin{bmatrix} \boldsymbol{A}_p & \boldsymbol{E}_p \boldsymbol{C}_w & 0 \\ 0 & \boldsymbol{A}_w & 0 \\ 0 & 0 & \boldsymbol{A}_v \end{bmatrix}, \boldsymbol{B} = \begin{bmatrix} \boldsymbol{B}_p & \boldsymbol{E}_p D_w & \boldsymbol{0} \\ \boldsymbol{0} & \boldsymbol{B}_w & \boldsymbol{0} \\ \boldsymbol{0} & \boldsymbol{0} & \boldsymbol{B}_v \end{bmatrix}$$

$$\boldsymbol{C} = [\,\boldsymbol{C}_p \quad 0 \quad \boldsymbol{C}_v\,], \quad \boldsymbol{D} = [\,D_p \quad E_p D_w \quad D_v\,]$$

References

[1] Ljung, L. (1999) *System Identification: Theory for the User*, Prentice-Hall, New Jersey.
[2] Soderstrom, T. and Stoica, P. (1989) *System Identification*, Prentice Hall, New Jersey.

Further Readings

Brown, R.G. and Hwang, P.Y. (1997) *Introduction to Random Signals and Applied Kalman Filtering*, John Wiley and Sons.

Ding, S. (2008) *Model-based Fault Diagnosis Techniques: Design Schemes*, Springer-Verlag.

Goodwin, G.C., Graeb, S.F., and Salgado, M.E. (2001) *Control System Design*, Prentice Hall, New Jersey.
Haykin, S. (2001) *Adaptive Filter Theory*, Prentice Hall, New Jersey.
Kay, S.M. (1988) *Modern Spectral Estimation: Theory and Application*, Prentice Hall, New Jersey.
Mitra, S.K. (2006) *Digital Signal Processing: A Computer-Based Approach*, McGraw Hill Higher Education, Boston.
Moon, T.K. and Stirling, W.C. (2000) *Mathematical Methods and Algorithms for Signal Processing*, Prentice Hall, New Jersey.
Oppenheim, A.V. and Schafer, R.W. (2010) *Discrete-Time Signal Processing*, Prentice-Hall, New Jersey.
Proakis, J.G. and Manolakis, D.G. (2007) *Digital Signal Processing: Principles, Algorithms and Applications*, Prentice Hall, New Jersey.

2

Characterization of Signals: Correlation and Spectral Density

2.1 Introduction

A very important characterization of signals is correlation and its frequency domain counterpart: power spectral density. It is a measure of interdependence or similarity between the two waveforms.

The operation of correlation has wide applications in science and engineering including detection of known signals buried in noise, pattern recognition, estimation of time delay (e.g., in radar and sonar for determining the range), identification of an unknown model of a system, synthesis, processing and compression of speech, and medical diagnosis [1–14]. It is employed to compute the measure of similarity or dissimilarity between two waveforms and enjoys the following properties of similarities and dissimilarities. Auto-correlation is a special case of correlation when the two signals are the same, that is, the auto-correlation is the correlation of the signal with itself. Auto-correlation is also a measure of similarity between a given time series and a lagged version of itself over successive time intervals. It is the same as calculating the correlation between two different time series, except that the same time series is used twice – once in its original form and once lagged by one or more time periods. For example the auto-correlation of a zero mean white noise is a delta function, the auto-correlation of a sum of signal and a zero mean white noise process is the auto-correlation of the signal for all lag values except at zero lag value.

Given a waveform with a rational spectrum, the waveform and its auto-correlation share the same characteristics. For example the auto-correlation function of a constant, exponential, sinusoid exponentially damped sinusoid is respectively a constant, exponential, and a sinusoid and exponentially damped sinusoid. The auto-correlation function of a periodic signal is periodic of the same period while the cross-correlation of two signals of different periods is zero. Auto-correlation finds application in many areas, including extraction of the features from the signal, estimation of its model, detection of non-randomness in data, and the periodicity in a waveform.

Cross-correlation of a zero mean process and a deterministic waveform is zero, and that between two similar waveforms is accentuated. These properties are exploited to attenuate noise and enhance a known signal corrupted by noise. The delay between two similar waveforms can be found as the lag value (or time) at which their cross-correlation is maximum.

The power spectral density is the Fourier transform of the auto-correlation function, while the cross-power spectral density is the Fourier transform of the cross-correlation. The power and cross-spectral densities are widely used as they are computationally fast efficient implements, since the Fast Fourier

Identification of Physical Systems: Applications to Condition Monitoring, Fault Diagnosis, Soft Sensor and Controller Design, First Edition. Rajamani Doraiswami, Chris Diduch and Maryhelen Stevenson.

Transform (FFT) algorithm is used to implement them. Further, they provide better insight into the properties of auto- and cross-correlation. Correlation of two signals is essentially a measure of spectral overlap between them. A cross-power spectral density indicates how the energy/power is distributed over frequency. The power and cross-power spectral analysis finds application in many fields including medical diagnosis, speech recognition, performance monitoring, fault diagnosis, and market analysis to name a few. In medical diagnosis, the biological signals/images such as the electro-encephalogram (EEG), the electrocardiogram (ECG), the phonocardiogram (PCG), the myo-electrical signals, and so on, are classified using the power spectral density. The power spectral density serves as a feature to distinguish one class from the other. In speech processing, power spectral density plays a central role in systems involving the synthesis, recognition, and compression of speech. In radar and sonar the spectral content is analyzed to determine the location and the type of target, such as a plane, ship, or other. In seismology, spectral analysis of the signals gives useful information on the movement of the ground. The spectral analysis is used to monitor the health of mechanical system-rotating machinery, vibratory systems, and so on, where the spectral signature is used to determine the wear and tear, breakage, or malfunction of mechanical parts. In control systems, the spectral analysis is used to design a controller and also help in performance monitoring and fault diagnosis. In economics, meteorology, and stock market analysis, spectral analysis may reveal trends –hidden periodicities reveal certain recurring or cyclic phenomena.

The relation between cross- and auto-power spectral densities of the input and the output are derived. It is shown that the frequency response of a system may easily be computed from the cross-spectral density of the input and the output, and the power spectral density of the input. The cross-correlation between the input and the output can be used to estimate the time delay.

The magnitude squared coherence spectrum, which measures similarity or dissimilarity between two signals, is defined. This measure is the frequency domain counterpart of the correlation coefficient. The coherence spectrum is widely used in many areas including medical diagnosis, performance monitoring, and fault diagnosis. If the spectral bandwidth of interest is known, then only coherence over the frequency of interest is considered and the rest of the spectral regions are ignored as they contain noise and other artifacts.

2.2 Definitions of Auto- and Cross-Correlation (and Covariance)

If the two signals are identical, that is $x(k) = y(k)$, then the correlation of $x(k)$ with itself, denoted $r_{xx}(m)$, is called the *auto-correlation* of $x(k)$, and if they are different, that is $x(k) \neq y(k)$, then $r_{xy}(m)$ is termed the *cross-correlation* of $x(k)$ and $y(k)$. In the definition of correlation function, the index, m, is not a time index, rather, it is a time-lag, a time lead, or a time-shift index: m can take either positive or negative integer values. In general, $r_{xy}(m) \neq r_{yx}(m)$.

The correlation function of two signals of $x(k)$ and $y(k)$, is denoted by $r_{xy}(m)$, and depending upon the characteristics of $x(k)$ and $y(k)$, it is defined as follows:

- if $x(k)$ and $y(k)$ are real-valued energy signals, then we need to consider two cases:
  - the signals are of infinite duration

$$r_{xy}(m) = \lim_{N\to\infty} \sum_{k=-N}^{N} x(k)y(k-m) \tag{2.1}$$

  - the signals are of finite duration

$$r_{xy}(m) = \sum_{k=0}^{N-1} x(k)y(k-m) \tag{2.2}$$

- if $x(k)$ and $y(k)$ are real-valued power signals, then

$$r_{xy}(m) = \lim_{N\to\infty} \frac{1}{2N+1} \sum_{k=-N}^{N} x(k)y(k-m) \tag{2.3}$$

- if $x(k)$ and $y(k)$ are power signals and periodic with period M, then

$$\begin{aligned} x(k) &= x(k+\ell M) \\ y(k) &= y(k+\ell M) \quad \ell = \pm 1, \pm 2, \pm 3, \ldots \end{aligned} \tag{2.4}$$

$$r_{xy}(m) = \frac{1}{M} \sum_{k=0}^{M-1} x(k)y(k-m) \tag{2.5}$$

- if $x(k)$ and $y(k)$ are real-valued jointly wide-sense stationary random signals, then

$$r_{xy}(m) = E[x(k)y(k-m)] = \int_{-\infty}^{\infty}\int_{-\infty}^{\infty} x(k)y(k-m)f_{xy}(x(k), y(k-m))dx(k)dy(k-m) \tag{2.6}$$

  where $E[(.)]$ is the usual expectation operator indicating an ensemble average of the random processes (.). Assuming that the random process is ergodic in the mean and the correlation, their estimates are obtained from a single realization

$$\mu_x = E[x(k)] = \lim_{N\to\infty} it \frac{1}{2N+1} \sum_{k=-N}^{N} x(k) \tag{2.7}$$

The mean $\mu_y = E[y(k)]$ can be estimated similarly. The correlation is estimated as:

$$r_{xy}(m) = E[x(k)y(k-m)] = \lim_{N\to\infty} it \frac{1}{2N+1} \sum_{k=-N}^{N} x(k)y(k-m) \tag{2.8}$$

Thanks to the assumption of wide-sense stationarity, the correlation is only a function of the lag m and not of time index k. Since a random signal is a power signal, the averaging operation is used.

Computation of Correlation Using Truncated Signals

In practice, the correlation of infinitely long signals (energy or power) is estimated from truncated versions of the signals. If $x(k)$ and $y(k)$, $k = 0, 1, 2, \ldots, N-1$, are truncated signals, then the correlation $r_{xy}(m)$ is:

$$r_{xy}(m) = \sum_{k=0}^{N-1} x(k)y(k-m) \tag{2.9}$$

In this case the number of samples N is chosen to be sufficiently large, so that $r_{xy}(m)$ captures completely the characteristics of the signal including its steady-state and transient behaviors. For example, if it is a transient signal, N should be much larger than its settling time, and if periodic N should contain several periods of the signals. Expanding the summation operation we get

$$r_{xy}(m) = \begin{cases} \sum_{k=0}^{m-1} x(k)y(k-m) + \sum_{k=m}^{N} x(k)y(k-m) & m \geq 0 \\ \sum_{k=0}^{N-m} x(k)y(k-m) + \sum_{k=N-m+1}^{N} x(k)y(k-m) & m < 0 \end{cases} \tag{2.10}$$

All the data required for computing the correlation function $r_{xy}(m)$ are not available, as can be deduced from the first summation term $\sum_{k=0}^{m-1} x(k)y(k-m)$ for $m \geq 0$, and the second summation term $\sum_{k=N-m+1}^{N} x(k)y(k-m)$ for $m \geq 0$. The unavailable data for $m \geq 0$ are $y(k) \; : k = -1, -2, \dots, -m$, and $y(k) \; : k = N+1, N+2, \dots, N+m$ for $m < 0$. To overcome this problem, the values for the unavailable data may be assumed equal to: zero, the mean value of the signal, or to a periodic extension of the signal.

The truncated signals may be treated either as power or energy signals, although they are commonly treated as an energy signals.

Summary of the Definitions of Correlation

Considering the signal characterization, the definition of correlation may be expressed compactly as:

$$r_{xy}(m) = \begin{cases} \lim_{N\to\infty} \sum_{k=-N}^{N} x(k)y(k-m) & \text{infinite duration energy signals} \\ \sum_{k=0}^{N-1} x(k)y(k-m) & \text{finite duration energy signals} \\ \lim_{N\to\infty} \frac{1}{2N+1} \sum_{k=-N}^{N} x(k)y(k-m) & \text{power signal} \\ r_{xy}(m) = \frac{1}{M} \sum_{k=0}^{M-1} x(k)y(k-m) & \text{M-periodic power signal} \\ \lim_{N\to\infty} \frac{1}{2N+1} \sum_{k=-N}^{N} x(k)y(k-m) & \text{ergodic random (power) signals} \end{cases} \tag{2.11}$$

Comments

- *If $x(k)$ and $y(k)$ are complex-valued signals, then $y(k)$ should be replaced by its complex conjugate $y^*(k)$, in the definition of $r_{xy}(m)$ above.*
- *Correlation of energy and finite duration signals involves summing operations, while the power signals, such as the periodic and the random signals, involve averaging operations.*

Alternative Expression of Correlation

A change of variables allows one to express the correlation $r_{xy}(m)$ in the alternative form shown below.

Change the variable of summation from k to $\bar{k}$ where

$$k = \bar{k} + m \tag{2.12}$$

$$r_{xy}(m) = \lim_{N\to\infty} \frac{1}{2N+1} \sum_{\bar{k}=-N}^{N} x(\bar{k}+m)y(\bar{k}) = \lim_{N\to\infty} it \frac{1}{2N+1} \sum_{k=-N}^{N} x(k+m)y(k) \tag{2.13}$$

It important to emphasize that the time shift m is the difference between the time indices associated with the first signal $x(k)$ and that of the second signal $y(k)$ in the correlation sum.

Covariance Function

Similar to correlation, one may define a covariance function of two stationary random signals $x(k)$ and $y(k)$. The covariance function is essentially the correlation function of two zero-mean signals, which are obtained from signals $x(k)$ and $y(k)$ by subtracting their means. The covariance of $x(k)$ and $y(k)$ is denoted by $c_{xy}(m)$. It is defined by

$$c_{xy}(m) = E[(x(k) - \mu_x)(y(k-m) - \mu_y)] \tag{2.14}$$

where $\mu_x = E[x(k)]$ and $\mu_y = E[y(k)]$ are the mean values of $x(k)$ and $y(k)$. Expanding and simplifying we get

$$c_{xy}(m) = E[x(k)y(k-m)] - \mu_x\mu_y = r_{xy}(m) - \mu_x\mu_y \tag{2.15}$$

2.2.1 Properties of Correlation

Except when noted, we derive all the properties of correlation for the case of energy signals whose definitions are given by Eq. (2.11); it can be shown that, when applicable, the resulting properties may be extended to correlation of other signal types.

Correlation and Convolution

Proposition The correlation $x(k)$ and $y(k)$ is the same as the convolution of $x(k)$ and $y(-k)$

A convolution of $x(k)$ and $y(k)$ denoted $conv(x(k), y(k))$ is defined as

$$conv(x(k), y(k)) = \sum_{k=-\infty}^{\infty} x(k)y(m-k) \tag{2.16}$$

Proof: Consider the expression for the correlation of the energy signal given in Eq. (2.11)

$$r_{xy}(m) = \sum_{k=-\infty}^{\infty} x(k)y(k-m) \tag{2.17}$$

where N may be finite or infinite. Let $s(k) = y(-k)$. Rewriting $y(k-m) = y(-(m-k)) = s(m-k)$ we get

$$r_{xy}(m) = \sum_{k=-\infty}^{\infty} x(k)y(-(m-k)) = \sum_{k=-\infty}^{\infty} x(k)s(m-k) = conv(x(k), y(-k)) \tag{2.18}$$

where $conv(x(k), y(-k))$ denotes the convolution of $x(k)$ and $y(-k)$.

The correlation complex valued signals $x(k)$ and $y(k)$ are the same as the convolution of complex values $x(k)$ and $y^*(-k)$. A convolution of complex valued $x(k)$ and $y(k)$ is defined as

$$conv(x(k), y^*(k)) = \sum_{k=-\infty}^{\infty} x(k)y^*(m-k) \tag{2.19}$$

Thus $r_{xy}(m)$ is the convolution of $x(k)$ and $y(-k)$. Note that $y(-k)$ is the time reversed sequence of $y(k)$. This means that the same algorithm may be used to compute correlation and convolution simply by reversing one of the sequences. Consequently, if one of the two sequences possesses symmetry such that the reversed sequence is a time shifted version of the original sequence (as is the case with constants and sinusoids), then correlation of the two sequences will be equal to the time shifted versions of the convolutions of the two sequences. As shown below, the frequency domain provides additional insight regarding the relationship between correlation and convolution.

The frequency domain provides additional insight regarding the relationship between correlation and convolution. Letting $\Im(.)$ denote Fourier transform of (.), we have

$$\Im\{r_{xy}(m)\} = X(f)Y^*(f) = |X(f)|\,|Y(f)|\,\angle(\varphi_x(f) - \varphi_y(f)) \tag{2.20}$$

$$\Im\{conv(x(k), y(k))\} = X(f)Y(f) = |X(f)|\,|Y(f)|\,\angle(\varphi_x(f) + \varphi_y(f)) \tag{2.21}$$

Where $X(f) = |X(f)| \angle\left(\varphi_x(f)\right)$; $Y(f) = |Y(f)| \angle\left(\varphi_y(f)\right)$; $\varphi_x(f)$,$\varphi_y(f)$, $|X(f)|$ and $|Y(f)|$ are the phase and the magnitude spectra of $X(f)$ and $Y(f)$. Comparing the expressions (2.20) and (2.21) we can deduce that the magnitudes of the Fourier transforms of correlation and convolution are identical while their phase spectra differ. Note that convolution is a filtering operation while correlation is a pattern matching operation.

Relation between $r_{xy}(m)$ and $r_{yx}(-m)$

For complex values signals it can be deduced

$$r_{xy}(m) = r^*_{yx}(-m) \tag{2.22}$$

Proof: Follows from Eqs. (2.1) and (2.12).

Symmetry Property of the Auto-Correlation Function

$$r_{xx}(m) = r^*_{xx}(-m) \tag{2.23}$$

Proof: It is special case of Eq. (2.22) when $x(k) = y(k)$.

This shows that the auto-correlation function is Hermitian. In case of real-valued signal, the autocorrelation is real and even.

Maximum Values of Auto- and Cross-Correlations

The absolute value of the correlation is bounded as follows:

$$r_{xx}(0) \geq |r_{xx}(m)| \tag{2.24}$$

$$\sqrt{r_{xx}(0)r_{yy}(0)} \geq |r_{xy}(m)| \tag{2.25}$$

$$|r_{xy}(m) + r^*_{xy}(m)| \leq r_{xx}(0) + r_{yy}(0) \tag{2.26}$$

Proof: Consider the following identity

$$\lim_{N\to\infty}\frac{1}{2N+1}\sum_{k-N}^{N}(ax(k)+by(k-m))\left(a^{*}x^{*}(k)+b^{*}y^{*}(k-m)\right)\geq 0 \tag{2.27}$$

where a and b are complex scalars. Recall the following definitions:

$$r_{xy}(m)=\lim_{N\to\infty}\frac{1}{2N+1}\sum_{k-N}^{N}x(k)y^{*}(k-m) \tag{2.28}$$

$$r_{xx}(0)=\lim_{N\to\infty}\frac{1}{2N+1}\sum_{k-N}^{N}x(k)x^{*}(k) \tag{2.29}$$

$$r_{yy}(0)=\lim_{N\to\infty}\frac{1}{2N+1}\sum_{k-N}^{N}y(k-m)y^{*}(k-m) \tag{2.30}$$

Expanding and using the above definitions, the expression (2.27) becomes

$$a^{2}r_{xx}(0)+b^{2}r_{yy}(0)+2abr_{xy}(m)\geq 0 \tag{2.31}$$

Expressing in a quadratic form we get

$$\begin{bmatrix}a & b\end{bmatrix}\begin{bmatrix}r_{xx}(0) & r_{xy}(m)\\ r_{xy}^{*}(m) & r_{yy}(0)\end{bmatrix}\begin{bmatrix}a^{*}\\ b^{*}\end{bmatrix}\geq 0 \text{ for all } a \text{ and } b \tag{2.32}$$

This shows that the center matrix in Eq. (2.32) is Hermitian and positive semi-definite, which implies that its determinant is real-valued. Furthermore, due to the inequality above, the determinant must be greater than or equal to zero:

$$r_{xx}(0)r_{yy}(0)\geq |r_{xy}(m)|^{2} \tag{2.33}$$

Setting $x(k)=y(k)$ we get

$$r_{xx}^{2}(0)\geq |r_{xx}(m)|^{2} \tag{2.34}$$

Since $r_{xx}(0)\geq 0$ and $r_{yy}(0)\geq 0$, the inequalities (2.24) and (2.25) hold. Setting $a=1$ and $b=-1$ in the inequality (2.32) we get Eq. (2.26).

Time Delay Between Two Signals

Let $x(k)$ and $y(k)$ denote two complex-valued signals where $y(k)$ is a delayed version of $x(k)$:$y(k)=x(k-\tau)$ and τ is the time delay. An estimate of the time delay is given by

$$\hat{\tau}=\arg\max_{m}\{r_{yx}(m)+r_{yx}^{*}(m)\} \tag{2.35}$$

where the cost function $r_{yx}(m)+r_{yx}^{*}(m)=2real\{r_{yx}(m)\}$

The real-valued signal estimate of the time delay becomes

$$\hat{\tau}=\arg\max_{m}\{r_{yx}(m)\} \tag{2.36}$$

Proof: Replacing $y(k)$ by $x(k-\tau)$ in the definition of the correlation $r_{yx}(m)$ we find

$$r_{yx}(m) = \lim_{N\to\infty} \sum_{k=-N}^{N} y(k)x^*(k-m) = \lim_{N\to\infty} \sum_{k=-N}^{N} x(k-\tau)x^*(k-m) \tag{2.37}$$

From Eq. (2.37), it follows that

$$r_{yx}(m) = r_{xx}(m-\tau) \tag{2.38}$$

Using the maximum value property of auto-correlation from Eq. (2.26), and in conjunction with Eq. (2.38), we see that

$$|r_{xy}(m) + r_{xy}^*(m)| = |r_{xx}(m-\tau) + r_{xx}^*(m-\tau)| \le r_{xx}(0) + r_{yy}(0) \tag{2.39}$$

From which it follows that the maximum value of $|r_{xy}(m) + r_{xy}^*(m)|$ occurs at $m = \tau$.

In the case that $x(k)$ and $y(k)$ are real-valued signals, using the maximum value property of auto-correlation from Eq. (2.25), and in conjunction with Eq. (2.38), we see that

$$|r_{yx}(m)| = |r_{xx}(m-\tau)| \le r_{xx}(0) \tag{2.40}$$

From which it follows that the maximum value of $|r_{yx}(m)|$ occurs at $m = \tau$.

Additive Property

Consider a signal $y(k) = \sum_{i=1}^{M} x_i(k)$ which is formed of a sum of M signals $x_i(k)$. The auto-correlation function of $y(k)$ is the sum of the cross- and auto-correlation functions $r_{x_i x_j}(m)$ of $x_i(k)$ and $x_j(k)$ given by

$$r_{yy}(m) = \sum_{i=1}^{M}\sum_{j=1}^{M} r_{x_i x_j}(m) = \sum_{i=1}^{M} r_{x_i x_i}(m) + \sum_{i=1}^{M}\sum_{\substack{j=1 \\ j\ne i}}^{M} r_{x_i x_j}(m) \tag{2.41}$$

Proof: Consider the expression for $r_{yy}(m)$ for the case that $y(k)$ is an energy signal. Substituting $y(k) = \sum_{i=1}^{M} x_i(k)$ we get

$$r_{yy}(m) = \lim_{N\to\infty} \sum_{k=-N}^{N} y(k)y(k-m) = \lim_{N\to\infty} \sum_{k=-N}^{N} \left(\sum_{i=1}^{M} x_i(k)\right)\left(\sum_{j=1}^{M} x_j(k-m)\right) \tag{2.42}$$

Changing the order of summation, using the product of the two sums' formulae given in the Appendix, and simplifying we get

$$r_{yy}(m) = \sum_{i=1}^{M}\sum_{j=1}^{M}\left(\lim_{N\to\infty} \sum_{k=-N}^{N} x_i(k)x_j(k-m)\right) \tag{2.43}$$

Using the definition of correlation we get

$$r_{yy}(m) = \sum_{i=1}^{M}\sum_{j=1}^{M} r_{x_i x_j}(m) = \sum_{i=1}^{M} r_{x_i x_i}(m) + \sum_{i=1}^{M}\sum_{\substack{j=1\\ j\neq i}}^{M} r_{x_i x_j}(m) \tag{2.44}$$

Comments *The auto-correlation function of a sum of waveforms is not equal to sum of their auto-correlation; instead, it is a sum of the auto- and cross-correlations of the various component waveforms. In the case when the waveforms are uncorrelated, the cross-correlations will be zero, and hence the auto-correlation of a sum of waveforms is the sum of their auto-correlations.*

If $x(k)$, $y(k)$ or both $x(k)$, $y(k)$ are periodic with period M, their correlation $r_{xy}(m)$ will also periodic with a same period M, that is:

$$r_{xy}(m) = r_{xy}(m + \ell M): \quad \ell = \pm 1, \pm 2, \pm 3, \ldots \tag{2.45}$$

Proof: We will assume that $y(k)$ is periodic. Let us consider the definition Eq. (2.5) of a correlation function of $x(k)$ and $y(k)$. Substituting $m = m + \ell M$ we get

$$r_{xy}(m + \ell M) = \frac{1}{M}\sum_{k=0}^{M-1} x(k)y(k - m - \ell M) \tag{2.46}$$

Since $y(k) = y(k + \ell M)$ $\ell = \pm 1, \pm 2, \pm 3, \ldots$, we get $y(k - m) = y(k - m - \ell M)$

$$r_{xy}(m + \ell M) = \frac{1}{M}\sum_{k=0}^{M-1} x(k)y(k - m) = r_{xy}(m) \tag{2.47}$$

Hence we have verified Eq. (2.45). Similarly we can prove that $r_{xy}(m)$ is periodic if $x(k)$ or both $x(k)$ and $y(k)$ are periodic.

Noise Annihilation Property

We will first show that the cross-correlation of a deterministic signal, $s(k)$, and a zero mean Wide-Sense Stationary (WSS) white noise process,$v(k)$,is zero. The cross-correlation $s(k)$ and $v(k)$ is given by

$$E\left[s(k)v(k - m)\right] = 0 \tag{2.48}$$

Proof: Consider the cross-correlation of $s(k)$ and $v(k)$ where $E\left[v(k)\right] = 0$. Simplifying we get

$$E\left[s(k)v(k - m)\right] = s(k)E\left[v(k - m)\right] = 0 \tag{2.49}$$

Noise annihilation

Consider a noisy signal $x(k) = s(k) + v(k)$ where $s(k)$ is a deterministic signal and $v(k)$ is a zero-mean WSS random process. The cross-correlation of $x(k)$ and $s(k)$ is

$$r_{xs}(m) = r_{ss}(m) \tag{2.50}$$

Proof:

$$r_{xs}(k) = E[x(k)s(k-m)] = E[s(k)s(k-m)] + E[v(k)s(k-m)] \tag{2.51}$$

Since the second term on the right is zero in view of Eq. (2.48), we get

$$r_{xs}(k) = E[x(k)s(k-m)] = E[s(k)s(k-m)] = r_{ss}(m) \tag{2.52}$$

The noise annihilating property of cross-correlation with a known deterministic signal is widely used for detecting a known signal in noise. A system which performs this operation is usually termed a correlator or matched filter. Consider now the auto-correlation noisy signal $x(k) = s(k) + v(k)$

$$r_{xx}(k) = E[x(k)x(k-m)] \tag{2.53}$$

Expanding yields

$$r_{xx}(k) = E[s(k)s(k-m)] + E[s(k)v(k-m)] + E[v(k)s(k-m)] + E[v(k)v(k-m)] \tag{2.54}$$

Using the noise annihilation property (2.48) we get

$$r_{xx}(k) = E[s(k)s(k-m)] + E[v(k)v(k-m)] \tag{2.55}$$

Thus, assuming the noise, v(k), to be a zero-mean WSS random process, the auto-correlation of a noisy signal, $x(k) = s(k) + v(k)$, will be equal to the sum of the auto-correlations of the signal and the noise. Generally, the auto-correlation of the noise will be dominant over an interval surrounding the lag value $m = 0$ but will decay as $|m|$ increases. Assuming the signal to be correlated over a longer interval than the noise (this assumption will generally be valid for low-frequency signals as well as for periodic signals), the auto-correlation of $s(k)$ will dominate for larger $|m|$. Thus, an inspection of the auto-correlation function of a noisy signal for large $|m|$ will help to determine the characteristics of any buried signals and will reveal the presence of any buried periodic signals

Generalization of the Properties of Correlation

Except when noted, we derived all the properties of correlation for the case of energy signals. In general, these properties will hold for the other types of signals. The exceptions are fairly obvious; for example, the periodicity property applies only to the case when at least one of the signals is periodic and the noise annihilation property applies only to the case when one of the signals is a zero-mean WSS random process.

2.2.2 *Normalized Correlation and Correlation Coefficient*

The normalized correlation of two signals, $x(k)$ and $y(k)$, is denoted by $\gamma_{xy}(m)$ and defined as:

$$\gamma_{xy}(m) = \frac{r_{xy}(m)}{\sqrt{r_{xx}(0)\, r_{yy}(0)}} \tag{2.56}$$

Using the inequality (2.25), we find that the normalized cross-correlation is restricted to values between -1 and $+1$:

$$-1 \le \gamma_{xy}(m) \le 1 \tag{2.57}$$

The correlation coefficient is defined as

$$\rho_{xy} = \frac{c_{xy}(0)}{\sqrt{c_{xx}(0)\, c_{yy}(0)}} \tag{2.58}$$

where $c_{xy}(m)$ is the covariance of $x(k)$ and $y(k)$ given in Eq. (2.14). If $x(k)$ and $y(k)$ are zero-mean signals, then the correlation coefficient is equal to the normalized correlation at the lag index of $m = 0$. The correlation coefficient $x(k)$ and $y(k)$ satisfies:

- $-1 \le \rho_{xy} \le 1$
- If $x(k)$ and $y(k)$ are uncorrelated then $\rho_{xy} = 0$.
- $\rho_{xy} = 1$ if and only if $y(k) = ax(k) + b$ where $a > 0$.
- $\rho_{xy} = -1$ if and only if $y(k) = -ax(k) + b$ where $a > 0$.

Correlation coefficient ρ_{xy} is a measure of linear dependence between the signals $x(k)$ and $y(k)$:$\rho_{xy} = 1$ indicates a positive (increasing) linear relationship, $\rho_{xy} = -1$ indicates decreasing (negative) linear relationship, and some value between -1 and 1 indicates a lesser degree of linear dependence between the variables. If ρ_{xy} approaches zero, there is less dependence, and the closer the coefficient is to either -1 or 1, the stronger the dependence between the variables.

2.3 Spectral Density: Correlation in the Frequency Domain

In practice, the frequency domain representation of a correlation function plays an important role. The correlation of two waveforms is essentially a measure of the degree of overlap in their spectral content. Generally, when the data record is long, the correlation can be computed faster when implemented in the frequency domain rather than directly in the time domain. This is due mainly to the fast DFT computation using the FFT algorithm. Frequency domain implementation requires $O\left(N \log_2 N\right)$ while time domain requires $O\left(N^2\right)$ arithmetic operations.

If $x(k)$ and $y(k)$ are two signals, then the Discrete Time Fourier Transform (DTFT) of their cross-correlation function is termed the *cross-energy spectral density*; it is denoted as $P_{xy}(f)$ and defined as:

$$P_{xy}(f) = \lim_{N \to \infty} \sum_{m=-N}^{N} r_{xy}(m) e^{-j2\pi fm} \tag{2.59}$$

The definition of spectral density is the same for the energy or the power signals. The thing that changes with signal type (energy or power) is the expression for the cross-correlation function $r_{xy}(m)$. If the signals are energy signals, then the spectral density is termed *energy spectral density*, and if they are power signals, then the spectral density is termed *power spectral density*.

Rayleigh Theorem

The energy of the signal, denoted E_x, may be expressed in terms of the energy spectral density

$$E_x = \sum_{k=-\infty}^{\infty} |x(k)|^2 = \int_0^1 |X(f)|^2\, df \tag{2.60}$$

If $X(\ell)$ is a discrete Fourier transform of $x(k)$ then

$$\sum_{k=0}^{N-1} x^2(k) = \frac{1}{N}\sum_{\ell=0}^{N-1} |X(\ell)|^2 \tag{2.61}$$

Comment *In many cases, it is simpler to obtain the correlation function by first obtaining $R_{xy}(z) = \Im\{r_{xy}(m)\}$, the z-domain equivalent of the power spectral density, and then using the inverse z-transform to obtain the correlation function $r_{xy}(m)$. An example demonstrating this approach will be provided later.*

It is convenient to express the energy and power spectral densities in terms of the complex-valued z-transform variable, z, instead of directly in terms of the Fourier transform variable, f; the use of the variable, z, emphasizes the discrete-time nature of the associated time-domain. Once the z-transform of a function is found, the DTFT may be obtained by evaluating the z-transform on the unit circle (i.e., replacing z by $z = e^{j2\pi f}$). Since $z = e^{j2\pi f}$ is a periodic function of f (with period equal to 1), so will be the DTFT, and thus the DTFT need only be specified on an interval of length 1 (e.g., $-1 \le f \le 1$ or $-0.5 \le f \le 0.5$); this is in contrast to the continuous-time Fourier transform which must be specified for all frequencies. We have expressed the DTFT in terms of z, ω, or f as all of arguments are commonly employed.

If $x(k)$, $y(k)$, or both are periodic, then their cross-correlation will also be periodic with this period. We generally don't talk about the DTFT of periodic functions because they are not absolutely summable; however, we can define a generalized DTFT for periodic functions which will consist of a sum of delta functions. For periodic signals, the power spectrum is more commonly defined as a line spectrum where the power at each frequency is the squared magnitude of the associated Fourier Series Coefficient. The generalized DTFT simply replaces the delta functions in the power spectrum by delta Functions that have area equal to the power at a particular frequency.

Figure 2.1 shows the complex vector, $e^{j\omega}$, sweeping out a unit circle in the z-plane as the frequency variable, omega, varies between -pi $-\pi$ to π. The frequency, ω, is the angle of the vector $e^{j\omega}$ as measured in the counter-clockwise direction from the positive real axis.

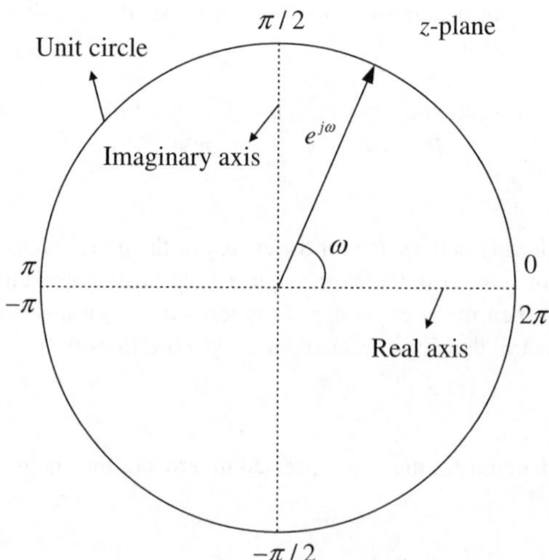

Figure 2.1 Z-plane, unit circle, and frequency

Similarly, $e^{j2\pi f}$ sweeps out the unit circle in the z-plane as f varies from 0 to 1; the value of f associated with any given point on the unit circle indicates the fraction of the way around the unit circle (when starting at the positive real axis and traveling the circle in the counter-clockwise direction) to the point's location. For example, the point where the unit circle intercepts the positive imaginary axis corresponds to a frequency $f = 1/4$, the point where the unit circle intercepts the negative real axis corresponds to $f = 1/2$, the point where the unit circle intercepts the negative imaginary axis corresponds to a frequency $f = 3/4$, which is also equivalent to a frequency of $f = -1/4$. Recall that the DTFT, when expressed as a function of the discrete-time frequency variable, f, is periodic with period 1; hence the DTFT will have the same value at $f = 3/4$ as it has at $f = -1/4$. When characterizing the frequency content of a signal, it is easiest to focus on the interval $-0.5 \leq f \leq 0.5$. Signals whose frequency components are confined to the interval $0 \leq |f| \leq 0.25$ are considered low-pass signals, those with frequency content confined to the interval $0.25 \leq |f| \leq 0.5$ are considered high-pass signals, and those whose frequency content is contained within the interval are considered band-pass signals.

In the following section, we will derive expressions for $R_{xy}(z)$, the z-transform of the correlation function, $r_{xy}(m)$, in terms of the individual z-transforms, $X(z)$ and $Y(z)$. The cross-spectral density $P_{xy}(z)$ can then be found by evaluating $R_{xy}(z)$ on the unit circle.

2.3.1 Z-transform of the Correlation Function

As explained in the previous section, finding $R_{xy}(z)$ is sometimes a convenient first step in finding the cross-correlation function and/or the cross-spectral density. In this section we start by defining $R_{xy}(z)$ as the z-transform of $r_{xy}(m)$. We then make use of various z-transform properties to express $R_{xy}(z)$ in terms of $X(z)$ and $Y(z)$. The case of energy signals, power signals, and random signals are treated separately due to differences in the definition of $r_{xy}(m)$ for these cases.

Case 1 Energy signals
Theorem: Let $x(k)$ and $y(k)$ be energy signals whose correlation function is given by Eq. (2.1).

$$R_{xy}(z) = X(z)\,Y^*((z^*)^{-1}) \tag{2.62}$$

In the case of real-values signals:

$$R_{xy}(z) = X(z)\,Y(z^{-1}) \tag{2.63}$$

Proof: Since $r_{xy}(m)$ is the convolution of $x(k)$ and $y^*(-k)$ we may use the *z-transforms* properties to find that:

$$R_{xy}(z) = X(z)Y^*((z^*)^{-1}) \tag{2.64}$$

In the case of real-valued signals $r_{xy}(m)$ is the convolution of $x(k)$ and $y(-k)$ and hence we get:

$$R_{xy}(z) = X(x)Y(z^{-1}) \tag{2.65}$$

Case 2 Power signals
Theorem: Let $x(k)$ and $y(k)$ be deterministic power signals whose correlation function $r_{xy}(m)$ is given by Eq. (2.11). Then the z-transform of the correlation function is denoted $R_{xy}(z)$ and given by:

$$R_{xy}(z) = \lim_{N\to\infty}\left(\frac{1}{2N+1}X_N(z)\,Y_N^*((z^*)^{-1})\right) \tag{2.66}$$

where $X_N(z) = \sum\limits_{k=-N}^{N} x(k)z^{-k}$ and $Y_N(z) = \sum\limits_{k=-N}^{N} y(k)z^{-k}$ are truncated z-transforms of $x(k)$ and $y(k)$ respectively. In the case of real signals:

$$R_{xy}(z) = \lim_{N\to\infty}\left(\frac{1}{2N+1}X_N(z)\,Y_N(z^{-1})\right) \tag{2.67}$$

Proof: In the case of deterministic power signals, the cross-correlation, $r_{xy}(m)$, is given by Eq. (2.3). Taking the z-transform yields:

$$R_{xy}(z) = \lim_{N\to\infty}\frac{1}{2N+1}\sum_{m=-\infty}^{\infty}\left(\sum_{k=-N}^{N} x(k)y(k-m)\right)z^{-m} \tag{2.68}$$

Let $x_N(k)$ and $y_N(k)$ denote truncated versions of $x(k)$ and $y(k)$ as defined below:

$$x_N(k) = \begin{cases} x(k) & -N \le k \le N \\ 0 & else \end{cases}$$

$$y_N(k) = \begin{cases} y(k) & -N \le k \le N \\ 0 & else \end{cases}$$

Using the truncated signals, the expression for $R_{xy}(z)$ becomes

$$R_{xy}(z) = \lim_{N\to\infty}\frac{1}{2N+1}\sum_{m=-N}^{N}\left(\sum_{k=-\infty}^{\infty} x_N(k)y_N(k-m)\right)z^{-m} \tag{2.69}$$

Since the summation term inside the bracket is the convolution of $x(k)$ and $y(-k)$ we get

$$R_{xy}(z) = \lim_{N\to\infty}\frac{1}{2N+1}X_N(z)\,Y_N(z^{-1}) \tag{2.70}$$

For the case of complex-valued signals we get

$$R_{xy}(z) = \lim_{N\to\infty}\frac{1}{2N+1}X_N(z)\,Y_N^*((z^*)^{-1}) \tag{2.71}$$

Case 3 Random signals
Let $x(k)$ and $y(k)$ be jointly WSS processes. Then assuming that the cross-correlation $r_{xy}(m)$ is jointly summable we may find

$$R_{xy}(z) = \lim_{N\to\infty}\frac{1}{2N+1}E[X_N(z)\,Y_N(z^{-1})] \tag{2.72}$$

Where similar to case 2, $X_N(z) = \sum\limits_{k=-N}^{N} x(k)z^{-k}$ and $Y_N\left(z^{-1}\right) = \sum\limits_{k=-N}^{N} y(k)z^{k}$. If the random signal is jointly WSS and ergodic in correlation, then $R_{xy}(z)$ may be computed from a single realization of the joint process:

$$R_{xy}(z) = \lim_{N\to\infty}\frac{X_N(z)\,Y_N(z^{-1})}{2N+1} \tag{2.73}$$

For complex-values random signals we get

$$R_{xy}(z) = \lim_{N\to\infty} \frac{X_N(z)\, Y_N^*((z^*)^{-1})}{2N+1} \tag{2.74}$$

Using Eqs. (2.63), (2.67), and (2.72), the z-transform of the correlation $R_{xy}(z)$ for the real-valued energy, power, and the random signals may be expressed compactly as:

$$R_{xy}(z) = \begin{cases} X(z)\,Y(z^{-1}) & \text{energy signal} \\ \lim_{N\to\infty} \left(\dfrac{X_N(z)\,Y_N(z^{-1})}{2N+1}\right) & \text{power signal} \\ \lim_{N\to\infty} \left(\dfrac{E[X_N(z)\,Y_N(z^{-1})]}{2N+1}\right) & \text{random signal} \end{cases} \tag{2.75}$$

Comment *Regardless of whether a signal is classified as an energy signal, a deterministic power signal, or a random power signal the expression for and the properties of the correlation function are essentially the same; the main difference is that the summation operation required to compute the correlation of the energy signal is replaced by an averaging operation when computing the correlation of the power signal.*

For pedagogical reason, correlation functions of energy signals are often analyzed to conceptual understanding of the spectral density.

In practice, it is impractical to acquire an infinite number of samples and, thus, the power signals are treated as energy signals (i.e., truncated) for the purpose of computing their correlations and spectral densities. In order for the estimate the correlation/spectral density to be accurate, the record length must be larger than the period of the signal's lowest frequency, and the sampling period must be smaller than the half period of the signal's highest frequency component.

2.3.2 Expressions for Energy and Power Spectral Densities

As previously stated, the cross-spectral density $P_{xy}(f)$ is defined as the DTFT of $R_{xy}(z)$ on the unit circle $z = e^{j2\pi f}$:

$$P_{xy}(f) = R_{xy}(e^{j2\pi f}) \tag{2.76}$$

In the case that $x(k)$ and $y(k)$ are real-valued energy signals, it was shown that $R_{xy}(z) = X(z)Y\left(z^{-1}\right)$ and hence,

$$P_{xy}(f) = R_{xy}(e^{j2\pi f}) = X(e^{j2\pi f})Y(e^{-j2\pi f}) \tag{2.77}$$

Furthermore, letting $X(f)$ and $Y(f)$ denote the DTFTs $x(k)$ and $y(k)$ so that $X(f) = X\left(e^{j2\pi f}\right)$ and $Y(f) = Y\left(e^{j2\pi f}\right)$, we may rewrite the previous equation as:

$$P_{xy}(f) = X(f)\,Y(-f) \tag{2.78}$$

Taking the inverse DTFT of the previous relationship reveals the familiar relationship that

$$Y(-f) = Y^*(f) \tag{2.79}$$

In the case that $x(k)$ and $y(k)$ are complex-valued energy signals, $R_{xy}(z) = X(z)Y((z^*)^{-1})$ and hence

$$P_{xy}(f) = R_{xy}(e^{j2\pi f}) = X(e^{j2\pi f})Y^*(e^{-j2\pi f}) \tag{2.80}$$

Letting $X_f(f)$ and $Y_f(f)$ denote the DTFTs $x(k)$ and $y(k)$ we may rewrite the previous equation as:

$$P_{xy}(f) = X(f)\,Y^*(f) \tag{2.81}$$

Note that the order of the subscripts is important. The DTFT of the signal is associated with the second subscript whose DTFT is conjugated in the expression for the cross-spectral density. Taking the inverse DTFT of the previous relationship yields the previously discovered relationship that shows that the cross-correlation function $r_{xy}(k)$ can be expressed as the convolution of $x(k)$ and $y^*(-k)$,

$$r_{xy}(k) = conv\,(x(k), y^*(-k)) \tag{2.82}$$

where in finding the inverse DTFT of the relation above, we used the DTFT property that multiplication in frequency is equivalent to convolution in time as well as the DTFT property: $y^*(-k) \Leftrightarrow Y^*(f)$.

Energy and Energy Spectral Density

If $x(k)$ is an energy signal, its energy spectral density is $P_{xx}(f)$ and is defined as the DTFT of the signal's auto-correlation function. Just as the auto-correlation function is a special case of the cross-correlation function, the energy spectral density is a special case of the cross-spectral density. Setting $y(k) = x(k)$ in Eq. (2.81), we find:

$$P_{xx}(f) = X(f)\,X^*(f) \tag{2.83}$$

The energy spectral density, $P_{xx}(f)$ shows how the energy of the signal, $x(k)$, is distributed over frequency.

According to Rayleigh's theorem, the energy, E_x, of the signal $x(k)$ can be found either in the time domain or in the frequency domain, as shown by the Eq. (2.60). Note that the sum in the middle expression above is $r_{xx}(0)$ and the integrand on the rightmost expression is the energy spectral density, $P_{xx}(f)$. Hence the equation above may be rewritten as:

$$E_x = r_{xx}(0) = \int_0^1 P_{xx}(f)\,df \tag{2.84}$$

This last relationship illustrates a well-known property of DTFT pairs, namely that the value of the discrete-time signal at sample no. 0 is equal to the area under its DTFT on the interval $[\,0 \quad 1\,]$.

Power and Power Spectral Density

The development above will hold similarly for power signals. The same limiting and averaging operations that were used to define $R_{xx}(z)$ in terms of the individual z-transforms may be used to express $P_{xx}(f)$ in terms of the individual DTFTs.

In the case of periodic signals, the power spectrum will be non-zero only at harmonically related discrete frequencies. In particular, if $x(k)$ is periodic with period N, it may be expressed in terms of its discrete-time Fourier series expansion as:

$$x(k) = \sum_{n=0}^{N-1} c_n e^{j2\pi nk/N} \tag{2.85}$$

where $c_n = \dfrac{1}{N}\sum_{k=0}^{N-1} x(k)e^{-j2\pi kn/N}$.

Similar to Rayleigh's theorem for energy signals, the average power,P_x, of the signal may be expressed in terms of the discrete-time Fourier series coefficients as:

$$P_x = r_{xx}(0) = \frac{1}{N}\sum_{k=0}^{N-1} x(k)x^*(k) = \sum_{k=0}^{N-1} c_n c_n^* \tag{2.86}$$

Despite the fact that the power spectrum of a periodic signal is a line spectrum (i.e., discrete in frequency), use of the delta function allows it to be represented as a continuous-time spectrum. For example, noting that the coefficients $\{c_n\}$ are periodic with period N so that $c_{n+N} = c_n$ and that the coefficients c_n is associated with discrete-time frequency $f = \frac{n}{N}$, we may define the power spectral density, $P_{xx}(f)$, as

$$P_{xx}(f) = \sum_{n=-\infty}^{\infty} c_n c_n^* \delta\left(f - \frac{n}{N}\right) \tag{2.87}$$

With $P_{xx}(f)$ defined this way, we note that the expressions for the average power P_x of the periodic signal $x(k)$ may be written as:

$$P_x = r_{xx}(0) = \int_0^1 P_{xx}(f)\, df \tag{2.88}$$

As the Discrete Fourier Transform (DFT) algorithm is often used as a frequency analysis tool, it is useful to know that the discrete-time Fourier series coefficients are identically equal to a scaled-version of the N-point DFT; in particular, if $\{X(n)\}$ denotes the N-point DFT of the N-periodic signal, $x(k)$, then the Fourier series coefficients, $\{c_n\}$, may be expressed in terms of the DFT coefficients as:

$$c_n = \frac{1}{N} X(n) \tag{2.89}$$

We may thus rewrite the expressions for the average power of the periodic signal $x(k)$ as:

$$P_x = \frac{1}{N}\sum_{k=0}^{N-1} x(k)x^*(k) = \frac{1}{N^2}\sum_{n=0}^{N-1} X(n)X^*(n) \tag{2.90}$$

Summarizing expressions for energy E_x for the energy signal, the average power P_x for power signals and the average power for periodic (power) signals may be expressed compactly as:

$$r_{xx}(0) = \int_0^1 P_{xx}(f)\, df = \begin{cases} E_x & \textit{energy signal} \\ P_x & \textit{power signal} \\ \sum_{k=0}^{N-1} c_n c_n^* & \textit{periodic signal} \end{cases} \tag{2.91}$$

Cross-Spectral Density and Degree of Similarity/Dissimilarity

Two signals are said to be uncorrelated if there is no spectral overlap between them (i.e., if $P_{xy}(f) = 0$) for all f; otherwise, we say they are correlated.

$$P_{xy}(f) = \begin{cases} 0 & \text{for all } f \quad \text{uncorrelated} \\ \neq 0 & \text{for some } f \quad \text{correlated} \end{cases} \tag{2.92}$$

For example, two sinusoids of different frequency are uncorrelated since there is no overlap in their spectra (whereever one spectrum has a non-zero value, the other has a value of 0).

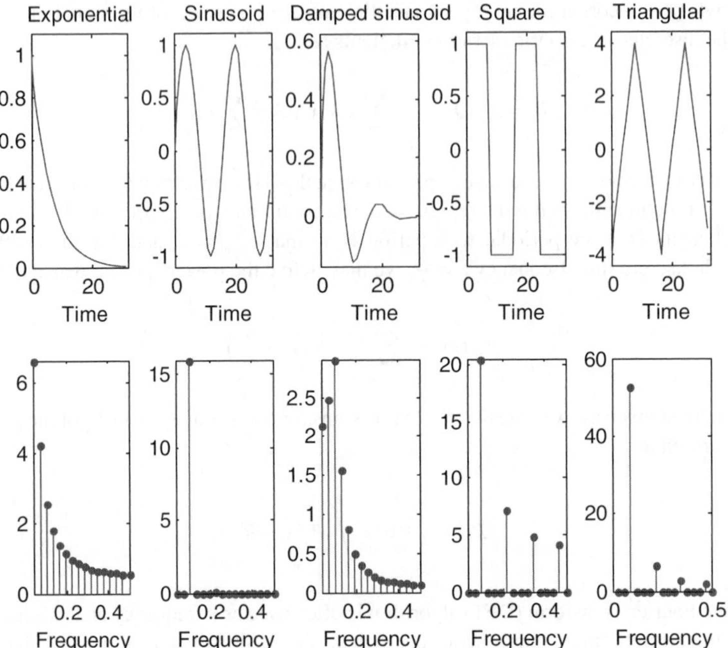

Figure 2.2 Correlated signals

For example, a sine wave, a square wave, and a triangular wave with identical fundamental frequency f_0, are correlated since their spectra have all have non-zero values at $f = f_0$.

Figure 2.2 shows examples of correlated signals. Note that the time domain signals (a decaying exponential, a sinusoid, a damped sinusoid, a square wave, and a triangular wave) are shown along the top row with the corresponding spectra shown in the bottom row. Since all spectra have a non-zero value at $f = 1/16$, the signals are said to be correlated.

Figure 2.3 shows examples of uncorrelated signals; the signals include a discrete-time sinusoid with fundamental frequency $f_0 = 1/32$, a square wave with fundamental frequency $f_0 = 1/16$, and a triangular wave with fundamental frequency $f_0 = 1/8$. Since the even harmonics of both the square wave and the triangular wave are all equal to zero, and since the square and triangular waves do not have any odd harmonics in common (i.e., there are no odd integers m and k such that $m/16 = k/8$), there is no frequency for which the spectra of the square and triangular waves both have non-zero values. Furthermore, the spectra of the square and triangular waves both have a value of zero at f=1/16 which is the only value for which the spectrum of the sinusoid is non-zero. Hence there is no overlap between the spectra of the signals shown in Figure 2.3.

Figure 2.3 shows examples of uncorrelated signals. A sinusoid, a square wave, and a triangular wave are shown at the top while their spectra are shown at the bottom. Note that there is no frequency where there is a spectral overlap.

A normalized measure of correlation, termed coherence, is introduced in the next section.

2.4 Coherence Spectrum

The coherence spectrum provides a frequency-dependent measure of similarity (or dissimilarity) between two stationary random signals and is derived from the normalized cross-spectral density. The coherence

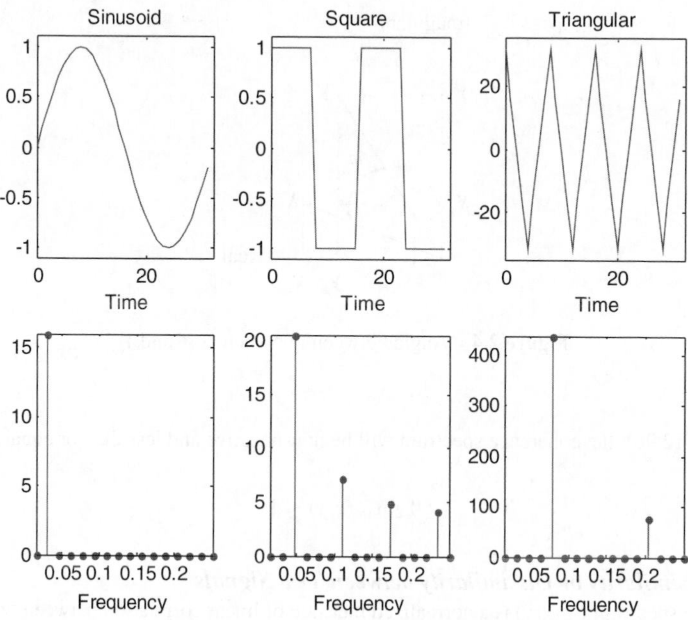

Figure 2.3 Uncorrelated signals

is a measure of the correlation between two signals, at each frequency. In this work the coherence spectral measure is defined as the cosine-squared of an angle between the Fourier transforms of the two signals. This measure is used widely, including by industries that specialize in test and measurement devices for monitoring the status of a system.

The normalized cross-spectral density of signals $x(k)$ and $y(k)$, denoted $P_{xy}^{N}(f)$, is given by

$$P_{xy}^{N}(f) = \frac{P_{xy}(f)}{\sqrt{P_{xx}(f)P_{yy}(f)}} = \frac{X(f)Y^{*}(f)}{|X(f)|\,|Y(f)|} \tag{2.93}$$

The coherence spectrum, denoted $C_h(x, y)$, is derived from the normalized cross-spectral density $P_{xy}^{N}(f)$ so that it is geared for applications including fault diagnosis and performance monitoring [1], It is defined as:

$$C_h(x, y) = \frac{|real\,(X(f)\,Y^{*}(f))|^2}{|X(f)|^2\,|Y(f)|^2} \tag{2.94}$$

Expressing $X(f) = X_r(f) + jX_i(f)$, $Y(f) = Y_r(f) + jY_i(f)$ where $X_r(f) = real\,\{X(f)\}$, $X_i(f) = image\,\{X(f)\}$, $Y_r(f) = real\,\{Y(f)\}$; $Y_i(f) = image\,\{Y(f)\}$, we get:

$$C_h(x, y) = \frac{(X_r(f)Y_r(f) + X_i(f)Y_i(f))^2}{(X_r^2(f) + X_i^2(f))(Y_r^2(f) + Y_i^2(f))} \tag{2.95}$$

It is shown in the Appendix that the expression for the coherence spectrum given by Eq. (2.94) is equal to the $\cos^2(\varphi)$ where φ is the angle between $X(f)$ and $Y(f)$ as shown in Figure 2.4:

$$C_h(x, y) = \cos^2(\varphi) \tag{2.96}$$

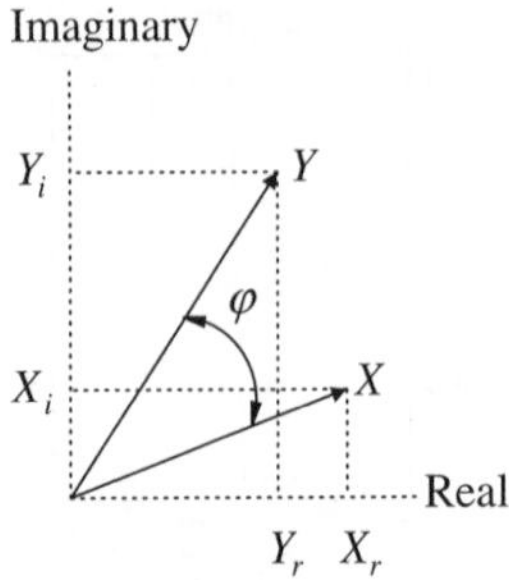

Figure 2.4 Angle between two vectors X and Y

In view of Eq. (2.96), the coherence spectrum will be non-negative and less than or equal to unity:

$$0 \le C_h(x, y) \le 1 \tag{2.97}$$

Measure of Similarity or Dissimilarity between Two Signals

The coherence spectrum $C_h(x, y)$ is a normalized measure of linear correlation between two signals $x(k)$ and $y(k)$. It is essentially the degree of spectral overlap between spectra $X(f)$ and $Y(f)$. The coherence spectrum of two signals will be unity at all frequencies, that is, $C_h(x, y) = 1$ for all $f \in [0\,0.5]$ if $x(k)$ and $y(k)$ are identical or collinear, that is, $y(k) = \alpha x(k)$ (or equivalently $Y(f) = \alpha X(f)$) where α is a scaling factor.

If $X(f)$ and $Y(f)$ are not identical or collinear, then the coherence spectrum will be less than unity at some frequencies

$$C_h(x, y) < 1 \text{ for some or all } f \tag{2.98}$$

The coherence spectrum is widely used in many areas, including medical diagnosis, performance monitoring, and fault diagnosis [1–3]. Further, it can detect if the measurement is contaminated by correlated noise or external signal sources, additional inputs present in the system, or system nonlinearity. If the spectral bandwidth of interest is known, then only coherence over the frequency of interest is considered and the rest of the spectral regions are ignored as they contain noise and other artifacts.

2.5 Illustrative Examples in Correlation and Spectral Density

2.5.1 Deterministic Signals: Correlation and Spectral Density

Auto-Correlation

Example 2.1 Constant

Compute the auto-correlation function of $s(k) = A, \quad k \in (-\infty\ +\infty)$

Solution: Since $s(k)$ is a power signal, its auto-correlation function is given by

$$r_{ss}(m) = \underset{N\to\infty}{\lim it}\left\{\frac{1}{(2N+1)}\sum_{n=-N}^{N} s(k)s(k-m) = \frac{1}{(2N+1)}\sum_{k=-N}^{N} A^2\right\} = A^2$$

Hence the auto-correlation of a constant is the square of the constant. Furthermore, since $r_{ss}(m) = A^2$ the power spectral density is the Fourier transform of the auto-correlation function, we find the power spectral density of the constant A is a delta function with area A^2 positioned at $f = 0$.

$$P_{ss}(f) = A^2\delta(f)$$

Example 2.2 Sinusoid
Find the auto-correlation function of

$$s(k) = A\sin(2\pi f_0 k + \varphi),\ \ k \in (-\infty \quad +\infty),\quad f_0 = 1/M$$

where $s(k) = s(k + M)$.

Solution: Since $s(k)$ is M-periodic (i.e., $s(k) = s(k + M)$), its auto-correlation function:

$$r_{ss}(m) = \frac{1}{M}\sum_{k=0}^{M-1} s(k)s(k-m) = \frac{1}{M}\sum_{k=0}^{M-1} A^2\sin(2\pi f_0 k + \phi)\sin(2\pi f_0(k-m) + \phi)$$

Using the identity $2\sin A\sin B = \cos(A - B) - \cos(A - B)$ we find:

$$r_{ss}(m) = \frac{1}{2M}\sum_{k=0}^{M-1} A^2\cos\left(2\pi f_0 m\right) - \frac{1}{2M}\sum_{k=0}^{M-1} A^2\cos[2\pi f_0\,(2k - m) + 2\varphi]$$

Note that the second term on the right is zero since it is a sum of a sinusoid over two periods. Furthermore the summand of the first term is constant (does not depend upon the time k). Hence

$$r_{ss}(m) = \frac{A^2}{2}\cos(2\pi f_o m)$$

This example illustrates that the auto-correlation of a sinusoid of frequency f_0, amplitude A, and arbitrary phase φ, $s(k) = A\sin(2\pi f_0 k + \varphi)$ is a sinusoid of the same frequency, f_0, and amplitude A^2; the phase information of the original sinusoid is not preserved by the auto-correlation function,$r_{ss}(m) = (A^2/2)\cos(2\pi f_o m)$, as per the properties of the auto-correlation function, the maximum value of A^2 will occur at a lag of 0. Taking the Fourier transform of $r_{xx}(m)$, we find the power spectral density of a sinusoid with frequency f_0 and amplitude A is:

$$P_{ss}(f) = \frac{A^2}{4}[\delta(f - f_0) + \delta(f + f_0)]$$

Example 2.3 Exponential
Determine the auto-correlation of $s(k)$ where

$$s(k) = \begin{cases} \rho^k & 0 \le k < \infty \\ 0 & k < 0 \end{cases}$$

where $|\rho| < 1$.

Solution: The signal $s(k)$ is an energy signal. The auto-correlation $r_{ss}(m)$ is given by

$$r_{ss}(m) = \sum_{k=0}^{\infty} s(k)s(k-m)$$

We will use the frequency-domain approach. The energy spectral density is given by

$$E_{ss}(z) = S(z)\,S(z^{-1})$$

where

$$S(z) = \frac{1}{1 - \rho z^{-1}}$$

Substituting for $S(z)$ and $S(z^{-1})$ we get

$$E_{ss}(z) = \frac{1}{(1-\rho z^{-1})(1-\rho z)}$$

Using partial fraction expansion we get

$$E_{ss}(z) = \frac{\alpha_0}{1-\rho z^{-1}} + \frac{\alpha_0}{1-\rho z} + \beta_0$$

where α_0 and β_0 are the coefficients of the expansion. The equation relating energy spectral density and its partial fraction expansion becomes

$$\frac{1}{(1-\rho z^{-1})(1-\rho z)} = \frac{\alpha_0}{1-\rho z^{-1}} + \frac{\alpha_0}{1-\rho z} + \beta_0$$

Note that the first term on the right is the z-transform of $r_{ss}(m)$ for $m \geq 0$, and the sum of the second and the third term is the z-transform of $r_{ss}(m)$ for $m < 0$. Simplifying yields

$$1 = \alpha_0 - \alpha_0\rho z + \alpha_0 - \alpha_0\rho z^{-1} + \beta_0 - \beta_0\rho z - \beta_0\rho z^{-1} + \beta_0\rho^2$$

Collecting terms in z^{-1}, z, and constant, and equating them to zero, the unknowns α_0 and β_0 are

$$\alpha_0 = \frac{1}{1-\rho^2}$$
$$\beta_0 = -\alpha_0$$

Taking the inverse z-transform, the correlation function $r_{ss}(m)$ becomes

$$r_{ss}(m) = \frac{\rho^{|m|}}{1-\rho^2}$$

Figure 2.5 shows the exponential signal $s(k)$, the magnitude of its Fourier transform $|S(f)|$, the auto-correlation $r_{ss}(m)$, and the energy spectral density $E_{ss}(f) = |S(f)|^2$.

Example 2.4 Damped sinusoid
Let us compute the auto-correlation of a damped exponential signal which is defined for $(0 \quad \infty)$ by

$$s(k) = \rho^k \sin 2\pi f_0 k$$

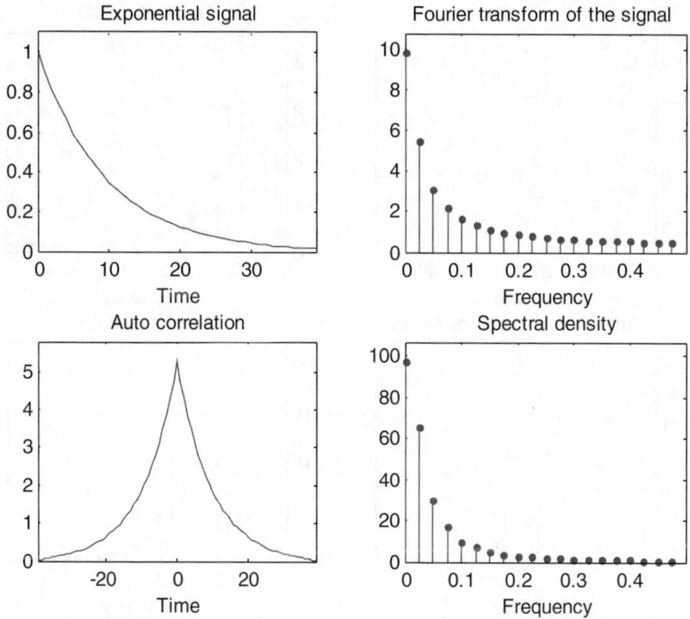

Figure 2.5 An exponential signal, its Fourier transform, auto-correlation, and spectral density

Solution: It is easier to use the frequency-domain rather than the time-domain approach to obtain the correlation.

$$E_{ss}(z) = S(z)S(z^{-1})$$

where

$$S(z) = \frac{b_1 z^{-1}}{D_s(z)};\ D_s(z) = 1 + a_1 z^{-1} + a_2 z^{-2};\ a_1 = -2\rho\cos\omega_0;\ a_2 = \rho^2;\ b_1 = \rho\sin\omega_0$$

Substituting for $S(z)$ and $S(z^{-1})$ we get

$$E_{ss}(z) = \frac{\rho^2 b_1^2}{(1 + a_1 z^{-1} + a_2 z^{-2})(1 + a_1 z + a_2 z^2)}$$

Using the partial fraction expansion given in the Appendix we get

$$\frac{b_1^2}{D_s(z)D_s(z^{-1})} = \frac{\alpha_1 z^{-1} + \alpha_0}{D_s(z)} + \frac{\alpha_1 z + \alpha_0}{D_s(z^{-1})} + \beta_0$$

where α_1 α_0 and β_0 are the coefficients of partial expansion; the first term on the right is the z-transform of $r_{ss}(m)$ for $m \geq 0$, and the sum of the second and the third term is the z-transform of $r_{ss}(m)$ for $m < 0$. The partial fraction coefficients α_1, α_0, and β_0 are determined from

$$\begin{bmatrix} b_1^2 \\ 0 \end{bmatrix} = \begin{bmatrix} 2a_1 & 1 - a_1^2 - a_2^2 \\ 1 + a_2 & -a_1 a_2 \end{bmatrix} \begin{bmatrix} \alpha_1 \\ \alpha_0 \end{bmatrix}$$

$$\beta_0 = -\alpha_0$$

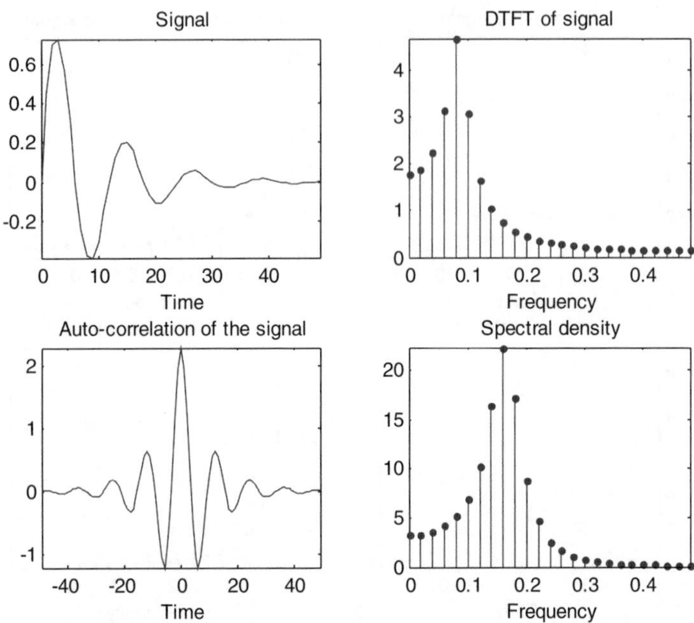

Figure 2.6 A damped sinusoid, its Fourier transform, and auto-correlation spectral density

Taking the inverse z-transform of $E_{ss}(z)$ yields the auto-correlation function $r_{ss}(m)$ of $s(k)$. Thus *the auto-correlation of a damped sinusoid* $s(k) = \rho^k \sin 2\pi f_0 k$ *is a damped sinusoid* given by

$$r_{ss}(m) = \alpha_0 \rho^{|m|} \cos 2\pi f_0 m$$

The correlation does not preserve the phase information of the sinusoid. Figure 2.6 shows the exponential signal $s(k)$, the magnitude of its Fourier transform $|S(f)|$, the auto-correlation $r_{ss}(m)$, and energy spectral density $P_{ss}(f) = |S(f)|^2$.

Cross-correlation

Example 2.5 Cross-correlation of two sinusoids
Compute the cross-correlation of two sinusoids $s_1(k)$ and $s_2(k)$ of two different frequencies f_1 and f_2, and phases φ_1 and φ_2 which are defined for $k \in (-\infty \quad \infty)$

$$s_1(k) = \sin(2\pi f_1 k + \varphi_1)$$
$$s_2(k) = \sin(2\pi f_2 k + \varphi_2)$$

Solution: We will use frequency domain approach by computing the DTFTs of the two signals, The DTFTs of $s_1(k)$ and $s_2(k)$ are

$$S_1(f) = \frac{\cos\varphi_1}{2j}[\delta(f - f_1) - \delta(f + f_1)] + \frac{\sin\varphi_1}{2}[\delta(f - f_1) + \delta(f + f_1)]$$

$$S_2(f) = \frac{\cos\varphi_2}{2j}[\delta(f - f_2) - \delta(f + f_2)] + \frac{\sin\varphi_2}{2}[\delta(f - f_2) + \delta(f + f_2)]$$

The cross-power spectral density $P_{12}(f)$ of $s_1(k)$ and $s_2(k)$ is

$$P_{12}(f) = \lim_{N\to\infty} it\left(\frac{S_{N1}(f)S^*_{N2}(f)}{N}\right)$$

Since there is no overlap in the spectral frequencies $S_1(f)$ and $S_2(f)$ we get

$$P_{12}(f) = 0$$

Thus the cross-spectral density, and hence the cross-correlation of two sinusoids with different frequencies, is zero.

Example 2.6 Sinusoid and square wave
Compute the cross-correlation of a sinusoid $s_1(k)$ and symmetric square wave $s_2(k)$, which are defined for $k \in (-\infty \quad \infty)$, and assuming that their periods are identical:

$$s_1(k) = \sin(2\pi f_0 k)$$
$$s_2(k) = square(2\pi f_0 k)$$

Solution: Expanding the symmetric square wave using Fourier series yields

$$s_2(k) = \frac{4}{\pi}\sum_{n=1,3,5,\dots}^{\infty}\left(\frac{1}{n}\right)\sin 2\pi n f_0 k$$

The cross-correlation of $s_1(k)$ and $s_2(k)$ is

$$r_{12}(m) = \frac{1}{M}\sum_{k=0}^{M-1} s_1(k)s_2(k-m) = \frac{4}{\pi M}\sum_{k=0}^{M-1}\sin(2\pi f_0 k)\left(\sum_{n=1,3,5,\dots}^{\infty}\left(\frac{1}{n}\right)\sin[2\pi n f_0(k-m)]\right)$$

Since the cross-correlation of two sinusoids of different frequencies is zero, all product terms other than $n = 1$ are zero, yielding

$$r_{12}(m) = \frac{4}{\pi M}\sum_{k=0}^{M-1}\sin(2\pi f_0 k)\sin[2\pi f_0(k-m)]$$

Hence we get

$$r_{12}(m) = \frac{2}{\pi}\cos 2\pi f_0 m$$

The cross-spectral density $P_{12}(f)$ is given by

$$P_{12}(f) = \frac{2}{\pi^2}[\delta(f - f_0) + \delta(f + f_0)]$$

Example 2.7 Waveform and its delayed version
Compute the cross-correlation of the two waveforms $x(k)$ and $y(k)$ where $y(k)$ is a delayed version of $x(k)$ and determine the time delay τ between the two rectangular pulse waveforms where $c_k = 1$, for all k:

$$x(k) = \begin{cases} c_k & 0 \le k \le N-1 \\ 0 & else \end{cases}$$

$$y(k) = x(k-\tau) = \begin{cases} c_{k-\tau} & \tau \le k \le N-1+\tau \\ 0 & else \end{cases}$$

Solution: We will use the frequency-domain approach and consider a general case of a finite duration signal, and then obtain the solution for the case of a rectangular pulse. The cross-correlation function is related to the auto-correlation function as follows

$$r_{xy}(m) = \sum_{k=0}^{N-1} x(k)y(k-m) = \sum_{k=0}^{N-1} x(k)x(k-\tau-m) = r_{xx}(m+\tau)$$

Using the frequency-domain approach, we will compute $r_{xy}(m)$ by computing first $r_{xx}(m)$ and then $r_{xy}(m) = r_{xx}(m+\tau)$.

Computation of $r_{xx}(m)$
The energy spectral density $E_{xx}(z)$

$$E_{xx}(z) = X(z)X(z^{-1}) = \sum_{i=0}^{N-1}\sum_{j=0}^{N-1} c_i c_j z^{-i+j}$$

Expanding using $j = i + m$, and since $c_i = 0$ for $i \ge N$ and $i < 0$, we get

$$E_{xx}(z) = \sum_{m=0}^{N-1}\sum_{i=0}^{N-1-m} c_i c_{i+m} z^{-m} + \sum_{m=0}^{N-1}\sum_{i=0}^{N-1-m} c_i c_{i+m} z^{m} - \sum_{i=0}^{N-1} c_i^2$$

We will consider the special case when $x(k)$ is a rectangular pulse. In this case since $c_k = 1$ for all k we get

$$E_{xx}(z) = \sum_{m=0}^{N-1}\sum_{i=0}^{N-1-m} z^{-m} + \sum_{m=0}^{N-1}\sum_{i=0}^{N-1-m} z^{m} - \sum_{i=0}^{N-1} 1$$

Simplifying we get

$$E_{xx}(z) = \sum_{m=0}^{N-1} (N-m)\, z^{-m} + \sum_{m=0}^{N-1} (N-m)\, z^{m} - N$$

The first term on the right is the z-transform of $r_{xx}(m)$ for $m \ge 0$, the second term is for $m \le 0$, and the last term is for $m = 0$.

Taking the inverse z-transform the auto-correlation function $r_{xx}(m)$ becomes

$$r_{xx}(m) = \begin{cases} N - |m| & 0 \le |m| \le N-1 \\ 0 & |m| \ge N \end{cases}$$

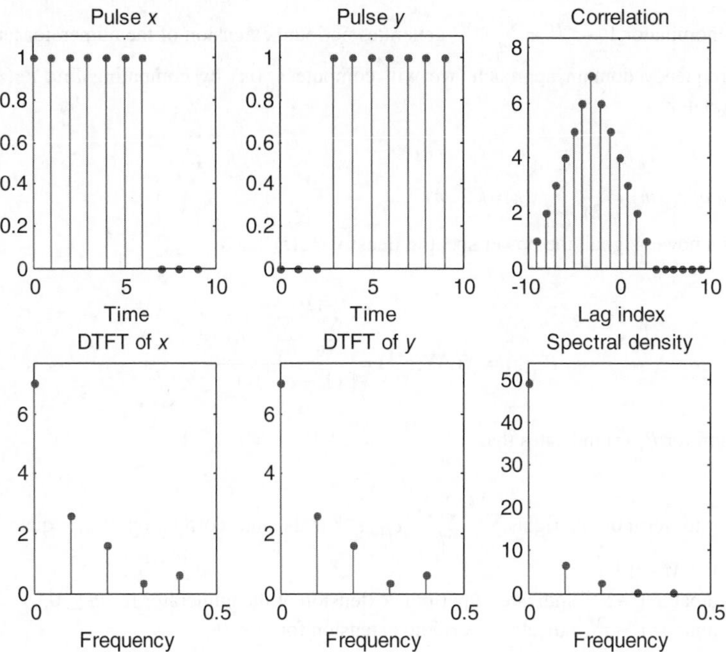

Figure 2.7 A pulse, its delayed version, the cross-correlation, and their magnitude spectra

Hence the cross-correlation function $r_{xy}(m) = r_{xx}(m + \tau)$ becomes

$$r_{xy}(m) = r_{xx}(m+\tau) = \begin{cases} N - |m+\tau| & 0 \le |m+\tau| \le N-1 \\ 0 & |m| \ge N \end{cases}$$

The maximum of the cross-correlation occurs when $m = -\tau$ and the maximum value is N.

The maximum value of the cross-correlation occurs at $m = -\tau = -3$, which is the time delay between the two waveforms. Figure 2.7 shows a pulse $x(k)$ of duration $N = 7$ and its delayed version $y(k) = x(k - \tau)$ where the delay $\tau = 3$. The cross-correlation $r_{xx}(m)$ is maximum at $m = -\tau$, indicating that the cross-correlation may be used to determine the time delay between two signals. The spectra of the signals and the energy spectral density are shown at the bottom of the figure. Note that *the cross-correlation of two pulse waveforms is a triangular pulse*.

Example 2.8 Square wave and its delayed version
We will use the frequency domain approach and consider a general case of an M-periodic wave form and then obtain the solution for the case of a square wave. The z-transform of a M-periodic signal $x(k)$ is

$$X(z) = \frac{\sum_{i=0}^{M-1} c_i z^{-i}}{1 - z^{-M}}$$

where the denominator $1 - z^{-M} = \sum_{i=0}^{\infty} z^{-Mi}$ generates periodic extension of the numerator term $\sum_{i=0}^{M-1} c_i z^{-i}$. Using the frequency domain approach, we will compute $r_{xy}(m)$ by computing first $r_{xx}(m)$ and then $r_{xy}(m) = r_{xx}(m + \tau)$.

Computation of $r_{xx}(m) = \frac{1}{M} \sum_{k=0}^{N-1} x(k)x(k-m)$

Since $x(k)$ is a power signal, the power spectral density $P_{xx}(z)$

$$P_{xx}(z) = X(z)X(z^{-1}) = \frac{\sum_{m=0}^{M-1} \sum_{i=0}^{M-1-m} c_i c_{i+m} z^{-m}}{(1 - z^{-M})(1 - z^{M})}$$

The expression for $P_{xx}(z)$ indicates that

- The numerator term on the right $\sum_{m=0}^{M-1} \sum_{i=0}^{M-1-m} c_i c_{i+m} z^{-m}$ is the auto-correlation of the square wave in the interval $[\,0 \quad M-1\,]$.
- The denominator $1 - z^{-M}$ indicates a periodic extension of the numerator for $m \geq 0$.
- The denominator $1 - z^{M}$ indicates a periodic extension for $m < 0$.

Let us consider a square wave which has an odd symmetry given by

$$c_i = \begin{cases} 1 & 0 \leq i \leq M_1 - 1 \\ -1 & M_1 \leq i \leq M - 1 \end{cases}$$

where $M_1 = M/2$. For this case it can be shown that the expression for the numerator term is

$$\sum_{m=0}^{M-1} \sum_{i=0}^{M-1-m} c_i c_{i+m} z^{-m} = \sum_{m=0}^{M-1} M - 4\,|m|: \quad -M_1 \leq m \leq M_1 - 1$$

Taking the inverse z-transform of $P_{xx}(z)$ we get

$$r_{xx}(m) = \begin{cases} N - 4\,|m| & M_1 \leq m \leq M_1 - 1 \\ r_{xx}(m + \ell M) & \ell = \pm 1, \pm 2, \pm 3, \ldots, \end{cases}$$

That is, *the auto-correlation of an odd symmetric square wave is an* M-*periodic triangular wave.* Hence the cross-correlation $r_{xy}(m) = r_{xx}(m + \tau)$ of a square wave $x(k)$ and its delayed version $y(k) = x(k - \tau)$ is

$$r_{xy}(m) = \begin{cases} N - 4\,|m + \tau| & M_1 \leq m + \tau \leq M_1 - 1 \\ r_{xy}(m + \ell M) & \ell = \pm 1, \pm 2, \pm 3, \ldots, \end{cases}$$

Figure 2.8 shows a square wave, its delayed version, and their cross-correlation on the top and their spectra at the bottom. The cross-correlation $r_{xx}(m)$ is maximum at $m = -\tau$, indicating that the cross-correlation may be used to determine the time delay between two signals. Further *the cross-correlation of two square waves is a triangular wave.*

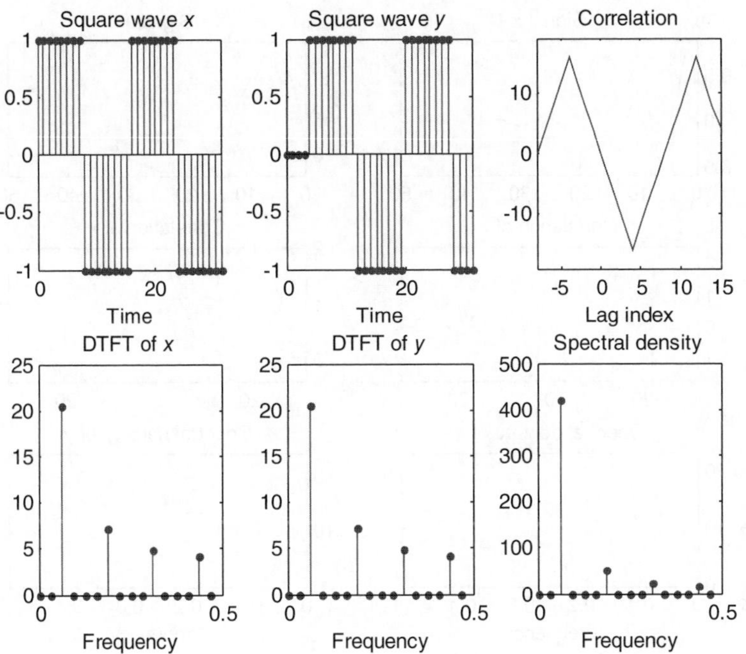

Figure 2.8 A square wave, its delayed version, their cross-correlation, and their spectra

Coherence Spectrum

Example 2.9 Two damped sinusoids
Let $x(k)$, and $y(k)$ be signals defined for $k \in (0 \quad \infty)$ given by

$$x(k) = \rho_1^k \sin \omega_1 k$$

$$y(k) = \rho_2^k \sin \omega_2 k$$

Compute the coherence spectrum of $x(k)$ and $y(k)$.

Solution:

$$X(z) = \frac{\rho_1 z^{-1} \sin \omega_1}{1 - 2\rho_1 z^{-1} \cos \omega_1 + \rho_1^2 z^{-2}}$$

$$Y(z) = \frac{\rho_2 z^{-1} \sin \omega_2}{1 - 2\rho_2 z^{-1} \cos \omega_2 + \rho_2^2 z^{-2}}$$

Figure 2.9 shows the damped sinusoidal signals $x(k)$ and $y(k)$, auto-correlation functions $r_{xx}(m)$ and $r_{yy}(m)$, and their energy spectral densities $E_{xx}(f)$ and $E_{yy}(f)$.

The coherence spectrum $C_h(x, y)$ and the energy spectral density $E_{xy}(f)$ are shown in Figure 2.10. The two signals differ in the frequency interval $(0 \quad 0.2)$, since $C_h(x, y) < 1$, and are identical over the rest.

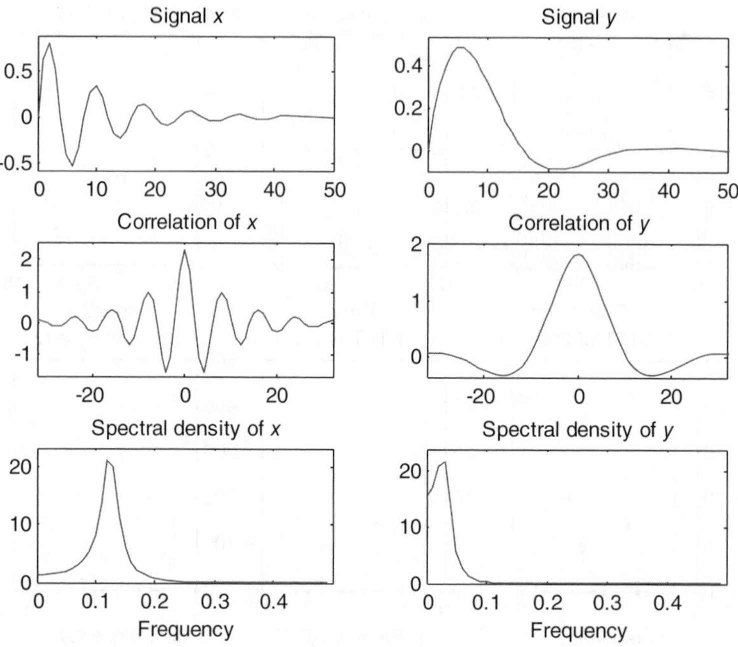

Figure 2.9 Signals, auto-correlations, and energy spectral densities

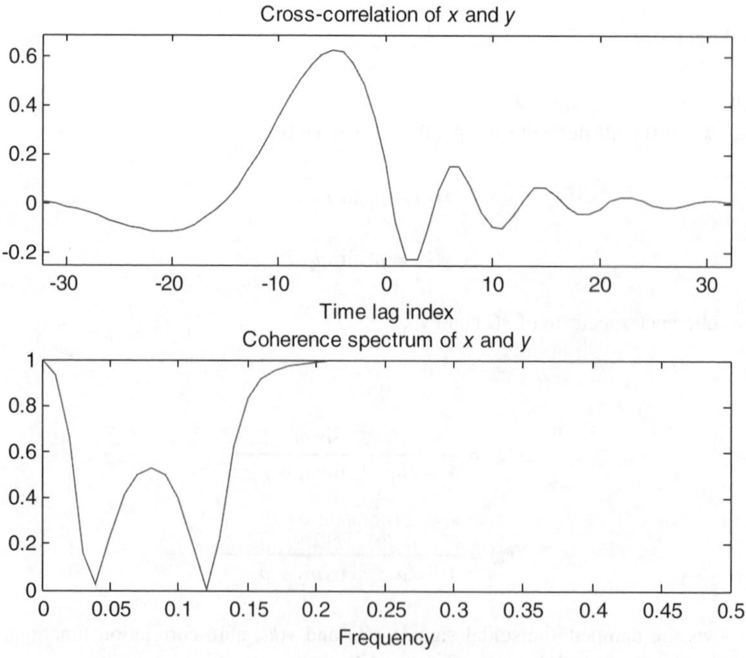

Figure 2.10 The cross-correlation and the coherence spectrum of the signals

2.5.2 Random Signals: Correlation and Spectral Density

Example 2.10 Noisy signals
Let $x_0(k)$ and $x(k)$ be signals and $v(k)$ be a zero-mean white noise process corrupting $x(k)$ defined for $k \in (-\infty \quad \infty)$:

$$x_0(k) = sin2\pi f_0 k$$
$$x(k) = x_0(k) + v(k)$$

where $v(k)$ is a zero-mean white noise process with variance σ_v^2. Using the property that the auto-correlation of a sum of two uncorrelated signal is the sum of their auto-correlations, as shown in Eq. (2.41), the auto-correlation of the noisy signal is the sum of the auto-correlation of the noise-free signal and the auto-correlation of the white noise given by

$$r_{xx}(m) = r_{x_0x_0}(m) + r_{vv}(m)$$

Since $v(k)$ is a white noise, its auto-correlation is

$$r_{vv}(m) = \sigma_v^2 \delta(m)$$

The auto-correlation of the noisy signal is the auto-correlation of the noise free signal except at zero lag. At zero lag the auto-correlation is the sum of auto-correlation of the noise free signal at zero lag and the variance of the white noise:

$$r_{xx}(m) = \begin{cases} r_{x_0x_0}(0) + \sigma_v^2 & m = 0 \\ r_{x_0x_0}(m) & m \neq 0 \end{cases}$$

Figure 2.11 shows the noise free signal $x_0(k)$, noisy signal $x(k)$, the noise $v(k)$, their auto-correlations $r_{x_0x_0}(m)$, $r_{xx}(m)$, and $r_{vv}(m)$ respectively, and their spectral densities $P_{x_0x_0}(m)$, $P_{xx}(m)$, and $P_{vv}(m)$. From the auto-correlation of the noisy signal, we can deduce the amplitude and the frequency (but not the phase of the sinusoid) of the noise free sinusoidal signal $x_0(k)$, and the variance of the white noise. The variance of the noise σ_v^2 is the difference between the correlation at zero lag $r_{xx}(0)$, and the amplitude of the sinusoid computed from $r_{xx}(m)$.

Signal Enhancing and Noise Attenuation Property

The signal enhancing and noise attenuating properties of correlation are widely used to determine (i) which of the known signals is present in a given noisy observation with applications including demodulation, fault isolation, and pattern classification, and (ii) estimated time delay between a noisy signal and a reference signal, with applications in system identification, biomedical signal processing, estimation of distance, and so on.

Example 2.11 Noisy signal with a known deterministic signal
Let $x_0(k)$,$x(k)$, and $y(k)$ be signals and $v(k)$ be a zero-mean white noise process corrupting $x(k)$ defined for $k \in (-\infty \quad \infty)$:

$$x_0(k) = sin(2\pi f_0 k + \varphi)$$
$$x(k) = x_0(k) + v(k)$$
$$y(k) = sin2\pi f_0 k$$

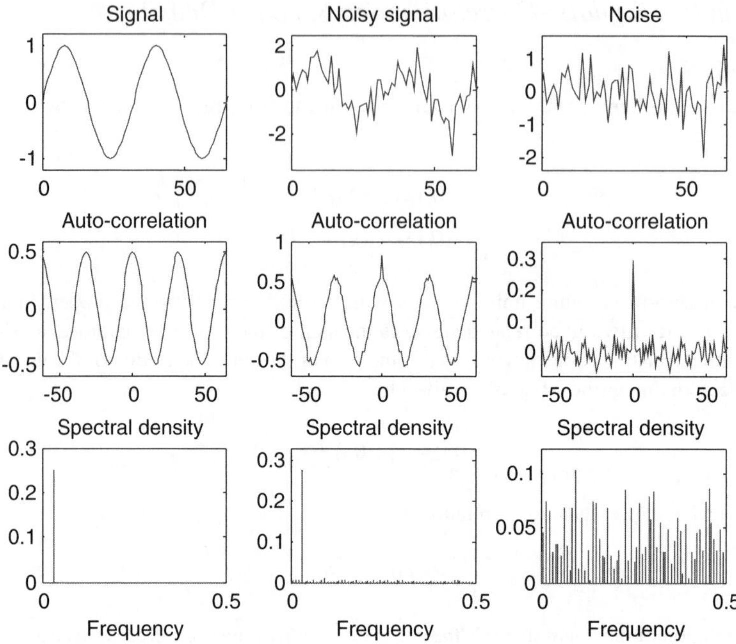

Figure 2.11 Signals, correlation, and power spectral densities

The signal $y(k)$ is highly correlated with the signal $x_0(k)$ and uncorrelated with the noise $v(k)$. The cross-correlation $r_{xy}(m)$ of $x(k)$ and $y(k)$ will enhance the signal $x_0(k)$ buried in $x(k)$ and attenuate the noise $v(k)$ corrupting $x(k)$. The cross-correlation of $x(k)$ and $y(k)$ is given by

$$r_{xy}(m) = r_{x_0y}(m) + r_{vy}(m)$$

Since $v(k)$ is uncorrelated with $y(k)$ we get

$$r_{xy}(m) = r_{x_0y}(m)$$

The cross-correlation $r_{x_0y}(m)$ is the noise free sinusoid

$$r_{xy}(m) = \frac{1}{2}\cos 2\pi f_0 m$$

Figure 2.12 shows the noisy signal $x(k)$, a signal $y(k)$ which is matched to $x_0(k)$, the cross-correlation $r_{xy}(m)$ which is matched to the filtering operation, the spectrum X, the spectrum Y, and the cross-power spectral density $P_{xy}(f)$.

It is clear from the figure that cross-correlating the noisy signal $x(k)$ by a signal $y(k)$ which is highly correlated with $x_0(k)$ and uncorrelated with the noise $v(k)$, produces a remarkable result. The cross-correlated signal $r_{xy}(m)$, also termed the matched filter output, is noise free and contains all the characteristics of the signal $x_0(k)$ except for its phase.

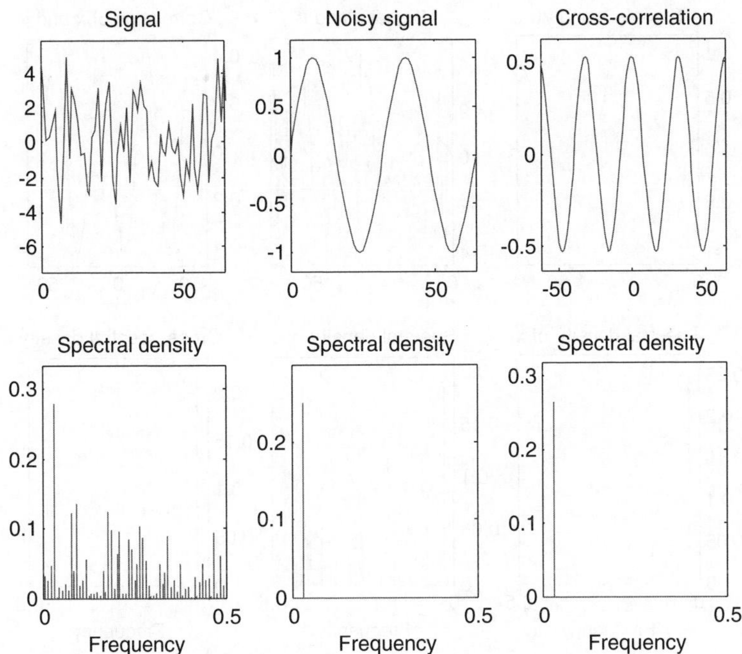

Figure 2.12 Matched filtering: a noisy signal and matching signal

Estimation of the Delay between Two Waveforms

Example 2.12 Two sinusoids
Let $x(k)$, $y_0(k)$, and $y(k)$ be signals and $v(k)$ be a zero-mean white noise process corrupting $y(k)$ defined for $k \in (-\infty \quad \infty)$ given by

$$x(k) = sin2\pi fk$$
$$y_0(k) = sin2\pi f_0 (k - \tau)$$
$$y(k) = y_0(k) + v(k)$$

Compute the time delay τ between the reference signal $x(k)$ and the noisy signal $y(k)$.

Solution: The reference signal $x(k)$ is highly correlated with the signal $y_0(k)$ and is uncorrelated with the noise $v(k)$. Consider the cross-correlation $r_{xy}(m)$ of $x(k)$ and $y(k)$

$$r_{xy}(m) = r_{xy_0}(m) + r_{xv}(m)$$

Since $v(k)$ is uncorrelated with $y(k)$ we get

$$r_{xy}(m) = r_{xy_0}(m)$$

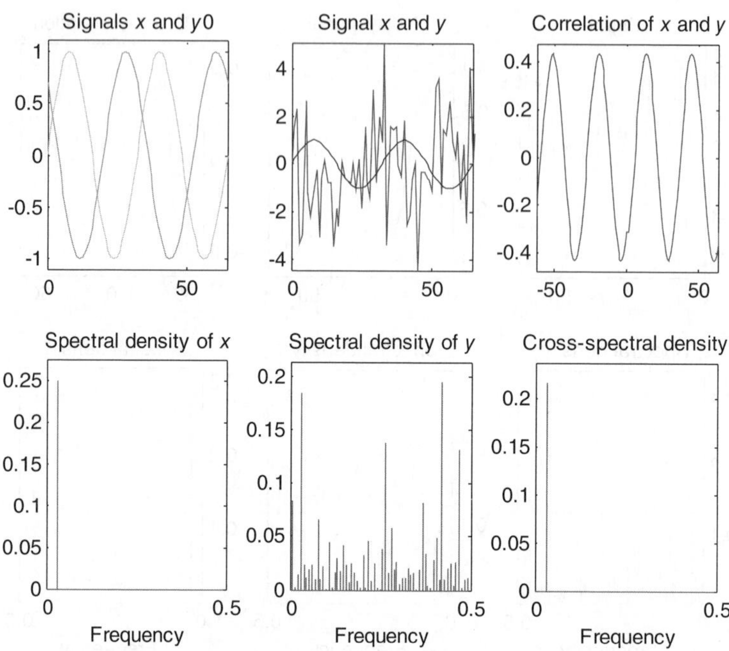

Figure 2.13 Signals, cross correlation, and spectral densities

Since $y_0(k)$ is the delayed version of $x(k)$, that is $y_0(k) = x(k - \tau)$, we get

$$r_{xy}(m) = r_{xx}(m - \tau)$$

Note that the time delay τ is the value of the time lag m where $r_{xy}(m)$ is maximum, that is $\tau = \max_m \{r_{xy}(m)\}$. Figure 2.13 shows the reference signal $x(k)$ and the noise-free signal $y_0(k)$, the noisy signal $y(k)$ which is a delayed and noise corrupted version of $x(k)$, the cross-correlation $r_{xy}(m)$ of $x(k)$ and $y(k)$, the power spectral density $P_{xx}(f)$ of $x(k)$, the power spectral density $P_{yy}(f)$ of $y(k)$, and the cross-power spectral density of $P_{xy}(f)$ of $x(k)$ and $y(k)$. Thanks to the noise attenuation property of the correlation operator, $r_{xy}(m)$ is practically a noise free sinusoid. An accurate estimate of the delay is given by the first maximum value of the correlation, namely $\tau = -20$, in spite of $y_0(k)$ being totally buried in the noise $v(k)$.

Comments *The results of simulation verify the following properties of correlation/spectral density:*

- *Similarity properties:*
  - *A waveform with a rational spectrum preserves its characteristics.*
    - *the auto-correlation of a constant is the square of the constant*
    - *an exponential is an exponential*
    - *a sinusoid is a sinusoid of the same frequency*
    - *an exponentially damped sinusoid is an exponentially damped sinusoid*
  - *A periodic signal is a periodic signal.*
  - *A square waveform is a triangular waveform.*
  - *A triangular waveform is a parabolic waveform.*

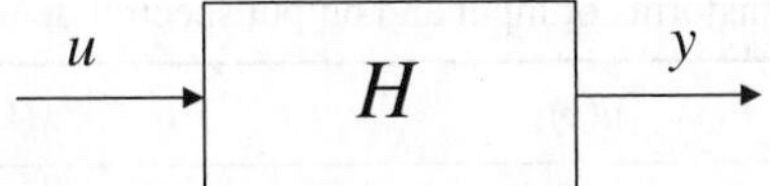

Figure 2.14 Input, output, and transfer function

- *Dissimilarity properties:*
  - *Cross-correlation of a deterministic waveform and a zero-mean stochastic process is zero.*
  - *Cross-correlation function of two waveforms with no spectral overlap is zero. For example. periodic waveforms of different periods are zero.*
  - *Cross-correlation function of stochastic processes, $x(k)$ and $y(k)$, is zero if the $E[X(f)Y^*(f)] = 0$.*
  - *Cross-correlation of periodic waveforms of different periods are zero.*

The properties of similarity/dissimilarity hinge upon the amount of spectral overlap between the two waveforms.

2.6 Input–Output Correlation and Spectral Density

In many applications one is faced with the problem of determining the cross-correlation between two waveforms that are related by some transfer function H, as shown in Figure 2.14 with an input u and an output y.

Input and Output Correlations and Spectral Densities

Consider a linear time invariant and asymptotically stable system whose impulse response is $h(k)$ and transfer function is $H(z)$, and with input $u(k)$ and output $y(k)$. It is assumed that the impulse response is causal, that is

$$h(k) = 0 \text{ for } k < 0 \tag{2.99}$$

The output $y(k)$ is a convolution of the input $u(k)$ and the impulse response $h(k)$ given by

$$y(k) = \sum_{i=0}^{k} h(i)u(k-i) \tag{2.100}$$

Taking the z-transform we get

$$Y(z) = H(z)U(z) \tag{2.101}$$

Input–output cross-correlation $r_{uy}(m) = E[u(k)y(k-m)]$ is given by

$$r_{uy}(m) = E\left[u(k)\left(\sum_{i=0}^{k} h(i)u(k-i-m)\right)\right] \tag{2.102}$$

Table 2.1 z-transforms of input and output spectral densities

	$u(k)$	$y(k)$
$u(k)$	$P_{uu}(z)$	$H\left(z^{-1}\right)P_{uu}(z)$
$y(k)$	$H(z)P_{uu}(z)$	$H(z)H\left(z^{-1}\right)P_{uu}(z)$

Note: The system $H(z)$ is causal whereas the system $H\left(z^{-1}\right)$ is anti-causal. The impulse response $h(k)$ of $H(z)$ is zero for $k < 0$, while the impulse response of $H\left(z^{-1}\right)$ is $h(-k)$ which is zero for $k > 0$. Expressing in terms of the frequency variable f using $z = e^{j2\pi f}$ we get.

Similarly cross-correlation $r_{yu}(m) = E[y(k)u(k-m)]$ becomes

$$r_{yu}(m) = E\left[\left(\sum_{i=0}^{k} h(i)u(k-i)\right)u(k-m)\right] \tag{2.103}$$

Using Eq. (2.101), the cross-spectral density $P_{uy}(z) = E[U(z)Y(z^{-1})]$ becomes

$$P_{uy}(z) = H(z^{-1})E[U(z)X(z^{-1})] = H(z^{-1})P_{uu}(z) \tag{2.104}$$

Similarly $P_{yu}(z) = E[Y(z)U(z^{-1})]$ and $P_{yy}(z) = E[Y(z)Y(z^{-1})]$ are given by

$$P_{yu}(z) = H(z)E[U(z)U(z^{-1})] = H(z)P_{uu}(z) \tag{2.105}$$

$$P_{yy}(z) = E[H(z)U(z)H(z^{-1})U(z^{-1})] = H(z)H(z^{-1})P_{uu}(z) \tag{2.106}$$

Tables 2.1 and 2.2 give the spectral densities relating all four combinations of U and Y in z-transform and Fourier transform for a general case when the input signal $u(k)$ may be an energy, a power, a finite duration, or a random signal by appropriately defining spectral density $P_{uu}(z)$.

2.6.1 Generation of Random Signal from White Noise

Applications of the expressions of spectral densities given by Eqs. (2.105) and (2.106) include system identification and modeling of a random signal as an output of a non-minimum phase linear system with white noise input. Let us choose the input signal to be a zero-mean white noise process $v(k)$ with flat power spectral density $P_{vv}(z) = \sigma_v^2$. Setting $u(k) = v(k)$ and substituting $P_{vv}(z) = \sigma_v^2$ in Eqs. (2.105) and (2.106) we get

$$P_{yv}(z) = \sigma_v^2 H(z) \tag{2.107}$$

$$P_{yy}(z) = \sigma_v^2 H(z)H(z^{-1}) \tag{2.108}$$

Table 2.2 Fourier transforms of input and output spectral densities

	$u(k)$	$y(k)$
$u(k)$	$P_{uu}(f)$	$H^*(f)P_{uu}(f)$
$y(k)$	$H(f)P_{uu}(f)$	$\lvert H(f)\rvert^2 P_{uu}(f)$

Expressing in terms of the frequency f by setting $z = e^{j2\pi f}$, we get

$$P_{yv}(f) = \sigma_v^2 H(f) \tag{2.109}$$

$$P_{yy}(f) = \sigma_v^2\, |H(f)|^2 \tag{2.110}$$

Let us express the power spectral density in terms of the coefficients of the transfer function

$$H(z) = \frac{\sum\limits_{i=0}^{n_b} b_i z^{-i}}{\sum\limits_{i=0}^{n_a} a_i z^{-i}} \tag{2.111}$$

where a_i and b_i are the denominator and the numerator coefficients with $a_0 = 1$, and n_a is the order of the system while n_b is order of the numerator. The power spectral density becomes

$$P_{yy}(z) = \sigma_v^2 H(z)H(z^{-1}) = \sigma_v^2 \frac{\sum\limits_{i=0}^{n_b} b_i z^{-i} \sum\limits_{j=0}^{n_b} b_j z^{j}}{\sum\limits_{i=0}^{n_a} a_i z^{-i} \sum\limits_{i=0}^{n_a} a_j z^{j}} = \sigma_v^2 \frac{\sum\limits_{i=0}^{n_b}\sum\limits_{j=0}^{n_b} b_i b_j z^{-(i-j)}}{\sum\limits_{i=0}^{n_a}\sum\limits_{j=0}^{n_a} a_i a_j z^{-(i-j)}} \tag{2.112}$$

Defining $m = i - j$ and simplifying we get

$$P_{yy}(z) = \sigma_v^2 \frac{\sum\limits_{m=0}^{n_b}\sum\limits_{i=0}^{n_b-m} b_i b_{i+m} (z^{-m} + z^m)}{\sum\limits_{m=0}^{n_a}\sum\limits_{i=0}^{n_a-m} a_i a_{i+m} (z^{-m} + z^m)} \tag{2.113}$$

The spectral density $P_{yy}(z)$ is a function of $z^{-m} + z^m$, which implies

$$P_{yy}(z) = P_{yy}\left(z^{-1}\right) \tag{2.114}$$

This shows that the poles and zeros of the $P_{yy}(z)$ have reciprocal symmetry: if z_i is a zero, then $1/z_i$ is also a zero, and if p_i is a pole, then $1/p_i$ is also a pole. Substituting $z = e^{j2\pi f}$ and noting that $z^{-m} + z^m = 2cos2\pi mf$ we get

$$P_{yy}(f) = \sigma_v^2 \frac{\sum\limits_{i=0}^{n_b} b_i^2 + 2\sum\limits_{m=0}^{n_b}\sum\limits_{i=0}^{n_b-m} b_i b_{i+m} \cos 2\pi mf}{\sum\limits_{i=0}^{n_a} a_i^2 + 2\sum\limits_{m=0}^{n_a}\sum\limits_{i=0}^{n_a-m} a_i a_{i+m} \cos 2\pi mf} \tag{2.115}$$

The spectral density $P_{yy}(f)$ expressed in terms of frequency f is a function of $\cos 2\pi mf$.

2.6.2 *Identification of Non-Parametric Model of a System*

A model of the system $H(z)$ may be modeled in (i) parametric form, as given Eq. (2.111) where the transfer function is expressed in the terms of the numerator coefficients $\{b_i\}$ and denominator coefficients $\{a_i\}$, the numerator order n_b and model order n_a; or (ii) non-parametric form expressed in terms of its frequency response $H(\omega)$ (or equivalently by its impulse response $h(k)$).

In practice, one is faced with the problem of identification using a single realization of the input–output data record. The expression for the transfer function $H(z)$ of a system given by Eq. (2.105) provides a solution:

$$H(z) = \frac{P_{yx}(z)}{P_{uu}(z)} \tag{2.116}$$

Since there are only a single realizations of $U(z)$ and $Y(z)$, the ensemble average (expectation operation) of a variable is replaced by the variable itself. Using the expression for spectral density in terms of $U(z)$ and $Y(z)$ we get

$$H(z) = \frac{Y(z)\,U\left(z^{-1}\right)}{X(z)\,U\left(z^{-1}\right)} = \frac{Y(z)}{U(z)} \tag{2.117}$$

Expressing in terms of the frequency f by setting $z = e^{j2\pi f}$, we get

$$H(f) = \frac{Y(f)}{U(f)} \tag{2.118}$$

Thus one can identify a non-parametric model $H(\omega)$ by (i) computing the Fourier transforms of the input $U(f)$ and that of the output $Y(f)$ and dividing $Y(f)$ by $U(f)$. The impulse response $h(k)$ of the system may be obtained by computing the inverse Fourier transform of $U(f)$. Let a random signal $y(k)$ be modeled as an output of a linear system $H(z)$ driven by a zero-mean white noise with variance σ_v^2 given by

$$Y(z) = H(z)V(z) \tag{2.119}$$

where $V(z)$ is the z-transform of the white noise process $v(k)$.The cross-power spectral density of $y(k)$ is $P_{yy}(z) = \sigma_v^2 H(z)H(z^{-1})$. In practice, however, the problem is to find a transfer function $H(z)$ given the power spectral density $P_{yy}(z)$(and not the other way round: given $H(z)$ to find $P_{yy}(z)$). The solution to the problem of determining $H(z)$ given $P_{yy}(z)$ is not unique. However, if we restrict ourselves to a class of all causal "minimum phase transfer functions" then there is a unique solution. If the power spectral density is given explicitly in the form $P_{yy}(z) = \sigma_v^2 H(z)H(z^{-1})$, a parametric model of $H(z)$ may be computed using a partial fraction of $P_{yy}(z)$, as shown in Examples 2.3, 2.4, and 2.8. In practice, however, the power spectral density has to be estimated from the data record. A popular approach is to employ an auto-correlation function $r_{yy}(m)$ of the output $y(k)$ to estimate the parametric model by expressing $H(z)$ in terms of the numerator coefficients $\{b_i\}$ and denominator coefficients $\left\{a_i\right\}$, the numerator order n_b and model order n_a. The auto-correlation function $r_{yy}(m)$ may be computed by taking the inverse Fourier transform $P_{yy}(\omega)$. In the case where the input to the system $x(k)$ is a zero mean white noise process, the auto-correlation $r_{yy}(m)$ is merely the auto-correlation of the impulse response of the system $h(k)$ scaled by σ_v– as can be deduced from Eq. (2.110). Identification of the parametric model of a system will be considered in a later chapter. In the next section, however, we will consider the identification of a signal model.

2.6.3 *Identification of a Parametric Model of a Random Signal*

We will consider all-pole (AR), pole-zero (ARMA), and all zero (MA) models and obtain a set of equations relating the coefficients of the model and the auto-correlation function of the random signals from which the coefficients may be estimated. First an AR, then ARMA, and finally an MA model are considered.

Identification of AR Coefficients

In many applications in modeling of colored noise (or disturbance), it is generally assumed that the colored noise is generated by filtering the white noise by an all-pole filter, that is the numerator of the filter transfer function is a constant. Consider the signal model given by Eq. (2.119) where the $H(z)$ is an all-pole transfer function given by

$$H(z) = \frac{Y(z)}{V(z)} = \frac{1}{1 + a_1 z^{-1} + a_2 z^{-2} + \cdots + a_{n_a} z^{-n_a}} \tag{2.120}$$

Expressing in the time domain we get

$$y(k) + \sum_{i=1}^{n_a} a_i y(k-i) = v(k) \tag{2.121}$$

The model is generally called an Auto-Regressive (AR) model of a signal. We will derive the model governing the auto-correlation of $y(k)$ from the above equation. Multiply both sides of the equation by $y(k-m)$ for $m \geq 0$, and take expectation. We get

$$E\left[y(k)y(k-m)\right] + \sum_{i=1}^{n_a} a_i E\left[y(k-i)y(k-m)\right] = E\left[v(k)y(k-m)\right] \tag{2.122}$$

Using the definition $r_{yy}(m) = E\left[y(k)y(k-m)\right]$ and $r_{vy}(m) = E\left[v(k)y(k-m)\right]$ we get

$$r_{yy}(m) + \sum_{i=1}^{n_a} a_i r_{yy}(m-i) = r_{vy}(m) \tag{2.123}$$

Consider the right-hand side of the above equation. The expression for $r_{vy}(m)$ is

$$r_{vy}(m) = \begin{cases} E\left[v(k)y(k)\right] & m = 0 \\ E\left[v(k)y(k-m)\right] & m > 0 \end{cases} \tag{2.124}$$

From Eq. (2.121), we deduce that the past values of $y(k)$ do not depend upon the present value of $v(k)$, that is $E\left[v(k)y(k-m)\right] = 0$ for $m > 0$. Hence Eq. (2.124) becomes

$$r_{vy}(m) = \begin{cases} E\left[v(k)y(k)\right] & m = 0 \\ 0 & m > 0 \end{cases} \tag{2.125}$$

Substituting for $y(k)$ using Eq. (2.121) we get

$$r_{vy}(m) = \begin{cases} E\left[v(k)y(k)\right] = -\sum_{i=1}^{n_a} a_i E\left[v(k)y(k-i)\right] + E\left[v^2(k)\right] & m = 0 \\ 0 & m > 0 \end{cases} \tag{2.126}$$

As $E\left[v(k)y(k-i)\right] = 0$ for $i > 0$ we get

$$r_{vy}(m) = \begin{cases} \sigma_v^2 & m = 0 \\ 0 & m > 0 \end{cases} \tag{2.127}$$

Using Eq. (2.127), Eq. (2.123) becomes

$$r_{yy}(m) + \sum_{i=1}^{n_a} a_i r_{yy}(m-i) = \begin{cases} \sigma_v^2 & m = 0 \\ 0 & m > 0 \end{cases} \tag{2.128}$$

Since $r_{yy}(m) = r_{yy}(-m)$ only $m \geq 0$ is considered. Given the auto-correlation functions $\{r_{yy}(m)\}$, we may now estimate the coefficients $\{a_i\}$ using the above equation. Given values $m = 1, 2, \dots n_a$ (considering only values of $m \geq 1$) Eq. (2.128) may be expressed as a linear matrix equation given by

$$\begin{bmatrix} r_{yy}(0) & r_{yy}(1) & r_{yy}(2) & . & r_{yy}(n_a-1) \\ r_{yy}(1) & r_{yy}(0) & r_{yy}(1) & . & r_{yy}(n_a-2) \\ r_{yy}(2) & r_{yy}(1) & r_{yy}(0) & . & r_{yy}(n_a-3) \\ . & . & . & . & . \\ r_{yy}(n_a) & r_{yy}(n_a-1) & r_{yy}(n_a-2) & . & r_{yy}(0) \end{bmatrix} \begin{bmatrix} a_1 \\ a_2 \\ a_3 \\ . \\ a_{n_a} \end{bmatrix} = - \begin{bmatrix} r_{yy}(1) \\ r_{yy}(1) \\ r_{yy}(1) \\ . \\ r_{yy}(n_a) \end{bmatrix} \tag{2.129}$$

The linear matrix equation relating the AR model parameters is called the *Yule–Walker* equation. The matrix formed of the auto-correlation functions $\{r_{yy}(m)\}$ is a Toeplitz matrix. The matrix has a special structure. The elements along the main and the off diagonals are all equal. It is a constant-diagonal matrix.

Coefficients of an ARMA Model

Unlike the AR case, a frequency domain rather than a time domain approach is employed to relate the ARMA coefficients and the auto-correlation functions, as frequency domain approach is simpler. The ARMA model of the random signal is

$$y(k) + \sum_{i=1}^{n_a} a_i y(k-i) = \sum_{i=0}^{n_b} b_i v(k-i) \tag{2.130}$$

Consider the signal model given by Eq. (2.119) where the $H(z)$ is given by

$$H(z) = \frac{Y(z)}{V(z)} = \frac{b_0 + b_1 z^{-1} + b_2 z^{-2} + \cdots + b_{n_b} z^{-n_b}}{1 + a_1 z^{-1} + a_2 z^{-2} + \cdots + a_{n_a} z^{-n_a}} \tag{2.131}$$

Cross-multiplying yields

$$D(z)Y(z) = N(z)V(z) \tag{2.132}$$

where $N(z) = b_0 + b_1 z^{-1} + b_2 z^{-2} + \cdots + b_{n_b} z^{-n_b}$ and $D(z) = 1 + a_1 z^{-1} + a_2 z^{-2} + \cdots + a_{n_a} z^{-n_a}$. Multiplying by $Y\left(z^{-1}\right)$ on both sides we get

$$D(z)Y(z)Y(z^{-1}) = N(z)V(z)Y(z^{-1}) \tag{2.133}$$

Substituting for $Y(z^{-1})$ using Eq. (2.131) and taking expectation yields

$$D(z)E[Y(z)Y(z^{-1})] = N(z)H(z^{-1})E[V(z)V(z^{-1})] \tag{2.134}$$

Taking inverse z-transform we get

$$r_{yy}(m) + \sum a_i r_{yy}(m-i) = \sigma_v^2 \sum_{i=0}^{n_b} b_i g(m-i) \tag{2.135}$$

Since $g(m) = h(-m)$(where $h(m)$ is the impulse response of $H(z)$) is anti-causal, $g(m) = 0$ for $m < 0$, we get

$$r_{yy}(m) + \sum_{i=1}^{n_a} a_i r_{yy}(m-i) = \begin{cases} \sigma_v^2 \sum_{i=0}^{n_b} b_i g(m-i) & 0 \le m \le n_b \\ 0 & m > n_b \end{cases} \tag{2.136}$$

Since $r_{yy}(m) = r_{yy}(-m)$ only $m \geq 0$ is considered.

Coefficients of an MA Model

As for the ARMA model, a frequency domain rather than a time domain approach is employed to relate the MA coefficients and the auto-correlation functions. The MA model of the random signal is

$$y(k) = \sum_{i=0}^{n_b} b_i v(k-i) \tag{2.137}$$

The signal model $H(z)$ is given by

$$H(z) = \frac{Y(z)}{V(z)} = b_0 + b_1 z^{-1} + b_2 z^{-2} + \cdots + b_{n_b} z^{-n_b} \tag{2.138}$$

Cross-multiplying and taking expectation gives

$$Y(z) = N(z)V(z) \tag{2.139}$$

Substituting for $Y\left(z^{-1}\right)$ using Eq. (2.138) and taking expectation yields

$$E[Y(z)Y(z^{-1})] = N(z)N(z^{-1})E[V(z)V(z^{-1})] \tag{2.140}$$

Taking the inverse z-transform and using the Appendix to compute $N(z)N\left(z^{-1}\right)$ we get

$$r_{yy}(m) = \begin{cases} \sigma_v^2 \sum_{i=m}^{n_b} b_i b_{i-m} & 0 \le m \le n_b \\ 0 & m > n_b \\ r_{yy}(-m) & m < 0 \end{cases} \tag{2.141}$$

Comments

- *Similarity and noise annihilation properties.*
  In all cases – namely the AR, the MA, and the ARMA models – the auto-correlation function $r_{yy}(m)$ generated as an output of the linear system $H(z)$ driven by a zero-mean white noise, and a deterministic

signal $y(k)$ generated as an output of the same system $H(z)$ driven by a delta function, satisfies an identical homogenous difference equation: the auto-correlation $r_{yy}(m)$ satisfies

$$r_{yy}(m) + \sum_{i=1}^{n_a} a_i r_{yy}(m-i) = 0,\ m > n_b$$

The deterministic signal $y(k)$ satisfies

$$y(k) + \sum_{i=1}^{n_a} a_i y(k-i) = 0, k > n_b$$

Since the auto-correlation function $r_{yy}(m)$ and $y(k)$ satisfy an identical homogenous difference equation, they both belong to the same class of signals, such as the class of polynomials, exponentials, sinusoids, weighted sinusoids, or their combinations. For example if $y(k)$ belongs to the class of exponential, then its auto-correlation $r_{yy}(m)$ will also belong to the same class of exponentials. Further, if a signal $s(k)$ is buried in a zero-mean white noise $v(k)$, then the auto-correlation $r_{yy}(m)$ of the noisy signal $y(k) = s(k) + v(k)$ belongs to the class of noise-free signal $s(k)$. This justifies the properties of noise annihilation and similarity properties of correlation in Section 2.2.2.

- *The estimated of the AR model using the Yule–alker equation is guaranteed to be stable and, thanks to Toeplitz structure of the correlation matrix, the Yule–Walker equation can be solved in $O\left(n_a^2\right)$ operations instead of $O\left(n_a^3\right)$. Hence it is used widely to estimate the model of a random as well as a deterministic signal, for example, speech processing, interference cancellation, echo cancellation, and identification.*
- *The MA coefficients of the ARMA and MA models are nonlinear in the coefficients of the coefficients of the transfer function $H(z)$.*

2.7 Illustrative Examples: *Modeling and Identification*

Example 2.13 Random signals
Model a random signal $y(k)$ as an output of an AR model excited by a zero-mean unity variance white noise. Assuming the order of the AR model to be (i) a first order (ii) a second order, and (ii) a third order, determine the AR coefficients for each case.

Solution:

- Consider an example of a first-order AR model of random signal

$$y(k) = a_1 y(k-1) + v(k-1)$$

The estimated $\hat{a}_1$ unknown coefficient a_1 may be determined using Eq. (2.129).

$$r_{yy}(1) = -\hat{a}_1 r_{yy}(0)$$

Simplifying we get

$$\hat{a}_1 = r_{yy}(1)/r_{yy}(0)$$

- Consider an example of a second-order AR model of random signal

$$y(k) = a_1 y(k-1) + a_2 y(k-2) + v(k-1)$$

Table 2.3 Estimated and true coefficients

	True coefficients	Estimated coefficients
First-order model	0.8	0.81
Second-order model	[−1.5588 0.8100]	[−1.5597 0.8088]
Third-order model	[−2.3588 2.0571 − 0.6480]	[−2.3353 2.0124 − 0.6244]

Estimates $\hat{a}_1$ and $\hat{a}_2$ of the unknown coefficients may be determined from

$$\begin{bmatrix} r_{yy}(0) & r_{yy}(1) \\ r_{yy}(1) & r_{yy}(0) \end{bmatrix} \begin{bmatrix} \hat{a}_1 \\ \hat{a}_2 \end{bmatrix} = -\begin{bmatrix} r_{yy}(1) \\ r_{yy}(2) \end{bmatrix}$$

- Consider an example of a second-order AR model of random signal.

$$y(k) = a_1 y(k-1) + a_2 y(k-2) + a_3 y(k-3) + v(k-1)$$

Estimates $\hat{a}_1$, $\hat{a}_2$, and $\hat{a}_3$ of the unknown coefficients may be determined from Eq. (2.129)

$$\begin{bmatrix} r_{yy}(0) & r_{yy}(1) & r_{yy}(2) \\ r_{yy}(1) & r(0) & r_{yy}(1) \\ r_{yy}(2) & r_{yy}(1) & r(0) \end{bmatrix} \begin{bmatrix} a_1 \\ a_2 \\ a_3 \end{bmatrix} = -\begin{bmatrix} r(1) \\ r(2) \\ r(3) \end{bmatrix}$$

Simulation Results

The coefficients of the first, second, and third were estimated by estimating the auto-correlation functions of the random signals. Only a single realization of the $N = 4096$ data samples were employed in computing the correlation estimates. The estimate and the true coefficients are given below in Table 2.3.

Examples of Modeling and Non-parametric Identification

Figure 2.15 shows the generation of low-pass, band-pass, and high-pass random signals by filtering a zero-mean white noise process using a low-pass filter, a band-pass filter, and a high-pass filter as given in Eq. (2.119). The input $u(k)$ is a zero-mean white noise process. The outputs of all the filters, and their estimated auto-correlations and spectral densities, are shown. Since only a single realization of the data is employed, true auto-correlations and power spectral densities are also plotted for comparison. Using the Fourier transforms of the input and the output of each of the filters, the non-parametric frequency response model is identified, and identified frequency responses are compared with the true ones. The top three figures show the outputs of low-pass, band-pass, and high-pass filters. The mid three figures show the estimated and true auto-correlation of the filter outputs, and the bottom three figures show the identified and true non-parametric models.

Modeling of Random Signals

Example 2.14 Random signals

Consider a stochastic process $y(k)$, which is generated using a zero-mean white noise process with unit variance, $v(k)$

$$y(z) = H(z)v(z) \quad where \quad H(z) = \frac{1 - 0.3z^{-1}}{1 - 0.5z^{-1}}$$

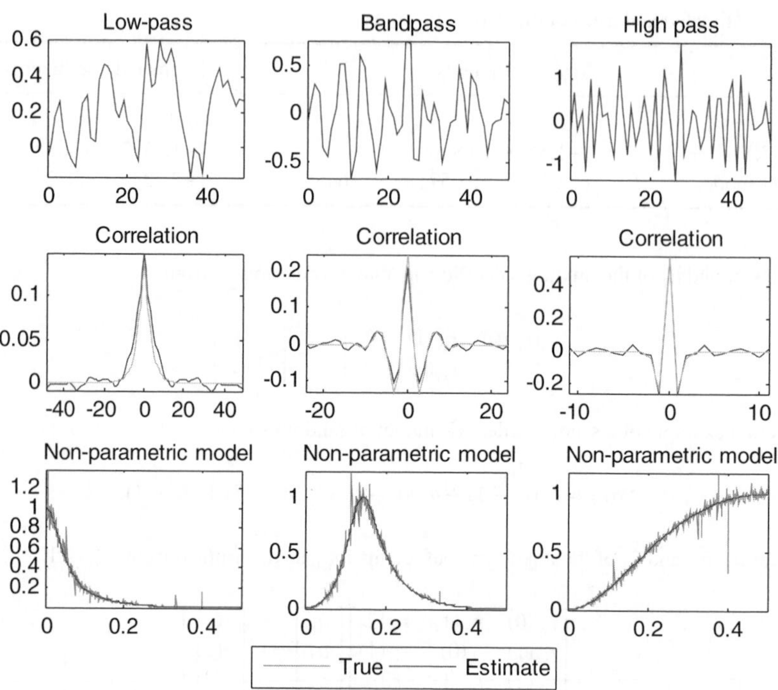

Figure 2.15 Non-parametric modeling and identification of random signals

Obtain expressions for the power spectral densities $P_{yy}(z)$, $P_{yv}(z)$, $P_{vy}(z)$, $P_{vv}(z)$.

Solution:

- $P_{vv}(z) = \sigma^2$, $P_{yv}(z) = \sigma^2 H(z)$, $P_{vy}(z) = \sigma^2 H(z^{-1})$, $P_{yy}(z) = \sigma^2 H(z)H(z^{-1})$
  Expanding the expression for $P_{yy}(z)$ we get

$$P_{yy}(z) = \sigma^2 \left(\frac{1 - 0.3z^{-1}}{1 - 0.5z^{-1}}\right)\left(\frac{1 - 0.3z}{1 - 0.5z}\right)$$

  Expanding we get

$$P_{yy}(z) = \sigma^2 \left(\frac{1 - 0.3\left(z^{-1} + z\right) + 0.09}{1 - 0.5\left(z^{-1} + z\right) + 0.25}\right) = \sigma^2 \left(\frac{1.09 - 0.3\left(z^{-1} + z\right)}{1.25 - 0.5\left(z^{-1} + z\right)}\right)$$

  Substituting $z = e^{j2\pi f}$ we get

$$P_{yy}(f) = \sigma^2 \left(\frac{1.09 - 0.6\cos 2\pi f}{1.25 - \cos 2\pi f}\right)$$

- The poles and the zeros of $P_{yy}(z)$: poles: 0.5 and 2; zeros: 0.3 and 3.333.
- The poles and zeros have reciprocal symmetrical: if p is pole then 1/p is also a pole, if z is a zero the 1/z is also a zero.

Example 2.15 Let $y(k)$ be some random process generated as an output of $H(z)$ excited by a zero-mean white noise process with unity variance. Compute the power spectral density $P_{yy}(z)$ if

$$H(z) = \frac{b_0 + b_1 z^{-1} + b_2 z^{-2}}{a_0 + a_1 z^{-1} + a_2 z^{-2}}$$

where the leading coefficient in the denominator $a_0 = 1$.

Solution: Power spectral density $P_{yy}(z)$ takes the form

$$P_{yy}(z) = \sigma_v^2 H(z)H(z^{-1})$$

Substituting for $H(z)$ yields

$$P_{yy}(z) = \sigma_v^2 \left(\frac{b_0 + b_1 z^{-1} + b_2 z^{-2}}{a_0 + a_1 z^{-1} + a_2 z^{-2}} \right) \left(\frac{b_0 + b_1 z^{1} + b_2 z^{2}}{a_0 + a_1 z^{1} + a_2 z^{2}} \right)$$

Simplifying we get

$$P_{yy}(z) = \sigma_v^2 \frac{(b_0^2 + b_1^2 + b_2^2) + (b_0 b_1 + b_1 b_2)(z + z^{-1}) + b_0 b_2 (z^2 + z^{-2})}{(a_0^2 + a_1^2 + a_2^2) + (a_0 a_1 + a_1 a_2)(z + z^{-1}) + a_0 a_2 (z^2 + z^{-2})}$$

Substituting $z = e^{j2\pi f}$ we get

$$z + z^{-1} = 2\cos 2\pi f$$

$$z^2 + z^{-2} = 2\cos 4\pi f$$

Hence

$$P_{yy}(\omega) = \sigma_v^2 \frac{(b_0^2 + b_1^2 + b_2^2) + 2(b_0 b_1 + b_1 b_2)\cos 2\pi f + 2b_0 b_2 cos 4\pi f}{(a_0^2 + a_1^2 + a_2^2) + 2(a_0 a_1 + a_1 a_2)\cos 2\pi f + 2a_0 a_2 \cos 4\pi f}$$

Example 2.16 Let $x(k)$ and $y(k)$ be two random signals modeled respectively as the outputs of $G(z)$ and $H(z)$ excited by zero-mean white noise processes with unit variance where

$$X(z) = G(z)v(z)$$
$$Y(z) = H(z)v(z)$$

where $G(z) = \dfrac{z^{-1}}{1 - \rho_1 z^{-1}}$; $H(z) = \dfrac{b_1 z^{-1}}{1 + a_1 z^{-1} + a_2 z^{-2}}$; $b_1 = \rho_2 \sin\omega_0$; $a_1 = -2\rho_2 \cos\omega_0$; $a_2 = \rho_2^2$; $\rho_1 = 0.8$; $\rho_2 = 0.9$ and $\omega_0 = \pi/6$. Determine the correlation functions $r_{xx}(m)$, $r_{yy}(m)$, and $r_{xy}(m)$, and the corresponding spectral densities $E_{xx}(f)$, $E_{yy}(f)$, and $E_{xy}(f)$.

Solution:

- The power spectral density of $x(k)$ is

$$E_{xx}(z) = \sigma^2 G(z)G(z^{-1}) = \frac{1}{(1-\rho_1 z^{-1})(1-\rho_1 z)}$$

Using the Appendix, the partial fraction expansion becomes

$$E_{xx}(z) = X(z)X(z^{-1}) = \frac{1}{1-\rho_1^2}\left(\frac{1}{1-\rho_1 z^{-1}} + \frac{1}{1-\rho_1 z} - 1\right)$$

Taking the inverse z-transform, yields

$$r_{xx}(m) = \frac{\rho_1^{|m|}}{1-\rho_1^2}$$

- The spectral density of $y(k)$ is

$$E_{yy}(z) = \sigma^2 H(z)H(z^{-1}) = \frac{b_1^2}{(1+a_1 z^{-1}+a_2 z^{-2})(1+a_1 z+a_2 z^2)}$$

Using the partial fraction expansion we get

$$E_{yy}(z) = \sigma^2 H(z)H(z^{-1}) = \frac{A_{yy}z^{-1}+B_{yy}}{1+a_1 z^{-1}+a_2 z^{-2}} + \frac{A_{yy}z+B_{yy}}{1+a_1 z+a_2 z^2} + C_{yy}$$

Using the Appendix, the coefficients A_{yy}, B_{yy}, and C_{yy} are obtained from

$$\begin{bmatrix} b_1^2 \\ 0 \end{bmatrix} = \begin{bmatrix} 2a_1 & 1-a_1^2-a_2^2 \\ 1+a_2 & -a_1 a_2 \end{bmatrix}\begin{bmatrix} B_{yy} \\ A_{yy} \end{bmatrix}$$

and

$$C_{xy} = -B_{xy}$$

Taking the inverse z-transform we get

$$r_{yy}(m) = B_{yy}\rho_2^{|m|}\cos\omega_0 m$$

- The spectral density of $x(k)$ and $y(k)$ is

$$E_{xy}(z) = X(z)Y(z^{-1}) = \sigma^2 G(z)H(z^{-1})$$

Using the partial fraction expansion, we get

$$E_{xy}(z) = \frac{A_{xy}z^{-1}}{(1-\rho_1 z^{-1})} + \frac{B_{xy}x + C_{xy}}{(1-2\rho_2\cos\omega_0 z+\rho_2^2 z^2)} + D_{xy}$$

Using the Appendix we get

$$\begin{aligned}
A_{xy} &= \frac{b_0\rho_1 + b_1\rho_1^2}{\rho_1^2 a_2 + \rho_1 a_1 + 1} \\
B_{xy} &= b_1 - A_{xy}a_2 \\
C_{xy} &= \frac{A_{xy}}{\rho_1} \\
D_{xy} &= 0
\end{aligned}$$

Taking the inverse z-transform we get

$$r_{xy}(m) = \begin{cases} A_{xy}\rho_1^m & m \geq 1 \\ C_{xy}\rho_2^m \cos\omega_0 m & m \leq 0 \end{cases}$$

Simulation Result

Figure 2.16 shows random signals $x(k)$ and $y(k)$ generated using a linear time invariant model driven by zero-mean white noise process. The auto-correlations $r_{xx}(m)$ and $r_{yy}(m)$, and cross-correlation $r_{xy}(m)$ are (i) estimated using a single realization of the random signal and (ii) computed analytically using the model which is termed as "true." Similarly, the spectral densities are computed. The top three figures

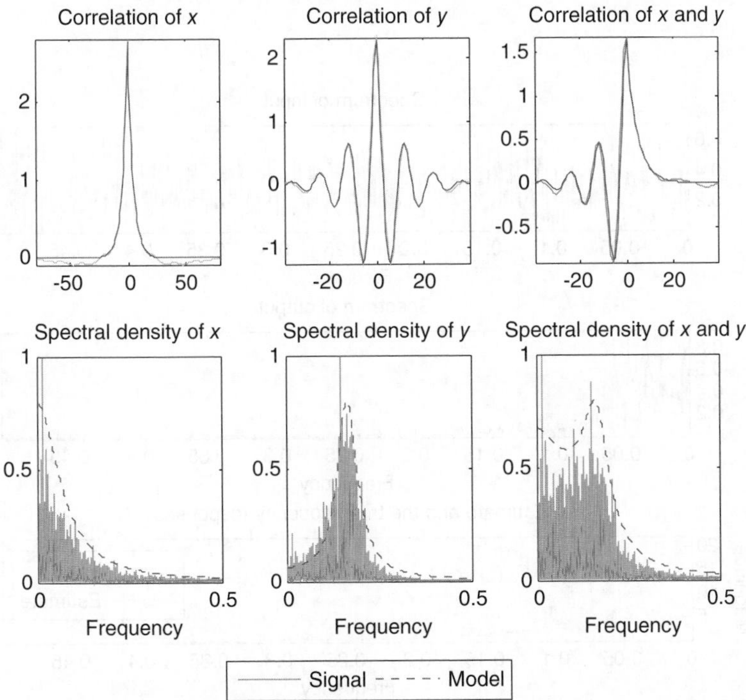

Figure 2.16 True and estimated correlations and spectral densities of $x(k)$ and $y(k)$

give true and estimated correlations of $x(k)$, $y(k)$, and the cross-correlation of $x(k)$ and $y(k)$. The bottom three figures give true and estimated spectral densities of $x(k)$, $y(k)$, and the cross-spectral density of $x(k)$ and $y(k)$.

System identification

Example 2.17 Estimate from the input and the output data, the non-parametric model in terms of frequency response of a system $H(z)$ given by

$$H(z) = \frac{z^{-1}}{1 - 1.7387z^{-1} + 0.81z^{-2}}$$

Solution: The input is a chosen to be a zero-mean white noise process. The Fourier transforms of the input and the output are computed and the frequency response is estimated from a single realization of the random input and the output. The estimate of the non-parametric model expressed in terms of the frequency response is

$$H(f) = \frac{Y(f)}{U(f)}$$

The estimate and the true frequency responses are given in Figure 2.17.

Spectral densities of input and noisy output of a system are shown in Figure 2.18, where u is the input, s is the noise-free output, v is the measurement noise, and y is the noisy output.

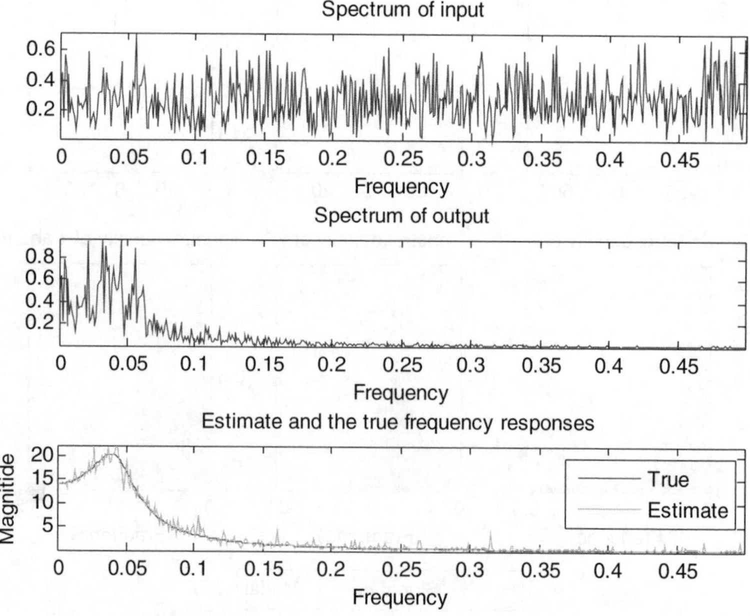

Figure 2.17 Estimated and true frequency responses

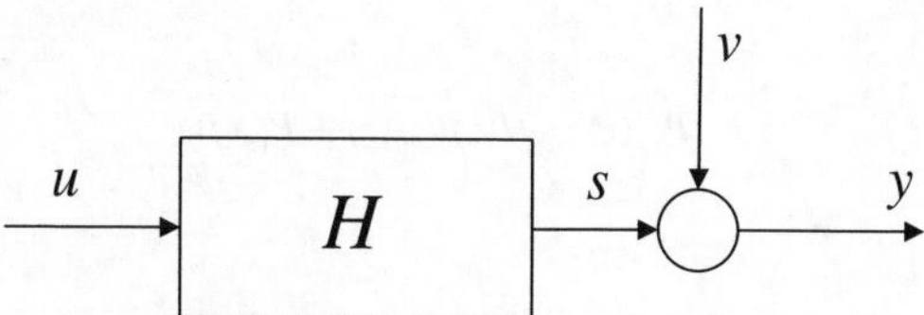

Figure 2.18 Input, noise-free, and noisy output

Example 2.18 Consider the following stochastic process $y(k)$ given by

$$y(k) = s(k) + v(k)$$
$$s(z) = H(z)u(z)$$

where $u(k)$ is the input and $v(k)$ is the zero-mean white noise process with variance σ_v^2.

1. Derive the expressions for the power spectral densities $P_{yy}(z)$, $P_{yv}(z)$, $P_{vy}(z)$, and $P_{vv}(z)$.
2. Derive the above expression if $u(k)$ and $v(k)$ are uncorrelated.
3. Is it possible to estimate $H(z)$ if $u(k)$ and $v(k)$ are uncorrelated?

Solution:

- Consider $P_{yy}(z)$
  Since $y(k) = s(k) + v(k)$

$$P_{yy}(z) = P_{ss}(z) + P_{vv}(z) + P_{sv}(z) + P_{vs}(z)$$

Consider $P_{ss}(z)$

$$P_{ss}(z) = H(z)H(z^{-1})P_{uu}(z)$$

Consider $P_{sv}(z)$

$$P_{sv}(z) = s(z)v(z^{-1}) = H(z)u(z)v(z^{-1}) = H(z)P_{uv}(z)$$

Consider $P_{vs}(z)$

$$P_{vs}(z) = v(z)s(z^{-1}) = v(z)H(z^{-1})u(z^{-1}) = H(z^{-1})P_{vu}(z)$$

Hence

$$P_{yy}(z) = H(z)H(z^{-1})P_{uu}(z) + P_{vv}(z) + H(z)P_{uv}(z) + H(z^{-1})P_{vu}(z)$$

- Consider $P_{yu}(z)$

$$P_{yu}(z) = P_{su}(z) + P_{vu}(z)$$

Since

$$P_{su}(z) = s(z)u(z^{-1}) = H(z)u(z)u(z^{-1}) = H(z)P_{uu}(z)$$

Hence

$$P_{yu}(z) = H(z)P_{uu}(z) + P_{vu}(z)$$

- Consider $P_{uy}(z)$

$$P_{uy}(z) = P_{us}(z) + P_{uv}(z)$$

Since

$$P_{us}(z) = u(z)s(z^{-1}) = H(z^{-1})u(z)u(z^{-1}) = H(z^{-1})P_{uu}(z)$$

Hence

$$P_{uy}(z) = H(z^{-1})P_{uu}(z) + P_{uv}(z)$$

- Substituting $P_{vv}(z) = \sigma_v^2$ and assuming that $u(k)$ and $v(k)$ are uncorrelated, which implies $P_{uv}(z) = 0$ and $P_{vu}(z) = 0$, we get
- $P_{yy}(z) = H(z)H(z^{-1})P_{uu}(z) + \sigma_v^2$
- $P_{yu}(z) = H(z)P_{uu}(z)$
- $P_{uy}(z) = H(z^{-1})P_{uu}(z)$

Expressing in terms of the frequency ω we get

- $P_{yy}(f) = |H(f)|^2 P_{uu}(f) + \sigma_v^2$
- $P_{yu}(f) = H(f)P_{uu}(f)$
- $P_{uy}(f) = H^*(f)P_{uu}(f)$
- Estimation of $H(z)$.

Consider the expression of $P_{yu}(z)$

$$P_{yu}(z) = H(z)P_{uu}(z)$$

It is possible to identify $H(z)$ knowing the power spectral densities $P_{yu}(z)$ and $P_{uu}(z)$

$$H(z) = \frac{P_{yu}(z)}{P_{uu}(z)}$$

The transfer function is estimated by dividing the cross-power spectral density of $y(k)$ and $u(k)$ by the auto-power spectral density of $u(k)$.

Identifying the Form of a Signal from a Noisy Measurement

Thanks to the properties of similarity and noise annihilation, the correlation and spectral density are widely used to determine the form (or class) of a signal buried in zero-mean uncorrelated noise.

Example 2.19 A random signal $y(k)$ is a sum of a deterministic signal $s(k)$ and a zero-mean white noise process $v(k)$ given by

$$y(k) = s(k) + v(k)$$

- Give expressions for the auto-correlation function and the power spectral density of $y(k)$.

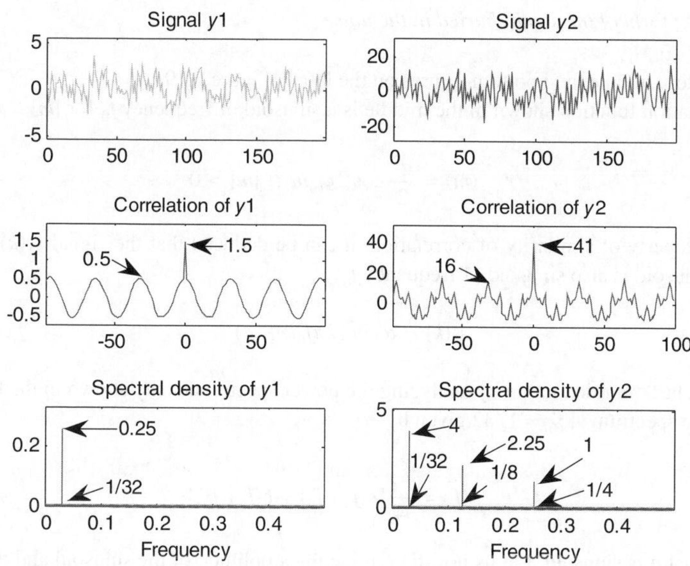

Figure 2.19 Two random signals, auto-correlations, power spectral densities

- Consider the random signals $y_1(k)$ and $y_2(k)$ containing signals $s_1(k)$ and $s_2(k)$ respectively.

$$y_i(k) = s_i(k) + v_i(k), \ i = 1, 2$$

Figure 2.19 shows random signals $y_1(k)$ and $y_2(k)$. The top, the middle, and the bottom figures on the right give the plots of the output $y_1(k)$, its auto-correlation, and its spectra density, and those on the left show $y_2(k)$, its auto-correlation, and its spectral density. Determine (i) the form $s_1(k)$ and the variance σ_1^2 of $v_1(k)$, and (ii) the form of $s_2(k)$ and the variance σ_2^2. The relevant values of auto-correlations and spectral densities are indicated.

Solution:

- The power spectral density of $y(k)$ takes the form.

$$P_{yy}(z) = P_{ss}(z) + P_{sv}(z) + P_{vs}(z) + P_{vv}(z)$$

Since $s(k)$ and $v(k)$ are uncorrelated, implying $P_{sv}(z) = 0$ and $P_{vs}(z) = 0$, we get

$$P_{yy}(z) = P_{ss}(z) + P_{vv}(z)$$

Further, as $v(k)$ is white noise we get

$$P_{yy}(z) = P_{ss}(z) + \sigma^2$$

That is

$$P_{yy}(\omega) = P_{ss}(\omega) + \sigma_v^2$$

Taking the inverse transform we get

$$r_{yy}(m) = r_{ss}(m) + \sigma_v^2 \delta(m)$$

Determination of form of the signal buried in the noise

Case 1 The top, middle, and bottom figures on the left of Figure 2.19.
The auto-correlation function shown in the middle is a sinusoid of frequency f_0 for $|m| > 0$.

$$r_{s_1 s_1}(m) = \frac{\alpha^2}{2} \cos 2\pi f_0 m \ : \ |m| > 0$$

Invoking the property of similarity of correlation, it can be deduced that the signal $s_1(k)$ buried in the noise is also sinusoid is also sinusoid of frequency f_0:

$$s_1(k) = \alpha \sin\left(2\pi f_0 k + \varphi\right)$$

We may cross-check our deduction by analyzing the power spectral density shown in the bottom figure. Since it is a line spectrum at $f_0 = 1/32$ given by

$$P_{y_1 y_1}(f) = \frac{\alpha_1^2}{4}[\delta(f - f_0) + \delta(f + f_0)]$$

this confirms that it is sinusoid. Let us now determine the amplitude of the sinusoid and the variance of the noise. From the auto-correlation function we get

$$r_{y_1 y_1}(0) = r_{s_1 s_1}(0) + \sigma_1^2 = \frac{\alpha_1^2}{2} + \sigma_1^2 = 1.5$$

as the amplitude of the auto-correlation, which is a sinusoid, is $\dfrac{\alpha_1^2}{2} = 0.5$ we get

$$\alpha_1 = 0.5$$

Hence

$$\sigma_1^2 = 1$$

We cannot, however, determine the phase of the sinusoid.

Case 2 The top, the middle and the bottom figures on the right of Figure 2.19
The auto-correlation function shown in the middle is periodic. Invoking the property of periodicity, it can be deduced that the signal $s_2(k)$ is a periodic signal buried in the noise. It is, however, difficult to determine additional characteristics of the signal. Let us analyze the power spectral density shown at the bottom. It is a line spectrum located at three frequencies $1/32$, $1/8$ and 1/4 with magnitude 4, 2.25 and 1 respectively given by

$$P_{y_2 y_2}(f) = 4\left[\delta\left(f - \frac{1}{32}\right) + \delta\left(f + \frac{1}{32}\right)\right] + \frac{9}{4}\left[\delta\left(f - \frac{1}{8}\right) + \delta\left(f - \frac{1}{8}\right)\right] + \left[\delta\left(f - \frac{1}{4}\right) + \delta\left(f - \frac{1}{4}\right)\right]$$

Hence the signal $s_2(k)$ is a sum of three sinusoids. The amplitudes of the sinusoids α_1, α_2, and α_3 are computed from the relations $\dfrac{\alpha_1^2}{4} = 4$, $\dfrac{\alpha_2^2}{4} = \dfrac{9}{4}$, and $\dfrac{\alpha_3^2}{4} = 1$. The general form of $s_2(k)$ is

$$s_2(k) = 4\sin(2\pi k/32 + \varphi_1) + 3\sin(2\pi k/32 + \varphi_2) + \sin(2\pi k/32 + \varphi_3)$$

Let us now determine variance of the noise. The auto-correlation of $s_2(k)$ is given by

$$r_{s_2s_2}(m) = 8\cos(2\pi m/32) + 6\cos(2\pi m/32) + 2\cos(2\pi m/32)$$

Using the data indicated we get

$$r_{y_2y_2}(0) = r_{s_2s_2}(0) + \sigma_2^2 = 8 + 6 + 2 + \sigma_2^2 = 16 + \sigma_2^2 = 41$$

Hence

$$\sigma_2^2 = 25$$

Comments *Correlations and the power spectral density provide a simple scheme to determine the form of the signal buried in uncorrelated noise thanks to the noise annihilation and similarity properties. In simple cases, such as a single sinusoid in zero-mean white noise, the auto-correlation function is sufficient to determine the form.*

In the presence of colored noise, it is comparatively difficult to determine the form of the signal unless the data record is very long.

2.8 Summary

Definitions of Correlation Function

$$r_{xy}(m) = \begin{cases} \lim\limits_{N\to\infty} \sum\limits_{k=-N}^{N} x(k)y(k-m) & \text{infinite duration energy signals} \\ \sum\limits_{k=0}^{N-1} x(k)y(k-m) & \text{finite duration energy signals} \\ \lim\limits_{N\to\infty} \dfrac{1}{2N+1} \sum\limits_{k=-N}^{N} x(k)y(k-m) & \text{power signal} \\ r_{xy}(m) = \dfrac{1}{M} \sum\limits_{k=0}^{M-1} x(k)y(k-m) & \text{M-periodic power signal} \\ \lim\limits_{N\to\infty} \dfrac{1}{2N+1} \sum\limits_{k=-N}^{N} x(k)y(k-m) & \text{ergodic random (power) signals} \end{cases}$$

Properties of Correlation Function

- *Correlation and convolution*

$$r_{xy}(m) = \sum_{k=-N}^{N} x(k)y(-(m-k)) = conv\,(x(k), y(-k))$$

$$\Im\{r_{xy}(m)\} = X(f)Y^*(f) = |X(f)|\,|Y(f)|\,\angle(\varphi_x(f) - \varphi_y(f))$$

$$\Im\,\{conv(x(k), y(k))\} = X(f)Y(f) = |X(f)|\,|Y(f)|\,\angle(\varphi_x(f) + \varphi_y(f))$$

- *Cross-correlation functions* $r_{xy}(m)$ and $r_{xy}(m)$*:*

$$r_{xy}(m) = r_{yx}(-m)$$

- *Symmetry property of auto-correlation*

$$r_{xx}(m) = r_{xx}(-m)$$

- *Maximum values of auto- and cross-correlations*

$$r_{xx}(0) \geq |r_{xx}(m)|$$
$$\sqrt{r_{xx}(0)r_{yy}(0)} \geq |r_{xy}(m)|$$
$$|r_{xy}(m) + r_{xy}^{*}(m)| \leq r_{xx}(0) + r_{yy}(0)$$

- *Time delay between two signals*
  $\hat{\tau} = \arg\max_{m}\{r_{xy}(m)\}$ for real signals
  $\hat{\tau} = \arg\max_{m}\{r_{yx}(m) + r_{yx}^{*}(m)\}$ for complex signals
- *Additive property*

$$r_{yy}(m) = \sum_{i,j} r_{x_i x_j}(m) = \sum_{i=1}^{M} r_{x_i x_i}(m) + \sum_{i \neq j} r_{x_i x_j}(m)$$

$$r_{yy}(m) = \sum_{i=1}^{M} r_{x_i x_i}(m) \text{ if } r_{x_i x_j}(m) = 0 \textit{ for } i \neq j$$

- *Periodicity property*
  If $x(k)$ and $y(k)$ are periodic: $x(k) = x(k + \ell M)$, then

$$r_{xy}(m) = r_{xy}(m + \ell M) \ : \quad \ell = \pm 1, \pm 2, \pm 3, \ldots$$

- *Noise annihilation property*

$$E\left[s(k)v(k-m)\right] = 0$$

If $x(k) = s(k) + v(k)$ then $r_{xs}(m) = r_{ss}(m)$

Normalized Correlation and Correlation Coefficient

$$\gamma_{xy}(m) = \frac{r_{xy}(m)}{\sqrt{r_{xx}(0)\, r_{yy}(0)}} \quad \text{where } -1 \leq \gamma_{xy}(m) \leq 1$$

The correlation coefficient

$$\rho_{xy} = \frac{c_{xy}(0)}{\sqrt{c_{xx}(0)\, c_{yy}(0)}}$$

The correlation coefficient $x(k)$ and $y(k)$ satisfies:

- $-1 \leq \rho_{xy} \leq 1$
- if $x(k)$ and $y(k)$ are uncorrelated then $\rho_{xy} = 0$
- $\rho_{xy} = 1$ if and only if $y(k) = ax(k) + b$ where $a > 0$
- $\rho_{xy} = -1$ if and only if $y(k) = -ax(k) + b$ where $a > 0$.

Energy and Power Spectral Densities

$$P_{xy}(f) = \lim_{N\to\infty} \sum_{m=-N}^{N} r_{xy}(m)e^{-j2\pi fm}$$

Z-transform of the Correlation Function

$$R_{xy}(z) = \begin{cases} X(z)\,Y\left(z^{-1}\right) & \text{energy signal} \\ \lim\limits_{N\to\infty}\left(\dfrac{X_N(z)\,Y_N\left(z^{-1}\right)}{2N+1}\right) & \text{power signal} \\ \lim\limits_{N\to\infty}\left(\dfrac{E\left[X_N(z)\,Y_N\left(z^{-1}\right)\right]}{2N+1}\right) & \text{random signal} \end{cases}$$

Cross-Spectral Density

$$P_{xy}(f) = X(f)\,Y(-f)$$

Note that the order of the subscripts is important. The DTFT of the signal is associated with the second subscript whose DTFT is conjugated in the expression for the cross-spectral density.

Spectral Density

$$P_{xx}(f) = |X(f)|^2$$

Energy and Average Power

$$r_{xx}(0) = \int_0^1 P_{xx}(f)df = \begin{cases} E_x & \textit{energy signal} \\ P_x & \textit{power signal} \\ \sum\limits_{k=0}^{N-1} c_n c_n^* & \textit{periodic signal} \end{cases}$$

Cross-spectral Density and Degree of Similarity/Dissimilarity

$$P_{xy}(f) = \begin{cases} 0 & \text{for all } f \quad \text{uncorrelated} \\ \neq 0 & \text{for some } f \quad \text{correlated} \end{cases}$$

Similarity properties:

- A waveform with a rational spectrum preserves its characteristics.
- A periodic signal is a periodic signal.
- A square waveform is a triangular waveform.
- A triangular waveform is a parabolic waveform.

Dissimilarity properties:

- Cross-correlation of a deterministic waveform and a zero-mean stochastic process is zero.
- Cross-correlation function of two waveforms with no spectral overlap is zero. For example periodic waveforms of different periods are zero.

- Cross-correlation function of stochastic processes, $x(k)$ and $y(k)$, is zero if the $E\left[X(f)Y^*(f)\right] = 0$.
- Cross-correlations of periodic waveforms of different periods are zero.

Coherence Spectrum

$$C_h(x,y) = \frac{|real(X(f)Y^*(f))|^2}{|X(f)|^2|Y(f)|^2} = \frac{(X_rY_r + X_iY_i)^2}{(X_rX_r + X_iX_i)^2 + (Y_rY_r + Y_iY_i)^2}$$

$$C_h\,(x,y) \leq 1 \quad \text{for some or all } f$$

Input–Output Correlation and Spectral Density

Generation of random signal from white noise

$$P_{yy}(z) = \sigma_v^2 H(z)H(z^{-1}) \text{ and } P_{yy}(f) = \sigma_v^2\,|H(f)|^2$$

Identification of non-parametric model of a system

$$H(z) = \frac{P_{yx}(z)}{P_{uu}(z)}$$

Identification of a parametric model of a random signal

- ARMA model

$$r_{yy}(m) + \sum_{i=1}^{n_a} a_i r_{yy}(m-i) = \begin{cases} \sigma_v^2 \sum\limits_{i=0}^{n_b} b_i g(m-i) & 0 \leq m \leq n_b \\ 0 & m > n_b \end{cases}$$

- AR model

$$\begin{bmatrix} r_{yy}(0) & r_{yy}(1) & r_{yy}(2) & . & r_{yy}(n_a-1) \\ r_{yy}(1) & r_{yy}(0) & r_{yy}(1) & . & r_{yy}(n_a-2) \\ r_{yy}(2) & r_{yy}(1) & r_{yy}(0) & . & r_{yy}(n_a-3) \\ . & . & . & . & . \\ r_{yy}(n_a) & r_{yy}(n_a-1) & r_{yy}(n_a-2) & . & r_{yy}(0) \end{bmatrix} \begin{bmatrix} a_1 \\ a_2 \\ a_3 \\ . \\ a_{n_a} \end{bmatrix} = -\begin{bmatrix} r_{yy}(1) \\ r_{yy}(1) \\ r_{yy}(1) \\ . \\ r_{yy}(n_a) \end{bmatrix}$$

- MA model

$$r_{yy}(m) = \begin{cases} \sigma_v^2 \sum\limits_{i=m}^{n_b} b_i b_{i-m} & 0 \leq m \leq n_b \\ 0 & m > n_b \\ r_{yy}(-m) & m < 0 \end{cases}$$

2.9 Appendix

Formula for the Product of Two Sums

$$\sum_{i=1}^{M} x_i \sum_{j=1}^{M} x_j = \sum_{i=1}^{M}\sum_{j=1}^{M} x_i x_j$$

Angle between Two Vectors

Proposition If X and Y are two complex functions $X = X_r + jX_i$ and $Y = Y_r + jY_i$, where X_r and X_i are the real and the imaginary components of X, and Y_r and Y_i are the real and the imaginary components of Y. Then

$$\cos^2 \varphi = \frac{|real\,\{XY^*\}|^2}{|X|^2\,|Y|^2} = \frac{(X_rY_r + X_iY_i)^2}{\left(X_r^2 + X_i^2\right)\left(Y_r^2 + Y_i^2\right)},$$

where φ is the angle between X and Y^*.

Proof: Using [4]

$$\cos \varphi = \frac{real\,(XY^*)}{|X|\,|Y|}$$

Expanding $real\{XY^*\}$, $|X|$ and $|Y|$ we get

$$\begin{aligned} real\{XY^*\} &= real\{(X_r + jX_i)(Y_r - jY_i)\} = X_rY_r + X_iY_i \\ |X| &= \sqrt{X_r^2 + X_i^2} \\ |Y| &= \sqrt{Y_r^2 + Y_i^2} \end{aligned}$$

Thus the proposition is established.

Partial Fraction Expansion of Spectral Density

First-order system

Consider an energy spectral density $E_{xx}(z) = X(z)X(z^{-1})$ where

$$X(z) = \frac{b_0}{1 - \rho z^{-1}}$$

The energy spectral density becomes

$$E_{xx}(z) = X(z)X(z^{-1}) = \frac{b_0^2}{(1 - \rho z^{-1})(1 - \rho z)}$$

The partial fraction expansion becomes

$$E_{xx}(z) = X(z)X(z^{-1}) = \frac{A}{1 - \rho z^{-1}} + \frac{A}{1 - \rho z} + C$$

Cross-multiplying we get

$$b_0^2 = A(1 - \rho z) + A(1 - \rho z^{-1}) + C(1 - \rho z)(1 - \rho z^{-1})$$

Collecting terms in z and z^{-1} constant we get

$$b_0^2 = 2A + C + C\rho^2 + (A\rho + C\rho)z - (A\rho + C\rho)z^{-1}$$

Equating to zero the constant, and the coefficients of z and z^{-1}, we get

$$C = -A$$

$$A = \frac{b_0^2}{(1-\rho^2)}$$

$$E_{xx}(z) = X(z)X(z^{-1}) = \frac{b_0^2}{1-\rho^2}\left(\frac{1}{1-\rho z^{-1}} + \frac{1}{1-\rho z} - 1\right)$$

Second-order system

Consider an energy spectral density $E_{xx}(z) = X(z)X(z^{-1})$ where

$$X(z) = \frac{b_1 z^{-1} + b_0}{1 + a_1 z^{-1} + a_2 z^{-2}}$$

The energy spectral density becomes

$$E_{xx}(z) = X(z)X(z^{-1}) = \frac{(b_1 z^{-1} + b_0)(b_1 z + b_0)}{(1 + a_1 z^{-1} + a_2 z^{-2})(1 + a_1 z + a_2 z^2)}$$

The partial fraction expansion becomes

$$E_{xx}(z) = X(z)X(z^{-1}) = \frac{Az^{-1} + B}{1 + a_1 z^{-1} + a_2 z^{-2}} + \frac{Az + B}{1 + a_1 z + a_2 z^2} + C$$

Cross-multiplying we get

$$\begin{aligned} b_1^2 + b_0^2 + b_1 b_0 z + b_1 b_0 z^{-1} &= (Az^{-1} + B)(1 + a_1 z + a_2 z^2) + (Az + B)(1 + a_1 z^{-1} + a_2 z^{-2}) \\ &\quad + C(1 + a_1 z + a_2 z^2)(1 + a_1 z^{-1} + a_2 z^{-2}) \end{aligned}$$

Consider the terms on the right-hand side

$$\begin{aligned} (Az^{-1} + B)(1 + a_1 z + a_2 z^2) &= (Aa_1 + B) + Az^{-1} + (Aa_2 + Ba_1)z + Ba_2 z^2 \\ (Az + B)(1 + a_1 z^{-1} + a_2 z^{-2}) &= (Aa_1 + B) + Az + (Aa_2 + Ba_1)z^{-1} + Ba_2 z^{-2} \\ (1 + a_1 z + a_2 z^2)(1 + a_1 z^{-1} + a_2 z^{-2}) &= \left(1 + a_1^2 + a_2^2\right) + (a_1 + a_1 a_2)z^{-1} + (a_1 + a_1 a_2)z + a_2 z^{-2} + a_1 z^2 \end{aligned}$$

The partial fraction expansion after cross-multiplication becomes

$$\begin{aligned} b_1^2 + b_0^2 + b_1 b_0 z + b_1 b_0 z^{-1} &= \left[2(Aa_1 + B) + C\left(1 + a_1^2 + a_2^2\right)\right] + [A(1 + a_2) + Ba_1 + C(a_1 + a_1 a_2)]z^{-1} \\ &\quad + [A(1 + a_2) + Ba_1 + C(a_1 + a_1 a_2)]z + [Ba_1 + Ca_1]z^{-2} + [Ba_2 + Ca_2]z^2 \end{aligned}$$

Setting the coefficients of z^i, z^{-i}, and the constant to zero we get

$$\begin{aligned} b_1^2 + b_0^2 &= 2(Aa_1 + B) + C(1 + a_1^2 + a_2^2) \\ b_1 b_0 &= A(1 + a_2) + Ba_1 + C(a_1 + a_1 a_2) \\ Ba_1 + Ca_1 &= 0 \\ Ba_2 + Ca_2 &= 0 \end{aligned}$$

Hence

$$\begin{bmatrix} b_0^2 + b_1^2 \\ b_0 b_1 \end{bmatrix} = \begin{bmatrix} 2a_1 & 1 - a_1^2 - a_2^2 \\ 1 + a_2 & -a_1 a_2 \end{bmatrix} \begin{bmatrix} A \\ B \end{bmatrix}$$

$$B = -C$$

Hence

$$E_{xx}(z) = X(z)X(z^{-1}) = \frac{Az^{-1} + B}{1 + a_1 z^{-1} + a_2 z^{-2}} + \frac{Az + B}{1 + a_1 z + a_2 z^2} - B$$

Cross-spectral density
The cross-spectral density is given by

$$E_{xy}(z) = X(z)Y(z^{-1}) = \frac{b_0 + b_1 z}{(1 - \rho_1 z^{-1})(1 + a_1 z + a_2 z^2)}$$

The partial fraction expansion becomes

$$\frac{b_0 + b_1 z}{(1 - \rho_1 z^{-1})(1 + a_1 z + a_2 z^2)} = \frac{Az^{-1}}{1 - \rho_1 z^{-1}} + \frac{Bz + C}{1 + a_1 z + a_2 z^2} + D$$

Cross-multiplying we get

$$\begin{aligned} b_1 z + b_0 &= Az^{-1} + Aa_1 + Aa_2 z + Bz + C - \rho_1 B - \rho_1 C z^{-1} \\ &\quad + D(1 + a_1 z + a_2 z^2 - \rho_1 z^{-1} - a_1 \rho_1 - a_2 \rho_1 z) \end{aligned}$$

Collecting terms in powers of z we get

$$\begin{aligned} b_1 z + b_0 &= Aa_1 + C - \rho_1 B + D - Da_1 \rho_1 + z(Aa_2 + B + Da_1 - Da_2 \rho_1) \\ &\quad - z^{-1}(A - \rho_1 C + \rho_1 D) + z^2 Da_2 \end{aligned}$$

Equating the coefficients of the powers of z to zero yields

$$D = 0$$

$$A = \rho_1 C$$

$$Aa_1 + C - B\rho_1 - b_0 = 0$$

$$Aa_2 + B - b_1 = 0$$

Simplifying we get

$$A = \frac{b_0 \rho_1 + b_1 \rho_1^2}{\rho_1^2 a_2 + \rho_1 a_1 + 1}$$

$$B = b_1 - Aa_2$$

$$C = \frac{A}{\rho_1}$$

The partial fraction expansion becomes

$$E_{xy}(z) = \frac{Az^{-1}}{1-\rho_1 z^{-1}} + \frac{Bz+C}{1+a_1 z + a_2 z^2}$$

References

[1] Neve, A. (2012) *Coherence Function: Theory and Visual Engineering Environment Applications*. Agilent Technologies: Education Corner.com Experiments.

[2] Binnie, C., Cooper, R., Mauguiere, F., *et al.* (eds) (2003) *Clinical Neurophysiology: EEG,Paediatric Neurophysiology, Special Techniques and Applications*, Vol. **2**, Elsevier Science BV, Amsterdam.

[3] Ropella, K., Sahakian, A., Baerman, J., and Swiryn, S. (1989) Coherence spectrum: a quantitative discriminator of fibrillatory and non-fibrillatory cardiac rythms. *Circulation: Journal of the American Heart Association*, **80**, 112–119.

[4] Kay, S.M. (1998) *Fundamentals of Signal Processing: Detection Theory*, Prentice Hall, New Jersey.

[5] Proakis, J.G. and Manolakis, D.G. (2007) *Digital Signal Processing: Principles, Algorithms and Applications*, Pearson Prentice Hall, New Jersey.

[6] Haykin, S. (2000) *Communication Systems*, John Wiley and Sons, New York.

[7] Oppenheim, A.V. and Schafer, R.W. (2010) *Discrete-Time Signal Processing*, Prentice-Hall, New Jersey.

[8] Kay, S.M. (1998) *Fundamentals of Signal Processing: Detection Theory*, Prentice Hall, New Jersey.

[9] Ljung, L. (1999) *System Identification: Theory for the User*, Prentice-Hall, New Jersey.

[10] Kay, S.M. (1988) *Modern Spectral Estimation: Theory and Application*, Prentice Hall, New Jersey.

[11] Mitra, S.K. (2006) *Digital Signal Processing: A Computer-Based Approach*, McGraw Hill Higher Education, Boston.

[12] Moon, T.K. and Stirling, W.C. (2000) *Mathematical Methods and Algorithms for Signal Processing*, Prentice Hall, New Jersey.

[13] Verhaegan, M. and Verdult, V. (2007) *Filtering and System Identification*, Cambridge University Press, Cambridge.

[14] Raol, J., Girija, G. and Singh, J. (2004) *Modeling and Parameter Estimation*, IEE Control Engineering Series 65, Institution of Electrical Engineers.

3

Estimation Theory

3.1 Overview

Estimation theory is a branch of statistics with wide application in science and engineering, and particularly forms the backbone of system identification. It deals with estimating an unknown parameter from measurements. The map relating the unknown parameter to the measurement may be described either by a mathematical model or by a probability density function. Mostly the model is assumed to be linear, or the PDF is assumed to be Gaussian, as the solution to the estimation problem is simplified. The estimate can be expressed in closed form, linear, and computationally efficient.

However, there are many practical cases where the above assumptions may not hold. In these cases it is assumed that the model is nonlinear or the PDF is non-Gaussian. A set of measurements, in general, contains a mixture of both "good data" and "bad data" points. Loosely speaking, good data are those which are close to the mean while bad data are located away from the mean. The bad data points are due to faulty instruments, unexpected failure of communication links, intermittent faults, and modeling errors, to name a few. As the bad data points are due to unpredictable causes it is difficult to characterize them statistically. Bad data can only be defined loosely as data containing errors that are worse than one normally would expect. Gaussian PDFs model random variables which are clustered close to their mean, while a non-Gaussian PDF, such as a Laplacian or Cauchy, models data which contains a mix of good and bad data. A zero-mean unity variance Gaussian PDF $f_y(y)$ has thin tails and approaches $\exp(-y^2/2\sigma_y^2)$ asymptotically. About 68% of values drawn from a Gaussian distribution are within one standard deviation σ_y away from the mean; about 95% of the values lie within two standard deviations; and about 99.7% are within three standard deviations. This justifies modeling good data by a Gaussian PDF. A non-Gaussian PDF with a thicker tail, such as a Laplacian or Cauchy, models bad data. A Laplacian PDF asymptotically approaches $\exp(-\sqrt{2}|y|/\sigma_y)$ while a Cauchy PDF approaches $\alpha/(\pi y^2)$. The thicker the tails the higher percentage of the data migrate towards the tails.

Figure 3.1 shows Gaussian, Laplace, and Cauchy PDFs and the random variables generated by these PDFS. On the left, subfigures (a), (b), and (c) show respectively the Gaussian, Laplace, and the Cauchy PDFs. The Gauss and Laplace PDFs have both unit mean and unit variance. The Cauchy has unit median and infinite variance. The subfigures D, E, and F on the right show respectively the Gaussian and Laplace and Cauchy random variables. The random variables which lie within an interval of $[-1 \quad 1]$ around the unit mean (or median) are assumed to be good data and those lying outside the interval are assumed to be dad data. The lines separating good data from the bad data are indicated in black at $y = 0$ and $y = 2$ while the mean is indicated by a line at $y = 1$.

From Figure 3.1 we can deduce that the Gaussian generates the largest percentage of good data while the Cauchy generates the smallest percentage.

Identification of Physical Systems: Applications to Condition Monitoring, Fault Diagnosis, Soft Sensor and Controller Design, First Edition. Rajamani Doraiswami, Chris Diduch and Maryhelen Stevenson.

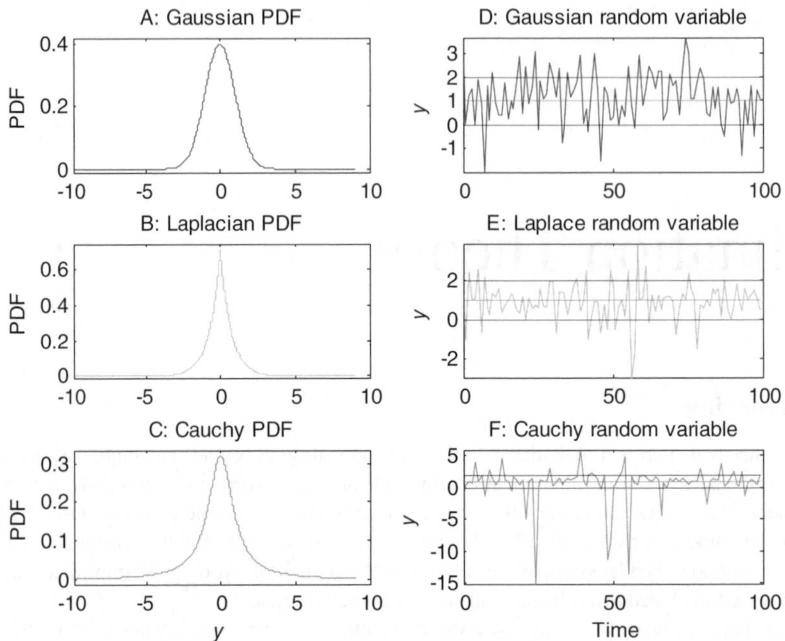

Figure 3.1 Gaussian, Laplace, and Cauchy random variables

In many practical cases, the PDF of the data may be unknown or partially known. An *ad hoc* or incorrect choice of PDF may result in a very poor performance of the estimator. For example, choosing a Gaussian PDF (assuming the data to be good) when the PDF is Cauchy (data is very bad), will mean the estimator performance will be catastrophic as the variance of the estimation error will be infinitely large. To deal with this statistical uncertainty, an estimator which gives an adequate performance over a class of PDFs is desirable. Such an estimator is said to be robust. A widely used approach to robust estimation is the so called "Min-Max" or game theoretic approach [1, 2]. Nature tries to maximize the covariance of the estimation error by choosing the worst-case PDF, while the engineer minimizes the error by choosing the best estimator. The engineer is thus forced to obtain an estimator by minimizing the estimation error for the worst-case PDF. An example of the worst-case PDF obtained using Min-Max is the "part Gauss-part Laplace" PDF. It is a thin-tailed Gaussian over the region containing good data and a thicker tailed Laplacian over the bad data region.

The choice of performance measure to obtain an estimate of the non-random parameter plays an important role. The intuitive and direct approach of choosing the measure to be the variance of the estimation error will yield an estimator to be a function of the estimator itself. This is meaningless, as the estimate of an unknown is a function of the unknown. Hence, alternative indirect schemes have been proposed so that the estimators have desirable properties such as unbiasedness and efficiency. An estimator is unbiased if the expectation of the estimate is equal to the true value of the estimated parameter. In other words if a number of experiments are performed, then the mean value of the estimates approach the true parameter value. The property of unbiasedness alone is not sufficient to ensure the good performance of an estimator. The variance of the estimator is an important measure. Ideally, the estimator must be unbiased and have the smallest possible variance.

The earliest stimulus for the development of estimation theory was apparently provided by astronomical studies in which the motion of planets and comets was studied using telescopic measurements. To solve this problem, Karl Gauss, around 1795, proposed the linear least-squares method, which is still the most popular approach. Least-squares estimation is a method of fitting the measurements (data or observation values) to a specified linear model. A best fit is obtained by minimizing the sum of the squares of the residual, where a residual is defined as an error between the measured value (data or observed value) and the value obtained using the model. The linear least-squares estimator is unbiased. However, it is not efficient unless the measurements are independent and identically (i.i.d.) random variables. The least-squares method may perform poorly when the PDF is non-Gaussian.

The question arises as to what is the lowest possible covariance of estimation error? The answer to this question may be obtained from the *Cramér–Rao lower bound*, which gives the lowest possible covariance that is achievable by an unbiased estimator. The Cramér–Rao lower bound (CRLB), named in honor of Herald Cramér and Calyampudi Radhakrishna Rao who were among the first to derive it, expresses a lower bound on the covariance of estimators of a deterministic parameter. The inverse of the lower bound is called the *Fisher information*. An unbiased estimator which achieves this lower bound is said to be efficient. An estimator which achieves the lowest possible mean squared error among all unbiased methods, is therefore the minimum variance unbiased (MVU) estimator. The next question arises as to how to determine an efficient estimator.

In 1922, R. A. Fisher introduced the method termed Maximum Likelihood Estimation (MLE) in the case when the PDF of the measurement is known. The unknown parameter is determined from the maximization of the probability that the observed occurs. It is based on maximizing a likelihood function, which is the PDF of the data expressed as a function of the parameter to be estimated. The maximum likelihood estimator is widely used, as the estimate gives the minimum estimation error covariance and serves as a gold standard for evaluating the performance of other estimators. It has the following properties:

Small sample property: If an efficient estimator exists, it is given by a maximum likelihood estimator.

Large sample properties: The maximum-likelihood estimator possesses a number of attractive and desirable properties. As the sample size increases to infinity, sequences of maximum-likelihood estimators have these properties:

- *Consistency*: subsequence of the sequence of MLEs converges in probability to the value being estimated.
- *Asymptotic normality*: as the sample size increases, the distribution of the MLE tends to the Gaussian distribution with a mean equal to the true mean, and the covariance matrix equal to the inverse of the Fisher information matrix.
- *Efficiency*: The covariance of the estimation attains the Cramér–Rao lower bound when the sample size tends to infinity.

In general, the estimates are implicit nonlinear functions of the parameter to be estimated and the estimate is obtained recursively. If the PDF of the measurement data is Gaussian, the maximum likelihood estimation method simplifies to a weighted least-squares method where the weight is the inverse of the noise covariance matrix.

3.2 Map Relating Measurement and the Parameter

3.2.1 Mathematical Model

Consider the problem of estimating a $Mx1$ parameter $\boldsymbol{\theta} = [\,\theta_1 \quad \theta_2 \quad \theta_3 \quad . \quad \theta_M\,]^T$ where the map relating the parameter to $Nx1$ measurement $y = [\,y(0) \quad y(1) \quad y(2) \quad . \quad y(N-1)\,]^T$ is given by some linear algebraic model

$$y = \boldsymbol{H}\boldsymbol{\theta} + \boldsymbol{v} \tag{3.1}$$

where $\boldsymbol{v}$ is zero-mean measurement noise with covariance $\boldsymbol{\Sigma}_v$; $\boldsymbol{H}$ is NxM matrix. Figure 3.2 shows the map relating the unknown parameter $\boldsymbol{\theta}$ and the measurement y where v is additive noise.

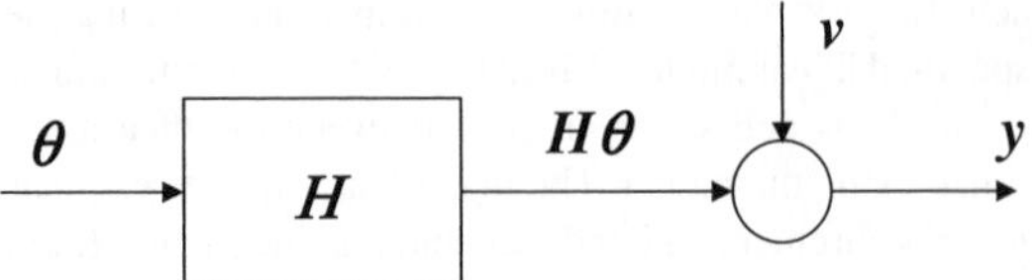

Figure 3.2 Map relating the unknown parameter and the measurement

3.2.2 Probabilistic Model

The measurement $\mathbf{y}$ is a random variable with distribution $f_y(\mathbf{y})$ which belongs to a family of PDFs parameterized by $\boldsymbol{\theta}$. In estimation theory the parameter $\boldsymbol{\theta}$ is unknown and the objective is to estimate $\boldsymbol{\theta}$ from y. To emphasize the probabilistic map relating y and $\boldsymbol{\theta}$ the PDF $f_y(\mathbf{y})$ governing the measurement y is denoted explicitly in terms of both y and $\boldsymbol{\theta}$ as:

$$f_y(\mathbf{y};\boldsymbol{\theta}) = f_y(y(0), y(1), y(2), \ldots, y(N-1); \boldsymbol{\theta}) \tag{3.2}$$

where $f_y(\mathbf{y};\boldsymbol{\theta})$ is the PDF governing the measurement y parameterized by $\boldsymbol{\theta}$. Examples of typical PDFs are given below:

3.2.2.1 Gaussian PDF: Denoted $f_g(\mathbf{y};\boldsymbol{\theta})$

The most common example is the Gaussian PDF given by

$$f_g(\mathbf{y};\boldsymbol{\theta}) = \frac{1}{\sqrt{(2\pi)^N\,|\Sigma_v|}} \exp\left\{-\frac{1}{2}(\mathbf{y}-\boldsymbol{\mu}_y)^T\boldsymbol{\Sigma}_v^{-1}(\mathbf{y}-\boldsymbol{\mu}_y)\right\} \tag{3.3}$$

The mean $\mu_y = E\,[\mathbf{y}]$ and the covariance of $\mathbf{y}$ denoted $\mathrm{cov}\,(y)$ are:

$$E\,[\mathbf{y}] = \mathbf{H}\boldsymbol{\theta} \tag{3.4}$$

$$\mathrm{cov}\,(\mathbf{y}) = \boldsymbol{\Sigma}_v \tag{3.5}$$

3.2.2.2 Uniform PDF: Denoted $f_u(y;\theta)$

A continuous uniform PDF for scalar y and scalar H = 1:

$$f_u(y;\theta) = \begin{cases} \dfrac{1}{b-a} & \textit{for}\; a \le y \le b \\ 0 & \textit{for}\; y < a \;\textit{or}\; y > b \end{cases} \tag{3.6}$$

The mean and the variance are:

$$E\,[y] = \mu_y = \theta = \frac{b+a}{2} \tag{3.7}$$

$$\mathrm{var}(y) = \sigma_y^2 = \frac{(b-a)^2}{12} \tag{3.8}$$

Treating unknown parameter θ as its mean, that is $\theta = \mu_y$, the uniform PDF may be expressed in terms of its mean and variance as

$$f_u(y;\theta) = \begin{cases} \dfrac{1}{2\sigma_y\sqrt{3}} & for\ -\sigma_y\sqrt{3} \le y - \mu_y \le \sigma_y\sqrt{3} \\ 0 & otherwise \end{cases} \tag{3.9}$$

The uniform PDF is used to model a random variable y when its variations around its mean are equally probable.

3.2.2.3 Laplacian PDF: Denoted $f_e(y;\theta)$

Laplace (also called double exponential) distribution is a continuous probability distribution named after Pierre-Simon Laplace and is called the double exponential because it may be interpreted as two exponential distributions spliced together back-to-back. It takes the form for a scalar case as:

$$f_e(y;\theta) = \frac{1}{\sqrt{2}\sigma_y}\exp\left(-\frac{\sqrt{2}|y-\mu_y|}{\sigma_y}\right) \tag{3.10}$$

The Laplace has thicker tails compared to that of the normal PDF and is used to model measurement y with larger random variations around its mean $\mu_y = H\theta$ and variance σ_y^2.

3.2.2.4 Worst-Case PDF: Part Gauss-Part Laplace Denoted $f_{ge}(y;\theta)$

Part Gauss-part Laplace is a worst-case PDF for obtaining a robust estimator, and plays an important role in parameter estimation when the PDF of the measurement is unknown, as will be seen later. The measurement space is divided into two regions Υ_{gd} and Υ_{bd}, where Υ_{gd} is a finite region around the mean μ_y of y and $\Upsilon_{bd} = \Re - \Upsilon_{gd}$ is the rest of the measurement space. Υ_{gd} and Υ_{bd} are defined as follows:

$$\Upsilon_{gd} = \{y : |y-\mu_y| \le a_{gd}\} \tag{3.11}$$

$$\Upsilon_{bd} = \{y : |y-\mu_y| > a_{gd}\} \tag{3.12}$$

where $a_{gd} > \mathbf{0}$ is some scalar separating the good from bad data points. The part Gauss-part Laplace PDF $f_y(y)$ is defined as

$$f_{ge}(y;\theta) = \begin{cases} \kappa\exp\left\{-\dfrac{1}{2\sigma_{ge}^2}(y-\mu_y)^2\right\} & y \in \Upsilon_{gd} \\ \kappa\exp\left\{\dfrac{a_{gd}^2}{2\sigma_{ge}^2}\right\}\exp\left\{-\dfrac{a_{gd}}{2\sigma_{ge}^2}|y-\mu_y|\right\} & y \in \Upsilon_{bd} \end{cases} \tag{3.13}$$

where $\mu_y = E[y] = H\theta$; κ and σ_{gd}^2 are determined such that (i) the integral of $f_y(y)$ is unity, ensuring thereby that $f_{ge}(y;\theta)$ is a PDF; and (ii) the constraint on the variance is met. Similar to the Laplacian PDF, the worst-case PDF has exponentially decaying tails.

3.2.2.5 Cauchy PDF: Denoted $f_c(y;\theta)$

Cauchy distribution is a continuous distribution. Its mean does not exist and its variance is infinity. Its median and mode are equal to the location parameter μ_y. The Cauchy has very thick tails compared to both the normal and Laplace PDFs and is used to model measurement y with very large random variations around its median μ_y.

$$f_c(y;\theta) = \frac{\alpha}{\pi[(y-\mu_y)^2+\alpha^2]} \tag{3.14}$$

where α is the scale parameter and α/π determines the peak amplitude.

3.2.2.6 Comparisons of the PDFs

Gaussian, Laplacian, part Gauss-part Laplace, Cauchy, and the uniform PDFs are shown in Figure 3.3. All the PDFs except the Cauchy have zero-mean, and Cauchy has zero-median. The variance of the Gaussian, the Laplacian, and part Gauss-part Laplace was 30, while Cauchy and the uniform have infinite variances. Note that the tails of the Gaussian PDF are the thinnest while that of the Cauchy are the thickest, with Laplacian and part Gauss-part Laplace thicker than Gaussian but thinner than Cauchy. The uniform is flat over the entire range of y.

3.2.3 Likelihood Function

The likelihood function plays a central role in an estimation method termed maximum likelihood estimation. A function closely related to the PDF, called "likelihood function," serves as a performance measure for the maximum likelihood estimation. A likelihood function of the random variable $\mathbf{y}$ is the

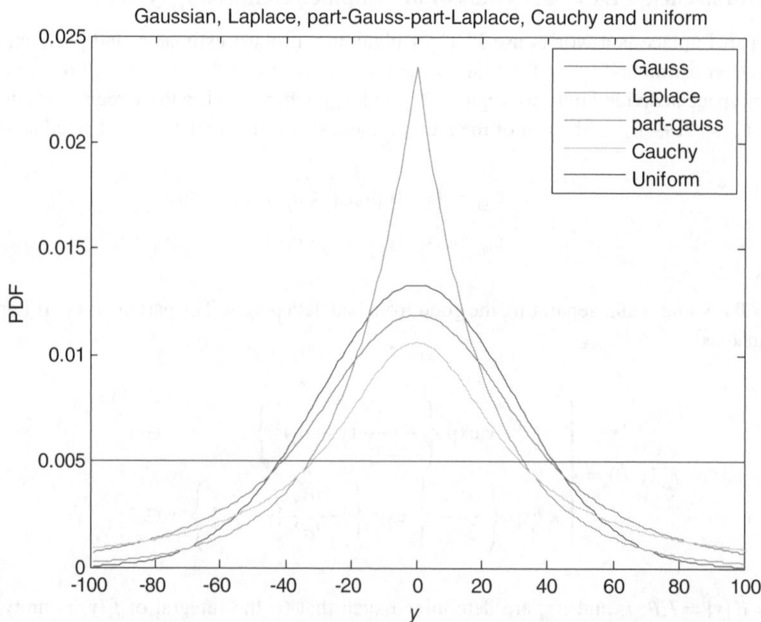

Figure 3.3 Gauss Laplace, part Gauss-part Laplace, Cauchy, and the uniform PDFs

PDF $f_y(y;\boldsymbol{\theta})$ of y expressed as a function of the parameter $\boldsymbol{\theta}$. It may be loosely defined as the conditional PDF of $\boldsymbol{\theta}$ given y with a subtle distinction between the two. In the case of the conditional PDF, denoted $f_y(\mathbf{y};\boldsymbol{\theta})$ for convenience, the parameter $\boldsymbol{\theta}$ is fixed and y is a random variable, while in the case of likelihood function, denoted $L(\boldsymbol{\theta}|y)$, the parameter $\boldsymbol{\theta}$ is a variable and the random variable y is fixed

$$L(\boldsymbol{\theta}|\mathbf{y}) = f_y(y;\boldsymbol{\theta}) \tag{3.15}$$

As commonly used PDFs are exponential functions of $\boldsymbol{\theta}$, it is convenient to work with the natural logarithm of the likelihood function, termed the log-likelihood function, rather than the likelihood function itself. The log-likelihood function, denoted $l(\boldsymbol{\theta}|\mathbf{y})$, is defined as

$$l(\boldsymbol{\theta}|y) = ln(L(\boldsymbol{\theta}|\mathbf{y})) \tag{3.16}$$

where ln (.) is the natural logarithm of (.). The log-likelihood function $l(\boldsymbol{\theta}|y)$ is the cost function used in obtaining the maximum likelihood estimator.

3.3 Properties of Estimators

3.3.1 Indirect Approach to Estimation

Let us formulate the problem of estimation of the unknown parameter $\boldsymbol{\theta}$ from the measurement y using the direct approach of minimizing the obvious performance metric, namely the covariance of the estimation error:

$$\min_{\hat{\theta}}\{E[(\boldsymbol{\theta}-\hat{\boldsymbol{\theta}})(\boldsymbol{\theta}-\hat{\boldsymbol{\theta}})^T]\} \tag{3.17}$$

where $\hat{\boldsymbol{\theta}}$ is an estimate of $\boldsymbol{\theta}$ and $\boldsymbol{\theta}-\hat{\boldsymbol{\theta}}$ is parameter estimation error. Since the covariance matrix is positive semi-definite, the minimum value of the covariance matrix is zero, and the optimal estimate $\hat{\boldsymbol{\theta}}$ is:

$$\hat{\boldsymbol{\theta}} = \boldsymbol{\theta} \tag{3.18}$$

This solution makes no sense as the estimate $\hat{\boldsymbol{\theta}}$ is expressed as a function of the unknown $\boldsymbol{\theta}$. We must seek an estimator which expresses the estimate as a function of the measurement. The estimate must be some function of the measurement:

$$\hat{\boldsymbol{\theta}} = \phi(\mathbf{y}) \tag{3.19}$$

We will consider two intuitive approaches to obtaining an estimate as a function of the measurement, based on assuming (i) the noise is zero-mean and (ii) the median of the noise is zero.

Noise is zero-mean: Let us now employ an intuitively simple approach to estimate θ using a scalar linear model:

$$y(k) = H\theta + v(k) \tag{3.20}$$

where $y(k)$, H, and $v(k)$ is a zero-mean random variable. Invoking the zero-mean property of the measurement noise $E[v(k)] = 0$, the estimate $\hat{\theta}$ yields:

$$E[y(k) - H\hat{\theta}] = 0 \tag{3.21}$$

Simplifying we get

$$\hat{\theta} = \frac{1}{H}E[y(k)] \tag{3.22}$$

Remarks The expression for the estimate $\hat{\theta}$ may be obtained by minimizing the 2-norm of the error $y(k) - H\hat{\theta}$ given by

$$\min_{\hat{\theta}} E[(y(k) - H\hat{\theta})^2] \tag{3.23}$$

Differentiating with respect to $\hat{\theta}$ and setting it to zero, the two-norm $E[(y(k) - H\hat{\theta})^2]$ yields Eq. (3.21). Let us verify the performance of the estimator by computing its mean $E\left[\hat{\theta}\right]$. Substituting for $y(k)$ in Eq. (3.22) using Eq. (3.20) and $E\left[v(k)\right] = 0$, we get:

$$E[\hat{\theta}] = \frac{E\left[H\theta + v\right]}{H} = \theta + E\left[v\right] = \theta \tag{3.24}$$

This shows that the performance of the estimate is ideal. However, the optimal estimate involves computation of an ensemble average of the measurement $E\left[y(k)\right]$. In practice, as only a finite and single realization of the measurement is available, an estimate of the ensemble average, namely, the time average of the measurement, is employed and the estimate θ becomes:

$$\hat{\theta} = \frac{1}{H}\left(\frac{1}{N}\sum_{k=0}^{N-1} y(k)\right) \tag{3.25}$$

where the time average $\frac{1}{N}\sum_{k=0}^{N-1} y(k)$ is an estimate of the ensemble average $E\left[y(k)\right]$; and N is the number of measurements.

The estimator $\hat{\theta}$ is a linear function of the measurement $y(k)$.

Let us analyze the asymptotic behavior of the estimator (3.25) in terms of its bias and the variance.

3.3.2 Unbiasedness of the Estimator

An estimator $\hat{\boldsymbol{\theta}}$ is an unbiased estimator of θ if

$$E[\hat{\boldsymbol{\theta}}] = \begin{cases} \boldsymbol{\theta} & \textit{if } \boldsymbol{\theta} \textit{ is not random} \\ E[\boldsymbol{\theta}] & \textit{if } \boldsymbol{\theta} \textit{ is random} \end{cases} \tag{3.26}$$

Let us now verify the unbiasedness of the estimator $\hat{\theta}$. Substituting for $y(k)$ in Eq. (3.25) using Eq. (3.20) we get

$$\hat{\theta} = \frac{1}{H}\left(\frac{1}{N}\sum_{k=0}^{N-1} H\theta\right) + \frac{1}{H}\left(\frac{1}{N}\sum_{k=0}^{N-1} v(k)\right) \tag{3.27}$$

Simplifying we get

$$\hat{\theta} = \theta + \frac{1}{H}\left(\frac{1}{N}\sum_{k=0}^{N-1} v(k)\right) \tag{3.28}$$

Taking expectation on both sides and invoking $E\left[v(k)\right] = 0$ yields:

$$E\left[\hat{\theta}\right] = \theta \tag{3.29}$$

This shows that the estimate is *unbiased*. The definition of an unbiased estimator is given by Eq. (3.26). We will derive a condition for the unbiasedness for a class of linear estimators given by

$$\hat{\theta} = \boldsymbol{F}\boldsymbol{y} \tag{3.30}$$

The following lemma gives the condition

Lemma 3.1 *Consider a linear model* $\boldsymbol{y} = \boldsymbol{H}\boldsymbol{\theta} + \boldsymbol{v}$. *A linear estimator,* $\hat{\theta} = \boldsymbol{F}\boldsymbol{y}$ *is unbiased if and only if*

$$\boldsymbol{F}\boldsymbol{H} = \boldsymbol{I} \tag{3.31}$$

Proof: Substituting for y in Eq. (3.30) using $\boldsymbol{y} = \boldsymbol{H}\boldsymbol{\theta} + \boldsymbol{v}$ we get

$$\hat{\theta} = \boldsymbol{F}\boldsymbol{H}\boldsymbol{\theta} + \boldsymbol{F}\boldsymbol{v} \tag{3.32}$$

Taking expectation on both sides and noting that $\boldsymbol{v}$ is a zero-mean random variable we get,

$$\boldsymbol{E}\left[\hat{\theta}\right] = \boldsymbol{F}\boldsymbol{H}\boldsymbol{E}[\theta] + \boldsymbol{F}\boldsymbol{E}[\boldsymbol{v}] = \boldsymbol{F}\boldsymbol{H}\boldsymbol{E}[\theta]$$

From the definition of unbiasedness, we conclude $\boldsymbol{F}\boldsymbol{H} = \boldsymbol{I}$. In the scalar example (3.20) the condition (3.31) holds.

3.3.3 Variance of the Estimator: Scalar Case

Let us now verify the variance of the estimator, as the unbiasedness alone is sufficient to deduce its performance. For simplicity we will assume that $v(k)$ is a zero-mean white noise with variance σ_v^2. The variance of the unbiased estimator $\hat{\theta}$ is obtained using Eq. (3.28). Squaring and taking expectation yields:

$$\boldsymbol{E}[(\hat{\theta} - \theta)^2] = \frac{1}{H^2} E\left[\left(\frac{1}{N}\sum_{k=0}^{N-1} v(k)\right)^2\right] = \frac{1}{N^2}\left(\frac{\sigma_v^2}{H^2}\right) \tag{3.33}$$

This shows that the variance of the estimator is a function of the "signal to noise ratio" $\dfrac{\sigma_v^2}{H^2}$ and inversely proportional to N^2. The asymptotic behavior of the variance is given by:

$$\lim_{N\to\infty} it\, E[(\hat{\theta} - \theta)^2] = 0 \tag{3.34}$$

The variance is asymptotically zero.

3.3.4 Median of the Data Samples

A median of a sample value is the numerical value that separates the higher half from the lower half of the sample value. In the case of a finite number of samples, the median is determined by sorting the sample values from the lowest to highest values and picking the middle one. Let $y = [\, y(0) \;\; y(1) \;\; y(2) \;\; . \;\; y(N-1)\,]^T$ be N data samples. Sort the data from the lowest to the highest values, $x_0 < x_1 < x_2 < x_3 < \cdots < x_{N-1}$ where $x_0 = \min_k y(k)$ and $x_{N-1} = \max_k y(k)$: $\{x_i\}$ is termed as order statistics of $\{y(k)\}$. Then, the median of y denoted *median*(y) is given by

$$\text{median}(\mathbf{y}) = \begin{cases} x_{\frac{N+1}{2}} & N\ odd \\ \dfrac{1}{2}\left(x_{\frac{N}{2}} + x_{\frac{N}{2}+1}\right) & N\ even \end{cases} \tag{3.35}$$

A median is also a central point that minimizes the average of the absolute deviations. A formal definition of a median is that it is the value that minimizes the 1-norm of the estimation error:

$$\min_{\hat{\theta}} \left\{ E|y - H\hat{\theta}| \right\} \tag{3.36}$$

Thus, when the noise is zero-mean the estimator is determined from the minimization of 2-norm (3.23), while for the case of zero-median the estimator is obtained from the minimization of the 1-norm (3.36). The question arises as to how to evaluate an estimator. This is addressed in the next section.

3.3.5 *Small and Large Sample Properties*

Often, the performance of an estimator is evaluated assuming that there are an infinite number of data samples. A theoretical evaluation of estimation error for small samples is very difficult. The question arises of whether there are measures that can evaluate the performance of an estimator when the sample size is small. The commonly used measures are

- unbiasedness
- efficiency

The property of unbiasedness is not good enough to ensure the good performance of an estimator. The variance of the estimator is an important measure. Ideally, the estimator must be unbiased with the smallest possible variance. This topic is addressed next in the section on Cramér–Rao lower bound. It will be shown later that an unbiased estimator is efficient if its covariance attains the Cramér–Rao lower bound.

3.3.6 *Large Sample Properties*

In the case of a non-Gaussian PDF, the estimates may not satisfy the desirable properties of unbiasedness and efficiency when the number of data samples is small. However, they may satisfy these properties when the number of samples is infinitely large. The behavior of an estimator for a large sample size is termed asymptotic behavior and is used as a measure of the performance of an estimator. Asymptotic behaviors include consistency, asymptotic unbiasedness, and asymptotic efficiency.

3.3.6.1 Consistent Estimator

In practice, a single estimator is constructed as a function of the number of data samples N, and a sequence of estimators is obtained as sample size N grows to infinity. If this sequence of estimators converges in probability to the true value, then the estimator is said to be consistent.

Let $\hat{\theta}_N$ be an estimator of θ from N data samples $y = [\,y(0) \quad y(1) \quad y(2) \quad . \quad y(N-1)\,]^T$. Then $\hat{\theta}_N$ is a consistent estimator of θ if $\hat{\theta}_N$ converges to θ in probability:

$$\lim_{N\to\infty} it\, P\{|\theta - \hat{\theta}| > \varepsilon\} = 0, \tag{3.37}$$

for every $\varepsilon > 0$.

3.3.6.2 Asymptotically Unbiased Estimator

An estimator $\hat{\theta}_N$ is said to be an asymptotically unbiased estimator of θ if the estimator is unbiased and if the number of data samples is infinite

$$\lim_{N\to\infty} it\, E\left[\hat{\theta}_N\right] = \begin{cases} \theta & \textit{if } \theta \textit{ is not random} \\ E\left[\theta\right] & \textit{if } \theta \textit{ is random} \end{cases} \tag{3.38}$$

3.3.6.3 Asymptotically Efficient Estimator

Ideally, the estimator must be unbiased with the smallest possible variance for a finite number of data samples. In general, this property may not hold for small sample sizes. If the property of efficiency holds for an infinitely large number of samples, N, then the estimator is said to be asymptotically efficient. It will be shown later that an estimator is asymptotically efficient if its covariance attains the Cramér–Rao lower bound.

3.4 Cramér–Rao Inequality

Definition *An unbiased estimator denoted* $\hat{\theta}^0$ *is said to be more efficient compared to any other unbiased estimator* $\hat{\theta} = \phi(y)$ *if the covariance of* $\hat{\theta}^0$ *is smaller than that of* $\hat{\theta} = \phi(y)$*:*

$$\text{cov}(\hat{\theta}^0) \le \text{cov}\left(\hat{\theta}\right) \tag{3.39}$$

where $\text{cov}(\hat{\theta}) = E[(\theta - \hat{\theta})(\theta - \hat{\theta})^T]$*; and if* $\boldsymbol{P}$ *and* $\boldsymbol{Q}$ *are two positive semi-definite matrices then* $\boldsymbol{P} \ge \boldsymbol{Q}$ *implies* $\boldsymbol{P} - \boldsymbol{Q}$ *is positive semi-definite.*

The *Cramér–Rao lower bound* gives the lowest possible covariance that is achievable by an unbiased estimator. The lower bound is equal to the inverse of the *Fisher information*. An unbiased estimator which achieves this lower bound is said to be efficient. An estimator which achieves the lowest possible mean squared error among all unbiased methods is therefore the minimum variance unbiased (MVU) estimator. We will first consider the scalar and extend the result to the vector case. We will assume the following:

- The derivation of Cramér–Rao lower bound assumes weak conditions on the PDF $f_y(y;\theta)$ and the estimator $\hat{\theta}(y)$.
- $f_y(y;\theta) > 0$.
- $\dfrac{\delta f_y(y;\theta)}{\delta\theta}$ exists and is finite.

- the PDF $f_y(y;\boldsymbol{\theta})$ is twice differentiable with respect $\boldsymbol{\theta}$ and satisfies the following regularity condition:

$$E\left[\frac{\delta l(\boldsymbol{\theta}|\mathbf{y})}{\delta\boldsymbol{\theta}}\right] = \mathbf{0} \text{ for all } \boldsymbol{\theta} \tag{3.40}$$

where $\frac{\delta l(\boldsymbol{\theta}|\mathbf{y})}{\delta\boldsymbol{\theta}}$ is the gradient of $l(\boldsymbol{\theta}|\mathrm{y})$ with respect to $\boldsymbol{\theta}$.

- The operations of integration with respect to $\mathbf{y}$ and differentiation with respect to $\boldsymbol{\theta}$ can be interchanged in $\mathrm{E}[\hat{\boldsymbol{\theta}}(\mathbf{y})]$.

$$\frac{\delta}{\delta\boldsymbol{\theta}}\int_{-\infty}^{\infty}\hat{\boldsymbol{\theta}}(\mathbf{y})f_y(\mathbf{y})d\mathbf{y} = \int_{-\infty}^{\infty}\hat{\boldsymbol{\theta}}(\mathbf{y})\frac{\delta f_y(\mathbf{y})}{\delta\boldsymbol{\theta}}d\mathbf{y} \tag{3.41}$$

This condition, that the integration and differentiation can be swapped, holds for all well-behaved PDFs which satisfy the following:

1. If $f_y(y;\boldsymbol{\theta})$ has bounded support in $\mathbf{y}$, and the bounds do not depend on $\boldsymbol{\theta}$.
2. If $f_y(y;\boldsymbol{\theta})$ has infinite support, is continuously differentiable, and the integral converges uniformly for all $\boldsymbol{\theta}$.
3. The estimate $\hat{\boldsymbol{\theta}}$ is unbiased, that is $E\left[\hat{\boldsymbol{\theta}}\right] = \boldsymbol{\theta}$.

Remark The uniform PDF does not satisfy the regularity condition and hence Cramér–Rao inequality cannot be applied.

3.4.1 Scalar Case: θ and $\hat{\theta}$ Scalars while y is a Nx1 Vector

Lemma 3.2 *The Cramér–Rao inequality of an unbiased estimator $\hat{\theta}$ is*

$$\mathrm{var}(\hat{\theta}) \geq \frac{1}{I_F} \tag{3.42}$$

where $\mathrm{var}(\hat{\theta}) = E[(\hat{\theta}-\theta)^2]$; *$I_F$ is the Fisher information given by*

$$I_F = E\left[\left(\frac{\partial l(\theta|\mathrm{y})}{\partial\theta}\right)^2\right] = -E\left[\frac{\partial^2 l(\theta|\mathrm{y})}{\partial\theta^2}\right] \tag{3.43}$$

$\frac{\delta l(\theta|\mathbf{y})}{\delta\theta}$ *is the partial derivative of $l(\theta|\mathbf{y})$ with respect to θ and is given by*

$$\frac{\delta l(\theta|\mathbf{y})}{\delta\theta} = \frac{H}{\sigma_v^2}\frac{f_y'(y;\theta)}{f_y(y;\theta)} \tag{3.44}$$

where $f_y'(y;\theta) = \frac{\delta f_y(y;\theta)}{\delta\theta}$ *and evaluated at the true value of θ. The Fisher information, I_F, is constant or a function of θ but it is not a function of the measurement* $\mathbf{y}$.

Corollary 3.2 *An estimator $\hat{\theta}$ attains the Cramér–Rao lower bound:*

$$\mathrm{var}(\hat{\theta}) = \frac{1}{I_F} \tag{3.45}$$

If and only if

$$\frac{\delta l(\theta|\mathbf{y})}{\delta\theta} = I_F(\hat{\theta} - \theta) \tag{3.46}$$

See Appendix for the proof.

3.4.2 Vector Case: θ is a Mx1 Vector

Lemma 3.3 *The Cramér–Rao inequality of an unbiased estimator $\hat{\boldsymbol{\theta}}$ is*

$$\mathrm{cov}(\hat{\boldsymbol{\theta}}) \geq \boldsymbol{I}_F^{-1} \tag{3.47}$$

where $\boldsymbol{P} \geq \boldsymbol{Q}$ is interpreted as $\boldsymbol{P} - \boldsymbol{Q}$ and is positive semi-definite; $\mathrm{cov}(\hat{\boldsymbol{\theta}}) = E[(\hat{\boldsymbol{\theta}} - \boldsymbol{\theta})(\hat{\boldsymbol{\theta}} - \boldsymbol{\theta})^T]$; $\boldsymbol{I}_F$ is an MxM Fisher information matrix given by

$$I_F = E\left[\left(\frac{\partial l(\boldsymbol{\theta}|\mathrm{y})}{\partial\boldsymbol{\theta}}\right)\left(\frac{\partial l(\boldsymbol{\theta}|\mathrm{y})}{\partial\boldsymbol{\theta}}\right)^T\right] = -E\left[\frac{\partial^2 l(\boldsymbol{\theta}|\mathrm{y})}{\partial\boldsymbol{\theta}^2}\right] \tag{3.48}$$

where $\dfrac{\partial^2 l(\boldsymbol{\theta}|\mathrm{y})}{\partial\boldsymbol{\theta}^2}$ is the Hessian matrix of the likelihood function $l(\boldsymbol{\theta}|\mathbf{y})$;

$$\frac{\delta l(\boldsymbol{\theta}|\mathbf{y})}{\delta\boldsymbol{\theta}} = \frac{\boldsymbol{H}^T\Sigma^{-1}f_y'(\mathbf{y};\boldsymbol{\theta})}{f_y(\mathbf{y};\boldsymbol{\theta})} \tag{3.49}$$

where $f_y'(\mathrm{y};\theta) = \dfrac{\delta f_y(\mathrm{y};\theta)}{\delta\theta}$. The Fisher information matrix $\boldsymbol{I}_F$ is constant or a function of $\boldsymbol{\theta}$ but it is not a function of the measurement $\mathbf{y}$. The element ij of $\boldsymbol{I}_F$ denoted $[\boldsymbol{I}_F]_{ij}$ is:

$$\left[\boldsymbol{I}_F\right]_{ij} = E\left[\left(\frac{\partial l(\boldsymbol{\theta}|\mathrm{y})}{\partial\theta_i}\right)\left(\frac{\partial l(\boldsymbol{\theta}|\mathrm{y})}{\partial\theta_j}\right)\right] = -E\left[\frac{\partial^2 l(\boldsymbol{\theta}|\mathrm{y})}{\partial\theta_i\partial\theta_j}\right] \tag{3.50}$$

Corollary 3.3 *An estimator $\hat{\boldsymbol{\theta}}$ attains the Cramér–Rao lower bound:*

$$\mathrm{cov}(\hat{\boldsymbol{\theta}}) = \boldsymbol{I}_F^{-1} \tag{3.51}$$

If and only if

$$\frac{\delta l(\boldsymbol{\theta}|\mathbf{y})}{\delta\boldsymbol{\theta}} = \boldsymbol{I}_F\left(\hat{\boldsymbol{\theta}} - \boldsymbol{\theta}\right) \tag{3.52}$$

See Appendix for the proof.

In view of the Cramér–Rao inequality and an estimator $\hat{\theta}$ we may define an efficient estimator given by Eq. (3.39) precisely as follows:

Definition *An estimator* $\hat{\theta}$ *is said to be* efficient *if its variance attains the Cramér–Rao lower bound (3.45) for the scalar case and Eq. (3.51) for the vector case or equivalently satisfies the condition (3.46) for the scalar and Eq. (3.52) for the vector case.*

The estimator is asymptotically efficient *if it attains the Cramér–Rao lower bound when the number of data samples approaches infinity.*

3.4.3 Illustrative Examples: Cramér–Rao Inequality

Example 3.1 i.i.d. Gaussian PDF

Single data sample: Let y be a measurement generated by the mathematical model

$$y = \theta + v \tag{3.53}$$

The probabilistic model is given by the PDF

$$f_g(y;\theta) = \frac{1}{\sqrt{2\pi\sigma^2}} \exp\left\{-\frac{1}{2\sigma_v^2}(y-\theta)^2\right\} \tag{3.54}$$

The log-likelihood function $l(\theta|y)$ is

$$l(\theta|y) = -\frac{1}{2}\ln(2\pi\sigma^2) - \frac{1}{2\sigma_v^2}(y-\theta)^2 \tag{3.55}$$

Taking partial derivative with respect to θ yields

$$\frac{\partial l(\theta|y)}{\partial\theta} = \frac{1}{\sigma_v^2}(y-\theta) \tag{3.56}$$

Comparing the right sides of Eqs. (3.56) and (3.46) we deduce the following:

$$\hat{\theta} = y \tag{3.57}$$

The Fisher information is constant:

$$I_F(\theta) = \frac{1}{\sigma_v^2} \tag{3.58}$$

It is reciprocal of the variance of the measurement noise σ_v^2.

N data samples: Let $y = [\,y(0) \quad y(1) \quad y(2) \quad . \quad y(N-1)\,]^T$ be $Nx1$ measurement where $y(k)$; $k = 0, 1, 2, \dots N-1$ are independent and identically distributed Gaussian random variables. The mathematical model that generates the measurements is:

$$y(k) = \theta + v(k), \quad k = 0, 1, 2, \dots, N-1 \tag{3.59}$$

Equivalently, the probabilistic model given by the PDF

$$f_g(\mathbf{y};\theta) = \frac{1}{\left(2\pi\sigma^2\right)^{\frac{N}{2}}} \exp\left\{-\frac{1}{2\sigma_v^2}\sum_{k=0}^{N-1}(y(k)-\theta)^2\right\} \tag{3.60}$$

The log-likelihood function $l(\theta|\mathbf{y})$ becomes

$$l(\theta|\mathbf{y}) = -\frac{N}{2}\ln(2\pi\sigma^2) - \frac{1}{2\sigma_v^2}\sum_{i=0}^{N-1}(y(k)-\theta)^2 \tag{3.61}$$

Taking the partial derivative yields

$$\frac{\partial l(\theta|\mathbf{y})}{\partial\theta} = \frac{1}{\sigma_v^2}\sum_{i=0}^{N-1}(y(k)-\theta) = \frac{N}{\sigma_v^2}\left(\frac{1}{N}\sum_{k=0}^{N-1}y(k)-\theta\right) \tag{3.62}$$

Let us verify whether an efficient estimator exists using Eq. (3.46):

$$\frac{\partial l(\theta|y)}{\partial\theta} = \frac{N}{\sigma_v^2}\left(\frac{1}{N}\sum_{k=0}^{N-1}y(k)-\theta\right) \tag{3.63}$$

Comparing the right-hand side of Eq. (3.46) with that of Eq. (3.63) we deduce the following:

$$\hat{\theta} = \frac{1}{N}\sum_{k=0}^{N-1}y(k) \tag{3.64}$$

Since $\{y(k)\}$ is i.i.d., the Fisher information from Eq. (3.62) is:

$$I_F = E\left[\left(\frac{\partial l(\theta|\mathbf{y})}{\partial\theta}\right)^2\right] = \frac{N}{\sigma_v^2} \tag{3.65}$$

The Fisher information is constant N times the reciprocal of the noise variance σ_v^2:

$$I_F = \frac{N}{\sigma_v^2} \tag{3.66}$$

Properties of the estimator: The estimator for the Gaussian PDF is the mean. Let us verify whether $\hat{\theta}$ is unbiased. Taking the expectation of Eq. (3.64) we get:

$$E\left[\hat{\theta}\right] = \frac{1}{N}\sum_{k=0}^{N-1}E\left[y(k)\right] \tag{3.67}$$

Using Eq. (3.59) we get

$$E\left[\hat{\theta}\right] = \frac{1}{N}\sum_{k=0}^{N-1}E\left[y(k)\right] = \frac{1}{N}\sum_{k=0}^{N-1}\theta + \frac{1}{N}\sum_{k=0}^{N-1}E\left[v(k)\right] = \theta \tag{3.68}$$

Since $E\left[\hat{\theta}\right] = \theta$, the estimator $\hat{\theta}$ is unbiased. The estimator $\hat{\theta}$ is efficient as it satisfies the Cramér–Rao equality condition (3.46). The estimator is unbiased and efficient for any sample size N.

Example 3.2 i.i.d. distributed Laplacian PDF

Single data sample: The mathematical and the probabilistic model are given by Eqs. (3.53) and (3.10). We will consider a simplified PDF given by

$$f_e(y;\theta) = \frac{1}{\sqrt{2\sigma_v^2}} \exp\left(-\frac{\sqrt{2}}{\sigma_v}|y-\theta|\right) \tag{3.69}$$

The likelihood function $l(\theta|y)$ is

$$l(\theta|y) = -\frac{1}{2}\ln(2\sigma_v^2) - \frac{\sqrt{2}}{\sigma_v}|y-\theta| \tag{3.70}$$

Expanding the definition of $|y(k)-\theta|$, we get

$$l(\theta|y) = \begin{cases} -\frac{1}{2}\ln(2\sigma_v^2) - \frac{\sqrt{2}}{\sigma_v}(y-\theta) \ for\, y > \theta \\ -\frac{1}{2}\ln(2\sigma_v^2) \ for\, y = \theta \\ -\frac{1}{2}\ln(2\sigma_v^2) + \frac{\sqrt{2}}{\sigma_v}(y-\theta)\, for\, y < \theta \end{cases} \tag{3.71}$$

It is differentiable everywhere except at $y = \theta$. Taking the partial derivative yields

$$\frac{\partial l(\theta|y)}{\partial\theta} = \frac{\sqrt{2}}{\sigma_v} sign\,(y-\theta) = \begin{cases} \frac{\sqrt{2}}{\sigma_v} & y > \theta \\ 0 & y = \theta \\ -\frac{\sqrt{2}}{\sigma_v} & y < \theta \end{cases} \tag{3.72}$$

The Fisher information is:

$$I_F = E\left[\left(\frac{\partial l(\theta|\mathbf{y})}{\partial\theta}\right)^2\right] = \frac{2}{\sigma_v^2} \tag{3.73}$$

The Cramér–Rao inequality is

$$\text{var}\left(\hat{\theta}\right) \geq \frac{\sigma_v^2}{2} \tag{3.74}$$

N data samples: Let $y = \begin{bmatrix} y(0) & y(1) & y(2) & . & y(N-1) \end{bmatrix}^T$ be $Nx1$ measurements where $y(k)$; $k = 0, 1, 2, \dots N-1$ are independent and identically distributed Laplacian random variables. The mathematical model that generates the measurements is given by Eq. (3.59). Equivalently, the probabilistic model is:

$$f_e(\mathbf{y};\theta) = \left(\frac{1}{\sqrt{2\sigma_v^2}}\right)^N \exp\left(-\frac{\sqrt{2}}{\sigma_v}\sum_{k=0}^{N-1}|y(k)-\theta|\right) \tag{3.75}$$

Taking the natural logarithm, the likelihood function $l(\theta|\mathbf{y})$ becomes

$$l(\theta|\mathbf{y}) = -\frac{N}{2}\ln(2\sigma_v^2) - \frac{\sqrt{2}}{\sigma_v}\sum_{k=0}^{N-1}|y(k) - \theta| \tag{3.76}$$

Taking the partial derivative yields

$$\frac{\partial l(\theta|\mathbf{y})}{\partial\theta} = \frac{\sqrt{2}}{\sigma_v}\sum_{k=0}^{N-1} sign\,(y(k) - \theta) \tag{3.77}$$

The partial derivative of the log-likelihood function is nonlinear, which is positive constant for $y > 0$ and negative constant for $y < 0$. This type of nonlinear function is termed a *hard-limiter*. The Fisher information is:

$$I_F = E\left[\left(\frac{\partial l(\theta|\mathbf{y})}{\partial\theta}\right)^2\right] = \frac{2N}{\sigma_v^2} \tag{3.78}$$

The Cramér–Rao inequality is

$$\text{var}\,(\theta) \geq \frac{\sigma_v^2}{2N} \tag{3.79}$$

The properties of the estimator are deferred to the next section on maximum likelihood estimation.

Example 3.3 Part Gauss-part Laplace PDF
Consider the PDF given by Eq. (3.13). Assuming $\mu_y = \theta$ the PDF becomes

$$f_{ge}(y;\theta) = \begin{cases} \kappa\exp\left\{-\frac{1}{2\sigma_{ge}^2}(y-\theta)^2\right\} & |y-\theta| \leq a_{gd} \\ \kappa\exp\left\{\frac{a_{gd}^2}{2\sigma_{ge}^2}\right\}\exp\left\{-\frac{1}{2\sigma_{ge}^2}|y-\theta|a_{gd}\right\} & |y-\theta| > a_{gd} \end{cases} \tag{3.80}$$

From the Appendix, the expression for the derivative of the log-likelihood function is

$$\frac{\delta l(\theta|y)}{\delta\theta} = \begin{cases} \frac{1}{\sigma_{ge}^2}(y-\theta) & |y-\theta| \leq a_{gd} \\ \frac{a_{gd}}{\sigma_{ge}^2} sign\,(y-\theta) & |y-\theta| > a_{gd} \end{cases} \tag{3.81}$$

The derivative of the log-likelihood function is nonlinear: it is linear over Υ_{gd} and is positive constant over $y - \theta > a_{gd}$ and negative constant over $y - \theta < a_{gd}$. This type of nonlinear function is termed a *soft limiter*.

The Fisher information is:

$$I_F = E\left[\left(\frac{\partial l(\theta|\mathbf{y})}{\partial\theta}\right)^2\right] = \frac{1}{\sigma_{ge}^4}\left(\sigma_{gd}^2 + a_{gd}^2\left(1 - \lambda_{gd}\right)\right) \tag{3.82}$$

where $\sigma_{gd}^2 = \int_{y=\theta-a_{gd}}^{y=\theta+a_{gd}} (y-\theta)^2 f_y(y)dy$ is called "partial variance" and $\lambda_{gd} = \int_{-a_{gd}}^{a_{gd}} f_y(y)dy$ is called "partial probability" of y over the region Υ_{gd}. The partial variance and partial probability are respectively the variance and the probability of the random variables restricted to a finite region around the mean, which in this case is the good data region.

Remarks The Part Gauss-part Laplace PDF is the worst-case PDF which maximizes the Fisher information for a class of all continuous PDFs which are constrained by the partial covariance.

$$f_{ge}(\mathbf{y};\theta) = \arg\min_{f_y(y)} \{I_F\} \text{ such that } \int_{y=\theta-a_{gd}}^{y=\theta+a_{gd}} (y-\theta)^2 f_{ge}(y)dy \le \sigma_{gd}^2 \tag{3.83}$$

One would have expected the worst-case PDF $f_{ge}(\mathbf{y};\theta)$ to have tails thicker than those of Cauchy. That is, one would expect the worst-case PDF to generate more bad data than the Cauchy PDF.

Example 3.4 i.i.d. Cauchy PDF

Single data sample: The mathematical and the probabilistic models are given by Eqs. (3.53) and (3.14) respectively. We will consider a simplified PDF given by

$$f_c(y;\theta) = \frac{\alpha}{\pi[(y-\theta)^2+\alpha^2]} \tag{3.84}$$

where α is the scale parameter and θ is median (or the mode or location parameter) of the PDF. The likelihood function $l(\theta|y)$ becomes

$$l(\theta|y) = \ln(\alpha) - \ln(\pi) - \ln(\pi[(y-\theta)^2+\alpha^2]) \tag{3.85}$$

The partial derivative of $l(\theta|y)$ is:

$$\frac{\delta l(\theta|y)}{\delta\theta} = \frac{2(y-\theta)}{[(y-\theta)^2+\alpha^2]} \tag{3.86}$$

The derivative of the log-likelihood function is nonlinear. It asymptotically decays as $1/|y|$.

The Fisher information using the Appendix yields:

$$I_F = E\left[\left(\frac{\partial l(\theta|y)}{\partial\theta}\right)^2\right] = \frac{1}{2\alpha^2} \tag{3.87}$$

N data samples: Let $\mathbf{y} = [\,y(0) \quad y(1) \quad y(2) \quad . \quad y(N-1)\,]^T$ be $Nx1$ measurements where $y(k)$; $k = 0, 1, 2, \ldots N-1$ are independent and identically distributed Cauchy random variables. The mathematical model that generates the measurements is given by Eq. (3.59). Equivalently, the probabilistic model is:

$$f_c(\mathbf{y};\theta) = \prod_{k=0}^{N-1}\left(\frac{\alpha}{\pi[(y(k)-\theta)^2+\alpha^2]}\right) \tag{3.88}$$

Taking the natural logarithm, the likelihood $l(\theta|\mathbf{y})$ s:

$$l(\theta|\mathbf{y}) = N\ln(\alpha) - N\ln(\pi) - \sum_{k=0}^{N-1} \ln(\pi[(y(k)-\theta)^2+\alpha^2]) \tag{3.89}$$

Taking the partial derivative yields

$$\frac{\delta l(\theta|\mathbf{y})}{\delta\theta} = \sum_{k=0}^{N-1}\left(\frac{2\,(y(k)-\theta)}{[(y(k)-\theta)^2+\alpha^2]}\right) \tag{3.90}$$

The Fisher information using the Appendix becomes

$$I_F = E\left[\left(\frac{\delta l(\theta|\mathbf{y})}{\delta\theta}\right)^2\right] = E\left[\left(\sum_{k=0}^{N-1}\left(\frac{2\,(y(k)-\theta)}{[(y(k)-\theta)^2+\alpha^2]}\right)\right)^2\right] \tag{3.91}$$

Using Eqs. (3.86) and (3.87) we get

$$I_F = E\left[\left(\frac{\delta\ln\left(f_y(y;\theta)\right)}{\delta\theta}\right)^2\right] = \frac{N}{2\alpha^2} \tag{3.92}$$

The properties of the estimator are deferred to the next section on maximum likelihood estimation.

Remarks Comparing Eqs. (3.58) and (3.66); (3.73) and (3.78); and (3.87) and (3.92) we deduce that the Fisher Information for N independent and identically distributed data samples $y = [\,y(0) \quad y(1) \quad y(2) \quad . \quad y(N-1)\,]^T$ is N times the Fisher information for a single data $y = \theta + v$.

The performance of an estimator depends upon the likelihood function. The log-likelihood function $l(\boldsymbol{\theta}|\mathbf{y})$ and its partial derivative $\dfrac{\delta l\,(\boldsymbol{\theta}|\mathbf{y})}{\delta\boldsymbol{\theta}}$ are shown in Figure 3.4. The top figures show the PDFs, the middle figures show the log-likelihood functions, and the bottom figures show the negatives of the partial derivative of the log-likelihood function for the Gaussian, the Laplacian, the part Gauss-part Laplace, and the Cauchy PDF.

Example 3.5 Multivariate Gaussian PDF
Consider the problem of estimation of the $Mx1$ parameter $\boldsymbol{\theta} = [\,\theta_1 \quad \theta_2 \quad \theta_3 \quad . \quad \theta_M\,]^T$ where the linear mathematical and the equivalent probabilistic models are:

$$\mathbf{y} = \boldsymbol{H}\boldsymbol{\theta} + \boldsymbol{v} \tag{3.93}$$

where $\boldsymbol{v}$ is zero-mean measurement Gaussian noise with covariance Σ_v; $\boldsymbol{H}$ is NxM matrix. The probabilistic model is

$$f_g(\mathbf{y};\boldsymbol{\theta}) = \frac{1}{\sqrt{(2\pi)^N\,|\boldsymbol{\Sigma}_v|}}\exp\left\{-\frac{1}{2}(\mathbf{y}-\boldsymbol{H}\boldsymbol{\theta})^T\,\boldsymbol{\Sigma}_v^{-1}\,(\mathbf{y}-\boldsymbol{H}\boldsymbol{\theta})\right\} \tag{3.94}$$

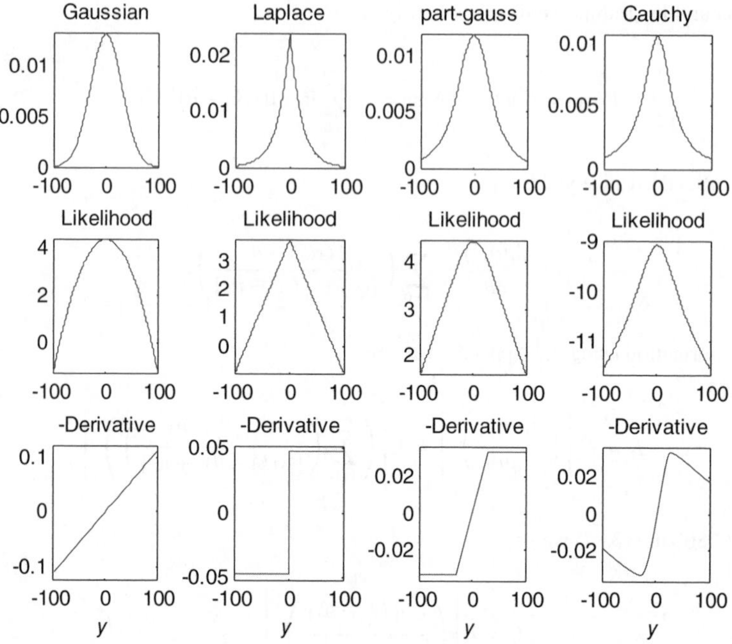

Figure 3.4 Log-likelihood function and its partial derivatives

Taking the natural logarithm likelihood function $l(\theta|y)$ becomes

$$l(\theta|\mathbf{y}) = -\frac{N}{2}\ln(2\pi) - \frac{1}{2}\ln|\boldsymbol{\Sigma}_v| - \frac{1}{2}(\mathbf{y} - \boldsymbol{H\theta})^T \boldsymbol{\Sigma}_v^{-1}(\mathbf{y} - \boldsymbol{H\theta}) \tag{3.95}$$

Expanding the right-hand side, considering only the terms in $\boldsymbol{\theta}$, the gradient vector becomes:

$$\frac{\delta l(\theta|\mathbf{y})}{\delta\theta} = \frac{1}{2}\frac{\delta}{\delta\theta}\left(\mathbf{y}^T\boldsymbol{\Sigma}\mathbf{y} - \boldsymbol{\theta}^T\boldsymbol{H}^T\boldsymbol{\Sigma}_v^{-1}\mathbf{y} - \mathbf{y}^T\boldsymbol{\Sigma}_v^{-1}\boldsymbol{H\theta} + \boldsymbol{\theta}^T\boldsymbol{H}^T\boldsymbol{\Sigma}_v^{-1}\boldsymbol{H\theta}\right) \tag{3.96}$$

Vector calculus for any $Mx1$ vector $\boldsymbol{p}$, $1xM$ vector s, and MxM matrix $\boldsymbol{R}$ is given by:

$$\frac{\partial\boldsymbol{\theta}^T\boldsymbol{p}}{\partial\boldsymbol{\theta}} = \boldsymbol{p}\ ;\ \frac{\partial s^T\boldsymbol{\theta}}{\partial\boldsymbol{\theta}} = s\ ;\ \frac{\partial}{\partial\boldsymbol{\theta}}\{\boldsymbol{\theta}^T R\boldsymbol{\theta}\} = 2R\boldsymbol{\theta} \tag{3.97}$$

Using the vector calculus, the expression for the gradient vector becomes

$$\frac{\delta l(\theta|\mathbf{y})}{\delta\theta} = -\boldsymbol{H}^T\boldsymbol{\Sigma}_v^{-1}\mathbf{y} + \boldsymbol{H}^T\boldsymbol{\Sigma}_v^{-1}\boldsymbol{H\theta} = \boldsymbol{H}^T\boldsymbol{\Sigma}_v^{-1}(\mathbf{y} - \boldsymbol{H\theta}) \tag{3.98}$$

The Fisher information matrix $\boldsymbol{I}_F$ given by Eq. (3.48) becomes

$$I_F = E\left[\left(\frac{\partial l(\theta|\mathbf{y})}{\partial\theta}\right)\left(\frac{\partial l(\theta|\mathbf{y})}{\partial\theta}\right)^T\right] = \boldsymbol{H}^T\Sigma_v^{-1}E[(\mathbf{y} - \boldsymbol{H\theta})(\mathbf{y} - \boldsymbol{H\theta})^T]\boldsymbol{\Sigma}_v^{-1}\boldsymbol{H} \tag{3.99}$$

From the model (3.93) we get

$$E[(y - H\theta)(y - H\theta)^T] = E[vv^T] = \Sigma_v \tag{3.100}$$

Using (3.100) the expression (3.99) for I_F becomes

$$I_F = H^T \Sigma_v^{-1} H \tag{3.101}$$

Let us verify whether an efficient estimator exists. Using Eqs. (3.98) and (3.101) the condition for the Cramér–Rao equality (3.52) becomes

$$H^T \Sigma_v^{-1} (y - H\theta) = H^T \Sigma_v^{-1} H(\hat{\theta} - \theta) \tag{3.102}$$

Equating both sides yields

$$H^T \Sigma_v^{-1} H\hat{\theta} = H^T \Sigma_v^{-1} y \tag{3.103}$$

The efficient estimator exists and is given by

$$\hat{\theta} = (H^T \Sigma_v^{-1} H)^{-1} H^T \Sigma_v^{-1} y \tag{3.104}$$

Remarks The log-likelihood function $l(\theta|y)$ for various PDFs is given below:
Gaussian: $l(\theta|y)$ is a quadratic in $y - \mu_y$ and its partial derivative $\frac{\delta}{\delta\theta} l(\theta|y)$ is linear in y
Laplacian: $l(\theta|y)$ is absolute value of $y - \mu_y$ and $\frac{\delta}{\delta\theta} l(\theta|y)$ is a hard limiter.
Part Gauss-part Laplace: $l(\theta|y)$ is part quadratic in $y - \mu_y \in \Upsilon_{gd}$ and part absolute value in $y - \mu_y \in \Upsilon_{bd}$; $\frac{\delta}{\delta\theta} l(\theta|y)$ is a soft limiter, that is, it is part linear in $y - \mu_y \in \Upsilon_{gd}$ and part constant for $y - \mu_y \in \Upsilon_{bd}$
Properties of the estimator: The estimator for Gaussian PDF is the mean. Let us verify whether $\hat{\theta}$ is unbiased. Taking the expectation of Eq. (3.104) we get:

$$E\left[\hat{\theta}\right] = \left(H^T \Sigma_v^{-1} H\right)^{-1} H^T \Sigma_v^{-1} E[y] \tag{3.105}$$

The expression for $E[y]$ using the linear model (3.93) is:

$$E[y] = HE[\theta] + E[v] \tag{3.106}$$

Since $E[v] = 0$ and θ is deterministic we get:

$$E[y] = H\theta \tag{3.107}$$

Substituting for $E[y]$ in (3.105) we get:

$$E\left[\hat{\theta}\right] = \left(H^T \Sigma_v^{-1} H\right)^{-1} H^T \Sigma_v^{-1} H\theta = \theta \tag{3.108}$$

Since $E[\hat{\theta}] = \theta$, $\hat{\theta}$ is unbiased. It is efficient as $\hat{\theta}$ satisfies the Cramér–Rao equality condition Eq. (3.52).

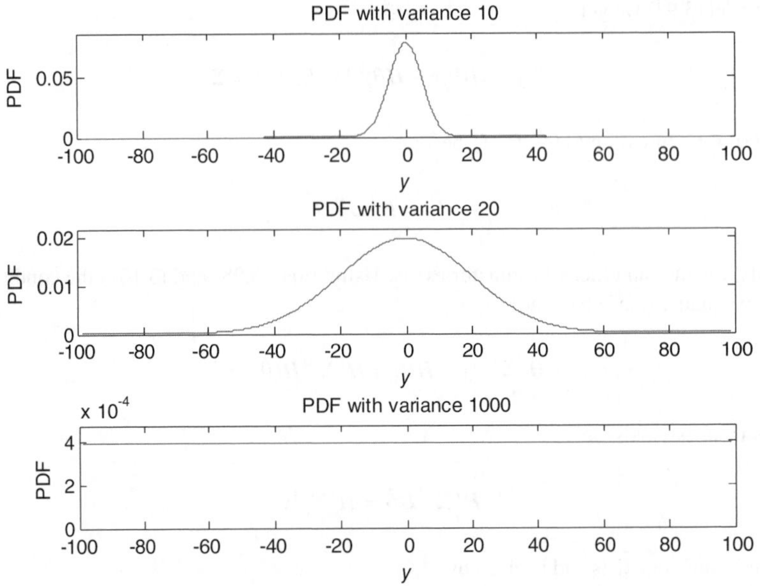

Figure 3.5 PDFs of different variances: small, medium, and large

3.4.4 Fisher Information

The denominator of the Cramér–Rao lower bound of the inequality (3.43) is called the *Fisher information*. The name "Fisher information" was given after the famous nineteenth-century statistician, Ronald Fisher. The Fisher information depends upon the rate of change of the PDF of y with respect to the parameter θ. The greater the expectation of the rate of change at a given value of θ, the larger will be the Fisher information: the greater the rates of change, the easier it will be distinguish θ from its neighboring values. The largest rate of change (infinite rate of change) will be when the PDF is a delta function located at $y = \theta$, while the smallest rate of change (zero rate of change) will occur if the PDF has very large variance. In general, if the PDF has a peak at θ, the Fisher information will be large, and it will be small if the PDF is flat around θ. Figure 3.5 shows zero-mean, $\theta = 0$, Gaussian PDFs with variances 5, 20, and 1000. It can be deduced that the smaller the variance, the greater is the rate of change of the PDF in the neighborhood of $\theta = 0$, implying the larger Fisher information, and the larger the variance, the less the rate of change of the PDF, implying the smaller the Fisher information.

3.4.4.1 Information Content

The Fisher information is an amount of information about the parameter θ contained in the measurement y. The Fisher information is non-negative and satisfies the additive properties of information measure. If there are N independent measurement i.i.d random variables containing information about the unknown parameter, then the Fisher information is N times the Fisher information when there is single measurement. See Example 3.1 and Example 3.2.

Lemma 3.4 *If $y(k)$, $k = 0, 1, 2, \ldots, N-1$ be N i.i.d. random variables containing information about θ and the PDF $f_y(y(k);\theta)$, $k = 0, 1, 2, \ldots, N-1$ satisfy regularity condition (3.40), then*

$$I_F^N = NI_F \tag{3.109}$$

where $I_F^N = E\left[\left(\frac{\partial l(\theta|\mathbf{y})}{\partial\theta}\right)^2\right]$ *and* $I_F = E\left[\left(\frac{\partial l(\theta|\mathbf{y})}{\partial\theta}\right)^2\right]$ *are respectively the Fisher information when N measurements* $y = [\,y(0)\quad y(1)\quad y(2)\quad .\quad y(N-1)\,]^T$ *and one measurement* $y(k)$ *are used.*

Proof: The PDF of y is $f_y(\mathbf{y};\theta)$. Since $\{y(k)\}$ are i.i.d. we get

$$l(\theta|\mathbf{y}) = \sum_{k=0}^{N-1} \ln(f_y(y(k);\theta)) \tag{3.110}$$

Invoking the statistical independence of the measurements and using the Appendix I_F^N is:

$$I_F^N = \sum_{k=0}^{N-1} E\left[\left(\frac{\partial l\,(\theta|\mathbf{y}(k))}{\partial\theta}\right)^2\right] \tag{3.111}$$

Since $\{y(k)\}$ is i.i.d. random variable, I_F is the same for all $y(k)$:

$$I_F = E\left[\left(\frac{\partial l\,(\theta|\mathbf{y}(k))}{\partial\theta}\right)^2\right] \quad \textit{for all } k \tag{3.112}$$

Hence Eq. (3.109) holds.

Remarks Thanks to the additive property (3.109), the larger the number of i.i.d data samples, the higher is the Fisher information, and as a result the lower is the Cramér–Rao lower bound.

Example 3.1 and Example 3.2 illustrate the additive property of the Fisher information.

3.5 Maximum Likelihood Estimation

There are two generally accepted methods of parameter estimation, namely the least-squares estimation and the maximum likelihood estimation. The maximum likelihood or ML estimator is widely used as the estimate gives the minimum estimation error covariance, and serves as a gold standard for evaluating the performance of other estimators. It is efficient as it achieves the Cramér–Rao lower bound if it exists. It is based on maximizing a likelihood function of the PDF of the data expressed as a function of the parameter to be estimated. In general the estimates are implicit nonlinear functions of the parameter to be estimated and the estimated is obtained recursively. If the PDF of the measurement data is Gaussian, the maximum likelihood estimation method simplifies to a weighted least-squares method where the weight is the inverse of the noise covariance matrix.

3.5.1 Formulation of Maximum Likelihood Estimation

Consider the problem of estimating a non-random parameter $\boldsymbol{\theta}$ from $y = \boldsymbol{H\theta} + \boldsymbol{v}$. The optimal estimate $\hat{\boldsymbol{\theta}}$ is obtained by maximizing the likelihood function $L\,(\boldsymbol{\theta}|y)$ of the random variable $\boldsymbol{y}$, which is essentially the PDF of $\boldsymbol{y}$ expressed as a function of the parameter to be estimated $\boldsymbol{\theta}$. It is the likelihood that the measurement $\boldsymbol{y}$ is characterized the parameter $\boldsymbol{\theta}$. The Maximum Likelihood (ML) estimate is obtained by maximizing the likelihood function. As commonly used PDFs are exponential functions of $\boldsymbol{\theta}$, a log-likelihood function $l(\boldsymbol{\theta}|y) = \ln f_y(\boldsymbol{y}; \boldsymbol{\theta})$, instead of the likelihood function $L(\boldsymbol{\theta}|y)$, is commonly employed. The ML estimate is obtained by maximizing the log-likelihood function

$$\hat{\theta} = \arg\max_{\theta} \{l(\boldsymbol{\theta}|y)\} \tag{3.113}$$

It is obtained by setting the partial derivatives of $l(\boldsymbol{\theta}|y)$ to zero with respect to $\boldsymbol{\theta}$:

$$\frac{\delta l(\boldsymbol{\theta}|y)}{\delta \boldsymbol{\theta}} = 0 \tag{3.114}$$

Lemma 3.5 *If a measurement model is (i) linear,* $\boldsymbol{y} = \boldsymbol{H\theta} + \boldsymbol{v}$, *(ii)* $\boldsymbol{v}$ *is a zero-mean multivariate Gaussian random variable with* $\text{cov}(\boldsymbol{v}) = \Sigma_v$, *and (iii) the data matrix* $\boldsymbol{H}$ *has full rank and is constant, then the maximum likelihood estimator is unbiased and efficient and is given by*

$$\hat{\theta} = (\boldsymbol{H}^T \boldsymbol{\Sigma}_v^{-1} \boldsymbol{H})^{-1} \boldsymbol{H}^T \boldsymbol{\Sigma}_v^{-1} \boldsymbol{y} \tag{3.115}$$

Proof: The PDF of $\boldsymbol{y}$ is

$$f_g(\boldsymbol{y}; \boldsymbol{\theta}) = \frac{1}{\sqrt{(2\pi)^N \det(\boldsymbol{\Sigma}_v)}} \exp\left\{ -\frac{(\boldsymbol{y} - \boldsymbol{H\theta})^T \Sigma_v^{-1} (\boldsymbol{y} - \boldsymbol{H\theta})}{2} \right\} \tag{3.116}$$

The log-likelihood function $l(\boldsymbol{\theta}|\boldsymbol{y})$ is

$$l(\boldsymbol{\theta}|\boldsymbol{y}) = -\frac{(\boldsymbol{y} - \boldsymbol{H\theta})^T \Sigma_v^{-1} (\boldsymbol{y} - \boldsymbol{H\theta})}{2} - \frac{N}{2} \ln 2\pi - \frac{1}{2} \ln \det \Sigma_v \tag{3.117}$$

The ML estimate is obtained from maximizing the log-likelihood function (3.113):

$$\hat{\theta} = \arg\max_{\theta} \left\{ -\frac{(\boldsymbol{y} - \boldsymbol{H\theta})^T \Sigma_v^{-1} (\boldsymbol{y} - \boldsymbol{H\theta})}{2} - \frac{N}{2} \ln 2\pi - \frac{1}{2} \ln \det \boldsymbol{\Sigma}_v \right\} \tag{3.118}$$

Equivalently differentiating $l(\boldsymbol{\theta}|y)$ with respect to $\boldsymbol{\theta}$ and setting to zero the ML estimate $\hat{\theta}(y)$ satisfies

$$\frac{\delta l(\boldsymbol{\theta}|y)}{\delta \boldsymbol{\theta}} = \boldsymbol{H}^T \Sigma_v^{-1} \boldsymbol{y} - \boldsymbol{H}^T \Sigma_v^{-1} \boldsymbol{H\theta} = \boldsymbol{0} \tag{3.119}$$

The ML estimator $\hat{\theta}$ is that value of $\boldsymbol{\theta}$ that satisfies the above equation:

$$\hat{\theta} = \left(\boldsymbol{H}^T \Sigma_v^{-1} \boldsymbol{H}\right)^{-1} \boldsymbol{H}^T \Sigma_v^{-1} \boldsymbol{y} \tag{3.120}$$

Note: The maximum likelihood estimator is identical to that obtained using the Cramér–Rao equality condition (3.52). The maximum likelihood estimator is both unbiased and efficient. From Eq. (3.46) we deduce that the estimate $\hat{\theta}$ is efficient with Fisher information

$$\boldsymbol{I}_F = \boldsymbol{H}^T \Sigma_v^{-1} \boldsymbol{H} \tag{3.121}$$

Remarks If the measurement model is linear and the measurement noise is Gaussian, then the maximum likelihood estimator is linear in the measurement, unbiased, and efficient.

When the model is linear and the noise is Gaussian, the cost function for the ML estimator is the log-likelihood function given by Eq. (3.117). For estimating the mean value $\boldsymbol{\theta}$, only the first term on the right-hand side is considered in the minimization as the rest of the two terms are not a function of $\boldsymbol{\theta}$. The cost function reduces to the following minimization problem after suppressing the negative sign and suppressing factor 2 in the denominator:

$$\hat{\theta} = \arg\min_{\theta}\{(y - \boldsymbol{H\theta})^T \boldsymbol{\Sigma}_v^{-1}(y - \boldsymbol{H\theta})\} \tag{3.122}$$

In the statistics, the term $\boldsymbol{y} - \boldsymbol{H\theta}$ is called residual: the residual is an error between the measurement and its estimate using the model that relates the measurement $\boldsymbol{y}$ and $\boldsymbol{\theta}$. The estimation scheme given by Eq. (3.122) is termed the *weighted least-squares method.*

The ML estimator for the case when the model is linear and the noise is zero-mean Gaussian is identical to the weighted least-squares estimator.

3.5.2 *Illustrative Examples: Maximum Likelihood Estimation of Mean or Median*

Example 3.6 i.i.d. Gaussian PDF

Consider the scalar version of Example 3.4 with $\boldsymbol{H} = [1 \quad 1 \quad 1 \quad . \quad 1]^T$ and $\boldsymbol{\Sigma}_v = \boldsymbol{I}\sigma_v^2$. The unknown parameter θ is scalar with $y = \begin{bmatrix} y(0) & y(1) & y(2) & . & y(N-1) \end{bmatrix}^T$:

$$y(k) = \theta + v(k), \quad k = 0, 1, 2, 3, \ldots, N-1 \tag{3.123}$$

We will consider two cases: (i) $v(k)$ is i.i.d. zero-mean Gaussian noise; (ii) $v(k)$ is zero-mean independent Gaussian noise but not identically distributed.

(i) $\boldsymbol{v}$ is independent and identically distributed zero-mean Gaussian noise with covariance $\boldsymbol{\Sigma}_v = \boldsymbol{I}\sigma_v^2$
Substituting $\boldsymbol{\Sigma}_v = \boldsymbol{I}\sigma_v^2$, the log-likelihood function (3.117) reduces to

$$l(\theta|\boldsymbol{y}) = -\frac{1}{2\sigma_v^2}\sum_{k=0}^{N-1}(y(k) - \theta)^2 - \frac{N}{2}\ln 2\pi - \frac{1}{2}\ln \sigma_v \tag{3.124}$$

The expression for the estimate $\hat{\theta}$ given by (3.118) becomes:

$$\hat{\theta} = \arg\max_{\theta}\left\{-\frac{1}{2\sigma_v^2}\sum_{k=0}^{N-1}(y(k) - \theta)^2 - \frac{N}{2}\ln 2\pi - \frac{1}{2}\ln \sigma_v\right\} \tag{3.125}$$

Differentiating and setting to zero yields:

$$\frac{\delta l(\theta|\boldsymbol{y})}{\delta\theta} = -\frac{1}{\sigma_v^2}\sum_{k=0}^{N-1}(y(k) - \theta) = 0 \tag{3.126}$$

Or, substituting $\boldsymbol{\Sigma}_v = \boldsymbol{I}\sigma_v^2$ in Eq. (3.120), the estimate $\hat{\theta}$ becomes

$$\hat{\theta} = \frac{1}{N}\sum_{i=1}^{N} y(i) \tag{3.127}$$

The Fisher information (3.121) becomes

$$\boldsymbol{I}_F(\theta) = \left(\boldsymbol{H}^T \boldsymbol{\Sigma}_v^{-1} \boldsymbol{H}\right) = \frac{N}{\sigma_v^2} \tag{3.128}$$

The variance of the estimation error

$$E\left[(\theta - \hat{\theta})^2\right] = \boldsymbol{I}_F^{-1} = \left(\boldsymbol{H}^T \boldsymbol{\Sigma}_v^{-1} \boldsymbol{H}\right)^{-1} = \left(\boldsymbol{H}^T \boldsymbol{H}\right)^{-1} \sigma_v^2 = \frac{\sigma_v^2}{N} \tag{3.129}$$

(ii) $\boldsymbol{v}$ is zero-mean independent but not identically distributed Gaussian noise with covariance $\Sigma_v = diag(\sigma_{v1} \quad \sigma_{v1} \quad \cdots \quad \sigma_{vn})$. The estimate $\hat{\theta}$ may easily be deduced.
The estimate and the variance of the estimation error are

$$\hat{\theta} = \left(\boldsymbol{H}^T \boldsymbol{\Sigma}_v^{-1} \boldsymbol{H}\right)^{-1} \boldsymbol{H}^T \boldsymbol{\Sigma}_v^{-1} \boldsymbol{y} = \left(\sum_{i=0}^{N-1} \frac{1}{\sigma_{vi}^2}\right)^{-1} \sum_{i=0}^{N-1} \frac{y(i)}{\sigma_{vi}^2} \tag{3.130}$$

The Fisher information is:

$$\boldsymbol{I}_F = \left(\boldsymbol{H}^T \boldsymbol{\Sigma}_v^{-1} \boldsymbol{H}\right) = \sum_{i=0}^{N-1} \frac{1}{\sigma_{vi}^2} \tag{3.131}$$

The estimator is unbiased and efficient.

Example 3.7 i.i.d. Laplacian PDF
Consider the linear measurement model (3.123), $v(k)$ is i.i.d. zero-mean random noise with a Laplacian PDF given by Eq. (3.75):

$$f_e(\boldsymbol{y};\theta) = \left(\frac{1}{\sqrt{2\sigma_v^2}}\right)^N \exp\left(-\frac{\sqrt{2}}{\sigma_v}\sum_{k=0}^{N-1} |y(k) - \theta|\right) \tag{3.132}$$

The log-likelihood function $l\,(\boldsymbol{\theta}|\boldsymbol{y})$ is

$$l\,(\boldsymbol{\theta}|\boldsymbol{y}) = -\frac{N}{2}\ln\left(2\sigma^2\right) - \frac{\sqrt{2}}{\sigma_v}\sum_{k=0}^{N-1} |y(k) - \theta| \tag{3.133}$$

The ML estimate is obtained from maximizing the log-likelihood function (3.113):

$$\hat{\theta} = \arg\max_{\theta}\left\{-\frac{N}{2}\ln\left(2\sigma^2\right) - \frac{\sqrt{2}}{\sigma_v}\sum_{k=0}^{N-1} |y(k) - \theta|\right\} \tag{3.134}$$

Equivalently, differentiating $\ln f_y(\mathbf{y})$ with respect to $\boldsymbol{\theta}$ and setting to zero the ML estimate is obtained. The estimator $\hat{\theta}$ is obtained by setting Eq. (3.77) to zero:

$$\frac{\delta l(\theta|\mathbf{y})}{\delta\theta} = \frac{\sqrt{2}}{\sigma_v}\sum_{k=0}^{N-1} sign\,(y(k)-\theta) = 0 \tag{3.135}$$

The estimator $\hat{\theta}$ is that value of θ that solves the above equation. The ML estimate $\hat{\theta}$ is that value θ that must satisfy Eq. (3.135), making the summation of all the signs of the terms $(y(k)-\theta)$ equal zero. If we choose the estimate to be the median of all the sample values, the number of positive and negative signs of the terms $(y(k)-\theta)$ will be equal, ensuring the sum of the signs is zero. Thus it may be deduced that the maximum likelihood estimator is the median of $\mathbf{y} = \begin{bmatrix} y(0) & y(1) & y(2) & . & y(N-1)\end{bmatrix}^T$:

$$\hat{\theta} = \text{median}([\,y(0) \quad y(1) \quad y(2) \quad . \quad y(N-1)\,]^T) \tag{3.136}$$

The median is determined by sorting the sample values from the lowest to highest values and picking the middle one so that the measurement data is split into two equal parts: in one part the data samples are all larger and in the other the data samples are all smaller than the median. Hence $sign\left(y(k)-\hat{\theta}\right)$, $k = 0, i, 2, \ldots, N-1$ will have an equal number of 1's and -1's with the result

$$\sum_{k=0}^{N-1} sign\left(y(k)-\hat{\theta}\right) = 0 \tag{3.137}$$

Hence

$$\frac{\sqrt{2}}{\sigma_v}\sum_{k=0}^{N-1} sign\left(y(k)-\hat{\theta}\right) = 0 \tag{3.138}$$

Consider the expression of the estimator (3.136). Expressing the median in terms of θ and $\mathbf{v}$ using the model (3.123) we get

$$\hat{\theta} = \theta + median([\,v(0) \quad v(1) \quad v(2) \quad . \quad v(N-1)\,]^T) \tag{3.139}$$

Using the definition of the median (3.35)

$$\hat{\theta} = \begin{cases} \theta + v(\ell) & N\ odd \\ \theta + \frac{1}{2}\,(v(N/2) + v(N/2+1)) & N\,even \end{cases} \tag{3.140}$$

where the index ℓ separates the lower half from the upper half while the index m is the center that divides the upper and the lower halves. Taking expectation on both sides, using the definition of the median (3.35), and since $v(k)$ is zero-mean, yields

$$E\left[\hat{\theta}\right] = \theta \tag{3.141}$$

Thus the median is an unbiased estimator.

It is not possible to show that the median satisfies the Cramér–Rao equality condition (3.46). It can be shown, however, that it asymptotically efficient.

Example 3.8 i.i.d. part Gauss-part Laplacian PDF
The expression for the part Gauss-part Laplacian PDF given by Eq. (3.80) becomes:

$$f_{ge}(\mathbf{y};\theta) = \begin{cases} \kappa^N \exp\left\{-\dfrac{1}{2\sigma_v^2}\sum_{k=0}^{N-1}(y(k)-\theta)^2\right\} & |y(k)-\theta| \le a_{gd} \\ \kappa^N \exp\left\{\dfrac{Na_{gd}^2}{2\sigma_v^2}\right\}\exp\left\{-\dfrac{a_{gd}}{2\sigma_v^2}\sum_{k=0}^{N-1}|y(k)-\theta|\right\} & |y(k)-\theta| > a_{gd} \end{cases} \tag{3.142}$$

The log-likelihood function $l(\theta|\mathbf{y})$ is

$$l(\theta|\mathbf{y}) = \begin{cases} \ln\left(\kappa^N\right) - \dfrac{1}{2\sigma_y^2}\sum_{k=0}^{N-1}(y(k)-\theta)^2 & |y(k)-\theta| \le a_{gd} \\ \ln\left(\kappa^N\right) + \dfrac{Na_{gd}^2}{2\sigma_y^2} - \dfrac{a_{gd}}{2\sigma_y^2}\sum_{k=0}^{N-1}|y(k)-\theta| & |y(k)-\theta| > a_{gd} \end{cases} \tag{3.143}$$

The ML estimate is obtained from maximizing the log-likelihood

$$\hat{\theta} = \arg\max_{\theta}\{l(\theta|\mathbf{y})\} \tag{3.144}$$

Equivalently the ML estimate is determined by finding the roots of the derivative of the log-likelihood function

$$\frac{\delta l(\theta|\mathbf{y})}{\delta\theta} = 0 \tag{3.145}$$

In this case there is no closed form solution. The estimate is computed recursively using Eq. (3.145). The maximum likelihood estimator although inefficient for finite samples, is asymptotically efficient.

Remark A robust estimator is obtained using the maximum likelihood method when the PDF is unknown using the Min-Max approach. A ML estimator is sought which minimizes the log-likelihood function for the worst-case PDF, namely $f_{ge}(\mathbf{y};\theta)$. It is interesting that the worst-case PDF is asymptotically Laplacian with exponentially decaying tails. Intuitively, one would have expected the worst-case PDF to have tails thicker than that of Cauchy. That is, the worst-case PDF is expected to generate more bad data than Cauchy.

Example 3.9 i.i.d. Cauchy PDF
Consider the linear measurement model (3.123) with $v(k)$ is i.i.d. zero-mean random noise with Cauchy PDF. The log-likelihood function is given by Eq. (3.89). The ML estimate is obtained from maximizing the log-likelihood function (3.113):

$$\hat{\theta} = \arg\max_{\theta}\left\{N\ln(\alpha) - N\ln(\pi) - \sum_{k=0}^{N-1}\ln\left(\pi[(y(k)-\theta)^2+\alpha^2]\right)\right\} \tag{3.146}$$

Equivalently the ML estimate is determined by finding the roots of Eq. (3.114) whose expression is given in Eq. (3.90):

$$\frac{\delta l(\theta|\mathbf{y})}{\delta\theta} = \sum_{k=0}^{N-1}\left(\frac{2\,(y(k)-\theta)}{\left[(y(k)-\theta)^2+\alpha^2\right]}\right) = 0 \tag{3.147}$$

There is no closed form solution and the estimate is computed recursively.

The Cauchy distribution is a "pathological distribution," which is often used to model a measurement data with large outliers. Its mean does not exist and its variance is infinite. But its median and mode exist and are well defined. Finding the roots of Eq. (3.147) is computationally burdensome. One simple and robust, although suboptimal, solution is to estimate the median value of the samples.

Remarks The cost function, denoted $c(\theta|\mathbf{y})$, for the ML estimate of Gaussian, Laplace, and Cauchy is derived from the $l(\theta|\mathbf{y})$ given respectively in Eqs. (3.60), (3.133), (3.143), and (3.89). The cost function is a convex function of $y - \mu_y$. Since the constant terms do not contribute to the constant, for comparing the cost functions associated with different PDFs, a constant term is subtracted to the likelihood function $l(\theta|\mathbf{y})$ so that the cost function is zero when $y = \mu_y$:

$$c(\theta|\mathbf{y}) = 0 \text{ when } y = \mu_y \tag{3.148}$$

The cost function for different PDFs when $\mu_y = \theta$ becomes:

$$c\,(\theta|\mathbf{y}) = \begin{cases} \dfrac{1}{2\sigma_v^2}\displaystyle\sum_{i=0}^{N-1}(y(k)-\theta)^2 & \textit{Gaussian} \\ \dfrac{\sqrt{2}}{\sigma_v}\displaystyle\sum_{k=0}^{N-1}|y(k)-\theta| & \textit{Laplacian} \\ \displaystyle\sum_{k=0}^{N-1}\ln(\pi[(y(k)-\theta)^2+\alpha^2]) - N\,ln(\pi\alpha^2) & \textit{Cauchy} \end{cases} \tag{3.149}$$

Similarly, the cost function for part Gauss-part Laplace using the expression (3.143) is:

$$c\,(\boldsymbol{\theta}|\mathbf{y}) = \begin{cases} \dfrac{1}{2\sigma_y^2}\displaystyle\sum_{k=0}^{N-1}(y(k)-\theta)^2 & |y(k)-\theta| \le a_{gd} \\ \dfrac{a_{gd}}{2\sigma_y^2}\displaystyle\sum_{k=0}^{N-1}|y(k)-\theta| & |y(k)-\theta| > a_{gd} \end{cases} \tag{3.150}$$

The ML estimator $\hat{\theta}$ of the mean θ is determined by minimizing the cost function with respect to θ:

$$\hat{\theta} = \arg\min_{\theta}\{c(\theta|y)\} \tag{3.151}$$

In the case when the PDF is not known, then the robust estimator of θ is found by minimizing the cost function for the worst-case PDF:

$$\hat{\theta} = \arg\min_{\theta}\left\{\max_{f_y\in\mathbb{C}}\{c(\theta|y)\}\right\} \tag{3.152}$$

Remarks The functional forms of the cost function $c(\theta|\mathbf{y})$ for different PDFs are as follows:

1. *Gaussian:* It is a L_2 metric (or distance function) of the $y - \theta$, that is it is a quadratic function of $y - \theta$.
2. *Laplace:* It is a L_1 metric of $y - \theta$, that is it is an absolute value function of $y - \theta$.
3. *A part Gauss-part Laplace*: It is a compromise between a L_2 and a L_1 of $y - \theta$, that is it is a quadratic function for small $y - \theta$ and an absolute function for large $y - \theta$.
4. *Cauchy:* It is a log-quadratic function of $y - \theta$.

The weightings of cost function depend upon the PDF.

1. All cost functions, except that of the worst-case (part Gauss-part Laplace), do not discriminate between the good and bad data. They weigh equally all the measurement deviation $y - \theta$.
2. The Gaussian cost function is a quadratic function of all $y - \theta$ including both good and bad data. The ML estimator using Gaussian PDF performs poorly in the presence of bad data and gives an acceptable performance when the measurement data is "good."
3. The part Gauss-part Laplace (worst-case) cost function is employed when the PDF of the measurement data is unknown or partially known. It discriminates between the good and the bad data, thereby ensuring adequate performance of the ML estimator in the face measurement data (including the worst case of bad data) generated by a class of all PDFs. It is a quadratic function of $y - \theta$ for all good data, $|y - \theta| \leq a_{gd}$, and an absolute function of $y - \theta$ for all bad data, $|y - \theta| > a_{gd}$.
4. Similar to the worst-case cost function, the Laplace cost function weights bad measurement data as an absolute function of $y - \theta$ for all $|y - \theta| > a_{gd}$. Hence the ML estimator using Laplacian PDF is robust to the presence of bad data.
5. The Cauchy cost function is a log-quadratic function of all $y - \theta$. It assigns the lowest weights, especially to the bad data. The ML estimator penalizes the bad data the most.
6. The ML estimates for Gaussian and Laplace are easy to compute. For the Gaussian PDF it is the mean of the data samples, while for the Laplace it is the median of the data samples. In the case when the PDF is not known, it is simpler to employ the mean when the data is believed to be "good" and the median when data is believed to be "bad."

The performance of the ML estimators for Gauss, Laplace, part Gauss-part Laplace, and Cauchy PDFs are illustrated in Figure 3.6. The ML estimate of $\theta = 1$ is obtained for a linear model $y(k) = \theta + v(k)$ for the cases when $v(k)$ zero-mean Gaussian, Laplacian, part Gauss-part Laplace, and Cauchy i.i.d. random variables. The variance for Gaussian, Laplacian, and part Gauss-part Laplace random variables is unity. The ML estimates are obtained for both the correct as well as the incorrect choice of PDFs. The leftmost subfigures (a), (f), (k), and (p) show the Gaussian, the Laplace, the part Gauss-part Laplace, and the Cauchy measurement random variables respectively. The ML estimates assuming Gaussian, Laplace, part Gauss-part Laplace. and Cauchy are shown in (i) the top subfigures (b), (c), (d), and (e) respectively when the true PDF is Gaussian, (ii) the middle subfigures (g), (h), (i), and (j) respectively when the true PDF is Laplace, (iii) the subfigures (l), (m), (n), and (o) respectively when the true PDF is part Gauss-part Laplace, and (iv) the bottom subfigures (q), (r), (s), and (t) when the true PDF is Cauchy. The subfigures (b), (h), and (o) show the ML estimate when the true PDF is assumed. The rest show the estimates with the incorrect choice of PDF. The captions "est G," "est L," "est P," and "est C" indicate that the estimates are computed assuming Gaussian, Laplacian, part Gauss-part Laplace, and Cauchy PDFs respectively. Similarly "data G," "data L," "data P," and "data C" indicate that measurement data is generated respectively assuming Gaussian, Laplace, part Gauss-part Laplace, and Cauchy PDFs. These cost functions associated with the PDFs are shown in Figure 3.7.

Table 3.1 shows the performance of the ML estimator. The ideal case, when the assumed PDF is equal to the true PDF, and the practical case, when the assumed PDF is different from the true PDF, are

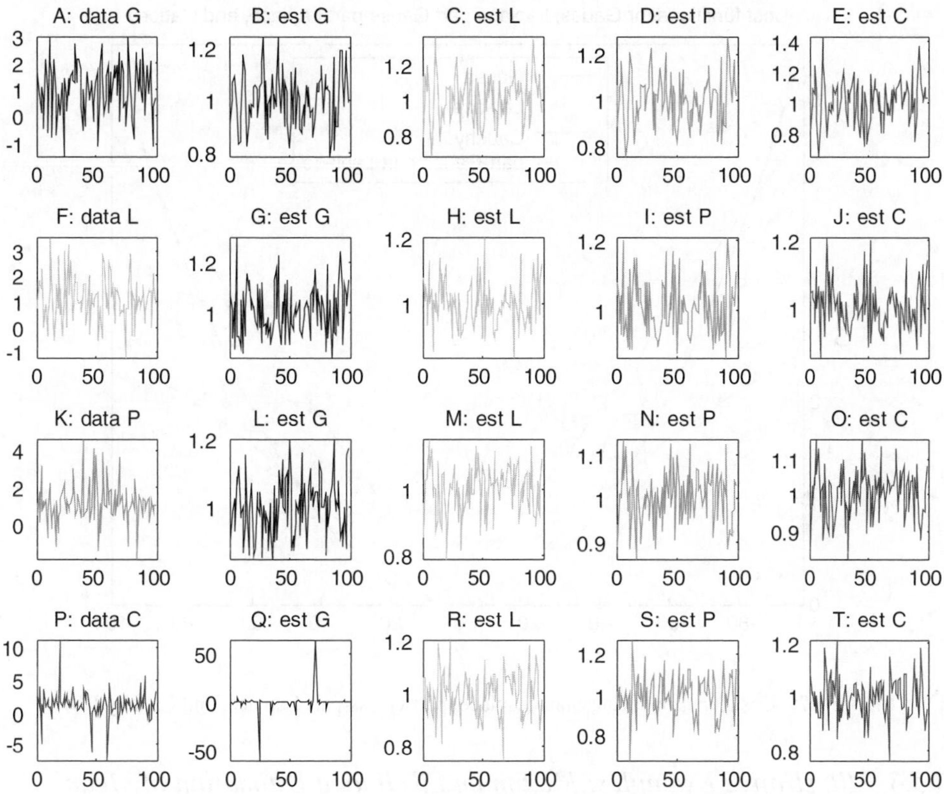

Figure 3.6 Performance of the ML estimators for various PDFs

considered. The ML estimates of the mean and the variance of the estimation errors are given when the measurement data comes from the Gaussian, the Laplacian, the part Gauss-part Laplace, and the Cauchy PDFs. The leftmost column indicates the true PDFs (Gaussian, Laplacian, part Gauss-part Laplace, or Cauchy). The rows are associated with the true PDF, and the columns with the assumed PDFs. The ML estimate of the mean and the variance of the estimate are given at the intersection of the row (denoting the true PDF) and the columns (denoting the assumed PDFs). For example, The ML estimation of the mean and variance of the estimation error when the true PDF is Gaussian and the assumed PDF is Cauchy are given respectively by the entries in the first row and fifth column, and the first row and the ninth column.

We can deduce the following from the Figure 3.6 and the Table 3.1:

1. The performance of the ML estimator is the best when the assumed PDF is equal to the true PDF.
2. In the case when the PDF is unknown, the ML estimator based on the worst-case PDF is robust to uncertainties in the PDF generating the measurement data. The estimator performs adequately over all measurement data PDFs.
3. The performance of the estimator assuming a Gaussian PDF when the data is bad, such as Cauchy distributed measurements, is catastrophic.
4. The performance of the estimator assuming Laplace PDF is acceptable.

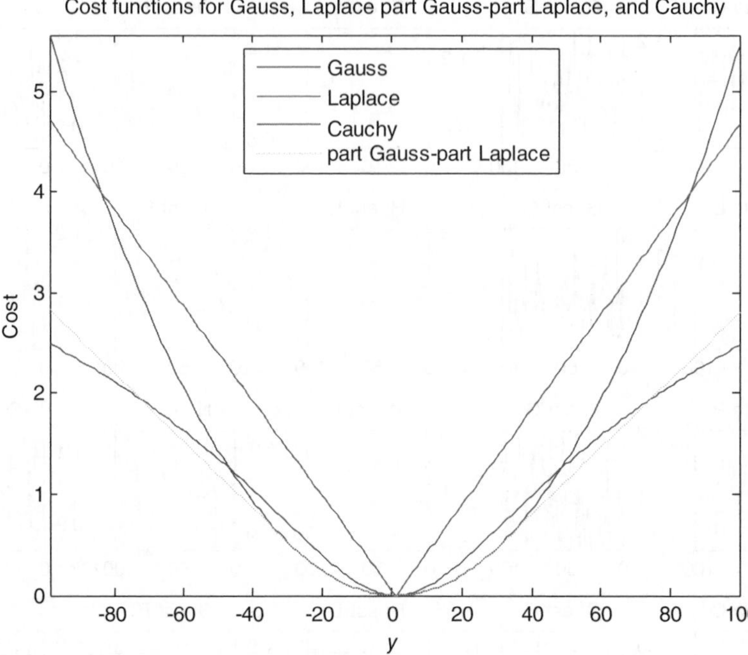

Figure 3.7 Cost functions: Gaussian, Laplace, part Gauss-part Laplace, and Cauchy PDFs

3.5.3 *Illustrative Examples: Maximum Likelihood Estimation of Mean and Variance*

So far we have concentrated on estimating the location parameter θ which is a mean or median or model of a PDF. However, there is a need to estimate the variance from the measurement data for determining the PDF and to evaluate the performance of the estimator.

Example 3.10 i.i.d. Gaussian PDF
Consider a scalar version of Example 3.4 with $\boldsymbol{H} = [\,1 \quad 1 \quad 1 \quad . \quad 1\,]^T$ and $\boldsymbol{\Sigma}_v = \boldsymbol{I}\sigma_v^2$. From Eq. (3.124), the log-likelihood function for the unknown parameters θ and the variance σ_v^2 denoted $l(\theta, \sigma_v^2|\mathbf{y})$ is:

$$l(\theta, \sigma_v^2|\mathbf{y}) = -\frac{1}{2\sigma_v^2}\sum_{k=0}^{N-1}(y(k) - \theta)^2 - \frac{N}{2}\ln 2\pi - \frac{N}{2}\ln\sigma_v^2\,, \tag{3.153}$$

Table 3.1 Performance of the ML estimator

	Estimate of the mean				Estimate of the variance			
	Gauss	Laplace	Part.	Cauchy	Gauss	Laplace	Part.	Cauchy
Gauss	1.0212	1.0169	1.0203	1.0125	0.0125	0.0164	0.0133	0.0203
Laplace	1.0045	1.0092	1.0063	1.0067	0.0104	0.0061	0.0080	0.0062
Part...	1.0007	1.0018	1.0043	1.0037	0.0072	0.0057	0.0036	0.0050
Cauchy	1.7102	1.0098	1.0044	1.0094	27.2594	0.0072	0.0103	0.0060

The ML estimates $\hat{\theta}$ and $\hat{\sigma}_v$ are obtained from maximizing the log-likelihood function $l\left(\theta, \sigma_v | \mathbf{y}\right)$:

$$\begin{bmatrix} \hat{\theta} \\ \hat{\sigma}_v^2 \end{bmatrix} = \arg\max_{\sigma_v, \theta} \left\{ -\frac{1}{2\sigma_v^2} \sum_{k=0}^{N-1} (y(k) - \theta)^2 - \frac{N}{2} \ln 2\pi - \frac{N}{2} \ln \sigma_v^2 \right\} \tag{3.154}$$

Partial differentiation of $l\left(\theta, \sigma_v^2 | \mathbf{y}\right)$ with respect to θ and σ_v we get

$$\frac{\delta}{\delta\theta} l\left(\theta, \sigma_v^2 | \mathbf{y}\right) = \frac{1}{\sigma_v^2} \sum_{k=0}^{N-1} (y(k) - \theta) \tag{3.155}$$

$$\frac{\delta}{\delta\sigma_v^2} l\left(\theta, \sigma_v^2 | \mathbf{y}\right) = \frac{1}{2\sigma_v^4} \sum_{k=0}^{N-1} (y(k) - \theta)^2 - \frac{N}{2\sigma_v^2} \tag{3.156}$$

Setting the partial derivatives (3.155) and (3.156) to zero, the ML estimates $\hat{\theta}$ and $\hat{\sigma}_v^2$ are those values of θ and σ_v^2 which satisfy the following equation:

$$\begin{bmatrix} \dfrac{\delta}{\delta\theta} l\left(\theta, \sigma_v^2 | \mathbf{y}\right) \\ \dfrac{\delta}{\delta\sigma_v^2} l\left(\theta, \sigma_v^2 | \mathbf{y}\right) \end{bmatrix} = \begin{bmatrix} \dfrac{1}{\sigma_v^2} \displaystyle\sum_{k=0}^{N-1} (y(k) - \theta) \\ \dfrac{1}{2\sigma_v^4} \displaystyle\sum_{k=0}^{N-1} (y(k) - \theta)^2 - \dfrac{N}{2\sigma_v^2} \end{bmatrix} = \begin{bmatrix} 0 \\ 0 \end{bmatrix} \tag{3.157}$$

The ML estimate $\hat{\theta}$ satisfies:

$$\frac{1}{\sigma_v^2} \sum_{k=0}^{N-1} \left(y(k) - \hat{\theta}\right) = 0 \tag{3.158}$$

From Eq. (3.158) the ML estimate $\hat{\theta}$ of θ is:

$$\hat{\theta} = \frac{1}{N} \sum_{i=1}^{N} y(i) \tag{3.159}$$

The variance of the estimator $\hat{\theta}$ from Eq. (3.129) is given by

$$\text{var}\left(\hat{\theta}\right) = \frac{\sigma_v^2}{N} \tag{3.160}$$

The ML estimate $\hat{\sigma}_v^2$ of σ_v^2 is:

$$\hat{\sigma}_v^2 = \frac{1}{N} \sum_{k=0}^{N-1} \left(y(k) - \hat{\theta}\right)^2 \tag{3.161}$$

It has been already established that $\hat{\theta}$ is unbiased and efficient in Example 3.4.

Let us verify the properties of $\hat{\sigma}_v^2$. We will show that $\hat{\sigma}_v^2$ is biased. Substituting $y(k) = \theta + v(k)$ in the expression $\hat{\theta} = \frac{1}{N}\sum_{i=1}^{N} y(i)$ in Eq. (3.161) yields

$$E\left[\hat{\sigma}_v^2\right] = \frac{1}{N}\sum_{k=0}^{N-1} E\left[\left(\theta + v(k) - \frac{1}{N}\sum_{i=0}^{N-1}(\theta + v(i))\right)^2\right] = \frac{1}{N}\sum_{k=0}^{N-1} E\left[\left(v(k) - \frac{1}{N}\sum_{i=0}^{N-1} v(i)\right)^2\right] \tag{3.162}$$

Expanding and using the fact that $\{v(i)\}$ is i.i.d. we get

$$E\left[\hat{\sigma}_v^2\right] = \frac{1}{N}\sum_{k=0}^{N-1} E\left[\left(\frac{1}{N}\left((N-1)\,v(k) - \sum_{i=0,i\neq k}^{N-1} v(i)\right)\right)^2\right] = \frac{N-1}{N}\sigma_v^2 \tag{3.163}$$

It has been shown that the variance of the estimator $\hat{\sigma}_v^2$ is given by

$$\text{var}\left(\hat{\sigma}_v^2\right) = \frac{2(N-1)}{N^2}\sigma_v^4 \tag{3.164}$$

Equation (3.163) shows that $\hat{\sigma}_v^2$ is "slightly" biased since $E\left[\hat{\sigma}_v^2\right] = \frac{N-1}{N}\sigma_v^2 \neq \sigma_v^2$. It is, however, asymptotically unbiased. We may fix the problem of bias by simply redefining the estimator (3.161) as:

$$\hat{\sigma}_{v\_unb}^2 = \frac{N}{N-1}\hat{\sigma}_v^2 = \frac{1}{N-1}\sum_{k=0}^{N-1}\left(y(k) - \hat{\theta}\right)^2 \tag{3.165}$$

where $\hat{\sigma}_{v\_unb}^2$ is the redefined estimator of the variance. Let us now verify whether it is unbiased. Taking the expectation and using Eq. (3.163) yields:

$$E\left[\hat{\sigma}_{v\_unb}^2\right] = \sigma_v^2 \tag{3.166}$$

Consider the Fisher information matrix (3.50):

From Eq. (3.155) $E\left[\dfrac{\partial^2 l\left(\theta,\sigma_v|y\right)}{\partial\theta^2}\right] = -\frac{N}{\sigma_v^2}$; $E\left[\dfrac{\partial^2 l\left(\theta,\sigma_v|y\right)}{\partial\theta\partial\sigma_v^2}\right] = -\frac{1}{\sigma_v^4}\sum_{k=0}^{N-1} E\left[(y(k)-\theta)\right] = 0$; and using Eq. (3.156) $E\left[\dfrac{\partial^2 l\left(\theta,\sigma_v|y\right)}{\partial\left(\sigma_v^2\right)^2}\right] = -\frac{1}{\sigma_v^6}\sum_{k=0}^{N-1} E\left[(y(k)-\theta)^2\right] + \frac{N}{2\sigma_v^4} = -\frac{N}{\sigma_v^4} + \frac{N}{2\sigma_v^4} = -\frac{N}{2\sigma_v^4}$.

Hence the Fisher information matrix becomes

$$\boldsymbol{I}_F = \begin{bmatrix} -E\left[\dfrac{\partial^2 l\left(\theta,\sigma_v|y\right)}{\partial\theta^2}\right] & -E\left[\dfrac{\partial^2 l\left(\theta,\sigma_v|y\right)}{\partial\theta\partial\sigma_v^2}\right] \\ -E\left[\dfrac{\partial^2 l\left(\theta,\sigma_v|y\right)}{\partial\theta\partial\sigma_v^2}\right] & -E\left[\dfrac{\partial^2 l\left(\theta,\sigma_v|y\right)}{\partial\left(\sigma_v^2\right)^2}\right] \end{bmatrix} = \begin{bmatrix} \dfrac{N}{\sigma_v^2} & 0 \\ 0 & \dfrac{N}{2\sigma_v^4} \end{bmatrix} \tag{3.167}$$

Let us verify the efficiency of the estimators $\hat{\theta}$ and $\hat{\sigma}_v^2$ by comparing their variances with the diagonal elements of the inverse of the Fisher information matrix $\boldsymbol{I}_F^{-1}$. From expressions of the variances (3.160) and (3.164) with diagonal elements of $\boldsymbol{I}_F^{-1} = diag\begin{pmatrix} \frac{\sigma_v^2}{N} & \frac{2\sigma_v^4}{N} \end{pmatrix}$, it can be deduced that (a) $\hat{\theta}$ is efficient whereas $\hat{\sigma}_v^2$ is asymptotically efficient.

Example 3.11 i.i.d. Gaussian PDF
We will use the results of the linear model $\boldsymbol{y} = \boldsymbol{H\theta} + \boldsymbol{v}$ in Lemma 3.5 with $\boldsymbol{\Sigma}_v = \boldsymbol{I}\sigma_v^2$. The log-likelihood function for the unknown parameters $\boldsymbol{\theta}$ and the variance σ_v^2 denoted $l\left(\boldsymbol{\theta}, \sigma_v^2 | \boldsymbol{y}\right)$ is:

$$l(\boldsymbol{\theta}|\boldsymbol{y}) = -\frac{(\boldsymbol{y} - \boldsymbol{H\theta})^T (\boldsymbol{y} - \boldsymbol{H\theta})}{2\sigma_v^2} - \frac{N}{2}\ln 2\pi - \frac{1}{2}\ln \sigma_v^2 \tag{3.168}$$

The ML estimates $\hat{\boldsymbol{\theta}}$ and $\hat{\sigma}_v$ are obtained from maximizing the log-likelihood function $l\left(\boldsymbol{\theta}, \sigma_v | \boldsymbol{y}\right)$:

$$\begin{bmatrix} \hat{\boldsymbol{\theta}} \\ \hat{\sigma}_v^2 \end{bmatrix} = \arg\max_{\sigma_v, \boldsymbol{\theta}} \left\{ -\frac{(\boldsymbol{y} - \boldsymbol{H\theta})^T (\boldsymbol{y} - \boldsymbol{H\theta})}{2\sigma_v^2} - \frac{N}{2}\ln 2\pi - \frac{N}{2}\ln \sigma_v^2 \right\} \tag{3.169}$$

Partial differentiation of $l\left(\theta, \sigma_v^2 | \boldsymbol{y}\right)$ with respect to θ and σ_v gives

$$\frac{\delta}{\delta \boldsymbol{\theta}} l(\boldsymbol{\theta}|\boldsymbol{y}) = \frac{1}{\sigma_v^2} \boldsymbol{H}^T (\boldsymbol{y} - \boldsymbol{H\theta}) \tag{3.170}$$

$$\frac{\delta}{\delta \sigma_v^2} l\left(\boldsymbol{\theta}, \sigma_v^2 | \boldsymbol{y}\right) = \frac{1}{2\sigma_v^4} (\boldsymbol{y} - \boldsymbol{H\theta})^T (\boldsymbol{y} - \boldsymbol{H\theta}) - \frac{N}{2\sigma_v^2} \tag{3.171}$$

Setting the partial derivatives (3.155) and (3.156) to zero, the ML estimates $\hat{\theta}$ and $\hat{\sigma}_v^2$ are those values of $\boldsymbol{\theta}$ and σ_v^2 which satisfy the following equation:

$$\begin{bmatrix} \dfrac{\delta}{\delta \boldsymbol{\theta}} l\left(\boldsymbol{\theta}, \sigma_v^2 | \boldsymbol{y}\right) \\ \dfrac{\delta}{\delta \sigma_v^2} l\left(\boldsymbol{\theta}, \sigma_v^2 | \boldsymbol{y}\right) \end{bmatrix} = \begin{bmatrix} \dfrac{1}{\sigma_v^2} \boldsymbol{H}^T (\boldsymbol{y} - \boldsymbol{H\theta}) \\ \dfrac{1}{2\sigma_v^4} (\boldsymbol{y} - \boldsymbol{H\theta})^T (\boldsymbol{y} - \boldsymbol{H\theta}) - \dfrac{N}{2\sigma_v^2} \end{bmatrix} = \begin{bmatrix} 0 \\ 0 \end{bmatrix} \tag{3.172}$$

The ML estimator $\hat{\boldsymbol{\theta}}$ is that value of $\boldsymbol{\theta}$ that satisfies the above equation:

$$\hat{\boldsymbol{\theta}} = \left(\boldsymbol{H}^T \boldsymbol{H}\right)^{-1} \boldsymbol{H}^T \boldsymbol{y} \tag{3.173}$$

The ML estimate $\hat{\sigma}_v^2$ of σ_v^2 is:

$$\hat{\sigma}_v^2 = \frac{1}{N} \left(\boldsymbol{y} - \boldsymbol{H}\hat{\boldsymbol{\theta}}\right)^T \left(\boldsymbol{y} - \boldsymbol{H}\hat{\boldsymbol{\theta}}\right) \tag{3.174}$$

Invoking Lemma 3.6 in the Appendix, the expectation of the estimate of the variance is:

$$E\left[\hat{\sigma}_v^2\right] = \frac{N-1}{N} \sigma_v^2 \tag{3.175}$$

This shows that $\hat{\sigma}_v^2$ is "slightly" *biased.* It is, however, asymptotically unbiased. We may fix the problem of bias by simply redefining the estimator (3.161) as:

$$\hat{\sigma}_{v\_unb}^2 = \frac{N}{N-1} \hat{\sigma}_v^2 = \frac{1}{N-1} \left(\boldsymbol{y} - \boldsymbol{H}\hat{\boldsymbol{\theta}}\right)^T \left(\boldsymbol{y} - \boldsymbol{H}\hat{\boldsymbol{\theta}}\right) \tag{3.176}$$

Example 3.12 Multivariate Gaussian PDF
Let $\boldsymbol{y} = \boldsymbol{H\theta} + \boldsymbol{v}$ be a linear model, $\boldsymbol{v}$ a zero-mean Gaussian random variable with $\text{cov}(\boldsymbol{v}) = \boldsymbol{\Sigma}_v$. The PDF of $\boldsymbol{y}$ is:

$$f_y(y) = \frac{1}{\sqrt{(2\pi)^N |\boldsymbol{\Sigma}_v|}} \exp\left\{-\frac{1}{2}(y - \boldsymbol{H\theta})^T \boldsymbol{\Sigma}_v^{-1} (y - \boldsymbol{H\theta})\right\} \tag{3.177}$$

Let $y^p = [\,y_1 \quad y_2 \quad y_3 \quad . \quad y_p\,]^T$ be p realizations of $\boldsymbol{y}$ obtained by performing p independent experiments. We will assume the sequence $\{y_i : i = 1, 2, 3, \dots, p\}$ to be i.i.d. Gaussian distributed random vectors. The problem is to estimate the mean $\boldsymbol{\theta}$ and the covariance matrix $\boldsymbol{\Sigma}_v$ from p realizations $\boldsymbol{y}_i$: $i = 1, 2, \dots, p$. The joint PDF of $\boldsymbol{y}^p$ is

$$f_y(y^p) = \frac{1}{\sqrt{(2\pi)^{Np} |\boldsymbol{\Sigma}_v|^p}} \exp\left\{-\frac{1}{2}\sum_{i=1}^{p}(y_i - \boldsymbol{H\theta})^T \boldsymbol{\Sigma}_v^{-1} (y_i - \boldsymbol{H\theta})\right\} \tag{3.178}$$

The log-likelihood function $l(\boldsymbol{\theta}, \boldsymbol{\Sigma}_v^{-1} | y^p)$ given in Eq. (3.117) becomes:

$$l(\boldsymbol{\theta}, \boldsymbol{\Sigma}_v^{-1} | y^p) = -\frac{1}{2}\sum_{i=1}^{p}(y_i - \boldsymbol{H\theta})^T \boldsymbol{\Sigma}_v^{-1} (y_i - \boldsymbol{H\theta}) - c - \frac{p}{2}\ln|\boldsymbol{\Sigma}_v| \tag{3.179}$$

where $c = \dfrac{Np}{2}\ln 2\pi$ is a constant term. The ML estimates $\hat{\boldsymbol{\theta}}$ and $\hat{\boldsymbol{\Sigma}}_v$ are those values of $\boldsymbol{\theta}$ and $\boldsymbol{\Sigma}_v$ that maximize the log-likelihood function $l\left(\boldsymbol{\theta}, \boldsymbol{\Sigma}_v^{-1} | y^p\right)$:

$$\begin{bmatrix} \hat{\boldsymbol{\theta}} \\ \hat{\boldsymbol{\Sigma}}_v \end{bmatrix} = \arg\max_{\boldsymbol{\Sigma}_v^{-1}, \boldsymbol{\theta}} \left\{-\frac{1}{2}\sum_{i=1}^{p}(y_i - \boldsymbol{H\theta})^T \Sigma_v^{-1} (y_i - \boldsymbol{H\theta}) - c - \frac{p}{2}\ln|\boldsymbol{\Sigma}_v|\right\} \tag{3.180}$$

Partial differentiation with respect to $\boldsymbol{\theta}$ and $\boldsymbol{\Sigma}_v^{-1}$ yields

$$\frac{\delta}{\delta\boldsymbol{\theta}} l(\boldsymbol{\theta}, \boldsymbol{\Sigma}_v^{-1} | y^p) = \sum_{i=1}^{p} \boldsymbol{H}^T \boldsymbol{\Sigma}_v^{-1} (y_i - \boldsymbol{H\theta}) \tag{3.181}$$

$$\frac{\delta}{\delta\boldsymbol{\Sigma}_v^{-1}} l\left(\boldsymbol{\theta}, \boldsymbol{\Sigma}_v^{-1} | y^p\right) = \frac{\delta}{\delta\boldsymbol{\Sigma}_v^{-1}} \left\{-\frac{1}{2}\sum_{i=1}^{p}(y_i - \boldsymbol{H\theta})^T \boldsymbol{\Sigma}_v^{-1} (y_i - \boldsymbol{H\theta}) - c - \frac{p}{2}\ln|\Sigma_v|\right\} \tag{3.182}$$

Expressing the quadratic term in Eq. (3.182) as a trace of the matrix product yields

$$-\frac{1}{2}\frac{\delta}{\delta\boldsymbol{\Sigma}_v^{-1}} \left(trace\left(\sum_{i=1}^{p}(y_i - \boldsymbol{H\theta})^T \boldsymbol{\Sigma}_v^{-1} (y_i - \boldsymbol{H\theta}) \right)\right) - \frac{\delta}{\delta\boldsymbol{\Sigma}_v^{-1}}\left(\frac{p}{2}\ln|\boldsymbol{\Sigma}_v|\right) \tag{3.183}$$

Using the properties of trace and determinant:
$trace\,\{ABC\} = trace\,\{BCA\}$; $|A| = \frac{1}{|A^{-1}|}$; $\ln|A| = -\ln|A^{-1}|$ yields

$$-\frac{1}{2}\frac{\delta}{\delta\boldsymbol{\Sigma}_v^{-1}} \left(trace\left(\boldsymbol{\Sigma}_v^{-1} \sum_{i=1}^{p}(y_i - \boldsymbol{H\theta})(y_i - \boldsymbol{H\theta})^T \right)\right) + \frac{p}{2}\frac{\delta}{\delta\boldsymbol{\Sigma}_v^{-1}}\left(\ln|\boldsymbol{\Sigma}_v^{-1}|\right) \tag{3.184}$$

Computing the partial differentiation of the trace and the logarithm terms with respect to Σ_v^{-1} using matrix calculus: $\frac{\delta}{\delta A} trace\{AB\} = \frac{\delta}{\delta A} trace\{BA\} = B^T; \frac{\delta}{\delta A}\ln(|A|) = (A^{-1})^T$ we get:

$$\frac{\delta}{\delta \boldsymbol{\Sigma}_v^{-1}} trace\left\{ \boldsymbol{\Sigma}_v^{-1} \frac{1}{2} \sum_{i=1}^{p} (\boldsymbol{y}_i - \boldsymbol{H\theta})(\boldsymbol{y}_i - \boldsymbol{H\theta})^T \right\} = \frac{1}{2} \sum_{i=1}^{p} (\boldsymbol{y}_i - \boldsymbol{H\theta})(\boldsymbol{y}_i - \boldsymbol{H\theta})^T \tag{3.185}$$

$$\frac{\delta \ln\left(|\boldsymbol{\Sigma}_v^{-1}|\right)}{\delta \boldsymbol{\Sigma}_v^{-1}} = \left(\boldsymbol{\Sigma}_v\right)^T = \boldsymbol{\Sigma}_v \tag{3.186}$$

Using the above expressions, the result of partial differentiation (3.182) becomes

$$\frac{\delta}{\delta \boldsymbol{\Sigma}_v^{-1}} l\left(\boldsymbol{\theta}, \boldsymbol{\Sigma}_v^{-1} | \boldsymbol{y}^p\right) = \frac{1}{2}\left(-\sum_{i=1}^{p} (\boldsymbol{y}_i - \boldsymbol{H\theta})(\boldsymbol{y}_i - \boldsymbol{H\theta})^T + \boldsymbol{\Sigma}_v p \right) \tag{3.187}$$

Setting the partial derivatives (3.181) and (3.187) to zero, the ML estimates $\hat{\boldsymbol{\theta}}$ and $\hat{\boldsymbol{\Sigma}}_v$ are those values of $\boldsymbol{\theta}$ and $\boldsymbol{\Sigma}_v$ which satisfy the following equation:

$$\begin{bmatrix} \frac{\delta}{\delta \boldsymbol{\theta}} l(\boldsymbol{\theta}, \Sigma_v^{-1} | \boldsymbol{y}^p) \\ \frac{\delta}{\delta \boldsymbol{\Sigma}_v^{-1}} l(\boldsymbol{\theta}, \boldsymbol{\Sigma}_v^{-1} | \boldsymbol{y}^p) \end{bmatrix} = \begin{bmatrix} \sum_{i=1}^{p} \boldsymbol{H}^T \boldsymbol{\Sigma}_v^{-1} (\boldsymbol{y}_i - \boldsymbol{H\theta}) \\ \frac{1}{2}\left(-\sum_{i=1}^{p} (\boldsymbol{y}_i - \boldsymbol{H\theta})(\boldsymbol{y}_i - \boldsymbol{H\theta})^T + \boldsymbol{\Sigma}_v p \right) \end{bmatrix} = \begin{bmatrix} \boldsymbol{0} \\ \boldsymbol{0} \end{bmatrix} \tag{3.188}$$

Solving for $\boldsymbol{\theta}$ and $\boldsymbol{\Sigma}_v$ yields

$$\begin{aligned} \hat{\boldsymbol{\theta}} &= \frac{1}{p} \sum_{i=1}^{p} \left(\boldsymbol{H}^T \boldsymbol{\Sigma}_v^{-1} \boldsymbol{H}^T\right)^{-1} \boldsymbol{H} \boldsymbol{\Sigma}_v^{-1} \boldsymbol{y}_i \\ \hat{\boldsymbol{\Sigma}}_v &= \frac{1}{p} \sum_{i=1}^{p} (\boldsymbol{y}_i - \boldsymbol{H}\hat{\boldsymbol{\theta}})(\boldsymbol{y}_i - \boldsymbol{H}\hat{\boldsymbol{\theta}})^T \end{aligned} \tag{3.189}$$

Example 3.13 i.i.d. Laplacian PDF
Consider the linear measurement model (3.123), $v(k)$ is i.i.d. zero-mean random noise with Laplacian PDF given by Eq. (3.75)

$$f_y(y; \theta) = \left(\frac{1}{\sqrt{2\sigma_v^2}} \right)^N \exp\left(-\frac{\sqrt{2}}{\sigma_v} \sum_{k=0}^{N-1} |y(k) - \theta| \right) \tag{3.190}$$

The log-likelihood function for estimating the mean θ and standard deviation σ_v, $l\left(\theta, \sigma_v | y\right)$ is

$$l\left(\theta, \sigma_v | y\right) = -\frac{N}{2} \ln\left(2\sigma_v\right) - \frac{\sqrt{2}}{\sigma_v} \sum_{k=0}^{N-1} |y(k) - \theta| \tag{3.191}$$

The ML estimates $\hat{\theta}$ and $\hat{\sigma}_v$ are obtained from maximizing the log-likelihood function $l(\theta, \sigma_v | y)$:

$$\begin{bmatrix} \hat{\theta} \\ \hat{\sigma}_v \end{bmatrix} = \arg\max_{\sigma_v, \theta} \left\{ -\frac{N}{2} \ln(2\sigma_v) - \frac{\sqrt{2}}{\sigma_v} \sum_{k=0}^{N-1} |y(k) - \theta| \right\} \tag{3.192}$$

Partial differentiation of $l(\theta, \sigma_v^2|y)$ with respect to θ and σ_v gives

$$\frac{\delta}{\delta\theta} l\left(\theta, \sigma_v^2|\mathbf{y}\right) = \frac{\sqrt{2}}{\sigma_v} \sum_{k=0}^{N-1} sign\left(y(k) - \theta\right) \tag{3.193}$$

$$\frac{\delta}{\delta\sigma_v} l\left(\theta, \sigma_v^2|\mathbf{y}\right) = \frac{\sqrt{2}}{\sigma_v^2} \sum_{k=0}^{N-1} |y(k) - \theta| - \frac{N}{2\sigma_v} \tag{3.194}$$

Setting the partial derivatives (3.193) and (3.194) to zero, the ML estimates $\hat{\theta}$ and $\hat{\sigma}_v$ are those values of θ and σ_v which satisfy the following equation:

$$\begin{bmatrix} \dfrac{\delta}{\delta\theta} l\left(\theta, \sigma_v^2|\mathbf{y}\right) \\ \dfrac{\delta}{\delta\sigma_v} l\left(\theta, \sigma_v^2|\mathbf{y}\right) \end{bmatrix} = \begin{bmatrix} \dfrac{\sqrt{2}}{\sigma_v} \displaystyle\sum_{k=0}^{N-1} sign\left(y(k) - \theta\right) \\ \dfrac{\sqrt{2}}{\sigma_v^2} \displaystyle\sum_{k=0}^{N-1} |y(k) - \theta| - \dfrac{N}{2\sigma_v} \end{bmatrix} = \begin{bmatrix} 0 \\ 0 \end{bmatrix} \tag{3.195}$$

The ML estimate $\hat{\theta}$ is given by

$$\hat{\theta} = \text{median}([\, y(0) \quad y(1) \quad y(2) \quad . \quad y(N-1)\,]^T) \tag{3.196}$$

The ML estimate $\hat{\sigma}_v$ is:

$$\hat{\sigma}_v = \frac{\sqrt{2}}{N} \sum_{k=0}^{N-1} |y(k) - \hat{\theta}| \tag{3.197}$$

3.5.4 Properties of Maximum Likelihood Estimator

1. The Cramér–Rao lower bound depends upon the PDF of the data, and one has to know the PDF either *a priori* or estimate it from the data before we can compute the lower bound. Hence in many practical cases it is not possible to compute the lower bound.
2. An efficient estimator may not always exist. However if the PDF is Gaussian, an efficient estimator always exists.
3. If the PDF is Gaussian, both the maximum likelihood and the best linear least-squares estimators are efficient.
4. If an efficient estimator exists, it is given by the maximum likelihood estimator.

3.6 Summary

Mathematical model

$$\mathbf{y} = \mathbf{H}\boldsymbol{\theta} + \mathbf{v}$$

Probabilistic model:

$$f_y(\mathbf{y}; \boldsymbol{\theta}) = f_y(y(0), y(1), y(2), \ldots, y(N-1); \boldsymbol{\theta})$$

Gaussian PDF: denoted $f_g(\mathbf{y};\boldsymbol{\theta})$

$$f_g(\mathbf{y};\boldsymbol{\theta}) = \frac{1}{\sqrt{(2\pi)^N|\Sigma_v|}}\exp\left\{-\frac{1}{2}(\mathbf{y}-\boldsymbol{\mu}_y)^T\boldsymbol{\Sigma}_v^{-1}(\mathbf{y}-\boldsymbol{\mu}_y)\right\}$$

$$\boldsymbol{\mu}_y = E[y];\ E[y] = H\theta;\ \operatorname{cov}(y) = \Sigma_v$$

Uniform PDF: denoted $f_u(y;\theta)$

$$f_u(y;\theta) = \begin{cases} \dfrac{1}{b-a} & \text{for } a \le y \le b \\ 0 & \text{for } y < a \text{ or } y > b \end{cases}$$

$$E[y] = \mu_y = \theta = \frac{b+a}{2};\ \operatorname{var}(y) = \sigma_y^2 = \frac{(b-a)^2}{12}$$

Laplacian PDF: denoted $f_e(y;\theta)$

$$f_e(y;\theta) = \frac{1}{\sqrt{2}\sigma_y}\exp\left(-\frac{\sqrt{2}|y-\mu_y|}{\sigma_y}\right)$$

Worst-case PDF: Part Gauss-part Laplace denoted $f_{ge}(y;\theta)$

$$f_{ge}(y;\theta) = \begin{cases} \kappa\exp\left\{-\dfrac{1}{2\sigma_{ge}^2}(y-\mu_y)^2\right\} & y \in \Upsilon_{gd} \\ \kappa\exp\left\{\dfrac{a_{gd}^2}{2\sigma_{ge}^2}\right\}\exp\left\{-\dfrac{a_{gd}}{2\sigma_{ge}^2}|y-\boldsymbol{\mu}_y|\right\} & y \in \Upsilon_{bd} \end{cases}$$

where $\Upsilon_{gd} = \{y : |y-\mu_y| \le a_{gd}\}$ and $\Upsilon_{bd} = \{y : |y-\mu_y| > a_{gd}\}$

Cauchy PDF: denoted $f_c(y;\theta)$

$$f_c(y;\theta) = \frac{\alpha}{\pi[(y-\mu_y)^2+\alpha^2]}$$

Likelihood function

$$L(\boldsymbol{\theta}|\mathbf{y}) = f_y(\mathbf{y};\boldsymbol{\theta})$$

The log-likelihood function

$$l(\boldsymbol{\theta}|\mathbf{y}) = \ln(L(\boldsymbol{\theta}|\mathbf{y}))$$

Properties of estimators
Scalar case:

$$y(k) = H\theta + v(k)$$
$$\min_{\hat{\theta}} E\left[(y(k) - H\hat{\theta})^2\right]$$
$$\hat{\theta} = \frac{1}{H}\left(\frac{1}{N}\sum_{k=0}^{N-1} y(k)\right)$$
$$E\left[\hat{\theta}\right] = \theta$$

Unbiasedness of the estimator

$$E[\hat{\boldsymbol{\theta}}] = \begin{cases} \boldsymbol{\theta} & \textit{if } \boldsymbol{\theta} \textit{ is non-random} \\ E[\boldsymbol{\theta}] & \textit{if } \boldsymbol{\theta} \textit{ is random} \end{cases}$$

A linear estimator, $\hat{\boldsymbol{\theta}} = \boldsymbol{F}\boldsymbol{y}$ is unbiased if and only if $\boldsymbol{F}\boldsymbol{H} = \boldsymbol{I}$

Variance of the Estimator: Scalar Case

$$E[(\hat{\theta} - \theta)^2] = \frac{1}{H^2}E\left[\left(\frac{1}{N}\sum_{k=0}^{N-1} v(k)\right)^2\right] = \frac{1}{N^2}\left(\frac{\sigma_v^2}{H^2}\right)$$

$$\lim_{N\to\infty} it\, E[(\hat{\theta} - \theta)^2] = 0$$

Median of the data samples

$$\text{median}(\boldsymbol{y}) = \begin{cases} x_{\frac{N+1}{2}} & N \textit{ odd} \\ \frac{1}{2}\left(x_{\frac{N}{2}} + x_{\frac{N}{2}+1}\right) & N \textit{ even} \end{cases}$$

A formal definition:

$$\min_{\hat{\theta}}\{E|y - H\hat{\theta}|\}$$

Small and large sample properties:
Consistent estimator
$\hat{\boldsymbol{\theta}}_N$ is a consistent estimator of θ if $\hat{\boldsymbol{\theta}}_N$ converges to $\boldsymbol{\theta}$ in probability:

$$\lim_{N\to\infty} it\, P\{|\boldsymbol{\theta} - \hat{\boldsymbol{\theta}}| > \varepsilon\} = 0, \text{ for every } \varepsilon > 0.$$

Asymptotically unbiased estimator:

$$\lim_{N\to\infty} it\, E\left[\hat{\boldsymbol{\theta}}_N\right] = \begin{cases} \boldsymbol{\theta} & \textit{if } \boldsymbol{\theta} \textit{ is random} \\ E\left[\boldsymbol{\theta}\right] & \textit{if } \boldsymbol{\theta} \textit{ is non-random} \end{cases}$$

Asymptotically efficient estimator
If the property of efficiency holds for infinitely large number samples N, then the estimator is said to be asymptotically efficient.

Cramér–Rao Inequality
The Cramér–Rao inequality of an unbiased estimator $\hat{\theta}$ is

$$\text{var}(\hat{\theta}) \geq \frac{1}{I_F}$$

I_F is the *Fisher information*, $I_F = E\left[\left(\frac{\partial l(\theta|y)}{\partial\theta}\right)^2\right] = -E\left[\frac{\partial^2 l(\theta|y)}{\partial\theta^2}\right]$.

$\text{var}(\hat{\theta}) = \frac{1}{I_F}$ if and only if $\frac{\delta l(\theta|y)}{\delta\theta} = I_F\left(\hat{\theta} - \theta\right)$

Vector Case: θ is a Mx1 Vector

$$\text{cov}(\hat{\boldsymbol{\theta}}) \geq \boldsymbol{I}_F^{-1}$$

where $\boldsymbol{I}_F = E\left[\left(\frac{\partial l(\boldsymbol{\theta}|\mathbf{y})}{\partial\boldsymbol{\theta}}\right)\left(\frac{\partial l(\boldsymbol{\theta}|\mathbf{y})}{\partial\boldsymbol{\theta}}\right)^T\right] = -E\left[\frac{\partial^2 l(\boldsymbol{\theta}|\mathbf{y})}{\partial\boldsymbol{\theta}^2}\right]$ is the Fisher information matrix $\text{cov}(\hat{\boldsymbol{\theta}}) = \boldsymbol{I}_F^{-1}$ if and only if $\frac{\delta l(\boldsymbol{\theta}|\mathbf{y})}{\delta\boldsymbol{\theta}} = \boldsymbol{I}_F\left(\hat{\boldsymbol{\theta}} - \boldsymbol{\theta}\right)$.

If $y(k)$, $k = 0, 1, 2, \ldots, N-1$ be N i.i.d. random variables then $I_F^N = NI_F$ where $I_F^N = E\left[\left(\frac{\partial l(\boldsymbol{\theta}|\mathbf{y})}{\partial\boldsymbol{\theta}}\right)^2\right]$ and $I_F = E\left[\left(\frac{\partial l(\boldsymbol{\theta}|\mathbf{y})}{\partial\boldsymbol{\theta}}\right)^2\right]$.

Maximum Likelihood Estimation

$$\hat{\boldsymbol{\theta}} = \left(\boldsymbol{H}^T\boldsymbol{\Sigma}_v^{-1}\boldsymbol{H}\right)^{-1}\boldsymbol{H}^T\boldsymbol{\Sigma}_v^{-1}\mathbf{y}$$

3.7 Appendix: Cauchy–Schwarz Inequality

$$\left(\int_{-\infty}^{\infty}(f(y))^2\,dy\right)\left(\int_{-\infty}^{\infty}(g(y))^2\,dy\right) \geq \left(\int_{-\infty}^{\infty}f(y)g(y)dy\right)^2 \tag{3.198}$$

where equality is achieved if and only if

$$f(\mathbf{y}) = cg(\mathbf{y}) \tag{3.199}$$

where c is a constant, which is it not a function of $\mathbf{y}$.

3.8 Appendix: Cramér–Rao Lower Bound

Regularity Condition: The derivation of the Cramér–Rao lower bound assumes two weak *regularity conditions*: one on the PDF $f_y(\mathbf{y})$ and the other on the estimator $\hat{\boldsymbol{\theta}}(\mathbf{y})$.

1. $f_y(\mathbf{y}) > 0$
2. $\frac{\delta f_y(\mathbf{y})}{\delta\theta}$ exists and is finite

3. The operations of integration with respect to $\boldsymbol{y}$ and differentiation with respect to $\boldsymbol{\theta}$ can be interchanged in $\boldsymbol{E}\left[\hat{\theta}\left(y\right)\right]$.

$$\frac{\delta}{\delta\theta}\int_{-\infty}^{\infty}\hat{\theta}(y)f_y(y)dy=\int_{-\infty}^{\infty}\hat{\theta}(y)\frac{\delta f_y(y)}{\delta\theta}dy$$

This conditions that the integration and differentiation can be swapped holds for all well-behaved PDFs:

1. If $f_y(\boldsymbol{y})$ has bounded support in x, and the bounds do not depend on θ.
2. If $f_y(\boldsymbol{y})$ has infinite support, is continuously differentiable, and the integral converges uniformly for all $\boldsymbol{\theta}$.

Cramér–Rao lower bound: We will first consider the scalar case and then extend the result to a vector case.

3.8.1 Scalar Case

Theorem 3.1 *Let* $y=[\,y(0)\quad y(1)\quad y(2)\quad .\quad y(N-1)\,]^T$ *is a* $N\mathrm{x}1$ *vector of measurements characterized by the probability density function (PDF)* $f_y(y;\boldsymbol{\theta})$*, the measurement y is a function of an unknown scalar parameter* θ*. Let* $\hat{\theta}$ *be an unbiased estimator of* θ *and only a function of y and not a function of the unknown parameter* θ*. If the regularity conditions hold then*

$$E[(\hat{\theta}(y)-\theta)^2]\geq\left(E\left[\left(\frac{\delta\ln f_y(y)}{\delta\theta}\right)^2\right]\right)^{-1}\tag{3.200}$$

where $I_F(\theta)=E\left[\left(\frac{\delta\ln f_y(y)}{\delta\theta}\right)^2\right]$ *is the Fisher information.*

Proof: Since $\hat{\theta}$ is an unbiased estimator of θ, the following identity holds for all θ

$$E\left[\hat{\theta}\right]-\theta=0\tag{3.201}$$

Expanding the expression of the expectation operator in terms of the PDF $f_y(\boldsymbol{y};\boldsymbol{\theta})$ we get

$$\int_{-\infty}^{\infty}\left(\hat{\theta}(\boldsymbol{y})-\theta\right)f_y(\boldsymbol{y};\boldsymbol{\theta})dy=0\tag{3.202}$$

The estimate $\hat{\theta}$ is a function only of $\boldsymbol{y}$ and not of θ and hence partial differentiation with respect to θ and invoking the regularity condition yields

$$\int_{-\infty}^{\infty}\left(\hat{\theta}\left(\boldsymbol{y}\right)-\theta\right)\frac{\delta f_y(\boldsymbol{y};\boldsymbol{\theta})}{\delta\theta}dy-\int_{-\infty}^{\infty}f_y(\boldsymbol{y};\boldsymbol{\theta})dy=0\tag{3.203}$$

Since $\int_{-\infty}^{\infty} f_y(y;\theta)dy = 1$ we get

$$\int_{-\infty}^{\infty} (\hat{\theta}(y) - \theta) \frac{\delta f_y(y;\theta)}{\delta\theta} dy = 1 \tag{3.204}$$

Dividing and multiplying the term inside the integral by $f_y(y;\theta)$, we get

$$\int_{-\infty}^{\infty} (\hat{\theta}(y) - \theta) \frac{\frac{\delta f_y(y;\theta)}{\delta\theta}}{f_y(y;\theta)} f_y(y;\theta)dy = 1 \tag{3.205}$$

Since $\frac{\delta \ln f_y(y;\theta)}{\delta\theta} = \frac{\frac{\delta f_y(y;\theta)}{\delta\theta}}{f_y(y;\theta)}$ we get

$$\int_{-\infty}^{\infty} (\hat{\theta}(y) - \theta) \frac{\delta \ln f_y(y;\theta)}{\delta\theta} f_y(y;\theta)dy = 1 \tag{3.206}$$

Expressing the integrand using $f_y(y;\theta) = \sqrt{f_y(y;\theta)}\sqrt{f_y(y;\theta)}$(as $f_y(y;\theta) \geq 0$) we get

$$\int_{-\infty}^{\infty} \left[(\hat{\theta}(y) - \theta) \sqrt{f_y(y;\theta)}\right] \left[\frac{\delta \ln f_y(y;\theta)}{\delta\theta} \sqrt{f_y(y;\theta)}\right] dy = 1 \tag{3.207}$$

Using the Cauchy–Schwarz inequality given in 3.7 Appendix, we get

$$\left(\int_{-\infty}^{\infty} (\hat{\theta}(y) - \theta)^2 f_y(y;\theta)dy\right)\left(\int_{-\infty}^{\infty} \left(\frac{\delta \ln f_y(y;\theta)}{\delta\theta}\right)^2 f_y(y;\theta)dy\right) \geq 1 \tag{3.208}$$

Expressing this in terms of the expectation we get

$$E[(\hat{\theta}(y) - \theta)^2]E\left[\left(\frac{\delta \ln f_y(y;\theta)}{\delta\theta}\right)^2\right] \geq 1 \tag{3.209}$$

Equivalently

$$E[(\hat{\theta}(y) - \theta)^2] \geq \frac{1}{I_F(\theta)} \tag{3.210}$$

Corollary 3.1 *The minimum variance unbiased estimator $\hat{\theta}$ (an estimator which achieves the lowest bound on the variance) is the one that satisfies Eq. (3.209) with equality. From Cauchy–Schwarz inequality this implies*

$$\hat{\theta}(y) - \theta = I_F \frac{\delta \ln f_y(y;\theta)}{\delta\theta} \tag{3.211}$$

3.8.2 Vector Case

Theorem 3.2 *Let* $\mathbf{y} = [\,y(0) \quad y(1) \quad y(2) \quad . \quad y(N-1)\,]^T$ *is a Nx1 vector of measurements characterized by the probability density function (PDF)* $f_y(\mathbf{y};\boldsymbol{\theta})$. *The measurement* $\mathbf{y}$ *is function of an unknown Mx1 parameter* $\boldsymbol{\theta}$. *Let* $\hat{\boldsymbol{\theta}}$ *be an unbiased estimator of* $\boldsymbol{\theta}$ *and only a function of* $\mathbf{y}$ *and not a function of the unknown parameter* $\boldsymbol{\theta}$. *If the regularity conditions hold then*

$$\mathrm{cov}(\hat{\boldsymbol{\theta}}) = E[(\hat{\boldsymbol{\theta}}(\mathbf{y})-\boldsymbol{\theta})(\hat{\boldsymbol{\theta}}(\mathbf{y})-\boldsymbol{\theta})^T] \geq \boldsymbol{I}_F^{-1}(\boldsymbol{\theta}) \tag{3.212}$$

where the MxM Fisher information matrix $\boldsymbol{I}_F(\boldsymbol{\theta}) = E\left[\left(\frac{\delta \ln f_y(\mathbf{y})}{\delta\boldsymbol{\theta}}\right)\left(\frac{\delta \ln f_y(\mathbf{y})}{\delta\boldsymbol{\theta}}\right)^T\right]$. *The inequality implies that the difference* $\mathrm{cov}\left(\hat{\boldsymbol{\theta}}\right) - \boldsymbol{I}_F^{-1}(\boldsymbol{\theta})$ *is positive semi-definite.*

Proof: [3] Since $\hat{\boldsymbol{\theta}}(y)$ is unbiased, $E[(\hat{\boldsymbol{\theta}}(y)-\boldsymbol{\theta})] = 0$. Partial differentiation with respect to $\boldsymbol{\theta}$ and invoking the regularity condition yields

$$\int_{-\infty}^{\infty} (\hat{\boldsymbol{\theta}}(\mathbf{y})-\boldsymbol{\theta})\left(\frac{\delta \ln f_y(\mathbf{y};\boldsymbol{\theta})}{\delta\boldsymbol{\theta}}\right)^T f_y(\mathbf{y};\boldsymbol{\theta})d\mathbf{y} = I \tag{3.213}$$

Pre-multiplying by $\boldsymbol{a}^T$ and post-multiplying by $\boldsymbol{b}$ where $\boldsymbol{a}$ and $\boldsymbol{b}$ are Mx1 vectors we get

$$\int_{-\infty}^{\infty} \boldsymbol{a}^T(\hat{\boldsymbol{\theta}}(\mathbf{y})-\boldsymbol{\theta})\left(\frac{\delta \ln f_y(\mathbf{y};\boldsymbol{\theta})}{\delta\boldsymbol{\theta}}\right)^T \boldsymbol{b} f_y(\mathbf{y};\boldsymbol{\theta})d\mathbf{y} = \boldsymbol{a}^T\boldsymbol{b} \tag{3.214}$$

Defining two scalars $h(\mathbf{y}) = \boldsymbol{a}^T(\hat{\boldsymbol{\theta}}(\mathbf{y})-\boldsymbol{\theta})$ and $g(\mathbf{y}) = \left(\frac{\delta \ln f_y(\mathbf{y};\boldsymbol{\theta})}{\delta\boldsymbol{\theta}}\right)^T \boldsymbol{b}$ Eq. (3.214) we get

$$\int_{-\infty}^{\infty} h(\mathbf{y})g(\mathbf{y})f_y(\mathbf{y};\boldsymbol{\theta})d\mathbf{y} = \boldsymbol{a}^T\boldsymbol{b} \tag{3.215}$$

Using the Cauchy–Schwarz inequality given in 3.7 Appendix

$$\left(\int_{-\infty}^{\infty} h^2(\mathbf{y})f_y(\mathbf{y})d\mathbf{y}\right)\left(\int_{-\infty}^{\infty} g^2(\mathbf{y})f_y(\mathbf{y})d\mathbf{y}\right) \geq \left(\boldsymbol{a}^T\boldsymbol{b}\right)^2 \tag{3.216}$$

where

$$\begin{aligned} h^2(\mathbf{y}) &= \boldsymbol{a}^T(\hat{\boldsymbol{\theta}}(\mathbf{y})-\boldsymbol{\theta})(\hat{\boldsymbol{\theta}}(\mathbf{y})-\boldsymbol{\theta})^T\boldsymbol{a} \\ g^2(\mathbf{y}) &= \boldsymbol{b}^T\left(\frac{\delta \ln f_y(\mathbf{y};\boldsymbol{\theta})}{\delta\boldsymbol{\theta}}\right)\left(\frac{\delta \ln f_y(\mathbf{y};\boldsymbol{\theta})}{\delta\boldsymbol{\theta}}\right)^T\boldsymbol{b} \end{aligned} \tag{3.217}$$

Expressing (3.216) in terms of $\mathrm{cov}\left(\hat{\boldsymbol{\theta}}\right)$, and $I_F(\boldsymbol{\theta})$ we get

$$\boldsymbol{a}^T\mathrm{cov}\left(\hat{\boldsymbol{\theta}}\right)\boldsymbol{a}\,\boldsymbol{b}^T\boldsymbol{I}_F\boldsymbol{b} \geq \left(\boldsymbol{a}^T\boldsymbol{b}\right)^2 \tag{3.218}$$

Since $\boldsymbol{b}$ is arbitrary choosing $\boldsymbol{b} = \boldsymbol{I}_F^{-1}(\boldsymbol{\theta})\boldsymbol{a}$ we get

$$\boldsymbol{a}^T \text{cov}(\hat{\boldsymbol{\theta}})\boldsymbol{a} \geq \boldsymbol{a}^T \boldsymbol{I}_F^{-1}(\boldsymbol{\theta})\boldsymbol{a} \tag{3.219}$$

Thus

$$\text{cov}(\hat{\boldsymbol{\theta}}) \geq \boldsymbol{I}_F^{-1}(\boldsymbol{\theta}) \tag{3.220}$$

Corollary 3.2 *The minimum variance unbiased estimator $\hat{\boldsymbol{\theta}}$ (an estimator which achieves the lowest bound on the variance) is the one that satisfies Eq. (3.220) with equality. From the Cauchy–Schwarz inequality this implies*

$$\hat{\boldsymbol{\theta}}(\boldsymbol{y}) - \boldsymbol{\theta} = \boldsymbol{I}_F(\boldsymbol{\theta}) \frac{\delta \ln f_y(\boldsymbol{y})}{\delta \boldsymbol{\theta}} \tag{3.221}$$

3.9 Appendix: Fisher Information: Cauchy PDF

$$I_F = E\left[\left(\frac{\partial \ln(f_y(y;\theta))}{\partial \theta}\right)^2\right] = \int_{-\infty}^{\infty} \left(\frac{\partial \ln(f_y(y;\theta))}{\partial \theta}\right)^2 f_y(y;\theta)dy \tag{3.222}$$

Substituting $\dfrac{\partial \ln(f_y(y;\theta))}{\partial \theta} = \dfrac{2y}{y^2+\alpha^2}; f_y(y;\theta) = \dfrac{\alpha}{\pi\left(\alpha^2+y^2\right)}$ we get

$$I_F = \frac{4\alpha}{\pi} \int_{-\infty}^{\infty} \frac{y^2}{\left(y^2+\alpha^2\right)^3} dy \tag{3.223}$$

The integral term is:

$$\int \frac{y^2}{(y^2+\alpha^2)^3} dy = \frac{1}{8\alpha^3} \tan\left(\frac{y}{\alpha}\right) + \frac{y}{8\alpha^2(y^2+\alpha^2)} - \frac{y}{4(y^2+\alpha^2)^2} \tag{3.224}$$

Substituting the upper limit of ∞ and the lower limit of integration $-\infty$ for y in the expression on the right-hand side of Eq. (3.224), the Fisher information given by Eq. (3.223) is:

$$I_F = \frac{1}{2\alpha^2} \tag{3.225}$$

3.10 Appendix: Fisher Information for i.i.d. PDF

The PDF of $\boldsymbol{y}$ is $f_y(\boldsymbol{y};\theta)$. Since $\{y(k)\}$ are i.i.d. we get

$$\ln\left(f_y(\boldsymbol{y};\theta)\right) = \sum_{k=0}^{N-1} \ln\left(f_y(y(k);\theta)\right) \tag{3.226}$$

The Fisher information I_F is:

$$I_F = E\left[\left(\sum_{k=0}^{N-1} \ln(f_y(y(k);\theta))\right)^2\right] \tag{3.227}$$

Expanding the summation term we get

$$I_F = \sum_{k=0}^{N-1} E\left[\left(\frac{\partial \ln(f_y(y(k);\theta))}{\partial\theta}\right)^2\right] + \sum_i \sum_{j\neq i} E\left[\left(\frac{\partial \ln(f_y(y(i);\theta))}{\partial\theta}\right)\left(\frac{\partial \ln(f_y(y(j);\theta))}{\partial\theta}\right)\right] \tag{3.228}$$

Expanding, and since $y(i)$ and $y(j)$ are independent yields, the product term becomes:

$$E\left[\left(\frac{\partial \ln(f_y(y(i);\theta))}{\partial\theta}\right)\left(\frac{\partial \ln(f_y(y(j);\theta))}{\partial\theta}\right)\right] = E\left[\left(\frac{\partial \ln(f_y(y(i);\theta))}{\partial\theta}\right)\right] E\left[\left(\frac{\partial \ln(f_y(y(j);\theta))}{\partial\theta}\right)\right] \tag{3.229}$$

Hence I_F becomes

$$I_F = \sum_{k=0}^{N-1} E\left[\left(\frac{\partial \ln(f_y(y(k);\theta))}{\partial\theta}\right)^2\right] + \sum_i \sum_{j\neq i} E\left[\left(\frac{\partial \ln(f_y(y(i);\theta))}{\partial\theta}\right)\right] E\left[\left(\frac{\partial \ln(f_y(y(j);\theta))}{\partial\theta}\right)\right] \tag{3.230}$$

Imposing the regularity condition (3.40), the product terms on the right-hand sides vanish. We get:

$$I_F = \sum_{k=0}^{N-1} E\left[\left(\frac{\partial \ln(f_y(y(k);\theta))}{\partial\theta}\right)^2\right] \tag{3.231}$$

3.11 Appendix: Projection Operator

Let $\boldsymbol{H}$ be some NxM matrix with $N \geq M$ and $rank(\boldsymbol{H}) = M_r$. The projection operator denoted $\boldsymbol{P}_r$ associated with $\boldsymbol{H}$ is defined as

$$\boldsymbol{P}_r = \boldsymbol{H}\boldsymbol{H}^{\dagger} = \boldsymbol{H}\left(\boldsymbol{H}^T\boldsymbol{H}\right)^{-1}\boldsymbol{H}^T \tag{3.232}$$

Properties of the projection operator are:
$\boldsymbol{P}_r = \boldsymbol{H}\boldsymbol{H}^{\dagger}$ has following properties:

1. $\boldsymbol{P}_r^T = P_r$ is symmetrical
2. $\boldsymbol{P}_r^2 = \boldsymbol{P}_r$ hence $\boldsymbol{P}_r^m = \boldsymbol{P}_r$ *for* $m = 1, 2, 3, \ldots,$
3. Eigenvalues of $\boldsymbol{I} - \boldsymbol{P}_r$ are only ones and zeros
   Let M_r be the rank of a NxM matrix $\boldsymbol{H}$
   - M_r eigenvalues will be ones
   - The rest of the $N - M_r$ eigenvalues will be zeros
   - $trace\left(\boldsymbol{I} - \boldsymbol{P}_r\right) = N - M_r$.
4. $\boldsymbol{I} - \boldsymbol{P}_r$ projects a vector on to a space perpendicular to the range space of the matrix $\boldsymbol{H}$
5. If $\boldsymbol{H}$ is non-singular square matrix then $\boldsymbol{P}_r = \boldsymbol{I}$; $\boldsymbol{I} - \boldsymbol{P}_r = \boldsymbol{0}$.

Lemma 3.6 *Let* $\boldsymbol{y} = \boldsymbol{H}\boldsymbol{\theta} + \boldsymbol{v}$ *and* $\boldsymbol{H}$ *have full rank:rank* $(\boldsymbol{H}) = M$. *The ML estimator* $\hat{\boldsymbol{\theta}}$ *of* $\boldsymbol{\theta}$ *is*

$$\hat{\boldsymbol{\theta}} = \left(\boldsymbol{H}^T\boldsymbol{H}\right)^{-1}\boldsymbol{H}^T\boldsymbol{y}, \tag{3.233}$$

and the ML estimator $\hat{\sigma}_v^2$ of σ_v is

$$\hat{\sigma}_v^2 = \frac{1}{N}\left(y - H\hat{\theta}\right)^T \left(y - H\hat{\theta}\right) \tag{3.234}$$

Then

$$E\left[\hat{\sigma}_v^2\right] = \left(1 - \frac{M}{N}\right)\sigma_v^2 \tag{3.235}$$

Proof: Substituting $\hat{\theta} = \left(\boldsymbol{H}^T\boldsymbol{H}\right)^{-1}\boldsymbol{H}^T\boldsymbol{y}$ and

$$\hat{\sigma}_v^2 = \frac{1}{N}\left(\boldsymbol{y} - \boldsymbol{H}\left(\boldsymbol{H}^T\boldsymbol{H}\right)^{-1}\boldsymbol{H}^T\boldsymbol{y}\right)^T\left(\boldsymbol{y} - \boldsymbol{H}\left(\boldsymbol{H}^T\boldsymbol{H}\right)^{-1}\boldsymbol{H}^T\boldsymbol{y}\right) \tag{3.236}$$

Substituting $y = \boldsymbol{H}\boldsymbol{\theta} + \boldsymbol{v}$ in the expression $\boldsymbol{H}\left(\boldsymbol{H}^T\boldsymbol{H}\right)^{-1}\boldsymbol{H}^T\boldsymbol{y}$ we get

$$\boldsymbol{y} - \boldsymbol{H}\left(\boldsymbol{H}^T\boldsymbol{H}\right)^{-1}\boldsymbol{H}^T\boldsymbol{y} = \boldsymbol{H}\boldsymbol{\theta} + \boldsymbol{v} - \boldsymbol{H}\left(\boldsymbol{H}^T\boldsymbol{H}\right)^{-1}\boldsymbol{H}^T\left(\boldsymbol{H}\boldsymbol{\theta} + \boldsymbol{v}\right) \tag{3.237}$$

Simplifying we get

$$\boldsymbol{y} - \boldsymbol{H}\left(\boldsymbol{H}^T\boldsymbol{H}\right)^{-1}\boldsymbol{H}^T\boldsymbol{y} = \boldsymbol{v} - \boldsymbol{H}\left(\boldsymbol{H}^T\boldsymbol{H}\right)^{-1}\boldsymbol{H}^T\boldsymbol{v} = \left(\boldsymbol{I} - \boldsymbol{H}\left(\boldsymbol{H}^T\boldsymbol{H}\right)^{-1}\boldsymbol{H}^T\right)\boldsymbol{v} \tag{3.238}$$

Using the definition of projection operator $\boldsymbol{P}_r$ (3.232) we get

$$\boldsymbol{y} - \boldsymbol{H}\left(\boldsymbol{H}^T\boldsymbol{H}\right)^{-1}\boldsymbol{H}^T\boldsymbol{y} == \left(\boldsymbol{I} - \boldsymbol{P}_r\right)\boldsymbol{v} \tag{3.239}$$

Using Eq. (3.239), the expression (3.236) for $\hat{\sigma}_v^2$ becomes:

$$\hat{\sigma}_v^2 = \frac{1}{N}\boldsymbol{v}^T\left(\boldsymbol{I} - \boldsymbol{P}_r\right)^T\left(\boldsymbol{I} - \boldsymbol{P}_r\right)\boldsymbol{v} \tag{3.240}$$

Using the properties of the projection operator yields

$$\hat{\sigma}_v^2 = \frac{1}{N}\boldsymbol{v}^T\left(\boldsymbol{I} - \boldsymbol{P}_r\right)\boldsymbol{v} \tag{3.241}$$

Taking expectation yields

$$E\left[\hat{\sigma}_v^2\right] = \frac{1}{N}E\left[\boldsymbol{v}^T\left(\boldsymbol{I} - \boldsymbol{P}_r\right)\boldsymbol{v}\right] \tag{3.242}$$

Since $\{v(k)\}$ is zero mean i.i.d. with variance σ_v^2, only the diagonal elements of $\boldsymbol{I} - \boldsymbol{P}$ will contribute to the non-zero values of the quadratic term on the right, and we get

$$E\left[\hat{\sigma}_v^2\right] = \frac{1}{N}trace\left(\boldsymbol{I} - \boldsymbol{P}_r\right)\sigma_v^2 \tag{3.243}$$

Using the property $trace\left(\boldsymbol{I} - \boldsymbol{P}_r\right) = N - M$ we get Eq. (3.235).

3.12 Appendix: Fisher Information: Part Gauss-Part Laplace

Consider the PDF given by Eq. (3.13). Assuming $\mu_y = \theta$ the PDF becomes

$$f_y(y) = \begin{cases} \kappa \exp\left\{-\frac{1}{2\sigma_y^2}(y-\theta)^2\right\} & -a_{gd} \le y-\theta \le a_{gd} \\ \kappa \exp\left\{\frac{a_{gd}^2}{2\sigma_y^2}\right\} \exp\left\{-\frac{1}{2\sigma_y^2}|y-\theta| a_{gd}\right\} & |y-\theta| > a_{gd} \end{cases} \tag{3.244}$$

Taking the logarithm of the PDF yields

$$\ln\left(f_y(y)\right) = \begin{cases} \ln(\kappa) - \frac{1}{2\sigma_y^2}(y-\theta)^2 & -a_{gd} \le y-\theta \le a_{gd} \\ \ln(\kappa) + \frac{a_{gd}^2}{2\sigma_y^2} - \frac{a_{gd}}{2\sigma_y^2}|y-\theta| & |y-\theta| > a_{gd} \end{cases} \tag{3.245}$$

Differentiating with respect to θ yields

$$\frac{\delta}{\delta\theta}\ln(f_y(y)) = \begin{cases} \frac{1}{\sigma_y^2}(y-\theta) & -a_{gd} \le y-\theta \le a_{gd} \\ \frac{a_{gd}}{\sigma_y^2} sign(y-\theta) & |y-\theta| > a_{gd} \end{cases} \tag{3.246}$$

The Fisher information is:

$$I_F = E\left[\left(\frac{\partial \ln(f_y(y;\theta))}{\partial\theta}\right)^2\right] = \int_{-\infty}^{\infty}\left(\frac{\partial \ln(f_y(y;\theta))}{\partial\theta}\right)^2 f_y(y)dy \tag{3.247}$$

Splitting the integration interval over Υ_{gd} and Υ_{bd} yields

$$I_F = \int_{-a_{gd}}^{a_{gd}}\left(\frac{\partial \ln(f_y(y;\theta))}{\partial\theta}\right)^2 f_y(y)dy + \int_{|y-\theta|>a_{gd}}\left(\frac{\partial \ln(f_y(y;\theta))}{\partial\theta}\right)^2 f_y(y)dy \tag{3.248}$$

Using Eq. (3.246) we get

$$I_F = \frac{1}{\sigma_y^4}\int_{-a_{gd}}^{a_{gd}}(y-\theta)^2 f_y(y)dy + \frac{a_{gd}^2}{\sigma_y^4}\int_{|y-\theta|>a_{gd}} f_y(y)dy \tag{3.249}$$

Since $\int_{-\infty}^{\infty} f_y(y)dy = 1$ we get

$$\int_{|y-\theta|>a_{gd}} f_y(y)dy = 1 - \int_{-a_{gd}}^{a_{gd}} f_y(y)dy \tag{3.250}$$

Hence the expression for Fisher information becomes

$$I_F = \frac{1}{\sigma_y^4} \int_{-a_{gd}}^{a_{gd}} (y-\theta)^2 f_y(y)dy + \frac{a_{gd}^2}{\sigma_y^4} \left(1 - \int_{-a_{gd}}^{a_{gd}} f_y(y)dy \right) \tag{3.251}$$

Simplifying we get

$$I_F = \frac{1}{\sigma_y^4} \left(\sigma_{gd}^2 + a_{gd}^2 \left(1 - \lambda_{gd} \right) \right) \tag{3.252}$$

where $\sigma_{gd}^2 = \int_{y=\theta-a_{gd}}^{y=\theta+a_{gd}} (y-\theta)^2 f_y(y)dy$ is called "partial variance" and $\lambda_{gd} = \int_{-a_{gd}}^{a_{gd}} f_y(y)dy$ is called "partial probability" of y over the region Υ_{gd}.

Problem

3.1 Consider the same example of Gaussian measurements. Verify whether the following estimator $\hat{\theta}$ is (i) biased, (ii) efficient, (iii) asymptotically biased and asymptotically efficient.

References

[1] Doraiswami, R. (1976) A decision theoretic approach to parameter estimation. *IEEE Transactions on Automatic Control*, **21**(6), 860–866.

[2] Huber, P.J. (1964). Robust estimation of location parameter. *Annals of Mathematical Statistics*, (35), 73–102.

[3] Kay, S.M. (1993) *Fundamentals of Signal Processing: Estimation theory*, Prentice Hall, New Jersey.

Further Readings

Haykin, S. (2001) *Adaptive Filter Theory*, Prentice Hall, New Jersey.

Mendel, J. (1995) *Lessons in Estimation Theory in Signal Processing, Communications and Control*, Prentice-Hall, New Jersey.

Mitra, S.K. (2006) *Digital Signal Processing: A Computer-Based Approach*, McGraw Hill Higher Education, Boston.

Moon, T.K. and Stirling, W.C. (2000) *Mathematical Methods and Algorithms for Signal Processing*, Prentice Hall, NJ.

Mix, D.F. (1995) *Random Signal Processing*, Prentice Hall, New Jersey.

Olofsson, P. (2005). *Probability, Statistics and Stochastic Process*, John Wiley and Sons, New Jersey.

Oppenheim, A.V. and Schafer, R.W. (2010) *Discrete-Time Signal Processing*, Prentice-Hall, New Jersey.

Proakis, J.G. and Manolakis, D.G. (2007) *Digital Signal Processing: Principles, Algorithms and Applications*, Prentice Hall, New Jersey.

4

Estimation of Random Parameter

4.1 Overview

In this chapter, estimation of a random parameter is considered. The measurement model relating the random parameter and the measurement is assumed to be linear. The measurement is corrupted by an additive zero-mean white noise. The estimate of the unknown random parameter is obtained by minimizing mean-square error (MSE), which is the expectation of the error between the parameter and its estimate in the mean-square error sense. The estimator is termed as the minimum mean-square (MMSE) estimator. The term MMSE specifically refers to estimation in a Bayesian setting with quadratic cost function. Unlike non-random parameter estimation schemes, such as the maximum likelihood, the Bayesian scheme deals with estimation of unknown parameters which are treated as random variables. The PDF including the mean values of the unknown parameters is assumed to be known *a priori*. If the PDF is assumed to be Gaussian, then the mean and the covariance of the unknown parameters are assumed to known. The Bayesian approach, based directly on Bayes' theorem, provides a probabilistic setting so that the *a priori* known PDF of the random parameters is incorporated into the estimator.

One may question the requirement of the *a priori* knowledge of the PDF, especially the mean. Consider the problem of estimating an unknown parameter which varies randomly. The unknown random parameters may be modeled (probabilistically) by a PDF. For example if the PDF is assumed to Gaussian, the mean and the covariance may be estimated by performing a number of experiments using, say, the maximum likelihood method. Using this *a priori* knowledge, the estimate of the actual parameter is obtained using a Bayesian-based MMSE approach. Thus the MMSE estimator incorporates the *a priori* knowledge of the probabilistic model and the *a posteriori* information provided by the measurement. This approach of fusing the information provided by the model and the measurement forms the backbone of the Kalman filter estimation scheme.

The derivation of MMSE and its properties is based on seminal works of [1, 2].

4.2 Minimum Mean-Squares Estimator (MMSE): Scalar Case

Let the measurement model be

$$y = ax + v \tag{4.1}$$

Identification of Physical Systems: Applications to Condition Monitoring, Fault Diagnosis, Soft Sensor and Controller Design, First Edition. Rajamani Doraiswami, Chris Diduch and Maryhelen Stevenson.

where x is a zero-mean Gaussian process with variance, σ_x^2, v is a zero-mean white Gaussian noise with variance, σ_v^2, a is some constant, and x, v, and y are all scalars. The stochastic processes x and v are uncorrelated. We will obtain the expressions for the following:

- The Minimum Mean-Squared Error (MMSE) estimate $\hat{x}$ of x.
- The estimation error.

Consider the problem of estimating the MMSE estimate $\hat{x}$ from the following minimization of the mean-squared estimation error (MMSE):

$$\min_{\hat{x}}\{E[(x-\hat{x})^2]\} \tag{4.2}$$

The solution to this problem is obviously $\hat{x}=x$ since with this choice the performance measure $E[(x-\hat{x})^2]=0$, which is the minimum value:

$$\hat{x}=x \tag{4.3}$$

However, the above solution is meaningless. If we know x why should we estimate it! In other words the problem is ill posed. Let us now restrict the MMSE estimator to be $\hat{x}=\alpha$ where α is a constant, which is to be determined from the minimization problem (4.2). Substituting $\hat{x}=\alpha$ the minimization problem becomes:

$$\min_{\alpha}\{E[(x-\hat{x})^2]=E[x^2]-2\alpha E[x]+\alpha^2\} \tag{4.4}$$

Differentiating J with respect to α and setting it to zero yields

$$\hat{x}=E\,[x] \tag{4.5}$$

This estimator makes sense if the measurement y is unavailable: with no other information about the unknown x available, the best estimate is $\hat{x}=E[x]$.

4.2.1 Conditional Mean: Optimal Estimator

Let us now consider the practical case where the measurement y is available so that the estimator is restricted as a function only of the measurement y. This restriction, namely $\hat{x}$ is a function only of y, may be imposed simply by modifying the performance measure to be a conditional expectation of the mean squared estimation error given y. The modified MMSE error problem becomes:

$$\min_{\hat{x}}\{E[(x-\hat{x}(y))^2|y]\} \tag{4.6}$$

Expanding the conditional expectation measure yields:

$$E[(x-\hat{x}(y))^2|y]=E[x^2+\hat{x}^2(y)-2x\hat{x}(y)|y] \tag{4.7}$$

Simplifying yields

$$E[(x-\hat{x}(y))^2|y]=E[x^2|y]+E[\hat{x}^2(y)|y]-2E[x\hat{x}(y)|y] \tag{4.8}$$

Since $\hat{x}(y)$ and $\hat{x}^2(y)$ are functions only of y, using $E[\varphi(y)|y] = \varphi(y)$ where $\varphi(y)$ is a function only of y, the performance measure simplifies to:

$$E[(x - \hat{x}(y))^2|y] = E[x^2|y] + \hat{x}^2(y) - 2\hat{x}(y)E[x|y] \tag{4.9}$$

The MMSE estimator $\hat{x}(y)$ is obtained by differentiating J with respect to $\hat{x}(y)$ and setting to zero:

$$\frac{\delta}{\delta x}\{E[x^2|y] + \hat{x}^2(y) - 2\hat{x}(y)E[x|y]\} = 0 \tag{4.10}$$

Differentiating and simplifying yields

$$E[\hat{x}(y)] = E[E[x|y]] = E[x] \tag{4.11}$$

$$\hat{x}(y) = E\,[x|\,y] \tag{4.12}$$

This shows that the conditional mean $E[x|y]$ is an optimal estimator $\hat{x}(y)$ of x when we restrict the estimator to be a class of all estimators which are a function of the measurement. Let us verify whether the MMSE estimator $\hat{x}(y)$ is unbiased. Taking the expectation on both sides of Eq. (4.12) and using $E[E[x|y]] = E[x]$ we get

$$E\,[\hat{x}(y)] = E\,[E\,[x|\,y]] = E\,[x] \tag{4.13}$$

Since $E[\hat{x}(y)] = E[x]$, the MMSE estimator $\hat{x}(y)$ is an unbiased estimator of x. Further the minimum mean-square error $E[(x - \hat{x}(y))^2|y]$ is the conditional variance $\hat{x}(y)$ of x given y.

4.3 MMSE Estimator: Vector Case

Linear Measurement Model

We will assume the measurement model relating the unknown $M \times 1$ parameter $\boldsymbol{x} = [x(0)\quad x(1)\quad x(2)\quad .\quad x(N-1)]^T$, and the $N \times 1$ measurement $\boldsymbol{y} = [y(0)\quad y(1)\quad y(2)\quad .\quad y(N-1)]^T$ to be linear given by:

$$\boldsymbol{y} = \boldsymbol{H}\boldsymbol{x} + \boldsymbol{v} \tag{4.14}$$

where $\boldsymbol{v}$ is $N \times 1$ zero-mean measurement noise with covariance $\boldsymbol{\Sigma}_v$ and $\boldsymbol{H}$ is a $N \times M$ matrix.

Performance Measure

Let $\hat{\boldsymbol{x}}$ be $M \times 1$ minimum mean-squared error estimator of $\boldsymbol{x}$ with a resulting estimation error $\boldsymbol{x} - \hat{\boldsymbol{x}}$. For notational convenience, the dependence on $\boldsymbol{y}$ in $\hat{\boldsymbol{x}}(\boldsymbol{y})$ is suppressed and is indicated by $\hat{\boldsymbol{x}}$. Taking a cue from the previous section to ensure that the estimator is a function only of the measurement $\boldsymbol{y}$, thereby avoiding an impractical solution, the performance measure is chosen to be a conditional mean-squared estimation error rather than a mean-squared estimation error. The conditional mean-squared estimation error measure is a $M \times M$ positive semi-definite matrix denoted $\boldsymbol{J}(\boldsymbol{y})$ given by

$$\boldsymbol{J}(\boldsymbol{y}) = E[(\boldsymbol{x} - \hat{\boldsymbol{x}})(\boldsymbol{x} - \hat{\boldsymbol{x}})^T|\boldsymbol{y}] \tag{4.15}$$

An estimate $\hat{\boldsymbol{x}}(\boldsymbol{y})$ is obtained from the minimization of $\boldsymbol{J}(\boldsymbol{y})$:

$$\min_{\hat{x}} \boldsymbol{J}(\boldsymbol{y}) \tag{4.16}$$

The definition of minimization of a positive definite matrix performance measure $\boldsymbol{J}(y)$ indicated by Eq. (4.16) is given below:

Definition *A positive definite matrix* $\boldsymbol{J}_0$ *is the minimum value of the positive definite matrix* $\boldsymbol{J}$ *if the following matrix inequality holds:*

$$\boldsymbol{J}_0 - \boldsymbol{J} \leq 0 \; for \; all \; \boldsymbol{J} \tag{4.17}$$

where the notation $\boldsymbol{J}_0 - \boldsymbol{J} \leq 0$ *implies that* $\boldsymbol{J}_0 - \boldsymbol{J}$ *is a negative semi-definite matrix.*

Expanding the conditional MMSE (4.15) we get

$$E[(\boldsymbol{x}-\hat{\boldsymbol{x}})(\boldsymbol{x}-\hat{\boldsymbol{x}})^T|\boldsymbol{y}] = E[\boldsymbol{x}\boldsymbol{x}^T|\boldsymbol{y}] + E[\hat{\boldsymbol{x}}\hat{\boldsymbol{x}}^T|\boldsymbol{y}] - E[\boldsymbol{x}\hat{\boldsymbol{x}}^T|\boldsymbol{y}] - E[\hat{\boldsymbol{x}}\boldsymbol{x}^T|\boldsymbol{y}] \tag{4.18}$$

Using the property of conditional expectation $E[\hat{\boldsymbol{x}}|\boldsymbol{y}] = \hat{\boldsymbol{x}}$, namely $E[\hat{\boldsymbol{x}}\boldsymbol{x}^T|\boldsymbol{y}] = \hat{\boldsymbol{x}}E[\boldsymbol{x}^T|\boldsymbol{y}]$; $E[\boldsymbol{x}\hat{\boldsymbol{x}}^T|\boldsymbol{y}] = E[\boldsymbol{x}|\boldsymbol{y}]\hat{\boldsymbol{x}}^T$ we get

$$E[(\boldsymbol{x}-\hat{\boldsymbol{x}})(\boldsymbol{x}-\hat{\boldsymbol{x}})^T|\boldsymbol{y}] = E[\boldsymbol{x}\boldsymbol{x}^T|\boldsymbol{y}] + \hat{\boldsymbol{x}}\hat{\boldsymbol{x}}^T - E[\boldsymbol{x}|\boldsymbol{y}]\hat{\boldsymbol{x}}^T - \hat{\boldsymbol{x}}E[\boldsymbol{x}^T|\boldsymbol{y}] \tag{4.19}$$

The problem of minimization of the conditional expectation Eq. (4.16) becomes

$$\min_{\hat{\boldsymbol{x}}}\{E[\boldsymbol{x}\boldsymbol{x}^T|\boldsymbol{y}] + \hat{\boldsymbol{x}}\hat{\boldsymbol{x}}^T - E[\boldsymbol{x}|\boldsymbol{y}]\hat{\boldsymbol{x}}^T - \hat{\boldsymbol{x}}E[\boldsymbol{x}^T|\boldsymbol{y}]\} \tag{4.20}$$

There are two approaches to minimization of the above expression:

- Completing the square approach.
- Vector calculus approach.

Completing the Square Approach

Let us first employ the complete square approach. The last three terms on the right-hand side is a vector version of $a^2 - 2ab$ where a and b respectively takes the role of $\hat{\boldsymbol{x}}$ and $E[\boldsymbol{x}|\boldsymbol{y}(E[\boldsymbol{x}^T|\boldsymbol{y}])$; a^2 and $2ab$ the role of $\hat{\boldsymbol{x}}\hat{\boldsymbol{x}}^T$ and $E[\boldsymbol{x}|\boldsymbol{y}]\hat{\boldsymbol{x}}^T + \hat{\boldsymbol{x}}E[\boldsymbol{x}^T|\boldsymbol{y}]$ respectively.

Adding and subtracting b^2 so as to complete the square we get, $a^2 - 2ab = a^2 - 2ab + b^2 - b^2 = (a-b)^2 - b^2$. We will use this approach to our problem.

Adding and subtracting $E[\boldsymbol{x}|\boldsymbol{y}]E[\boldsymbol{x}^T|\boldsymbol{y}]$ to the expression for $\boldsymbol{J}$ in Eq. (4.19) we get

$$\begin{aligned} \boldsymbol{J} = E[\boldsymbol{x}\boldsymbol{x}^T|\boldsymbol{y}] + \hat{\boldsymbol{x}}\hat{\boldsymbol{x}}^T - E[\boldsymbol{x}|\boldsymbol{y}]\hat{\boldsymbol{x}}^T - \hat{\boldsymbol{x}}E[\boldsymbol{x}^T|\boldsymbol{y}] \\ + E[\boldsymbol{x}|\boldsymbol{y}]E[\boldsymbol{x}^T|\boldsymbol{y}] - E[\boldsymbol{x}|\boldsymbol{y}]E[\boldsymbol{x}^T|\boldsymbol{y}] \end{aligned} \tag{4.21}$$

Completing the square similar to $((a-b)^2 - b^2)$ on the right-hand side we get

$$\boldsymbol{J} = E[\boldsymbol{x}\boldsymbol{x}^T|\boldsymbol{y}] + (\hat{\boldsymbol{x}} - E[\boldsymbol{x}|\boldsymbol{y}])(\hat{\boldsymbol{x}} - E[\boldsymbol{x}|\boldsymbol{y}])^T - E[\boldsymbol{x}|\boldsymbol{y}]E[\boldsymbol{x}^T|\boldsymbol{y}] \tag{4.22}$$

The minimization problem (4.20) reduces to finding an estimate $\hat{\boldsymbol{x}}$ which minimizes the expression on the right-hand side of Eq. (4.22). The performance measure $\boldsymbol{J}$ will be minimum if we choose $\hat{\boldsymbol{x}}$ so that the second term on the right-hand side is zero. Hence the MMSE estimate, $\hat{\boldsymbol{x}}$ is

$$\hat{\boldsymbol{x}} = E[\boldsymbol{x}|\boldsymbol{y}] \tag{4.23}$$

Vector Calculus Approach

Consider the minimization of the conditional covariance $\boldsymbol{J}$ given by Eq. (4.19). The optimal estimate $\hat{\boldsymbol{x}}$ is the solution of the partial derivative of $\boldsymbol{J}$:

$$\frac{\delta}{\delta\hat{\boldsymbol{x}}}\{\boldsymbol{J}\} = \frac{\delta}{\delta\hat{\boldsymbol{x}}}\{E[\boldsymbol{x}\boldsymbol{x}^T|\boldsymbol{y}] + \hat{\boldsymbol{x}}\hat{\boldsymbol{x}}^T - E[\boldsymbol{x}|\boldsymbol{y}]\hat{\boldsymbol{x}}^T - \hat{\boldsymbol{x}}E[\boldsymbol{x}^T|\boldsymbol{y}]\} = 0 \quad (4.24)$$

where a partial differentiation of a matrix using the notation $\frac{\delta}{\delta\hat{\boldsymbol{x}}}\{\boldsymbol{J}\} = 0$ implies

$$\frac{\delta}{\delta\hat{x}_i}\{\boldsymbol{J}\} = 0, \quad i = 1, 2, 3, \ldots, M \quad (4.25)$$

Using this definition yields

$$\frac{\delta}{\delta\hat{x}_i}\{\boldsymbol{J}\} = \frac{\delta}{\delta\hat{x}_i}\{\hat{\boldsymbol{x}}\hat{\boldsymbol{x}}^T\} - \frac{\delta}{\delta\hat{x}_i}\{E[\boldsymbol{x}|\boldsymbol{y}]\hat{\boldsymbol{x}}^T\} - \frac{\delta}{\delta\hat{x}_i}\{\hat{\boldsymbol{x}}E[\boldsymbol{x}^T|\boldsymbol{y}]\} = 0 \quad (4.26)$$

Simplifying we get

$$\frac{\delta}{\delta\hat{\boldsymbol{x}}}\{\boldsymbol{J}\} = 2\hat{\boldsymbol{x}} - 2E[\boldsymbol{x}|\boldsymbol{y}] = 0 \quad (4.27)$$

Hence the MMSE estimator is given by

$$\hat{\boldsymbol{x}} = E[\boldsymbol{x}|\boldsymbol{y}] \quad (4.28)$$

Unbiased Estimator

Let us verify whether $\hat{\boldsymbol{x}} = E[\boldsymbol{x}|\boldsymbol{y}]$ is unbiased. Taking expectation of Eq. (4.23) yields

$$E[\hat{\boldsymbol{x}}] = E[E[\boldsymbol{x}|\boldsymbol{y}]] \quad (4.29)$$

Since $E[E[\boldsymbol{x}|\boldsymbol{y}]] = E[\boldsymbol{x}]$ we get

$$E[\hat{\boldsymbol{x}}] = E[\boldsymbol{x}] \quad (4.30)$$

4.3.1 Covariance of the Estimation Error

Since $\hat{\boldsymbol{x}} = E[\boldsymbol{x}|\boldsymbol{y}]$ is unbiased, the performance measure $\boldsymbol{J} = E[(\boldsymbol{x} - \hat{\boldsymbol{x}})(\boldsymbol{x} - \hat{\boldsymbol{x}})^T|\boldsymbol{y}]$ is the conditional covariance of the estimation error, and its minimum value $\boldsymbol{J}_0$ is obtained by substituting $\hat{\boldsymbol{x}} = E[\boldsymbol{x}|\boldsymbol{y}]$ in the expression for $\boldsymbol{J}$ given in Eq. (4.22). The minimum conditional covariance is:

$$\boldsymbol{J}_0 = E[\boldsymbol{x}\boldsymbol{x}^T|\boldsymbol{y}] - E[\boldsymbol{x}|\boldsymbol{y}]E[\boldsymbol{x}^T|\boldsymbol{y}] \quad (4.31)$$

The covariance of the estimator $\hat{\boldsymbol{x}} = E[\boldsymbol{x}|\boldsymbol{y}]$, denoted $\text{cov}(\hat{\boldsymbol{x}}) = E[(\boldsymbol{x} - \hat{\boldsymbol{x}})(\boldsymbol{x} - \hat{\boldsymbol{x}})^T]$, is obtained by taking the expectation of $\boldsymbol{J}_0$. The covariance $\text{cov}(\hat{\boldsymbol{x}})$ becomes:

$$\text{cov}(\hat{\boldsymbol{x}}) = E[\boldsymbol{J}_0] = E[E[\boldsymbol{x}\boldsymbol{x}^T|\boldsymbol{y}]] - E[E[\boldsymbol{x}|\boldsymbol{y}]E[\boldsymbol{x}^T|\boldsymbol{y}]] \quad (4.32)$$

Simplifying using $E[E[\boldsymbol{x}\boldsymbol{x}^T|\boldsymbol{y}]] = E[\boldsymbol{x}\boldsymbol{x}^T]$ we get

$$\text{cov}(\hat{\boldsymbol{x}}) = E[\boldsymbol{x}\boldsymbol{x}^T] - E[E[\boldsymbol{x}|\boldsymbol{y}]E[\boldsymbol{x}^T|\boldsymbol{y}]] \quad (4.33)$$

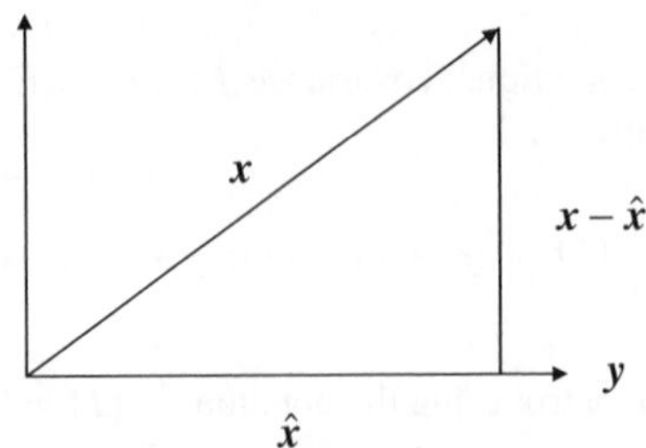

Figure 4.1 The orthogonal relation between the estimate and the estimation error

4.3.2 Conditional Expectation and Its Properties

The unconditional and conditional expectations satisfy the following relation [1]:

$$E[\mathbf{x}g^T(\mathbf{y})] = E[E[\mathbf{x}|\mathbf{y}]\mathbf{g}^T(\mathbf{y})] \tag{4.34}$$

For a scalar case we get

$$E[xg(y)] = E[E[x|y]g(y)] \tag{4.35}$$

Orthogonality Principle

The MMSE estimator $\hat{\mathbf{x}} = E[\mathbf{x}|\mathbf{y}]$ of $\mathbf{x}$ is orthogonal to the estimation error $x - E[\mathbf{x}|\mathbf{y}]$:

$$E[(\mathbf{x} - E[\mathbf{x}|\mathbf{y}])(E[\mathbf{x}|\mathbf{y}])^T] = 0 \tag{4.36}$$

Proof: Expanding Eq. (4.36) yields

$$E[(\mathbf{x} - E[\mathbf{x}|\mathbf{y}])(E[\mathbf{x}|\mathbf{y}])^T] = E[\mathbf{x}(E[\mathbf{x}|\mathbf{y}])^T] - E[E[\mathbf{x}|\mathbf{y}](E[\mathbf{x}|\mathbf{y}])^T] \tag{4.37}$$

Using Eq. (4.34) on the first term becomes

$$E[\mathbf{x}(E[\mathbf{x}|\mathbf{y}])^T] = E[E[\mathbf{x}|\mathbf{y}](E[\mathbf{x}|\mathbf{y}])^T] \tag{4.38}$$

Substituting Eq. (4.38) in Eq. (4.37) we get Eq. (4.36). For the scalar case the result Eq. (4.36) simplifies to the following orthogonal relation:

$$E[(x - E[x|y])E[x|y]] = 0 \tag{4.39}$$

The orthogonal relation between the MMSE estimator $\hat{\mathbf{x}}$ and the estimation error $\mathbf{x} - \hat{\mathbf{x}}$ is shown in Figure 4.1.

4.4 Expression for Conditional Mean

We will derive an expression for conditional mean when (i) the measurement is linear and (ii) $\mathbf{x}$ and $\mathbf{v}$ are independent. Consider the minimum mean-squared estimator $\hat{\mathbf{x}}$ given by Eq. (4.28). Expressing in terms of the conditional PDF yields:

$$\hat{\mathbf{x}} = E[\mathbf{x}|\mathbf{y}] = \int_{-\infty}^{\infty} \mathbf{x} f(\mathbf{x}|\mathbf{y})\, d\mathbf{x} \tag{4.40}$$

where $f(\boldsymbol{x}|\boldsymbol{y})$ can be expressed either in terms of (i) joint PDF $f(\boldsymbol{x},\boldsymbol{y})$ and marginal PDF $f_y(\boldsymbol{y})$, or (ii) the conditional PDF $f(\boldsymbol{y}|\boldsymbol{x})$ and the marginal PDFs $f_x(\boldsymbol{x})$ and $f_y(\boldsymbol{y})$ as:

$$f(\boldsymbol{x}|\boldsymbol{y}) = \frac{f(\boldsymbol{x},\boldsymbol{y})}{f_y(\boldsymbol{y})} = \frac{f(\boldsymbol{y}|\boldsymbol{x})f_x(\boldsymbol{x})}{f_y(\boldsymbol{y})} \tag{4.41}$$

Since x and v are independent, using the linear model (4.14) the expression for the conditional PDF $f(\boldsymbol{y}|\boldsymbol{x})$ and joint PDF $f(\boldsymbol{x},\boldsymbol{y})$ in terms of noise PDF $f_v(\boldsymbol{v})$ become

$$\begin{aligned} f(\boldsymbol{y}|\boldsymbol{x}) &= f_v(\boldsymbol{y} - \boldsymbol{H}\boldsymbol{x}) \\ f(\boldsymbol{x},\boldsymbol{y}) &= f_v(\boldsymbol{y} - \boldsymbol{H}\boldsymbol{x})f_x(\boldsymbol{x}) \end{aligned} \tag{4.42}$$

Substituting for $f(\boldsymbol{y}|\boldsymbol{x})$ in Eq. (4.41) the $f(\boldsymbol{x}|\boldsymbol{y})$ becomes

$$f(\boldsymbol{x}|\boldsymbol{y}) = \frac{f_v(\boldsymbol{y} - \boldsymbol{H}\boldsymbol{x})f_x(\boldsymbol{x})}{f_y(\boldsymbol{y})} \tag{4.43}$$

Substituting for $f(\boldsymbol{x}|\boldsymbol{y})$ in the expression for $\hat{\boldsymbol{x}}$ in Eq. (4.40) we get:

$$\hat{\boldsymbol{x}} = E[\boldsymbol{x}|\boldsymbol{y}] = \frac{\int_{-\infty}^{\infty} \boldsymbol{x} f_v(\boldsymbol{y} - \boldsymbol{H}\boldsymbol{x})f_x(\boldsymbol{x})d\boldsymbol{x}}{f_y(\boldsymbol{y})} \tag{4.44}$$

where the marginal PDF $f_y(\boldsymbol{y})$ of $\boldsymbol{y}$ is:

$$f_y(\boldsymbol{y}) = \int_{-\infty}^{\infty} f_x(\boldsymbol{x})f_v(\boldsymbol{y} - \boldsymbol{H}\boldsymbol{x})d\boldsymbol{x} \tag{4.45}$$

The determination of the MMSE estimator $\hat{\boldsymbol{x}}$ is computationally burdensome as it involves integration operations given by Eqs. (4.44) and (4.45). We will consider two special cases where it may be more convenient to compute the estimator:

- The PDF of the unknown parameter $\boldsymbol{x}$ is Gaussian.
- All the PDFs are Gaussian.

4.4.1 MMSE Estimator: Gaussian Random Variables

The Gaussian PDF models random variables which are clustered close to the mean as it has very thin tails representing "good data" with negligible outliers. A non-Gaussian PDF, such as a Laplacian or a Cauchy with thicker tails, models "bad data" with many outliers. The bad data points are due to faulty instruments, unexpected failure of communication links, intermittent faults, and modeling errors, to name a few.

The measurement $y = \boldsymbol{H}\boldsymbol{x} + \boldsymbol{v}$ is commonly used to model an unknown random parameter $\boldsymbol{x}$ clustered close to the mean ("good signal") corrupted by measurement noise $\boldsymbol{v}$ which contain outliers ("bad noise"). As a result the measurement random variable $\boldsymbol{y}$ will be "bad measurement." Hence in practice it may be reasonable to assume the PDF of $\boldsymbol{x}$ to be Gaussian and that of $\boldsymbol{y}$ to be non-Gaussian. The expression for the MMSE estimator, namely the conditional expectation of $\boldsymbol{x}$ given $\boldsymbol{y}$, is considerably simplified thanks to the property of a Gaussian PDF of $\boldsymbol{x}$ as the ratio between the gradient of the PDF and the PDF itself is linear function of $\boldsymbol{x}$. The MMSE estimator, as well as the covariance of the estimation error is only a function of the PDF of the measurement $\boldsymbol{y}$.

4.4.2 MMSE Estimator: Unknown is Gaussian and Measurement Non-Gaussian

4.4.2.1 Scalar Case

We will first consider the scalar version and then extend to the vector case. A scalar version for the expression for the conditional mean given by Eq. (4.44) is:

$$E[x|y] = \frac{\int\limits_{-\infty}^{\infty} x f_v(y - ax) f_x(x)\, dx}{f_y(y)} \tag{4.46}$$

The Gaussian PDF of x is:

$$f_x(x) = \frac{1}{\sigma_x \sqrt{2\pi}} \exp\left\{ -\frac{x^2}{2\sigma_x^2} \right\} \tag{4.47}$$

Differentiating with respect to x yields:

$$f_x'(x) = -\frac{x^2}{\sigma_x^2} \left(\frac{1}{\sigma_x \sqrt{2\pi}} \exp\left\{ -\frac{x^2}{2\sigma_x^2} \right\} \right) = -\frac{x}{\sigma_x^2} f_x(x) \tag{4.48}$$

where $f_x'(x) = \frac{d}{dx} f_x(x)$. Using Eq. (4.48), substituting for $x f_x(x)$ in the integrand of the expression on the right-hand side of Eq. (4.46) yields

$$E[x|y] = \frac{-\sigma_x^2 \int\limits_{-\infty}^{\infty} f_v(y - ax) f_x'(x)\, dx}{f_y(y)} \tag{4.49}$$

Lemma 4.1 *If (i) $y = hx + v$, (ii) x and v are zero-mean random variables, (iii) x and v are independent, (iv) the PDFs $f_x(x)$, $f_y(y)$, and $f_v(v)$ are symmetric about the mean and absolutely continuous, and x is a Gaussian random variable with PDF $f_x(x) = \frac{1}{\sqrt{2\pi\sigma_x^2}} \exp\left\{ -\frac{x^2}{2\sigma_x^2} \right\}$, then*

$$\hat{x} = E[x|y] = -\sigma_x^2 h \left(\frac{f_y'(y)}{f_y(y)} \right) \tag{4.50}$$

where $f_y'(y) = \frac{df_y(y)}{dy}$

Proof: The proof is given in the Appendix.

Variance of the MMSE Estimator

The conditional variance of the MMSE estimator $\hat{x}$ given by Eq. (4.31) becomes

$$\text{var}(\hat{x}|y) = E[(x - \hat{x})^2 | y] \tag{4.51}$$

Expanding using Eq. (4.31)

$$\operatorname{var}(\hat{x}|y) = E[x^2|y] - E[\hat{x}^2|y] \tag{4.52}$$

Since $\hat{x}$ is a function only of y, $E[\hat{x}^2|y] = \hat{x}^2$. Hence

$$\operatorname{var}(\hat{x}|y) = E[x^2|y] - \sigma_x^4 h^2 \left(\frac{f_y'(y)}{f_y(y)}\right)^2 \tag{4.53}$$

The variance of the MMSE estimator is obtained by taking expectation of the conditional variance (4.53) and is given by

$$\operatorname{var}(\hat{x}) = \sigma_x^2 - \sigma_x^4 h^2 E\left[\left(\frac{f_y'(y)}{f_y(y)}\right)^2\right] \tag{4.54}$$

We will express the $\operatorname{var}(\hat{x})$ in terms of the Fisher information I_F:

$$I_F = E\left[\left(\frac{\partial \ln\left(f_y(y;x)\right)}{\partial x}\right)^2\right] \tag{4.55}$$

Using the linear model $y = hx + v$ we get

$$I_F = h^2 E\left[\left(\frac{f_y'(y)}{f_y(y)}\right)^2\right] \tag{4.56}$$

Using the expression for I_F, $\operatorname{var}(\hat{x})$ can be expressed in terms of I_F

$$\operatorname{var}(\hat{x}) = \sigma_x^2 - \sigma_x^4 I_F \tag{4.57}$$

4.4.2.2 Vector Case

We will now consider the vector case

Lemma 4.2 *If (i)* $\mathbf{y} = \mathbf{H}\mathbf{x} + \mathbf{v}$, *(ii)* $\mathbf{x}$ *and* $\mathbf{v}$ *are zero-mean random variables, (iii)* $\mathbf{x}$ *and* $\mathbf{v}$ *are independent, (iv) the PDFs* $f_x(\mathbf{x})$, $f_y(\mathbf{y})$, *and* $f_v(\mathbf{v})$ *are symmetric about the mean and absolutely continuous, and* $\mathbf{x}$ *is a Gaussian random variable with PDF* $f_x(\mathbf{x}) = \dfrac{1}{\sqrt{(2\pi)^N|\mathbf{\Sigma}_x|}}\exp\left\{-\dfrac{1}{2}\mathbf{x}\,\mathbf{\Sigma}_x^{-1}\mathbf{x}^T\right\}$, *then*

$$E[\mathbf{x}|\mathbf{y}] = -\Sigma_x \mathbf{H}^T \left(\frac{f_y'(\mathbf{y})}{f_y(\mathbf{y})}\right) \tag{4.58}$$

where $f_y'(\mathbf{y}) = \dfrac{df_y(\mathbf{y})}{d\mathbf{y}} = \left[\dfrac{df_y(\mathbf{y})}{dy(0)} \quad \dfrac{df_y(\mathbf{y})}{dy(1)} \quad \dfrac{df_y(\mathbf{y})}{dy(2)} \quad . \quad \dfrac{df_y(\mathbf{y})}{dy(N-1)}\right]^T$

Proof: The proof is given in the Appendix.

Covariance of the MMSE Estimator

Using the expression for the covariance of the MMSE estimator (4.33) we get

$$\mathrm{cov}(\hat{\boldsymbol{x}}) = \boldsymbol{\Sigma}_x - \boldsymbol{\Sigma}_x \boldsymbol{H}^T E\left[\left(\frac{f_y'(\boldsymbol{y})}{f_y(\boldsymbol{y})}\right)\left(\frac{f_y'(\boldsymbol{y})}{f_y(\boldsymbol{y})}\right)^T\right]\boldsymbol{H}\boldsymbol{\Sigma}_x \tag{4.59}$$

We will express the $\mathrm{cov}(\hat{\boldsymbol{x}})$ in terms of the Fisher information matrix $\boldsymbol{I}_F$ given by

$$\boldsymbol{I}_F = E\left[\left(\frac{\partial \ln\left(f_y(\boldsymbol{y};\boldsymbol{x})\right)}{\partial \boldsymbol{x}}\right)\left(\frac{\partial \ln\left(f_y(\boldsymbol{y};\boldsymbol{x})\right)}{\partial \boldsymbol{x}}\right)^T\right] \tag{4.60}$$

Using the linear model $\boldsymbol{y} = \boldsymbol{H}\boldsymbol{x} + \boldsymbol{v}$ we get

$$\boldsymbol{I}_F = H^T E\left[\left(\frac{f_y'(\boldsymbol{y})}{f_y(\boldsymbol{y})}\right)\left(\frac{f_y'(\boldsymbol{y})}{f_y(\boldsymbol{y})}\right)^T\right] H \tag{4.61}$$

Expressing $\mathrm{cov}(\hat{\boldsymbol{x}})$ in terms of $\boldsymbol{I}_F$ we get

$$\mathrm{cov}(\hat{\boldsymbol{x}}) = \boldsymbol{\Sigma}_x - \boldsymbol{\Sigma}_x \boldsymbol{I}_F^{-1} \boldsymbol{\Sigma}_x \tag{4.62}$$

Remarks

1. The MMSE estimator $\hat{\boldsymbol{x}}$ is in general a nonlinear function of the measurement $\boldsymbol{y}$.
2. The estimator $\hat{\boldsymbol{x}}$ is a function of the PDF of the measurement $f_y(\boldsymbol{y})$ and the covariance of the random parameter $\boldsymbol{\Sigma}_x$.
3. The covariance of the MMSE estimator $\mathrm{cov}(\hat{\boldsymbol{x}})$ is a function of the PDF of the measurement $f_y(\boldsymbol{y})$ and the covariance of the random parameter $\boldsymbol{\Sigma}_x$.
4. The ratio of the gradient of the PDF to the PDF, $\dfrac{f_y'(\boldsymbol{y})}{f_y(\boldsymbol{y})}$, plays a crucial role in $\hat{\boldsymbol{x}}$ and its covariance $\mathrm{cov}(\hat{\boldsymbol{x}})$.
5. The covariance of the MMSE estimator is a function of the Fisher information.

4.4.3 The MMSE Estimator for Gaussian PDF

In most engineering applications, the PDFs of random variables are assumed to be Gaussian and the model is assumed to be linear.

Most physical systems are nonlinear. However, a linearized model about an operating point is employed successfully in many applications, such as mechatronic, process control, power, robotic, and aero-space systems. A linearized model captures nonlinear system behavior for small variations about an operating point.

A random process is a signal which is corrupted by an unwanted random waveform termed noise. The noise is generated in all electronic circuits and devices as a result of random fluctuations in current or voltage caused by the random movement of electrons. This noise is termed thermal noise or Johnson noise. The noise corrupting the signal at the macroscopic level is caused by the superposition of large numbers of thermal noise generated at the atomic level. In view of Central Limit Theorem, it may be justified to assume the noise corrupting the signal is Gaussian.

Thanks to advances in computers, microprocessors, digital signal processors, and dedicated processors, it is possible to store and process large data. In applications such as estimation, fault detection and isolation, and pattern classification, the sum of a large number of independent and identically distributed measurements are employed. The sum of i.i.d. random variables is Gaussian in view of Central Limit Theorem.

The other important reason for assuming a linear model and a Gaussian PDF is that solutions to many engineering problems in analysis, estimation, control, detection, and isolation become mathematically tractable. A closed form solution may be obtained, and in many cases the solution takes a linear form. For example, a controller, a filter, and an estimator are linear, while a discriminant function used in detection, isolation, and pattern classification is linear.

In this section we will give expressions of joint and conditional PDFs, expectation, conditional expectation, and conditional mean.

4.4.3.1 The Joint and Conditional Gaussian PDF

Let $\mathbf{x} = [x(0) \quad x(1) \quad x(2) \quad . \quad x(N-1)]^T$ and $\mathbf{y} = [\,y(0) \quad y(1) \quad y(2) \quad . \quad y(M-1)\,]^T$ be two $Nx1$ and $Mx1$ random variables. Let $\boldsymbol{\mu}_x = E[\mathbf{x}]$ be $Nx1$ mean vector and $\boldsymbol{\Sigma}_x = E[(\mathbf{x}-\boldsymbol{\mu}_x)(\mathbf{x}-\boldsymbol{\mu}_x)^T]$ be NxN covariance of $\mathbf{x}$; $\boldsymbol{\mu}_y = E[\mathbf{y}]$ is $Mx1$ mean and $\boldsymbol{\Sigma}_y = E[(\mathbf{y}-\boldsymbol{\mu}_y)(\mathbf{y}-\boldsymbol{\mu}_y)^T]$ is MxM covariance of $\mathbf{y}$; $\boldsymbol{\Sigma}_{xy} = E[(\mathbf{x}-\boldsymbol{\mu}_x)(\mathbf{y}-\boldsymbol{\mu}_y)^T]$ is NxM covariance and $\boldsymbol{\Sigma}_{yx} = \boldsymbol{\Sigma}_{xy}^T = E[(\mathbf{y}-\boldsymbol{\mu}_x)(\mathbf{x}-\boldsymbol{\mu}_x)^T]$ is MxN cross-covariance.

Let $\mathbf{z} = \begin{bmatrix} \mathbf{x} \\ \mathbf{y} \end{bmatrix}$ be a $(N+M)x1$ joint random variable formed of $\mathbf{x}$ and $\mathbf{y}$. The joint PDF of $\mathbf{z}$ denoted $f_{xy}(\mathbf{x},\mathbf{y})$ is:

$$f_{xy}(\mathbf{x},\mathbf{y}) = \frac{1}{\sqrt{(2\pi)^N |\boldsymbol{\Sigma}_z|}} \exp\left\{-\frac{1}{2}\left(\mathbf{z}-\boldsymbol{\mu}_z\right)^T \boldsymbol{\Sigma}_z^{-1}\left(\mathbf{z}-\boldsymbol{\mu}_z\right)\right\}, -\infty \le \mathbf{z} \le \infty \tag{4.63}$$

where

$$\boldsymbol{\mu}_z = \begin{bmatrix} \boldsymbol{\mu}_x \\ \boldsymbol{\mu}_y \end{bmatrix} \text{ is } (N+M)x1; \; \boldsymbol{\Sigma}_z = E\left[(\mathbf{z}-\boldsymbol{\mu}_z)(\mathbf{z}-\boldsymbol{\mu}_z)^T\right] \text{ is } (N+M)x(N+M); \; \boldsymbol{\Sigma}_z = \begin{bmatrix} \boldsymbol{\Sigma}_x & \boldsymbol{\Sigma}_{xy} \\ \boldsymbol{\Sigma}_{yx} & \boldsymbol{\Sigma}_y \end{bmatrix};$$

$$\boldsymbol{\Sigma}_z^{-1} = \begin{bmatrix} (\boldsymbol{\Sigma}_x - \boldsymbol{\Sigma}_{xy}\boldsymbol{\Sigma}_y^{-1}\boldsymbol{\Sigma}_{yx})^{-1} & -\boldsymbol{\Sigma}_x^{-1}\boldsymbol{\Sigma}_{xy}(\boldsymbol{\Sigma}_y - \boldsymbol{\Sigma}_{yx}\boldsymbol{\Sigma}_x^{-1}\boldsymbol{\Sigma}_{xy})^{-1} \\ -(\boldsymbol{\Sigma}_y - \boldsymbol{\Sigma}_{yx}\boldsymbol{\Sigma}_x^{-1}\boldsymbol{\Sigma}_{xy})^{-1}\boldsymbol{\Sigma}_{yx}\boldsymbol{\Sigma}_x^{-1} & (\boldsymbol{\Sigma}_y - \boldsymbol{\Sigma}_{yx}\boldsymbol{\Sigma}_x^{-1}\boldsymbol{\Sigma}_{xy})^{-1} \end{bmatrix}$$

$$|\boldsymbol{\Sigma}_z| = |\boldsymbol{\Sigma}_x||\boldsymbol{\Sigma}_y - \boldsymbol{\Sigma}_{yx}\boldsymbol{\Sigma}_x^{-1}\boldsymbol{\Sigma}_{xy}|$$

The conditional PDF $f(\mathbf{x}|\mathbf{y})$, which plays an important role in estimation of random parameter, may be expressed in terms of the joint and the marginal PDFs as:

$$f(\mathbf{x}|\mathbf{y}) = \frac{f_{xy}(\mathbf{x},\mathbf{y})}{f_y(\mathbf{y})} \tag{4.64}$$

The expression for $f(\mathbf{x}|\mathbf{y})$ becomes

$$f(\mathbf{x}|\mathbf{y}) = \frac{1}{\sqrt{(2\pi)^N \left|\boldsymbol{\Sigma}_{x|y}\right|}} \exp\left\{-\frac{1}{2}(\mathbf{x}-\boldsymbol{\mu}_{x|y})^T\boldsymbol{\Sigma}_{x|y}^{-1}(\mathbf{x}-\boldsymbol{\mu}_{x|y})\right\}, -\infty \le \mathbf{x} \le \infty \tag{4.65}$$

The conditional PDF $f(\mathbf{x}|\mathbf{y})$ is also multivariate Gaussian with mean $\boldsymbol{\mu}_{x|y}$ and conditional covariance $\boldsymbol{\Sigma}_{x|y}$.

4.4.3.2 Conditional Mean and Conditional Variance

The conditional mean is affine in $\mathbf{y}$ (linear with an additive constant term) and its closed form expression is given by

$$\hat{\mathbf{x}} = E[\mathbf{x}|\mathbf{y}] = \boldsymbol{\mu}_x + \boldsymbol{\Sigma}_{xy}\boldsymbol{\Sigma}_y^{-1}(\mathbf{y} - \boldsymbol{\mu}_y) \tag{4.66}$$

The conditional covariance $\mathrm{cov}(\hat{\mathbf{x}}|\mathbf{y})$ of the estimation error is

$$\mathrm{cov}(\hat{\mathbf{x}}|\mathbf{y}) = E[(\mathbf{x} - E[\mathbf{x}|\mathbf{y}])(\mathbf{x} - E[\mathbf{x}|\mathbf{y}])^T] = \boldsymbol{\Sigma}_x - \boldsymbol{\Sigma}_{xy}\boldsymbol{\Sigma}_y^{-1}\boldsymbol{\Sigma}_{yx} \tag{4.67}$$

Remarks In the case when $\boldsymbol{x}\,\boldsymbol{y}$ are Gaussian random variables, the following holds:

1. Conditional mean $\boldsymbol{E}[\boldsymbol{x}|\boldsymbol{y}]$ is a function of $\boldsymbol{y}$.
2. Conditional covariance $\mathrm{cov}(\hat{\boldsymbol{x}}|\boldsymbol{y})$ is not a function of $\boldsymbol{y}$, that is $\mathrm{cov}(\hat{\boldsymbol{x}}) = \mathrm{cov}(\hat{\boldsymbol{x}}|\boldsymbol{y})$.
3. Conditional PDF: bivariate Gaussian

$$E[x|y] = \mu_{x|y} = \mu_x + \frac{\mathrm{cov}\,(x,y)}{\sigma_y^2}\left(y - \mu_y\right) = \mu_x + \rho\frac{\sigma_x}{\sigma_y}\left(y - \mu_y\right) \tag{4.68}$$

$$\mathrm{var}\,(x) = \mathrm{var}\,(x|y) = \sigma_x^2\left(1 - \rho^2\right) = \sigma_x^2 - \frac{\mathrm{cov}^2\,(x,y)}{\sigma_y^2} \tag{4.69}$$

4.4.4 *Illustrative Examples*

Example 4.1 Scalar Gaussian random variables
We will assume a linear model $y = ax + v$ where x and v are zero-mean Gaussian random variables with variances σ_x^2 and σ_v^2 with PDFs

$$f_x(x) = \frac{1}{\sigma_x\sqrt{2\pi}}\exp\left\{-\frac{x^2}{\sigma_x^2}\right\} \tag{4.70}$$

$$f_v(v) = \frac{1}{\sigma_v\sqrt{2\pi}}\exp\left\{-\frac{v^2}{\sigma_v^2}\right\} \tag{4.71}$$

As the random variables are zero mean and Gaussian, the MMSE estimate $\hat{x}$ is given by

$$\hat{x} = E[x|y] = \alpha y \quad where \quad \alpha = \frac{E[xy]}{E[y^2]} \tag{4.72}$$

Substituting $y = ax + v$ in the expression for α, and noting that x and v are uncorrelated gives;

$$\hat{x} = \frac{E[xy]}{E[y^2]}y = \frac{E[x(ax+v)]}{E[(ax+v)^2]}y = \frac{aE[x^2]}{a^2E[x^2] + E[v^2]}y = \frac{a\sigma_x^2}{a^2\sigma_x^2 + \sigma_v^2}\mathbf{y} \tag{4.73}$$

Thus the estimate $\hat{x}$ becomes

$$\hat{x} = \left(\frac{a\sigma_x^2}{a^2\sigma_x^2 + \sigma_v^2}\right)y = \left(\frac{aSNR}{a^2SNR + 1}\right)y \quad where \quad SNR = \frac{\sigma_x^2}{\sigma_v^2} \tag{4.74}$$

The expression for the variance of the estimation error is

$$E[(x-\hat{x})^2] = E\left[(x-\alpha y)^2\right] \quad where \quad \alpha = \frac{a\sigma_x^2}{a^2\sigma_x^2+\sigma_v^2} \tag{4.75}$$

Substituting for y we get

$$E[(x-\hat{x})^2] = E\left[(x-\alpha\,(ax+v))^2\right] = \sigma_x^2(1-a\alpha)^2+\alpha^2\sigma_v^2 \tag{4.76}$$

Substituting for α yields

$$E[(x-\hat{x})^2] = (1-a\alpha)^2\sigma_x^2+\alpha^2\sigma_v^2 \quad where \quad \alpha = \frac{a\sigma_x^2}{a^2\sigma_x^2+\sigma_v^2} \tag{4.77}$$

Simplifying we get:

$$E[(x-\hat{x})^2] = \frac{\sigma_x^2\sigma_v^2}{a^2\sigma_x^2+\sigma_v^2} \tag{4.78}$$

Numerical example

$$a = 1;\ \sigma_x^2 = 0.2;\ \sigma_v^2 = 1$$

Performance of the MMSE estimator is shown in Figure 4.2. Subfigures (a), (b), (c), and (d) show respectively the measurements y, the random parameter x, the random parameter and its MMSE estimate

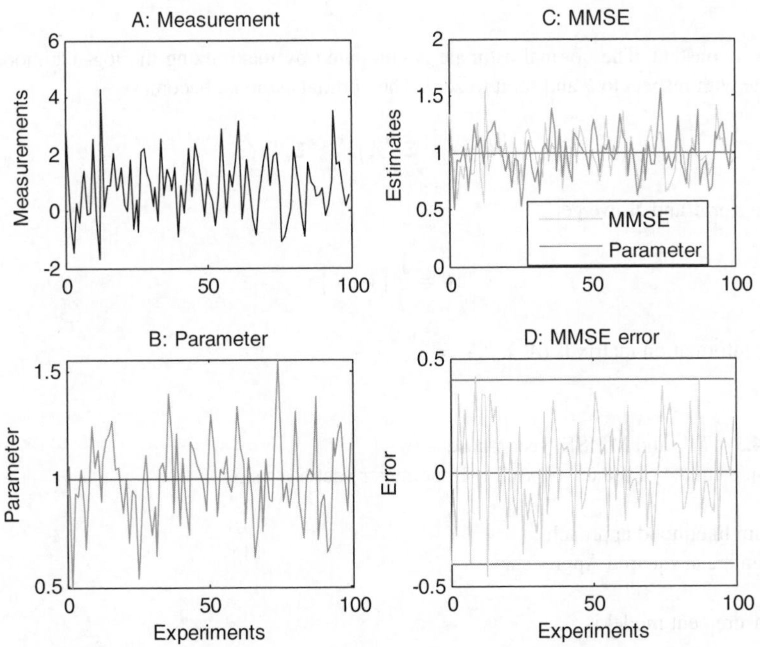

Figure 4.2 Performance of the MMSE estimator

$\hat{x}$, and the parameter estimation error $x - \hat{x}$. The theoretical mean and the standard deviations of the estimation error are shown. The estimation error variance is $E[(x - \hat{x})^2] = 0.1667$ and the standard deviation is 0.4082.

Example 4.2 ML estimate, scalar case
Obtain the maximum likelihood estimate of an unknown deterministic parameter θ given a measurement model

$$\boldsymbol{y} = \boldsymbol{A}\theta + \boldsymbol{v}$$

where $\boldsymbol{y}$ is a 2×1 vector of measurement θ is a scalar, $\boldsymbol{v}$ is a 2×1 zero-mean white Gaussian noise with covariance,

$$\boldsymbol{\Sigma}_v = \begin{bmatrix} 1 & 0 \\ 0 & 2 \end{bmatrix}$$

and $\boldsymbol{A}$ is a 2×1 vector given by

$$\boldsymbol{A} = \begin{bmatrix} 1 \\ 2 \end{bmatrix}$$

Solution

>The log-likelihood function $l(\boldsymbol{\theta}|y) = \ln f_y(\boldsymbol{y}; \boldsymbol{\theta})$ is given by

$$l(\theta|y) = \alpha - \frac{(y - \boldsymbol{A}\theta)^T \boldsymbol{\Sigma}_v^{-1} (\boldsymbol{y} - \boldsymbol{A}\theta)}{2}$$

where α is a constant. The optimal estimate is computed by maximizing the log-likelihood function: differentiate with respect to θ and set it to zero. The optimal estimate becomes

$$\hat{\theta} = \left(\boldsymbol{A}^T \boldsymbol{\Sigma}_v^{-1} \boldsymbol{A}\right)^{-1} \boldsymbol{A}^T \boldsymbol{\Sigma}_v^{-1} \boldsymbol{y}$$

Substituting for $\boldsymbol{A}$ and $\boldsymbol{\Sigma}_v$ we get

$$\hat{\theta} = \frac{1}{3} \begin{bmatrix} 1 & 1 \end{bmatrix} y$$

The Fisher information matrix is $(\boldsymbol{A}^T \boldsymbol{\Sigma}_v^{-1} \boldsymbol{A}) = 3$

Example 4.3 ML and MMSE, vector case
Compute an estimate of the unknown 2×1 parameter vector $\boldsymbol{\theta}$ using

- Maximum likelihood approach.
- Minimum mean squared Approach.

The measurement model is:

$$\boldsymbol{y} = \boldsymbol{A}\boldsymbol{\theta} + \boldsymbol{v} \tag{4.79}$$

where $\boldsymbol{y}$ is a 2×1 vector of measurement, θ is 2×1 random parameter with mean $\mu_\theta = 1$ with variance $\sigma_\theta = 0.2$, $\boldsymbol{v}$ is a 2×1 zero-mean white Gaussian noise with covariance $\sigma_v = \begin{bmatrix} 1 & 0 \\ 0 & 2 \end{bmatrix}$, and $\boldsymbol{A}$ is a 2×2 matrix given by $\boldsymbol{A} = \begin{bmatrix} 1 & 2 \\ 2 & 1 \end{bmatrix}$. We will assume that θ is uncorrelated with measurement noise $\boldsymbol{v}$.

a) *Maximum likelihood estimate*
   The log-likelihood function $l(\boldsymbol{\theta}|y)$ is given by

$$l(\boldsymbol{\theta}|\mathrm{y}) = \alpha - \frac{(\boldsymbol{y} - \boldsymbol{A\theta})^T \boldsymbol{\Sigma}_v^{-1} (\boldsymbol{y} - \boldsymbol{A\theta})}{2} \tag{4.80}$$

   where α is a constant. The optimal estimate is computed by maximizing the log-likelihood function: differentiate with respect to $\boldsymbol{\theta}$ and set it to zero. The optimal estimate $\hat{\boldsymbol{\theta}}_{ML}$ becomes:

$$\hat{\boldsymbol{\theta}}_{ML} = \left(\boldsymbol{A}^T \boldsymbol{\Sigma}_v^{-1} \boldsymbol{A}\right)^{-1} \boldsymbol{A}^T \boldsymbol{\Sigma}_v^{-1} \boldsymbol{y} \tag{4.81}$$

   The covariance of the estimation error is:

$$\mathrm{cov}(\boldsymbol{\theta}) = \boldsymbol{I}_F^{-1} \tag{4.82}$$

   Where $\boldsymbol{I}_F = \boldsymbol{H}^T \boldsymbol{\Sigma}_v^{-1} \boldsymbol{H}$.
   Substituting for A and Σ we get

$$\hat{\boldsymbol{\theta}}_{ML} = \frac{1}{3} \begin{bmatrix} -1 & 2 \\ 2 & -1 \end{bmatrix} \boldsymbol{y} \tag{4.83}$$

b) *Minimum Mean Square Error estimator*
   Using Eq. (4.66) the MMSE becomes

$$\hat{\boldsymbol{\theta}} = \boldsymbol{E}[\boldsymbol{\theta}|\boldsymbol{y}] = \boldsymbol{\mu}_\theta + \boldsymbol{\Sigma}_{\theta y} \boldsymbol{\Sigma}_y^{-1} (\boldsymbol{y} - \boldsymbol{\mu}_y) \tag{4.84}$$

   where $\boldsymbol{\mu}_y = \boldsymbol{A\mu}_\theta$. Using the measurement model (4.79) the cross-covariance matrix $\boldsymbol{\Sigma}_{\theta y}$ becomes

$$\boldsymbol{\Sigma}_{\theta y} = E[(\boldsymbol{\theta} - \boldsymbol{\mu}_\theta)(\boldsymbol{y} - \boldsymbol{\mu}_y)^T] = E[(\boldsymbol{\theta} - \boldsymbol{\mu}_\theta)(\boldsymbol{A\theta} + \boldsymbol{v} - \boldsymbol{\mu}_y)^T] \tag{4.85}$$

   Since $\boldsymbol{\mu}_y = \boldsymbol{A\mu}_\theta$, $\boldsymbol{\Sigma}_\theta = E[(\boldsymbol{\theta} - \boldsymbol{\mu}_\theta)(\boldsymbol{\theta} - \boldsymbol{\mu}_\theta)^T]$ and $E[\boldsymbol{\theta v}^T] = 0$ we get:

$$\boldsymbol{\Sigma}_{\theta y} = E[(\boldsymbol{\theta} - \boldsymbol{\mu}_\theta)(\boldsymbol{A}(\boldsymbol{\theta} - \boldsymbol{\mu}_\theta) + \boldsymbol{v})^T] = E[(\boldsymbol{\theta} - \boldsymbol{\mu}_\theta)(\boldsymbol{\theta} - \boldsymbol{\mu}_\theta)^T]\boldsymbol{A}^T = \boldsymbol{\Sigma}_\theta \boldsymbol{A}^T \tag{4.86}$$

   Consider the covariance $\boldsymbol{\Sigma}_y = E[(\boldsymbol{y} - \boldsymbol{\mu}_y)(\boldsymbol{y} - \boldsymbol{\mu}_y)^T]$. Using $E[\boldsymbol{\theta v}^T] = 0$ and $\boldsymbol{\mu}_y = \boldsymbol{A\mu}_\theta$ we get:

$$\boldsymbol{\Sigma}_y = E[(\boldsymbol{A\theta} + \boldsymbol{v} - \boldsymbol{\mu}_y)(\boldsymbol{A\theta} + \boldsymbol{v} - \boldsymbol{\mu}_y)^T] = \boldsymbol{A}E[(\boldsymbol{\theta} - \boldsymbol{\mu}_\theta)(\boldsymbol{\theta} - \boldsymbol{\mu}_\theta)^T]\boldsymbol{A}^T + E[\boldsymbol{vv}^T] \tag{4.87}$$

   Using the definition $\boldsymbol{\Sigma}_\theta = E[(\boldsymbol{\theta} - \boldsymbol{\mu}_\theta)(\boldsymbol{\theta} - \boldsymbol{\mu}_\theta)^T]$ and $\boldsymbol{\Sigma}_v = E[\boldsymbol{vv}^T]$ we get:

$$\boldsymbol{\Sigma}_y = \boldsymbol{A\Sigma}_\theta \boldsymbol{A}^T + \boldsymbol{\Sigma}_v \tag{4.88}$$

   Using Eqs. (4.85) and (4.86), the MMSE estimate Eq. (4.84) becomes:

$$\hat{\boldsymbol{\theta}} = E[\boldsymbol{\theta}|\boldsymbol{y}] = \boldsymbol{\mu}_\theta + \boldsymbol{\Sigma}_\theta \boldsymbol{A}^T (\boldsymbol{A\Sigma}_\theta \boldsymbol{A}^T + \boldsymbol{\Sigma}_v)^{-1} (\boldsymbol{y} - \boldsymbol{A\mu}_\theta) \tag{4.89}$$

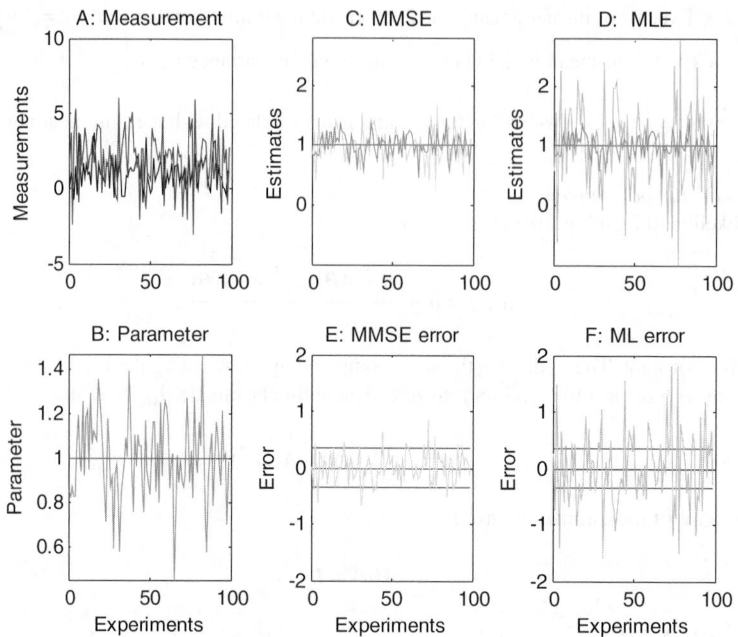

Figure 4.3 Minimum mean square error and maximum likelihood estimators.

Using Eqs. (4.86) and (4.88), the variance of the estimate becomes:

$$\text{cov}(\hat{\theta}) = \mathbf{\Sigma}_\theta - \mathbf{\Sigma}_\theta \mathbf{A}^T(\mathbf{A}\mathbf{\Sigma}_\theta \mathbf{A}^T + \mathbf{\Sigma}_v)^{-1}\mathbf{A}\mathbf{\Sigma}_\theta \tag{4.90}$$

Results of evaluation:

Performance of MMSE and ML estimators are shown in Figure 4.3. Subfigures (a), (b), (c), (d), (e), and (f) show respectively the measurements $y(1)$ and $y(2)$, the random parameter θ, parameter θ and its MMSE estimate $\hat{\theta}$, parameter θ and its ML estimate $\hat{\theta}_{ML}$, the parameter estimation error $\theta - \hat{\theta}$ with MMSE scheme, and the parameter estimation error $\theta - \hat{\theta}_{ML}$ with ML scheme. The theoretical mean and the standard deviations of the estimation error are shown. The standard deviations of the MMSE estimate and that of the ML estimate are given in Table 4.1. The standard deviations of the MMSE estimate and the ML estimate were computed using Eqs. (4.90) and (4.82).

Comments *The MMSE and the MLE schemes are designed to handle random and non-random unknown parameters respectively.*

The performance of the MMSE scheme is superior to that of the MLE approach when the unknown parameter is random.

Using ML estimation scheme to estimate a random parameter (assuming it to be a non-random parameter), gives a poor estimation error performance.

Table 4.1 Standard deviation

MMSE scheme	MLE scheme
0.3536	0.5774

The MMSE gives superior performance as it fuses the information provided by the model and the measurement data. The MMSE scheme requires, however, a priori *knowledge of the mean value of the parameter to be estimated.*

The performance MMSE estimator for the vector case, Example 4.3, is superior to that of the scalar case, Example 4.1. In the vector case the scalar random parameter is estimated using two measurements while a single measurement is employed in the scalar case. The standard deviations of the estimators for the vector and the scalar cases were respectively 0.3536 and 0.4082.

Example 4.4 True or false

State whether the following statements are true or false:

1. The maximum likelihood estimate will always be a linear function of the measurement independent of the PDF of the measurement noise.
2. The maximum likelihood estimate will be a linear function of the measurement if and only if the probability density function of measurement noise is Gaussian.
3. The minimum mean-squared error estimate of the unknown random parameter is always a linear function of the measurement independent of the probability density function of the measurement noise.
4. The minimum mean-squared error estimate of the unknown random parameter will be a linear function of the measurement if the probability density functions and the random parameter and measurement are both Gaussian.

Solution

1. False.
2. True.
3. False.
4. True.

4.5 Summary

The measurement model relating the random parameter and the measurement is assumed to be Minimum Mean Squares Estimator (MMSE):

Linear Measurement Model

$$\mathbf{y} = \mathbf{H}\mathbf{x} + \mathbf{v}$$

Performance Measure

$$J(\mathbf{y}) = E[(\mathbf{x} - \hat{\mathbf{x}})(\mathbf{x} - \hat{\mathbf{x}})^T | \mathbf{y}]$$

An estimate $\hat{\mathbf{x}}(\mathbf{y})$ is obtained from the minimization of $\mathbf{J}(\mathbf{y})$:

$$\min_{\hat{x}} \mathbf{J}(\mathbf{y})$$

$$\hat{x} = E[\mathbf{x}|\mathbf{y}]$$

Covariance of the Estimation Error

$$\text{cov}(\hat{\mathbf{x}}) = E[\mathbf{x}\mathbf{x}^T] - E[E[\mathbf{x}|\mathbf{y}]E[\mathbf{x}^T|\mathbf{y}]]$$

Conditional Expectation and its Properties

$$E[\boldsymbol{x}\boldsymbol{g}^T(\boldsymbol{y})] = E[E[\boldsymbol{x}|\boldsymbol{y}]\boldsymbol{g}^T(\boldsymbol{y})]$$

Orthogonality Principle

The MMSE estimator $\hat{\boldsymbol{x}} = E[\boldsymbol{x}|\boldsymbol{y}]$ of $\boldsymbol{x}$ is orthogonal to the estimation error $\boldsymbol{x} - E[\boldsymbol{x}|\boldsymbol{y}]$:

$$E[(\boldsymbol{x} - E[\boldsymbol{x}|\boldsymbol{y}])(E[\boldsymbol{x}|\boldsymbol{y}])^T] = 0$$

Expression for Conditional Mean

$$\hat{\boldsymbol{x}} = E[\boldsymbol{x}|\boldsymbol{y}] = \frac{\int_{-\infty}^{\infty} \boldsymbol{x} f_v(\boldsymbol{y} - \boldsymbol{H}\boldsymbol{x}) f_x(\boldsymbol{x}) d\boldsymbol{x}}{f_y(\boldsymbol{y})}$$

$$f_y(\boldsymbol{y}) = \int_{-\infty}^{\infty} f_x(\boldsymbol{x}) f_v(\boldsymbol{y} - \boldsymbol{H}\boldsymbol{x})\, d\boldsymbol{x}$$

MMSE Estimator: Gaussian Random Variables

If (i) $\boldsymbol{y} = \boldsymbol{H}\boldsymbol{x} + \boldsymbol{v}$, (ii) $\boldsymbol{x}$ and $\boldsymbol{v}$ are zero-mean random variables, (iii) $\boldsymbol{x}$ and $\boldsymbol{v}$ are independent, (iv) the PDFs $f_x(\boldsymbol{x}), f_y(\boldsymbol{y})$, and $f_v(\boldsymbol{v})$ are symmetric about the mean and absolutely continuous and (v) $\boldsymbol{x}$ is

$$f_x(\boldsymbol{x}) = \frac{1}{\sqrt{(2\pi)^N |\Sigma_x|}} \exp\left\{ -\frac{1}{2} \boldsymbol{x}\, \Sigma_x^{-1} \boldsymbol{x}^T \right\}, \text{ then}$$

$$E[\boldsymbol{x}|\boldsymbol{y}] = -\boldsymbol{\Sigma}_x \boldsymbol{H}^T \left(\frac{f_y'(\boldsymbol{y})}{f_y(\boldsymbol{y})} \right)$$

Covariance of the MMSE Estimator

$$\text{cov}(\hat{\boldsymbol{x}}) = \boldsymbol{\Sigma}_x - \boldsymbol{\Sigma}_x \boldsymbol{I}_F^{-1} \boldsymbol{\Sigma}_x$$

Conditional Mean and Conditional Variance

$$\hat{\boldsymbol{x}} = \boldsymbol{E}[\boldsymbol{x}|\boldsymbol{y}] = \boldsymbol{\mu}_x + \boldsymbol{\Sigma}_{xy} \boldsymbol{\Sigma}_y^{-1} (\boldsymbol{y} - \boldsymbol{\mu}_y)$$

$$\text{cov}(\hat{\boldsymbol{x}}|\boldsymbol{y}) = E[(\boldsymbol{x} - E[\boldsymbol{x}|\boldsymbol{y}])(\boldsymbol{x} - E[\boldsymbol{x}|\boldsymbol{y}])^T] = \boldsymbol{\Sigma}_x - \boldsymbol{\Sigma}_{xy} \boldsymbol{\Sigma}_y^{-1} \boldsymbol{\Sigma}_{yx}$$

4.6 Appendix: Non-Gaussian Measurement PDF

The random parameter $\boldsymbol{x}$ is Gaussian while the measurement $\boldsymbol{y}$ is non-Gaussian. We will assume that all the PDFs are symmetrical, and absolutely continuous. We will assume that all random variables are zero mean: $\boldsymbol{\mu}_x = 0$; $\boldsymbol{\mu}_y = 0$.

4.6.1 Expression for Conditional Expectation

The conditional expectation is:

$$\hat{\boldsymbol{x}} = E[\boldsymbol{x}|\boldsymbol{y}] = \int_{-\infty}^{\infty} \boldsymbol{x} f(\boldsymbol{x}|\boldsymbol{y})\, d\boldsymbol{x} \tag{4.91}$$

where $f(\boldsymbol{x}|\boldsymbol{y})$ can be expressed either in terms of (i) joint PDF $f(\boldsymbol{x},\boldsymbol{y})$ and marginal PDF $f_y(\boldsymbol{y})$, or (ii) the conditional PDF $f(\boldsymbol{y}|\boldsymbol{x})$ and the marginal PDFs $f_x(\boldsymbol{x})$ and $f_y(\boldsymbol{y})$ as:

$$f(\boldsymbol{x}|\boldsymbol{y}) = \frac{f(\boldsymbol{x},\boldsymbol{y})}{f_y(\boldsymbol{y})} = \frac{f(\boldsymbol{y}|\boldsymbol{x})f_x(\boldsymbol{x})}{f_y(\boldsymbol{y})} \tag{4.92}$$

where $\boldsymbol{x}$ and $\boldsymbol{y}$ are related by the linear measurement model is $\boldsymbol{y} = \boldsymbol{H}\boldsymbol{x} + \boldsymbol{v}$. Since $\boldsymbol{x}$ and $\boldsymbol{v}$ are independent, using the linear model, the expression for conditional PDF $f(\boldsymbol{y}|\boldsymbol{x})$ and joint PDF $f(\boldsymbol{x},\boldsymbol{y})$ in terms of noise PDF $f_v(\boldsymbol{v})$ become

$$f(\boldsymbol{y}|\boldsymbol{x}) = f_v(\boldsymbol{y} - \boldsymbol{H}\boldsymbol{x}) \tag{4.93}$$

$$f(\boldsymbol{x},\boldsymbol{y}) = f_v(\boldsymbol{y} - \boldsymbol{H}\boldsymbol{x})f_x(\boldsymbol{x}) \tag{4.94}$$

Substituting for $f(\boldsymbol{y}|\boldsymbol{x})$, $f(\boldsymbol{x}|\boldsymbol{y})$ becomes

$$f(\boldsymbol{x}|\boldsymbol{y}) = \frac{f_v(\boldsymbol{y} - \boldsymbol{H}\boldsymbol{x})f_x(\boldsymbol{x})}{f_y(\boldsymbol{y})} \tag{4.95}$$

Substituting for $f(\boldsymbol{x}|\boldsymbol{y})$ in the expression for $\hat{\boldsymbol{x}}$ we get:

$$E[\boldsymbol{x}|\boldsymbol{y}] = \frac{\int_{-\infty}^{\infty} \boldsymbol{x} f_v(\boldsymbol{y} - \boldsymbol{H}\boldsymbol{x})f_x(\boldsymbol{x})d\boldsymbol{x}}{f_y(\boldsymbol{y})} \tag{4.96}$$

where the marginal PDF $f_y(\boldsymbol{y})$ of $\boldsymbol{y}$ is:

$$f_y(\boldsymbol{y}) = \int_{-\infty}^{\infty} f_x(\boldsymbol{x})f_v(\boldsymbol{y} - \boldsymbol{H}\boldsymbol{x})d\boldsymbol{x} \tag{4.97}$$

4.6.2 Conditional Expectation for Gaussian x and Non-Gaussian y

We will derive the expression for the conditional expectation for the scalar $\boldsymbol{x}$ and $\boldsymbol{y}$ first and then generalize to the vector case $Mx1$ x and $Nx1y$.

Scalar x and y

Lemma 4.1 *If (i) $y = hx + v$, (ii) x and v are zero-mean random variables, (iii) x and v are independent, (iv) the PDFs $f_x(x)$, $f_y(y)$, and $f_v(v)$ are symmetric about the mean and absolutely continuous, and x is a Gaussian random variable with PDF $f_x(x) = \frac{1}{\sqrt{2\pi\sigma_x^2}}\exp\left\{-\frac{x^2}{2\sigma_x^2}\right\}$, then*

$$E[x|y] = -\sigma_x^2 h \frac{f_y'(y)}{f_y(y)} \tag{4.98}$$

where $f_y'(y) = \frac{df_y(y)}{dy}$

Proof: The conditional expectation for the scalar case is

$$E[x|y] = \frac{1}{f_y(y)}\int_{-\infty}^{\infty} x f_v(y - hx) f_x(x)\, dx \tag{4.99}$$

Differentiating the Gaussian PDF of the zero-mean random variable x, we get

$$\frac{df_x(x)}{dx} = \frac{d}{dx}\left(\frac{1}{\sqrt{2\pi\sigma_x^2}}\exp\left\{-\frac{x^2}{2\sigma_x^2}\right\}\right) = -\frac{x}{\sigma_x^2}f_x(x) \tag{4.100}$$

Hence

$$xf_x(x) = -\sigma_x^2 f_x'(x) \tag{4.101}$$

where $f_x'(x) = \dfrac{df_x(x)}{dx}$. Substituting $xf_x(x) = -\sigma_x^2 f_x'(x)$ in the integrand on the right-hand side of the expression for conditional expectation (4.99) we get

$$E[x|y] = -\frac{\sigma_x^2}{f_y(y)}\int_{-\infty}^{\infty} f_v(y-hx)f_x'(x)\,dx \tag{4.102}$$

The partial differentiation of the joint density function $f(\boldsymbol{x},\boldsymbol{y}) = f_v(y-Hx)f_x(x)$ gives

$$\frac{\delta f(\boldsymbol{x},\boldsymbol{y})}{\delta x} = f_v(y-hx)f_x'(x) + \frac{\delta f_v(y-hx)}{\delta x}f_x(x) \tag{4.103}$$

Since $\dfrac{\delta f_v(y-hx)}{\delta x} = -h\dfrac{\delta f_v(y-hx)}{\delta y}$ we get

$$\frac{\delta f(\boldsymbol{x},\boldsymbol{y})}{\delta x} = f_v(y-hx)f_x'(x) - h\frac{\delta f_v(y-hx)}{\delta y}f_x(x) \tag{4.104}$$

Substituting for $f_v(y-hx)f_x'(x)$ in the expression for conditional expectation Eq. (4.102) becomes

$$E[x|y] = -\frac{\sigma_x^2}{f_y(y)}\int_{-\infty}^{\infty}\left(\frac{\delta f(\boldsymbol{x},\boldsymbol{y})}{\delta x}\right)dx - \frac{\sigma_x^2 h}{f_y(y)}\int_{-\infty}^{\infty}\left(\frac{\delta f_v(y-hx)}{\delta y}f_x(x)\right)dx \tag{4.105}$$

Since $f(x,y)$ is even and symmetrical about the origin $x=0$ and $y=0$, $\dfrac{\delta f(\boldsymbol{x},\boldsymbol{y})}{\delta x}$ is symmetric and odd about the origin. Hence the first term on the right vanishes and we get

$$E[x|y] = -\frac{\sigma_x^2 h}{f_y(y)}\int_{-\infty}^{\infty}\left(\frac{\delta f_v(y-hx)}{\delta y}f_x(x)\right)dx \tag{4.106}$$

Since $f_y(\boldsymbol{y}) = \int_{-\infty}^{\infty} f_x(\boldsymbol{x})f_v(\boldsymbol{y}-\boldsymbol{Hx}\,d\boldsymbol{x}$, $\dfrac{df_y(\boldsymbol{y})}{dy} = \displaystyle\int_{-\infty}^{\infty}\frac{\delta f_v(\boldsymbol{y}-\boldsymbol{Hx})}{\delta y}f_x(\boldsymbol{x})d\boldsymbol{x}$, we get

$$E[x|y] = -\sigma_x^2 h\frac{f_y'(y)}{f_y(y)} \tag{4.107}$$

where $f'(y) = \dfrac{df_y(y)}{dy}$.

Lemma 4.2 *If (i)* $y = Hx + v$, *(ii)* x *and* v *are zero-mean random variables, (iii)* x *and* v *are independent, (iv) the PDFs* $f_x(x)$, $f_y(y)$, *and* $f_v(v)$ *are symmetric about the mean and absolutely continuous, and* x *is a Gaussian random variable with PDF* $f_x(x) = \frac{1}{\sqrt{(2\pi)^N|\Sigma_x|}}\exp\left\{-\frac{1}{2}x\Sigma_x^{-1}x^T\right\}$, *then*

$$E[x|y] = -\Sigma_x H^T \frac{f'_y(y)}{f_y(y)} \tag{4.108}$$

where $f'_y(y) = \frac{df_y(y)}{dy} = \left[\frac{df_y(y)}{dy(0)} \quad \frac{df_y(y)}{dy(1)} \quad \frac{df_y(y)}{dy(2)} \quad \cdot \quad \frac{df_y(y)}{dy(N-1)}\right]^T$

Proof: The conditional expectation is

$$E[x|y] = \frac{1}{f_y(y)}\int_{-\infty}^{\infty} x f_v(y - Hx) f_x(x) dx \tag{4.109}$$

Differentiating the Gaussian PDF of the zero-mean random variable x, we get

$$\frac{\delta f_x(x)}{\delta x} = \frac{\delta}{\delta x}\left(\frac{1}{\sqrt{(2\pi)^N|\Sigma_x|}}\exp\left\{-\frac{1}{2}x\Sigma_x^{-1}x^T\right\}\right) = -\Sigma_x^{-1}x f_x(x) \tag{4.110}$$

Hence

$$x f_x(x) = -\Sigma_x \frac{\delta f_x(x)}{\delta x} \tag{4.111}$$

where $\frac{\delta f_x(x)}{\delta x} = \left[\frac{\delta f_x(x)}{\delta x(0)} \quad \frac{\delta f_x(x)}{\delta x(1)} \quad \frac{\delta f_x(x)}{\delta x(2)} \quad \cdot \quad \frac{\delta f_x(x)}{\delta x(N-1)}\right]^T$.

Using Eq. (4.111), the integrand on the right-hand side of the expression for conditional expectation Eq. (4.109) becomes

$$E[x|y] = -\Sigma_x \frac{1}{f_y(y)}\int_{-\infty}^{\infty} f_v(y - Hx)\frac{\delta f_x(x)}{\delta x} dx \tag{4.112}$$

From the partial differentiation of the joint density function $f(x, y) = f_v(y - Hx) f_x(x)$ we get

$$\frac{\delta f(x, y)}{\delta x} = f_v(y - Hx)\frac{\delta f_x(x)}{\delta x} + \frac{\delta f_v(y - Hx)}{\delta x} f_x(x) \tag{4.113}$$

Since $\frac{\delta f_v(y - Hx)}{\delta x} = -H^T \frac{\delta f_v(y - Hx)}{\delta y}$, and rearranging we get

$$f_v(y - Hx)\frac{\delta f_x(x)}{\delta x} = \frac{\delta f(x, y)}{\delta x} + H^T \frac{\delta f_v(y - Hx)}{\delta y} f_x(x) \tag{4.114}$$

The expression for conditional expectation Eq. (4.112) becomes

$$E[\boldsymbol{x}|\boldsymbol{y}] = -\frac{\boldsymbol{\Sigma}_x}{f_y(\boldsymbol{y})}\int_{-\infty}^{\infty}\frac{\delta f(\boldsymbol{x},\boldsymbol{y})}{\delta x}d\boldsymbol{x} - \frac{\boldsymbol{\Sigma}_x\boldsymbol{H}^T}{f_y(\boldsymbol{y})}\int_{-\infty}^{\infty}\frac{\delta f_v(\boldsymbol{y}-\boldsymbol{H}\boldsymbol{x})}{\delta \boldsymbol{y}}f_x(\boldsymbol{x})\,dx \tag{4.115}$$

Since the joint multivariate PDF $f(\boldsymbol{x},\boldsymbol{y})$ is symmetric and even about the origin, its partial derivative $\frac{\delta f(\boldsymbol{x},\boldsymbol{y})}{\delta \boldsymbol{x}}$ is symmetric and an odd function of $\boldsymbol{x}$ about the origin. Hence the first term on the right vanishes and we get

$$E[\boldsymbol{x}|\boldsymbol{y}] = -\frac{\boldsymbol{\Sigma}_x\boldsymbol{H}^T}{f_y(\boldsymbol{y})}\frac{\delta}{\delta \boldsymbol{y}}\int_{-\infty}^{\infty} f_v(\boldsymbol{y}-\boldsymbol{H}\boldsymbol{x})f_x(\boldsymbol{x})dx \tag{4.116}$$

Using the expression for $f_y(\boldsymbol{y})$ using Eq. (4.97) we get

$$E[\boldsymbol{x}|\boldsymbol{y}] = -\boldsymbol{\Sigma}_x\boldsymbol{H}^T\frac{1}{f_y(\boldsymbol{y})}\frac{\delta f_y(\boldsymbol{y})}{\delta \boldsymbol{y}} \tag{4.117}$$

References

[1] Mendel, J. (1995) *Lessons in Estimation Theory in Signal Processing, Communications and Control*, Prentice Hall, New Jersey.

[2] Olofsson, P. (2005) *Probability, Statistics and Stochastic Process*, John Wiley and Sons, New Jersey.

Further Readings

Haykin, S. (2001) *Adaptive Filter Theory*, Prentice Hall, New Jersey.

Kay, S.M. (1993) *Fundamentals of Signal Processing: Estimation Theory*, Prentice Hall, New Jersey.

Mix, D.F. (1995) *Random Signal Processing*, Prentice Hall, New Jersey.

5

Linear Least-Squares Estimation

5.1 Overview

Linear least-squares estimation is a method of fitting the measurements (data or observation values) to a specified linear model. A best fit is obtained by minimizing the sum of the squares of the residual, where a residual is defined as an error between the measured value (data or observed value) and the value obtained using the model. Carl Friedrich Gauss invented the method of least squares in 1794 for predicting the planetary motion which is completely described by six parameters. A linear model that can best describe the measurement data is generally derived from physical laws. The model is set of algebraic equations governing the measurements and is completely described by a set of parameters termed herein as *feature vector*. In order to attenuate the effect of measurement errors on the estimation accuracy, the number of measurements is generally chosen to be much larger than the number of parameters to be estimated, resulting in what is commonly known as an *over determined* set of equations. Further the measurement data may not contain sufficient information about the parameters, resulting in an *ill conditioned* set of equations. The estimation of model parameters has wide application in many areas of science and engineering, including system identification, controller design, fault diagnosis, and condition monitoring. The least-squares method of Gauss is still widely used for estimating unknown parameters from the measurement data. The main reason is that it does require any probabilistic assumptions such as the underlying PDF of the measurement error, which is generally unknown *a priori* and is difficult to estimate *a posteriori*.

A generalized version of the least-squares method, popularly known as the weighted least-squares method, is developed. The least squares and the more general weighted least-squares estimate is unbiased, and is the best linear unbiased estimator. Most importantly, the optimal estimate is a solution of set a linear equations (assuming that the measurement model is linear) which can be efficiently computed using the Singular Value Decomposition (SVD). The least-squares estimation has a very useful geometric interpretation. The residual is orthogonal to the hyper plane generated by the columns of the matrix. This is called the orthogonality principle.

5.2 Linear Least-Squares Approach

The linear least-squares approach includes the following topics

- Linear algebraic model
- Objective function
- Least-squares estimation

Identification of Physical Systems: Applications to Condition Monitoring, Fault Diagnosis, Soft Sensor and Controller Design, First Edition. Rajamani Doraiswami, Chris Diduch and Maryhelen Stevenson.

- Normal equation: the equation governing the optimal estimate
- Properties of least-squares estimate
- Illustrative examples
- Cramér–Rao inequality
- Maximum likelihood method.

5.2.1 Linear Algebraic Model

We will restrict ourselves to the case where the set of algebraic equations relating the unknown parameters and the measurements is linear. A solution to a set of linear equations is important as it forms the backbone of estimation theory. A brief background is presented in this section. Consider a set of linear equations

$$\mathbf{y} = \mathbf{H}\boldsymbol{\theta} + \mathbf{v} \tag{5.1}$$

where $\mathbf{y}$ is a $(N \times 1)$ vector of measurements, $\mathbf{H} : \Re^M \to \Re^N$ is a $(N \times M)$ matrix, $\boldsymbol{\theta}$ is a $(M \times 1)$ deterministic parameter vector to be estimated, and $\mathbf{v}$ is an $(N \times 1)$ error vector. It is assumed that the matrix $\mathbf{H}$ has a full rank. In system identification, which is the focus of this chapter, the objective is to estimate the unknown vector, termed feature vector, which is formed of the numerator and the denominator coefficient system transfer function, from the input–output data using the measurement Eq. (5.1). In view of this application $\mathbf{H}$ and $\boldsymbol{\theta}$ are termed herein as *data matrix* and *feature vector*. The error term $\mathbf{v}$ is generally assumed to be a zero-mean white noise process or a zero-mean colored noise process. In the case when the error term is not zero-mean, the unknown mean may be included in the model by augmenting feature vector $\boldsymbol{\theta}$ and data matrix $\mathbf{H}$. Estimation of the unknown parameters depends upon the structure of the matrix including its dimension and its rank. The following cases are considered:

- If (i) the number of equations N is strictly less than the number of unknown parameters M, $N < M$ or (ii) when the number of equations is equal to the number of parameters $N = M$ but the resulting square matrix is singular, the set of equations are called *under determined* equations. In this case, a solution always exists. But the solution is not unique as there is an infinite number of solutions. In this case a constraint version of the least-squared method is employed to find a solution which (i) has a minimum norm, and (ii) gives the best fit between the measurement and its estimate.
- If the number of equations N is greater than the number of unknown parameters M, $N > M$ termed *over determined* equations, a solution may not exist. This case is the most common in system identification. In order to attenuate the effect of measurement errors on the parameter estimates, a larger number of measurements (or observations) compared to the number of unknown parameters is made. Due the presence of measurement errors the observation may not lie in the range space of the matrix of the observation model. The least-squares method is to obtain the best fit between the measurement and the estimate obtained using the model.

We will focus mainly on the over-determined set of equations.

5.2.2 Least-Squares Method

The term least-squares method describes a frequently used approach to solving over-determined or an inaccurate set of equations in an indirect sense. Instead of minimizing directly the parameter estimation error, the error between the measured value (data or observed value) and the value obtained using the model, termed *residual*, is minimized: a best fit between the measurement and its estimate is obtained by minimizing the sum of the squares of the weighted residual. A model that can best describe the measurement data is derived from physical laws.

5.2.3 *Objective Function*

5.2.3.1 Minimization Parameter Estimation Error

Let us first consider a direct approach to estimate the unknown parameter $\boldsymbol{\theta}$. Let $\hat{\boldsymbol{\theta}}$ be an estimate of $\boldsymbol{\theta}$. The optimal $\hat{\boldsymbol{\theta}}$ is obtained by minimizing the sum of the squares of the parameter estimation error $\boldsymbol{\theta} - \hat{\boldsymbol{\theta}}$:

$$\min_{\hat{\theta}}\{(\boldsymbol{\theta} - \hat{\boldsymbol{\theta}})^T(\boldsymbol{\theta} - \hat{\boldsymbol{\theta}})\} \tag{5.2}$$

Since the objective function is non-negative, the optimal estimate is that which will make the objective function equal to zero. Hence the optimal $\hat{\boldsymbol{\theta}}$ is given by

$$\hat{\boldsymbol{\theta}} = \boldsymbol{\theta} \tag{5.3}$$

Since we do not know $\boldsymbol{\theta}$, the optimal solution given by Eq. (5.3) is meaningless as it cannot be implemented. An alternative method of minimizing error in estimation of the measurement termed the residual is used instead of the parameter estimation error.

5.2.3.2 Minimization of the Residual

The estimate $\hat{\boldsymbol{\theta}}$ of the unknown parameter $\boldsymbol{\theta}$ is estimated by minimizing the sum of the squares of the residual, denoted $\boldsymbol{e}$, which is defined as

$$\boldsymbol{e} = \boldsymbol{y} - \hat{\boldsymbol{y}} \tag{5.4}$$

The residual is the difference between the measurement $\boldsymbol{y}$ and its estimate $\hat{\boldsymbol{y}} = \boldsymbol{H}\hat{\boldsymbol{\theta}}$ which is obtained by using the model $\boldsymbol{y} = \boldsymbol{H}\boldsymbol{\theta}$. The objective function $J = \left(\boldsymbol{y} - \boldsymbol{H}\hat{\boldsymbol{\theta}}\right)^T(\boldsymbol{y} - \boldsymbol{H}\hat{\boldsymbol{\theta}})$ and the least-squares estimation problem is formulated as

$$\min_{\hat{\theta}}\{(\boldsymbol{y} - \boldsymbol{H}\hat{\boldsymbol{\theta}})^T(\boldsymbol{y} - \boldsymbol{H}\hat{\boldsymbol{\theta}})\} \tag{5.5}$$

A more general form least-squares problem, termed the *weighted least-squares* problem, is to weight the measurements based on an *a priori* knowledge of the measurement accuracy so that $J = (\boldsymbol{y} - \boldsymbol{H}\hat{\boldsymbol{\theta}})^T\boldsymbol{W}(\boldsymbol{y} - \boldsymbol{H}\hat{\boldsymbol{\theta}})$:

$$\min_{\hat{\theta}}\{(\boldsymbol{y} - \boldsymbol{H}\hat{\boldsymbol{\theta}})^T\boldsymbol{W}(\boldsymbol{y} - \boldsymbol{H}\hat{\boldsymbol{\theta}})\} \tag{5.6}$$

where $\boldsymbol{W}$ is some NxN symmetric and positive definite matrix $\boldsymbol{W} = \boldsymbol{W}^T$. Since the measurement noise covariance $\boldsymbol{\Sigma}_v = E[\boldsymbol{v}\boldsymbol{v}^T]$ is a measure of an error in the measurement data $\boldsymbol{y}$, the weighting matrix W is chosen to be the inverse of the covariance matrix. The larger the $\boldsymbol{\Sigma}_v^{-1}$, the more accurate is the model and *vice versa*. More weight is given to those measurements for which the elements of $\boldsymbol{\Sigma}_v^{-1}$ is larger. The weight is chosen as

$$\boldsymbol{W} = \boldsymbol{\Sigma}_v^{-1} \tag{5.7}$$

5.2.3.3 Un-Weighted and Weighted Least-Squares

The weighted least squares may be posed as an un-weighted least-squares problem by "filtering the data." Consider the measurement model given by Eq. (5.1). Since $\boldsymbol{W}$ is a positive definite matrix, it can be expressed as a product of its square root,

$$\boldsymbol{W} = \boldsymbol{W}^{1/2}\boldsymbol{W}^{1/2} \tag{5.8}$$

Pre-multiplying by the filter operator $\boldsymbol{W}^{1/2}$ on both sides we get

$$\bar{\boldsymbol{y}} = \bar{\boldsymbol{H}}\boldsymbol{\theta} + \bar{\boldsymbol{v}} \tag{5.9}$$

where $\bar{\boldsymbol{y}} = \boldsymbol{W}^{1/2}\boldsymbol{y}$, $\bar{\boldsymbol{v}} = \boldsymbol{W}^{1/2}\boldsymbol{v}$ and $\bar{\boldsymbol{H}} = \boldsymbol{W}^{1/2}\boldsymbol{H}$. Substituting for $\boldsymbol{W}$ using Eq. (5.8), the weighted objective function becomes

$$J = \left(\boldsymbol{y} - \boldsymbol{H}\hat{\boldsymbol{\theta}}\right)^T \boldsymbol{W}^{1/2}\boldsymbol{W}^{1/2}\left(\boldsymbol{y} - \boldsymbol{H}\hat{\boldsymbol{\theta}}\right) = \left(\bar{\boldsymbol{y}} - \bar{\boldsymbol{H}}\hat{\boldsymbol{\theta}}\right)^T\left(\bar{\boldsymbol{y}} - \bar{\boldsymbol{H}}\hat{\boldsymbol{\theta}}\right) \tag{5.10}$$

Thus the weighted objective function can be expressed as an un-weighted objective function. The least-squares problem Eq. (5.5) becomes

$$\min_{\hat{\theta}}\{(\bar{\boldsymbol{y}} - \bar{\boldsymbol{H}}\hat{\boldsymbol{\theta}})^T(\bar{\boldsymbol{y}} - \bar{\boldsymbol{H}}\hat{\boldsymbol{\theta}})\} \tag{5.11}$$

In view of this, there is no loss of generality to consider an un-weighted least-squares problem with $\boldsymbol{W} = \boldsymbol{I}$ by filtering the data with $\boldsymbol{W}^{1/2}$.

Comments *Objective functions other than the 2-norm of the residual may be used, such as the 1-norm and the infinity norm of the residuals. They are as follows:*

- *1-norm of the residuals: the objective is to minimize the sum of the absolute values of the residuals* $\{e(i)\}$ *where* $e(i)$ *is the* i*th element of the* Nx1 *residual vector* $\boldsymbol{e}$*:*

$$\min_{\theta}\left\{\sum_{i=1}^{N}|e(i)|\right\} \tag{5.12}$$

  It is computationally more difficult than the least-squares. However, it is more robust as the estimated parameters are less sensitive to the presence of spurious data points or outliers.
- *Infinity norm of the residual: the objective is to minimize the largest residual*

$$\min_{\theta}\max_{i}\{e(i)\} \tag{5.13}$$

  This is also known as a Chebyshev fit (instead of least-squares fit) of the data. It is frequently used in the design of digital filters and in the development of approximations.

  Both 1-norm and infinity-norm optimizations are reformulated as linear programming problems.

5.2.4 Optimal Least-Squares Estimate: Normal Equation

Consider the weighted least-squares problem formulated in Eq. (5.6). The optimal least-squares estimate is obtained by differentiating the objective function J with respect to each element of $\boldsymbol{\theta}$. For notational simplicity we will use the results of vector calculus. The optimal solution is

$$\frac{dJ}{d\hat{\theta}} = 0 \tag{5.14}$$

Expanding the objective function J yields

$$J = y^T \boldsymbol{W}\boldsymbol{y} - \boldsymbol{y}^T \boldsymbol{W}\boldsymbol{H}\hat{\boldsymbol{\theta}} - \hat{\boldsymbol{\theta}}^T \boldsymbol{H}^T \boldsymbol{W}\boldsymbol{y} + \hat{\boldsymbol{\theta}}^T \boldsymbol{H}^T \boldsymbol{W}\boldsymbol{H}\hat{\boldsymbol{\theta}} \tag{5.15}$$

Differentiating the scalar J with respect to the vector $\boldsymbol{\theta}$ and setting it equal to zero using vector calculus, namely $\frac{d}{d\boldsymbol{x}}\{\boldsymbol{a}^T\boldsymbol{x}\} = \boldsymbol{a}$ and $\frac{d}{d\boldsymbol{x}}\{\boldsymbol{x}^T\boldsymbol{P}\boldsymbol{x}\} = 2\boldsymbol{P}\boldsymbol{x}$, we get

$$\frac{dJ}{d\hat{\boldsymbol{\theta}}} = -\boldsymbol{H}^T\boldsymbol{W}\boldsymbol{y} - \boldsymbol{y}^T\boldsymbol{H}\boldsymbol{W} + 2\boldsymbol{H}^T\boldsymbol{W}\boldsymbol{H}\hat{\boldsymbol{\theta}} = -2\boldsymbol{H}^T\boldsymbol{W}\boldsymbol{y} + 2\boldsymbol{H}^T\boldsymbol{W}\boldsymbol{H}\hat{\boldsymbol{\theta}} = \boldsymbol{0} \tag{5.16}$$

The optimal estimate of $\boldsymbol{\theta}$ denoted $\hat{\boldsymbol{\theta}}$ is the solution of the following equation given by

$$\boldsymbol{H}^T\boldsymbol{W}\boldsymbol{H}\hat{\boldsymbol{\theta}} = \boldsymbol{H}^T\boldsymbol{W}\boldsymbol{y} \tag{5.17}$$

This equation is often called a *normal equation*. If $\boldsymbol{H}$ has a full rank, $rank\{\boldsymbol{H}\} = \min(M, N) = M$ or equivalently $\boldsymbol{H}^T\boldsymbol{W}\boldsymbol{H}$ is invertible, then the optimal least-squares estimate has a closed form solution given by

$$\hat{\boldsymbol{\theta}} = \left(\boldsymbol{H}^T\boldsymbol{W}\boldsymbol{H}\right)^{-1}\boldsymbol{H}^T\boldsymbol{W}\boldsymbol{y} \tag{5.18}$$

The expression $(\boldsymbol{H}^T\boldsymbol{W}\boldsymbol{H})^{-1}\boldsymbol{H}^T\boldsymbol{W}$ is called pseudo-inverse, denoted $\boldsymbol{H}^{\dagger}$

$$\boldsymbol{H}^{\dagger} = (\boldsymbol{H}^T\boldsymbol{W}\boldsymbol{H})^{-1}\boldsymbol{H}^T\boldsymbol{W} \tag{5.19}$$

The estimate of the measurement $\hat{\boldsymbol{y}} = \boldsymbol{H}\hat{\boldsymbol{\theta}}$ is given by

$$\hat{\boldsymbol{y}} = \boldsymbol{H}\hat{\boldsymbol{\theta}} = \boldsymbol{H}\left(\boldsymbol{H}^T\boldsymbol{W}\boldsymbol{H}\right)^{-1}\boldsymbol{H}^T\boldsymbol{W}\boldsymbol{y} \tag{5.20}$$

One of the important advantages of the least-squares method is that it has closed form solution and the optimal solution is a linear function of the measurements thanks to the fact that (i) the model is linear and (ii) the objective function is quadratic. In the case when data matrix $\boldsymbol{H}$ is rank deficient, that is $rank\{\boldsymbol{H}\} < M$, and consequently $\boldsymbol{H}^T\boldsymbol{W}\boldsymbol{H}$ is not invertible, SVD (Singular Value Decomposition) of the data matrix is employed to compute the optimal least-squares estimate $\hat{\boldsymbol{\theta}}$. The model $\boldsymbol{y} = \boldsymbol{H}\boldsymbol{\theta} + \boldsymbol{v}$ and the estimate $\hat{\boldsymbol{\theta}} = (\boldsymbol{H}^T\boldsymbol{H})^{-1}\boldsymbol{H}^T\boldsymbol{y}$ are shown in Figure 5.1.

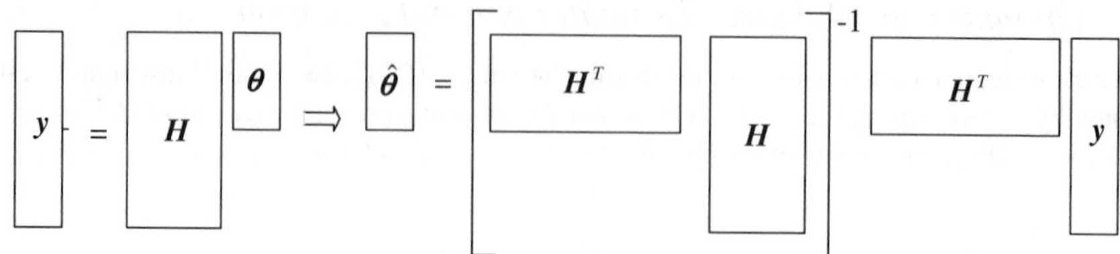

Figure 5.1 Visual representation of the linear model and the optimal least-squares estimate

5.2.5 Geometric Interpretation of Least-Squares Estimate: Orthogonality Principle

We will show that the when the least-squares estimate is optimal the residual is orthogonal to the column vectors of the data matrix. For notational convenience let us assume $\boldsymbol{W} = \boldsymbol{I}$. Rewriting the normal equation (5.17) we get

$$\boldsymbol{H}^T(\boldsymbol{y} - \hat{\boldsymbol{y}}) = \boldsymbol{0} \tag{5.21}$$

This shows that the residual is orthogonal to the data matrix. Expressing the data matrix explicitly in terms of its M column vectors $\boldsymbol{H} = \begin{bmatrix} \boldsymbol{h}_1 & \boldsymbol{h}_2 & \cdots & \boldsymbol{h}_M \end{bmatrix}$ the orthogonality condition (5.21) becomes

$$\boldsymbol{h}_i^T\left(\boldsymbol{y} - \boldsymbol{H}\hat{\boldsymbol{\theta}}\right) = 0 \quad i = 1, 2, 3, \dots, M \tag{5.22}$$

A geometrical interpretation of Eq. (5.22) is that the residual is orthogonal to each of the column vectors of the data matrix. $\boldsymbol{y} \in \Re^N$, $\hat{\boldsymbol{\theta}} \in \Re^M$, $\hat{\boldsymbol{y}} \in \Re^N$, $\hat{\boldsymbol{y}} = \boldsymbol{H}\hat{\boldsymbol{\theta}} \in span(\boldsymbol{h}_1 \quad \boldsymbol{h}_2 \quad \cdots \quad \boldsymbol{h}_M)$ and the residual $\boldsymbol{y} - \boldsymbol{H}\hat{\boldsymbol{\theta}}$ which is perpendicular to the span of $\boldsymbol{H}$ form a right-angled triangle. This is the well-known *orthogonality principle*. The residual represents the part of the measurement $\boldsymbol{y}$ that is not modeled by $\boldsymbol{y} = \boldsymbol{H\theta}$ as $M < N$. An expression for the residual given in Eq. (5.4) plays an important role in evaluating the performance of the least-squares estimator. Substituting $\hat{\boldsymbol{y}} = \boldsymbol{H}\hat{\boldsymbol{\theta}}$ we get

$$\boldsymbol{e} = \boldsymbol{y} - \boldsymbol{H}\hat{\boldsymbol{\theta}} \tag{5.23}$$

Using Eq. (5.18) and the definition of pseudo-inverse yields

$$\boldsymbol{e} = \boldsymbol{y} - \boldsymbol{H}(\boldsymbol{H}^T\boldsymbol{W}\boldsymbol{H})^{-1}\boldsymbol{H}^T\boldsymbol{W}\boldsymbol{y} = \boldsymbol{y} - \boldsymbol{H}\boldsymbol{H}^{\dagger}\boldsymbol{y} \tag{5.24}$$

This has an interesting geometric interpretation. Define an operator $\boldsymbol{P}_r$ given by

$$\boldsymbol{P}_r = \boldsymbol{H}(\boldsymbol{H}^T\boldsymbol{W}\boldsymbol{H})^{-1}\boldsymbol{H}^T\boldsymbol{W} = \boldsymbol{H}\boldsymbol{H}^{\dagger} \tag{5.25}$$

Using Eq. (5.20), $\hat{\boldsymbol{y}}$ and $\boldsymbol{e}$ in terms of the projection operator $\boldsymbol{P}_r$ become

$$\hat{\boldsymbol{y}} = \boldsymbol{P}_r\boldsymbol{y} \tag{5.26}$$

$$\boldsymbol{e} = \boldsymbol{y} - \hat{\boldsymbol{y}} = (\boldsymbol{I} - \boldsymbol{P}_r)\boldsymbol{y} \tag{5.27}$$

Thus $\boldsymbol{P}_r$ projects the measurement $\boldsymbol{y}$ onto the range space of data matrix $\boldsymbol{H}$ so that $\hat{\boldsymbol{y}} = \boldsymbol{P}_r\boldsymbol{y}$ while $\boldsymbol{I} - \boldsymbol{P}_r$ is called the *orthogonal complement projector* which projects the measurement $\boldsymbol{y}$ so that $\boldsymbol{e} = (\boldsymbol{I} - \boldsymbol{P}_r)\boldsymbol{y}$: $\boldsymbol{e}$

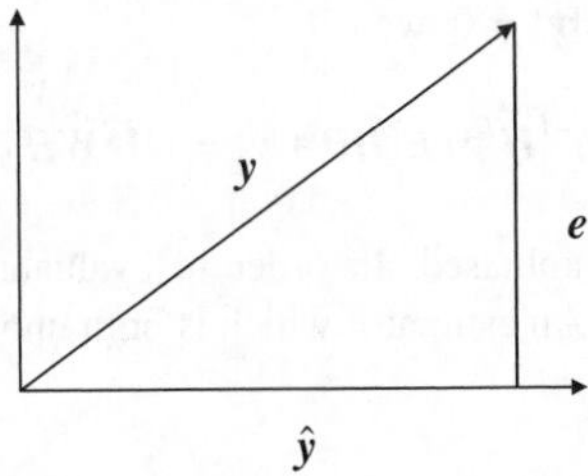

Figure 5.2 The measurement, its estimate and the residual form a right-angled triangle

is orthogonal to the $\hat{y}$ null space shown in Figure 5.2. Substituting for y in Eq. (5.27) and using Eq. (5.1) we get

$$e = (I - P_r)(H\theta + v) = H\theta + v - P_r H\theta - P_r v = (I - P_r)v \tag{5.28}$$

Consider the inner product $\hat{y}^T e$. Using Eqs. (5.26), and (5.27), and $P_r^2 = P$ we get

$$e^T \hat{y} = y^T P_r (I - P_r) y = y^T (P_r - P_r^2) y = 0 \tag{5.29}$$

Taking expectation we get

$$E\left[e^T \hat{y}\right] = 0 \tag{5.30}$$

The residual is thus orthogonal to the estimate. See Figure 5.2.

5.3 Performance of the Least-Squares Estimator

We will evaluate the performance of the least-squares estimator considering unbiasedness, the covariance of the estimation error, the mean-squared residual error, and comparison with other linear unbiased estimators.

5.3.1 *Unbiasedness of the Least-Squares Estimate*

The estimate is a random variable since the measurement is corrupted by a noise term. The question arises whether the expectation of the estimate of a parameter will be equal to its true value, that is, will the average of all the estimates obtained from an infinite number of experiments be equal to their true value. If so, then the estimate is said to be unbiased. Unbiasedness of an estimator is important as it indicates whether the model and/or the measurement are subject to a systemic error. The formal definition of an unbiased estimator is

$$E[\hat{\theta}] = \theta \tag{5.31}$$

We will show that the least-squares estimate is unbiased if (i) the data matrix H is deterministic and (ii) the noise v is a zero-mean random variable $E[v] = 0$. Substituting for $\hat{\theta}$ in the definition and using Eq. (5.18) yields

$$E[\hat{\theta}] = E[(H^T W H)^{-1} H^T W y] = (H^T W H)^{-1} H^T W E[y] \tag{5.32}$$

Substituting $y = H\theta + v$ and since $E[v] = 0$, we get

$$E[\hat{\theta}] = (H^T WH)^{-1} H^T WE[H\theta + v] = (H^T WH)^{-1} H^T WH\theta = \theta \quad (5.33)$$

Thus the least-squares estimate is unbiased. In order to evaluate the accuracy of an estimator the unbiasedness alone is not sufficient. An estimator which is both unbiased and has low covariance of the estimation error is desirable.

5.3.1.1 Illustrative Example: Unbiasedness, Whiteness, and Orthogonality

An illustrative example is given to compare the performance of the estimator with a zero-mean white noise process with different variances.

Example 5.1 The measurement model (5.1) where $H = \begin{bmatrix} 1 & 1 & . & 1 \end{bmatrix}^T$ is a $N \times 1$ vector, the true parameter $\theta = 1$, and v is $N \times 1$ measurement noise vector, $N = 500$ samples. The estimates of the parameters were computed using N data samples and 100 experiments were performed. Figure 5.3 shows the (i) estimate $\hat{\theta}$, (ii) the orthogonality of the residual e and the measurement estimate $\hat{y}$, and (iii) the whiteness of the residual for the noise standard deviation, $\sigma_v = 1$ and $\sigma_v = 0.25$. For the case when $\sigma_v = 1$, the subfigures (a), (b), and (c) show respectively the estimate vs. experiments, and the inner product of the residual and the estimate vs. experiments (for verifying the property of orthogonality), and the auto-correlation of the residual vs. time lag. Similarly the subfigures (d), (e), (f) show the performance when $\sigma_v = 0.25$.

Subfigures (a) and (d) show that the estimates are unbiased. The larger the variance of the noise, the larger is the variance of the estimator. Hence the measure of performance of the estimator should

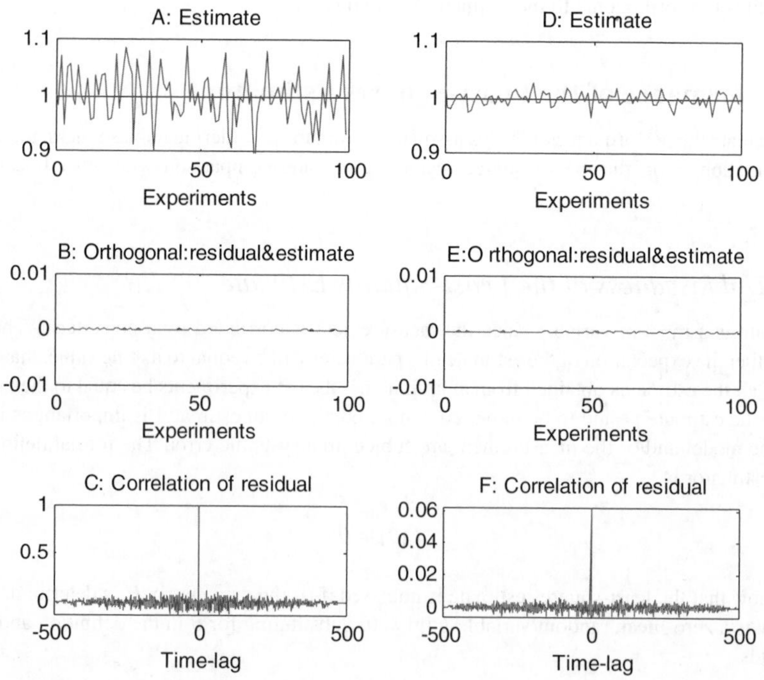

Figure 5.3 Unbiasedness of the estimate, whiteness of the residual and orthogonality

include the variance of the estimator besides the unbiasedness. The auto-correlation of the residual is a good visual indicator of the whiteness of the measurement noise as shown in subfigures (c) and (f). The maximum value of the auto-correlation, which occurs at zero lag, indicates the variance of the noise. Orthogonality of the residual and the estimated measurement (computed using time average) verifies the theoretical result (5.30) derived using ensemble average.

5.3.2 *Covariance of the Estimation Error*

The main measure of performance of an estimator is the covariance of the estimation error, denoted $\text{cov}(\hat{\theta})$ given by

$$\text{cov}(\hat{\theta}) = E[(\theta - \hat{\theta})(\theta - \hat{\theta})^T] \tag{5.34}$$

Consider the expression for $\hat{\theta}$ given in Eq. (5.18). Substituting $\boldsymbol{y} = \boldsymbol{H}\boldsymbol{\theta} + \boldsymbol{v}$ we get

$$\hat{\theta} = (\boldsymbol{H}^T\boldsymbol{W}\boldsymbol{H})^{-1}\boldsymbol{H}^T\boldsymbol{W}\,(\boldsymbol{H}\boldsymbol{\theta} + \boldsymbol{v}) = \boldsymbol{\theta} + \left(\boldsymbol{H}^T\boldsymbol{W}\boldsymbol{H}\right)^{-1}\boldsymbol{H}^T\boldsymbol{W}\boldsymbol{v} \tag{5.35}$$

Substituting Eq. (5.35) in Eq. (5.34) we get

$$cov(\hat{\theta}) = (\boldsymbol{H}^T\boldsymbol{W}\boldsymbol{H})^{-1}\boldsymbol{H}^T\boldsymbol{W}\boldsymbol{\Sigma}_v\boldsymbol{W}\boldsymbol{H}(\boldsymbol{H}^T\boldsymbol{W}\boldsymbol{H})^{-1} \tag{5.36}$$

Let us analyze the special case when $\boldsymbol{v}$ is a zero-mean white noise process with covariance $\boldsymbol{\Sigma}_v$ and $\boldsymbol{W} = \boldsymbol{\Sigma}_v^{-1}$. In this case we get

$$cov(\hat{\theta}) = (\boldsymbol{H}^T\boldsymbol{\Sigma}_v^{-1}\boldsymbol{H})^{-1} \tag{5.37}$$

5.3.2.1 Asymptotic Behavior of the Parameter Estimate

Let us now analyze the asymptotic behavior of the estimate $\hat{\theta}$ when the number of data samples N is asymptotically large. Let us consider the expression of $\hat{\theta}$ given in Eq. (5.35). Dividing and multiplying the second term on the right by N we get

$$\hat{\theta} = \theta + \left(\frac{\boldsymbol{H}^T\boldsymbol{W}\boldsymbol{H}}{N}\right)^{-1}\frac{\boldsymbol{H}^T\boldsymbol{W}\boldsymbol{v}}{N} \tag{5.38}$$

We will assume the following [1]

$$\lim_{N\to\infty} it \left(\frac{\boldsymbol{H}^T\boldsymbol{W}\boldsymbol{H}}{N}\right)^{-1} = \boldsymbol{\Sigma}_H^{-1} < \infty \tag{5.39}$$

$$\lim_{N\to\infty} it\,\frac{\boldsymbol{H}^T\boldsymbol{W}\boldsymbol{v}}{N} = 0 \tag{5.40}$$

The condition (5.39) merely states that the "sample covariance" $\boldsymbol{\Sigma}_H$ of the weighted data matrix $\boldsymbol{W}^{1/2}\boldsymbol{H}$ exists and is non-singular:

$$\lim_{N\to\infty} it \left(\frac{\boldsymbol{H}^T\boldsymbol{W}\boldsymbol{H}}{N}\right) = \boldsymbol{\Sigma}_H > 0 \tag{5.41}$$

Consider the condition (5.40). Expanding the left-hand side yields

$$\lim_{N\to\infty} it \frac{\boldsymbol{H}^T \boldsymbol{W} \boldsymbol{v}}{N} = \lim_{N\to\infty} it \left(\frac{1}{N} \sum_{i=1}^{N} \bar{\boldsymbol{h}}_j \boldsymbol{v}_j \right) = 0 \tag{5.42}$$

where $\bar{\boldsymbol{h}}_j$ is the jth column vector of $\boldsymbol{H}^T\boldsymbol{W}$.This condition essentially states that the average value of linear combinations of a zero-mean random variable is asymptotically zero. Both these conditions are not restrictive to our problems as we assume that the data matrix has full rank and the noise is a zero-mean random variable. In view of these two assumptions Eq. (5.38) becomes

$$\hat{\theta} \to \theta \; as \; N \to \infty \tag{5.43}$$

5.3.3 Properties of the Residual

An expression for the mean-squared value, asymptotic value, and the auto-correlation of the residual are considered

5.3.3.1 Mean-Squared Residual

Consider the expression of the minimum mean-squared residual, denoted $\sigma_{res}^2 = E[\boldsymbol{e}^T\boldsymbol{e}/N]$. Using the expression (5.28) we get

$$\sigma_{res}^2 = E[\boldsymbol{v}^T(\boldsymbol{I} - \boldsymbol{P}_r)^T(\boldsymbol{I} - \boldsymbol{P}_r)\boldsymbol{v}/N] \tag{5.44}$$

Using the property of the projection operator $\boldsymbol{P}_r^2 = \boldsymbol{P}_r$ yields

$$\sigma_{res}^2 = E[\boldsymbol{v}^T(\boldsymbol{I} - \boldsymbol{P}_r)\boldsymbol{v}/N] \tag{5.45}$$

Using the property the trace of a matrix $trace\{AB\} = trace\ \{BA\}$ and using $\boldsymbol{\Sigma}_v = E[\boldsymbol{v}\boldsymbol{v}^T]$

$$\sigma_{res}^2 = trace\{(\boldsymbol{I} - \boldsymbol{P}_r)\boldsymbol{\Sigma}_v\}/N \tag{5.46}$$

Let us analyze the minimum mean-squared residual when v is a zero-mean white noise with covariance $\boldsymbol{\Sigma}_v = \sigma_v^2\boldsymbol{I}$. In this case we get

$$\sigma_{res}^2 = \sigma_v^2 trace\{(\boldsymbol{I} - \boldsymbol{P}_r)/N\} \tag{5.47}$$

Assuming that the data matrix $\boldsymbol{H}$ has full rank and using the property of the projection operator $trace\{(\boldsymbol{I} - \boldsymbol{P}_r)\} = N - M$, we get

$$\sigma_{res}^2 = (1 - M/N)\,\sigma_v^2 \tag{5.48}$$

It is interesting to note that the minimum mean-squared residual depends only upon the number of unknown parameters M and the number of data samples N and the noise variance σ_v^2 and not on the data matrix $\boldsymbol{H}$. Consider the expression for the mean-squared residual when the number of data samples N is infinitely large. Using Eq. (5.48) we get

$$\lim_{N\to\infty} it\, \sigma_{res}^2 = \sigma_v^2 \tag{5.49}$$

5.3.3.2 Covariance and the Mean-Squared Value of the Residual

We will assume that the weighting matrix is optimal, $\boldsymbol{W} = \boldsymbol{\Sigma}_v^{-1}$. Consider the expression of the covariance of the residual, denoted cov(*res*) $= E[\boldsymbol{e}\boldsymbol{e}^T]$. Using the expression $\boldsymbol{e} = (\boldsymbol{I} - \boldsymbol{P}_r)\boldsymbol{v}$ given by Eq. (5.28) we get:

$$\text{cov}(\boldsymbol{res}) = E[\boldsymbol{e}\boldsymbol{e}^T] = E[(\boldsymbol{I} - \boldsymbol{P}_r)\boldsymbol{v}\boldsymbol{v}^T(\boldsymbol{I} - \boldsymbol{P}_r)^T] = (\boldsymbol{I} - \boldsymbol{P}_r)\boldsymbol{\Sigma}_v(\boldsymbol{I} - \boldsymbol{P}_r)^T \tag{5.50}$$

Simplifying using the property of the projection operator $\boldsymbol{P}_r^2 = \boldsymbol{P}_r$ yields

$$\text{cov}(\boldsymbol{res}) = (\boldsymbol{I} - \boldsymbol{P}_r)\boldsymbol{\Sigma}_v(\boldsymbol{I} - \boldsymbol{P}_r)^T = \boldsymbol{\Sigma}_v - \boldsymbol{\Sigma}_v\boldsymbol{P}_r^T - \boldsymbol{P}_r\boldsymbol{\Sigma}_v + \boldsymbol{P}_r\boldsymbol{\Sigma}_v\boldsymbol{P}_r^T \tag{5.51}$$

Substituting $\boldsymbol{P}_r = \boldsymbol{H}(\boldsymbol{H}^T\boldsymbol{\Sigma}_v^{-1}\boldsymbol{H})^{-1}\boldsymbol{H}^T\boldsymbol{\Sigma}_v^{-1}$ and $\boldsymbol{P}_r^T = \boldsymbol{\Sigma}_v^{-1}\boldsymbol{H}(\boldsymbol{H}^T\boldsymbol{\Sigma}_v^{-1}\boldsymbol{H})^{-1}\boldsymbol{H}^T$ yields:

$$\text{cov}(\boldsymbol{res}) = \boldsymbol{\Sigma}_v - \boldsymbol{H}\left(\boldsymbol{H}^T\boldsymbol{\Sigma}_v^{-1}\boldsymbol{H}\right)^{-1}\boldsymbol{H}^T \tag{5.52}$$

Let us compute the mean-squared residual $\sigma_{res}^2 = E[\boldsymbol{e}^T\boldsymbol{e}/N]$. Taking the trace of the matrices on both sides of Eq. (5.52), and dividing by N we get

$$\sigma_{res}^2 = E[\frac{\boldsymbol{e}^T\boldsymbol{e}}{N}] = \frac{1}{N}trace\left\{\boldsymbol{\Sigma}_v\right\} - \frac{1}{N}trace\left\{\boldsymbol{H}\left(\boldsymbol{H}^T\boldsymbol{\Sigma}_v^{-1}\boldsymbol{H}\right)^{-1}\boldsymbol{H}^T\right\} \tag{5.53}$$

Using the property of the trace $trace\{\boldsymbol{ABC}\} = trace\{\boldsymbol{CBA}\}$ we get:

$$\sigma_{res}^2 = \frac{1}{N}trace\{\boldsymbol{\Sigma}_v\} - \frac{1}{N}trace\left\{\boldsymbol{H}^T\boldsymbol{H}\left(\boldsymbol{H}^T\boldsymbol{\Sigma}_v^{-1}\boldsymbol{H}\right)^{-1}\right\} \tag{5.54}$$

Dividing and multiplying the second term on the right by N we get

$$\sigma_{res}^2 = \frac{1}{N}trace\left\{\boldsymbol{\Sigma}_v\right\} - \frac{1}{N}trace\left\{\frac{\boldsymbol{H}^T\boldsymbol{H}}{N}\left(\frac{\boldsymbol{H}^T\boldsymbol{\Sigma}_v^{-1}\boldsymbol{H}}{N}\right)^{-1}\right\} \tag{5.55}$$

Invoking the finiteness assumption (5.39), $\lim\limits_{N\to\infty}\dfrac{\boldsymbol{H}^T\boldsymbol{H}}{N} < \infty$ and $\lim\limits_{N\to\infty}\dfrac{\boldsymbol{H}^T\boldsymbol{\Sigma}_v^{-1}\boldsymbol{H}}{N} < \infty$, we get:

$$\lim_{N\to\infty}\sigma_{res}^2 = \sigma_v^2 \tag{5.56}$$

5.3.3.3 Asymptotic Expression of the Residual

Consider the expression of the residual (5.28). Expanding the expression we get

$$\boldsymbol{e} = (\boldsymbol{I} - \boldsymbol{P}_r)\boldsymbol{v} = (I - \boldsymbol{H}(\boldsymbol{H}^T\boldsymbol{W}\boldsymbol{H})^{-1}\boldsymbol{H}^T\boldsymbol{W})\boldsymbol{v} \tag{5.57}$$

Similar to Eq. (5.38), dividing and multiplying by N

$$\boldsymbol{e} = \boldsymbol{v} - \boldsymbol{H}\left(\frac{\boldsymbol{H}^T\boldsymbol{W}\boldsymbol{H}}{N}\right)^{-1}\left(\frac{\boldsymbol{H}^T\boldsymbol{W}\boldsymbol{v}}{N}\right) \tag{5.58}$$

Taking the limit and using the condition (5.40) we get

$$e \to v \text{ as } N \to \infty \tag{5.59}$$

It interesting to note that the larger the number of data samples N, the closer the residual is to the noise. Let us compute the auto-correlation of the residual for the case when N is large. Clearly the auto-correlation of the residual will approach that of the noise as N is large.

$$E\left[e(n)e(n-m)\right] \to E\left[v(n)v(n-m)\right] \text{ as } N \to \infty \tag{5.60}$$

where $e(n)$ and $v(n)$ are the nth element of $\boldsymbol{e}$ and $\boldsymbol{v}$ respectively. If the noise $\boldsymbol{v}$ is a zero-mean white noise the auto-correlation of the residual is a delta function

$$E[e(n)e(n-m)] = \delta(n-m) = \begin{cases} E[e^2(n)] & m = n \\ 0 & m \neq n \end{cases} \tag{5.61}$$

We say that the residual is white if the auto-correlation is a delta function as given in Eq. (5.61). There is simple test, termed the whiteness test, to verify whether the residual is white [1].

5.3.3.4 Illustrative Example: Performance with White and Colored Noise

An illustrative example is given to compare the performance of the estimator in the presence of white and colored measurement noise processes.

Example 5.2 The measurement model (5.1) where $\boldsymbol{H} = \begin{bmatrix} 1 & 1 & . & 1 \end{bmatrix}^T$ is a $Nx1$ vector, the true parameter $\boldsymbol{\theta} = 1$, and v is $Nx1$ measurement noise vector, $N = 2000$ samples. Two cases of measurement noise are considered:

Case 1: $\boldsymbol{v}$ is a zero-mean with unit covariance, $\boldsymbol{\Sigma}_v = I$, white noise process.

Case 2: $\boldsymbol{v}$ is a zero-mean colored noise process generated as an output of a filter $H_v(z) = (1-a)\dfrac{z^{-1}}{1-az^{-1}}$ with $a = 0.98$.

Figure 5.4 shows the asymptotic behavior of the parameter estimate, the residual, and its auto-correlation for case 1 and case 2 when v is a zero-mean white noise process and zero-mean colored noise. The estimates of the parameters were computed using N data samples and 100 experiments were performed. For case 1, subfigures (a), (b), and (c) show respectively (i) the true parameter and its estimates vs. the experiments, (ii) a single realization of the residual and the measurement noise, and (iii) auto-correlation of the residual and the measurement noise. Similarly, subfigures (d), (e), and (f) show for case 2 when the noise is colored.

In both the cases when the data samples are large (i) the residual and the auto-correlation of the residual are close to those of the noise, and (ii) the parameter estimates are close to the true value. The variance of the parameter estimation error is $cov(\hat{\theta}) = 0.0011$ for the white noise case, while for the colored noise the covariance is larger, $cov(\hat{\theta}) = 0.0688$. The auto-correlation of the residual is a delta function only for the zero-mean white noise case as it should be. With colored noise, a larger number of data samples N are required to achieve the same desired asymptotic behavior obtained with a white noise.

The residual and the measurement noise, and the auto-correlation of the residual and that of the noise, are not distinguishable in subfigures (b), (c), (e), and (f). The simulation results confirm the theoretical results (based on ensemble average) (5.43), (5.59), and (5.60).

Pathological case:
An interesting pathological case occurs when the data matrix $\boldsymbol{H}$ is square and non-singular with $N = M$, and the noise is a zero-mean white noise process. From Eq. (5.48) it can be deduced that the mean-squared residual is zero even if the noise variance is non-zero and the number of data samples is finite and merely

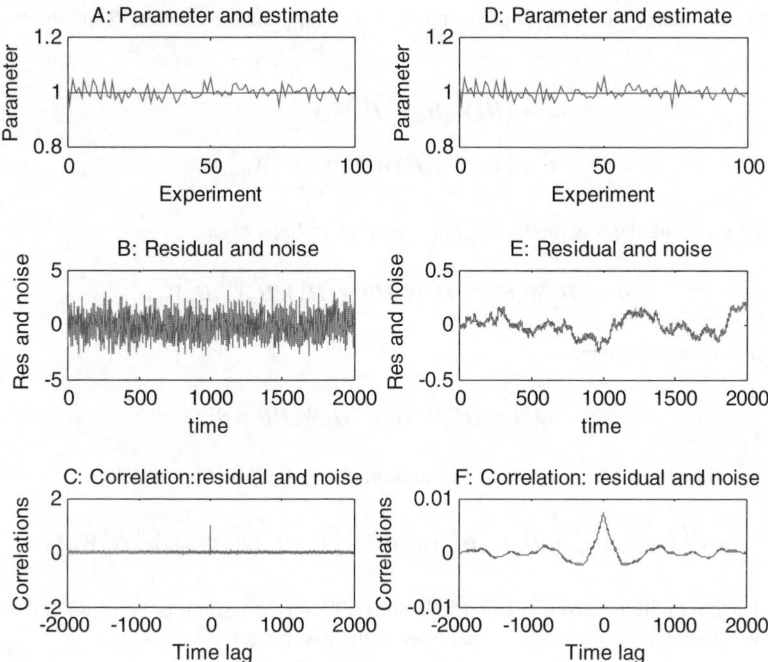

Figure 5.4 Asymptotic performance of the estimator with white and colored noise

equal to the number of unknown parameters. However, the $\text{cov}(\hat{\theta})$ is non-zero as can be deduced from Eq. (5.36).

5.3.4 *Model and Systemic Errors: Bias and the Variance Errors*

It is important to emphasize that the properties of the least-squares estimator hold if the linear model given by Eq. (5.1), and the statistics of the measurement noise are both accurate. In practice, the above assumptions may not hold. The structure of an assumed model, termed the nominal model and denoted $\boldsymbol{H}_0$, and the true model $\boldsymbol{H}$, generally differ as the structure of the true model, characterized by the order of the numerator, the order of the denominator order, and the delay, is unknown unless extreme care is taken to validate the model. For example, the structure of the model may be derived from physical laws and its validation by an analysis of the residual. Further, there will be systemic errors such as those resulting from inaccurate mean and covariance of the measurement noise, and external inputs contaminating the measurements including low frequency disturbances, drifts, offsets, trends, and non-zero-mean noise.

There will be a *bias error* due to model mismatch and both *bias* and *variance error* due to the measurement noise and disturbance. If, however, the noise and the disturbance are zero-mean, there will be only variance error.

5.3.4.1 Model Error

Let us consider the effect of modeling error on the covariance of the estimation error and the residual. The true data matrix $\boldsymbol{H}$ of the measurement model (5.1) is not known accurately, and is assumed to be $\boldsymbol{H}_0$. The assumed or nominal model is given by

$$\boldsymbol{y} = \boldsymbol{H}_0\boldsymbol{\theta} + \boldsymbol{v} \tag{5.62}$$

The nominal data matrix model $\boldsymbol{H}_0$ is then employed in computing the estimate $\hat{\theta}$ and the residual $\boldsymbol{e}$ we get

$$\hat{\theta} = \left(\boldsymbol{H}_0^T \mathbf{W}_0 \boldsymbol{H}_0\right)^{-1} \boldsymbol{H}_0^T \mathbf{W}_0 \boldsymbol{y} \tag{5.63}$$

$$\boldsymbol{e} = \left(\boldsymbol{I} - \boldsymbol{H}_0 \left(\boldsymbol{H}_0^T \mathbf{W}_0 \boldsymbol{H}_0\right)^{-1} \boldsymbol{H}_0^T \mathbf{W}_0\right) \boldsymbol{y} \tag{5.64}$$

Substituting for $\boldsymbol{y}$ using the true model $\boldsymbol{y} = \boldsymbol{H}\boldsymbol{\theta} + \boldsymbol{v}$ in Eq. (5.63) we get

$$\hat{\theta} = \left(\boldsymbol{H}_0^T \mathbf{W}_0 \boldsymbol{H}_0\right)^{-1} \boldsymbol{H}_0^T \mathbf{W}_0 \boldsymbol{H}\boldsymbol{\theta} + \left(\boldsymbol{H}_0^T \mathbf{W} \boldsymbol{H}_0\right)^{-1} \boldsymbol{H}_0^T \mathbf{W}_0 \boldsymbol{v} \tag{5.65}$$

The estimate is biased given by

$$E[\hat{\theta}] = \left(\boldsymbol{H}_0^T \mathbf{W}_0 \boldsymbol{H}_0\right)^{-1} \boldsymbol{H}_0^T \mathbf{W}_0 \boldsymbol{H}\boldsymbol{\theta} \neq \boldsymbol{\theta} \tag{5.66}$$

Consider the residual given by Eq. (5.64). Substituting for $\boldsymbol{y}$ using the true model

$$\boldsymbol{e} = \left(\boldsymbol{I} - \boldsymbol{H}_0 \left(\boldsymbol{H}_0^T \mathbf{W}_0 \boldsymbol{H}_0\right)^{-1} \boldsymbol{H}_0^T \mathbf{W}_0\right) \boldsymbol{H}\boldsymbol{\theta} + \left(\boldsymbol{I} - \boldsymbol{H}_0 \left(\boldsymbol{H}_0^T \mathbf{W}_0 \boldsymbol{H}_0\right)^{-1} \boldsymbol{H}_0^T \mathbf{W}_0\right) \boldsymbol{v} \tag{5.67}$$

Assuming that the conditions given by Eqs. (5.39) and (5.40) on the asymptotic behavior of the covariance matrix of the estimation error and the noise respectively hold we get

$$\boldsymbol{e} \to \left(\boldsymbol{I} - \boldsymbol{H}_0 \left(\boldsymbol{H}_0^T \mathbf{W}_0 \boldsymbol{H}_0\right)^{-1} \boldsymbol{H}_0^T \mathbf{W}_0\right) \boldsymbol{H}\boldsymbol{\theta} + \boldsymbol{v} \; as \; N \to \infty \tag{5.68}$$

This shows that the residual asymptotically approaches noise plus a bias term.

The mean of the residual $\boldsymbol{e}$ is asymptotically non-zero given by

$$E\left[\boldsymbol{e}\right] \to \left(\boldsymbol{I} - \boldsymbol{H}_0 \left(\boldsymbol{H}_0^T \mathbf{W}_0 \boldsymbol{H}_0\right)^{-1} \boldsymbol{H}_0^T \mathbf{W}_0\right) \boldsymbol{H}\boldsymbol{\theta} \; as \; N \to \infty \tag{5.69}$$

$\left(\boldsymbol{I} - \boldsymbol{H}_0 \left(\boldsymbol{H}_0^T \mathbf{W}_0 \boldsymbol{H}_0\right)^{-1} \boldsymbol{H}_0^T \mathbf{W}_0\right) \boldsymbol{H}\boldsymbol{\theta}$ is termed as*bias error*

As result of the bias error, the auto-correlation of the residual will not be a delta function.

5.3.4.2 Systemic Error

In this case there is systemic error but no modeling error, that is $\boldsymbol{H} = \boldsymbol{H}_0$. The system error may be modeled as an unknown additive term μ

$$\boldsymbol{y} = \boldsymbol{H}\boldsymbol{\theta} + \boldsymbol{\mu} + \boldsymbol{v} \tag{5.70}$$

Using Eq. (5.18) and substituting for $\boldsymbol{y}$ and simplifying the estimate $\hat{\theta}$ becomes

$$\hat{\theta} = \boldsymbol{\theta} + (\boldsymbol{H}^T \mathbf{W} \boldsymbol{H})^{-1} \boldsymbol{H}^T \mathbf{W} \boldsymbol{\mu} + \left(\boldsymbol{H}^T \mathbf{W} \boldsymbol{H}\right)^{-1} \boldsymbol{H}^T \mathbf{W} \boldsymbol{v} \tag{5.71}$$

The estimate is biased given by

$$E[\hat{\theta}] = \boldsymbol{\theta} + \left(\boldsymbol{H}^T \mathbf{W} \boldsymbol{H}\right)^{-1} \boldsymbol{H}^T \mathbf{W} \boldsymbol{\mu} \neq \boldsymbol{\theta} \tag{5.72}$$

Assuming $\boldsymbol{W} = \boldsymbol{\Sigma}_v^{-1}$ the covariance of the estimation is

$$cov(\hat{\theta}) = \left(\boldsymbol{H}^T \boldsymbol{\Sigma}_v^{-1} \boldsymbol{H}\right)^{-1} \tag{5.73}$$

Consider the expression for the residual given by Eq. (5.24). Substituting for $\boldsymbol{y}$ we get

$$\boldsymbol{e} = \boldsymbol{\mu} + \boldsymbol{v} - \boldsymbol{H}\left(\boldsymbol{H}^T \boldsymbol{W} \boldsymbol{H}\right)^{-1} \boldsymbol{H}^T \boldsymbol{W} \boldsymbol{v} \tag{5.74}$$

Assuming that the conditions given by Eqs. (5.39) and (5.40) on the asymptotic behavior of the covariance matrix of the estimation error and the noise respectively hold we get

$$\boldsymbol{e} \to \boldsymbol{\mu} + \boldsymbol{v} \text{ as } N \to \infty \tag{5.75}$$

This shows that the residual asymptotically approaches noise plus a bias term. Due to the presence of systemic error there will be bias error $\boldsymbol{\mu}$ and the covariance error $\left(\boldsymbol{H}^T \boldsymbol{\Sigma}_v^{-1} \boldsymbol{H}\right)^{-1}$. If, however, the noise and the disturbance are zero-mean, there will be only covariance error. The auto-correlation of the residual will not be a delta function.

Example 5.3 Illustrative example: performance with model and systemic errors
Consider the model (5.62). An example of a second-order system is considered with 4*x*2 data matrix $\boldsymbol{H}$. First the modeling error is simulated by selecting a wrong structure by the assuming the model order is 1 instead of the true order 2 with $\boldsymbol{H}_0$ a 4*x*1 vector. The actual and the assumed matrices were $\boldsymbol{H} = \begin{bmatrix} 2 & 1 \\ 3 & 4 \\ 2 & 5 \\ 4 & 5 \end{bmatrix}$ and $\boldsymbol{H}_0 = \begin{bmatrix} 2 \\ 3 \\ 2 \\ 4 \end{bmatrix}$.

Then a systemic error was introduced by introducing a bias term in the model simulating a constant disturbance in the true model. In this case it was assumed that the true and the assumed data matrices are equal $\boldsymbol{H}_0 = \boldsymbol{H}$. Figure 5.5 shows the results of simulation when there are modeling and systemic errors. Subfigures (a) and (b) show respectively the true parameters and the estimate (note that due to the modeling error resulting from assuming that the model order is 1 instead of 2, there is only one estimate and two unknown parameters) and the auto-correlation of the residual and the auto-correlation of the noise. Similarly subfigures (c) and (d) show the true parameters and their estimates and the auto-correlations when there is a systemic error resulting from an additive bias term μ.

Comment *When there is a modeling and/or or a systemic error, there will be a bias in the parameter estimates and the auto-correlation of the residual will not be a delta function. The auto-correlation of the residual is a good indicator of a systemic error or model error or both.*

Lemma 5.1 *Let $\breve{\theta}$ be some linear unbiased estimate of θ given by*

$$\breve{\theta} = \boldsymbol{F}y \tag{5.76}$$

Then

$$\boldsymbol{cov}(\hat{\theta}) \leq cov(\breve{\theta}) \tag{5.77}$$

where from (5.37), $cov(\hat{\theta}) = (\boldsymbol{H}^T \boldsymbol{\Sigma}_v^{-1} \boldsymbol{H})^{-1}$.

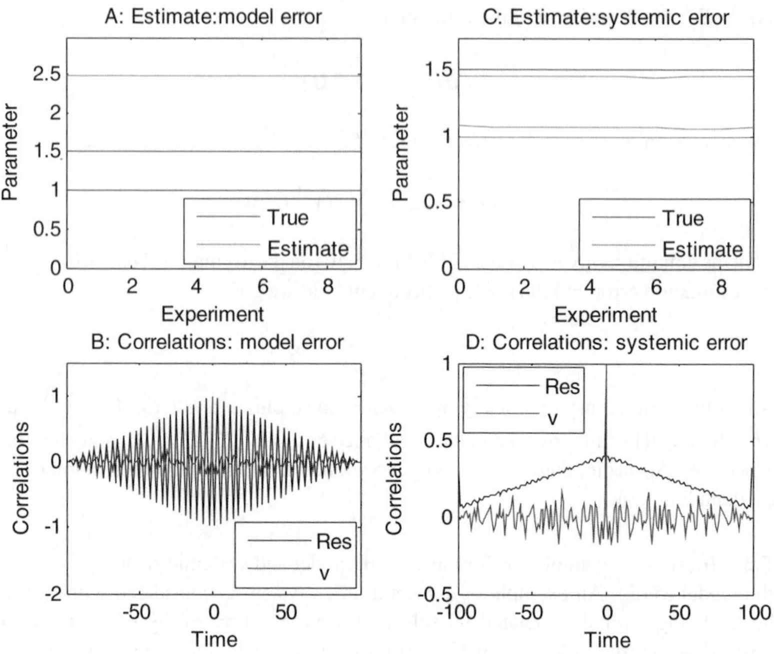

Figure 5.5 Estimates and correlations with model and systemic errors

Proof: Using Eq. (5.1) we get

$$\breve{\boldsymbol{\theta}} = \boldsymbol{FH\theta} + \boldsymbol{Fv} \tag{5.78}$$

Taking expectation we get

$$E[\breve{\boldsymbol{\theta}}] = \boldsymbol{FH\theta} \tag{5.79}$$

Since $\breve{\boldsymbol{\theta}}$ is unbiased, that is $E[\breve{\boldsymbol{\theta}}] = \boldsymbol{\theta}$ we get:

$$\boldsymbol{FH} = \boldsymbol{I} \tag{5.80}$$

Substituting Eq. (5.80) in Eq. (5.78) yields:

$$\breve{\boldsymbol{\theta}} = \boldsymbol{\theta} + \boldsymbol{Fv} \tag{5.81}$$

The covariance of the parameter estimation error $cov(\breve{\boldsymbol{\theta}})$ becomes

$$cov\left(\breve{\boldsymbol{\theta}}\right) = \boldsymbol{FE}\left[\boldsymbol{vv}^T\right]\boldsymbol{F}^T = \boldsymbol{F}\boldsymbol{\Sigma}_v\boldsymbol{F}^T \tag{5.82}$$

Recall the covariance of the least-squares estimate (5.37), $cov(\hat{\theta}) = \left(H^T \Sigma_v^{-1} H\right)^{-1}$. Define the following positive semi-definitive matrix, denoted P_{FH}

$$P_{FH} = \left(F - \left(H^T W H\right)^{-1} H^T W\right) \Sigma_v \left(F - \left(H^T W H\right)^{-1} H^T W\right)^T \tag{5.83}$$

where $W = \Sigma_v^{-1}$. Simplifying using $W = \Sigma_v^{-1}$

$$P_{FH} = F\Sigma_v F^T + \left(H^T W H\right)^{-1} - FH\left(H^T W H\right)^{-1} - \left(H^T W H\right)^{-1} H^T F^T \tag{5.84}$$

Using Eq. (5.80) we get

$$P_{FH} = F\Sigma_v F^T - (H^T W H)^{-1} \tag{5.85}$$

Since P_{FH} is positive semi-definite $P_{FH} \geq 0$ from Eq. (5.83) we deduce

$$F\Sigma_v F^T \geq (H^T W H)^{-1} \tag{5.86}$$

Comparing the expressions for $cov(\hat{\theta})$ and $cov(\breve{\theta})$ from Eqs. (5.37) and (5.82) and using the inequality (5.86) we conclude

$$cov(\hat{\theta}) \leq cov(\breve{\theta}) \tag{5.87}$$

Comment *The weighted least-squares estimator with* $W = \Sigma_v^{-1}$ *is the best linear unbiased estimator. In a later section on Cramér–Rao inequality, we will show that the weighted least-squares estimator is best if the PDF of the noise is Gaussian.*

5.4 Illustrative Examples

Example 5.4 Scalar model

$$y = H\theta + v \tag{5.88}$$

where H and θ are scalars.

Solution: $H^{\dagger} = \left(H^T W H\right)^{-1} H^T W = \frac{1}{H}$, $P_r = HH^{\dagger} = 1$, $1 - P_r = 0$

$$\hat{\theta} = H^{\dagger} y = \frac{y}{H} \tag{5.89}$$

$$\hat{y} = P_r y = y \tag{5.90}$$

$$e = y - \hat{y} = \left(1 - P_r\right) y = 0 \tag{5.91}$$

$$\text{cov}(\hat{\theta}) = \left(H^T H\right)^{-1} \sigma_v^2 = \frac{\sigma_v^2}{H^2} \tag{5.92}$$

Comment *It is a pathological case. The residual is zero even though the output is corrupted by noise. However, the covariance of the parameter estimation is not zero.*

Example 5.5 Vector model with scalar parameter

$$y = H\theta + v \tag{5.93}$$

where $\boldsymbol{H}$ is $Nx1$ vector of all ones, $\boldsymbol{H} = [1 \quad 1 \quad . \quad 1]^T, \boldsymbol{y} = \begin{bmatrix} y(1) & y(2) & \cdots & y(N) \end{bmatrix}^T$ is a $N \times 1$ vector.

Solution:

$$\boldsymbol{H}^{\dagger} = \left(\boldsymbol{H}^T \boldsymbol{W} \boldsymbol{H}\right)^{-1} \boldsymbol{H}^T \boldsymbol{W}, \quad \boldsymbol{P}_r = \boldsymbol{H}\boldsymbol{H}^{\dagger}$$

a. $\boldsymbol{W} = \boldsymbol{I}$

The error term $\boldsymbol{v}$ is independent and identically distributed zero-mean white noise with variance $E[\boldsymbol{v}\boldsymbol{v}^T] = \boldsymbol{I}\sigma_v^2$.

$$\boldsymbol{H}^{\dagger} = \left(\boldsymbol{H}^T\boldsymbol{H}\right)^{-1}\boldsymbol{H}^T, \boldsymbol{H}^T\boldsymbol{H} = N \text{ and } \boldsymbol{H}^T\boldsymbol{y} = \sum_{i=1}^{N} y(i) \text{ and } \boldsymbol{P}_r = \boldsymbol{H}\boldsymbol{H}^{\dagger} = \left(\frac{1}{N}\right)\boldsymbol{I}_1, \boldsymbol{I} - \boldsymbol{P}_r = \boldsymbol{I} - \left(\frac{1}{N}\right)\boldsymbol{I}_1$$

where $1 = \begin{bmatrix} 1 & 1 & \cdots & 1 \end{bmatrix}^T$ is a $Nx1$ vector of all ones, and $\boldsymbol{I}_1$ is a NxN matrix whose elements are all one. The estimate $\hat{\theta}$ is

$$\hat{\theta} = \boldsymbol{H}^{\dagger}\boldsymbol{y} = \left(\boldsymbol{H}^T\boldsymbol{H}\right)^{-1}\boldsymbol{H}^T\boldsymbol{y} = \frac{1}{N}\sum_{i=1}^{N} y(i) \tag{5.94}$$

$$\hat{y} = \boldsymbol{P}_r\boldsymbol{y} = 1\left(\frac{1}{N}\sum_{i=1}^{N} y(i)\right) \tag{5.95}$$

The covariance of the estimation error is

$$\text{cov}(\hat{\theta}) = \sigma_v^2(\boldsymbol{H}^T\boldsymbol{H})^{-1} = \frac{\sigma_v^2}{N} \tag{5.96}$$

The residual is given by

$$\boldsymbol{e} = (\boldsymbol{I} - \boldsymbol{P}_r)\boldsymbol{v} = \boldsymbol{v} - \boldsymbol{P}_r\boldsymbol{v} = \left(\boldsymbol{I} - \frac{1}{N}I_1\right)\boldsymbol{v} \tag{5.97}$$

The covariance of the residual Eq. (5.52) becomes:

$$E[\boldsymbol{e}\boldsymbol{e}^T] = \left(\boldsymbol{I} - \frac{1}{N}I_1\right)\boldsymbol{\Sigma}_v \tag{5.98}$$

Consider the covariance of the estimation error given by Eq. (5.96). We get

$$\text{cov}(\hat{\theta}) = 0 \text{ } as \text{ } N \to \infty \tag{5.99}$$

The asymptotic condition (5.39) is clearly satisfied as

$$\lim_{N\to\infty} it \left(\frac{\boldsymbol{H}^T\boldsymbol{H}}{N}\right)^{-1} = 1 \tag{5.100}$$

Further as the noise is zero-mean white noise, it satisfies Eq. (5.40)

$$\lim_{N\to\infty} it \frac{1}{N}\sum_{i=1}^{N} v(i) = 0 \tag{5.101}$$

Hence the residual approaches the noise asymptotically

$$\lim_{N\to\infty} it\, \boldsymbol{e} = \boldsymbol{v} \tag{5.102}$$

a. $\boldsymbol{W} = \boldsymbol{\Sigma}_v^{-1} = diag\left(1/\sigma_{v1} \quad 1/\sigma_{v1} \quad \cdots \quad 1/\sigma_{vn}\right)$

The error term $\boldsymbol{v}$ is a zero-mean independent but not identically distributed white noise with covariance $\boldsymbol{\Sigma}_v = diag\left(\sigma_{v1} \quad \sigma_{v1} \quad \cdots \quad \sigma_{vn}\right)$,

$$\boldsymbol{W} = \boldsymbol{\Sigma}_v^{-1} = diag\left(1/\sigma_{v1} \quad 1/\sigma_{v1} \quad \cdots \quad 1/\sigma_{vn}\right), \text{ and } \left(\boldsymbol{H}^T\boldsymbol{W}\boldsymbol{H}\right)^{-1} = \left(\sum_{i=1}^{N}\frac{1}{\sigma_{vi}^2}\right)^{-1} \text{ and}$$

$$\boldsymbol{H}^{\dagger} = \left(\boldsymbol{H}^T\boldsymbol{W}\boldsymbol{H}\right)^{-1}\boldsymbol{H}^T\boldsymbol{W} = \left(\sum_{i=1}^{N}\frac{1}{\sigma_{vi}^2}\right)^{-1}\begin{bmatrix}\frac{1}{\sigma_{v1}^2} & \frac{1}{\sigma_{v2}^2} & \cdots & \frac{1}{\sigma_{vN}^2}\end{bmatrix}$$

$$\boldsymbol{P}_r = \boldsymbol{H}\left(\boldsymbol{H}^T\boldsymbol{\Sigma}_v^{-1}\boldsymbol{H}\right)^{-1}\boldsymbol{H}^T\boldsymbol{\Sigma}_v^{-1}$$

The weighted least-squares estimate $\hat{\theta} = \boldsymbol{H}^{\dagger}\boldsymbol{y}$ is

$$\hat{\theta} = \left(\boldsymbol{H}^T\boldsymbol{\Sigma}_v^{-1}\boldsymbol{H}\right)^{-1}\boldsymbol{H}^T\boldsymbol{\Sigma}_v^{-1}\boldsymbol{y} = \left(\sum_{i=1}^{n}\frac{1}{\sigma_{vi}^2}\right)^{-1}\sum_{i=1}^{n}\frac{y(i)}{\sigma_{vi}^2} \tag{5.103}$$

The covariance of the estimation error is

$$\text{cov}(\hat{\theta}) = \left(\sum_{i=1}^{N}\frac{1}{\sigma_{vi}^2}\right)^{-1} \tag{5.104}$$

The residual is given by

$$\boldsymbol{e} = \left(\boldsymbol{I} - \boldsymbol{P}_r\right)\boldsymbol{v} = \boldsymbol{v} - H\left(\boldsymbol{H}^T\boldsymbol{\Sigma}_v^{-1}\boldsymbol{H}\right)^{-1}\boldsymbol{H}^T\boldsymbol{\Sigma}_v^{-1}\boldsymbol{v} = \left(\boldsymbol{I} - \left(\sum_{i=1}^{N}\frac{1}{\sigma_{vi}^2}\right)^{-1}\boldsymbol{I}_1\boldsymbol{\Sigma}_v^{-1}\right)\boldsymbol{v} \tag{5.105}$$

Covariance of the residual Eq. (5.52) becomes:

$$\text{cov}(\boldsymbol{e}) = \boldsymbol{\Sigma}_v - \boldsymbol{H}\left(\boldsymbol{H}^T\boldsymbol{\Sigma}_v^{-1}\boldsymbol{H}\right)^{-1}\boldsymbol{H}^T = \boldsymbol{\Sigma}_v - \left(\sum_{i=1}^{N}\frac{1}{\sigma_{vi}^2}\right)^{-1}\boldsymbol{I}_1 \tag{5.106}$$

Consider the covariance of the estimation error given by Eq. (5.104). We get

$$\text{cov}(\hat{\boldsymbol{\theta}}) = \left(\sum_{i=1}^{N}\frac{1}{\sigma_{vi}^2}\right)^{-1} = 0 \text{ as } N \to \infty \tag{5.107}$$

The asymptotic condition (5.39) is clearly satisfied as

$$\lim_{N\to\infty} it \frac{\boldsymbol{H}^T\boldsymbol{\Sigma}_v^{-1}\boldsymbol{H}}{N} = \lim_{N\to\infty} it \frac{1}{N}\sum_{i=1}^{n}\frac{1}{\sigma_{vi}^2} < \infty \tag{5.108}$$

Further as the noise is zero-mean white noise, it satisfies Eq. (5.40)

$$\lim_{N\to\infty} it \frac{1}{N}\sum_{i=1}^{N}\frac{v(i)}{\sigma_{vi}^2} = 0 \tag{5.109}$$

Hence the residual approaches the noise asymptotically

$$\lim_{N\to\infty} it\, \boldsymbol{e} = \boldsymbol{v} \tag{5.110}$$

b. $\boldsymbol{W} = \boldsymbol{\Sigma}_v^{-1}$;
The error term $\boldsymbol{v}$ is a colored noise with non-diagonal covariance $\boldsymbol{\Sigma}_v$ matrix. The pseudo-inverse and the projection matrix are $\boldsymbol{H}^\dagger = \left(\boldsymbol{H}^T\boldsymbol{\Sigma}_v^{-1}\boldsymbol{H}\right)^{-1}\boldsymbol{H}^T\boldsymbol{\Sigma}_v^{-1}$; $\boldsymbol{P}_r = \boldsymbol{H}\left(\boldsymbol{H}^T\boldsymbol{\Sigma}_v^{-1}\boldsymbol{H}\right)^{-1}\boldsymbol{H}^T\boldsymbol{\Sigma}_v^{-1}$. The information matrix $\boldsymbol{H}^T\boldsymbol{\Sigma}_v^{-1}\boldsymbol{H}$ is a positive scalar as $\boldsymbol{H} = \begin{bmatrix}1 & 1 & . & 1\end{bmatrix}^T$.

$$\hat{\theta} = \boldsymbol{H}^\dagger\boldsymbol{y} = \left(\boldsymbol{H}^T\boldsymbol{\Sigma}_v^{-1}\boldsymbol{H}\right)^{-1}\boldsymbol{H}^T\boldsymbol{\Sigma}_v^{-1}\boldsymbol{y} \tag{5.111}$$

The covariance of the estimation error is

$$\text{cov}(\hat{\boldsymbol{\theta}}) = \left(\boldsymbol{H}^T\boldsymbol{\Sigma}_v^{-1}\boldsymbol{H}\right)^{-1} \tag{5.112}$$

The residual is given by

$$\boldsymbol{e} = \boldsymbol{v} - \boldsymbol{P}_r\boldsymbol{v} = \boldsymbol{v} - \boldsymbol{H}\left(\boldsymbol{H}^T\boldsymbol{\Sigma}_v^{-1}\boldsymbol{H}\right)^{-1}\boldsymbol{H}^T\boldsymbol{\Sigma}_v^{-1}\boldsymbol{v} \tag{5.113}$$

Consider the covariance of the estimation error given by Eq. (5.112). Let $\bar{\boldsymbol{h}}_i$ be the ith columnvector of $\bar{\boldsymbol{H}} = \boldsymbol{\Sigma}_v^{-1/2}\boldsymbol{H}$ so that $\boldsymbol{H}^T\boldsymbol{\Sigma}_v^{-1}\boldsymbol{H} = \sum_{i=1}^{N}\bar{\boldsymbol{h}}_i^T\bar{\boldsymbol{h}}_i$. Since the terms $\{\bar{\boldsymbol{h}}_i^T\bar{\boldsymbol{h}}_i\}$ are positive and $\boldsymbol{H}^T\boldsymbol{\Sigma}_v^{-1}\boldsymbol{H} = \sum_{i=1}^{N}\bar{\boldsymbol{h}}_i^T\bar{\boldsymbol{h}}_i$ is a sum of positive values, the expression for the covariance Eq. (5.112) becomes:

$$\text{cov}(\hat{\boldsymbol{\theta}}) = \left(\sum_{i=1}^{N}\bar{\boldsymbol{h}}_i^T\bar{\boldsymbol{h}}_i\right)^{-1} = 0 \; as \; N \to \infty \tag{5.114}$$

Since $\boldsymbol{H}^T\boldsymbol{\Sigma}_v^{-1}\boldsymbol{H}$ is positive definite matrix, using the Appendix, the asymptotic condition (5.39) is clearly satisfied as

$$0 < \lim_{N\to\infty} it \left(\frac{\boldsymbol{H}^T\boldsymbol{\Sigma}_v^{-1}\boldsymbol{H}}{N}\right)^{-1} < \infty \tag{5.115}$$

Further as the noise is a zero-mean white noise, it satisfies Eq. (5.40)

$$\lim_{N\to\infty} it \left\{\frac{\boldsymbol{H}^T\boldsymbol{W}\boldsymbol{v}}{N}\right\} = 0 \tag{5.116}$$

Hence the residual approaches the noise asymptotically

$$\lim_{N\to\infty} it\, \boldsymbol{e} = \boldsymbol{v} \tag{5.117}$$

5.4.1 *Non-Zero-Mean Measurement Noise*

In some applications there is a bias error due to calibration error, an offset or other external input effect on the measurements. If the bias error is an unknown constant, we may club the unknown bias error with the desired parameter to be estimated, and estimate an augmented unknown parameter formed of the desired parameter and the bias parameter.

Example 5.6 Non-zero-mean error term

$$E[\boldsymbol{v}] = \boldsymbol{\mu} \tag{5.118}$$

The linear model can be rewritten with an error term which is zero-mean

$$\boldsymbol{y} = \boldsymbol{H\theta} + \boldsymbol{\mu} + \bar{\boldsymbol{v}} \tag{5.119}$$

where $\bar{\boldsymbol{v}}$ is a zero-mean term. Define augmented feature vector $\boldsymbol{\theta}_{ag} = [\,\boldsymbol{\theta} \quad \boldsymbol{\mu}\,]^T$, and the augmented data matrix $\boldsymbol{H}_{ag} = [\,\boldsymbol{H} \quad \mathbf{1}\,]$ where $\mathbf{1}$ is a $Nx1$ vector of all ones. Then the estimate of the augmented vector $\hat{\boldsymbol{\theta}}_{ag}$ becomes

$$\hat{\boldsymbol{\theta}}_{ag} = \left(\boldsymbol{H}_{ag}^T \boldsymbol{W} \boldsymbol{H}_{ag}\right)^{-1} \boldsymbol{H}_{ag} \boldsymbol{W} \boldsymbol{y} \tag{5.120}$$

Remarks

- The over-determined set of equations attenuate the effect of noise.
- The larger the number of data samples N compared to the number of unknown parameters M, the smaller the covariance of the estimation error performance of the least-squares estimator in the presence of noise as can be deduced from Eqs. (5.96), (5.107), and (5.114).
- The larger the number of data samples N compared to the number of unknown parameters M, the smaller the mean-squared residual as shown in Eqs. (5.98), (5.105). and (5.113).

5.5 Cramér–Rao Lower Bound

The most desirable property of an estimator is that it is unbiased and has the lowest possible parameter error covariance. In Section 5.3 we showed that the least-squares estimator is unbiased and has the lowest possible error covariance among the class of all linear unbiased estimators. A question arises as to how the linear least-squares compares with a class of all (instead of a restricted class of linear) unbiased estimators. The CRLB gives the theoretically minimal variance for class of all unbiased estimators of a deterministic parameter. It is widely used to measure the efficiency of an estimator. A measure of efficiency is the ratio of the theoretically minimal variance given by CRLB to the actual variance of the estimator. This measure is less than or equal to 1. An estimator with efficiency 1.0 is said to be an "efficient estimator." In many applications, for mathematical tractability, an estimator is derived by making simplifying assumptions. The performance of the estimator is evaluated using the measure of efficiency. For example, the least-squares estimator is efficient if the PDF is Gaussian and the weighting matrix is chosen to be the inverse of the covariance matrix of the measurement. The derivation of the CRLB is simple. The estimator is assumed to be unbiased and the PDF is assumed to be absolutely continuous. Differentiating the expression for an unbiased estimator with respect to the parameter, the CRLB is derived using Cauchy–Schwartz inequality. The lower bound is shown to be equal to the *Fisher information*. Although there exists a lower bound, an estimator which achieves the lower bound may not exist. A condition for the existence of the lower bound is given. It is shown that, if there exists a lower bound, then it is given by the maximum likelihood estimator.

Let $\mathbf{y} = [\,y(1) \quad y(2) \quad \cdots \quad y(N)\,]^T$ is a $Nx1$ vector of measurements characterized by the probability density function (PDF) $f_y(\mathbf{y})$. The measurement $\mathbf{y}$ is function of an unknown scalar parameter $\boldsymbol{\theta}$. Let $\hat{\boldsymbol{\theta}}$ be an unbiased estimator of θ that is only a function of $\mathbf{y}$ and not a function of the unknown parameter $\boldsymbol{\theta}$. Then

$$\text{cov}(\hat{\boldsymbol{\theta}}) = E[(\hat{\boldsymbol{\theta}} - \boldsymbol{\theta})(\hat{\boldsymbol{\theta}} - \boldsymbol{\theta})^T] \geq \boldsymbol{I}_F^{-1} \tag{5.121}$$

where the MxM Fisher information matrix

$$\boldsymbol{I}_F(\theta) = E\left[\left(\frac{\delta \ln f_y(\mathbf{y})}{\delta \boldsymbol{\theta}}\right)\left(\frac{\delta \ln f_y(\mathbf{y})}{\delta \boldsymbol{\theta}}\right)^T\right].$$

5.6 Maximum Likelihood Estimation

The maximum likelihood estimator is widely used as the estimate gives the minimum estimation error covariance, and serves as a gold standard for evaluating the performance of other estimators. It is efficient as it achieves the Cramér–Rao lower bound if it exists. It is based on maximizing a likelihood function of the PDF of the data expressed as a function of the parameter to be estimated. In general the estimates are implicit nonlinear functions of the parameter to be estimated and the estimate is obtained recursively. If the PDF of the measurement data is Gaussian, the maximum likelihood estimation method simplifies to a weighted least-squares method where the weight is the inverse of the noise covariance matrix. The ML estimate is obtained by maximizing the log-likelihood function

$$\hat{\boldsymbol{\theta}} = \arg\max_{\theta} \left\{ \log f_y(\mathbf{y}) \right\} \tag{5.122}$$

5.6.1 Illustrative Examples

Example 5.7 Efficiency of the least-squares estimate

Consider the measurement model $\mathbf{y} = \boldsymbol{H\theta} + \mathbf{v}$ where $\mathbf{v}$ is a zero-mean Gaussian PDF with covariance $\boldsymbol{\Sigma}_v$. The PDF of $\mathbf{y}$ is

$$f_y(\mathbf{y}) = \frac{1}{\sqrt{(2\pi)^N \det\left(\boldsymbol{\Sigma}_v\right)}} \exp\left\{ -\frac{(\mathbf{y} - \boldsymbol{H\theta})^T\, \boldsymbol{\Sigma}_v^{-1}\, (\mathbf{y} - \boldsymbol{H\theta})}{2} \right\} \tag{5.123}$$

The log-likelihood function $\ln f_y(\boldsymbol{\theta}|\mathbf{y})$ becomes

$$\ln f_y(\mathbf{y}) = -\frac{(\mathbf{y} - \boldsymbol{H\theta})^T\, \boldsymbol{\Sigma}_v^{-1}\, (\mathbf{y} - \boldsymbol{H\theta})}{2} - \frac{N}{2} \ln 2\pi - \frac{1}{2} \ln \det \boldsymbol{\Sigma} \tag{5.124}$$

Differentiating $\ln f_y(\mathbf{y})$ with respect to $\boldsymbol{\theta}$ and setting to zero the ML estimate $\hat{\boldsymbol{\theta}}(\mathbf{y})$ satisfies

$$\frac{\delta}{\delta \boldsymbol{\theta}} \ln f_y(\mathbf{y}) = \boldsymbol{H}^T\, \boldsymbol{\Sigma}_v^{-1} (\mathbf{y} - \boldsymbol{H\theta}) \tag{5.125}$$

The Fisher information is given by:

$$I_F(\theta) = E\left[\left(\frac{\delta \ln f_y(y)}{\delta\theta}\right)\left(\frac{\delta \ln f_y(y)}{\delta\theta}\right)^T\right] = E\left[H^T \Sigma_v^{-1}(y - H\theta)(y - H\theta)^T \Sigma_v^{-1} H\right] \quad (5.126)$$

Using $y = H\theta + v$ we get:

$$I_F = H^T \Sigma_v^{-1} H \quad (5.127)$$

Using the expression of the covariance of the estimation error given by Eq. (5.37), we get

$$\text{cov}(\hat{\theta}) = \left(H^T \Sigma_v^{-1} H\right)^{-1} = I_F^{-1}(\theta) \quad (5.128)$$

We can deduce from the inequality (5.121) that the covariance of the estimation error is efficient and the Fisher information is:

$$I_F = H^T \Sigma_v^{-1} H \quad (5.129)$$

Example 5.8 Maximum likelihood and the least-squares approach
Consider the Gaussian PDF given in the Example 5.7. Differentiating $\ln f_y(y)$ with respect to θ and setting to zero the ML estimate $\hat{\theta}_{ML}$ satisfies

$$\frac{\delta}{\delta\theta} \ln f_y(y) = H^T \Sigma_v^{-1}(y - H\theta) = 0$$

Assuming that H has a full rank $\left(H^T \Sigma_v^{-1} H\right)$ is non-singular. Multiplying by $\left(H^T \Sigma_v^{-1} H\right)^{-1}$

$$\left(H^T \Sigma_v^{-1} H\right)^{-1} \frac{\delta}{\delta\theta} \ln f_y(y) = \left(H^T \Sigma_v^{-1} H\right)^{-1} H^T \Sigma_v^{-1} y - \theta = 0 \quad (5.130)$$

The ML estimator $\hat{\theta}_{ML}$ is that value of θ that satisfies the above equation:

$$\hat{\theta}_{ML} = \left(H^T \Sigma_v^{-1} H\right)^{-1} H^T \Sigma_v^{-1} y \quad (5.131)$$

This shows that weighted least-squares estimate and the maximum likelihood are identical if $W = \Sigma_v^{-1}$.

Comments *If the PDF of the noise* **v** *is a zero-mean Gaussian then*

- *The weighted least-squares with weight* $W = \Sigma_v^{-1}$ *and the ML estimators are both efficient as they attain the Cramér–Rao lower bound. They both yield the minimum variance unbiased estimator.*
- *The Cramér–Rao lower bound for both ML and the weighted least-squares method is*

$$I_F^{-1}(\theta) = \left(H^T \Sigma_v^{-1} H\right)^{-1}$$

5.6.1.1 Illustrative Example

Consider the measurement model with scalar unknown parameters but vector measurements $\boldsymbol{y} = \boldsymbol{H}\boldsymbol{\theta} + \boldsymbol{v}$ given in Example 5.7

a) The error term $\boldsymbol{v}$ is an independent and identically distributed zero-mean white noise with variance $E[\boldsymbol{v}^T\boldsymbol{v}] = \sigma_v^2$. Choose $\boldsymbol{W} = \boldsymbol{I}$. The estimate is $\hat{\theta} = \frac{1}{N}\sum_{i=1}^{N} y(i)$ and the variance of the estimation error is $\text{var}(\hat{\boldsymbol{\theta}}) = \frac{\sigma_v^2}{N}$. Let us verify whether the estimate is efficient. The Cramér–Rao lower bound is given by

$$\boldsymbol{I}_F^{-1}(\theta) = \left(\boldsymbol{H}^T\boldsymbol{\Sigma}_v^{-1}\boldsymbol{H}\right)^{-1} = \frac{\sigma_v^2}{N} \tag{5.132}$$

It is efficient as the covariance of the estimator equals the Cramér–Rao lower bound. The Fisher information is

$$\boldsymbol{I}_F(\theta) = \left(\boldsymbol{H}^T\boldsymbol{\Sigma}_v^{-1}\boldsymbol{H}\right) = \frac{N}{\sigma_v^2} \tag{5.133}$$

b) The error term $\boldsymbol{v}$ is a zero-mean independent but not identically distributed white noise with covariance $\boldsymbol{\Sigma}_v = diag(\,\sigma_{v1} \quad \sigma_{v1} \quad \cdots \quad \sigma_{vn}\,)$, $\boldsymbol{W} = \boldsymbol{\Sigma}_v^{-1} = diag\left(1/\sigma_{v1} \quad 1/\sigma_{v1} \quad \cdots \quad 1/\sigma_{vn}\right)$. The estimate and the variance of the estimation error are

$$\hat{\theta} = \left(\sum_{i=1}^{n}\frac{1}{\sigma_{vi}^2}\right)^{-1}\sum_{i=1}^{n}\frac{y(i)}{\sigma_{vi}^2} \quad \text{and} \quad \text{var}(\hat{\boldsymbol{\theta}}) = \left(\sum_{i=1}^{N}\frac{1}{\sigma_{vi}^2}\right)^{-1}$$

The Cramér–Rao lower bound is given by

$$\boldsymbol{I}_F^{-1}(\boldsymbol{\theta}) = \left(\boldsymbol{H}^T\boldsymbol{\Sigma}_v^{-1}\boldsymbol{H}\right)^{-1} = \left(\sum_{i=1}^{N}\frac{1}{\sigma_{vi}^2}\right)^{-1} \tag{5.134}$$

It is efficient and the Fisher information is

$$\boldsymbol{I}_F(\theta) = \left(\boldsymbol{H}^T\boldsymbol{\Sigma}_v^{-1}\boldsymbol{H}\right) = \sum_{i=1}^{N}\frac{1}{\sigma_{vi}^2} \tag{5.135}$$

5.7 Least-Squares Solution of Under-Determined System

If (i) the number of unknowns M is less than the number of equations N or (ii) the number of unknowns M is equal to the number of equations N but the resulting square matrix is singular, the system of equation is called *under-determined*. In this case

- A solution will always exist as the observation is always in the range space of the matrix.
- There are infinite solutions.

Since there are infinite solutions, it is preferable to choose a solution from the infinite set of solutions, the onewhich has a minimum norm. Let $\boldsymbol{H}$ be a $N \times M$ data matrix with $N < M$. Assuming $\boldsymbol{H}$ has a full rank, which implies $\boldsymbol{H}\boldsymbol{H}^T$ is invertible, the optimal solution $\hat{\boldsymbol{\theta}}$ is

$$\hat{\theta} = \boldsymbol{H}^T(\boldsymbol{H}\boldsymbol{H}^T)^{-1}\boldsymbol{y} \tag{5.136}$$

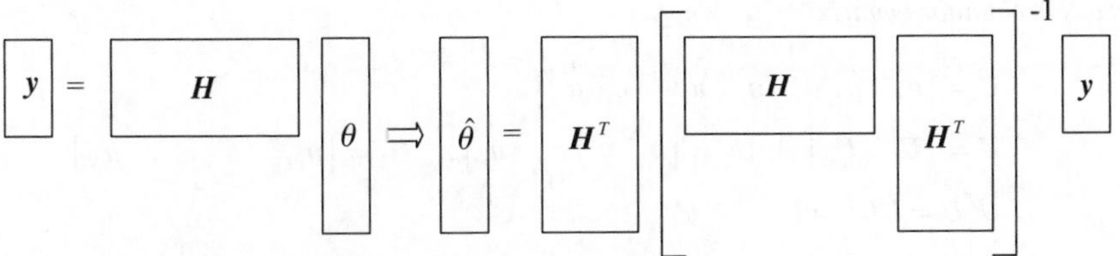

Figure 5.6 Visual representation of the under-determined system

Similarly to the over-determined case the projection matrix is given by

$$P_r = H^T(HH^T)^{-1}H \tag{5.137}$$

As an under-determined system has infinite solution, the general solution is given by

$$\hat{\theta} = H^T(HH^T)^{-1}y + (I - P_r)\theta_0 \tag{5.138}$$

where θ_0 is arbitrary. It can be verified that $\hat{\theta}$ is a solution of $H\hat{\theta} = y$. The estimate $\hat{y}$ of y becomes

$$\hat{y} = H\hat{\theta} = HH^T(HH^T)^{-1}y = y \tag{5.139}$$

Hence residual e

$$e = 0 \tag{5.140}$$

Comments *One cannot compare θ and $\hat{\theta}$. In general $\hat{\theta} \neq \theta$ since $\hat{\theta}$ is a minimum norm solution whereas θ is one of the infinite solutions of $H\theta = y$.*

The visual representation of the linear equation $y = H\theta$ and its solution $\hat{\theta} = H^T(HH^T)^{-1}y$ is given in the Figure 5.6. See the Appendix for the proof of the least-squares estimate.

5.8 Singular Value Decomposition

The SVD is one of the most elegant algorithms in numerical algebra for providing quantitative information about the structure of the system of linear equations. The SVD provides a robust solution of both the over-determined and under-determined least-squares problem, matrix approximation, conditioning of ill-conditioned matrices, and principal component analysis. It is employed in a variety of signal processing applications including spectrum analysis, filter design, system identification, model order reduction, estimation, image compression, and data reduction. It is the basis for MIMO robustness analysis using SVD plots of various frequency response plots. It also appears in many of the standard algorithms of robust control, such as H_∞ and H_2 synthesis and state-space balancing. It can be used to improve SNR for large arrays by identifying the directions or linear combinations of sensors which have the greatest sensitivity to the system's important parameters, such as modes of vibrations. It has an important role in least-squares solutions with application to signal processing, estimation, and system identification.

Theorem *Every NxM matrix H can be decomposed as*

$$H = USV^T \tag{5.141}$$

where U is $N \times N$ real matrix of rank $r \leq \min(M, N)$ where

- *U is a $N \times N$ unitary matrix*

$$U = \begin{bmatrix} u_1 & u_2 & . & u_r & u_{r+1} & . & u_N \end{bmatrix}$$
$$U = \begin{bmatrix} U_1 & U_2 \end{bmatrix}, \quad U_1 = \begin{bmatrix} u_1 & u_2 & . & u_r \end{bmatrix}, \quad U_2 = \begin{bmatrix} u_{r+1} & u_{r+2} & . & u_N \end{bmatrix}$$
$$U^T U = UU^T = I$$

where u_i is Nx1 vector and U is called the left singular matrixof H

- *V is a MxM unitary matrix*

$$V = \begin{bmatrix} v_1 & v_2 & . & v_r & v_{r+1} & . & v_M \end{bmatrix}$$
$$V = \begin{bmatrix} V_1 & V_2 \end{bmatrix}, \quad V_1 = \begin{bmatrix} v_1 & v_2 & . & v_r \end{bmatrix}, \quad V_2 = \begin{bmatrix} v_{r+1} & v_{r+2} & . & v_M \end{bmatrix}$$
$$V = \begin{bmatrix} V_1 & V_2 \end{bmatrix}, \quad V_1 = \begin{bmatrix} v_1 & v_2 & . & v_r \end{bmatrix}, \quad V_2 = \begin{bmatrix} v_{r+1} & v_{r+2} & . & v_M \end{bmatrix}$$
$$VV^T = V^T V = I$$

where v_i is a $M \times 1$ vector, V is called the right singular matrix of H.

- *S is a $N \times M$ rectangular matrix given by.*

$$S = \begin{bmatrix} \Sigma & 0 \\ 0 & 0 \end{bmatrix}, \quad \Sigma = diag \begin{bmatrix} \sigma_1 & \sigma_2 & \sigma_3 & . & \sigma_r \end{bmatrix}$$

$\{\sigma_i\}$ are the singular values of H which are positive

$$\sigma_1 \geq \sigma_2 \geq \sigma_3 \geq \cdots \geq \sigma_r > 0 \quad \text{and} \quad \sigma_{r+1} = \sigma_{r+2} = \cdots = \sigma_M = 0$$

The matrix H is decomposed into r *NxM matrices $\{u_i v_i^T\}$ of unity rank:*

$$H = \sum_{i=1}^{r} \sigma_i u_i v_i^T \tag{5.142}$$

Thus $H^T H$ and HH^T have the same eigenvalues $\{\sigma_i^2\}$ but with $\{v_i\}$ and $\{u_i\}$ respectively as eigenvectors.

The choice of $H^T H$ or HH^T in the definition of singular values is arbitrary. While these two matrices have different sizes and, therefore, a different number of eigenvalues (M eigenvalues for $H^T H$ and N for HH^T), they have similar non-zero eigenvalues, the positive roots of which are called the singular values of matrix A. Finally, it is obvious that the singular values of H and H^T are the same. The singular values of $H(H^T)$ the positive square roots of the eigenvalues of $H^T H$ (HH^T). A pictorial representation of SVD namely $H = USV^T$ is shown in Figure 5.7.

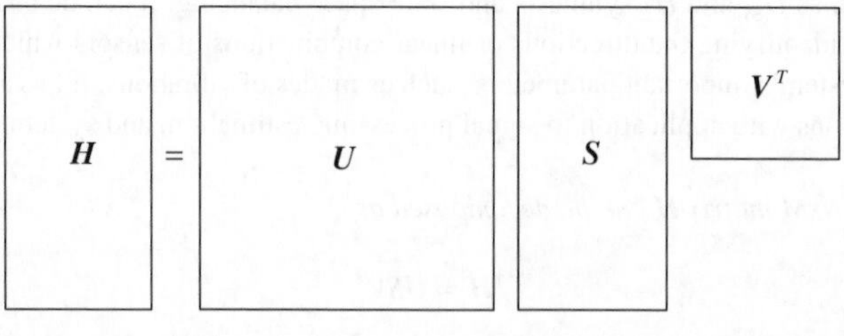

H = U S V^T

Figure 5.7 Pictorial representation of Singular Value Decomposition

5.8.1 *Illustrative Example: Singular and Eigenvalues of Square Matrices*

Example 5.9 Non-symmetric square matrix
Consider a square matrix, $\boldsymbol{H}$

$$\boldsymbol{H} = \begin{bmatrix} 1 & 2 \\ 3 & 4 \end{bmatrix}$$

The eigenvalue-eigenvector decomposition of $\boldsymbol{H}$ is

$$\begin{bmatrix} 1 & 2 \\ 3 & 4 \end{bmatrix} = \begin{bmatrix} -0.8246 & -0.4160 \\ 0.5658 & -0.9094 \end{bmatrix} \begin{bmatrix} -0.3723 & 0 \\ 0 & 5.3723 \end{bmatrix} \begin{bmatrix} -0.9231 & 0.4222 \\ -0.5743 & -0.8370 \end{bmatrix}$$

The singular value decomposition of $\boldsymbol{H}$ is

$$\begin{bmatrix} 1 & 2 \\ 3 & 4 \end{bmatrix} = \begin{bmatrix} -0.4046 & -0.9145 \\ -0.9145 & 0.4046 \end{bmatrix} \begin{bmatrix} 5.4650 & 0 \\ 0 & 0.3660 \end{bmatrix} \begin{bmatrix} -0.5760 & 0.8174 \\ -0.8174 & -0.5760 \end{bmatrix}$$

Example 5.10 Symmetric square matrix

$$\boldsymbol{H} = \begin{bmatrix} 1 & 2 \\ 2 & 3 \end{bmatrix}$$

The eigenvalue-eigenvector decomposition of $\boldsymbol{H}$ is

$$\begin{bmatrix} 1 & 2 \\ 2 & 3 \end{bmatrix} = \begin{bmatrix} -0.8507 & 0.5257 \\ 0.5257 & 0.8507 \end{bmatrix} \begin{bmatrix} -0.2361 & 0 \\ 0 & 4.2361 \end{bmatrix} \begin{bmatrix} -0.8507 & 0.5257 \\ 0.5257 & 0.8507 \end{bmatrix}$$

The singular value decomposition of $\boldsymbol{H}$ is

$$\begin{bmatrix} 1 & 2 \\ 2 & 3 \end{bmatrix} = \begin{bmatrix} -0.5257 & -0.8507 \\ -8507 & 0.5257 \end{bmatrix} \begin{bmatrix} 4.2361 & 0 \\ 0 & 0.2361 \end{bmatrix} \begin{bmatrix} -0.5257 & 0.8507 \\ -8507 & -0.5257 \end{bmatrix}$$

Comment *The SVD and eigenvalue decomposition for a square matrix are in general different. If, however, the matrix is symmetric, then singular values are the absolute values of the eigenvalues.*

Example 5.11 Rectangular matrix

$$\boldsymbol{H} = \begin{bmatrix} 1 & 2 \\ 2 & 4 \\ 3 & 6 \end{bmatrix}$$

The SVD of $\boldsymbol{H}$ is given by

$$\boldsymbol{U} = \begin{bmatrix} -0.2673 & 0.9562 & 0.1195 \\ -0.5345 & -0.0439 & -0.8440 \\ -0.8018 & -0.2895 & 0.5228 \end{bmatrix}, \quad \boldsymbol{V} = \begin{bmatrix} -0.4472 & -0.8944 \\ -0.8944 & 0.4472 \end{bmatrix}, \quad S = \begin{bmatrix} 8.3666 & 0 \\ 0 & 0 \\ 0 & 0 \end{bmatrix}$$

Its rank is 1 as it has only one non-zero singular value. The columns of $\boldsymbol{H}$ are linearly dependent.

5.8.2 Computation of Least-Squares Estimate Using the SVD

The least-squares problem was formulated and solved assuming that the data matrix $\boldsymbol{H}$ is of full rank. Thanks to the SVD decomposition, the full rank restriction may be lifted. Consider the linear measurement model (5.1) where the data matrix $\boldsymbol{H}$ need not be of full rank. Its rank is $r \leq \min(M, N)$.

5.8.2.1 Un-Weighted Least-Squares Estimate

The least-squares estimate $\hat{\boldsymbol{\theta}}$ is given by

$$\hat{\boldsymbol{\theta}} = \boldsymbol{V} \begin{bmatrix} \boldsymbol{\Sigma}^{-1} & \boldsymbol{0} \\ \boldsymbol{0} & \boldsymbol{0} \end{bmatrix} \boldsymbol{U}^T \boldsymbol{y} \tag{5.143}$$

where $\boldsymbol{\Sigma}^{-1} = diag\left(\sigma_1^{-1} \ \sigma_2^{-1} \ \cdots \ \sigma_r^{-1} \right)$ $\boldsymbol{U}$, $\boldsymbol{V}$, and $\boldsymbol{\Sigma}$ are the SVD of the weighted matrix $\boldsymbol{H}$

$$\boldsymbol{H} = \boldsymbol{U} \begin{bmatrix} \boldsymbol{\Sigma} & \boldsymbol{0} \\ \boldsymbol{0} & \boldsymbol{0} \end{bmatrix} \boldsymbol{V}^T \tag{5.144}$$

Compare the estimate $\hat{\boldsymbol{\theta}}$ for the case when $\boldsymbol{H}$ is of full rank given by Eq. (5.18). The pseudo-inverse becomes

$$\boldsymbol{H}^{\dagger} = \boldsymbol{V} \begin{bmatrix} \boldsymbol{\Sigma}^{-1} & \boldsymbol{0} \\ \boldsymbol{0} & \boldsymbol{0} \end{bmatrix} \boldsymbol{U}^T \tag{5.145}$$

The pseudo-inverse $\boldsymbol{H}^{\dagger}$ may be expressed as linear combinations of r rank-1 matrices

$$\boldsymbol{H}^{\dagger} = \sum_{i=1}^{r} \frac{1}{\sigma_i} \boldsymbol{u}_i \boldsymbol{v}_i^T \tag{5.146}$$

5.8.2.2 Weighted Least-Squares Estimate

In the case of the weighted least-squares method the transformation given in Eq. (5.9) is employed where $\boldsymbol{W} = \boldsymbol{W}^{1/2}\boldsymbol{W}^{1/2}$, $\bar{\boldsymbol{y}} = \boldsymbol{W}^{1/2}\boldsymbol{y}$, $\bar{\boldsymbol{v}} = \boldsymbol{W}^{1/2}\boldsymbol{v}$, and $\bar{\boldsymbol{H}} = \boldsymbol{W}^{1/2}\boldsymbol{H}$. The transformed variables $\bar{\boldsymbol{y}} = \boldsymbol{W}^{1/2}\boldsymbol{y}$ and $\bar{\boldsymbol{H}} = \boldsymbol{W}^{1/2}\boldsymbol{H}$ are employed instead of $\boldsymbol{y}$ and $\boldsymbol{H}$ respectively in the SVD computations given in Eqs. (5.143) and (5.144)

- The bounds on the covariance of the estimation error are

$$\|cov(\hat{\boldsymbol{\theta}})\| \leq \begin{cases} \dfrac{\sigma_{\max}^2\left(\boldsymbol{\Sigma}_v\right)}{\sigma_{\min}^2\left(\boldsymbol{\Sigma}_v^{-1/2}\boldsymbol{H}\right)} & if\ E\left[\boldsymbol{v}\boldsymbol{v}^T\right] = \boldsymbol{\Sigma}_v \\ \dfrac{\sigma_v^2}{\sigma_{\min}^2(\boldsymbol{H})} & if\ E\left[\boldsymbol{v}\boldsymbol{v}^T\right] = \boldsymbol{I}\sigma_v^2 \end{cases} \tag{5.147}$$

- The bounds on the residual.

$$\sigma_{res}^2 \leq \begin{cases} \dfrac{1}{N} trace\left(\boldsymbol{\Sigma}_v\right) & if\ E\left[\boldsymbol{v}\boldsymbol{v}^T\right] = \boldsymbol{\Sigma}_v \\ \sigma_v^2 & if\ E\left[\boldsymbol{v}\boldsymbol{v}^T\right] = \sigma_v^2\boldsymbol{I} \end{cases} \tag{5.148}$$

Comment *If data matrix $\boldsymbol{H}$ is not of maximal rank $r < \min(M,N)$ the SVD gives a least-squares solution $\hat{\theta} = \boldsymbol{H}^{\dagger}\boldsymbol{y}$ such that $\|\hat{\theta}\|$ has a minimum norm.*

5.8.2.3 Ill-Conditioned Matrices

Matrices in which small errors in the matrix elements are substantially amplified to produce large deviations in the solution are referred to as *ill-conditioned matrices*. Ill-conditioned matrices are those which are close to being rank deficient. Associated with each matrix is a number called the condition number, which indicates the degree to which it is ill-conditioned. The condition number of the matrix $\boldsymbol{H}$ denoted $\kappa(\boldsymbol{H})$ is defined in terms of the norms of the matrix and the pseudo-inverse matrix. Since the norm of the matrix and its inverse are respectively the maximum and minimum singular values of the matrix, the condition number is

$$\kappa(\boldsymbol{H}) = \|\boldsymbol{H}\|\|\boldsymbol{H}^{\dagger}\| = \frac{\sigma_{\max}(\boldsymbol{H})}{\sigma_{\min}(\boldsymbol{H})} \tag{5.149}$$

The condition number $\kappa(\boldsymbol{H})$ gives a measure of how much the error is in the data matrix $\boldsymbol{H}$ and the measurement $\boldsymbol{y}$ may be magnified in computing the solution of $\boldsymbol{y} = \boldsymbol{H\theta}$. Thus it is desirable to have $\kappa(\boldsymbol{H})$ close to 1. A poorly conditioned data matrix $\boldsymbol{H}$ will have "small" singular values. As a consequence, components of $\boldsymbol{H}^{\dagger}$ with small singular values in Eq. (5.146) will be amplified. Hence the pseudo-inverse will be very sensitive to any the small changes in the singular vectors $\boldsymbol{u}_i$, $\boldsymbol{v}_i$ or both. To avoid this problem, set the small singular values to zero and obtain a truncated $\boldsymbol{\Sigma}$ and compute the pseudo-inverse (5.145) using the truncated $\boldsymbol{\Sigma}$. However, determining which of the singular values may be deemed small is problem dependant and requires a good judgment.

Example 5.12 Ill-conditioned matrix

Consider the measurement model $\boldsymbol{y} = \boldsymbol{H\theta} + \boldsymbol{v}$ where $\boldsymbol{H} = \begin{bmatrix} 1+\varepsilon & 2 \\ 2+\varepsilon & 4 \\ 3+\varepsilon & 6 \end{bmatrix}$ where $\varepsilon = 0.0001$, $\boldsymbol{\theta} = \begin{bmatrix} 1 \\ 2 \end{bmatrix}$, $y = \begin{bmatrix} 5 \\ 10 \\ 15 \end{bmatrix}$. The data matrix is ill-conditioned as the column vectors are almost linearly dependent. The singular values of $\boldsymbol{H}$ are 8.3673 and 0.0005. The least-squares estimate $\hat{\theta}$ is given by

$$\hat{\theta} = \boldsymbol{V}\begin{bmatrix} \boldsymbol{\Sigma}^{-1} & \boldsymbol{0} \\ \boldsymbol{0} & \boldsymbol{0} \end{bmatrix}\boldsymbol{U}^T\boldsymbol{y} = \begin{bmatrix} -1.6666 \\ 3.3333 \end{bmatrix}$$

$$\hat{y} = \begin{bmatrix} 5.0019 \\ 10.0005 \\ 14.9990 \end{bmatrix}$$

The SVD of $\boldsymbol{H}$ is

$$\boldsymbol{U} = \begin{bmatrix} -0.2674 & 0.4081 & -0.8729 \\ -0.5345 & -0.8166 & -0.2180 \\ -0.8018 & 0.4083 & 0.4365 \end{bmatrix} \quad \boldsymbol{V} = \begin{bmatrix} -0.4473 & 0.8944 \\ -0.8944 & -0.4473 \end{bmatrix}, \quad \boldsymbol{S} = \begin{bmatrix} 8.3673 & 0 \\ 0 & 0.0005 \\ 0 & 0 \end{bmatrix}$$

The condition number is $\kappa(\boldsymbol{H}) = \dfrac{\sigma_{\max}(\boldsymbol{H})}{\sigma_{\min}(\boldsymbol{H})}$

Since the smallest singular value 0.0005 is very small compared to the largest 8.3673, let us zero the small singular value and compute the inverse using the truncated $\boldsymbol{\Sigma} = 8.3673$.

$$\hat{\theta} = \boldsymbol{V}\begin{bmatrix} 1/8.3673 & \mathit{0} \\ \mathit{0} & \mathit{0} \end{bmatrix}\boldsymbol{U}^T\boldsymbol{y} = \begin{bmatrix} 1 \\ 2 \end{bmatrix}$$

$$\hat{y} = \begin{bmatrix} 5.0003 \\ 9.9999 \\ 15.0 \end{bmatrix}$$

The estimate is very accurate with truncation. However, extreme care is required to determine when the singular value is small.

$$\|cov(\hat{\theta})\| \le \begin{cases} \dfrac{\sigma_{\max}^2\left(\boldsymbol{\Sigma}_v\right)}{\sigma_{\min}^2(\boldsymbol{H})} & \text{if } E\left[\boldsymbol{v}\boldsymbol{v}^T\right] = \boldsymbol{\Sigma}_v \\ \dfrac{\sigma_v^2}{\sigma_{\min}^2(\boldsymbol{H})} & \text{if } E\left[\boldsymbol{v}\boldsymbol{v}^T\right] = \sigma_v^2 \end{cases}$$

5.9 Summary

Model

$$\boldsymbol{y} = \boldsymbol{H}\boldsymbol{\theta} + \boldsymbol{v}$$

Objective Function

$$\min_{\hat{\theta}}\{(\boldsymbol{y} - \boldsymbol{H}\hat{\theta})^T\boldsymbol{W}(\boldsymbol{y} - \boldsymbol{H}\hat{\theta})\}$$

Least-Squares Estimate

- under-determined, over-determined and non-singular matrix $\boldsymbol{H}$

$$\hat{\theta} = \begin{cases} \boldsymbol{H}^T\left(\boldsymbol{H}\boldsymbol{H}^T\right)^{-1}\boldsymbol{y} & \text{if } N \le M \\ \boldsymbol{H}^{-1}\boldsymbol{y} & \text{if } N = M;\ \det(\boldsymbol{H}) \ne 0 \\ \hat{\theta} = \left(\boldsymbol{H}^T\boldsymbol{W}\boldsymbol{H}\right)^{-1}\boldsymbol{H}^T\boldsymbol{W}\boldsymbol{y} & \text{if } N > M \end{cases}$$

- The least-squares estimator is unbiased

$$E[\hat{\theta}] = \boldsymbol{\theta}$$

- The least-squares estimator is the best linear unbiased estimator.
- Orthogonality condition

$$E[\boldsymbol{e}^T\hat{\boldsymbol{y}}] = \mathit{0}$$

- Covariance of the estimation error

$$cov(\hat{\theta}) = \begin{cases} \left(\boldsymbol{H}^T\boldsymbol{\Sigma}_v^{-1}\boldsymbol{H}\right)^{-1} & \text{if } E\left[\boldsymbol{v}\boldsymbol{v}^T\right] = \boldsymbol{\Sigma}_v \\ \sigma_v^2\left(\boldsymbol{H}^T\boldsymbol{H}\right)^{-1} & \text{if } E\left[\boldsymbol{v}\boldsymbol{v}^T\right] = \sigma_v^2\boldsymbol{I} \end{cases}$$

- Residual

$$\sigma_{res}^2 = \begin{cases} trace\left\{\frac{1}{N}\left(\boldsymbol{I}-\boldsymbol{P}_r\right)\boldsymbol{\Sigma}_v\right\} & if\; E\left[\boldsymbol{v}\boldsymbol{v}^T\right]=\boldsymbol{\Sigma}_v \\ \sigma_{res}^2 = \left(1-\frac{M}{N}\right)\sigma_v^2 & if\; E\left[\boldsymbol{v}\boldsymbol{v}^T\right]=\sigma_v^2\boldsymbol{I} \end{cases}$$

- *Asymptotic properties:* $N \to \infty$

$$\hat{\boldsymbol{\theta}} \to \boldsymbol{\theta} \;as\; N \to \infty$$
$$\lim_{N\to\infty} it\, \sigma_{res}^2 = \sigma_v^2$$
$$\boldsymbol{e} \to \boldsymbol{v} \;as\; N \to \infty$$
$$E\left[\boldsymbol{e}(n)\boldsymbol{e}(n-m)\right] \to E\left[v(n)v(n-m)\right] \;\; as\; N \to \infty$$

If v is zero-mean white noise then

$$E[\boldsymbol{e}(n)\boldsymbol{e}(n-m)] = \delta(n-m) = \begin{cases} \sigma_v^2 & m=n \\ 0 & m \neq n \end{cases}$$

Un-Weighted and Weighted Least-Squares

$$\bar{\boldsymbol{y}} = \bar{\boldsymbol{H}}\boldsymbol{\theta} + \bar{\boldsymbol{v}}$$

where $\boldsymbol{W} = \boldsymbol{W}^{1/2}\boldsymbol{W}^{1/2}$, $\bar{\boldsymbol{y}} = \boldsymbol{W}^{1/2}\boldsymbol{y}$, $\bar{\boldsymbol{v}} = \boldsymbol{W}^{1/2}\boldsymbol{v}$ and $\bar{\boldsymbol{H}} = \boldsymbol{W}^{1/2}\boldsymbol{H}$

Model and Systemic Errors

$$E[\hat{\boldsymbol{\theta}}] = \boldsymbol{\theta} + (\boldsymbol{H}^T\boldsymbol{W}\boldsymbol{H})^{-1}\boldsymbol{H}^T\boldsymbol{W}\boldsymbol{\mu} \neq \boldsymbol{\theta}$$
$$\boldsymbol{e} = \boldsymbol{\mu} + \boldsymbol{v} - \boldsymbol{H}(\boldsymbol{H}^T\boldsymbol{W}\boldsymbol{H})^{-1}\boldsymbol{H}^T\boldsymbol{W}\boldsymbol{v}$$
$$\boldsymbol{e} \to \boldsymbol{\mu} + \boldsymbol{v} \;as\; N \to \infty$$

Hence the auto-correlation of the residual will not be delta function.

Augment the Parameter Vector for Systemic Error

$$\boldsymbol{\theta}_{ag} = [\boldsymbol{\theta} \quad \boldsymbol{\mu}]^T, \boldsymbol{H}_{ag} = [\boldsymbol{H} \quad \boldsymbol{1}]$$
$$\hat{\boldsymbol{\theta}}_{ag} = (\boldsymbol{H}_{ag}^T\boldsymbol{W}\boldsymbol{H}_{ag})^{-1}\boldsymbol{H}_{ag}\boldsymbol{W}\boldsymbol{y}$$

Cramér–Rao Lower Bound

Scalar case

$$\mathrm{var}(\hat{\theta}) = E[(\hat{\theta}(\boldsymbol{y}) - \theta)^2] \geq \left(E\left[\left(\frac{\delta \ln f_y(\boldsymbol{y})}{\delta\theta}\right)^2\right]\right)^{-1}$$

Vector case

$$\text{cov}(\hat{\theta}) = E[(\hat{\theta}(y) - \theta)(\hat{\theta}(y) - \theta)^T] \geq I_F^{-1}(\theta)$$

Efficient estimator

An unbiased estimator, which achieves The Cramér–Rao lower bound is said to be efficient.

Maximum Likelihood Estimation

$$\hat{\theta} = \arg\max_{\theta}\{\log f_y(y)\}$$

If the PDF of the noise $\boldsymbol{v}$ is zero0mean Gaussian then

- The weighted least-squares with weight $\boldsymbol{W} = \boldsymbol{\Sigma}_v^{-1/2}$ and the ML estimators are both efficient as they attain the Cramér–Rao lower bound. They both yield the minimum variance unbiased estimator.
- The Cramér–Rao lower bound for both ML and the weighted least-squares method is

$$I_F^{-1}(\theta) = \left(\boldsymbol{H}^T \boldsymbol{\Sigma}_v^{-1} \boldsymbol{H}\right)^{-1}$$

Least-Squares Estimates for Under-Determined System

$$\hat{\theta} = \boldsymbol{H}^T \left(\boldsymbol{H}\boldsymbol{H}^T\right)^{-1} \boldsymbol{y}$$
$$\hat{y} = \boldsymbol{H}\hat{\theta} = \boldsymbol{y}$$
$$\boldsymbol{e} = \boldsymbol{0}$$
$$\boldsymbol{H}^{\dagger} = \boldsymbol{H}^T \left(\boldsymbol{H}\boldsymbol{H}^T\right)^{-1}$$

Singular Value Decomposition

Every $N \times M$ matrix $\boldsymbol{H}$ can be decomposed as

$$\boldsymbol{H} = \boldsymbol{U}\boldsymbol{S}\boldsymbol{V}^T$$

The matrix $\boldsymbol{H}$ is decomposed into r rank-1 NxM matrices $\{\boldsymbol{u}_i \boldsymbol{v}_i^T\}$

$$\boldsymbol{H} = \sum_{i=1}^{r} \sigma_i \boldsymbol{u}_i \boldsymbol{v}_i^T$$

The condition number $\kappa(\boldsymbol{H})$ gives a measure of how close the matrix is to being rank deficient. $\kappa(\boldsymbol{H}) = \|\boldsymbol{H}\|\|\boldsymbol{H}^{\dagger}\| = \dfrac{\sigma_{\max}(\boldsymbol{H})}{\sigma_{\min}(\boldsymbol{H})}$

Computation of the Least-Squares Estimate Using SVD

$$\hat{\theta} = \boldsymbol{H}^{\dagger}\boldsymbol{y}$$

where

$$\boldsymbol{H}^{\dagger} = \boldsymbol{V}\begin{bmatrix} \boldsymbol{\Sigma}^{-1} & \boldsymbol{0} \\ \boldsymbol{0} & \boldsymbol{0} \end{bmatrix}\boldsymbol{U}^{T}$$

$\boldsymbol{\Sigma}^{-1} = diag\left(\sigma_1^{-1} \quad \sigma_2^{-1} \quad \cdots \quad \sigma_r^{-1}\right)$ $\boldsymbol{U}$, $\boldsymbol{V}$ and $\boldsymbol{\Sigma}$ are the SVD of the weighted matrix $\boldsymbol{H}$

$$\boldsymbol{H} = \boldsymbol{U}\begin{bmatrix} \boldsymbol{\Sigma} & \boldsymbol{0} \\ \boldsymbol{0} & \boldsymbol{0} \end{bmatrix}\boldsymbol{V}^{T}$$

The pseudo-inverse $\boldsymbol{H}^{\dagger}$ may be expressed as linear combination r rank-1 matrices

$$\boldsymbol{H}^{\dagger} = \sum_{i=1}^{r} \frac{1}{\sigma_i}\boldsymbol{u}_i\boldsymbol{v}_i^T$$

The bounds on the covariance of the estimation error are

$$\|cov(\hat{\boldsymbol{\theta}})\| \leq \begin{cases} \dfrac{1}{\sigma_{\min}^2\left(\boldsymbol{\Sigma}_v^{-1/2}\boldsymbol{H}\right)} & if\ E\left[\boldsymbol{v}\boldsymbol{v}^T\right] = \boldsymbol{\Sigma}_v \\ \dfrac{\sigma_v^2}{\sigma_{\min}^2(\boldsymbol{H})} & if\ E\left[\boldsymbol{v}\boldsymbol{v}^T\right] = \sigma_v^2 \end{cases}$$

The bounds on the residual

$$\sigma_{res}^2 \leq \begin{cases} \frac{1}{N}trace\left(\boldsymbol{\Sigma}_v\right) & if\ E\left[\boldsymbol{v}\boldsymbol{v}^T\right] = \boldsymbol{\Sigma}_v \\ \sigma_v^2 & if\ E\left[\boldsymbol{v}\boldsymbol{v}^T\right] = \sigma_v^2\boldsymbol{I} \end{cases}$$

5.10 Appendix: Properties of the Pseudo-Inverse and the Projection Operator

5.10.1 Over-Determined System

The pseudo-inverse satisfies one property, $\boldsymbol{H}^{\dagger}\boldsymbol{H} = \boldsymbol{I}$, but not the other property of an inverse $\boldsymbol{H}\boldsymbol{H}^{\dagger} \neq \boldsymbol{I}$. In fact $\boldsymbol{H}\boldsymbol{H}^{\dagger} = \boldsymbol{P}_r$ is a projection operator of $\boldsymbol{H}$.

$\boldsymbol{P}_r = \boldsymbol{H}\boldsymbol{H}^{\dagger}$ has the following properties

- $\boldsymbol{P}_r^T = \boldsymbol{P}_r$ is symmetrical
- $\boldsymbol{P}_r^2 = \boldsymbol{P}_r$ hence $\boldsymbol{P}_r^m = \boldsymbol{P}_r$ *for* $m = 1, 2, 3, \ldots,$
- Eigenvalues of $\boldsymbol{I} - \boldsymbol{P}_r$ are only ones and zeros.
- Let M_r be the rank of a *NxM* matrix $\boldsymbol{H}$
  - M_r eigenvalues will be ones
  - The rest of the $N - M_r$ eigenvalues will be zeros
  - $trace\left(\boldsymbol{I} - \boldsymbol{P}_r\right) = N - M_r$.
- $\boldsymbol{I} - \boldsymbol{P}_r$ projects a vector on to a space perpendicular to the range space of the matrix $\boldsymbol{H}$
- If $\boldsymbol{H}$ is non-singular square matrix then $\boldsymbol{P}_r = \boldsymbol{I}$; $\boldsymbol{I} - \boldsymbol{P}_r = 0$.

Weighted projection matrix:
In a weighted least squares method, the projection matrix is defined as

$$P_r = H\left(H^T W H\right)^{-1} H^T W$$

It is not a symmetric matrix $P_r^T \neq P_r$. However, the weighted projection operator satisfies all the properties listed above except the symmetry property.

5.10.2 Under-Determined System

The pseudo-inverse is given by

$$H^{\dagger} = H^T (HH^T)^{-1}$$

The pseudo-inverse satisfies one property, $HH^{\dagger} = I$, but not the other property of an inverse $H^{\dagger}H \neq I$. In fact $H^{\dagger}H$ is a projection operator of H. The projection operator is given by

$$P_r = H^{\dagger}H = H^T (HH^T)^{-1} H$$

The projection operator of the under-determined system also satisfies all the properties of the over-determined system.

5.11 Appendix: Positive Definite Matrices

- If A is a positive definite matrix, then A^{-1} is also a positive matrix.

Lemma 5.2 *If A is a positive definite matrix, then*

$$0 < \lim_{N\to\infty} it \frac{x^T A x}{N} < \infty \tag{5.150}$$

Proof: Since A is positive definite

$$0 < \lambda_{\min}(A) x^T x < x^T A x \leq \lambda_{\max}(A) x^T x \textit{ for all } x \neq 0 \tag{5.151}$$

where $\lambda_{\min}(A)$ and $\lambda_{\min}(A)$ are minimum and maximum eigenvalues of A. Let $m_{\min} = \min_i\{x_i^2\}$ and $m_{\max} = \max_i\left\{x_i^2\right\}$ be respectively the minimum and the maximum positive values of x_i^2. Since $x \neq 0$ and hence $x_i^2 \neq 0$ for all i, we get

$$m_{\min} N \leq x^T x = \sum_{i=1}^{N} x_i^2 \leq m_{\max} N \tag{5.152}$$

Hence we get

$$0 < N m_{\min} \lambda_{\min}(A) < x^T A x \leq N m_{\max} \lambda_{\max}(A) \tag{5.153}$$

Dividing by N and taking the limit yields

$$0 < m_{\min} \lambda_{\min}(A) < \lim_{N\to\infty} it \frac{x^T A x}{N} \leq m_{\max} \lambda_{\max}(A) \tag{5.154}$$

Hence we conclude $0 < \lim_{N\to\infty} it \frac{x^T A x}{N} < \infty$.

5.12 Appendix: Singular Value Decomposition of a Matrix

Theorem 5.2 *Every $N \times M$ matrix $\boldsymbol{H}$ can be decomposed as*

$$\boldsymbol{H} = \boldsymbol{U}\boldsymbol{S}\boldsymbol{V}^T \tag{5.155}$$

where $\boldsymbol{U}$ is $N \times N$ real matrix of rank $r \le \min(M,N)$ where
$\boldsymbol{U}$ is a $N \times N$ unitary matrix

$$\begin{aligned}
&\boldsymbol{U} = [\,\boldsymbol{u}_1 \quad \boldsymbol{u}_2 \quad . \quad \boldsymbol{u}_r \quad \boldsymbol{u}_{r+1} \quad . \quad \boldsymbol{u}_N\,]\\
&\boldsymbol{U} = [\,\boldsymbol{U}_1 \quad \boldsymbol{U}_2\,], \quad \boldsymbol{U}_1 = [\,\boldsymbol{u}_1 \quad \boldsymbol{u}_2 \quad . \quad \boldsymbol{u}_r\,], \quad \boldsymbol{U}_2 = [\,\boldsymbol{u}_{r+1} \quad \boldsymbol{u}_{r+2} \quad . \quad \boldsymbol{u}_N\,]\\
&\boldsymbol{U}^T\boldsymbol{U} = \boldsymbol{U}\boldsymbol{U}^T = \boldsymbol{I}
\end{aligned}$$

$\boldsymbol{u}_i$ is $N \times 1$ vector. $\boldsymbol{U}$ is called left singular matrix of $\boldsymbol{H}$
$\boldsymbol{V}$ is a $M \times M$ unitary matrix

$$\begin{aligned}
&\boldsymbol{V} = [\,\boldsymbol{v}_1 \quad \boldsymbol{v}_2 \quad . \quad \boldsymbol{v}_r \quad \boldsymbol{v}_{r+1} \quad . \quad \boldsymbol{v}_M\,]\\
&\boldsymbol{V} = [\,\boldsymbol{V}_1 \quad \boldsymbol{V}_2\,], \quad \boldsymbol{V}_1 = [\,\boldsymbol{v}_1 \quad \boldsymbol{v}_2 \quad . \quad \boldsymbol{v}_r\,], \quad \boldsymbol{V}_2 = [\,\boldsymbol{v}_{r+1} \quad \boldsymbol{v}_{r+2} \quad . \quad \boldsymbol{v}_M\,]\\
&\boldsymbol{V} = [\,\boldsymbol{V}_1 \quad \boldsymbol{V}_2\,], \quad \boldsymbol{V}_1 = [\,\boldsymbol{v}_1 \quad \boldsymbol{v}_2 \quad . \quad \boldsymbol{v}_r\,], \quad \boldsymbol{V}_2 = [\,\boldsymbol{v}_{r+1} \quad \boldsymbol{v}_{r+2} \quad . \quad \boldsymbol{v}_M\,]\\
&\boldsymbol{V}\boldsymbol{V}^T = \boldsymbol{V}^T\boldsymbol{V} = \boldsymbol{I}
\end{aligned}$$

$\boldsymbol{v}_i$ is a $M \times 1$ vector, V is called right singular matrix of $\boldsymbol{H}$.
$\boldsymbol{S}$ is a $N \times M$ rectangular matrix given by

$$\boldsymbol{S} = \begin{bmatrix} \boldsymbol{\Sigma} & \boldsymbol{0} \\ \boldsymbol{0} & \boldsymbol{0} \end{bmatrix}, \quad \boldsymbol{\Sigma} = diag\begin{bmatrix}\sigma_1 & \sigma_2 & \sigma_3 & . & \sigma_r\end{bmatrix}$$

where σ_i is the ith singular values of $\boldsymbol{H}$ which are positive

$$\sigma_1 \ge \sigma_2 \ge \sigma_3 \ge \cdots \ge \sigma_r > 0 \text{ and } \sigma_{r+1} = \sigma_{r+2} = \cdots = \sigma_M = 0$$

The matrix $\boldsymbol{H}$ is decomposed into r *rank-1 $N \times M$ matrices $\{\boldsymbol{H}_i\}$*

$$\boldsymbol{H} = \sum_{i=1}^{r} \sigma_i \boldsymbol{u}_i \boldsymbol{v}_i^T \tag{5.156}$$

A brief outline of the proof is given below:

Proof: Consider the *MxM* matrix $\boldsymbol{H}^T\boldsymbol{H}$. It is a symmetric and positive semi-definite matrix. From linear algebra we know the following:

- Every symmetric and positive semi-definite matrix $\boldsymbol{H}^T\boldsymbol{H}$ can be diagonalized by a unitary matrix with diagonal elements which are real and non-negative eigenvalues. If r is the rank, then there will be r non-zero eigenvalues and the rest will be zero.

Hence

$$H^T H v_i = \sigma_i^2 v_i \tag{5.157}$$

where $\{v_i\}$ form a set of orthonormal vectors and σ_i^2 is the ith eigenvalue of $H^T H$. Let V be MxM matrix formed of $Mx1$ orthonormal eigenvectors $\{v_i\}$, $V = [v_1 \quad v_2 \quad . \quad v_r \quad v_{r+1} \quad . \quad v_M]$ As V is a unitary matrix $V^T V = I$. Expressing Eq. (5.157) in a matrix form we get

$$H^T H V = V\Gamma \tag{5.158}$$

where $\Gamma = \begin{bmatrix} \Sigma^2 & 0 \\ 0 & 0 \end{bmatrix}$, $\Sigma^2 = diag\begin{bmatrix} \sigma_1^2 & \sigma_2^2 & . & \sigma_r^2 \end{bmatrix}$

Pre-multiplying by V^T and noting that V is unitary we get

$$V^T H^T H V = \Gamma \tag{5.159}$$

Substituting $V = \begin{bmatrix} V_1 & V_2 \end{bmatrix}$, $V_1 = \begin{bmatrix} v_1 & v_2 & . & v_r \end{bmatrix}$, $V_2 = \begin{bmatrix} v_{r+1} & v_{r+2} & . & v_M \end{bmatrix}$ we get

$$\begin{bmatrix} V_1^T \\ V_2^T \end{bmatrix} H^T H \begin{bmatrix} V_1 & V_2 \end{bmatrix} = \begin{bmatrix} V_1^T H^T H V_1 & V_1^T H^T H V_2 \\ V_2^T H^T H V_1 & V_2^T H^T H V_2 \end{bmatrix} = \begin{bmatrix} \Sigma^2 & 0 \\ 0 & 0 \end{bmatrix} \tag{5.160}$$

Hence we get

$$V_2^T H^T H V_2 = 0 \tag{5.161}$$

This implies

$$HV_2 = 0 \tag{5.162}$$

$$V_1^T H^T H V_1 = \Sigma^2 \tag{5.163}$$

$$V_2^T H^T H V_2 = V_2^T H^T H V_1 = 0 \tag{5.164}$$

Consider $U = [U_1 \quad U_2]$, $U_1 = [u_1 \quad u_2 \quad . \quad u_r]$, $U_2 = [u_{r+1} \quad u_{r+2} \quad . \quad u_N]$. Define

$$U_1 = HV_1\Sigma^{-1} \tag{5.165}$$

Using Eqs. (5.163) and (5.165) we get

$$U_1^T U_1 = \Sigma^{-1} V_1^T H^T H V_1 \Sigma^{-1} = \Sigma^{-1}\Sigma^2\Sigma^{-1} = I \tag{5.166}$$

Hence U_1 is unitary matrix. Create a matrix U_2 so that $U_1^T U_2 = U_2^T U_1 = 0$, $U_2^T U_2 = I$ and hence

$$U^T U = \begin{bmatrix} U_1^T U_1 & U_1^T U_2 \\ U_2^T U_1 & U_2^T U_2 \end{bmatrix} = \begin{bmatrix} I & 0 \\ 0 & I \end{bmatrix} \tag{5.167}$$

Now examine

$$U^T HV = \begin{bmatrix} U_1^T \\ U_2^T \end{bmatrix} H \begin{bmatrix} V_1 & V_2 \end{bmatrix} = \begin{bmatrix} U_1^T HV_1 & U_1^T HV_2 \\ U_2^T HV_1 & U_2^T HV_2 \end{bmatrix} \tag{5.168}$$

Using Eqs. (5.162), (5.165), and (5.167) we get

$$U^T H V = \begin{bmatrix} \Sigma^{-1} V_1^T H^T H V_1 & 0 \\ U_2^T U_1 \Sigma & 0 \end{bmatrix} = \begin{bmatrix} \Sigma^{-1} \Sigma^2 & 0 \\ U_2^T U_1 \Sigma & 0 \end{bmatrix} = \begin{bmatrix} \Sigma & 0 \\ 0 & 0 \end{bmatrix} = S \tag{5.169}$$

Thus

$$U^T H V = S \tag{5.170}$$

Pre-multiplying with U and post-multiplying with V^T we get

$$H = U S V^T \tag{5.171}$$

5.12.1 SVD and Eigendecompositions

Lemma 5.3 *If the H is a symmetric and non-negative definite $M \times M$ matrix of rank* r, *then its singular value decomposition and the eigendecomposition will be the same.*

Proof: Let q_i be a ith eigenvector of H associated with the ith eigenvalue λ_i:

$$H q_i = \lambda_i q_i \tag{5.172}$$

Pre-multiplying by H^T and noting that H and H^T have same eigenvalue, we get

$$H^T H q_i = \lambda_i H^T q_i = \lambda_i^2 q_i \tag{5.173}$$

That is, q_i is an eigenvector of $H^T H$ associated with λ_i^2. Thus, if q_i is normalized, we will have:

$$U = V = [\, q_1 \quad q_2 \quad . \quad q_M \,] = [\, v_1 \quad v_2 \quad . \quad v_r \quad v_{r+1} \quad . \quad v_M \,] \tag{5.174}$$

Moreover, since H is also assumed to be a non-negative definite matrix, its eigenvalues are non-negative real numbers. Hence

$$H = U S U^T \tag{5.175}$$

where $S = diag(\, \sigma_1 \quad \sigma_3 \quad \cdots \quad \sigma_r \quad 0 \quad \cdots \quad 0\,)$. Therefore

$$\lambda(H) = \sigma(H) \tag{5.176}$$

where $\lambda(H)$ and $\sigma(H)$ are respectively eigenvalues and the singular values of H. In general

$$\sigma_i(H) = \sqrt{\lambda_i \left(H^T H\right)} \tag{5.177}$$

5.12.2 Matrix Norms

The SVD can be used to compute the 2-norm and the Frobenius norm. The 2-norm is the maximum singular value of the matrix while the Frobenius norm is the sum of the squares of the singular values

$$\|\boldsymbol{H}\|_2 = \sigma_{\max}(\boldsymbol{H}) \tag{5.178}$$

If $\boldsymbol{H}$ is invertible, then

$$\|\boldsymbol{H}^{-1}\|_2 = \frac{1}{\sigma_{\min}(\boldsymbol{H})} \tag{5.179}$$

$$\|\boldsymbol{H}\|_F^2 = \sum_{i=1}^{r} \sigma_i^2 \tag{5.180}$$

5.12.3 Least Squares Estimate for Any Arbitrary Data Matrix H

Theorem 5.3 *Let be a* $\boldsymbol{H}$ *be a NxM matrix of rank* $r \le \min(M, N)$*. Then the least-squares estimate* $\hat{\theta} = arg\{\min_{\theta}(\boldsymbol{y} - \boldsymbol{H}\hat{\theta})^T \boldsymbol{W}(\boldsymbol{y} - \boldsymbol{H}\hat{\theta})\}$ *is given by*

$$\hat{\theta} = \boldsymbol{V} \begin{bmatrix} \boldsymbol{\Sigma}^{-1} & \boldsymbol{0} \\ \boldsymbol{0} & \boldsymbol{0} \end{bmatrix} \boldsymbol{U}^T \boldsymbol{y} \tag{5.181}$$

where $\mathbf{W}^{1/2}\boldsymbol{H} = \boldsymbol{USV}^T$ *is*

Proof:

- We will first consider the un-weighted least-squares problem.

  Substituting the SVD decomposition of $\boldsymbol{H}$ we get

$$J = (\boldsymbol{y} - \boldsymbol{USV}^T\hat{\theta})^T(\boldsymbol{y} - \boldsymbol{USV}^T\hat{\theta}) \tag{5.182}$$

where $\boldsymbol{S} = \begin{bmatrix} \boldsymbol{\Sigma} & \boldsymbol{0} \\ \boldsymbol{0} & \boldsymbol{0} \end{bmatrix}$, $\quad \boldsymbol{\Sigma} = diag\begin{bmatrix} \sigma_1 & \sigma_2 & \sigma_3 & . & \sigma_r \end{bmatrix}$

Insert $\boldsymbol{UU}^T$ to create a "weighted" inner product as it were. Since $\boldsymbol{UU}^T = \boldsymbol{I}$, the cost function J is unaffected and we get after simplification

$$J = (\boldsymbol{y} - \boldsymbol{USV}^T\hat{\theta})^T\boldsymbol{UU}^T(\boldsymbol{y} - \boldsymbol{USV}^T\hat{\theta}) = (\boldsymbol{U}^T\boldsymbol{y} - \boldsymbol{SV}^T\hat{\theta}_v)^T(\boldsymbol{U}^T\boldsymbol{y} - \boldsymbol{SV}^T\hat{\theta}_v) \tag{5.183}$$

Denoting $\boldsymbol{y}_u = \boldsymbol{U}^T\boldsymbol{y}$, $\hat{\theta}_v = \boldsymbol{V}^T\hat{\theta}$ we get

$$J = (\boldsymbol{y}_u - \boldsymbol{S}\hat{\theta}_v)^T(\boldsymbol{y}_u - \boldsymbol{S}\hat{\theta}_v) \tag{5.184}$$

Using definitions $\boldsymbol{S} = \begin{bmatrix} \boldsymbol{\Sigma} & \boldsymbol{0} \\ \boldsymbol{0} & \boldsymbol{0} \end{bmatrix}$, $\boldsymbol{y}_u = \begin{bmatrix} \boldsymbol{y}_{u1} \\ \boldsymbol{y}_{u2} \end{bmatrix}$, $\hat{\theta}_v = \begin{bmatrix} \hat{\theta}_{v1} \\ \hat{\theta}_{v2} \end{bmatrix}$ we get

$$J = \begin{pmatrix} \boldsymbol{y}_{u1} - \boldsymbol{\Sigma}\hat{\theta}_{v1} \\ \boldsymbol{y}_{u2} \end{pmatrix}^T \begin{pmatrix} \boldsymbol{y}_{u1} - \boldsymbol{\Sigma}\hat{\theta}_{v1} \\ \boldsymbol{y}_{u2} \end{pmatrix} = (\boldsymbol{y}_{u1} - \boldsymbol{\Sigma}\hat{\theta}_{v1})^T(\boldsymbol{y}_{u1} - \boldsymbol{\Sigma}\hat{\theta}_{v1}) + \boldsymbol{y}_{u2}^T\boldsymbol{y}_{u2} \tag{5.185}$$

Since the cost function J is not a function of $\hat{\theta}_{v2}$, we may choose any value for $\hat{\theta}_{v2}$ without affecting J. The least-squares estimate $\hat{\theta}_{v1}$ is given by

$$\hat{\theta}_{v1} = \begin{bmatrix} \Sigma^{-1} & 0 \\ 0 & 0 \end{bmatrix} y_u \tag{5.186}$$

The solution is not unique as $\hat{\theta}_{v2}$ is arbitrary. The general solution $\hat{\theta}_v$ is obtained by appending $\hat{\theta}_{v2}$ which is arbitrary to the solution $\hat{\theta}_{v1}$.

$$\hat{\theta}_v = \begin{bmatrix} \hat{\theta}_{v1} \\ \hat{\theta}_{v2} \end{bmatrix} \tag{5.187}$$

The norm of $\hat{\theta}_v$ is $\|\hat{\theta}_v\|^2 = \|\hat{\theta}_{v1}\|^2 + \|\hat{\theta}_{v2}\|^2$ [3]. A solution $\hat{\theta}_v$ which has minimum norm is obtained by setting $\hat{\theta}_{v2} = 0$ and the minimum norm solution becomes

$$\hat{\theta}_v = \hat{\theta}_{v1} = \begin{bmatrix} \Sigma^{-1} & 0 \\ 0 & 0 \end{bmatrix} y_u \tag{5.188}$$

Substituting for $\hat{\theta}_v$ and y_u using $y_u = U^T y$, $\hat{\theta}_v = V^T \hat{\theta}$, and noting $VV^T = I$ we get

$$\hat{\theta} = V \begin{bmatrix} \Sigma^{-1} & 0 \\ 0 & 0 \end{bmatrix} U^T y \tag{5.189}$$

Let us consider the weighted least-squares problem.

We will employ the transformed measurement model

$$\bar{y} = \bar{H}\theta + \bar{v} \text{ where } W = W^{1/2}W^{1/2},\ \bar{y} = W^{1/2}y,\ \bar{v} = W^{1/2}v \text{ and } \bar{H} = W^{1/2}H$$

With this transformation the weighted least-squares is converted to the un-weighted least-squares problem. For notational convenience of the SVD of $\bar{H} = W^{1/2}H$ and that of H are denoted by the same singular matrices, namely U, S, and V

$$\bar{H} = USU^T \tag{5.190}$$

The least-squares estimate of the weighted least-squares problem becomes

$$\hat{\theta} = V \begin{bmatrix} \Sigma^{-1} & 0 \\ 0 & 0 \end{bmatrix} U^T \bar{y} = V \begin{bmatrix} \Sigma^{-1} & 0 \\ 0 & 0 \end{bmatrix} U^T W^{1/2} y \tag{5.191}$$

Comment *If the data matrix H is not of maximal rank, $r < \min(M, N)$, then the solution is not unique. It has infinite solutions and the least-squares solution using the SVD gives a minimum norm solution. In other words the solution $\hat{\theta} = H^{\dagger} y$ is such that $\|\hat{\theta}\|$ is minimum.*

5.12.4 Pseudo-Inverse of Any Arbitrary Matrix

Theorem 5.4 *Let $W^{1/2}\bar{H} = USU^T$. The pseudo-inverse of a NxM matrix H of rank $r \leq \min(M,N)$ is*

$$H^{\dagger} = V\begin{bmatrix} \Sigma^{-1} & 0 \\ 0 & 0 \end{bmatrix} U^T \tag{5.192}$$

Proof: Consider the symmetric, square matrix, and positive semi-definite *MXM* matrix $H^T WH$

$$H^T WH = VS^T U^T USV^T = VS^T SV^T = V\begin{bmatrix} \Sigma^2 & 0 \\ 0 & 0 \end{bmatrix} V^T \tag{5.193}$$

Hence

$$\left(H^T WH\right)^{-1} = V\begin{bmatrix} \Sigma^{-2} & 0 \\ 0 & 0 \end{bmatrix} V^T \tag{5.194}$$

The pseudo-inverse $H^{\dagger}$ becomes

$$H^{\dagger} = \left(H^T WH\right)^{-1} H^T W = \left(H^T WH\right)^{-1} = V\begin{bmatrix} \Sigma^{-2} & 0 \\ 0 & 0 \end{bmatrix} V^T VSU^T = V\begin{bmatrix} \Sigma^{-1} & 0 \\ 0 & 0 \end{bmatrix} U \tag{5.195}$$

5.12.5 Bounds on the Residual and the Covariance of the Estimation Error

Consider the covariance of the estimation error $cov(\hat{\theta}) = (H^T \Sigma_v^{-1} H)^{-1}$. Substituting the SVD of H and using $\|A^{-1}\| = \dfrac{1}{\sigma_{\min}(A^{-1})}$ we get

$$cov(\hat{\theta}) \leq \frac{1}{\sigma_{\min}^2\left(\Sigma_v^{-1/2} H\right)} \tag{5.196}$$

Hence

$$\|cov(\hat{\theta})\| \leq \begin{cases} \dfrac{1}{\sigma_{\min}^2\left(\Sigma_v^{-1/2} H\right)} & \text{if } E\left[vv^T\right] = \Sigma_v \\ \dfrac{\sigma_v^2}{\sigma_{\min}^2 (H)} & \text{if } E\left[vv^T\right] = \sigma_v^2 \end{cases} \tag{5.197}$$

5.13 Appendix: Least-Squares Solution for Under-Determined System

Since there are infinite solutions, it is preferable to choose a solution from the infinite set of solutions, the one which has a minimum norm. The problem is formulated as follows:

$$\min_{\theta}\left\{\frac{1}{2}\|\theta\|^2\right\} \quad \text{such that} \quad H\theta = y \tag{5.198}$$

where $\|\theta\|^2 = \theta^T\theta$ is the 2-norm of θ. The cost function $\|\theta\|^2$ is divided by 2 for notational convenience so that it is canceled out on differentiation. It is a constrained minimization problem, which may be solved using the Lagrange multiplier technique: convert the constrained least-squares optimization into unconstrained least-squares optimization using Lagrange multipliers:

$$\min_{\theta,\lambda}\left\{\frac{\hat{\theta}^T\hat{\theta}}{2} - \lambda^T\left(H\hat{\theta} - y\right)\right\} \tag{5.199}$$

Differentiating with respect to $\hat{\theta}$ and λ yields

$$\hat{\theta} - H^T\lambda = 0 \tag{5.200}$$

$$H\hat{\theta} = y \tag{5.201}$$

Substituting for $\hat{\theta}$ in Eq. (5.201) using Eq. (5.200) and assuming H has a full rank (which implies HH^T is invertible), we get

$$\lambda = \left(HH^T\right)^{-1} y \tag{5.202}$$

Expressing $\hat{\theta}$ as a function of y by substituting for λ in Eq. (5.200) we get

$$\hat{\theta} = H^T\left(HH^T\right)^{-1} y \tag{5.203}$$

The estimate $\hat{y}$ of y becomes

$$\hat{y} = H\hat{\theta} = HH^T\left(HH^T\right)^{-1} y = y \tag{5.204}$$

5.14 Appendix: Computation of Least-Squares Estimate Using the SVD

The least-squares estimate is computed using a procedure similar to the over-determined case.

Let be a H be a NxM with $N < M$, and of rank $r \le \min(M, N)$. Then the least-squares estimate $\hat{\theta} = arg\{\min_{\hat{\theta}}(y - H\hat{\theta})^T W(y - H\hat{\theta})\}$ is given by

$$\hat{\theta} = V\begin{bmatrix} \Sigma^{-1} & 0 \\ 0 & 0 \end{bmatrix} U^T y \tag{5.205}$$

where $W^{1/2}H = USV^T$

Proof: Proof It is identical to the over-determined case.

References

[1] Ljung, L. (1999) *System Identification: Theory for the User*, Prentice-Hall, New Jersey.

[2] Mendel, J. (1995) *Lessons in Estimation Theory in Signal Processing, Communications and Control*, Prentice Hall, New Jersey.

[3] Haykin, S. (2001) *Adaptive Filter Theory*, Prentice Hall, New Jersey.

Further Readings

Doraiswami, R. (1976) A decision theoretic approach to parameter estimation. *IEEE Transactions on Automatic Control*, **21**(6), 860–866.

Kay, S.M. (1993) *Fundamentals of Signal Processing: Estimation Theory*, Prentice Hall, New Jersey.

Moon, T.K. and Stirling, W.C. (2000) *Mathematical Methods and Algorithms for Signal Processing*, Prentice Hall, New Jersey.

6

Kalman Filter

6.1 Overview

The Kalman filter is widely used in a plethora of science and engineering applications, including tracking, navigation, fault diagnosis, condition-based maintenance, performance (or health or product quality) monitoring, soft sensor, estimation of a signal of a known class buried in noise, speech enhancement, and controller implementation [1–12]. It also plays a crucial in system identification as the structure of the identification model is chosen to be the same as that of the Kalman filter.

It is a recursive estimator of the states of a dynamical system. The system is modeled in a state-space form driven by a zero-mean white noise process. The Kalman filter consists of two sets of equations: a static (or algebraic) and dynamic equation. The dynamic equation is driven by the input of the system and the residual. There is an additional input, termed disturbance, which may represent any input that affects the system, but unlike the system input, it cannot be accessed or manipulated by the user. The disturbances represent either the actual disturbance affecting the system or a mathematical artifice to give a measure of belief in the model. The actual disturbances may include effects such as the gravity load, electrical power demand, fluctuations in the flow in a fluid system, wind gusts, bias, power frequency signal, dc offset, crew or passenger load in vehicles such as space-crafts, ships, helicopters, and planes, and faults. An additive noise in the algebraic equation represents the measurement noise. A Kalman filter is designed to estimate the states of the system using the system input and output measurements.

There are two approaches to deriving the Kalman filter: one relies on the stochastic estimation theory [3, 11, 12] and the other on the deterministic theory [4]. A deterministic approach is adopted herein. The structure of the Kalman filter is determined using the *internal model principle* which establishes the necessary and sufficient condition for the tracking of the output of a dynamical system [13]. In accordance with this principle, the Kalman filter consists of (i) a copy of the system model driven by the residuals, and (ii) a gain term, termed the Kalman gain to stabilize the filter. The Kalman gain is determined from the minimization of the covariance of the state estimation error. The Riccati equation governing the evolution of the estimation error covariance is derived, from which the Kalman gain is then computed.

The internal model principle provides a mathematical justification for the robustness of the Kalman filter to noise and disturbances and for the high sensitivity (i.e., a lack of robustness) of the mean of the residuals to model mismatch. This property is exploited in designing a Kalman filter for applications such as performance monitoring and fault diagnosis. The structure of observers, low-, band-, and high-pass filters, and so on can all be justified from an "internal model principle" viewpoint.

The Kalman filter computes the estimates by fusing the *a posteriori* information provided by the measurement, and the *a priori* information contained in the model which governs the evolution of

Identification of Physical Systems: Applications to Condition Monitoring, Fault Diagnosis, Soft Sensor and Controller Design, First Edition. Rajamani Doraiswami, Chris Diduch and Maryhelen Stevenson.

the measurement. The covariance of the measurement noise, and the covariance of the plant noise, quantifies the degree of belief associated with the measurement and model information, respectively. These covariances play a crucial role in the performance of the Kalman filter. The estimate of the state is obtained as the best compromise between the estimates generated by the model and those obtained from the measurement, depending upon the plant noise and the measurement noise covariances. For example, when the plant noise covariance is very small, implying that the model is highly reliable, the estimate is computed essentially using the model. The Kalman gain is zero and the measurements are simply ignored in this case. On the other hand, when the measurement noise covariance is very small, implying that the measurements are highly reliable, the state estimate is computed from the measurements, ignoring the plant model. It is therefore very important to use the correct values for the covariances that reflect correctly the belief in the model. If the plant noise covariance (or the measurement noise covariance) is chosen to be small when the model is not reliable (or when the measurements are not reliable) the performance of the Kalman filter will significantly degrade. In practice an online tuning of the covariances may be preferable to avoid such adverse situations.

In applications such as condition-based monitoring, soft sensing, and fault diagnosis, a steady-state Kalman filter, instead of an optimal Kalman filter, is employed, where the Kalman gain is derived from the steady-state Riccati equation. Unlike the optimal Kalman filter, the steady-state Kalman filter is preferred in practice as it is time invariant, and does not require frequent restarts of the Kalman filter equations every time the Riccati equation has reached the steady-state.

The orthogonality relationships governing the residual, the state estimation error, the estimate, and the system output are derived and illustrated by examples. It is shown that

- Residual is a zero-mean white noise process.
- Estimation error is orthogonal to both the estimate of the state and the output.

These properties, especially the white noise properties of the residual, are exploited in many applications.

A number of simulated examples are given to illustrate the procedures for designing the Kalman filter. The role of the plant covariance and noise covariance, which give respectively a measure of belief in the model and the measurements, is also explained. The comparison of the optimal and the suboptimal Kalman filters is also discussed. The use of a suboptimal Kalman filter is justified in some applications where its performance is close to that of the optimal filter.

In the following sections, we shall develop the basic theory behind this novel approach and then test it using simulated data. Further, this novel approach is also successfully tested on a physical process control system represented by a lab-scale two-tank system.

This chapter sections are organized as follows. The Kalman filter based on the internal model principle is developed starting from the design of a controller so that the system output tracks a given reference signal in spite of the disturbance affecting the output. The reference and the disturbances are assumed to be deterministic signals. Then, duality between a controller and an estimator design problem is established. The problem of estimating a signal given its model is considered with applications in signal, speech, and biological signal processing. The design of an observer to estimate the states of a system from the input and the output of the system is developed to set the stage for deriving the Kalman filter. The internal model principle approach to Kalman filter design evolves from the design of controller, the estimation deterministic signal, and the design of an observer to estimate the state. The Kalman filter generalizes the problem of estimation of the states of a system when the output of the system is a random process. The internal model approach is generalized to handle stochastic processes. The disturbances and the measurement noise are assumed to be a sum of deterministic and random processes. The deterministic and random processes are modeled respectively as the outputs of linear time invariance system driven by delta function and a white noise process respectively.

The power and effectiveness of the Kalman filter in detecting the presence of model mismatch, that is the actual model of the system as it deviates from its nominal model (nominal model is used in the design of the Kalman filter), is demonstrated. The property of the Kalman filter residual, namely the residual is a zero-mean white noise process if and only if there is no model mismatch, is exploited judiciously in many applications including fault diagnosis and status monitoring. Thanks to the minimal variance of the estimation error, a small or incipient fault from the input–output data buried in the noise may be captured. Further, it is also used to verify the performance of the Kalman filter performance as well as to tune its gain.

A summary of the chapter is given. In the Appendices the derivation of the Kalman gain, the Riccati equation governing the covariance of the estimation error, the orthogonality properties of the Kalman filter, and the expression for the residual when there is a model mismatch are given.

6.2 Mathematical Model of the System

The model of the system is expressed in state-space form, which is a vector-matrix description of the dynamical system. The state-space form is convenient in that it handles multivariate and time-varying systems and lends itself to a powerful analysis, design, and realization (or implementation) of filters and controllers. The state of a system is defined as the minimal set of information required to completely characterize this system. For the sake of completeness, the state-space model of the system given in Chapter 1 is repeated here. The state-space model of the system is obtained by combining the models of the plant, the disturbance, the measurement noise, and the deterministic components of the noise and the disturbance.

6.2.1 *Model of the Plant*

The plant model in state-space form $(\boldsymbol{A}_p, \boldsymbol{B}_p, \boldsymbol{C}_p)$ is:

$$\begin{aligned} \boldsymbol{x}_p(k+1) &= \boldsymbol{A}_p \boldsymbol{x}_p(k) + \boldsymbol{B}_p r(k) + \boldsymbol{E}_p w_p(k) \\ y(k) &= \boldsymbol{C}_p \boldsymbol{x}_p(k) + v_p(k) \end{aligned} \tag{6.1}$$

where $\boldsymbol{x}_p(k)$ is the $n_p x1$ state vector, $y(k)$ the measured output, $r(k)$ the input to the plant, $w_p(k)$ is the disturbance or plant noise, and $v_p(k)$ the measurement noise. The input $r(k)$, the output $y(k)$, the disturbance $w_p(k)$, and the measurement noise $v_p(k)$ are assumed to be scalars. When the measurement or plant noise or both are small, the estimation error diverges due to numerical errors resulting from ill-conditioned matrices. To avoid this problem of divergence, a fictitious plant white noise $w_p(k)$, is added to the state-space (or dynamical) equation. In many cases, the disturbance term $w_p(k)$ is conveniently included as a mathematical artifice to represent our measure of belief in the model and to provide an additional parameter, in terms of its covariance, in designing a filter. The covariance term provides a trade-off between faster filter response and better noise attenuation in the filter estimates.

It is assumed that the direct transmission term, "D" (also called the feed-through or feed-forward term) is zero. The discretized model for many physical systems, such as mechatronics, process control system, and power systems may be modeled assuming "D = 0." Further, it introduces unnecessary complications in tasks such as closed loop system identification, controller and filter design.

6.2.2 *Model of the Disturbance and Measurement Noise*

The disturbance $w_p(k)$, and the measurement noise $v_p(k)$ are assumed to be a sum of a deterministic and random process. The random component of (i) the disturbance is modeled as an output of a linear time invariant system $(\boldsymbol{A}_w, \boldsymbol{B}_w, \boldsymbol{C}_w, D_w)$ driven by a zero-mean white noise process $w(k)$ and (ii) the measurement noise is modeled as an output of $(\boldsymbol{A}_v, \boldsymbol{B}_v, \boldsymbol{C}_v, D_v)$ driven by a zero-mean white noise process

$v(k)$. The deterministic components of the disturbance and the measurement noise are modeled as outputs of a linear time invariant system $(\boldsymbol{A}_d, \boldsymbol{C}_d)$, where $\boldsymbol{C}_d = \begin{bmatrix} \boldsymbol{C}_{wd} \\ \boldsymbol{C}_{vd} \end{bmatrix}$, driven by the initial condition.

6.2.2.1 State-Space Model of the Noise and Disturbance

$$\begin{aligned} w_p(k) &= w_r(k) + w_d(k) \\ v_p(k) &= v_r(k) + v_d(k) \end{aligned} \tag{6.2}$$

where $w_d(k)$ and $w_r(k)$, and $v_d(k)$, and $v_r(k)$ are respectively deterministic and zero-mean random components of the disturbance $w_p(k)$ and the measurement noise $v_p(k)$.

The state-space model $(\boldsymbol{A}_w, \boldsymbol{B}_w, \boldsymbol{C}_w, D_w)$ of the random disturbance $w_r(k)$ is given by:

$$\begin{aligned} \boldsymbol{x}_w(k+1) &= \boldsymbol{A}_w \boldsymbol{x}_w(k) + \boldsymbol{B}_w w(k) \\ w_r(k) &= \boldsymbol{C}_w \boldsymbol{x}_w(k) + D_w w(k) \end{aligned} \tag{6.3}$$

The state-space model $(\boldsymbol{A}_v, \boldsymbol{B}_v, \boldsymbol{C}_v, D_v)$ of the random measurement noise $v_r(k)$ is

$$\begin{aligned} \boldsymbol{x}_v(k+1) &= \boldsymbol{A}_v \boldsymbol{x}_v(k) + \boldsymbol{B}_v v(k) \\ v_r(k) &= \boldsymbol{C}_v \boldsymbol{x}_v(k) + D_v v(k) \end{aligned} \tag{6.4}$$

The combined state-space model $(\boldsymbol{A}_d, \boldsymbol{C}_d)$ of the deterministic disturbance $w_d(k)$ and the measurement noise $v_d(k)$ is:

$$\begin{aligned} \boldsymbol{x}_d(k+1) &= \boldsymbol{A}_d \boldsymbol{x}_d(k) \\ \begin{bmatrix} w_d(k) \\ v_d(k) \end{bmatrix} &= \boldsymbol{C}_d \boldsymbol{x}_d(k) \end{aligned} \tag{6.5}$$

6.2.3 Integrated Model of the System

The integrated model of the system is formed by combining the models (i) the plant $(\boldsymbol{A}_p, \boldsymbol{B}_p, \boldsymbol{C}_p)$, (ii) the random disturbance $(\boldsymbol{A}_w, \boldsymbol{B}_w, \boldsymbol{C}_w, D_w)$, and (iii) the random measurement noise $(\boldsymbol{A}_v, \boldsymbol{B}_v, \boldsymbol{C}_v, D_v)$ with $D_v = 1$, and the deterministic components $(\boldsymbol{A}_d, \boldsymbol{C}_d)$. Figure 6.1 shows the three state-space models of the plant, the disturbance, and the measurement noise.

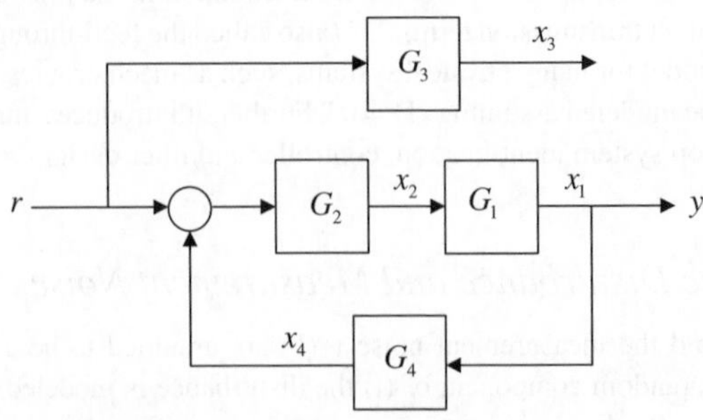

Figure 6.1 Example of an unobservable state (x_3)

The state-space model of the system is given by:

$$\begin{aligned} \boldsymbol{x}(k+1) &= \boldsymbol{A}\boldsymbol{x}(k) + \boldsymbol{B}r(k) + \boldsymbol{E}_w w(k) + \boldsymbol{E}_v v(k) \\ y(k) &= \boldsymbol{C}\boldsymbol{x}(k) + v(k) \end{aligned} \tag{6.6}$$

where

$$\boldsymbol{x} = \begin{bmatrix} \boldsymbol{x}_p \\ \boldsymbol{x}_w \\ \boldsymbol{x}_v \\ x_d \end{bmatrix}, \ \boldsymbol{u} = \begin{bmatrix} u_p \\ w \\ v \end{bmatrix}, \ \boldsymbol{A} = \begin{bmatrix} \boldsymbol{A}_p & \boldsymbol{E}_p\boldsymbol{C}_w & 0 & \boldsymbol{E}_p\boldsymbol{C}_{wd} \\ 0 & \boldsymbol{A}_w & 0 & 0 \\ 0 & 0 & \boldsymbol{A}_v & 0 \\ 0 & 0 & 0 & \boldsymbol{A}_d \end{bmatrix}, \ \boldsymbol{B} = \begin{bmatrix} \boldsymbol{B}_p \\ 0 \\ 0 \\ 0 \end{bmatrix}, \ \boldsymbol{E}_w = \begin{bmatrix} \boldsymbol{E}_p D_w \\ \boldsymbol{B}_w \\ 0 \\ 0 \end{bmatrix}, \ \boldsymbol{E}_v = \begin{bmatrix} 0 \\ 0 \\ \boldsymbol{B}_v \\ 0 \end{bmatrix}$$

$$\boldsymbol{C} = [\boldsymbol{C}_p \quad 0 \quad \boldsymbol{C}_v \quad \boldsymbol{C}_{vd}], \ \boldsymbol{D} = [0 \quad 0 \quad 1]$$

$\boldsymbol{x}(k)$ is the $nx1$ state vector, $y(k)$ the measured output, $r(k)$ the input to the system, $w(k)$ is the disturbance or plant noise, and $v(k)$ the measurement noise; $\boldsymbol{A}$ is nxn matrix, B is $nx1$, $\boldsymbol{C}$ is $1xn$ a, $\boldsymbol{E}_w$ is $nx1$ vectors, and D is a scalar. The vector matrix $\boldsymbol{E}_w$ is usually termed a disturbance distribution matrix and it is assumed that $(\boldsymbol{A}, \boldsymbol{E}_w)$ is controllable so that the disturbance affects all the states, and $(\boldsymbol{A}, \boldsymbol{C})$ is observable so that knowing the input and the output the states may be estimated.

The disturbance $w(k)$ and the measurement noise $v(k)$ are uncorrelated zero-mean white noise processes. The covariances of $w(k)$ and $v(k)$ are given by:

$$E[w^2] = Q \geq 0 \quad \text{and} \quad E[v^2] = R \geq 0 \tag{6.7}$$

6.2.4 Expression for the Output of the Integrated System

The system (6.6) is a three input and one output MIMO system. The inputs are the reference input $r(k)$, the disturbance $w(k)$, and the measurement noise $v(k)$ where $w(k)$ and $v(k)$ are inaccessible zero-mean white noise processes.

$$D(z)y(z) = N(z)r(z) + \upsilon(z) \tag{6.8}$$

where

$$\frac{N(z)}{D(z)} = \boldsymbol{C}(z\boldsymbol{I} - \boldsymbol{A})^{-1}\boldsymbol{B}; \ \frac{N_w(z)}{D(z)} = \boldsymbol{C}(z\boldsymbol{I} - \boldsymbol{A})^{-1}\boldsymbol{E}_w; \ \frac{N_v(z)}{D(z)} = \boldsymbol{C}(z\boldsymbol{I} - \boldsymbol{A})^{-1}\boldsymbol{E}_v + 1;$$

$$\upsilon(z) = N_w(z)w(z) + N_v(z)v(z)$$

6.2.5 Linear Regression Model

A linear regression model may be derived from the state-space model (6.6) using Eq. (6.8):

$$y(k) = \boldsymbol{\psi}^T(k)\boldsymbol{\theta} + \upsilon(k) \tag{6.9}$$

where $\psi^T(k) = [-y(k-1) \ -y(k-2) \ \cdots \ -y(k-n) r(k-1) \ r(k-2) \ \cdots \ r(k-n)]$ and the feature vector is $\theta = [a_1 \ a_2 \ \cdots \ a_n \ b_1 \ b_2 \ \cdots \ b_n]^T$.

Comments *The integrated model, formed of the plant and the disturbance and the measurement noise models, play a crucial role in the design of the Kalman filter and its applications as it gives a complete picture of the map relation of the input to the system to the measure output. The main focus of the*

Kalman filter is to estimate the state $\mathbf{x}(k)$ of the integrated system from the output $y(k)$. From the estimate of the state $\mathbf{x}(k)$, estimates of the other states, namely the plant $\mathbf{x}_p(k)$, the state $\mathbf{x}_w(k)$, and $\mathbf{x}_d(k)$ of the disturbance model and the state $\mathbf{x}_v(k)$ of the measurement model are derived. We will use the following terminologies

- *The output $y(k)$ and the state of the integrated model $\mathbf{x}(k)$ is termed respectively as the* actual output *and the* actual state *of the system.*
- *The output and the state when the disturbance $w(k) = 0$ and the measurement noise $v(k) = 0$ are termed the* true output *and the* true state.

6.2.6 *Observability*

In order to derive the structure of the Kalman filter, we need to assume that the system enjoys some necessary properties, including observability. Let us recall that the prime objective of the Kalman filter is to estimate the states of a dynamical system uniquely from the input and the output data. As it turns out, there are situations where some system states cannot unfortunately be estimated from the available input–output data as shown in the ensuing example. For example, in the system shown in Figure 6.1, it is not possible to estimate the state x_3 from the inputs $\{r(k)\}$ and the outputs $\{y(k)\}$.

The ability to extract the states from the input–output data depends upon the structure of the matrices $\boldsymbol{A}$ and $\boldsymbol{C}$. This ability is formally described, in system-theoretic terms, as the system *observability* property and is mathematically defined as:

A system $(\boldsymbol{A}, \boldsymbol{C})$ is observable if and only if the observability test matrix has a full rank

$$rank \left\{ \begin{bmatrix} C \\ CA \\ CA^2 \\ . \\ CA^{n-1} \end{bmatrix} \right\} = n \tag{6.10}$$

where n is the order of the system, or the number of states. If the system is observable then all the states may be estimated from the input–output data. Equivalently the observability condition may be stated in the frequency domain in terms of system matrix as:

$$rank \left\{ \begin{bmatrix} z\boldsymbol{I} - \boldsymbol{A} \\ \boldsymbol{C} \end{bmatrix} \right\} = n \text{ for all } z \in \mathbb{C} \tag{6.11}$$

where $\mathbb{C}$ is the complex z-plane. For a given transfer function there is an infinite number of state-space models $(\boldsymbol{A}, \boldsymbol{B}, \boldsymbol{C}, D)$ depending upon the choice of state variables. In some applications, the choice of state variables is physically motivated whereas in others an observer canonical form is preferred as it ensures that the system is observable and the number of parameters is minimal.

6.3 Internal Model Principle

The concept of an "internal model of a signal" is widely employed in the design of control systems. Although proportional integral (PI), and proportional integral and derivative (PID) type controllers have been widely used, their theoretical justification and extension were only provided much later in a seminal paper by [13] where the famous *internal model principle* was proposed. It gives the necessary structure of the controller so as to track a given reference signal independent of disturbances and perturbations in the plant. The principle states that the structure of a controller should contain (i) an internal model of the reference and the disturbance signals driven by the tracking error and (ii) a stabilizer to stabilize the closed loop formed of the plant and the internal model.

The internal model principle (IMP) was successfully used in the control system for tracking a reference signal and rejecting disturbances [13–21]. Our aim here is to extend this successful strategy employed in designing controllers to the design of a Kalman filter for applications including fault diagnosis, performance monitoring, estimation of the states of a dynamical system (observer), and the estimation of a signal of a known class buried in noise.

The internal model principle is extended to the design of the Kalman filter where the design objective is twofold: (i) the estimate of the system output generated by the Kalman filter must track the plant output and (ii) the variance of the estimation error is minimized. Since the output of a system is generally random in nature, the tracking objective is defined in statistical terms. An output is said to track a given reference input if the expectation is that the tracking error is asymptotically zero.

6.3.1 Controller Design Using the Internal Model Principle

A brief outline of the internal model principle applied to controller design is outlined with a view to obtain a structure of the Kalman filter. It is shown that the problem of designing an asymptotically unbiased estimator is equivalent to the designing a controller to ensure asymptotic tracking. In this section (i) the concept of an internal model of a deterministic signal is given, (ii) the internal model principle for control design is stated, (iii) a simple example is given to illustrate the design procedure, and (iv) the similarity between the design of a controller and an estimator is shown.

6.3.2 Internal Model (IM) of a Signal

An internal model of a signal is the model of the signal. The zero-input response of the internal model is a copy of the signal, that is, the zero-input response belongs to the same class as the signal. The zero-input response will depend upon the initial conditions of the internal model. For example, if the signal is a constant then the signal generated by the internal model is a class of all constants, and if the signal is a sinusoid, the zero-input response is a sinusoid of the same frequency but different amplitude and phase. A simple way to determine the transfer functions of the internal model is to determine the z-transform of the signal. Then the denominator of the transfer function of the internal model will be equal to the denominator polynomial of the z-transform of the signal.

Let the model of the reference input $r(k)$ be

$$G_r(z) = \frac{N_r(z)}{D_r(z)} \tag{6.12}$$

Then the IM of the signal $r(k)$ may simply be defined as:

$$G_{IM}(z) = \frac{N_{IM}(z)}{D_{IM}(z)} \text{ where } D_{IM}(z) = D_r(z) = 1 + \sum_{i=1}^{n_{IM}} a_{IMi} z^{-i} \tag{6.13}$$

where n_{IM} is the order and $\{a_{IMi}\}$ are the coefficients, and $N_I(z)$ is some arbitrary polynomial. The zero-input response of the internal model satisfies the following difference equation

$$y_{IM}(k) + \sum^{n_{IM}} a_{IMi} y_{IM}(k-i) = 0 \tag{6.14}$$

The zero-input response belongs to the class of all reference input signals to which $r(k)$ belongs. A particular member of this class will depend upon the initial condition of the IM. If the initial condition of the internal model (6.13) equals that of the model of the signal $r(k)$ given by Eq. (6.12), then the zero-input response of IM will be equal to the signal $r(k)$. Since the zero-input response of the internal model generates a class of reference input signal, we may for simplicity choose $N_I(z)$ to be some arbitrary polynomial, for example $N_I(z) = z^{-1}$, so that

$$G_{IM}(z) = \frac{z^{-1}}{D_{IM}(z)} \tag{6.15}$$

For example, an internal model of a class of all constants, a class of ramps, and class of sinusoids are given in Table 6.1 below.

Comment *The signal and its internal model represent, as it were, two sides of the same coin. A signal is completely characterized by its internal model. The internal model is a compact representation of a signal, as the number of parameters of the internal model is generally much less than the number of samples. This property is exploited in many applications including data compression and speech communication where the internal mode rather than the signal is employed. All signals may not have an internal model in the sense that they can be generated by a linear time invariant system. A signal, which has an internal model, is generally referred to as a signal with* rational spectrum *implying that its Fourier transform can be expressed as a ratio of a numerator and a denominator polynomial. In fact, the model of a signal is its z-transform. There are two approaches to signal processing, one based on the signal and its Fourier transform, and the other based on a model of the signal. The model based approach (if the model is available) is increasingly employed in many applications in recent times, including speech processing, controller design, communication, and spectral estimation [22–25].*

6.3.3 Controller Design

Consider a plant given by

$$y(z) = G_p(z)u(z) \text{ where } G_p(z) = \frac{N_p(z)}{D_p(z)} \tag{6.16}$$

Table 6.1 Class of signal and their internal models

Signal	State-space model	Denominator polynomial
$y(k) = \alpha$	$x(k+1) = x(k)$ $y(k) = x(k)$	$D_{IM}(z) = 1 - z^{-1}$
$y(k) = \beta_0 + \beta_1 k + \beta_2 k^2$	$\boldsymbol{x}(k+1) = \begin{bmatrix} 2 & 1 \\ -1 & 0 \end{bmatrix} \boldsymbol{x}(k)\frac{\pi}{3}$ $y(k) = [1 \quad 0]\boldsymbol{x}(k)$	$D_{IM}(z) = 1 - 2z^{-1} + z^{-2}$
$y(k) = a\sin(\omega k + \varphi)$	$\boldsymbol{x}(k+1) = \begin{bmatrix} 2\cos\omega & 1 \\ -1 & 0 \end{bmatrix} \boldsymbol{x}(k)$ $y(k) = [1 \quad 0]\boldsymbol{x}(k)$	$D_{IM}(z) = 1 - 2\cos\omega z^{-1} + z^{-2}$

The problem is to design a controller $G_c(z)$ to meet the following objectives:

- The closed loop system formed of the controller and the plant is asymptotically stable.
- The output of the plant $y(k)$ tracks the reference input $r(k)$

$$\lim_{k\to\infty} it\{y(k) - r(k)\} = 0 \tag{6.17}$$

The denominator polynomial of the internal model of $r(z)$ is $D_{IM}(z) = D_r(z)$.

The following *lemma* gives the necessary and sufficient conditions to meet the control objectives.

Assumptions *We will assume that there are no pole-zero cancellations between the zeros of the plant $G_p(z)$ and the poles of the reference input model $G_r(z)$. That is, there are no common factors between $N_p(z)$ and $D_r(z)$.*

Lemma 6.1 *The robust asymptotic tracking:*

$$\lim_{k\to\infty} it\{y(k) - r(k)\} = 0 \tag{6.18}$$

if and only if

- *The state-space model of the plant $(\boldsymbol{A}_p, \boldsymbol{B}_p, \boldsymbol{C}_p)$ is controllable and observable. That is, there is no pole-zero cancellations in the plant transfer function.*
- *A model of the system that has generated the input $r(k)$, termed "an internal model of $r(k)$," is included in the controller and the internal model is driven by the tracking error $e(k) = r(k) - y(k)$.*
- $\begin{vmatrix} zI - \boldsymbol{A}_p & \boldsymbol{B}_p \\ \boldsymbol{C}_p & 0 \end{vmatrix} \neq 0$ *for all* $\{z : |z\boldsymbol{I} - \boldsymbol{A}_{IM}| = 0\}$ *That is, there are no pole-zero cancellations between the poles of the internal model and the zeros of the plant.*
- *The internal model is driven by the tracking error $e(k) = r(k) - y(k)$.*
- *The resulting closed-loop system formed of the controller and the plant is stabilized.*

The asymptotic tracking (6.18) holds independent of all variation of the plant and the stabilizer parameters as long as (i) the closed system stable is asymptotically stable, and (ii) the internal model parameters are not perturbed

Proof: The proof of the necessary and sufficient conditions is given in [13] using a geometric approach.

A simple proof of sufficiency is outlined herein. For clarity, let us express the controller as a cascade combination of the internal model and a stabilizer:

$$G_c(z) = G_{IM}(z)G_s(z) \tag{6.19}$$

We may express the controller in terms of its numerator and the denominator polynomials as

$$G_c(z) = \frac{N_s(z)}{D_{IM}(z)D_s(z)} \tag{6.20}$$

Figure 6.2 shows the closed-loop system formed of the plant and controller and driven by the tracking error. The controller is comprised of an internal model and a stabilizer.

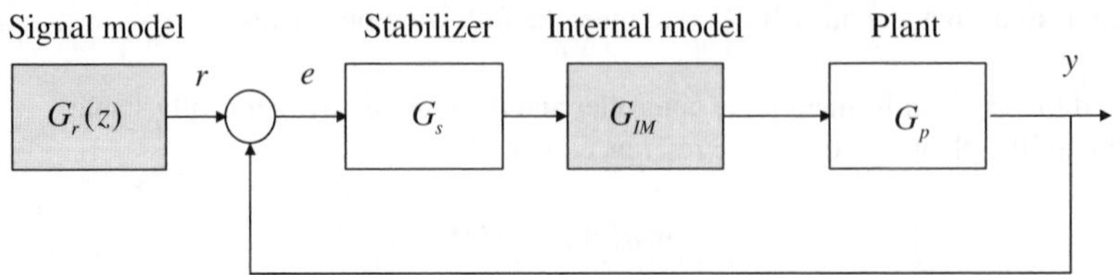

Figure 6.2 Controller design using the internal model principle

The tracking error $e(z)$ is given by:

$$e(z) = S(z)r(z) \tag{6.21}$$

where $S(z) = \dfrac{1}{1 + G_c(z)G_p(z)}$ is the sensitivity function. Expanding using Eq. (6.20) we get:

$$e(z) = \frac{D_{IM}(z)D_s(z)D_p(z)}{N_s(z)N_p(z) + D_{IM}(z)D_s(z)D_p(z)} r(z) \tag{6.22}$$

Expressing $r(z)$ in terms of its model (6.12) yields:

$$e(z) = \left(\frac{D_{IM}(z)D_s(z)D_p(z)}{N_s(z)N_p(z) + D_{IM}(z)D_s(z)D_p(z)} \right) \frac{N_r(z)}{D_r(z)} \tag{6.23}$$

Using the relation (6.15) between the internal model and the signal model, namely $D_r(z) = D_{IM}(z)$, we get:

$$e(z) = \left(\frac{D_{IM}(z)D_s(z)D_p(z)}{N_s(z)N_p(z) + D_{IM}(z)D_s(z)D_p(z)} \right) \frac{N_r(z)}{D_{IM}(z)} \tag{6.24}$$

Since there is a pole-zero cancellation between the poles of the transfer function of the reference input and the zeros of the sensitivity function (transfer function between the tracking error and the reference input), after canceling the common factor, $D_{IM}(z)$ yields:

$$e(z) = \frac{D_s(z)D_p(z)N_r(z)}{N_s(z)N_p(z) + D_{IM}(z)D_s(z)D_p(z)} \tag{6.25}$$

The denominator polynomial $N_s(z)N_p(z) + D_{IM}(z)D_s(z)D_p(z)$ of $e(z)$ is also the closed loop characteristic polynomial. Since the closed loop system is asymptotically stable the tracking error $e(k)$ is zero asymptotically:

$$\lim_{k \to \infty} it\{e(k)\} = 0 \tag{6.26}$$

The asymptotic tracking will hold in spite of the perturbations in the parameters of the plant and the stabilizer. However, if there is perturbation in the parameters of internal model $D_{IM}(z)$, there will be no common factor between $D_{IM}(z)$ and $D_r(z)$. Since the internal model is generally unstable with poles on the unit circle, the poles of the tracking error $e(z)$ will be unstable, and the asymptotic tracking will not hold. It is crucial that $D_r(z) = D_{IM}(z)$ to ensure cancellation of the unstable poles $e(z)$.

Comments *The internal model principle provides a necessary and sufficient structure for a controller to ensure asymptotic tracking. An intuitive proof of the necessity may be obtained by analyzing the structure of the controller shown in Figure 6.2.*

- *Assume that the tracking error $e(z)$ is zero, then the input to the forward loop transfer function (relating $e(z)$ and $y(z)$) $G_{IM}(z)G_s(z)G_p(z)$ is zero. As the zero input response of $G_{IM}(z)$ includes $r(k)$, the asymptotic tracking will occur.*
- *If there is no internal model in the forward loop, asymptotic tracking will not be possible.*
- *If, however, there is a common factor between $D_{IM}(z)$ and the numerator of the plant $N_p(z)$, there will be unstable pole-zero cancelation, and hence there will not be an asymptotic tracking.*
  Note: *The pole- zero cancellation of unstable poles of the exogenous signal model and the unstable zeros of a subsystem is permitted. But the pole-zero cancelation of the unstable poles and zeros of subsystems is not permissible as it will not ensure* internal stability.
- *If $D_{IM}(z)$ is a factor of the denominator of the plant $D_p(z)$, then there is no need to duplicate an internal model in the controller. For example, in practice many plants, such as dc motors relating the voltage input to the armature and the angular position output, have an integral term. In this case, to ensure asymptotic tracking of a constant reference input, there is no need to include an internal model in the controller.*

6.3.4 Illustrative Example: Controller Design

An example to illustrate the internal model principle for the controller design is given using a state-space model mainly to show the duality between the controller and estimator design approaches.

6.3.4.1 Problem Formulation

Design a controller for a plant to track a constant reference input which is a sum of a constant signal and a sinusoid of frequency ω_0 given by

$$r(k) = \alpha \cos(2\pi\omega_0 k) + r_0 \tag{6.27}$$

where $\alpha = 1$ is the amplitude of the sinusoid and $r_0 = 1.5$, $\omega_0 = \pi/8$.

The Plant Model (A_p, B_p, C_p)

$$\begin{aligned} \boldsymbol{x}_p(k+1) &= \boldsymbol{A}_p\boldsymbol{x}_p(k) + \boldsymbol{B}_p u_p(k) \\ y(k) &= \boldsymbol{C}_p\boldsymbol{x}_p(k) \end{aligned} \tag{6.28}$$

where $\boldsymbol{A}_p = \begin{bmatrix} 2\rho\cos\omega_p & 1 \\ -\rho^2 & 0 \end{bmatrix}, \omega_p = \dfrac{\pi}{16}, \rho = 0.9;\ \boldsymbol{B}_p = \begin{bmatrix} 1 \\ 0 \end{bmatrix};\ \boldsymbol{C}_p = [1 \quad 0]$.

Internal Model (A_{im}, B_{im}) of $r(k)$

$$\boldsymbol{x}_{im}(k+1) = \boldsymbol{A}_{im}\boldsymbol{x}_{im}(k) + \boldsymbol{B}_{im}e(k) \tag{6.29}$$

where $\boldsymbol{A}_{im} = \begin{bmatrix} 1 & 0 & 0 \\ 0 & 2\cos\omega_0 & -1 \\ 0 & 1 & 0 \end{bmatrix};\ \boldsymbol{B}_{im} = \begin{bmatrix} 1 \\ 1 \\ 0 \end{bmatrix}$; $\boldsymbol{x}_p$ is $n_p x1$ with $n_p = 2$ and $x_{im}(k)$ is $n_r x1$ state vectors; $n_r = 3$; $u_p(k)$ is the control input, and $y(k)$ is the output to be regulated.

Augmented Plant Model

The augmented plant formed of the plant (6.28) and the internal model (6.29) is given by:

$$\boldsymbol{x}_a(k+1) = \boldsymbol{A}_a\boldsymbol{x}_a(k) + \boldsymbol{B}_a u_p(k) \tag{6.30}$$

where $\boldsymbol{x}_a(k) = \begin{bmatrix} \boldsymbol{x}_p(k) \\ \boldsymbol{x}_{im}(k) \end{bmatrix}$, $\boldsymbol{A}_a = \begin{bmatrix} \boldsymbol{A}_p & 0 \\ -\boldsymbol{B}_{im}\boldsymbol{C}_p & \boldsymbol{A}_{im} \end{bmatrix}$, $\boldsymbol{B}_a = \begin{bmatrix} \boldsymbol{B}_p \\ 0 \end{bmatrix}$, $\boldsymbol{x}_a(k)$ is $nx1$ state with $n = n_p + n_r$.

Stabilizer: State Feedback

The closed loop is stabilized using the state feedback of the plant and the internal model states. The stabilizer is given by

$$u_p(k) = -\boldsymbol{K}_{im}\boldsymbol{x}_{im}(k) - \boldsymbol{K}_p\boldsymbol{x}_p(k) \tag{6.31}$$

where $\boldsymbol{K} = [\boldsymbol{K}_p \quad \boldsymbol{K}_{im}]$ is a $1xn$ state feedback gain with $\boldsymbol{K}_p = [1.9730 \quad 0.9989]$ and $\boldsymbol{K}_{im} = [-1.1422 \quad 0.8763]$ obtained by stabilization of the augmented plant (6.30). Since it is assumed that the plant $(\boldsymbol{A}_p, \boldsymbol{B}_p, \boldsymbol{C}_p)$ is controllable and observable and there are no pole-zero cancellations between the poles of the internal model and the zeros of the plant, there exists a state feedback gain $\boldsymbol{K}$ to ensure stability of the closed-loop system. Using linear quadratic optimization or pole placement or other approaches, the closed-loop poles may be located anywhere inside the unit circle:

$$|\lambda(\boldsymbol{A}_a - \boldsymbol{B}_a\boldsymbol{K})| < 1 \tag{6.32}$$

The closed-loop system using the augmented plant (6.30) and the state feedback stabilizer (6.31) becomes:

$$\begin{aligned} \boldsymbol{x}_{cl}(k+1) &= \boldsymbol{A}_{cl}\boldsymbol{x}_{cl}(k) + \boldsymbol{B}_{cl}r(k) \\ y(k) &= \boldsymbol{C}_{cl}\boldsymbol{x}_{cl}(k) \end{aligned} \tag{6.33}$$

where $\boldsymbol{x}_{cl}(k) = \begin{bmatrix} \boldsymbol{x}_p(k) \\ \boldsymbol{x}_{im}(k) \end{bmatrix}$; $\boldsymbol{A}_{cl} = \begin{bmatrix} \boldsymbol{A}_p - \boldsymbol{B}_p\boldsymbol{K}_p & -\boldsymbol{B}_p\boldsymbol{K}_{im} \\ -\boldsymbol{B}_{im}\boldsymbol{C}_p & \boldsymbol{A}_{im} \end{bmatrix}$, $\boldsymbol{B}_{cl} = \begin{bmatrix} 0 \\ \boldsymbol{B}_{im} \end{bmatrix}$, $\boldsymbol{C}_{cl} = [\boldsymbol{C}_p \quad 0]$; $\boldsymbol{x}_{cl}(k)$ is $nx1$ state. The closed-loop system (6.33) is thus formed of an internal model driven by the tracking error, the plant, and the stabilizer. The internal model (6.29) using Eq. (6.31) becomes

$$\begin{aligned} \boldsymbol{x}_{im}(k+1) &= \boldsymbol{A}_{im}\boldsymbol{x}_{im}(k) + \boldsymbol{B}_{im}e(k) \\ y_{im}(k) &= -\boldsymbol{K}_{im}\boldsymbol{x}_{im}(k) \end{aligned} \tag{6.34}$$

where $y_{im}(k)$ is the output of the IM and an input to the plant. The control input to the plant $u_p(k)$ becomes

$$u_p(k) = y_{im}(k) - \boldsymbol{K}_p\boldsymbol{x}_p(k) \tag{6.35}$$

The plant model after stabilization becomes

$$\begin{aligned} \boldsymbol{x}_p(k+1) &= (\boldsymbol{A}_p - \boldsymbol{B}_p\boldsymbol{K}_p)\boldsymbol{x}_p(k) + \boldsymbol{B}_p y_{im}(k) \\ y(k) &= \boldsymbol{C}_p\boldsymbol{x}_p(k) \end{aligned} \tag{6.36}$$

In order to set the stage for showing the duality between the controller and an estimator we may use an equivalent representation of the state-space model of the internal model. The transfer function of the

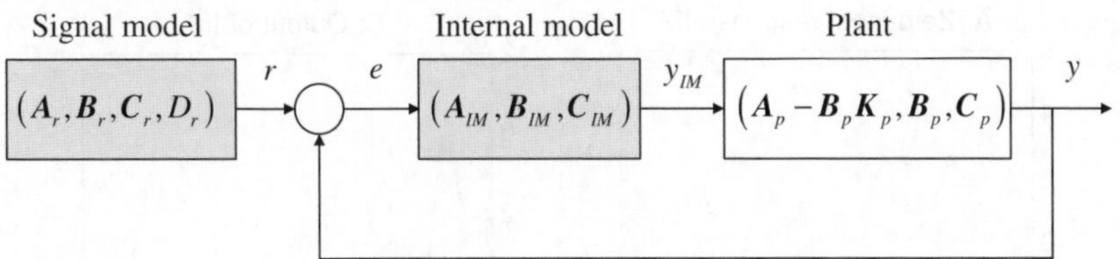

Figure 6.3 Necessary structure of a controller

internal model (6.34) is given by

$$G_{im}(z) = -K_{im}(zI - A_{im})^{-1}B_{im} \tag{6.37}$$

Since the transfer function is scalar, $G_{im}(z) = G_{im}^T(z)$, taking the transpose of Eq. (6.37) we get

$$-K_{im}(zI - A_{im})^{-1}B_{im} = -B_{im}^T\left(zI - A_{im}^T\right)^{-1}K_{im}^T \tag{6.38}$$

as the transfer function of the state-space model (6.34) is unchanged by substituting $A_{IM} = A_{im}^T$, $K_0 = K_{im}^T$, $C_{IM} = B_{im}^T$. Replacing $x_{im}(k)$, by $x_{IM}(k)$, an equivalent internal model (A_{IM}, K_0, C_{IM}) becomes

$$\begin{aligned} x_{IM}(k+1) &= A_{IM}x_{IM}(k) + K_0e(k) \\ y_{IM}(k) &= C_{IM}x_{IM}(k) \end{aligned} \tag{6.39}$$

where $y_{IM}(k) = y_{im}(k)$. The reference input signal $r(k)$ is modeled as an output of a LTI system driven by a delta function given by

$$\begin{aligned} x_r(k+1) &= A_rx_r(k) + B_r\delta(k) \\ r(k) &= C_rx_r(k) \end{aligned} \tag{6.40}$$

where $x_r(k)$ is $n_rx1, A_r = A_{IM} = \begin{bmatrix} 1 & 0 & 0 \\ 0 & 2\cos\omega_0 & 1 \\ 0 & -1 & 0 \end{bmatrix}; B_r = \begin{bmatrix} r_0 \\ \sin\omega_0 \\ 1 \end{bmatrix}; C_r = C_{IM} = [r_0 \quad \alpha \quad 0]; D_r = 0.$

The zero input response of internal model (6.39) will generate a class of all signals $r(k)$. The closed loop formed of the plant $(A_p - B_pK_p, B_p, C_p)$ given by Eq. (6.36) and the internal model (6.39) is shown in Figure 6.3. The necessary structure of the controller is formed of the internal model of the reference signal driven by the tracking error, and a state feedback stabilizer.

6.3.4.2 Simulation Results

Results of the simulation of the control system designed using the internal model principle are given in Figure 6.4. The zero-input response of the internal model of the reference input (6.39) when $e(k) \equiv 0$ is obtained for various initial conditions is shown in subfigure (a). The generated zero responses include the reference input shown by the dotted line. The plant output and the reference input are shown in subfigure (b) indicating asymptotic tracking. The output $y_{IM}(k)$ of the internal model is shown in subfigure (c). Note that the internal model generates a copy of the reference input when the tracking is zero asymptotically. This clearly demonstrates the necessity of the internal model embedded in the closed loop. Finally, the tracking error is shown in subfigure (d) which is asymptotically zero.

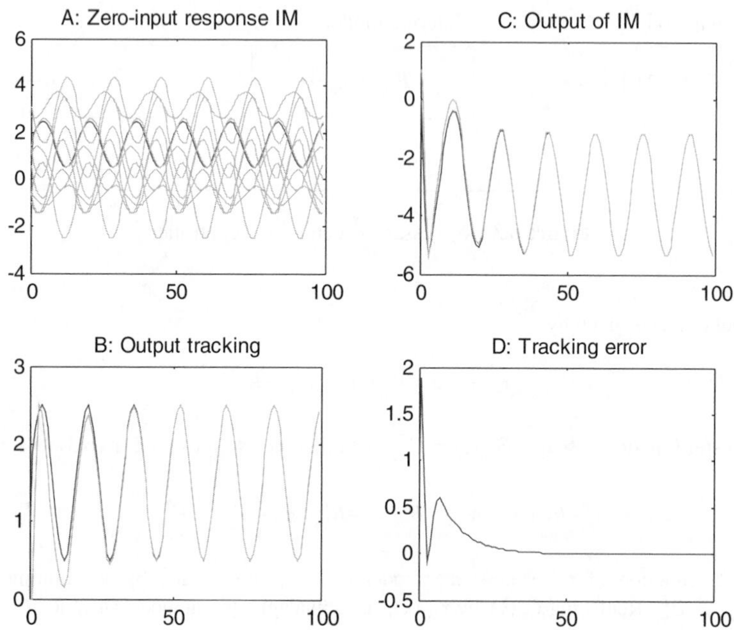

Figure 6.4 The internal model driven by tracking error: asymptotic tracking

6.4 Duality Between Controller and an Estimator Design

The estimator is a dual of controller design. An estimator design problem may be formulated as follows:

6.4.1 Estimation Problem

Let $y(k)$ be a deterministic signal. Design an estimator such that the estimation error $e(k) = y(k) - \hat{y}(k)$, where $\hat{y}(k)$ is the estimate of $y(k)$, is zero asymptotically:

$$\lim_{k\to\infty} it\{e(k)\} = 0 \tag{6.41}$$

Assumptions *The deterministic signal* $y(k)$ *is an output of a LTI system driven by the delta function:*

$$\begin{aligned} \boldsymbol{x}(k+1) &= \boldsymbol{A}\boldsymbol{x}(k) + \boldsymbol{B}\delta(k) \\ y(k) &= \boldsymbol{C}\boldsymbol{x}(k) \end{aligned} \tag{6.42}$$

where $\boldsymbol{x}(k)$ *is* nx1 *state,* $\boldsymbol{A}$ *is* nxn *matrix,* $\boldsymbol{B}$ *is* nx1 *input vector,* C *is* 1xn *output vector, and* $(\boldsymbol{A}, \boldsymbol{C})$ *is observable.*

6.4.2 Estimator Design

The estimator design problem is a dual of the controller design as can be deduced from Section 6.3. The reference input $r(k)$ and the control system output $y(k)$ in controller design take the place of $y(k)$ and $\hat{y}(k)$ respectively, $(\boldsymbol{A}, \boldsymbol{B}, \boldsymbol{C})$ replaces $(\boldsymbol{A}_r, \boldsymbol{B}_r, \boldsymbol{C}_r)$, and $(\boldsymbol{A}_{IM}, \boldsymbol{K}_0, \boldsymbol{C}_{IM})$ is the internal model of $(\boldsymbol{A}, \boldsymbol{B}, \boldsymbol{C})$. Finally, there is a fictitious plant $(\boldsymbol{A}_p, \boldsymbol{B}_p, \boldsymbol{C}_p)$ which is a static system with unit gain $G_p(z) = 1$. With

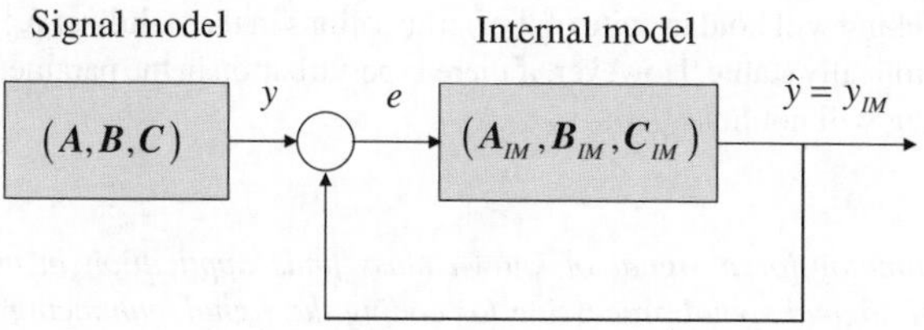

Figure 6.5 Necessary structure of an estimator

these substitutions, the necessary structure of the controller shown in Figure 6.3 becomes the necessary structure of the estimator as shown in Figure 6.5. The estimator consists of an internal model of the signal driven by the estimation error.

Lemma 6.2 *Let $y(k)$ be some deterministic output of a LTI system $(\boldsymbol{A}, \boldsymbol{B}, \boldsymbol{C})$ driven by a delta function given by Eq. (6.42). Assume that $(\boldsymbol{A}, \boldsymbol{C})$ is observable. Let $\hat{y}(k)$ be an estimate of $y(k)$. Then the state estimation error $\tilde{\boldsymbol{x}}(k) = \boldsymbol{x}(k) - \hat{\boldsymbol{x}}(k)$ is zero asymptotically*

$$\lim_{k\to\infty} it\{\tilde{x}(k)\} = 0 \tag{6.43}$$

if and only if *an internal model of $y(k)$ driven by estimation error is included, and the closed-loop system formed of the internal model is stabilized.*

Proof: Let the internal model of $y(k)$ be given by Eq. (6.39). Changing the notations of the internal model we get

$$\begin{aligned} \hat{\boldsymbol{x}}(k+1) &= \boldsymbol{A}\hat{\boldsymbol{x}}(k) + \boldsymbol{K}_0 e(k) + \boldsymbol{B}\delta(k) \\ \hat{y}(k) &= \boldsymbol{C}\hat{\boldsymbol{x}}(k) \end{aligned} \tag{6.44}$$

Substituting $e(k) = y(k) - \hat{y}(k)$ and using $y(k) = \boldsymbol{C}\boldsymbol{x}(k)$, and $\hat{y}(k) = \boldsymbol{C}\hat{\boldsymbol{x}}(k)$, and re-arranging we get

$$\begin{aligned} \hat{\boldsymbol{x}}(k+1) &= (\boldsymbol{A} - \boldsymbol{K}_0\boldsymbol{C})\hat{\boldsymbol{x}}(k) + \boldsymbol{K}_0\boldsymbol{C}\boldsymbol{x}(k) + \boldsymbol{B}\delta(k) \\ \hat{y}(k) &= \boldsymbol{C}\hat{\boldsymbol{x}}(k) \end{aligned} \tag{6.45}$$

Subtracting Eq. (6.45) from Eq. (6.42) we get

$$\begin{aligned} \tilde{\boldsymbol{x}}(k+1) &= (\boldsymbol{A} - \boldsymbol{K}_0\boldsymbol{C})\tilde{\boldsymbol{x}}(k) \\ e(k) &= \boldsymbol{C}\tilde{\boldsymbol{x}}(k) \end{aligned} \tag{6.46}$$

Since $(\boldsymbol{A}, \boldsymbol{C})$ is observable, there exists a state feedback $\boldsymbol{K}_0$ to ensure stability of the closed-loop system. Using linear quadratic optimization or pole placement or other approaches, the closed-loop poles may be located anywhere inside the unit circle:

$$|\lambda(\boldsymbol{A} - \boldsymbol{K}_0\boldsymbol{C})| < 1$$

Hence Eq. (6.43) holds. Since $e(k) = \boldsymbol{C}\tilde{\boldsymbol{x}}(k)$, Eq. (6.41) holds. The proof of necessity follows from the internal model principle.

The asymptotic tracking will hold in spite of the perturbations in the stabilizer $\boldsymbol{K}_0$ as long as the closed-loop system is asymptotically stable. However, if there is perturbation in the parameters of internal model, the asymptotic tracking will not hold.

Comment *The estimator for a signal of known class finds application in many areas including signal, speech, and biological signal processing for coding the signal, enhancing the signal, and noise annihilation.*

6.5 Observer: Estimator for the States of a System

An *observer* is an algorithm for estimating the state variables of a dynamic system. The estimation of a signal, assuming that it is an output of a LTI system $(\boldsymbol{A}, \boldsymbol{B}, \boldsymbol{C})$ driven by a delta function, is extended to the case of an output of a system driven by some deterministic input which is accessible, and $y(k)$ is an output of a system driven by an input $r(k)$ given by

$$\begin{aligned} \boldsymbol{x}(k+1) &= \boldsymbol{A}\boldsymbol{x}(k) + \boldsymbol{B}r(k) \\ y(k) &= \boldsymbol{C}\boldsymbol{x}(k) \end{aligned} \tag{6.47}$$

This represents a linearized model of a large class of physical systems.

6.5.1 *Problem Formulation*

Let $\hat{\boldsymbol{x}}(k)$ be the estimate of $\boldsymbol{x}(k)$. Design an estimator such that state estimation error $\tilde{\boldsymbol{x}}(k)$ is asymptotically zero

$$\lim_{k\to\infty} it\{\tilde{\boldsymbol{x}}(k)\} = 0 \tag{6.48}$$

Assumptions *It is assumed that the plant model* $(\boldsymbol{A}, \boldsymbol{B}, \boldsymbol{C})$ *is known,* $(\boldsymbol{A}, \boldsymbol{C})$ *is observable, and the input* $r(k)$ *is accessible*

6.5.2 *The Internal Model of the Output*

The internal model is simply a copy of the model that has generated $y(k)$ given by Eq. (6.47)

$$\begin{aligned} \hat{\boldsymbol{x}}(k+1) &= \boldsymbol{A}\hat{\boldsymbol{x}}(k) + \boldsymbol{B}r(k) + \boldsymbol{u}(k) \\ \hat{y}(k) &= \boldsymbol{C}\hat{\boldsymbol{x}}(k) \end{aligned} \tag{6.49}$$

where $\hat{\boldsymbol{x}}(k)$ is the states of the IM, $\boldsymbol{u}(k)$ is a $nx1$ input. It is an internal model of the system since when $\boldsymbol{u}(k) \equiv 0$ it is equal to the model (6.47) that has generated $y(k)$ given by

$$\begin{aligned} \hat{\boldsymbol{x}}(k+1) &= \boldsymbol{A}\hat{\boldsymbol{x}}(k) + \boldsymbol{B}r(k) \\ \hat{y}(k) &= \boldsymbol{C}\hat{\boldsymbol{x}}(k) \end{aligned} \tag{6.50}$$

When the initial conditions are identical $\hat{\boldsymbol{x}}(0) = \boldsymbol{x}(0)$ and $\boldsymbol{u}(k) \equiv 0$, then the outputs are identical: $y(k) \equiv \hat{y}(k)$. The model (6.50) is an *open loop-observer* which is used in some applications where the initial condition matching is ensured.

Lemma 6.3 *Let $y(k)$ be some deterministic output of a LTI system $(\boldsymbol{A}, \boldsymbol{B}, \boldsymbol{C})$ driven by a deterministic and an accessible input $r(k)$ given by Eq. (6.47). Assume that $(\boldsymbol{A}, \boldsymbol{C})$ is observable. Then the state estimation error $\tilde{\boldsymbol{x}}(k) = \boldsymbol{x}(k) - \hat{\boldsymbol{x}}(k)$ is zero asymptotically*

$$\lim_{k\to\infty} it\{\tilde{x}(k)\} = 0 \tag{6.51}$$

if and only if *the estimator contains an internal model of $y(k)$ given by Eq. (6.50) and is driven by estimation error and the closed-loop system formed of the internal model is stabilized. The stabilizer is simply a constant gain feedback of the estimation error $e(k) = y(k) - \hat{y}(k)$*

$$\boldsymbol{u}(k) = \boldsymbol{K}_0 e(k) \tag{6.52}$$

where $\boldsymbol{K}_0$ is a nx1 feedback gain. The estimator takes the following form

$$\begin{aligned} \hat{\boldsymbol{x}}(k+1) &= \boldsymbol{A}\hat{\boldsymbol{x}}(k) + \boldsymbol{B}r(k) + \boldsymbol{K}_0 e(k) \\ \hat{y}(k) &= \boldsymbol{C}\hat{\boldsymbol{x}}(k) \end{aligned} \tag{6.53}$$

Proof: An outline of the proof of the sufficient condition is given as the necessity follows from the internal model principle. We will now show that the state estimation error $\tilde{\boldsymbol{x}}(k) = \boldsymbol{x}(k) - \hat{\boldsymbol{x}}(k)$ is zero asymptotically. The expression for the estimation error in terms of the state estimation error is

$$e(k) = y(k) - \hat{y}(k) = \boldsymbol{C}\boldsymbol{x}(k) - \boldsymbol{C}\hat{\boldsymbol{x}}(k) = \boldsymbol{C}\tilde{\boldsymbol{x}}(k) \tag{6.54}$$

Subtracting the estimator model (6.53) from the plant model (6.47) and simplifying using Eq. (6.55) the state-space model governing the state estimation error becomes

$$\begin{aligned} \tilde{\boldsymbol{x}}(k+1) &= (\boldsymbol{A} - \boldsymbol{K}_0\boldsymbol{C})\tilde{\boldsymbol{x}}(k) \\ e(k) &= \boldsymbol{C}\tilde{\boldsymbol{x}}(k) \end{aligned} \tag{6.55}$$

Since $(\boldsymbol{A}, \boldsymbol{C})$ there exists a state-feedback gain $\boldsymbol{K}_0$ to ensure that the eigenvalues of $\boldsymbol{A} - \boldsymbol{K}_0\boldsymbol{C}$ can be located anywhere inside the unit circle as

$$\lambda(\boldsymbol{A} - \boldsymbol{K}_0\boldsymbol{C}) < 1 \tag{6.56}$$

Since the estimation error model (6.55) is asymptotically stable the estimation error is asymptotically zero, and thus meets the state estimator objective (6.48).

The asymptotic tracking will hold in spite of the perturbations in the stabilizer $\boldsymbol{K}_0$ as long as the closed loop system is asymptotically stable. However, if there is perturbation in the parameters of internal model, the asymptotic tracking will not hold.

The necessary structure of the observer is shown in Figure 6.6. The observer is a copy of the plant model driven by the output estimation error, and the resulting closed-loop system is stabilized using constant feedback gain.

6.5.3 *Illustrative Example: Observer with Internal Model Structure*

An example of the design of an observer to estimate the states of a system is given. The observer design is based on the necessary and sufficient conditions established in Lemma 6.3. The application of the internal model principle to the observer design is illustrated.

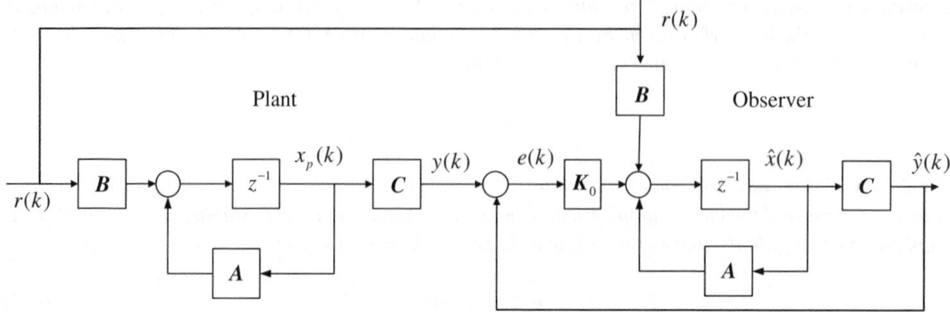

Figure 6.6 Necessary structure of an observer

6.5.3.1 Problem Formulation

An example of a plant is analyzed when plant output is subjected to random and deterministic disturbance and a random measurement noise (with no deterministic measurement noise component).

The Plant Model

The state-space model of the plant $(\boldsymbol{A}_p, \boldsymbol{B}_p, \boldsymbol{C}_p)$ given by

$$\begin{aligned} \boldsymbol{x}_p(k+1) &= \begin{bmatrix} 1.7654 & 1 \\ -0.81 & 0 \end{bmatrix} \boldsymbol{x}_p(k) + \begin{bmatrix} 1 \\ 0 \end{bmatrix} r(k) + \begin{bmatrix} 1 \\ 0 \end{bmatrix} w_p(k) \\ y(k) &= [1 \quad 0]\boldsymbol{x}_p(k) + v_p(k) \end{aligned} \tag{6.57}$$

State-Space Model of the Disturbance

The disturbance is a sum of random and deterministic components. The state-space model $(\boldsymbol{A}_w, \boldsymbol{B}_w, \boldsymbol{C}_w, \boldsymbol{D}_w)$ of the random disturbance $w_r(k)$ is given by:

$$\begin{aligned} x_w(k+1) &= 0.7x_w(k) + w(k) \\ w_r(k) &= 0.3x_w(k) \end{aligned} \tag{6.58}$$

The state-space model $(\boldsymbol{A}_d, \boldsymbol{C}_d)$ of a constant deterministic component is given by:

$$\begin{aligned} x_d(k+1) &= x_d(k) \\ w_d(k) &= x_d(k) \end{aligned} \tag{6.59}$$

The disturbance $w_p(k)$ is a sum of random and deterministic components.

State-Space Model of the Measurement Noise

The state-space model $(\boldsymbol{A}_v, \boldsymbol{B}_v, \boldsymbol{C}_v, \boldsymbol{D}_v)$ of the random measurement noise $v_p(k)$ is

$$\begin{aligned} x_v(k+1) &= -0.7x_w(k) + v(k) \\ v_p(k) &= 1.7x_v(k) \end{aligned} \tag{6.60}$$

Integrated System Model
Combining the plant model $(\boldsymbol{A}_p, \boldsymbol{B}_p, \boldsymbol{C}_p)$, the disturbance models $(\boldsymbol{A}_w, \boldsymbol{B}_w, \boldsymbol{C}_w, \boldsymbol{D}_w)$ and $(\boldsymbol{A}_d, \boldsymbol{C}_d)$, and the measurement noise model $(\boldsymbol{A}_v, \boldsymbol{B}_v, \boldsymbol{C}_v, \boldsymbol{D}_v)$, the integrated system model $(\boldsymbol{A}, \boldsymbol{B}, \boldsymbol{C}, D)$ given by Eq. (6.6) is

$$\boldsymbol{x}(k+1) = \begin{bmatrix} 1.7654 & 1 & 0.3 & 1 & 0 \\ -0.81 & 0 & 0 & 0 & 0 \\ 0 & 0 & 0.7 & 0 & 0 \\ 0 & 0 & 0 & -0.7 & 0 \\ 0 & 0 & 0 & 0 & 1 \end{bmatrix} \boldsymbol{x}(k) + \begin{bmatrix} 1 \\ 0 \\ 0 \\ 0 \\ 0 \end{bmatrix} r(k) + \begin{bmatrix} 0 \\ 0 \\ 1 \\ 0 \\ 0 \end{bmatrix} w(k) + \begin{bmatrix} 0 \\ 0 \\ 0 \\ 1 \\ 0 \end{bmatrix} v(k) \tag{6.61}$$
$$y(k) = [0.0427 \quad 0 \quad 0 \quad 1.7 \quad 0]\boldsymbol{x}(k)$$

Objective
Design an observer such that the state estimation error is zero asymptotically

$$\lim_{k\to\infty} it\{\tilde{\boldsymbol{x}}(k)\} = 0 \tag{6.62}$$

where $\tilde{\boldsymbol{x}}(k) = \boldsymbol{x}(k) - \hat{\boldsymbol{x}}(k)$, $\hat{\boldsymbol{x}}(k)$ is the estimate of $\boldsymbol{x}(k)$ generated by an observer.

Observer Design
The internal model of the plant output $y(k)$ and the observer are given respectively by Eqs. (6.49) and (6.53). The observer is the internal model driven by scaled tracking errors $\boldsymbol{K}_0 e(k)$. The integrated system and the necessary structure of an observer are shown in Figure 6.6. The feedback gain $\boldsymbol{K}_0$ is obtained using linear quadratic optimization given by

$$\boldsymbol{K}_0 = [6.6437 \quad -4.5545 \quad 0.0074 \quad 0.3957 \quad -0.2485]^T \tag{6.63}$$

The eigenvalues $\lambda(\boldsymbol{A} - \boldsymbol{K}_0\boldsymbol{C})$ are $-0.1641, 0.8036 \pm j0.2555, 0.7674$.

6.5.3.2 Simulation Results

The performance of the observer is analyzed by simulations including the following:

- The asymptotic tracking performance (6.62) or the lack of it in the face of the perturbation in stabilizer and the IM model of the observer.
- The behavior in the presence of disturbance and measurement noise.

Asymptotic Tracking Performance
Results of the simulation of the observer, which contains a copy of the plant model driven by the tracking error, are given in Figure 6.7. Subfigures (a) and (b) show the behavior of the observer under an ideal condition when there is no mismatch between the plant model and the internal model encapsulated in the observer, while subfigures (c), (d), (e), and (f) show the robustness or otherwise of the observer under stabilizer and plant model perturbations. Subfigure (a) shows the plant and the observer states. The state estimation errors are asymptotically zero. Subfigure (b) shows the zero input response of the internal model of the plant under various initial conditions. When the initial conditions of the plant and the internal model are identical, the corresponding states are the same. Subfigure (c) shows that asymptotic tracking holds in spite of the variations in the stabilizer gain $\boldsymbol{K}_0$ as long as the closed-loop stability is maintained. There are, of course, changes in the transient behavior under variations in the stabilizer. Subfigures (d), (e), and (f) show that the observer *is not robust* to perturbations in the internal model $\boldsymbol{A}$,

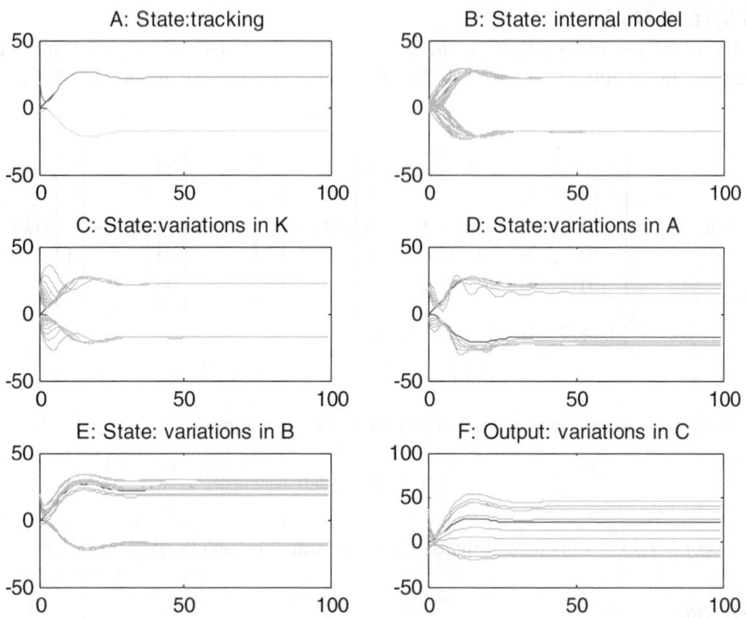

Figure 6.7 Observer: robust to stabilizer and not to the plant model

$\boldsymbol{B}$, and $\boldsymbol{C}$. The tracking objective (6.62) does not hold under any perturbation of the internal model. Note that under perturbation in $\boldsymbol{C}$ of the internal model, only the output $\hat{y}(k)$ is affected.

Effect of Disturbance and Measurement Noise

Figure 6.8 shows the output and the states of the plant when subjected to disturbance and measurement noise. Subfigure (a) shows the noise-free states, the actual states, and the estimated states. The correlation of the residual $e(k) = y(k) - \hat{y}(k)$ is shown in subfigure (b). Note that the residual observer is not a white noise process (the correlation function is not an approximate delta function).

Comments *An observer ensures tracking in spite of variations in the stabilizer thanks to the necessary structure. The observer consists of an internal model of the noise free plant output and is driven by the tracking error (or residual).*

However its performance in the presence of the noise and disturbance is poor. The residual of the observer is not a zero-mean white noise process, indicating that the residual contains some dynamics of the system and the noise models: the estimate has not completely extracted all the information from the input and the output of the system. This problem is addressed in the next section on Kalman filters. It is shown that the stabilizer gain should be determined optimally by taking into account the variances of the disturbances of the disturbance and the measurement noise beside the system model (and not the noise free model alone as is done in this example).

6.6 Kalman Filter: Estimator of the States of a Stochastic System

The Kalman filter may be referred to as a stochastic version of an observer. The Kalman filter has a structure identical to that of an observer, namely an internal model of the system output driven by the residual. The design of the Kalman filter and the observer are the same except that the stabilizer is designed optimally taking into account the statistics of the disturbance and noise.

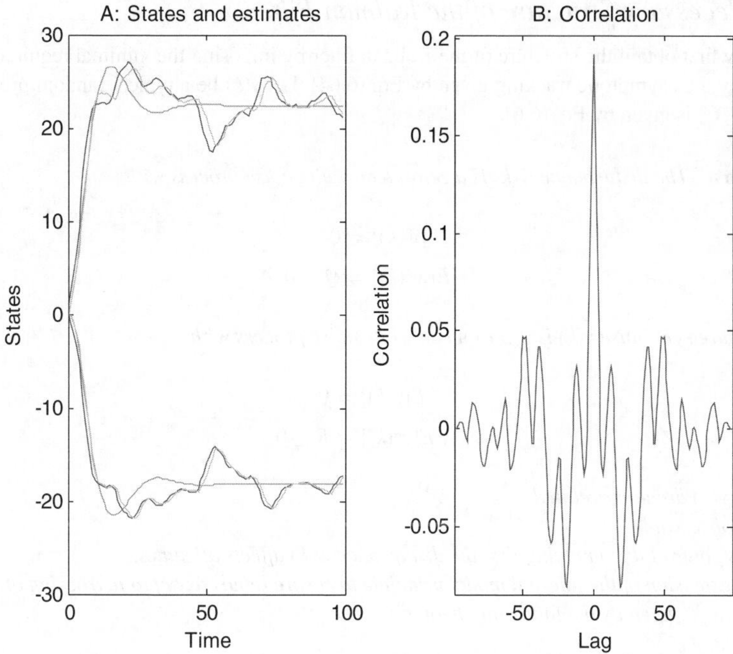

Figure 6.8 The effect of disturbance and the measurement noise

Similarly to an observer, the main objective of the Kalman filter is to estimate the states of a dynamical system from its input and the output.

6.6.1 Objectives of the Kalman Filter

The Kalman filter is designed to estimate the states of a system in the presence of noise and disturbance affecting the system output. The states, the output, and their estimates using the Kaman filter will also be random processes and not deterministic signals. Hence the design objectives are expressed in statistical terms. We will assume the random processes to be Gaussian so that the derivation of the Kalman filter is mathematically tractable, and further the mean and covariance completely characterize the random process. The performance of the Kalman filter is expressed in terms of the mean and covariance of the state estimation errors. Let the model of the system that generates the random processes, namely the state $\boldsymbol{x}(k)$ and the output $y(k)$, be the integrated model $(\boldsymbol{A}, \boldsymbol{B}, \boldsymbol{C})$ given by Eq. (6.6). Let $\hat{\boldsymbol{x}}(k)$ and $\hat{y}(k)$ be respectively the estimates of the state $\boldsymbol{x}(k)$ and the output $y(k)$. The design objectives may be stated as follows:

- The mean of the estimation error $\tilde{\boldsymbol{x}}(k) = \hat{\boldsymbol{x}}(k) - \boldsymbol{x}(k)$ is zero asymptotically, or equivalently the estimation error is asymptotically unbiased:

$$\lim_{k \to \infty} E[\tilde{x}(k)] = 0 \tag{6.64}$$

- The mean squared rrror of the state estimate is minimal:

$$\min_{\hat{x}} \left\{ E\left[(\boldsymbol{x}(k) - \hat{\boldsymbol{x}}(k))^T (\boldsymbol{x}(k) - \hat{\boldsymbol{x}}(k)) \right] \right\} \tag{6.65}$$

The expression of the mean squared error is the sum of the variances of the elements of the estimate.

6.6.2 Necessary Structure of the Kalman Filter

We will now first obtain the structure of the Kalman filter by imposing the minimal requirement on the filter, namely the asymptotic tracking given by Eq. (6.64). Let $y(k)$ be a typical random process whose model $(\boldsymbol{A}, \boldsymbol{B}, \boldsymbol{C})$ is given by Eq. (6.6).

Assumptions *The disturbance $w(k)$ is a zero-mean white noise process with*

$$\begin{aligned} E[w(k)] &= 0 \\ E[w^2(k)] &= Q > 0 \end{aligned} \tag{6.66}$$

- *The measurement noise $v(k)$ is a zero-mean white noise process with*

$$\begin{aligned} E[v(k)] &= 0 \\ E[v^2(k)] &= R \geq 0 \end{aligned} \tag{6.67}$$

- *$w(k)$ and $v(k)$ are uncorrelated.*
- *$(\boldsymbol{A}, \boldsymbol{C})$ is observable.*
- *$(\boldsymbol{A}, \boldsymbol{E}_w)$ is controllable implying that the disturbance $w(k)$ affects all states.*
- *A simple extension of the internal model principle to ensure robust asymptotic tracking of the random processes $\mathbf{x}(k)$ given by the following theorem:*

Theorem 6.1 *Let $\mathbf{x}(k)$ be an output of a LTI system $(\boldsymbol{A}, \boldsymbol{B}, \boldsymbol{C})$ driven by a deterministic and an accessible input $r(k)$ and assumptions given above hold. The state estimation error $\tilde{\mathbf{x}}(k) = \mathbf{x}(k) - \hat{\mathbf{x}}(k)$ is asymptotically zero in the statistical sense*

$$\lim_{k\to\infty} it\{E[\tilde{\mathbf{x}}(k)]\} = 0 \tag{6.68}$$

if and only if *the estimator contains (i) an internal model of $E[y(k)]$ and is driven by output estimation error (or residual) $e(k) = y(k) - \hat{y}(k)$ and (ii) the closed-loop system formed of the internal model is stabilized. The stabilizer is simply a feedback of the residual given by*

$$\hat{\mathbf{u}}(k) = \boldsymbol{K}e(k) \tag{6.69}$$

where $\hat{\mathbf{u}}(k)$ is the input of the internal model and $\boldsymbol{K}$ is the state feedback gain.

Proof: Follows from the internal model principle for controller design [1] and Lemmas 6.1, 6.2, and especially 6.3.

$E[y(k)]$ and $E[x(k)]$ are the *true output* and the *true state* of the system respectively. Theorem 6.1 gives a necessary and sufficient condition for the robust asymptotic tracking of the true state of the system. The necessary structure of the Kalman filter is identical to that of the observer, as shown in Figure 6.6. In the case of the optimal Kalman filter, stabilizer $\boldsymbol{K}$ is time-varying. However, we will use a suboptimal constant gain feedback similar to the observer.

6.6.3 Internal Model of a Random Process

An internal model of $E[y(k)]$ is derived from the integrated state-space model $(\boldsymbol{A}, \boldsymbol{B}, \boldsymbol{C})$ by taking expectation on both sides of the dynamic and the algebraic equations given by Eq. (6.6) or equivalently

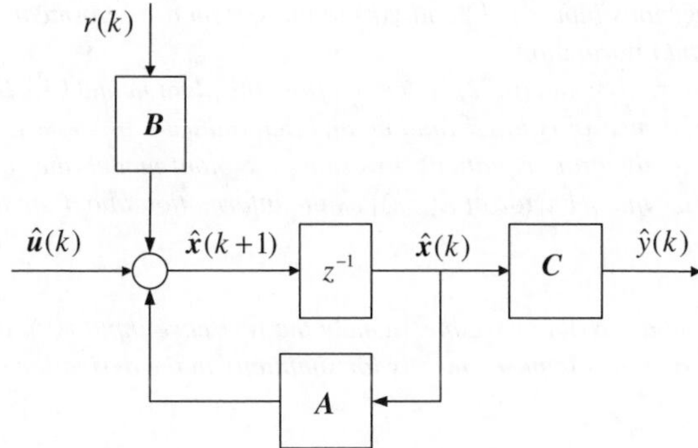

Figure 6.9 Internal model: the expected value of the output

by simply setting to zero the inaccessible zero mean random variables $w(k)$ and $v(k)$. The internal model of $E[y(k)]$ is given by

$$\begin{aligned} \hat{\boldsymbol{x}}(k+1) &= \boldsymbol{A}\hat{\boldsymbol{x}}(k) + \boldsymbol{B}r(k) + \hat{\boldsymbol{u}}(k) \\ \hat{y}(k) &= \boldsymbol{C}\hat{\boldsymbol{x}}(k) \end{aligned} \tag{6.70}$$

where $\hat{\boldsymbol{u}}(k)$ is an $nx1$ input vector. The internal model generates a class of all expected system outputs $\{E[y(k)]\}$ when the input $\hat{\boldsymbol{u}}(k) \equiv 0$, and they will be identical to the expected values of system states and output if their initial conditions are equal:

$$\begin{aligned} \hat{\boldsymbol{x}}(k) &= E[\boldsymbol{x}(k)] \\ \hat{y}(k) &= E[y(k)] \end{aligned} \quad if \quad \hat{\boldsymbol{u}}(k) \equiv 0 \; and \; \hat{\boldsymbol{x}}(0) = E[\boldsymbol{x}(0)] \tag{6.71}$$

The "new input" to the internal model $\hat{\boldsymbol{u}}(k)$ is introduced to set the stage for developing the Kalman filter. Figure 6.9 shows the internal model. It maps $\hat{\boldsymbol{u}}(k)$ and $r(k)$ to the output.

Comments *The internal model (6.70) is an open-loop estimator of the states of system as long as the initial conditions of the system and the internal model are identical and the integrated model is asymptotically stable. The open-loop estimator is used in some applications in view of its simplicity.*

The main objective is to estimate the states of the plant $\boldsymbol{x}_p(k)$. The internal model of $E[y(k)]$ derived from the plant model by taking expectation yields

$$\begin{aligned} \hat{\boldsymbol{x}}_p(k+1) &= \boldsymbol{A}_p\hat{\boldsymbol{x}}_p(k) + \boldsymbol{B}_p r(k) + \hat{\boldsymbol{u}}_p(k) \\ \hat{y}(k) &= \boldsymbol{C}_p\hat{\boldsymbol{x}}_p(k) \end{aligned} \tag{6.72}$$

The reasons for using an augmented model (6.70) whose state $\boldsymbol{x}(k)$ contains $\boldsymbol{x}_p(k)$ and the states of the noise and the disturbances, instead of a simpler state-space model of the plant $(\boldsymbol{A}_p, \boldsymbol{B}_p, \boldsymbol{C}_p)$ given by Eq. (6.72) which only contains the desired state $\boldsymbol{x}_p(k)$, are given below:

- *The state-space internal model $(\boldsymbol{A}, \boldsymbol{B}, \boldsymbol{C}, D)$ combines the models of the the plant $(\boldsymbol{A}_p, \boldsymbol{B}_p, \boldsymbol{C}_p)$, random disturbance $(\boldsymbol{A}_w, \boldsymbol{B}_w, \boldsymbol{C}_w, D_w)$, the measurement noise $(\boldsymbol{A}_v, \boldsymbol{B}_v, \boldsymbol{C}_v, D_v)$, and deterministic disturbance and the deterministic measurement noise $(\boldsymbol{A}_d, \boldsymbol{C}_d)$. Thus all the relevant information about the expected value of state $E[\boldsymbol{x}(k)]$ output $E[y(k)]$ has been completely captured by the augmented state-space model.*

The inaccessible exogenous inputs $w(k)$ and $v(k)$ to the system are zero-mean white noise processes and do not carry useful information.

- *On the other hand, internal model (6.72) is derived from the plant model $(\boldsymbol{A}_p, \boldsymbol{B}_p, \boldsymbol{C}_p)$ driven by $w_p(k)$ and $v_p(k)$, which are formed of colored random and deterministic processes. By taking expectation of the dynamic and the algebraic equations governing the plant model, the resulting model will not completely characterize the behavior of $E[y(k)]$ as the information about the dynamics of $w_p(k)$ and $v_p(k)$ are lost.*

We have assumed that the accessible variable, namely the reference input $r(k)$, is deterministic so that $r(k) = E[r(k)]$. This restriction is imposed merely for simplicity in the derivation.

6.6.4 Illustrative Example: Role of an Internal Model

An example to illustrate the role of IM is given. It is shown that the zero-input response of the internal model contains the true state. The example clearly shows that the IM model derived from the integrated model (6.70) containing the models of the noise and disturbance generates the true states of the system, while a simpler IM (6.72) derived from the plant model neglecting the dynamics of the noise and disturbance does not. The models of the plant, the disturbance, and the measurement noise are given in the Illustrative example: observer with internal model structure. The models of the plant $(\boldsymbol{A}_p, \boldsymbol{B}_p, \boldsymbol{C}_p)$, the disturbance $(\boldsymbol{A}_w, \boldsymbol{B}_w, \boldsymbol{C}_w, \boldsymbol{D}_w)$, the deterministic disturbance $(\boldsymbol{A}_d, \boldsymbol{C}_d)$, and the measurement noise $(\boldsymbol{A}_v, \boldsymbol{B}_v, \boldsymbol{C}_v, \boldsymbol{D}_v)$ are given respectively by Eqs. (6.57), (6.58), (6.59), and (6.60). Taking expectation of the integrated model $(\boldsymbol{A}, \boldsymbol{B}, \boldsymbol{C}, \boldsymbol{D})$ given in Eq. (6.61) we get:

$$E[\boldsymbol{x}(k+1)] = \begin{bmatrix} 1.7654 & 1 & 0.3 & 1 & 0 \\ -0.81 & 0 & 0 & 0 & 0 \\ 0 & 0 & 0.7 & 0 & 0 \\ 0 & 0 & 0 & -0.7 & 0 \\ 0 & 0 & 0 & 0 & 1 \end{bmatrix} E[\boldsymbol{x}(k)] + \begin{bmatrix} 1 \\ 0 \\ 0 \\ 0 \\ 0 \end{bmatrix} r(k) \tag{6.73}$$

$$E[y(k)] = [0.0427 \quad 0 \quad 0 \quad 1.7 \quad 0] E[\boldsymbol{x}(k)]$$

The internal model of the output $y(k)$ using Eq. (6.70) is given by

$$\hat{x}(k+1) = \begin{bmatrix} 1.7654 & 1 & 0.3 & 1 & 0 \\ -0.81 & 0 & 0 & 0 & 0 \\ 0 & 0 & 0.7 & 0 & 0 \\ 0 & 0 & 0 & -0.7 & 0 \\ 0 & 0 & 0 & 0 & 1 \end{bmatrix} \hat{x}(k) + \begin{bmatrix} 1 \\ 0 \\ 0 \\ 0 \\ 0 \end{bmatrix} r(k) + \boldsymbol{u}(k) \tag{6.74}$$

$$\hat{y}(k) = [0.0427 \quad 0 \quad 0 \quad 1.7 \quad 0] \hat{\boldsymbol{x}}(k)$$

6.6.4.1 Simulation Results

The true state $E[\boldsymbol{x}(k)]$ of the system (6.73) and the states $\hat{\boldsymbol{x}}(k)$ of the IM (6.74) when $\boldsymbol{u}(k) \equiv 0$ for different randomly selected initial conditions $\hat{\boldsymbol{x}}(0)$ are shown in Figure 6.10. Subfigure (a) shows the true state, and the states of the plant affected by noise and disturbances. Subfigure (b) shows the zero-input response of the state $\hat{\boldsymbol{x}}(k)$ of the integrated IM (6.74) for different initial conditions. It can be seen that one of the family of responses $\{\hat{\boldsymbol{x}}(k)\}$ of the internal model is equal to $E[\boldsymbol{x}(k)]$ when $\hat{\boldsymbol{x}}(0) = E[\boldsymbol{x}(0)]$. Subfigure (c)

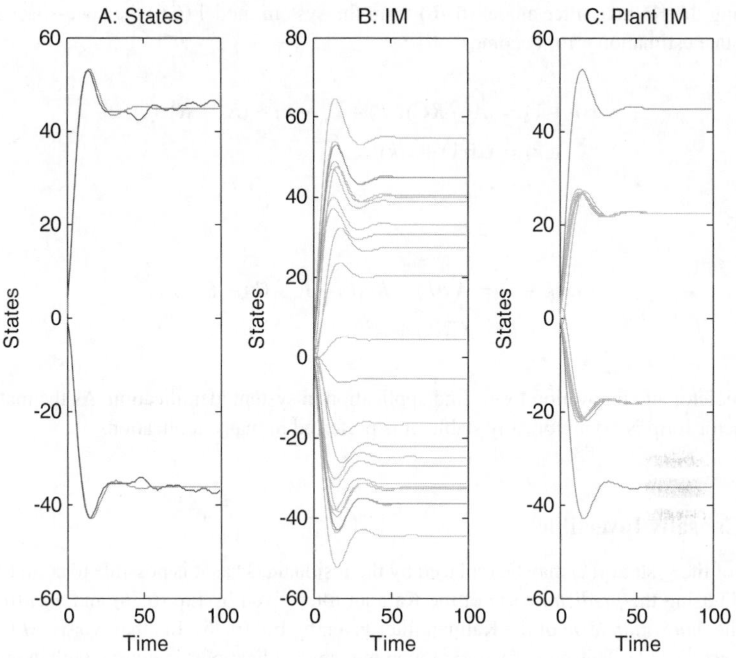

Figure 6.10 Comparison of the integrated and the plant internal models

shows the zero-input response of the state $\hat{\boldsymbol{x}}_p(k)$ of the plant IM (6.72) for different initial conditions. Note that the zero-input responses of the plant IM do not contain the true response. In view of this, the integrated model, and not the plant internal model, is employed in the Kalman filter.

6.6.5 *Model of the Kalman Filter*

Using the expression for the stabilizer (6.69) and the internal model (6.70), the state-space model of the Kalman filter becomes

$$\begin{aligned}\hat{\boldsymbol{x}}(k+1) &= \boldsymbol{A}\hat{\boldsymbol{x}}(k) + \boldsymbol{B}r(k) + \boldsymbol{K}e(k)\\ \hat{y}(k) &= \boldsymbol{C}\hat{\boldsymbol{x}}(k)\\ y(k) &= e(k) + \hat{y}(k)\end{aligned} \tag{6.75}$$

The Kalman filter model (6.75) is termed the *innovation form*.

Substituting for the residual $e(k) = y(k) - \boldsymbol{C}\hat{\boldsymbol{x}}(k)$ we get

$$\begin{aligned}\hat{\boldsymbol{x}}(k+1) &= (\boldsymbol{A} - \boldsymbol{K}\boldsymbol{C})\hat{\boldsymbol{x}}(k) + \boldsymbol{B}r(k) + \boldsymbol{K}y(k)\\ \hat{y}(k) &= \boldsymbol{C}\hat{\boldsymbol{x}}(k)\\ e(k) &= y(k) - \hat{y}(k)\end{aligned} \tag{6.76}$$

This model is often referred to as *predictor* form.

Subtracting the Kalman filter model (6.76) from the system model (6.6) the state-space model governing the state estimation error becomes

$$\begin{aligned} \tilde{\boldsymbol{x}}(k+1) &= (\boldsymbol{A} - \boldsymbol{KC})\tilde{\boldsymbol{x}}(k) + \boldsymbol{E}_w w(k) + (\boldsymbol{E}_v - \boldsymbol{K})v(k) \\ e(k) &= \boldsymbol{C}\tilde{\boldsymbol{x}}(k) + v(k) \end{aligned} \tag{6.77}$$

Equivalently

$$\begin{aligned} \tilde{\boldsymbol{x}}(k+1) &= \boldsymbol{A}\tilde{\boldsymbol{x}}(k) - \boldsymbol{K}e(k) + \boldsymbol{E}_w w(k) + \boldsymbol{E}_v v(k) \\ e(k) &= \boldsymbol{C}\tilde{\boldsymbol{x}}(k) + v(k) \end{aligned} \tag{6.78}$$

Both the predictor and innovation forms find application in system identification. As the matrix $\boldsymbol{A} - \boldsymbol{KC}$ of the predictor form is asymptotically stable, it is preferred in many applications.

6.6.5.1 Causally Invertible

The output of the system $y(k)$ may be replaced by the residual $e(k)$ as it is possible to compute $e(k)$ from $y(k)$ and $r(k)$ using the *predictor form* of the Kalman filter given by Eq. (6.76) and $y(k)$ from $e(k)$ and $r(k)$ using the *innovation form* of the Kalman filter given by Eq. (6.75). In other words $y(k)$ and $e(k)$ are *causally invertible* [11]. This property is exploited in many applications, including fault diagnosis, as the residual $e(k)$ is a zero-mean white noise process (as will be shown in later sections). The Kalman filter may be viewed as a *whitening filter*: the input to the Kalman filter is $y(k)$ and $r(k)$ and its output is $e(k)$, which is a zero-mean white noise process.

6.6.6 Optimal Kalman Filter

Having obtained the necessary structure of the Kalman filter to ensure the asymptotic unbiasedness condition (6.64), the next step is to derive an optimal state feedback gain $\boldsymbol{K}$, termed Kalman gain, by minimizing the mean-squared estimation error at each time instant given by Eq. (6.65). Since the estimate $\tilde{\boldsymbol{x}}(k+1)$ is a function of the Kalman gain $\boldsymbol{K}$ in view of state-space model (6.77), the mean squared error minimization problem may be posed as:

$$\min_{\boldsymbol{K}} E[\tilde{\boldsymbol{x}}^T(k+1)\tilde{\boldsymbol{x}}(k+1)] \tag{6.79}$$

In 6.9 Appendix, substituting for $\tilde{\boldsymbol{x}}(k+1)$ using Eq. (6.77), the expression of the covariance of estimation error $\boldsymbol{P}(k) = E[\tilde{\boldsymbol{x}}(k)\tilde{\boldsymbol{x}}^T(k)]$ is derived and is given by

$$\begin{aligned} \boldsymbol{P}(k+1) = \boldsymbol{AP}(k)\boldsymbol{A}^T + \boldsymbol{K}(k)(\boldsymbol{CP}(k)\boldsymbol{C}^T + R)\boldsymbol{K}^T(k) - \left(\boldsymbol{AP}(k)\boldsymbol{C}^T + \boldsymbol{E}_v R\right)\boldsymbol{K}^T \\ - \boldsymbol{K}(k)\left(\boldsymbol{CP}(k)\boldsymbol{A}^T + \boldsymbol{E}_v^T R\right) + \boldsymbol{E}_w \boldsymbol{E}_w^T Q + \boldsymbol{E}_v \boldsymbol{E}_v^T R \end{aligned} \tag{6.80}$$

6.6.7 Optimal Scalar Kalman Filter

To gain an insight into the parameters that affect the mean squared error, let us first derive the necessary condition assuming all the variables are scalars. Replacing vectors and matrices by scalars, the scalar version of the system model (6.6) and the Kalman filter (6.76) and the scalar version of estimation error equation (6.77) we get

$$\begin{aligned} x(k+1) &= Ax(k) + Br(k) + E_w w(k) + E_v v(k) \\ y(k) &= Cx(k) + v(k) \end{aligned} \tag{6.81}$$

$$\begin{aligned} \hat{x}(k+1) &= (A - K(k)C)\hat{x}(k) + Br(k) + K(k)y(k) \\ \hat{y}(k) &= C\hat{x}(k) \end{aligned} \tag{6.82}$$

Equivalently

$$\begin{aligned} \hat{x}(k+1) &= A\hat{x}(k) + Br(k) + K(k)e(k) \\ \hat{y}(k) &= C\hat{x}(k) \end{aligned} \tag{6.83}$$

$$\tilde{x}(k+1) = A\tilde{x}(k) - K(k)e(k) + E_w w(k) + E_v v(k) \tag{6.84}$$

Squaring both sides and taking expectation yields

$$P(k+1) = E\left[(A\tilde{x}(k) - K(k)e(k) + E_w w(k) + E_v v(k))^2\right] \tag{6.85}$$

Simplifying we get

$$P(k+1) = A^2P(k) + K^2(k)(C^2P(k) + R) - 2K(k)APC - 2K(k)E_v + E_w^2Q + E_v^2R \tag{6.86}$$

where, A, B, C, E_w, E_v, and K are respectively the scalar versions of $\mathbf{A}$, $\mathbf{B}$, $\mathbf{C}$, $\mathbf{E}_w$, $\mathbf{E}_v$, and $\mathbf{K}$. We will minimize the covariance matrix $P(k+1)$ with respect to K using the "completing the square approach" given in 6.9 Appendix for the vector case. The scalar version of the proof is given below.

Adding and subtracting a term $(APC + E_v)^2(C^2P(k) + R)^{-1}$ on the right-hand side of Eq. (6.86) and re-arranging we get

$$\begin{aligned} P(k+1) = A^2P(k) &- (APC + E_vR)^2(C^2P(k) + R)^{-1} + E_w^2Q + E_v^2R \\ &K^2(k)(C^2P(k) + R) - 2K(k)(APC + E_vR) + (APC + E_vR)^2(C^2P(k) + R)^{-1} \end{aligned} \tag{6.87}$$

Expressing the last three terms as a quadratic function of K yields

$$\begin{aligned} P(k+1) = A^2P(k) &- (APC + E_vR)^2(C^2P(k) + R)^{-1} + E_w^2Q + E_v^2R \\ &+(C^2P(k) + R)\left(K(k) - (APC + E_vR)(C^2P(k) + R)^{-1}\right)^2 \end{aligned} \tag{6.88}$$

The covariance matrix $P(k+1)$ is minimum with respect to $K(k)$. The quadratic term in $K(k)$ on the right-side is set equal to zero and the optimal gain becomes

$$K(k) = (AP(k)C + E_vR)(C^2P(k) + R)^{-1} \tag{6.89}$$

Substituting the expression for the optimal gain $K(k)$ the expression for the optimal covariance $P(k+1)$ becomes:

$$P(k+1) = A^2P(k) - (AP(k)C + E_vR)^2(C^2P(k) + R)^{-1} + E_w^2Q + E_v^2R \tag{6.90}$$

The above equations governing the covariance of the estimation error are a nonlinear first-order matrix equation termed the *Riccati equation.*

6.6.7.1 Expression in Terms of the Ratio of the Variance

It is convenient to express the Kalman gain $K(k)$ and the estimation error variance $P(k)$ in terms of the ratio of the variance of the disturbance and the variance of measurement noise, namely $Q_\rho = Q/R$. Dividing the expression for the covariance $P(k+1)$ by the measurement variance R we get

$$P_\rho(k+1) = A^2P_\rho(k) - (AP_\rho(k)C + E_v)^2(C^2P_\rho(k) + 1)^{-1} + E_w^2Q_\rho + E_v^2 \tag{6.91}$$

Dividing the numerator and the denominator expressions for the Kalman gain by R we get

$$K(k) = (AP_\rho(k)C + E_v)(C^2P_\rho(k) + 1)^{-1} \tag{6.92}$$

where $P_\rho = P/R$.

6.6.7.2 Kalman Gain: Variations in the Variance

We will make the following assumptions to handle the extreme values when the ratio of the variances assume $Q_\rho \to \infty$ or $Q_\rho \to 0$.

Lemma 6.4 *If the following assumptions hold*

- *The system model (6.81) is asymptotically stable,* $|A| < 1$ *if* $Q_\rho \to \infty$.
- *The system model (6.81) satisfies* $|A - E_vC| < 1$ *if* $Q_\rho \to 0$.

Then

$$P(k) \to \begin{cases} 0 & \text{if } Q_\rho \to 0 \\ E_w^2Q_\rho \to \infty & \text{if } Q_\rho \to \infty \end{cases} \tag{6.93}$$

$$K(k) \to \begin{cases} E_v & \text{if } Q_\rho \to 0 \\ A/C & \text{if } Q_\infty \to \infty \end{cases} \tag{6.94}$$

Proof:
Case 1: $Q_\rho \to 0$
It can be deduced that $P_\rho(k) \to 0$ is a solution of the Riccati equation (6.91) if $|A| < 1$. Since $P_\rho = P/R$, we get $P(k) \to 0$ Consider the expression for the Kalman gain (6.92).

Since $P_\rho(k) \to 0$, we get $K(k) \to E_v$.

Case 2: $Q_\rho \to \infty$
It can easily be deduced that $P_\rho(k) \to E_w^2Q_\rho \to \infty$ is a solution of the Riccati equation (6.91) if the system (6.81) is asymptotically stable. Since $P_\rho = P/R$, we get $P(k) \to \infty$. Consider the expression for the Kalman gain (6.93). Since $P_\rho(k) \to \infty$ we get $K(k) \to A/C$.

Consider Case I: $Q_\rho \to 0$, In this case the Kalman gain $K(k) \to E_v$ and the Kalman filter asymptotically approaches

$$\begin{aligned}\hat{x}(k+1) &= (A - E_v C)\hat{x}(k) + Br(k) + E_v y(k) \\ \hat{y}(k) &= C\hat{x}(k)\end{aligned} \tag{6.95}$$

It is asymptotically stable in view of the assumption $|A - E_v C| < 1$.

Consider Case II: $Q_\rho \to \infty$. The Kalman filter model asymptotically approaches with gain $K(k) \to \dfrac{A}{C}$ given by

$$\begin{aligned}\hat{x}(k+1) &= Br(k) + \tfrac{A}{C} y(k) \\ \hat{y}(k) &= C\hat{x}(k)\end{aligned} \tag{6.96}$$

The Kaman filter has the fastest response, termed finite settling time or dead-beat response.

Comment *The dead-beat design wherein all the eigenvalues of the Kalman filter matrix are located at the origin is highly desirable in certain applications where the model of the system is partially or totally unknown and the signal to noise ratio is large. The state estimate settles accurately in a finite number of sample periods and not asymptotically as is the case with all other design strategies. The number of sample periods to settle exactly at most equal the order of the system. In controller applications, however, the dead-beat design is constrained by a large control input signal and inter-sample oscillations [14–16].*

6.6.7.3 Steady-State Kalman Filter

The optimal Kalman gain is a function of time k. In practice, a steady-state constant gain is preferable since the Kalman filter has to re-start whenever it reaches the steady-state even though the Kalman filter will be suboptimal when the optimal time varying is substituted by the steady-state gain. The steady-state Kalman filter is obtained by setting $P(k+1) = P(k) = P$. The steady-state Kalman gain K becomes:

$$K = (APC + E_v R)(C^2 P + R)^{-1} \tag{6.97}$$

The equation governing the steady-state covariance P obtained from Eq. (6.91) is

$$P = A^2 P - (APC + E_v R)^2 (C^2 P + R)^{-1} + E_w^2 Q + E_v^2 R \tag{6.98}$$

The steady-state expressions for the Kalman gain and the variance of the estimation error in terms of the ratio of variance $Q_\rho = Q/R$ may be derived by dividing the expressions (6.98) and (6.99) by the variance of the measurement noise R:

$$K = (AP_\rho C + E_v)(C^2 P_\rho + R)^{-1} \tag{6.99}$$

$$P_\rho = A^2 P_\rho - (AP_\rho C + E_v)^2 (C^2 P_\rho + 1)^{-1} + E_w^2 Q_\rho + E_v^2 \tag{6.100}$$

where $P_\rho = \dfrac{P}{R}$.

6.6.8 Optimal Kalman Gain

In 6.9 Appendix, expressions for the optimal time varying Kalman gain are derived from the expression (6.80) using the "completing the square of the terms in the Kalman gain approach." The optimal gain and covariance are given by

$$\boldsymbol{K}(k) = \left(\boldsymbol{A}\boldsymbol{P}(k)\boldsymbol{C}^T + \boldsymbol{E}_v R\right)\left(\boldsymbol{C}\boldsymbol{P}(k)\boldsymbol{C}^T + R\right)^{-1} \tag{6.101}$$

$$\boldsymbol{P}(k+1) = \boldsymbol{A}\boldsymbol{P}(k)\boldsymbol{A} - \boldsymbol{K}(k)\left(\boldsymbol{C}\boldsymbol{P}(k)\boldsymbol{A}^T + \boldsymbol{E}_v^T R\right) + \boldsymbol{E}_w \boldsymbol{E}_w^T Q + \boldsymbol{E}_v \boldsymbol{E}_v^T R \tag{6.102}$$

The above equations governing the covariance of the estimation error are a nonlinear first-order matrix equation termed the *Riccati equation.* An alternate expression for the Riccati equation without explicating the Kalman gain is given by

$$\boldsymbol{P}(k+1) = \boldsymbol{A}\boldsymbol{P}(k)\boldsymbol{A} - \left(\boldsymbol{A}\boldsymbol{P}(k)\boldsymbol{C}^T + \boldsymbol{E}_v R\right)\left(\boldsymbol{C}\boldsymbol{P}(k)\boldsymbol{C}^T + R\right)^{-1}\left(\boldsymbol{A}\boldsymbol{P}(k)\boldsymbol{C}^T + \boldsymbol{E}_v R\right)^T + \boldsymbol{E}_w \boldsymbol{E}_w^T Q + \boldsymbol{E}_v \boldsymbol{E}_v^T R \tag{6.103}$$

The *Kalman gain and the error covariance: a function of* $Q_\rho = Q/R$.

From 6.10 Appendix, the expressions for the covariance (6.103) and the Kalman gain (6.101) may be expressed in terms of the ratio $Q_\rho = Q/R$ of the variance of the plant noise $w(k)$ and the measurement noise $v(k)$ are given by

$$\boldsymbol{P}_\rho(k+1) = \boldsymbol{A}\boldsymbol{P}_\rho(k)\boldsymbol{A}^T - \left(\boldsymbol{A}\boldsymbol{P}_\rho(k)\boldsymbol{C}^T + \boldsymbol{E}_v\right)\left(\boldsymbol{C}\boldsymbol{P}_\rho(k)\boldsymbol{C}^T + 1\right)^{-1}\left(\boldsymbol{A}\boldsymbol{P}_\rho(k)\boldsymbol{C}^T + \boldsymbol{E}_v\right)^T + \boldsymbol{E}_w \boldsymbol{E}_w^T Q_\rho + \boldsymbol{E}_v \boldsymbol{E}_v^T \tag{6.104}$$

$$\boldsymbol{K}(k) = \left(\boldsymbol{A}\boldsymbol{P}_\rho(k)\boldsymbol{C}^T + \boldsymbol{E}_v\right)\left(\boldsymbol{C}\boldsymbol{P}_\rho(k)\boldsymbol{C}^T + 1\right)^{-1} \tag{6.105}$$

where $\boldsymbol{P}_\rho(k) = \dfrac{\boldsymbol{P}_\rho(k)}{R}$.

6.6.9 Comparison of the Kalman Filters: Integrated and Plant Models

It is interesting to compare the equations governing Kalman gains and the covariance of the estimation error when (i) the system model $(\boldsymbol{A}, \boldsymbol{B}, \boldsymbol{C}, \boldsymbol{D})$ is subjected to correlated random and deterministic disturbance and measurement noise given by Eq. (6.6), and (ii) when the plant model $(\boldsymbol{A}_p, \boldsymbol{B}_p, \boldsymbol{C}_p)$ is given by Eq. (6.1). We will assume, however, that disturbance and measurement noise are both zero-mean uncorrelated white noise processes and there are no deterministic disturbance and measurement noise components. The Kalman filter equations in this case are given by:

$$\boldsymbol{K}_p(k) = \boldsymbol{A}_p \boldsymbol{P}_p(k) \boldsymbol{C}_p \left(\boldsymbol{C}_p \boldsymbol{P}_p(k) \boldsymbol{C}_p^T + R_p\right)^{-1} \tag{6.106}$$

$$\boldsymbol{P}_p(k+1) = \boldsymbol{A}_p \boldsymbol{P}_p(k) \boldsymbol{A}_p^T - \boldsymbol{A}_p \boldsymbol{P}_p(k) \boldsymbol{C}_p^T \left(\boldsymbol{C}_p \boldsymbol{P}_p(k) \boldsymbol{C}_p^T + R_p\right)^{-1} \boldsymbol{C}_p \boldsymbol{P}_p(k) \boldsymbol{A}_p^T + \boldsymbol{E}_p \boldsymbol{E}_p^T Q_p \tag{6.107}$$

where the subscript "p" is used distinguish the parameters associated with the model $(\boldsymbol{A}_p, \boldsymbol{B}_p, \boldsymbol{C}_p, \boldsymbol{D}_p)$ including the Kaman gain and covariance matrix.

Comparing the expressions for the Kalman gain and the Riccati equation (6.106) and (6.107) for Eq. (6.1) with those of given by Eqs. (6.101) and (6.103) for the integrated model (6.6). The Kalman gain Eq. (6.101) has an additional term $\boldsymbol{E}_v R$ and the Riccati equation (6.103) has an additional driving term $\boldsymbol{E}_v\boldsymbol{E}_v^T R$. The optimal gain and the Riccati equations for Eq. (6.1) may simply be derived from those of Eqs. (6.101) and (6.105) by setting $\boldsymbol{E}_v = 0$.

6.6.10 Steady-State Kalman Filter

In many practical applications the steady-state, which is suboptimal, instead of the optimal Kalman filter, is employed as the degradation in the performance is offset by the simplicity of implementation. The suboptimal Kalman filter employs the steady-state Kalman gain and the steady-state covariance of the estimation error, which are obtained by setting $\boldsymbol{P}(k+1) = \boldsymbol{P}(k) = \boldsymbol{P}$ in Eq. (6.103)

$$\boldsymbol{P} = \boldsymbol{APA} - \left(\boldsymbol{APC}^T + \boldsymbol{E}_v R\right)(\boldsymbol{CPC}^T + R)^{-1}\left(\boldsymbol{APC}^T + \boldsymbol{E}_v R\right)^T + \boldsymbol{E}_w\boldsymbol{E}_w^T Q + \boldsymbol{E}_v\boldsymbol{E}_v^T R \qquad (6.108)$$

This is a nonlinear algebraic equation which is a quadratic function of $\boldsymbol{P}$. The expression of the steady-state Kalman gain by setting $\boldsymbol{K}(k) = \boldsymbol{K}$ and $\boldsymbol{P}(k) = \boldsymbol{P}$ becomes

$$\boldsymbol{K} = \left(\boldsymbol{APC}^T + \boldsymbol{E}_v R\right)(\boldsymbol{CPC}^T + R)^{-1} \qquad (6.109)$$

An alternative expression for the covariance matrix equation derived from Eq. (6.102) is

$$\boldsymbol{P} = \boldsymbol{APA} - \boldsymbol{K}\left(\boldsymbol{APC}^T + \boldsymbol{E}_v R\right)^T + \boldsymbol{E}_w\boldsymbol{E}_w^T Q + \boldsymbol{E}_v\boldsymbol{E}_v^T R \qquad (6.110)$$

The steady-state expressions in terms of the ratio $Q_\rho = Q/R$ may be derived from Eqs. (6.109) and (6.110) by dividing by R:

$$\boldsymbol{K} = \left(\boldsymbol{AP}_\rho\boldsymbol{C}^T + \boldsymbol{E}_v\right)\left(\boldsymbol{CP}_\rho\boldsymbol{C}^T + 1\right)^{-1} \qquad (6.111)$$

$$\boldsymbol{P}_\rho = \boldsymbol{AP}_\rho\boldsymbol{A} - \boldsymbol{K}\left(\boldsymbol{AP}_\rho\boldsymbol{C}^T + \boldsymbol{E}_v\right)^T + \boldsymbol{E}_w\boldsymbol{E}_w^T Q_\rho + \boldsymbol{E}_v\boldsymbol{E}_v^T \qquad (6.112)$$

where $\boldsymbol{P}_\rho = \dfrac{\boldsymbol{P}}{R}$.

6.6.11 Internal Model and Statistical Approaches

It is important to point out at this juncture the key difference between the internal model (IM) approach, which is in essence a deterministic one, and the classical statistical approach in the design of a Kalman filter. The IM approach is based on first deriving a necessary and sufficient structure for this filter, and then determining the Kalman gain $\boldsymbol{K}$ by minimizing the covariance of the estimation error. On the other hand, the classical statistical approach is based on minimizing the covariance of the estimation error in order to determine both the structure and the gain assuming a Gaussian PDF. The classical statistical approach is based on the following optimization problem: $\min_{\hat{x}} E[(\boldsymbol{x}(k+1) - \hat{\boldsymbol{x}}(k+1))(\boldsymbol{x}(k+1) - \hat{\boldsymbol{x}}(k+1))^T]$ such that (6.6) holds: Though the above two approaches are differ, they lead to identical results. The structure and the gain obtained using (i) the classical statistical approach in a statistical setting based on a Gaussian PDF for the measurements, and (ii) the proposed internal-model-based approach in a deterministic setting, are in fact identical. Both approaches yield an *optimal Kalman filter* with a time-varying Kalman gain. The inclusion of an internal model ensures that the expected value of the estimation error will asymptotically vanish to zero for all variations in the Kalman gain as long as the stability of the filter is ensured. Further, since the internal model structure is a necessary and sufficient condition for the asymptotic tracking, the error will not be zero asymptotically if the internal model is not an exact copy of the system model. This lack of asymptotic tracking due to model mismatch is in fact judiciously exploited in a fault detection scheme based on a Kalman filter.

The internal model approach employed in deriving the structure of the Kalman filter is simple, intuitive, powerful, and provides a unifying framework to handle a large class of control and filtering problems, as explained next:

- In control applications, the structure of a controller needed to track a reference signal and reject a disturbance can be derived from the internal model principle.
- The design of conventional filters such as low-pass, band-pass, high-pass filters and the like used to filter out respectively a low-pass, band-pass. and high-pass noisy waveforms, and the like, may also be formulated using the IM principle.
- Applications of the Kalman filter based on the IM include signal enhancing applications in speech, biological signal processing, global positioning systems for navigation, tracking position of a moving objects, and phase-locked loop in receivers. It is used in practically all satellite navigation equipment and smart phones.
- An observer to reconstruct states from the system's output may also be derived using the IM principle.

The Kalman filter is the best linear filter that minimizes the mean-squared estimation error (a 2-norm of the expected value of the estimation error) at each time step. However, there may be a nonlinear filter that may outperform the Kalman filter if the plant and/or the measurement noise are non-Gaussian (especially PDFs with heavier tails implying a higher probability of the noise assuming large values). For the Gaussian case, however, the Kalman filter is the best minimum variance filter. Further, if the PDF of the measurement noise is not known except its variance, then the Kalman filter is a robust filter that minimizes the worst-case estimation error.

The Kalman filter performance depends upon the *a priori* knowledge of the system model $(\boldsymbol{A}, \boldsymbol{B}, \boldsymbol{C})$, the plant noise variance Q, and the measurement noise variance R.

The performance of the filter will deteriorate if an incorrect value of the plant model, the plant noise variance, or measurement noise variance is employed.

6.6.12 Optimal Information Fusion

The Kalman filter computes the state estimates by fusing the *a posteriori* information provided by the residual $e(k)$, and the *a priori* information contained in the model $(\boldsymbol{A}, \boldsymbol{B}, \boldsymbol{C})$, which governs the evolution of the measurements. See the equations governing the Kalman filter estimates (6.83) for the scalar case and Eq. (6.75) for the vector case. The covariance of the measurement noise, R, and the covariance of the plant noise Q, quantifies the degree of belief associated with the measurement and model information, respectively. These covariances play a crucial role in the performance of the Kalman filter, as explained next.

6.6.13 Role of the Ratio of Variances

In general, the state estimate is obtained as the best compromise between the estimates generated by both the model and measurements. The performance of the filter, and hence the quality of the estimate, really does not depend upon the influence of the individual matrices Q and R, but rather on their "ratio." This is best illustrated if we consider the case when the plant and the measurement noise are independent and identically distributed (i.i.d.) white noise processes. It is shown in 6.10 Appendix that the covariance of the estimation error and the Kalman gain depends on the ratio of the plant and the measurement noise variances, namely Q/R, rather than the individual covariance matrices themselves, which plays an important role in "tweaking" the performance level of the Kalman filter. It is for that reason that the ratio Q/R is referred to as the tuning parameter of the Kalman filter. This is to be expected since, if we view the Kalman filter as a decision-making device having to deal with two sources of uncertainty, that is, the system model and the measurement process, that are "characterized" by the covariance matrices Q and

R, respectively, it has to decide on which of these two sources it should place its belief (or reliability) when optimally estimating the system's state. The performance of the Kalman filter when the ratio Q/R is varied is considered, taking a cue from the scalar example of Section 6.6.7.

Case 1 The ratio Q/R is very small
When Q/R is very small, implying the covariance of the measurement noise, R, is very large compared to plant noise covariance Q, the information provided by the measurement will therefore not be reliable. This implies that the measurement $y(k)$ will be given less importance than the model, in computing the state estimate.

Scalar case:
It is shown for the scalar case that when Q/R is very small, P will be small and consequently the Kalman gain K will also be very small. This implies that the measurement $y(k)$ will be given less importance than the model in computing the state estimate. The Kalman filter will be asymptotically an open-loop system, and estimates are generated by the zero-input response of the IM.

Case 2 The ratio Q/R is very large
When Q/R is large, implying the covariance of the measurement noise, R, is very small compared to plant noise covariance Q, the information provided by the measurement will therefore be highly reliable, indicating that the thrust of the trust will be put upon the measurement process and not on the model. Let us now analyze the performance of the Kalman filter in this case.

Scalar case:
It is shown for the scalar case that if the ratio of the plant to the measurement noise covariance Q/R is very large, then all the eigenvalues of the Kalman filter matrix $A - KC$ will be located at the origin. In this case the Kalman filter is a dead-beat observer, also called a finite settling time observer as the time to reach the steady-state estimate is equal to the order of the system. This indicates that the estimates are computed mainly from the measurement model and not the dynamic model of the system $(\boldsymbol{A}, \boldsymbol{B})$.

Case 3 The ratio $0 < Q/R < \infty$
When the ratio of the plant to the measurement noise variance do not assume an extreme value of zero or infinity, indicating that the information provided by the dynamic model and the measurement need to be factored in the computation of the estimate. In this case, the weights given to the measurement model and the dynamic model of the system will depend upon the *a priori* information provided by the ratio Q/R.

- The larger the ratio Q/R, the larger will be the thrust on the measurement model and the smaller will be the thrust on the dynamic model.
- The smaller the ratio Q/R, the smaller will be the importance given to the measurement and the larger will be the importance given to the dynamic model.

6.6.14 *Fusion of Information from the Model and the Measurement*

The Kalman filter estimates the system states by fusing the information provided by the residual $e(k)$ and the *a priori* information contained in the model $(\boldsymbol{A}, \boldsymbol{B})$. As explained above, this fusion is controlled by the Kalman gain which itself is computed based on the *a priori* information of the plant embedded in the system model and the ratio of the noise covariances Q and R.

If the ratio Q/R is large the Kalman filter will rely on the measurement to generate the residual. The Kalman filter in this case will be robust (or less sensitive) to model uncertainties. If on the other hand the ratio Q/R is small, the Kalman filter will rely on the model to generate the residual. The Kalman filter in this case will be robust (or less sensitive) to the measurement noise.

If the actual model of the system and the assumed model of the Kalman filter are appreciably different from each other, then the covariance matrix Q should be chosen to be sufficiently large to reflect the model uncertainty. In such cases, any incorrect choice of a small value for Q will result in a poor performance of the Kalman filter. The filter's estimate is likely to end up tracking the output from a wrong model, a phenomenon usually encountered in system under-modeling or over-modeling.

In applications including fault diagnosis, performance monitoring, condition-based maintenance, and estimation of a signal of a known class (a signal whose internal model is known) buried in noise, it is desirable that the Kalman filter be highly sensitive (or least robust) to model uncertainties. To achieve this, the ratio Q/R must be very small, so that the dynamic model (A, B) is given the maximum importance compared to the measurement. In practice this requirement is met if the nominal model of the system is identified accurately. The requirement of high sensitivity (or the lack of robustness) of the Kalman filter to model uncertainty in the above stated applications is in striking contrast to that of a controller which has to be highly insensitive (highly robust) to model uncertainties.

6.6.15 Illustrative Example: Fusion of Information

Examples of a vector and scalar models are given to illustrate the fusion of information by the Kalman filter algorithm. The Kalman filter computes the state estimates by fusing the *a posteriori* information provided by the residual $e(k)$, and the *a priori* information contained in the model $(\boldsymbol{A}, \boldsymbol{B})$. The ratio variance of the plant and the measurement noise Q/R quantifies the degree of belief associated with the model and the measurement thereby decides the amount of weighting to be given to the model and the measurement by choosing the appropriate values of Kalman gains.

Analytical expressions for the covariance of the estimation error, the Kalman gain, and the eigenvalues of the closed-loop matrix of the Kalman filter are difficult to obtain for the vector model. However, an overall behavior of the Kalman filter for the vector case may be deduced from the simple scalar example. Hence the scalar model is used to provide insight on the roles of the Kalman gain and the covariance as Q/R is varied.

The behavior of the steady-state Kalman filter and the steady-state covariance of the estimation error for the integrated system model (6.61) and that of the scalar system model (6.81) are compared as the variance ratio Q/R is varied. The integrated system model is a fifth-order system formed of the plant, disturbance, and the measurement noise model developed in Section 6.6.3 while the scalar system model (6.81) which is developed in Section 6.6.7 is chosen to be $A = B = C = E_w = 1$, $D = 0$, $E_v = 0$ and $R = 1$. The steady-state Kalman gain $\boldsymbol{K}$ given and the steady-state covariance of the estimation error $\boldsymbol{P} = R\boldsymbol{P}_p$ are given respectively by Eqs. (6.111) and (6.112). For the scalar case, the steady-state Kalman gain K and the steady-state estimation error variance $P = RP_p$ are respectively given by Eqs. (6.99) and (6.100).

Figure 6.11 shows the comparison of the steady-state Kalman gain and the covariance of the estimation error as Q/R is varied. Subfigures (a) and (b) show respectively the variance of the steady-state estimation error P and steady-state Kalman gain for the scalar case when variance ratio Q/R is varied from 0 to 0.5. Subfigures (c) and (d) show respectively the trace of the steady-state estimation error covariance $trace\{\boldsymbol{P}\}$ (which is a sum of the steady-state variances of the estimation error of each of the states) and the five steady-state Kalman gains as the variance ratio Q/R in the vector case was varied from 0 to 50.

Table 6.2 shows the steady-state variance estimation error P for the scalar model, the sum of the variance of the steady-state estimation error $trace\{\boldsymbol{P}\}$ for the vector model, and the steady-state Kalman gains for both the scalar and the vector model when $Q/R = 0$ and $Q/R = 50$.

Comments *The steady-state variances of the estimation errors for (i) the vector model and (ii) the steady-state variance of scalar model increase monotonically as the ratio Q/R increases. This may be*

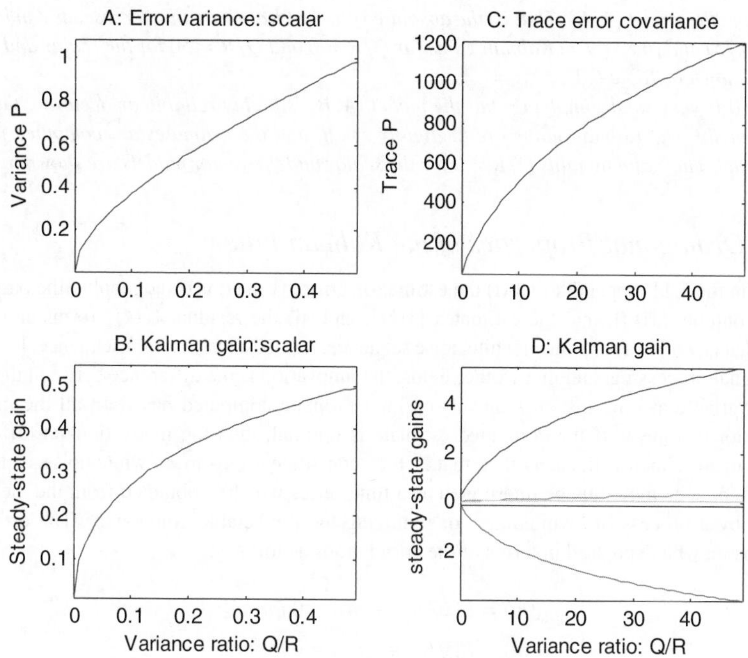

Figure 6.11 The steady-state variance and the gain as Q/R is varied

analytically justified from the expressions (6.100) and (6.112). An intuitive explanation for the increase in the estimation error variance is that, as Q/R increases the reliability or the accuracy of model $(\boldsymbol{A}, \boldsymbol{B})$ decreases. As a consequence, the inaccuracies in the estimates of the state computed from the model will increase with an increase in Q/R. The variance of the estimation errors for $Q/R = 0$ and $Q/R = 50$ for the scalar and the vector models are shown in Table 6.2

The steady-state Kalman gain increases as Q/R increases for the scalar case. For the vector case the elements of the steady-state Kalman gain may monotonically increase or decrease with an increase in Q/R so that the eigenvalues of $\boldsymbol{A} - \boldsymbol{KC}$ move closer to the origin and the Kalman filter response is faster, as can be deduced from Eqs. (6.101) and (6.102). An intuitive explanation is that as Q/R increases the reliability or the accuracy of model $(\boldsymbol{A}, \boldsymbol{B})$ decreases. Hence the residual $e(k)$ is weighted more than the

Table 6.2 Variances and the Kalman gains when $Q/R = 0$ and $Q/R = 50$

	Scalar model		Vector model	
	$Q/R = 0$	$Q/R = 50$	$Q/R = 0$	$Q/R = 50$
variance	$P = 0.0050$	$P = 0.9866$	$trace\{\boldsymbol{P}\} = \mathbf{21}.2684$	$trace\{\boldsymbol{P}\} = \mathbf{1}185$
Kalman gain	$K = 0.005$	$K = 0.4966$	$\boldsymbol{K} = \begin{bmatrix} 0.1052 \\ -0.0852 \\ -0.0000 \\ 0.0047 \\ -0.1661 \end{bmatrix}$	$\boldsymbol{K} = \begin{bmatrix} 5.3454 \\ -3.9019 \\ 0.3502 \\ 0.0140 \\ -0.1415 \end{bmatrix}$

estimates from the model $(\boldsymbol{A}, \boldsymbol{B})$. That is, the absolute values of Kalman gain increase, as can be deduced from Eqs. (6.75) and (6.83). The Kalman gains for $Q/R = 0$ and $Q/R = 50$ for the scalar and the vector models are shown in Table 6.2

When Q/R is very small, implying that the model $(\boldsymbol{A}, \boldsymbol{B})$ is highly reliable or accurate, the model is weighted heavily. The Kalman gains are negligibly small, and the estimates are computed practically from the model. The Kalman gains $Q/R = 0$ for the scalar and the vector models are shown in Table 6.2.

6.6.16 Orthogonal Properties of the Kalman Filter

It is shown in the 6.11 Appendix that (i) the estimation error $\tilde{x}(k+m)$ is orthogonal to the past residuals $\{e(k)\}$, the outputs $\{y(k)\}$, and the estimates $\{\hat{x}(k)\}$, and (ii) the residual $\{e(k)\}$ forms an orthogonal sequence, that is, $e(k)$ is a zero-mean white noise sequence, termed an *innovation* sequence. In time series analysis, signal processing, and many other fields, the innovation is the difference between the observed value of a variable at time instant k and its optimal estimate computed based on all the information available prior to time k. If the computed estimate is optimal, then the innovations at different time instants are uncorrelated with each other, that is, they constitute a zero-mean white noise sequence. The innovation sequence may thus be interpreted as a time series which is obtained from the measurement time series by a process of "whitening," or removing the predictable component. The orthogonality relationships may be expressed in terms of the correlations as follows:

$$r_{\tilde{x}e}(m) = E[\tilde{x}(k)e(k-m)] = 0 \text{ for } m \geq 1 \tag{6.113}$$

$$r_{\tilde{x}y}(m) = E[\tilde{x}(k)y(k-m)] = 0 \text{ for } m \geq 1 \tag{6.114}$$

$$r_{\overleftarrow{x}\hat{x}}(m) = E[\tilde{x}(k)\hat{x}^T(k-m)] = 0 \text{ for } m \geq 0 \tag{6.115}$$

The residual $e(k)$ is an innovation sequence and its auto-correlation is given by

$$r_{ee}(m) = E[e(k+m)e(k)] = \begin{cases} CP(k)C^T + R & m = 0 \\ 0 & m \neq 0 \end{cases} \tag{6.116}$$

Summarizing, the orthogonality relationships may be stated as follows:

- The present state estimation error $\tilde{\boldsymbol{x}}(k)$ is perpendicular to
  - the span of all the past residuals $e(k-m) : m \geq 1$
  - the span of all the past outputs $y(k-m) : m \geq 1$
  - the span of all the present and the past estimates $\tilde{\boldsymbol{x}}(k-m) : m \geq 1$.
- The present residual is perpendicular to the span of all the past and the future except the present residuals.

The span of a set of vectors is defined Eq. (6.60) in the Appendix.

Figure 6.12 shows a geometric representation of the orthogonality relationships governing the estimation error $\tilde{x}(k)$, the residual $e(k)$, the output $y(k)$, and the estimate $\hat{x}(k)$ given by Eqs. (6.113)–(6.116). The normalized cross-correlations of estimation error with (i) the residual and (ii) the estimate and the output are shown in subfigures (a), (b), and (c). Subfigure (d) shows the normalized auto-correlation of the residual for $m \neq 0$ while subfigure (e) shows that the state, its estimate, and the state estimation error form a right-angled triangle: $\boldsymbol{x} = \hat{\boldsymbol{x}} + \tilde{\boldsymbol{x}}$.

6.6.17 Ensemble and Time Averages

The orthogonal and correlation properties have been derived using the expectation operator $E[(.)]$, which is an ensemble average of an infinite number of realizations of the stochastic processes. One cannot

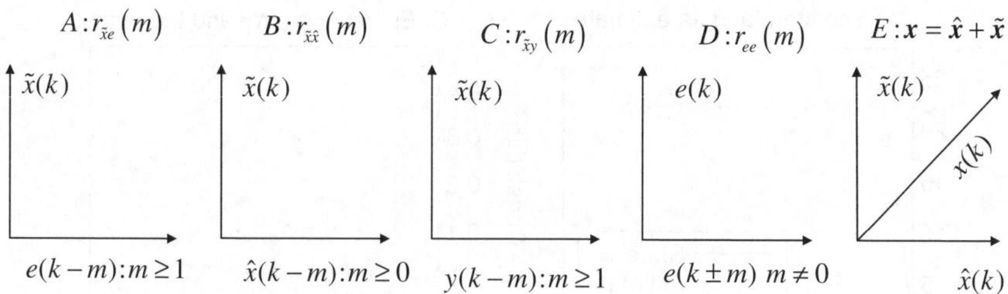

Figure 6.12 Geometric representation of cross and auto correlations

employ an ensemble average if a single realization from an experiment is available. If, however, the single realization is large enough to capture the stochastic behavior of an ensemble of realizations, then one can substitute the ensemble average by a time average. In other words, the ensemble is computed by taking a time average from a single, albeit infinitely large realization. The stochastic process where the time average equals the ensemble average when the data length is infinitely large is termed an *ergodic process*. The system and the Kalman filter variables such as the state, the output, and their estimates are stationary after the transients have decayed, and we may assume in this case that all the random processes are ergodic.

6.6.18 Illustrative Example: Orthogonality Properties of the Kalman Filter

The orthogonality properties of the Kalman filter are verified using a simulated example of the integrated model (6.61). It is shown that (i) the residual is a zero-mean white noise sequence (innovation sequence), (ii) the estimation error is uncorrelated with the past outputs, and (iii) the estimation error is uncorrelated with the present and past estimates. The correlations are computed using a time (and not an ensemble) average assuming all the random variables are ergodic processes. Figure 6.13 shows the orthogonality properties of the Kalman filter. Subfigure (a) shows the state of the plant and its estimate; subfigure (b) shows that the auto-correlation of the residual is a delta function; Subfigures (c) and (d) show respectively that the estimation error is uncorrelated with the output and the estimate.

Comment *It is to be emphasized that the orthogonality properties hold if and only if the internal model employed in the design of the Kalman filter is identical to the system model, and the Kalman gain is optimal. The property that the residual is a zero-mean white noise process if there is no mismatch between the system and internal model is widely used in many applications of the Kalman filter, especially in fault diagnosis.*

6.7 The Residual of the Kalman Filter with Model Mismatch and Non-Optimal Gain

The Kalman filter residual is employed for monitoring the status of the system including fault detection and isolation since residual $e(k)$ is a zero-mean white noise process (an innovation sequence) and its auto-correlation is a delta function (6.116) if and only if the internal model employed in the design of Kalman filter (6.75) is identical to that of the system model (6.6). In this section it will be shown that the residual will not be an innovation sequence if the internal model is not identical to the system model, but it will be a function of the model mismatch expressed in terms of (i) the deviation between the nominal and the actual system transfer function and then (ii) in terms of the deviation between the nominal and the actual feature vectors, and are established respectively in Lemma 6.5 and Lemma 6.6 respectively.

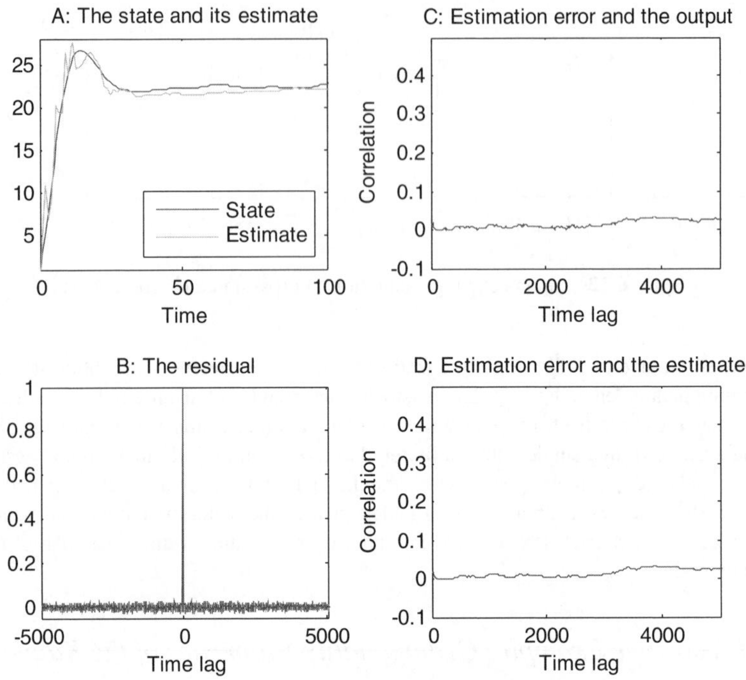

Figure 6.13 Orthogonality properties of the Kalman filter

Let the model of the Kalman filter be

$$\begin{aligned}\hat{\boldsymbol{x}}(k+1) &= \boldsymbol{A}_0\hat{\boldsymbol{x}}(k) + \boldsymbol{B}_0 r(k) + \boldsymbol{K}_0(y(k) - \boldsymbol{C}_0\hat{\boldsymbol{x}}(k)) \\ \hat{y}(k) &= \boldsymbol{C}_0\hat{\boldsymbol{x}}(k)\end{aligned} \tag{6.117}$$

where $(\boldsymbol{A}_0, \boldsymbol{B}_0, \boldsymbol{C}_0)$ the internal model is employed in the Kalman filter and $\boldsymbol{K}_0$ is the optimal Kalman gain for the assumed system model $(\boldsymbol{A}_0, \boldsymbol{B}_0, \boldsymbol{C}_0)$. Let $G_0(z) = \dfrac{N_0(z)}{D_0(z)}$ be a nominal version of transfer function $G(z) = \dfrac{N(z)}{D(z)}$ relating the output $y(z)$ to $r(z)$.

The deviation of the system model $(\boldsymbol{A}, \boldsymbol{B}, \boldsymbol{C})$ (or $G(z)$) from its nominal model $(\boldsymbol{A}_0, \boldsymbol{B}_0, \boldsymbol{C}_0)$ (or $G_0(z)$) occurs as a result of a variation in the normal operating regime, faults, or an error in the identification of the system model (for example to obtain an internal model in the design of the Kalman filter) or other causes. A fault may include variations in the sensor, the actuator, and the plant models.

6.7.1 *State Estimation Error with Model Mismatch*

The state-space model of the estimation error (6.78) for the case when there is a model mismatch may be derived by subtracting the Kalman filter model $(\boldsymbol{A}_0, \boldsymbol{B}_0, \boldsymbol{C}_0)$ given by Eq. (6.117) from the system model $(\boldsymbol{A}, \boldsymbol{B}, \boldsymbol{C})$ given by Eq. (6.6). Let the perturbations in $\boldsymbol{A}$, $\boldsymbol{B}$, and $\boldsymbol{C}$ be defined as $\Delta\boldsymbol{A} = \boldsymbol{A} - \boldsymbol{A}_0$, $\Delta\boldsymbol{B} = \boldsymbol{B} - \boldsymbol{B}_0$, and $\Delta\boldsymbol{C} = \boldsymbol{C} - \boldsymbol{C}_0$. Expressing in terms of the perturbations the system model becomes:

$$\begin{aligned}\boldsymbol{x}(k+1) &= (\boldsymbol{A}_0 + \Delta\boldsymbol{A})\boldsymbol{x}(k) + (\boldsymbol{B}_0 + \Delta\boldsymbol{B})r(k) + \boldsymbol{E}_w w(k) + E_v v(k) \\ y(k) &= (\boldsymbol{C}_0 + \Delta\boldsymbol{C})\boldsymbol{x}(k) + v(k)\end{aligned} \tag{6.118}$$

Subtracting Eq. (6.117) from Eq. (6.118) yields

$$\begin{aligned}\tilde{\boldsymbol{x}}(k+1) &= (\boldsymbol{A}_0 - \boldsymbol{K}_0\boldsymbol{C}_0)\tilde{\boldsymbol{x}}(k) + (\Delta\boldsymbol{A} - \boldsymbol{K}_0\Delta C)\boldsymbol{x}(k) + \Delta\boldsymbol{B}r(k) + \boldsymbol{E}_w w(k) + (\boldsymbol{E}_v - \boldsymbol{K}_0)v(k) \\ e(k) &= \boldsymbol{C}_0\tilde{\boldsymbol{x}}(k) + \Delta\boldsymbol{C}\boldsymbol{x}(k) + v(k)\end{aligned} \tag{6.119}$$

6.7.1.1 Model Mismatch Indicator

Lemma 6.5 *If the system and the Kalman filter models are given by Eqs. (6.6) and (6.117), then the frequency-domain expression of the residual, that is, e(z), is given, in terms of the plant output y(z) and reference input r(z), by:*

$$e(z) = \frac{D_0(z)}{F_0(z)}y(z) - \frac{N_0(z)}{F_0(z)}r(z) \tag{6.120}$$

where

$$\begin{aligned}\frac{D_0(z)}{F_0(z)} &= 1 - \boldsymbol{C}_0(z\boldsymbol{I} - \boldsymbol{A}_0 + \boldsymbol{K}_0\boldsymbol{C}_0)^{-1}\boldsymbol{K}_0; \\ \frac{N_0(z)}{F_0(z)} &= 1 - \boldsymbol{C}_0(z\boldsymbol{I} - \boldsymbol{A}_0 + \boldsymbol{K}_0\boldsymbol{C}_0)^{-1}\boldsymbol{B}_0;\; F_0(z) = |z\boldsymbol{I} - \boldsymbol{A}_0 + \boldsymbol{K}_0\boldsymbol{C}_0|\end{aligned}$$

Proof: See 6.12 Appendix.

Deviation in the Transfer Function

We will show that the residual is a function of the deviation in the system model $(\boldsymbol{A}, \boldsymbol{B}, \boldsymbol{C})$ and nominal system model $(\boldsymbol{A}_0, \boldsymbol{B}_0, \boldsymbol{C}_0)$. Substituting the expression for $y(z)$ given in Eq. (6.8) in Eq. (6.120), we get

$$e(z) = \frac{D_0(z)}{F_0(z)}\left(\Delta G(z)r(z) + \frac{\upsilon(z)}{D(z)}\right) \tag{6.121}$$

where $G(z) = \dfrac{N(z)}{D(z)}$ and $G_0(z) = \dfrac{N_0(z)}{D_0(z)}$ are respectively the actual and the nominal system models and the difference $\Delta G(z) = G(z) - G_0(z)$ is the model mismatch.

Deviation in the Feature Vector

Let $\boldsymbol{\theta} = [a_1 \;\; a_2 \;\; \cdots \;\; a_n \;\; b_1 \;\; b_2 \;\; \cdots \;\; b_n]^T$ is the actual feature vector of the system, and $\boldsymbol{\theta}^0 = [a_{10} \;\; a_{20} \;\; \cdots \;\; a_{n0} \;\; b_{10} \;\; b_{20} \;\; \cdots \;\; b_{n0}]^T$ is the assumed or nominal feature vector (the feature vector of the assumed internal model employed in the design of the Kalman filter). That is, $\boldsymbol{\theta}$ is the feature vector of the actual transfer function of the system, $G(z) = \dfrac{N(z)}{D(z)}$ and $\boldsymbol{\theta}^0$ is the feature vector of the nominal version: $G_0(z) = \dfrac{N_0(z)}{D_0(z)}$ with $D_0(z) = 1 + \sum_{i=1}^{n} a_{i0}z^{-i}; N_0(z) = \sum_{i=1}^{n} b_{i0}z^{-i}$. By letting $\Delta a_i = a_i - a_{i0}$; $\Delta b_i = b_i - b_{i0}$, we can then write:

$$\Delta\boldsymbol{\theta} = \boldsymbol{\theta} - \boldsymbol{\theta}^0 = [\Delta a_1 \;\; \Delta a_2 \;\; . \;\; \Delta a_n \;\; \Delta b_1 \;\; \Delta b_2 \;\; . \;\; \Delta b_n]^T \tag{6.122}$$

Lemma 6.6 *If the linear regression of the system is Eq. (6.9), then*

$$e(z) = \frac{1}{F_0(z)}(\boldsymbol{\psi}^T(z)\Delta\boldsymbol{\theta} + \upsilon(z)) \tag{6.123}$$

where $\boldsymbol{\psi}(z)$ is the z-transform of $\psi(k)$, given by:

$$\boldsymbol{\psi}^T(z) = [-z^{-1}y(z) \quad -z^{-2}y(z) \quad \cdots \quad -z^{-n}y(z) \quad z^{-1}r(z) \quad z^{-2}r(z) \quad \cdots \quad z^{-n}r(z)] \tag{6.124}$$

Proof: See 6.12 Appendix.
It is interesting to note that Eq. (6.123) reveals an interesting fact, namely that the residual of the Kalman filter $e(k)$ can be expressed as a sum of a fault indicator component $e_f(k)$ and a zero-mean noise component $e_0(k)$.

$$e(z) = e_f(z) + e_0(z) \tag{6.125}$$

where $e_f(z) = \dfrac{\boldsymbol{\psi}^T(z)\Delta\boldsymbol{\theta}}{F_0(z)}$; $e_0(z) = \dfrac{\upsilon(z)}{F_0(z)}$.

6.7.1.2 Whiteness of the Residual

We shall now state and prove the key results of this subsection in the form of two theorems:

Theorem 6.2 *The Kalman filter residual is a zero-mean random process if and only if there is no model mismatch.*

$$\lim_{k\to\infty} E[e(k)] = 0 \text{ if and only if } \Delta\theta = 0 \tag{6.126}$$

Proof: Note that the assumption that $\upsilon(k)$ is a zero-mean random process leads to $E[\upsilon(z)] = 0$ and, from (6.125), $E[e_0(k)] = 0$. Taking the expectation of (6.125) leads directly to:

$$E[e(z)] = E[e_f(z)] = \frac{E[\boldsymbol{\psi}^T(z)]\Delta\boldsymbol{\theta}}{F_0(z)} \tag{6.127}$$

Since $F_0(z)$ is an asymptotically stable polynomial, we conclude that (6.126) holds.

The next theorem (Theorem 6.3) characterizes the Kalman filter's residual even further than Theorem 6.2 by bringing out the crucial importance of the optimality of the Kalman filter gain.

Theorem 6.3 *The Kalman filter residual is a zero-mean white noise process if and only if there is no model mismatch and the Kalman Filter is optimal (i.e., time-varying). In such a case, the auto-correlation of the residual is a delta function*

$$E[e(k)e(k-m)] = \sigma_e^2\delta(m) \tag{6.128}$$

where $\sigma_e^2 = E[e^2(k)]$ if and only if $\Delta\theta = 0$ (i.e., no mismatch) and the Kalman gain K is optimal.

Comment *We have demonstrated the power and effectiveness of the Kalman filter in unifying the important dual aim of detecting the presence of model mismatch as indicated by Eqs. (6.126) and*

(6.128), and if there is a mismatch quantifying it in terms of the deviation in the feature vector between the actual and the assumed system models as shown in Eq. (6.127). The property of the Kalman filter residual, namely the residual is a zero-mean white noise process if and only if there is no model mismatch stated in Eq. (6.128) is exploited judiciously in many applications including fault diagnosis and status monitoring. Thanks to the minimal variance of the estimation error, a small or incipient fault from the input–output data buried in the noise may be captured. Further it is also used to verify the performance of the Kalman filter performance as well as to tune its gain.

6.7.2 Illustrative Example: Residual with Model Mismatch and Non-Optimal Gain

The state-space model of the plant $(\boldsymbol{A}, \boldsymbol{B}, \boldsymbol{C})$ is given by:

$$\begin{aligned} \boldsymbol{x}(k+1) &= \begin{bmatrix} 1.7654 & 1 \\ -0.81 & 0 \end{bmatrix} \boldsymbol{x}(k) + \begin{bmatrix} 1 \\ 0 \end{bmatrix} r(k) + \begin{bmatrix} 1 \\ 0 \end{bmatrix} w(k) \\ y(k) &= [1 \quad 0]\boldsymbol{x}(k) + v(k) \end{aligned} \tag{6.129}$$

The performance of the Kalman filter was evaluated by analyzing the auto-correlation of the residual for the following cases:

Case 1: The Kalman gain was suboptimal. The steady-state optimal Kalman gain was employed. $\boldsymbol{K}_0 = [1.0910 \quad 0.7382]^T$

Case 2: A mismatch between the true and the estimated variance. The estimated measurement noise variance was different from the true variance $R_0 = 100R$. The variance of the disturbance, however, was estimated correctly: $Q_0 = Q$. Recall that the performance depends only on the ratio of the variances Q/R.

Case 3: The Kalman gain was computed using a pole-placement approach to ensure merely stability without regard to the performance in the presence of noise and disturbance. The optimal steady-state Kalman gain was $\boldsymbol{K}_0 = [1.0910 \quad 0.7382]^T$. But the non-optimal gain obtained from the pole-placement approach was $\boldsymbol{K}_0 = [0.1305 \quad 0.1587]^T$.

Case 4: The system was subject to an "actuator fault" by choosing $\boldsymbol{B} = 0.5\boldsymbol{B}_0$.

Case 5: The system was subject to a "sensor fault" by choosing $\boldsymbol{C} = 0.7\boldsymbol{C}_0$.

Case 6: The system was subject to a "plant fault" by choosing $\boldsymbol{A} = 0.9\boldsymbol{A}_0$

Figure 6.14 shows the auto-correlation function of the residual for the above cases. Subfigures (a), (b), and (c), show respectively the autocorrelations for Case 1, Case 2, and Case 3 when the Kalman gain was not optimal but there was no model mismatch. Subfigures (d), (e), and (f) show respectively the autocorrelation functions for Case 4, Case 5, and Case 6 when model mismatches were introduced. However the Kalman gain was optimal. The auto-correlation for the optimal fault free case is shown for comparison in all the subfigures. The auto-correlations were computed assuming the residual process was ergodic using time average $\hat{r}_{xx}(m) = \sum_{k=0}^{N-1} x(k)x(k-m)$ where $N = 1000$.

Results of the Evaluation

The variance of the residual for all the cases is given in Table 6.3. The variance of the residual is the value of the auto-correlation at the zero lag, that is, it is maximum value of the auto-correlation.

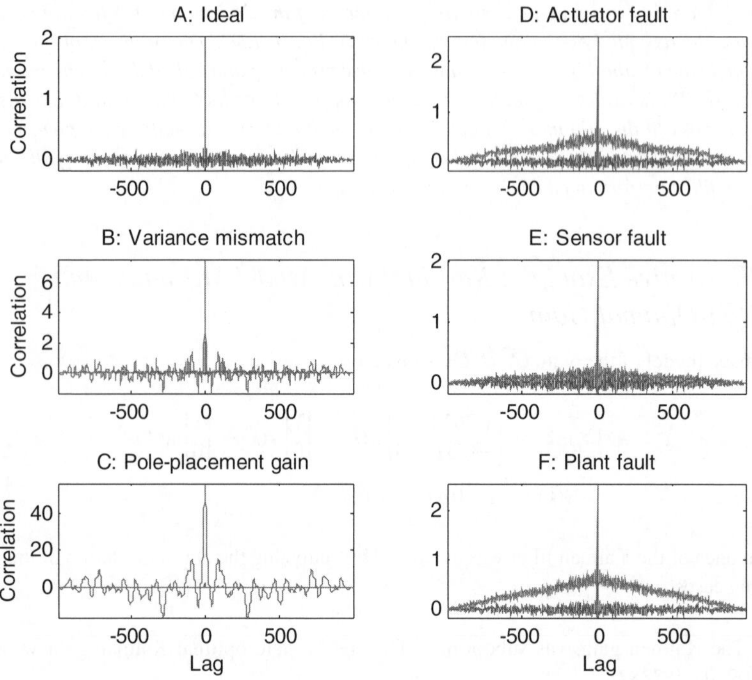

Figure 6.14 Auto-correlation with model mismatch and non-optimal gains

The auto-correlation of the residual computed using the time average is close to a delta function (6.128) (which will be case if an ensemble average is used) if and only if (i) the Kalman gain is optimal and (i) there is no model mismatch. See subfigure (a).

The optimal Kalman gain $\boldsymbol{K}_0$ is computed by fusing the *a posteriori* information provided by the measurement $y(k)$, and the *a priori* information contained in the model $(\boldsymbol{A}_0, \boldsymbol{B}_0, \boldsymbol{C}_0)$ employed in the Kalman filter. The covariance of the measurement noise R, and the covariance of the plant noise Q, quantifies the degree of belief associated with the measurement and model information, respectively. Hence the accuracy of the model $(\boldsymbol{A}_0, \boldsymbol{B}_0, \boldsymbol{C}_0)$ and the variances Q and R play a crucial role in the performance of the Kalman filter.

Non-Optimal Kalman Gain

The auto-correlations of the residual deviate from the ideal case when the Kalman gain is non-optimal resulting from (i) computing the gain using a wrong estimate of the variance Q or R, (ii) determining the gain purely for stabilization without consideration of the noise variances, or (iii) using (constant)

Table 6.3 Variance of the residual

	Ideal	Suboptimal gain	Variance mismatch	Pole-placement	Actuator fault	Sensor fault	Plant fault
Variance	2.0421	2.0429	7.6190	55.9842	2.5187	1.7666	2.5463

steady-state gain instead of the optimal time-varying gain. With steady-state optimal time invariant Kalman gain, however, its behavior is close to that of the optimal case. However, when the gain is computed purely for stabilization without consideration of the noise variances, the behavior may deviate considerably from that of the optimal case. From Table 6.3, the variances of the estimation error for the ideal (optimal), the optimal steady-state, variance mismatch resulting from an error in the estimation of the variance of the disturbance or measurement noise, and pole-placement based gain are respectively 2.0421, 2.0429, 7.6190, and 55.9842. The variance with pole-placement based gain without noise considerations is significantly larger, while with steady-state gain the difference in the variance compared to the case of the optimal gain is not significant, See subfigures (c) and (a). The residual is rather robust to uncertainties in the estimation of Q and R as can be deduced from subfigure (b).

Model Mismatch

The auto-correlation of the residual deviates from the ideal delta function when there is a model match as a result of $(\boldsymbol{A}_0, \boldsymbol{B}_0, \boldsymbol{C}_0)$ being not equal to $(\boldsymbol{A}, \boldsymbol{B}, \boldsymbol{C})$. See subfigures (d), (e), and (f).

Comments *The estimate of the state is obtained as the best compromise between the estimates generated by the model and those obtained from the measurement, depending upon the plant noise and the measurement noise covariances. For example, when the plant noise covariance is very small, implying that the model is highly reliable, the estimate is computed essentially using the model. The Kalman gain is zero and the measurements are simply ignored in this case. On the other hand when the measurement noise covariance is very small, implying that the measurements are highly reliable, the state estimates are computed from the measurements ignoring the plant model. It is therefore very important to use the correct values for the covariances that reflect correctly the belief in the model. If the plant noise covariance (or the measurement noise covariance) is chosen to be small when the model is not reliable (or when the measurements are not reliable) the performance of the Kalman filter will significantly degrade.*

In practice an online tuning of the covariances and the Kalman filter model may be preferable to avoid such adverse situations by exploiting the orthogonality properties of the Kalman filter estimates if there is no model mismatch as shown in Eq. (6.119), and the ratio of the plant to the noise variances are correct. That is:

- *Residual is a zero-mean white noise process as shown in Eq. (6.128).*
- *Estimation error is orthogonal to both the estimate of the state and the output as given by Eqs. (6.114) and (6.115).*

These properties, especially the white noise properties of the residual, are exploited in adapting the plant model and the measurement noise covariances [9]. If the residual is non-white, has non-zero mean, or has a higher variance then the model the plant, or the ratio of plant to the measurement noise variance is incorrect.

In physical systems, the plant and measurement noise covariance are generally unknown a priori *and are difficult to estimate* a posteriori *from the data. An online tuning technique is employed to determine the ratio of the disturbance and the measurement noise variances Q/R. The optimal choice is obtained from minimizing the mean-squared residual*

$$\min_{Q/R} \left\{ \frac{1}{N} \sum_{k=0}^{N-1} e^2(k) \right\} \tag{6.130}$$

N *is the number of data samples.*

6.8 Summary

State-Space Model of the Plant

The plant model in state-space form $(\boldsymbol{A}_p, \boldsymbol{B}_p, \boldsymbol{C}_p, D_p)$ is:

$$\begin{aligned}\boldsymbol{x}_p(k+1) &= \boldsymbol{A}_p\boldsymbol{x}_p(k) + \boldsymbol{B}_p r(k) + \boldsymbol{E}_p w_p(k) \\ y(k) &= \boldsymbol{C}_p\boldsymbol{x}_p(k) + D_p r(k) + v_p(k)\end{aligned}$$

State-Space Model of the Noise and Disturbance

$$\begin{aligned}w_p(k) &= w_r(k) + w_d(k) \\ v_p(k) &= v_r(k) + v_d(k)\end{aligned}$$

The state-space model $(\boldsymbol{A}_w, \boldsymbol{B}_w, \boldsymbol{C}_w, D_w)$ of the random disturbance $w_r(k)$ is given by:

$$\begin{aligned}\boldsymbol{x}_w(k+1) &= \boldsymbol{A}_w\boldsymbol{x}_w(k) + B_w w(k) \\ w_r(k) &= C_w\boldsymbol{x}_w(k) + D_w w(k)\end{aligned}$$

The state-space model $(\boldsymbol{A}_v, \boldsymbol{B}_v, \boldsymbol{C}_v, D_v)$ of the random measurement noise $v_r(k)$ is

$$\begin{aligned}\boldsymbol{x}_v(k+1) &= \boldsymbol{A}_v\boldsymbol{x}_v(k) + \boldsymbol{B}_v v(k) \\ v_r(k) &= \boldsymbol{C}_v\boldsymbol{x}_v(k) + D_v v(k)\end{aligned}$$

The combined state-space model $(\boldsymbol{A}_d, \boldsymbol{C}_d)$ of the deterministic disturbance $w_d(k)$ and the measurement noise $v_d(k)$ is:

$$\begin{aligned}\boldsymbol{x}_d(k+1) &= \boldsymbol{A}_d\boldsymbol{x}_d(k) \\ \begin{bmatrix} w_d(k) \\ v_d(k) \end{bmatrix} &= \begin{bmatrix} \boldsymbol{C}_{wd} \\ \boldsymbol{C}_{vd} \end{bmatrix} \boldsymbol{x}_d(k)\end{aligned}$$

Integrated State-Space Model of the System

Combining the plant model $(\boldsymbol{A}_p, \boldsymbol{B}_p, \boldsymbol{C}_p)$, the disturbance model $(\boldsymbol{A}_w, \boldsymbol{B}_w, \boldsymbol{C}_w, D_w)$ and the measurement noise model $(\boldsymbol{A}_v, \boldsymbol{B}_v, \boldsymbol{C}_v, D_v)$, and the determinist noise model $(\boldsymbol{A}_d, \boldsymbol{C}_d)$, the integrated model $(\boldsymbol{A}, \boldsymbol{B}, \boldsymbol{C}, \boldsymbol{D})$ becomes

$$\begin{aligned}\boldsymbol{x}(k+1) &= \boldsymbol{A}\boldsymbol{x}(k) + \boldsymbol{B}r(k) + \boldsymbol{E}_w w(k) + \boldsymbol{E}_v v(k) \\ y(k) &= \boldsymbol{C}\boldsymbol{x}(k) + \boldsymbol{D}r(k) + v(k)\end{aligned}$$

where

$$\boldsymbol{x} = \begin{bmatrix} \boldsymbol{x}_p \\ \boldsymbol{x}_w \\ \boldsymbol{x}_v \\ \boldsymbol{x}_d \end{bmatrix}, \ \boldsymbol{u} = \begin{bmatrix} u_p \\ w \\ v \end{bmatrix}, \ \boldsymbol{A} = \begin{bmatrix} \boldsymbol{A}_p & \boldsymbol{E}_p\boldsymbol{C}_w & 0 & \boldsymbol{E}_p\boldsymbol{C}_{wd} \\ 0 & \boldsymbol{A}_w & 0 & 0 \\ 0 & 0 & \boldsymbol{A}_v & 0 \\ 0 & 0 & 0 & \boldsymbol{A}_d \end{bmatrix}, \ \boldsymbol{B} = \begin{bmatrix} \boldsymbol{B}_p \\ 0 \\ 0 \\ 0 \end{bmatrix}, \ \boldsymbol{E}_w = \begin{bmatrix} \boldsymbol{E}_p D_w \\ \boldsymbol{B}_w \\ 0 \\ 0 \end{bmatrix}, \ \boldsymbol{E}_v = \begin{bmatrix} 0 \\ 0 \\ \boldsymbol{B}_v \\ 0 \end{bmatrix}$$

$$\boldsymbol{C} = [\boldsymbol{C}_p \quad 0 \quad \boldsymbol{C}_v \quad \boldsymbol{C}_{vd}], \ \boldsymbol{D} = [0 \quad 0 \quad 1]$$

Internal Model Principle

Controller design using the internal model principle

If (i) the state-space model of the plant $(\boldsymbol{A}_p, \boldsymbol{B}_p, \boldsymbol{C}_p)$ is controllable and observable, that is there is no pole-zero cancellations in the plant transfer function, (ii) a model of the system that has generated the input $r(k)$, termed "an internal model of $r(k)$," is included in the controller and the internal model is driven by the tracking error $e(k) = r(k) - y(k)$, (iii) the closed-loop system formed of the controller and the plant is asymptotically stable, then the tracking error is zero asymptotically.

Duality between Controller and an Estimator Design

Observer: Estimator for the States of a System

$$\begin{aligned}\hat{\boldsymbol{x}}(k+1) &= \boldsymbol{A}\hat{\boldsymbol{x}}(k) + \boldsymbol{B}r(k) + \boldsymbol{K}_0 e(k)\\ \hat{y}(k) &= \boldsymbol{C}\hat{\boldsymbol{x}}(k) + Dr(k)\end{aligned}$$

Kalman Filter: Estimator of the States of a Stochastic System

Objectives of the Kalman Filter

$$\begin{aligned}&\lim_{k\to\infty} E[\tilde{x}(k)] = 0\\ &\min_{\hat{x}}\{E[(\boldsymbol{x}(k)-\hat{\boldsymbol{x}}(k))^T(\boldsymbol{x}(k)-\hat{\boldsymbol{x}}(k))]\}\end{aligned}$$

Model of the Kalman Filter

Predictor form

$$\begin{aligned}\hat{\boldsymbol{x}}(k+1) &= \boldsymbol{A}\hat{\boldsymbol{x}}(k) + \boldsymbol{B}r(k) + \boldsymbol{K}e(k)\\ \hat{y}(k) &= \boldsymbol{C}\hat{\boldsymbol{x}}(k)\\ y(k) &= e(k) + \hat{y}(k)\end{aligned}$$

Innovation form

$$\begin{aligned}\hat{\boldsymbol{x}}(k+1) &= (\boldsymbol{A} - \boldsymbol{K}\boldsymbol{C})\hat{\boldsymbol{x}}(k) + \boldsymbol{B}r(k) + \boldsymbol{K}y(k)\\ \hat{y}(k) &= \boldsymbol{C}\hat{\boldsymbol{x}}(k)\\ e(k) &= y(k) - \hat{y}(k)\end{aligned}$$

Optimal Kalman gain

$$\begin{aligned}&\min_{\boldsymbol{K}(k)} E[\tilde{\boldsymbol{x}}^T(k+1)\tilde{\boldsymbol{x}}(k+1)]\\ &\boldsymbol{K}(k) = \left(\boldsymbol{A}\boldsymbol{P}(k)\boldsymbol{C}^T + \boldsymbol{E}_v R\right)\left(\boldsymbol{C}\boldsymbol{P}(k)\boldsymbol{C}^T + R\right)^{-1}\\ &\boldsymbol{P}(k+1) = \boldsymbol{A}\boldsymbol{P}(k)\boldsymbol{A} - \boldsymbol{K}(k)\left(\boldsymbol{C}\boldsymbol{P}(k)\boldsymbol{A}^T + \boldsymbol{E}_v^T R\right) + \boldsymbol{E}_w\boldsymbol{E}_w^T Q + \boldsymbol{E}_v\boldsymbol{E}_v^T R\end{aligned}$$

The Kalman gain and the error covariance: a function of $Q_\rho = Q/R$

$$\begin{aligned}\boldsymbol{P}_\rho(k+1) &= \boldsymbol{A}\boldsymbol{P}_\rho(k)\boldsymbol{A}^T - \left(\boldsymbol{A}\boldsymbol{P}_\rho(k)\boldsymbol{C}^T + \boldsymbol{E}_v\right)\left(\boldsymbol{C}\boldsymbol{P}_\rho(k)\boldsymbol{C}^T + 1\right)^{-1}\left(\boldsymbol{A}\boldsymbol{P}_\rho(k)\boldsymbol{C}^T + \boldsymbol{E}_v\right)^T + \boldsymbol{E}_w\boldsymbol{E}_w^T Q_\rho + \boldsymbol{E}_v\boldsymbol{E}_v^T\\ \boldsymbol{K}(k) &= \left(\boldsymbol{A}\boldsymbol{P}_\rho(k)\boldsymbol{C}^T + \boldsymbol{E}_v\right)\left(\boldsymbol{C}\boldsymbol{P}_\rho(k)\boldsymbol{C}^T + 1\right)^{-1}\end{aligned}$$

Steady-State Kalman Filter

$$\begin{aligned}
\boldsymbol{P} &= \boldsymbol{APA} - \left(\boldsymbol{APC}^T + \boldsymbol{E}_v R\right)(\boldsymbol{CPC}^T + R)^{-1}\left(\boldsymbol{APC}^T + \boldsymbol{E}_v R\right)^T + \boldsymbol{E}_w \boldsymbol{E}_w^T Q + \boldsymbol{E}_v \boldsymbol{E}_v^T R \\
\boldsymbol{K} &= \left(\boldsymbol{APC}^T + \boldsymbol{E}_v R\right)(\boldsymbol{CPC}^T + R)^{-1}
\end{aligned}$$

Orthogonality Properties of the Kalman Filter

$$\begin{aligned}
r_{\tilde{x}e}(m) &= E[\tilde{x}(k)e(k-m)] = 0 \; for \; m \geq 1 \\
r_{\tilde{x}y}(m) &= E[\tilde{x}(k)y(k-m)] = 0 \; for \; m \geq 1 \\
r_{\tilde{x}\hat{x}}(m) &= E[\tilde{x}(k)\hat{x}^T(k-m)] = 0 \; for \; m \geq 0 \\
r_{ee}(m) &= E[e(k+m)e(k)] = \begin{cases} CP(k)C^T + R & m = 0 \\ 0 & m \neq 0 \end{cases}
\end{aligned}$$

State Estimation Error with Model Mismatch

$$\begin{aligned}
\boldsymbol{x}(k+1) &= (\boldsymbol{A}_0 + \Delta\boldsymbol{A})\boldsymbol{x}(k) + (\boldsymbol{B}_0 + \Delta\boldsymbol{B})r(k) + \boldsymbol{E}_w w(k) + \boldsymbol{E}_v v(k) \\
y(k) &= (\boldsymbol{C}_0 + \Delta\boldsymbol{C})\boldsymbol{x}(k) + v(k) \\
\tilde{\boldsymbol{x}}(k+1) &= (\boldsymbol{A}_0 - \boldsymbol{K}_0\boldsymbol{C}_0)\tilde{\boldsymbol{x}}(k) + (\Delta\boldsymbol{A} - \boldsymbol{K}_0\Delta\boldsymbol{C})\boldsymbol{x}(k) + \Delta\boldsymbol{B}r(k) + \boldsymbol{E}_w w(k) + (\boldsymbol{E}_v - \boldsymbol{K}_0)v(k) \\
e(k) &= \boldsymbol{C}_0\tilde{\boldsymbol{x}}(k) + \Delta\boldsymbol{C}\boldsymbol{x}(k) + v(k)
\end{aligned}$$

Model Mismatch Indicator

$$e(z) = \frac{D_0(z)}{F_0(z)}y(z) - \frac{N_0(z)}{F_0(z)}r(z)$$

Deviation in the Transfer Function

$$e(z) = \frac{D_0(z)}{F_0(z)}\left(\Delta G(z)r(z) + \frac{\upsilon(z)}{D(z)}\right)$$

Deviation in the Feature Vector

$$\begin{aligned}
e(z) &= \frac{1}{F_0(z)}(\boldsymbol{\psi}^T(z)\Delta\boldsymbol{\theta} + \upsilon(z)) \\
e(z) &= e_f(z) + e_0(z)
\end{aligned}$$

where $e_f(z) = \dfrac{\boldsymbol{\psi}^T(z)\Delta\boldsymbol{\theta}}{F_0(z)}$; $e_0(z) = \dfrac{\upsilon(z)}{F_0(z)}$.

Whiteness of the Residual

$$\begin{aligned}
\lim_{k\to\infty} E[e(k)] &= 0 \text{ if and only if } \Delta\theta = 0 \\
E[e(k)e(k-m)] &= \sigma_e^2\delta(m)
\end{aligned}$$

where $\sigma_e^2 = E[e^2(k)]$ if and only if $\Delta\theta = 0$ (i.e., no mismatch) and the Kalman gain K is optimal.

6.9 Appendix: Estimation Error Covariance and the Kalman Gain

Lemma 6.7 *Consider the following linear time invariant state-space model of a system:*

$$\begin{aligned} \mathbf{x}(k+1) &= \mathbf{A}\mathbf{x}(k) + \mathbf{B}r(k) + \mathbf{E}_w w(k) + E_v v(k) \\ y(k) &= \mathbf{C}\mathbf{x}(k) + v(k) \end{aligned} \tag{6.131}$$

which satisfies the following properties:

- *The system $(\mathbf{A}, \mathbf{C})$ is detectable, that is all the states $\mathbf{x}(k)$ can be observed from the measurement $y(k)$ (this is a reasonable assumption since if the system is not detectable, then this would imply that there are states whose response cannot be seen (detected) in the output).*
- *The system $(\mathbf{A}, \mathbf{E}_w)$ is stabilizable. This may be interpreted as all the states $\mathbf{x}(k)$ are perturbed by the plant noise $w(k)$.*

Note here that the state-space model is not assumed to be asymptotically stable, that is the eigenvalues of $\mathbf{A}$ *are not assumed be strictly inside the unit circle.*

Then the optimal covariance matrix of the estimation error $\mathbf{P}(k)$ satisfies the following Riccati *equation:*

$$\mathbf{P}(k+1) = \mathbf{A}\mathbf{P}(k)\mathbf{A}^T - \left(\mathbf{A}\mathbf{P}(k)\mathbf{C}^T + \mathbf{E}_v R\right)(\mathbf{C}\mathbf{P}(k)\mathbf{C}^T + R)^{-1}\left(\mathbf{C}\mathbf{P}(k)\mathbf{A}^T + \mathbf{E}_v^T R\right) + \mathbf{E}_w \mathbf{E}_w^T Q + \mathbf{E}_v \mathbf{E}_v^T R \tag{6.132}$$

Then the steady-state $\mathbf{P} = \lim_{k\to\infty} \mathbf{P}(k)$ exists for any initial condition $\mathbf{P}(0)$ and satisfies the following steady-state version of the time-varying Riccati equation derived by setting $\mathbf{P}(k) = \mathbf{P}(k-1) = \mathbf{P}$:

$$\mathbf{P} = \mathbf{A}\mathbf{P}\mathbf{A}^T - \left(\mathbf{A}\mathbf{P}\mathbf{C}^T + \mathbf{E}_v R\right)(\mathbf{C}\mathbf{P}\mathbf{C}^T + R)^{-1}\left(\mathbf{C}\mathbf{P}\mathbf{A}^T + \mathbf{E}_v^T R\right) + \mathbf{E}_w \mathbf{E}_w^T Q + \mathbf{E}_v \mathbf{E}_v^T R \tag{6.133}$$

This equation is termed the steady-state (or algebraic) Riccati equation. Further, the steady-state Kalman gain takes the form:

$$\mathbf{K} = \left(\mathbf{A}\mathbf{P}\mathbf{C}^T + \mathbf{E}_v R\right)(\mathbf{C}\mathbf{P}\mathbf{C}^T + R)^{-1} \tag{6.134}$$

The steady-state Kalman filter becomes:

$$\hat{\mathbf{x}}(k+1) = \mathbf{A}\hat{\mathbf{x}}(k) + \mathbf{B}r(k) + \mathbf{K}(y(k) - \mathbf{C}\hat{\mathbf{x}}(k)) \tag{6.135}$$

The steady-state solution has the following key properties:

- *The solution $\mathbf{P}$ is the only non-negative definite solution.*
- *The closed-loop matrix $\mathbf{A} - \mathbf{K}\mathbf{C}$, where $\mathbf{K}$ is the steady-state Kalman gain, is Hurwitz, that is: $|\lambda(\mathbf{A} - \mathbf{K}\mathbf{C})| < 1$.*
- *The estimation error equation given by $\tilde{\mathbf{x}}(k+1) = (\mathbf{A} - \mathbf{K}\mathbf{C})\tilde{\mathbf{x}}(k)$ is asymptotically stable. Note that even though the states of the system may not be asymptotically stable, the estimation error equation is asymptotically stable.*

Proof: The proof is an extension of a simpler system with disturbance and measurement noise which are zero-mean uncorrelated white noise processes (as a consequence there is no measurement noise driving the state equations) and no direct transmission term given by

$$\begin{aligned} \boldsymbol{x}(k+1) &= \boldsymbol{A}\boldsymbol{x}(k) + \boldsymbol{B}r(k) + \boldsymbol{E}_w w(k) \\ y(k) &= \boldsymbol{C}\boldsymbol{x}(k) + v(k) \end{aligned} \tag{6.136}$$

See [11] for a complete proof of this plant model. The state estimation $\tilde{\boldsymbol{x}}(k)$ satisfies the following difference equation

$$\tilde{\boldsymbol{x}}(k+1) = \boldsymbol{A}\tilde{\boldsymbol{x}}(k) - \boldsymbol{K}(k)e(k) + \boldsymbol{E}_w w(k) + \boldsymbol{E}_v v(k) \tag{6.137}$$

The *Riccati equation* governing the covariance of estimation error $\boldsymbol{P}(k) = E[\tilde{\boldsymbol{x}}(k)\tilde{\boldsymbol{x}}^T(k)]$ is:

$$\boldsymbol{P}(k+1) = E[\boldsymbol{A}\tilde{\boldsymbol{x}}(k) - \boldsymbol{K}(k)e(k) + \boldsymbol{E}_w w(k) + \boldsymbol{E}_v v(k)][\boldsymbol{A}\tilde{\boldsymbol{x}}(k) - \boldsymbol{K}(k)e(k) + \boldsymbol{E}_w w(k) + \boldsymbol{E}_v v(k)]^T \tag{6.138}$$

Simplifying yields

$$\begin{aligned} \boldsymbol{P}(k+1) = {} & \boldsymbol{A}\boldsymbol{P}(k)\boldsymbol{A}^T + \boldsymbol{K}(k)(\boldsymbol{C}\boldsymbol{P}(k)\boldsymbol{C}^T + R)\boldsymbol{K}^T(k) - \left(\boldsymbol{A}\boldsymbol{P}(k)\boldsymbol{C}^T + \boldsymbol{E}_v R\right)\boldsymbol{K}^T \\ & - \boldsymbol{K}(k)\left(\boldsymbol{C}\boldsymbol{P}(k)\boldsymbol{A}^T + \boldsymbol{E}_v^T R\right) + \boldsymbol{E}_w \boldsymbol{E}_w^T Q + \boldsymbol{E}_v \boldsymbol{E}_v^T R \end{aligned} \tag{6.139}$$

We will minimize the covariance matrix $\boldsymbol{P}(k+1)$ with respect to $\boldsymbol{K}(k)$ using the "completing the square" approach.

Adding and subtracting a term $(\boldsymbol{A}\boldsymbol{P}(k)\boldsymbol{C}^T + \boldsymbol{E}_v R)(\boldsymbol{C}\boldsymbol{P}(k)\boldsymbol{C}^T + R)^{-1}(\boldsymbol{A}\boldsymbol{P}(k)\boldsymbol{C}^T + \boldsymbol{E}_v R)$ and re-arranging we get

$$\begin{aligned} \boldsymbol{P}(k+1) = {} & \boldsymbol{A}\boldsymbol{P}(k)\boldsymbol{A}^T + \boldsymbol{E}_w \boldsymbol{E}_w^T Q + \boldsymbol{E}_v \boldsymbol{E}_v^T R - \left(\boldsymbol{A}\boldsymbol{P}(k)\boldsymbol{C}^T + \boldsymbol{E}_v R\right)\left(\boldsymbol{C}\boldsymbol{P}(k)\boldsymbol{C}^T + R\right)^{-1}\left(\boldsymbol{A}\boldsymbol{P}(k)\boldsymbol{C}^T + \boldsymbol{E}_v R\right)^T \\ & + \boldsymbol{K}(k)(\boldsymbol{C}\boldsymbol{P}(k)\boldsymbol{C}^T + R)\boldsymbol{K}^T(k) + \left(\boldsymbol{A}\boldsymbol{P}(k)\boldsymbol{C}^T + \boldsymbol{E}_v R\right)\left(\boldsymbol{C}\boldsymbol{P}(k)\boldsymbol{C}^T + R\right)^{-1}\left(\boldsymbol{A}\boldsymbol{P}(k)\boldsymbol{C}^T + \boldsymbol{E}_v R\right)^T \\ & - \left(\boldsymbol{A}\boldsymbol{P}(k)\boldsymbol{C}^T + \boldsymbol{E}_v R\right)\boldsymbol{K}^T - \boldsymbol{K}(k)\left(\boldsymbol{C}\boldsymbol{P}(k)\boldsymbol{A}^T + \boldsymbol{E}_v^T R\right) \end{aligned} \tag{6.140}$$

Expressing as a quadratic function yields

$$\begin{aligned} \boldsymbol{P}(k+1) = {} & \boldsymbol{A}\boldsymbol{P}(k)\boldsymbol{A}^T + \boldsymbol{E}_w \boldsymbol{E}_w^T Q + \boldsymbol{E}_v \boldsymbol{E}_v^T R - \left(\boldsymbol{A}\boldsymbol{P}(k)\boldsymbol{C}^T + \boldsymbol{E}_v R\right)\left(\boldsymbol{C}\boldsymbol{P}(k)\boldsymbol{C}^T + R\right)^{-1}\left(\boldsymbol{A}\boldsymbol{P}(k)\boldsymbol{C}^T + \boldsymbol{E}_v R\right)^T \\ & \times \left[\boldsymbol{K}(k) - \left(\boldsymbol{A}\boldsymbol{P}(k)\boldsymbol{C}^T + \boldsymbol{E}_v R\right)\left(\boldsymbol{C}\boldsymbol{P}(k)\boldsymbol{C}^T + R\right)^{-1}\right]\left(\boldsymbol{C}\boldsymbol{P}(k)\boldsymbol{C}^T + R\right) \\ & \times \left[\boldsymbol{K}(k) - \left(\boldsymbol{A}\boldsymbol{P}(k)\boldsymbol{C}^T + \boldsymbol{E}_v R\right)\left(\boldsymbol{C}\boldsymbol{P}(k)\boldsymbol{C}^T + R\right)^{-1}\right]^T \end{aligned} \tag{6.141}$$

The covariance $\boldsymbol{P}(k+1)$ is minimized with respect to the Kalman gain $\boldsymbol{K}(k)$ if the quadratic term in $\boldsymbol{K}(k)$ on the right-hand side is set equal to zero, and we get:

$$\boldsymbol{K}(k) = \left(\boldsymbol{A}\boldsymbol{P}(k)\boldsymbol{C}^T + \boldsymbol{E}_v R\right)\left(\boldsymbol{C}\boldsymbol{P}(k)\boldsymbol{C}^T + R\right)^{-1} \tag{6.142}$$

Using Eq. (6.142) and after algebraic manipulation the expression (6.141) becomes

$$\boldsymbol{P}(k+1) = \boldsymbol{A}\boldsymbol{P}(k)\boldsymbol{A}^T - \left(\boldsymbol{A}\boldsymbol{P}(k)\boldsymbol{C}^T + \boldsymbol{E}_v R\right)\left(\boldsymbol{C}\boldsymbol{P}(k)\boldsymbol{C}^T + R\right)^{-1}\left(\boldsymbol{A}\boldsymbol{P}(k)\boldsymbol{C}^T + \boldsymbol{E}_v R\right)^T + \boldsymbol{E}_w \boldsymbol{E}_w^T Q + \boldsymbol{E}_v \boldsymbol{E}_v^T R \tag{6.143}$$

Comments *It is interesting that the Kalman gain and the Riccati equation have additional terms* $E_v R$ *and* $E_v E_v^T R$ *respectively when the disturbance and measurement noise are correlated random and deterministic processes, compared to the case when they are zero-mean white noise processes [11].*

The optimal gain and the Riccati equations, when the disturbance and the measurement noise are zero-mean white noise processes, are obtained simply by setting $E_v = 0$ *in (6.141) and (6.142):*

$$K = APC^T(CPC^T + R)^{-1} \tag{6.144}$$

$$P(k+1) = AP(k)A^T - AP(k)C^T(CP(k)C^T + R)^{-1}CP(k)A + E_w E_w^T Q \tag{6.145}$$

6.10 Appendix: The Role of the Ratio of Plant and the Measurement Noise Variances

Lemma 6.8 *The Kalman gain and the Riccati equation can be expressed in terms of the ratio of the the plant variance Q and the measurement noise variance* R *as follows*

$$K(k) = \left(AP_\rho(k)C^T + E_v\right)(CP_\rho(k)C^T + 1)^{-1} \tag{6.146}$$

$$P_\rho(k+1) = AP_\rho(k)A^T - \left(AP_\rho(k)C^T + E_v\right)\left(CP_\rho(k)C^T + 1\right)^{-1}\left(AP_\rho(k)C^T + E_v\right)^T + E_w E_w^T Q_\rho + E_v E_v^T \tag{6.147}$$

where $Q_\rho = \dfrac{Q}{R}$ *is the ratio of the plant and the measurement noise variances, and* $P_\rho = \dfrac{P}{R}$.

Proof: The proof follows by dividing the expression for the Kalman gain (6.16) and Riccati equation (6.17).

Comment *The performance of the Kalman filter depends only on the ratio* Q/R *of the plant and the measurement noise variance and not on* Q *and* R *separately. In practice, one does not know the variances* Q *and* R. *One may tune online* Q/R *so that the mean-squared residual is minimum.*

6.11 Appendix: Orthogonal Properties of the Kalman Filter

Definition *A stochastic process* $x(k)$ *is* stationary *if the following condition holds*

- *The mean of* $x(k)$ *is constant and bounded:* $E[x(k)] = \mu < \infty$ μ *is some constant.*
- *The correlation function of* $x(k)$ *denoted* $R_{xx}(k,m) = E[x^T(k)x(k-m)]$ *is bounded.*
- *The correlation function of is function only of time lag* m $R_{\tilde{x}\tilde{x}}(m) = E[\tilde{x}(k)\tilde{x}^T(k-m)] = 0, \quad m \neq 0.$

Definition *[26] A stochastic process* $x(k)$ *is* quasi-stationary *if the following condition holds*

- $E[x(k)] = \mu(k)$ *where* $\mu(k)$ *may in general be time varying and bounded.*
- $\lim_{k\to\infty} it\, R_{xx}(k,m) = R_{xx}(m)$ where $R_{xx}(k,m) = E[x(k)x(k-m)]$.

Comments *If a stochastic process is stationary, it is quasi-stationary.*

In this work all the stochastic processes, namely including the state $x(k)$, *its estimate* $\hat{\mathbf{x}}(k)$ *and estimation error* $\tilde{\mathbf{x}}(k)$, *are assumed to be quasi-stationary if the system* $(\boldsymbol{A}, \boldsymbol{B}, \boldsymbol{C})$ *is asymptotically stable, and controllable so that the system state* $\boldsymbol{x}(k)$ *and the output* $y(k)$ *are bounded (assuming that the input*

$r(k)$ is bounded). Assuming the system is observable, the Kalman filter state $\hat{\mathbf{x}}(k)$ and estimation error $\tilde{\mathbf{x}}(k)$ are also bounded. If the input to the system $r(k)$ is constant, the system will be asymptotically stationary. To ensure quasi-stationarity, the system and the Kalman filter must be in steady-state operating regime.

Theorem 6.4 *If the optimal Kaman filter equations are*

$$\begin{aligned}\hat{\mathbf{x}}(k+1) &= \boldsymbol{A}\hat{\mathbf{x}}(k) + \boldsymbol{B}r(k) + \boldsymbol{K}(k)e(k)\\ \hat{y}(k) &= \boldsymbol{C}\hat{\mathbf{x}}(k)\end{aligned} \tag{6.148}$$

$$\begin{aligned}\tilde{\mathbf{x}}(k+1) &= \boldsymbol{A}\tilde{\mathbf{x}}(k) - \boldsymbol{K}(k)e(k) + \boldsymbol{E}_w w(k) + \boldsymbol{E}_v v(k)\\ e(k) &= \boldsymbol{C}\tilde{\mathbf{x}}(k) + v(k)\end{aligned} \tag{6.149}$$

$$\boldsymbol{K}(k) = \left(\boldsymbol{A}\boldsymbol{P}_\rho(k)\boldsymbol{C}^T + \boldsymbol{E}_v R\right)\left(\boldsymbol{C}\boldsymbol{P}_\rho(k)\boldsymbol{C}^T + R\right)^{-1} \tag{6.150}$$

Then

a) The estimation error $\tilde{\mathbf{x}}(k)$ is orthogonal to the past residuals $\{e(k-m):\ m \geq 1\}$.

$$E[\tilde{\mathbf{x}}(k+m)e(k)] = 0 \textit{ for all } m \geq 1 \tag{6.151}$$

b) The residual $e(k+m)$ is orthogonal to the past residuals $e(k)$ for $m \geq 1$.

$$E[e(k+m)e(k)] = \begin{cases}\boldsymbol{C}\boldsymbol{P}(k)\boldsymbol{C}^T + R & m = 0\\ 0 & m \neq 0\end{cases} \tag{6.152}$$

Proof:

a) We will first prove (6.151). Consider the equation governing $\tilde{\mathbf{x}}(k)$ given by Eq. (6.149). Multiplying by $e(k)$ and taking expectation yields:

$$E[\tilde{\mathbf{x}}(k+1)e(k)] = \boldsymbol{A}E[\tilde{x}(k)e(k)] - \boldsymbol{K}(k)E[e^2(k)] + \boldsymbol{E}_w E[w(k)e(k)] + \boldsymbol{E}_v E[v(k)e(k)] \tag{6.153}$$

Substituting $e(k) = \boldsymbol{C}\tilde{\mathbf{x}}(k) + v(k)$ on the right-hand sides and simplifying using $E[v(k)w(k)] = 0$; $E[w(k)e(k)] = 0$; $E[v(k)e(k)] = R$, yields

$$E[\tilde{\mathbf{x}}(k+1)e(k)] = \boldsymbol{A}\boldsymbol{P}(k)\boldsymbol{C}^T - \boldsymbol{K}(k)(\boldsymbol{C}\boldsymbol{P}(k)\boldsymbol{C}^T + R) + \boldsymbol{E}_v R \tag{6.154}$$

Substituting for $\boldsymbol{K}(k)$ using Eq. (6.142) yields

$$E[\tilde{\mathbf{x}}(k+1)e(k)] = \mathbf{0} \tag{6.155}$$

Deriving the expression for $\tilde{\mathbf{x}}(k+2)$ using Eq. (6.149) and multiplying by $e(k)$, and taking expectation yields:

$$\begin{aligned}E[\tilde{\mathbf{x}}(k+2)e(k)] &= \boldsymbol{A}E[\tilde{\mathbf{x}}(k+1)e(k)] - \boldsymbol{K}(k+1)E[e(k+1)e(k)]\\ &\quad + \boldsymbol{E}_w E[w(k+1)e(k)] + \boldsymbol{E}_v E[v(k+1)e(k)]\end{aligned} \tag{6.156}$$

Using Eq. (6.77), $e(k+1) = \mathbf{C}\tilde{\mathbf{x}}(k+1) + v(k+1)$ and noting that the future values of zero-mean white noise $w(k)$ and $v(k)$ are uncorrelated with the present residual $e(k)$:

$$\begin{aligned} E[e(k)w(k+m)] &= 0 \\ E[e(k)v(k+m)] &= 0 \end{aligned} \quad \text{for all } m \geq 1 \tag{6.157}$$

Generalizing the result by repeated application of Eqs. (6.146), (6.156), and (6.157), we get

$$E[\tilde{\mathbf{x}}(k+m)e(k)] = \mathbf{0} \quad m \geq 1 \tag{6.158}$$

b) We will now prove $E[e(k+m)e(k)] = 0$ first for $m \geq 1$
Expressing the residual $e(k)$ in terms of $\tilde{x}(k)$ and $v(k)$ we get

$$e(k) = y(k) - \hat{y}(k) = \mathbf{C}\tilde{\mathbf{x}}(k) + v(k) \tag{6.159}$$

Substituting for $e(k)$ in the expression $E[e(k+m)e(k)]$ we get

$$E[e(k+m)e(k)] = E[(\mathbf{C}\tilde{\mathbf{x}}(k+m) + v(k+m))e(k)] \tag{6.160}$$

Using Eq. (6.158) the expression (6.160) becomes

$$E[e(k+m)e(k)] = E[v(k+m)e(k)] \, for \; all \; m \geq 1 \tag{6.161}$$

Using Eq. (6.157), the expression (6.161) becomes

$$E[e(k+m)e(k)] = 0 \, for \; all \; m \geq 1 \tag{6.162}$$

We will now extend the proof $E[e(k+m)e(k)] = 0$ for $m \leq -1$.

This problem is equivalent to the proof of $E[e(k-m)e(k)] = 0 \; for \; all \; m \geq 1$. The expression $E[e(k-m)e(k)]$ becomes

$$E[e(k-m)e(k)] = E[e(k-m)(\mathbf{C}\tilde{\mathbf{x}}(k) + v(k))] + \mathbf{C}E[\tilde{\mathbf{x}}(k)e(k-m)] + E[e(k-m)v(k)] \tag{6.163}$$

Using Eq. (6.158) the expression (6.163) becomes

$$E[e(k-m)e(k)] = E[e(k-m)v(k)] \, for \; m \geq 1 \tag{6.164}$$

Since the present measurement noise $v(k)$ is uncorrelated with the past residual $e(k-m)$, we get

$$E[e(k-m)e(k)] = 0 \, for \; m \geq 1 \tag{6.165}$$

Equivalently

$$E[e(k+m)e(k)] = 0 \, for \; m \leq -1 \tag{6.166}$$

From Eqs. (6.158) and (6.166) we deduce

$$E[e(k+m)e(k)] = 0 \, for \; m \neq 0 \tag{6.167}$$

Consider the case when $m = 0$

$$E[e^2(k)] = E\left[(\boldsymbol{C}\tilde{\boldsymbol{x}}(k) + v(k))(\boldsymbol{C}\tilde{\boldsymbol{x}}(k) + v(k))^T\right] = \boldsymbol{C}E[\tilde{\boldsymbol{x}}(k)\tilde{\boldsymbol{x}}^T(k)]\boldsymbol{C}^T + E[v^2(k)] \tag{6.168}$$

Using the definition $\boldsymbol{P} = E[\tilde{\boldsymbol{x}}(k)\tilde{\boldsymbol{x}}^T(k)]$ the expression (6.38) becomes

$$E[e^2(k)] = \boldsymbol{C}\boldsymbol{P}(k)\boldsymbol{C}^T + R \tag{6.169}$$

Thus Eq. (6.152) is proved.

Theorem 6.5

a) The estimation error $\tilde{\boldsymbol{x}}(k+m)$ is orthogonal to the past estimates $\hat{\boldsymbol{x}}(k)$ for $m \geq 0$:

$$E[\tilde{\boldsymbol{x}}(k+m)\hat{\boldsymbol{x}}^T(k)] = 0 \textit{ for } m \geq 0 \tag{6.170}$$

b) The estimation error $\tilde{\boldsymbol{x}}(k+m)$ forms an orthogonal to the measurement $y(k)$ for $m \geq 1$:

$$E[\tilde{\boldsymbol{x}}(k+m)y(k)] = 0 \textit{ for } m \geq 1 \tag{6.171}$$

Proof:

a) We will first prove $E[\tilde{\boldsymbol{x}}(k)\hat{\boldsymbol{x}}^T(i)] = 0,\ i = 1, 2,$ *and* 3 for $k \geq i$ and then generalize the results to $E[\tilde{\boldsymbol{x}}(k+m)\hat{\boldsymbol{x}}^T(k)] = 0\ m \geq 0$.

The following orthogonal properties are employed in the derivations:

- $E[\tilde{\boldsymbol{x}}(k+m)e(k)] = 0 \quad m \geq 1$ established in Theorem 6.4
- $E[\tilde{\boldsymbol{x}}(k)r(i)] = 0$ since $\tilde{\boldsymbol{x}}(k)$ is an uncorrelated zero-mean process, and is not a function of $r(i)$ as can be deduced from the Eq. (6.149) governing $\tilde{\boldsymbol{x}}(k)$.
- $E[\tilde{\boldsymbol{x}}(k)\hat{\boldsymbol{x}}^T(0)] = 0$ as $\hat{\boldsymbol{x}}(0)$ is assumed to be an uncorrelated zero-mean random variable.

In other words the estimation error $\tilde{\boldsymbol{x}}(k)$ is orthogonal to the residual $e(k-i)$ *for* $i \geq 1$, the reference input $r(i)$ and the initial condition $\hat{\boldsymbol{x}}(0)$.

- The equations governing $\hat{\boldsymbol{x}}(k)$ and $\tilde{\boldsymbol{x}}(k)$ are given by Eqs. (6.148) and (6.149).
    - Consider the estimate $\hat{x}(1)$ given by

$$\hat{\boldsymbol{x}}(1) = \boldsymbol{A}\hat{\boldsymbol{x}}(0) + \boldsymbol{K}(0)e(0) + \boldsymbol{B}r(0) \tag{6.172}$$

Then the expression for $E[\tilde{\boldsymbol{x}}(k)\hat{\boldsymbol{x}}^T(1)]$

$$E[\tilde{\boldsymbol{x}}(k)\hat{\boldsymbol{x}}^T(1)] = E[\tilde{\boldsymbol{x}}(k)(\boldsymbol{A}\hat{x}(0) + \boldsymbol{K}(0)e(0) + \boldsymbol{B}r(0))^T] \tag{6.173}$$

Since for $k \geq 1$, $\tilde{\boldsymbol{x}}(k)$ is orthogonal to $e(0)$ the reference input $r(0)$ and the initial condition $\hat{\boldsymbol{x}}(0)$

$$E[\tilde{\boldsymbol{x}}(k)\hat{\boldsymbol{x}}^T(1)] = 0 \text{ for } k \geq 1 \tag{6.174}$$

- Consider the estimate $\hat{\boldsymbol{x}}(2)$ given by

$$\hat{\boldsymbol{x}}(2) = \boldsymbol{A}\hat{\boldsymbol{x}}(1) + \boldsymbol{K}(1)e(1) + \boldsymbol{B}r(1) \tag{6.175}$$

Substituting for $\hat{\boldsymbol{x}}(1)$ using Eq. (6.172) we get

$$\hat{\boldsymbol{x}}(2) = \boldsymbol{A}^2\hat{\boldsymbol{x}}(0) + \boldsymbol{AK}(0)e(0) + \boldsymbol{AB}r(0) + \boldsymbol{K}(1)e(1) + \boldsymbol{B}r(1) \tag{6.176}$$

Substituting for $\hat{x}(2)$ we get

$$E[\tilde{\boldsymbol{x}}(k)\hat{\boldsymbol{x}}^T(2)] = E\left[\tilde{\boldsymbol{x}}(k)(\boldsymbol{A}^2\hat{\boldsymbol{x}}(0) + \boldsymbol{AK}(0)e(0) + \boldsymbol{AB}r(0) + \boldsymbol{K}(1)e(1) + \boldsymbol{B}r(1))^T\right] \tag{6.177}$$

Since for $k \geq 2$, $\tilde{\boldsymbol{x}}(k)$ is orthogonal to $e(0)$, $r(0)$, $e(1)$, $r(1)$ and $\hat{\boldsymbol{x}}(0)$ we get

$$E[\tilde{\boldsymbol{x}}(k)\hat{\boldsymbol{x}}^T(2)] = 0 \text{ for } k \geq 2 \tag{6.178}$$

- Consider the estimate $\hat{\boldsymbol{x}}(3)$ given by

$$\hat{\boldsymbol{x}}(3) = \boldsymbol{A}\hat{\boldsymbol{x}}(2) + \boldsymbol{K}(2)e(2) + \boldsymbol{B}r(2) \tag{6.179}$$

Substituting for $\hat{\boldsymbol{x}}(2)$ we get

$$\hat{\boldsymbol{x}}(3) = \boldsymbol{A}^3\hat{\boldsymbol{x}}(0) + \boldsymbol{A}^2\boldsymbol{K}(0)e(0) + \boldsymbol{A}^2\boldsymbol{B}r(0) + \boldsymbol{AK}(1)e(1) + \boldsymbol{AB}r(1) + \boldsymbol{K}(2)e(2) + \boldsymbol{B}r(2) \tag{6.180}$$

Substituting for $\hat{\boldsymbol{x}}(3)$ we get

$$\begin{aligned} E[\tilde{\boldsymbol{x}}(k)\hat{\boldsymbol{x}}^T(3)] &= E[\tilde{\boldsymbol{x}}(k)(\boldsymbol{A}^3\hat{\boldsymbol{x}}(0) + \boldsymbol{A}^2\boldsymbol{K}(0)e(0) + \boldsymbol{A}^2\boldsymbol{B}r(0) + \boldsymbol{AK}(1)e(1) \\ &\quad + \boldsymbol{AB}r(1) + \boldsymbol{K}(2)e(2) + \boldsymbol{B}r(2))^T] \end{aligned}$$

Since for $k \geq 3$, $\tilde{\boldsymbol{x}}(k)$ is uncorrelated with $e(0)$, $e(1)$, $e(2)$, $r(0)$, $r(1)$, $r(2)$, and $\hat{x}(0)$ we get

$$E[\tilde{\boldsymbol{x}}(k)\hat{\boldsymbol{x}}^T(3)] = 0 \quad k \geq 3 \tag{6.181}$$

Generalizing the result to the estimate $\hat{x}(k-m)$ using Eq. (6.148)

$$\hat{\boldsymbol{x}}(k-m) = \boldsymbol{A}^{k-m}\boldsymbol{x}(\hat{0}) + \sum_{i=1}^{k-m}\boldsymbol{A}^{k-m-i}\boldsymbol{B}r(i-1) + \sum_{i=1}^{k-m}\boldsymbol{A}^{k-m-i}\boldsymbol{K}(i-1)e(i-1) \tag{6.182}$$

Substituting for $\hat{\boldsymbol{x}}(k-m)$ in the expression $E[\tilde{\boldsymbol{x}}(k)\hat{\boldsymbol{x}}^T(k-m)]$ we get

$$E[\tilde{x}(k)\hat{x}^T(k-m)] = E\left[\tilde{\boldsymbol{x}}(k)\left(\boldsymbol{A}^{k-m}\hat{\boldsymbol{x}}(0) + \sum_{i=1}^{k-m}\boldsymbol{A}^{k+m-i}\boldsymbol{B}r(i-1) + \sum_{i=1}^{k-m}\boldsymbol{A}^{k-m-i}\boldsymbol{K}(i-1)e(i-1)\right)^T\right] \tag{6.183}$$

Simplifying we get

$$E[\tilde{x}(k)\hat{x}^T(k-m)] = E[\tilde{x}(k)\hat{x}^T(0)](A^T)^{k-m} + \sum_{i=1}^{k-m} E[\tilde{x}(k)e(i-1)]K^T(i-1)(A^T)^{k-m-i}$$

$$+ \sum_{i=1}^{k-m} E[\tilde{x}(k)r(i-1)]B^T(A^T)^{k-m-i} \tag{6.184}$$

Since the estimation error $\tilde{x}(k)$ is orthogonal to the residuals $\{e(i-1)\}$, the reference input $\{r(i-1)\}$ and the initial condition $\hat{x}(0)$ we get

$$E[\tilde{x}(k)\hat{x}^T(k-m)] = 0 \; for \; m \geq 0 \tag{6.185}$$

Note that the estimation error at any time instant also is orthogonal to the estimate at the same time instant, that is $E[\tilde{x}(k)\hat{x}^T(k)] = 0$

- We will now prove $E[\tilde{x}(k+m)y(k)] = 0 \; for \; m \geq 1$.
  Expressing the output $y(k) = Cx(k) + v(k)$ as a function of the state estimate $\hat{x}(k)$ and the residual $e(k)$ we get

$$y(k) = C\hat{x}(k) + e(k) \tag{6.186}$$

Using Eq. (6.186) the expression $E[\tilde{x}(k+m)y(k)]$ becomes:

$$E[\tilde{x}(k+m)y(k)] = E[\tilde{x}(k+m)C\hat{x}(k)] + E[\tilde{x}(k+m)e(k)] \tag{6.187}$$

As $C\hat{x}$ is a scalar, $C\hat{x}(k) = \hat{x}^T(k)C^T$. Re-writing the first term on the right we get

$$E[\tilde{x}(k+m)y(k)] = E[\tilde{x}(k+m)\hat{x}^T(k)C^T] + E[\tilde{x}(k+m)e(k)] \tag{6.188}$$

Using the orthogonality conditions (6.170) and (6.151) we get

$$E[\tilde{x}(k+m)y(k)] = 0 \; m \geq 1 \tag{6.189}$$

6.11.1 Span of a Matrix

Definition *Let X = $\{x_i : i = 1, 2, \dots\}$ be a subset of a vector space and S_x be the span of X. Then every vector $s \in S_x$ can be expressed as a linear combination of vectors in X, given by*

$$S_x = span\{x_1 \quad x_2 \quad x_3 \quad \dots \quad .\} = \left\{ s : s = \sum_i \alpha_i x_i, \quad \alpha_i \in R \right\} \tag{6.190}$$

where $\{x_i : i = 1, 2, \dots\}$ is the basis of S_x.

6.11.2 Transfer Function Formulae

Let (A, B, C, D) be the state-space model of the system where A, B, C, and D are respectively nxn matrix, $nx1$ column vector, $1xn$ row vector. and a scalar respectively. The transfer function $G(z)$ is given by

$$G(z) = \frac{N(z)}{D(z)} = C(zI - A)^{-1}B + D = \frac{CAdj(zI - A)B}{|zI - A|} + D = \frac{CAdj(zI - A)B + D|zI - A|}{|zI - A|} \tag{6.191}$$

We may express the numerator polynomial $N(z)$ in terms of the system matrix $\begin{bmatrix} zI - A & B \\ -C & D \end{bmatrix}$ as:

$$\begin{aligned} N(z) &= CAdj(zI - A)B + D|zI - A| = \left|\begin{bmatrix} zI - A & B \\ -C & D \end{bmatrix}\right| \\ D(z) &= |zI - A| \end{aligned} \tag{6.192}$$

6.12 Appendix: Kalman Filter Residual with Model Mismatch

Lemma 6.9 *If the system and the Kalman filter models are given by Eqs. (6.6) and (6.118), then the frequency-domain expression of the residual, that is, $e(z)$, is given, in terms of the plant output $y(z)$ and reference input $r(z)$, by:*

$$e(z) = \frac{D_0(z)}{F_0(z)} y(z) - \frac{N_0(z)}{F_0(z)} r(z)$$

where

$$\frac{D_0(z)}{F_0(z)} = 1 - C_0(zI - A_0 + K_0C_0)^{-1}K_0;$$

$$\frac{N_0(z)}{F_0(z)} = C_0(zI - A_0 + K_0C_0)^{-1}B_0; F_0(z) = |zI - A_0 + K_0C_0|$$

Proof: [4, 14] The Kalman filter when the internal model of the Kalman filter $(A_0, B_0\ C_0)$ is not equal to the system model (A, B, C) is:

$$\hat{x}(k+1) = (A_0 - K_0C_0)\hat{x}(k) + Br(k) + K_0y(k) \tag{6.193}$$

Taking z-transform yields:

$$(zI - A_0 + K_0C_0)\hat{x}(z) = Br(z) + K_0y(z) \tag{6.194}$$

Inverting, the expression for $\hat{x}(z)$ becomes

$$\hat{x}(z) = (zI - A_0 + K_0C_0)^{-1}Br(z) + (zI - A_0 + K_0C_0)^{-1}K_0y(z) \tag{6.195}$$

Substituting the expression for the residual $e(z) = y(z) - \hat{y}(z)$ becomes

$$e(z) = \left(1 - C_0(zI - A_0 + K_0C_0)^{-1}K_0\right) y(z) - C_0(zI - A_0 + K_0C_0)^{-1}Br(z) \tag{6.196}$$

Using the transfer function formulae (6.191) and (6.192) the first term on the right of Eq. (6.196) is:

$$1 - C_0(zI - A_0 + K_0C_0)^{-1}K_0 = \frac{\left|\begin{bmatrix} zI - A_0 + K_0C_0 & K_0 \\ C_0 & I \end{bmatrix}\right|}{|zI - A_0 + K_0C_0|} \tag{6.197}$$

Consider the square bracket on the upper right-hand side of Eq. (6.197). As the determinant value does not change by elementary operation on its rows and columns, post-multiplying the second row by $\boldsymbol{K}_0$ and subtracting the result from the first row yields:

$$\frac{\left|\begin{bmatrix} z\boldsymbol{I}-\boldsymbol{A}_0+\boldsymbol{K}_0\boldsymbol{C}_0 & \boldsymbol{K}_0 \\ \boldsymbol{C}_0 & 1 \end{bmatrix}\right|}{|z\boldsymbol{I}-\boldsymbol{A}_0+\boldsymbol{K}_0\boldsymbol{C}_0|} = \frac{\left|\begin{bmatrix} z\boldsymbol{I}-\boldsymbol{A}_0 & 0 \\ \boldsymbol{C}_0 & 1 \end{bmatrix}\right|}{|z\boldsymbol{I}-\boldsymbol{A}_0+\boldsymbol{K}_0\boldsymbol{C}_0|} = \frac{|z\boldsymbol{I}-\boldsymbol{A}_0|}{|z\boldsymbol{I}-\boldsymbol{A}_0+\boldsymbol{K}_0\boldsymbol{C}_0|} \tag{6.198}$$

Using the transfer function formulae (6.191) and (6.192) the second term on the right of Eq. (6.196) is:

$$\boldsymbol{C}_0(z\boldsymbol{I}-\boldsymbol{A}_0+\boldsymbol{K}_0\boldsymbol{C}_0)^{-1}\boldsymbol{B}_0 = \frac{\left|\begin{bmatrix} z\boldsymbol{I}-\boldsymbol{A}_0+\boldsymbol{K}_0\boldsymbol{C}_0 & \boldsymbol{B}_0 \\ -\boldsymbol{C}_0 & 0 \end{bmatrix}\right|}{|z\boldsymbol{I}-\boldsymbol{A}_0+\boldsymbol{K}_0\boldsymbol{C}_0|} \tag{6.199}$$

Here too, post-multiplying the second row by $\boldsymbol{K}_0$ and subtracting the result from the first row yields:

$$\boldsymbol{C}_0(z\boldsymbol{I}-\boldsymbol{A}_0+\boldsymbol{K}_0\boldsymbol{C}_0)^{-1}\boldsymbol{B}_0 = \frac{\left|\begin{bmatrix} z\boldsymbol{I}-\boldsymbol{A}_0 & \boldsymbol{B}_0 \\ -\boldsymbol{C}_0 & 0 \end{bmatrix}\right|}{|z\boldsymbol{I}-\boldsymbol{A}_0+\boldsymbol{K}_0\boldsymbol{C}_0|} \tag{6.200}$$

Let the transfer function of the actual model $(\boldsymbol{A}, \boldsymbol{B}, \boldsymbol{C})$ and assumed system model or the internal model $(\boldsymbol{A}_0, \boldsymbol{B}_0, \boldsymbol{C}_0)$ employed in the design of the Kalman filter relating input $r(z)$ and the output $y(z)$ respectively be $G(z) = \dfrac{N(z)}{D(z)}$ and $G_0(z) = \dfrac{N_0(z)}{D_0(z)}$. Using the formulae (6.62) we get

$$N(z) = \left|\begin{bmatrix} z\boldsymbol{I}-\boldsymbol{A} & \boldsymbol{B} \\ -\boldsymbol{C} & 0 \end{bmatrix}\right|;\ D(z) = |z\boldsymbol{I}-\boldsymbol{A}|;$$

$$N_0(z) = \left|\begin{bmatrix} z\boldsymbol{I}-\boldsymbol{A}_0 & \boldsymbol{B}_0 \\ -\boldsymbol{C}_0 & 0 \end{bmatrix}\right|;\ D_0(z) = |z\boldsymbol{I}-\boldsymbol{A}_0|.$$

Using the above expressions, Eqs. (6.198) and (6.200). Eq. (6.196) may be written compactly as:

$$e(z) = \frac{D_0(z)}{F_0(z)}y(z) - \frac{N_0(z)}{F_0(z)}r(z) \tag{6.201}$$

Thus the lemma is established.

Lemma 6.10 *If the linear regression model of the system is:*

$$y(k) = \boldsymbol{\psi}^T(k)\boldsymbol{\theta} + \upsilon(k) \tag{6.202}$$

where $\boldsymbol{\psi}^T(k) = [-y(k-1) \quad -y(k-2) \quad \cdots \quad -y(k-n) \quad r(k-1) \quad r(k-2) \quad \cdots \quad r(k-n)]$ *and the feature vector is* $\boldsymbol{\theta} = [a_1 \quad a_2 \quad \cdots \quad a_n \quad b_1 \quad b_2 \quad \cdots \quad b_n]^T$. *Then*

$$e(z) = \frac{1}{F_0(z)}(\boldsymbol{\psi}^T(z)\Delta\boldsymbol{\theta} + \upsilon(z))$$

Proof: [2] Consider the expression of the residual

$$e(z) = \frac{D_0(z)}{F_0(z)}y(z) - \frac{N_0(z)}{F_0(z)}r(z) \tag{6.203}$$

Substituting for $D_0(z) = 1 + \sum_{i=1}^{n} a_{i0}z^{-i}$ and $N_0(z) = \sum_{i=1}^{n} b_{i0}z^{-i}$, we get

$$F_0(z)e(z) = y(z) + \sum_{i=1}^{n} a_{i0}z^{-i}y(z) - \sum_{i=1}^{n} b_{i0}z^{-i}r(z) \tag{6.204}$$

The z-transform $\psi^T(z)$ of $\psi^T(k)$ is:

$$\psi^T(z) = [-z^{-1}y(z) \quad -z^{-2}y(z) \quad \cdots \quad -z^{-n}y(z) \quad z^{-1}r(z) \quad z^{-2}\quad r(z) \quad \cdots \quad z^{-n}r(z)] \tag{6.205}$$

Using Eq. (6.201) the expression Eq. (6.204) becomes

$$F_0(z)e(z) = y(z) - \boldsymbol{\psi}^T(z)\boldsymbol{\theta}^0 \tag{6.206}$$

Substituting the z-transform $y(z) = \boldsymbol{\psi}^T(z)\boldsymbol{\theta} + \boldsymbol{v}(z)$ of $y(k) = \boldsymbol{\psi}^T(k)\boldsymbol{\theta} + \boldsymbol{v}(k)$ given by Eq. (6.202), we get:

$$F_0(z)e(z) = \boldsymbol{\psi}^T(z)\Delta\boldsymbol{\theta} + \boldsymbol{v}(z) \tag{6.207}$$

where $\Delta\boldsymbol{\theta} = \boldsymbol{\theta} - \boldsymbol{\theta}^0 = [\Delta a_1 \quad \Delta a_2 \quad \cdot \quad \Delta a_n \quad \Delta b_1 \quad \Delta b_2 \quad \cdot \quad \Delta b_n]^T$.

References

Kalman Filter

[1] Doraiswami, R. and Cheded, L. (2013) Fault diagnosis of a sensor network: a distributed filtering approach. *Journal of Dynamic Systems, Measurement and Control*, **135**(5).

[2] Doraiswami, R. and Cheded, L. (2013) A unified approach to detection and isolation of parametric faults using a Kalman filter residuals. *Journal of Franklin Institute*, **350**(5), 938–965.

[3] Faragher, R. (2012) Understanding the basis of the Kalman filter via simple and intuitive derivation. *IEEE Signal Processing Magazine*, 128–132.

[4] Doraiswami, R. and Cheded, L. (2012) Kalman filter for fault detection: an internal model approach. *IET Control Theory and Applications*, **6**(5): 1–11.

[5] Groves, P. (2008) *Principles of GNSS, Inertial and Multisensor Integrated Navigation Systems*, Artech House, Norwood, MA.

[6] Oikonomou, V., Tzallas, A. and Fotiadis, D. (2007) A Kalman filter based methodology for EEG spike enhancement. *Computer Methods and Programs in Medicine, Elsevier*, 101–108.

[7] Anderson, B. and Moore, J. (2005) *Optimal Filtering*, Dover, New York.

[8] Doraiswami, R. and Price, R.S. (1998) A robust position estimation scheme using sun sensor. *IEEE Transactions on Instrumentation and Measurement*, **47**(2) 595–603.

[9] Brown, R.G. and Hwang, P.Y. (1997) *Introduction to Random Signals and Applied Kalman Filtering*, John Wiley and Sons.

[10] Doraiswami, R. (1996) A novel Kalman filter-based navigation using beacons. *IEEE Transactions on Aerospace and Electronic Systems*, **32**(2) 830–840.

[11] Mendel, J. (1995) *Lessons in Estimation Theory in Signal Processing, Communications and Control*, Prentice-Hall.

[12] Kay, S.M. (1993) *Fundamentals of Signal Processing: Estimation Theory*, Prentice-Hall PTR, New Jersey.

Internal Model Principle

[13] Francis, B.F. and Wonham, W.M. (1976) The internal model principle of control theory. *Automatica*, **12**: 457–465.

[14] Goodwin, G.C., Graeb, S.F., and Salgado, M.E. (2001) *Control System Design*, Prentice Hall, New Jersey.

[15] Doraiswami, R. and Bordry, F. (1987) Robust two-time level control strategy for sampled data servomechanism problem. *International Journal of System Sciences*, **18**(12), 2261–2277.

[16] Diduch, C. and Doraiswami, R. (1987) Robust servomechanism controller design for digital implementation. *IEEE Transactions on Industrial Electronics*, **34**(2), 172–179.

[17] Doraiswami, R. and Sharaf, A.M. (1985) A digital controller design for enhancing steady-state stability of interconnected power systems. *The International Journal of Electrical power and Energy Systems*, **7**(4): 210–214.

[18] Doraiswami, R. and Gulliver, A. (1984) A control strategy for computer numerical control machine exhibiting precision and rapidity. *Transactions of the ASME Journal of Dynamic Systems, Measurement and Control*, **106**(1): 56–62.

[19] Doraiswami, R. (1983) Robust control of a class of nonlinear servomechanism problem. *IEEE Proceedings*, **130**(2): 63–71.

[20] Doraiswami, R. (1982) Robust control strategy for a linear time-invariant multivariable sampled data servomechanism problem. *IEEE Proceedings: Part D*, **129**(6), 283–292.

[21] Doraiswami, R. (1978) A nonlinear load-frequency control design. *IEEE Transactions on Power*.

Signal Processing

[22] Proakis, J.G.and Manolakis, D.G. (2007) *Digital Signal Processing: Prnnciples, Algorithms and Applications*, Prentice Hall, New Jersey.

[23] Mitra, S.K. (2006) *Digital Signal Processing: A Computer-based Approach*, McGraw Hill Higher Education, Boston.

[24] Moon, T.K. and Stirling, W.C. (2000) *Mathematical Methods and Algorithms for Signal Processing*, Prentice Hall, New Jersey.

[25] Kay, S.M. (1988) *Modern Spectral Estimation: Theory and Application*, Prentice Hall, New Jersey.

Identification

[26] Ljung, L. (1999) *System Identification: Theory for the User*, Prentice-Hall, New Jersey.

7

System Identification

7.1 Overview

Identification of a physical system poses a great challenge as its model is complex and nonlinear, and the assumed linear model that is widely used is at best an approximation. The identification scheme must be tailored for a specific application. A number of identification schemes are presented to meet the requirements of various applications including fault diagnosis, condition monitoring, condition-based maintenance, soft sensor, controller design and simulation- based prediction [1–14]. A model termed the *diagnostic model* is identified for fault diagnosis, which completely captures the static and the dynamic behavior of the nominal fault-free and fault-bearing systems [1–3]. In the case of soft sensor applications, a set of models in the neighborhood of a given operating point is generated by performing a number of experiments. A model termed the *optimal nominal model* is identified, which is the best least-squares fit that the set of models thus obtained. The identification of the diagnostic and the optimal model are treated in later chapters.

Identification of a dynamical system at a given operating point is presented in this chapter. The identification schemes include the widely used prediction error, the subspace method, and the least squares methods. The Kalman filter plays a key role in the identification and its application [5].

The structure of the identification model relating the input and the output of the system is the same as that of the Kalman filter of the system. The Kalman filter-based structure will ensure that the equation error is a zero-mean white noise process. Most of the popular identification schemes use this structure to ensure that the estimates are unbiased and efficient.

The least-squares method represents an important cornerstone in the approximation and estimation theories and is widely used to identify a system as it is simple, numerically efficient, and yields a closed-form solution to parameter estimation problems. If the equation error is a zero-mean white noise process, the estimate is then unbiased and efficient. However, if the equation error is a colored noise, the estimate will be biased as the colored noise will be correlated with the data vector. To overcome this problem, approaches such as the prediction error method (PEM) and high-order least squares (HOLS) method [6, 8, 12, 13] have been proposed.

A high-order model is used in various applications, including the non-parameteric identification of impulse response, estimation of Markov parameters in subspace identification and in model predictive control, identification of a signal model, and in system identification. The use of a high-order model for the identification of a signal model is inspired by the seminal paper by [12] for the accurate estimation of the parameters of an impulse response from measurements corrupted by an additive white noise. It is shown via simulation that the variance of the parameter estimation error approaches the Cramér–Rao

Identification of Physical Systems: Applications to Condition Monitoring, Fault Diagnosis, Soft Sensor and Controller Design, First Edition. Rajamani Doraiswami, Chris Diduch and Maryhelen Stevenson.

lower bound (CRLB). It is shown analytically in [13] that using a higher-order model (with an order several times larger than the true order) improves significantly the accuracy of the parameter estimates. The higher-order method had not received much attention in system identification although it has been mentioned as an alternative scheme to the PEM [4,7] and has been successfully employed in identification for fault diagnosis in [1–3,6].

Most of the identification methods, including the LS, the PEM, and subspace method, use a Kalman filter model to derive the structure of the model set used in identification, [5]. The residual of the Kalman filter takes over the combined role of the disturbance and the measurement noise affecting the system. As the Kalman filter residual, termed also an innovation process, is a zero-mean white process, one can develop an appropriate identification scheme using the vast literature on the statistical estimation theory for parameter estimation in a filtered white noise process (or colored noise) [9].

A higher-order LS method is developed here starting with the derivation of the linear regression model based on the Kalman filter structure. The residual of the Kalman filter is expressed in terms of the numerator and denominator polynomials of the system, and the *Kalman polynomial* (which is the characteristic polynomial of the Kalman filter) [15]. Using this residual model, a linear regression model relating the system input, the system output, and the residual is obtained. The resulting equation error is a colored noise which is modeled as an output of a MA model (formed of the Kalman polynomial) driven by the Kalman filter residual. This residual, or innovation process, is a zero-mean white noise process. It is interesting to note that the linear regression model derived from the PEM is also the same. However, unlike in the PEM case, in the HOLS method, the regression model is whitened by dividing both the denominator and numerator polynomials by the Kalman polynomial. As the Kalman polynomial is stable, the coefficients of the polynomials resulting from the division operation are truncated to some finite, but large, number of terms. Although the equation error in the PEM is a zero-mean white noise process, it is not linear in the unknown parameters (formed of the coefficients of the numerator, the denominator, and the Kalman polynomials). In the subspace method, a higher-order model is similarly derived from the Kalman filter state-space model by truncating the impulse response, which is expressed in terms of the Markov parameters, to some finite but large number of terms.

In many applications, such as fault diagnosis, performance monitoring, and controller design, a reduced-order model is desired. The commonly-used approaches for obtaining a reduced-order model include the balanced realization approach and the frequency-weighted least-squares estimator approach. The balanced realization approach hinges on the equality of the controllability and observability Grammians, whose diagonal entries are the singular values of the Hankel matrix. The ratio of the maximum to the minimum singular values of the Hankel matrix will be the minimum and a reduced-order model is obtained based on this minimal ratio of the singular values. On the other hand, in the frequency-weighted LS estimator method, an estimate of the true system model, termed here the reduced-order model, is derived from the higher-order model by minimizing a the frequency-weighted residual. This approach, explained in the Appendix, is direct and is the preferred one in many applications where the best fit between the high-order and the true model over a specified frequency band, generally over the bandwidth of the system, is more desirable than over the entire frequency range including that of the noise[6, 8]. In the ideal noise-free case, thanks to the pole-zero cancellations, the high-order transfer function model will be identical to the true system model. In the presence of noise, however, the two transfer functions will not be equal because of the noise artifact. In the noisy case, the frequency response of the high-order model over the bandwidth of the system is analyzed. It is assumed that (i) the frequency response of the system is smooth in contrast to that of the noise artifact which is jagged and (ii) the dominant peaks in the high-order frequency response come from that of the system. Using an energy measure (instead of the amplitude of the frequency response at some frequency), the selected frequencies are those at the peaks, and the smoothness of the regions of frequency response where there are no peaks is then exploited. An algorithm was specifically developed to automatically select the frequencies at which peaks occur in the high-order frequency response.

A prediction error method is developed using the linear regression model. However, unlike the high-order least-squares method, the equation error is not whitened by polynomial division. The residual is nonlinear in the coefficients of the numerator and the denominator polynomials of the system transfer function, as well as the coefficients of Kalman polynomial. The sum of the squares of the residual is minimized. If the true system is contained in the chosen model structure, and if the prediction error sequence is a white noise, then the estimator will be consistent and asymptotically efficient. The estimates are obtained iteratively using the Newton–Raphson, and simpler Pseudo-Linear Regression (PLR) and Least Mean Squares (LMS) schemes are presented.

The subspace identification has received a lot of attention in recent years, as it is numerically efficient and robust, and requires minimal *a priori* information, such as the structure of the system, that is the model order of the numerator and the denominator polynomials and the delay[4, 5]. The only design parameter is the threshold values for truncation of the singular values. The subspace system identification, unlike other methods including prediction error methods, can handle seamlessly both multi-input–multi-output and single-input–single-output system identification without a need for choosing model structures and corresponding parameterization. This method estimates directly the state-space model for the system and is well suited for applications including filtering and controller design, especially the model predictive controller used widely in the process industry. However, although the identified state-space model is *similar* to that of the system (that is the rank, the eigenvalues, the determinant, and the trace are identical), the states of the system and those associated with the identified model may not have the same physical meaning. It is easier for a practitioner to implement this scheme as there are a small number of design parameters to be chosen. The subspace algorithm does not require nonlinear searches in the parameter space but is based on computationally reliable tools such as the SVD. Subspace identification is non-recursive based on robust SVD-based numerical methods and avoids problems with optimization and possible local minima. The model order selection process is simple, as it is based merely on truncating the "low singular values" of the estimated Hankel matrix. Two versions of the subspace method, namely the prediction and the innovation forms, are given. The prediction and the innovation forms use respectively the predictor model structure and the innovation model structure of the Kalman filter. The predictor form is numerically stable when the system model is poorly damped or the system is close to being unstable. The innovation form of the subspace model is derived from the predictor form.

7.2 System Model

A system may be described by a state-space model and an input–output model, such as the transfer function, the difference equation, or the linear regression model.

7.2.1 State-Space Model

The state-space model $(\boldsymbol{A}, \boldsymbol{B}, \boldsymbol{E}_w, \boldsymbol{C})$, termed a *process form*, has a single input and single output with a disturbance and a measurement noise given by

$$\begin{aligned} \boldsymbol{x}(k+1) &= \boldsymbol{A}\boldsymbol{x}(k) + \boldsymbol{B}r(k) + \boldsymbol{E}_w w(k) \\ y(k) &= \boldsymbol{C}\boldsymbol{x}(k) + v(k) \end{aligned} \tag{7.1}$$

where $\boldsymbol{x}(k)$ is a $(nx1)$ vector of states given by $\boldsymbol{x}(k) = \begin{bmatrix} x_1(k) & x_2(k) & x_3(k) & \dots & x_n(k) \end{bmatrix}^T$, $r(k)$, $w(k)$, and $v(k)$ are scalars representing the input, the disturbance (system noise), and measurement noise, respectively; $\boldsymbol{A}$ is a (nxn) transition matrix, $\boldsymbol{B}$ is a $(nx1)$ input vector, $\boldsymbol{E}_w$ is a $(nx1)$ disturbance entry vector, and $\boldsymbol{C}$ is a $(1xn)$ output vector. The direct transmission term, that is, the "D-term," is assumed to be zero in the state-space model of Eq. (7.1). It is also assumed that the system is controllable and observable.

7.2.2 Assumptions

7.2.2.1 Controllability and Observability

Minimal realization: The state-space model is controllable and observable so that its input–output behavior is identical to that of its transfer function model. The controllability and observability of the state-space model ensures that it is a minimal realization of the transfer function model. The realization is called minimal because it describes the system with the minimum number of states. The order of the system, which is the minimum number of the state variable, will also be the minimum. For a single-input–single-output system, the transfer function derived from the state-space model will have no pole-zero cancellations.

Input-state interaction: It is controllable if the inputs affect all the states. The states can be steered along any given trajectory in finite time by manipulating the inputs.

Output-state relationship: It is observable if we can "see," "hear," and "smell" what happens inside the system. The initial state of the system can be determined from finite input and the output data. The above assumptions can be precisely stated as follows:

- $(\boldsymbol{A}, \boldsymbol{B})$ and $(\boldsymbol{A}, \boldsymbol{E}_w)$ are controllable so that input $r(k)$ and the disturbance $w(k)$ affect the state $\boldsymbol{x}(k)$.
- $(\boldsymbol{A}, \boldsymbol{C})$ is observable so that $\boldsymbol{x}(k)$ can be determined from the output $y(k+i)$, $r(k+i)$ for $i = 0, 1, 2, \dots, n-1$ in the absence of the disturbance and the measurement noise, that is when $w(k)$ and $v(k)$ are both zero.

7.2.3 Frequency-Domain Model

The frequency-domain expression relating the input $r(z)$ and output $y(z)$ is given by

$$y(z) = G(z)r(z) + \vartheta(z) \tag{7.2}$$

where $G(z) = \boldsymbol{C}(z\boldsymbol{I} - \boldsymbol{A})^{-1}\boldsymbol{B} = \dfrac{N(z)}{D(z)}$ is the transfer function,

$$D(z) = |(z\boldsymbol{I} - \boldsymbol{A})| = 1 + \sum_{i=1}^{n} a_i z^{-i}$$

$$N(z) = \left\| \begin{bmatrix} z\boldsymbol{I} - \boldsymbol{A} & \boldsymbol{B} \\ -\boldsymbol{C} & 0 \end{bmatrix} \right\| = \sum_{i=1}^{n} b_i z^{-i}; \quad \frac{N_w(z)}{D(z)} = \frac{\sum\limits_{i=0}^{n} b_{wi} z^{-i}}{1 + \sum\limits_{i=0}^{n} a_i z^{-i}} = \boldsymbol{C}(z\boldsymbol{I} - \boldsymbol{A})^{-1}\boldsymbol{E}_w,$$

and the output error $\vartheta(z)$ is given by:

$$\vartheta(z) = \boldsymbol{C}(z\boldsymbol{I} - \boldsymbol{A})^{-1}\boldsymbol{E}_w w(z) + v(z) = \frac{N_w(z)w(z) + D(z)v(z)}{D(z)} \tag{7.3}$$

The ideal noise-free output of the system is given by:

$$y^0(z) = G(z)r(z) \tag{7.4}$$

Multiplying both sides of Eq. (7.2) by $D(z)$ and using the expression $G(z) = \frac{N(z)}{D(z)}$, yields:

$$D(z)y(z) = N(z)r(z) + D(z)\vartheta(z) \tag{7.5}$$

Rewriting so that $y(z)$ appears on the left and the rest on the right yields:

$$y(z) = (1 - D(z))\,y(z) + N(z)r(z) + \upsilon(z) \tag{7.6}$$

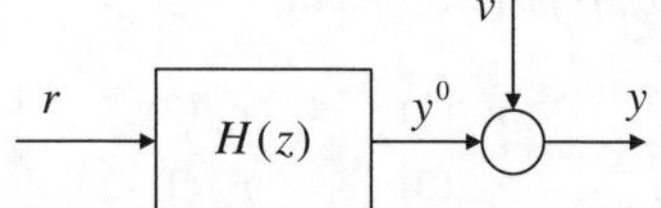

Figure 7.1 Frequency domain model

where $\upsilon(z) = D(z)\vartheta(z)$. Figure 7.1 shows the frequency-domain model.

7.2.3.1 Equation Error

The error term $\upsilon(z)$, termed the *equation error*, is a sum of two random processes, namely the colored disturbance and measurement noise. It is the combined effects of the disturbance $w(z)$ and the measurement noise $v(z)$ on the output $y(z)$ given by

$$\upsilon(z) = N_w(z)w(z) + D(z)v(z) \tag{7.7}$$

Using (7.6) the error term $\upsilon(z)$ may be expressed as follows:

$$\upsilon(z) = D(z)y(z) - N(z)r(z). \tag{7.8}$$

7.2.3.2 Difference Equation Model

The difference equation model derived from the transfer function model (7.6) is:

$$y(k) = -\sum_{i=1}^{n} a_i y(k-i) + \sum_{i=1}^{n} b_i r(k-i) + \upsilon(k) \tag{7.9}$$

7.2.4 *Input Signal for System Identification*

In system identification, the input to the system determines whether we can identify the system and if so how good the identified model is. Typical signals used in identification include a step function, a square wave, sum of sinusoids, a binary random signal, a white or a colored noise process, and a Pseudo-Random Binary Signal (PRBS) signal. The most important requirement is that the input signal must be persistently exciting.

7.2.4.1 Persistence of Excitation

It is assumed that the input signal $r(k)$ is persistently exciting. The persistently exciting input signal implies that it is "rich" so that all the modes of the system can be excited. The persistently exciting condition ensures that the complete dynamic behavior is of the system is completely captured by the output. This condition is crucial for many applications including the system identification, condition monitoring, and simulation-based method to predict the behavior of the system under various operating scenarios.

Definition *A signal $\{r(k)\}$ with power spectral density $P_{rr}(f)$ is said to be persistently exciting of order M_r if*

- *The $M_r x M_r$ correlation matrix, $\mathbf{r}_{rr}$ is positive definite*

$$\mathbf{r}_{rr} = \begin{bmatrix} r_{rr}(0) & r_{rr}(1) & r_{rr}(2) & . & r_{rr}(M_r-1) \\ r_{rr}(1) & r_{rr}(0) & r_{rr}(1) & . & r_{rr}(M_r-2) \\ r_{rr}(2) & r_{rr}(1) & r_{rr}(0) & . & r_{rr}(M_r-3) \\ . & . & . & . & . \\ r_{rr}(M_r-1) & r_{rr}(M_r-2) & r_{rr}(M_r-3) & . & r_{rr}(0) \end{bmatrix} > 0 \quad (7.10)$$

where the correlation is defined as $r_{rr}(m) = \lim_{N\to\infty} \frac{1}{N} \sum_{k=0}^{N} r(k)r(k-m)$

- *The frequency-domain equivalent condition (7.10) is that the power spectral density $P_{rr}(f)$ is non-zero at least at M_r distinct frequencies:*

$$P_{rr}(f) \neq 0 \text{ for } \{f_i : i = 1, 2, \ldots, M_r\} \quad (7.11)$$

Proof: The power spectral density of the Fourier transform of the correlation function $P_{rr}(f) = \Im\left(r_{rr}(m)\right)$. Taking the Fourier transform of the correlation matrix $\mathbf{r}_{rr}$ it is shown that $P_{rr}(f)$ satisfies the non-zero condition (7.11) [11].

- If $r(k)$ is persistently exciting of the order M_r, then $y(k)$ which is an output given in the frequency domain by $y(z) = G(z)r(z)$ is persistently exciting of the order M_y given by:

$$M_r - n_{zeros} \leq M_y \leq M_r \quad (7.12)$$

where n_{zeros} is the number of zeros of $G(z)$ on the unit circle.

Proof: The power spectral density $y(k)$ is $P_{yy}(f) = |G(f)|^2 P_{rr}(f)$. Using the equivalent definition of persistent excitation (7.11), the inequality governing the order M_y of persistence of excitation condition (7.12) follows.

7.2.4.2 Illustrative Examples: Persistency of a Signal

Typical signals, such as a delta function, step function, sum of sinusoids, pseudo-random binary sequence, and colored noise process are analyzed.

Example 7.1 An impulse function
The signal $r(k)$ is a delta function given by:

$$r(k) = \delta(k) = \begin{cases} 1 & k = 0 \\ 0 & else \end{cases} \quad (7.13)$$

The correlation of the delta function is:

$$r_{rr}(m) = \lim_{N\to\infty} \frac{1}{N} \sum_{k=0}^{N} r(k)r(k-m) = 0 \quad (7.14)$$

Since the correlation is zero, delta function is not a persistently exciting signal.

Example 7.2 Step function
The signal $r(k)$ is given by:

$$r(k) = \begin{cases} 1 & k \geq 0 \\ 0 & else \end{cases}$$

The correlation function is:

$$r_{rr}(m) = 1 \text{ for all } m \tag{7.15}$$

The power spectral density is:

$$P_{rr}(f) = \delta(f) \tag{7.16}$$

The correlation matrix (7.10) will have a unity rank, and the power spectral density is non-zero only at $f = 0$. Therefore the step function is persistent of order 1.

Example 7.3 Sum of sinusoids
The signal $r(k)$ is given by:

$$r(k) = \sum_{i=1}^{5} \sin(2\pi f_i k)$$

where $f_i = i/64 \quad : \quad i = 1, 2, 3, 4, 5$. The correlation function is:

$$r_{rr}(m) = \frac{1}{2} \sum_{i=1}^{5} \cos(2\pi f_i m) \tag{7.17}$$

The power spectral density is:

$$P_{rr}(f) = \frac{1}{4} \sum_{i=1}^{5} [\delta(f - f_i) + \delta(f + f_i)] \tag{7.18}$$

The correlation matrix (7.10) will have a rank 5, and the power spectral density is non-zero at five distinct frequencies $\{f_i : i = 1, 2, 3, 4, 5\}$. The sum of five sinusoids is a persistently exciting signal of order 5.

Example 7.4 Colored noise process
A colored noise is generated as an output of a low-pass filter driven by a zero-mean white noise process $v(k)$ with variance $\sigma^2 = 0.01$ given by:

$$r(z) = H(z)v(z) \tag{7.19}$$

where $H(f) = \dfrac{bz^{-1}}{1 - az^{-1}}$, $a = 0.7$ and $b = 0.3$. The correlation function $r_{rr}(m)$ is

$$r_{rr}(m) = \left(\frac{b^2}{1 - a^2} \right) a^{|m|} \tag{7.20}$$

The power spectral density is

$$P_{rr}(f) = \sigma^2 |H(f)|^2 \tag{7.21}$$

where $H(f) = H(z = e^{j2\pi f})$.

The colored noise is persistently exciting of any order as the power spectral density is non-zero for an infinite number of distinct frequencies.

Example 7.5 Pseudo-random sequence (PRBS)
The PRBS is a random signal in the sense that it enjoys the properties of a white noise sequence. It is "pseudo" because it is deterministic and periodic, unlike a random white noise sequence. A PRBS with values +1 and -1 is generated using the state space mode given by:

$$\begin{gathered} \boldsymbol{x}(k+1) = \begin{bmatrix} 0 & 0 & 1 & 0 & 1 \\ 1 & 0 & 0 & 0 & 0 \\ 0 & 1 & 0 & 0 & 0 \\ 0 & 0 & 0 & 0 & 0 \\ 0 & 0 & 0 & 1 & 0 \end{bmatrix} \boldsymbol{x}(k) \\ r(k) = \begin{bmatrix} 0 & 0 & 0 & 0 & 2 \end{bmatrix} \boldsymbol{x}(k) - 1 \end{gathered} \tag{7.22}$$

where the dynamic equation governing the states are computed using binary (modulo-2) arithmetic. It is a periodic signal with a period $M = 2^5 - 1 = 31$. The auto-correlation function $r_{rr}(m)$, and the power spectral density $P_{rr}(f)$ of the waveform $r(k)$, over one period M are similar to that of a white noise sequence given by:

$$r_{rr}(m) = \begin{cases} 1 & \text{if } m = 0 \\ -\frac{1}{M} & \text{otherwise} \end{cases} \tag{7.23}$$

$$P_{rr}(f) = \frac{1}{M^2}\delta(f) + \frac{M+1}{M^2}\sum_{\ell=1}^{M-1} \delta\left(f - \frac{\ell}{M}\right) \tag{7.24}$$

The power spectral density is non-zero over M distinct frequencies. Hence it is persistently exciting of order M.

Figure 7.2 shows the persistent excitation properties of a sum of sinusoids, a PRBS, and a colored noise process. Subfigures (a), (d), and (g) show respectively the input signals sum of sinusoids, the PRBS, and the colored noise, subfigures (b), (e), and (h) show respectively their correlation functions, and subfigures (c), (f), and (i) show their respective power spectral densities. The non-zero power spectral density is a quick indicator of the order of the persistent excitation. From the auto-correlation function it not easy to determine visually the order of persistency. It shows clearly that the sum of five sinusoids has the order of 5, the PRBS has an order of 31, and the colored noise has an infinitely large order.

Comments *The persistent excitation condition is important for ensuring that the estimates of the parameter of a system are consistent, especially when the input–output data is noisy. In the case of noisy data the number of data samples must be large enough to average out the noise effects on the estimates. For an* n*th order system, it has been shown that the input signal must be persistently exciting of order* 2n.

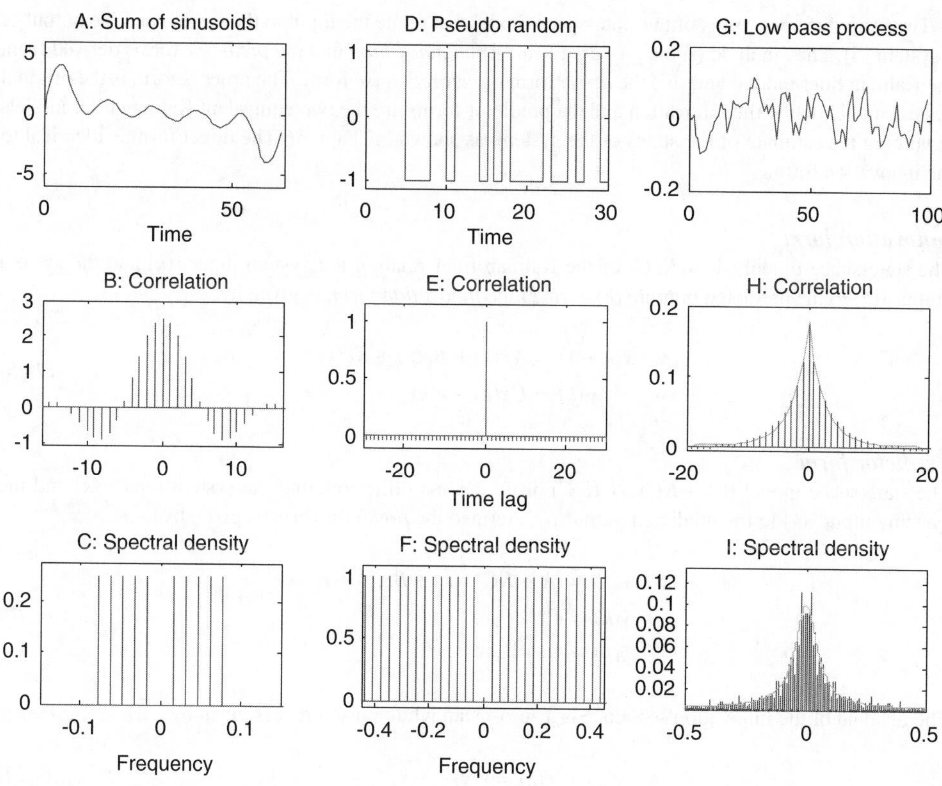

Figure 7.2 Examples of persistently exciting signals

A quantitative measure which complements the order of persistently excitation, is the condition number of the correlation matrix (7.10). The condition number is also indicative of the computational efficiency or the robustness of the estimation algorithms such as the least-squares method.

In the ideal noise free case (or large signal to noise ratio case), the system parameters may be identified from finite data samples. In the ideal case, it is not necessary that the input signal is persistently exciting. For example, one can identify the nth *order system from* 2n *data points by exciting a system with an input which is not persistently exciting or persistently exciting of an order of less than* 2n. *For example, impulse and step input function are widely used to determine the system parameters such as time delay, static gain, overshoot, and natural frequency.*

7.3 Kalman Filter-Based Identification Model Structure

The error term (7.7) in the input–output model (7.6) is a sum of two colored noises representing the effect of the disturbance and the measurement noise on the output.

Most algorithms for system identification, such as the least-squares method and the prediction error method require that the error term in the input–output model is a zero-mean white noise process. Hence a different, albeit indirect, approach to obtain the input–output model satisfying this requirement is addressed.

There are different forms of the state-space model that relate the input to the system and the output of a system [5]. They include (i) the *process* form, (ii)the *innovation* and the *predictor* forms derived from the Kalman filter model, and (iii) the *direct* form or *colored noise* form. The process form is the original system model, while the innovation and the predictor forms are the two equivalent Kalman filter models governing the estimate of the states of the system, as shown inChapter 6 The direct form is identical to the innovation form.

Innovation form

The state-space model $(\boldsymbol{A}, \boldsymbol{B}, \boldsymbol{K}, \boldsymbol{C})$ of the Kalman filter relating the system input $r(k)$ and the system output $y(k)$ to the predicted output $\hat{y}(k)$, termed the *innovation form*, is given by:

$$\begin{aligned} \hat{\boldsymbol{x}}(k+1) &= \boldsymbol{A}\hat{\boldsymbol{x}}(k) + \boldsymbol{B}r(k) + \boldsymbol{K}e(k) \\ y(k) &= \boldsymbol{C}\hat{\boldsymbol{x}}(k) + e(k) \end{aligned} \tag{7.25}$$

Predictor form

The state-space model $((\boldsymbol{A} - \boldsymbol{KC}), \boldsymbol{B}, \boldsymbol{K}, \boldsymbol{C})$ of the Kalman filter relating the system input $r(k)$ and the system output $y(k)$ to the predicted output $\hat{y}(k)$, termed the *predictor form*, is given by

$$\begin{aligned} \hat{\boldsymbol{x}}(k+1) &= (\boldsymbol{A} - \boldsymbol{KC})\hat{\boldsymbol{x}}(k) + \boldsymbol{B}r(k) + \boldsymbol{K}y(k) \\ \hat{y}(k) &= \boldsymbol{C}\hat{\boldsymbol{x}}(k) \\ y(k) &= \hat{y}(k) + e(k) \end{aligned} \tag{7.26}$$

The residual of the innovation sequence is a zero-mean white noise process given by

$$e(k) = y(k) - \hat{y}(k) \tag{7.27}$$

where $\hat{\boldsymbol{x}}(k)$ and the predictor $\hat{y}(k)$ are respectively the best estimate of $\boldsymbol{x}(k)$ and the measurement $y(k)$ given all the input–output data until the (recent past) time instant $k-1$, that is, the outputs $\{y(i) : i = 0, 1, 2, \dots k-1\}$ and the inputs $\{r(i) : i = 0, 1, 2, \dots k-1\}$, $e(k)$ is the residual which is a zero-mean white noise process, termed an *innovation* process. In other words $\hat{y}(k)$ is the conditional mean of $y(k)$, given all the past measurements

$$\hat{y}(k) = E\left[y(k)\,|y(0), y(1), \dots, y(k-1)\right] \tag{7.28}$$

7.3.1 Expression for the Kalman Filter Residual

Figure 7.3 shows the generation of the Kalman filter residual from the input–output system model. The frequency-domain expression relating inputs $r(z)$ and $y(z)$ to the residual $e(z)$ is given by (see Chapter 6):

$$e(z) = \frac{D(z)}{F(z)}y(z) - \frac{N(z)}{F(z)}r(z) \tag{7.29}$$

The above model (7.29) is termed the *residual model.*

Adding and subtracting $y(z)$ on the right-hand sides of Eq. (7.29), and rearranging so that $y(z)$ appears on the left and the rest of the terms on the right-hand side, we get

$$y(z) = \left(1 - \frac{D(z)}{F(z)}\right)y(z) + \frac{N(z)}{F(z)}r(z) + e(z) \tag{7.30}$$

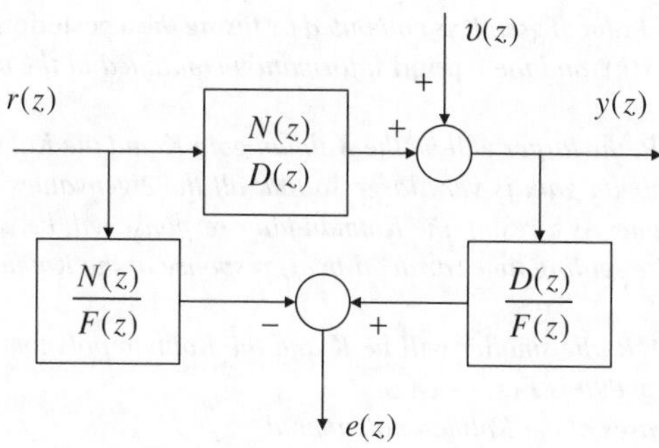

Figure 7.3 Generation of the Kalman filter residual

Multiplying (7.29) by $F(z)$ and rearranging yields:

$$D(z)y(z) = N(z)r(z) + F(z)e(z) \tag{7.31}$$

where the Kalman polynomial $F(z) = |\boldsymbol{A} - \boldsymbol{KC}| = 1 + \sum_{i=1}^{n} c_i z^{-i}$, which is the characteristic polynomial of the Kalman filter, $\{c_i\}$ are the coefficients of the denominator of the Kalman filter.

Subtracting both sides of Eq. (7.31) from $y(z)$ and rearranging gives:

$$y(z) = (1 - D(z))\,y(z) + N(z)r(z) + F(z)e(z) \tag{7.32}$$

Taking the inverse transform of Eq. (7.32), we get

$$y(k) = -\sum_{i=1}^{n} a_i y(k-i) + \sum_{i=1}^{n} b_i r(k-i) + e(k) + \sum_{i=1}^{n} c_i e(k-i) \tag{7.33}$$

Substituting $e(k) = y(k) - \hat{y}(k)$ and simplifying yields the predictor equation

$$\hat{y}(k) = -\sum_{i=1}^{n} c_i \hat{y}(k-i) + \sum_{i=1}^{n} (c_i - a_i)y(k-i) + \sum_{i=1}^{n} b_i r(k-i) \tag{7.34}$$

7.3.1.1 Kalman Filter Residual and the Equation Error

Comparing the expressions for the equation error term $\upsilon(z)$ (7.8) for the process form (7.6) and the Kalman filter residual $e(z)$ given in Eq. 7.29 it can be deduced that:

$$\upsilon(z) = N_w(z)w(z) + D(z)v(z) = F(z)e(z) \tag{7.35}$$

From Eq. (7.35), it is clear that the use of a Kalman filter whitens the equation error of the process form. Moreover, the required whitening process consists simply of dividing the equation error $\upsilon(z)$ by the stable Kalman polynomial $F(z)$, or equivalently inverse-filtering $\upsilon(z)$ by the stable filter $F(z)$. This, in fact, provides the prime motivation for using the Kalman filter-based, rather than the process form (the actual system model).

Comments *The optimal Kalman gain* $\mathbf{K}$ *is computed by fusing the* a posteriori *information provided by the residual* $e(k) = y(k) - \hat{y}(k)$, *and the* a priori *information contained in the model* $(\mathbf{A}, \mathbf{B}, \mathbf{C})$ *employed in the Kalman filter.*

The larger the ratio Q/R, *the larger will be the Kalman gain* $\mathbf{K}$ *and the Kalman polynomial* $F(z)$ *will be closer to unity. The Kalman gain is very large so that all the eigenvalues will migrate towards the origin of the complex z-plane. As a result, the Kalman filter response will be very fast. We will consider an extreme case of the finite settling time (or dead-beat) response of the Kalman filter corresponding to the case when* $F(z) \to 1$.

The smaller the ratio Q/R, *the smaller will be* $\mathbf{K}$ *and the Kalman polynomial* $F(z)$ *will be closer to the system polynomial* $D(z)$, *that is* $F(z) \to D(z)$.

Consider two extreme cases of the Kalman polynomial

$$F(z) = |\mathbf{A} - \mathbf{KC}| = 1 + \sum_{i=1}^{n} c_i z^{-i} \text{ namely } F(z) \to D(z) \text{ and } F(z) \to 1.$$

Case 1 $F(z) \to D(z)$

When the variance Q *of the disturbance* $w(k)$ *is very small compared to the variance* R *of the measurement noise* $v(k)$, *then the Kalman gain is very small and the Kaman filter response will be sluggish. In this case*

$$F(z) \to D(z) \text{ when } Q/R \to 0 \tag{7.36}$$

From Eqs. (7.30) and (7.35) we deduce the following:

$$y(z) = \frac{N(z)}{D(z)} r(z) + e(z) \tag{7.37}$$

$$\upsilon(z) \to D(z)e(z) \tag{7.38}$$

Case 2 $F(z) \to 1$

In this case all the eigenvalues of the Kalman filter matrix $\mathbf{A} - \mathbf{KC}$ *are zero, that is the coefficients of* $F(z)$ *are all zero, that is* $c_i = 0$ *for all i. It can be deduced from Eq. (7.30) that the Kalman filter model becomes:*

$$y(z) = (1 - D(z))\, y(z) + N(z)r(z) + e(z) \tag{7.39}$$

From Eq. (7.34) we get

$$\upsilon(z) \to e(z) \tag{7.40}$$

7.3.2 Direct Form or Colored Noise Form

Instead of using the state-space model (with the disturbance driving the dynamical equations and the additive measurement noise entering the measurement equation), taking a cue from the Kalman filer approach, we may simply assume that the error term is a colored noise generated by filtering a zero-mean white noise process. Then a whitening filter, which is the inverse of the colored noise filter, is used to generate a predictor of the output. This direct approach is used [7] as it is simple to grasp, intuitive, and does not require the in-depth analysis of the Kalman filter. 7.3.2.1 Input–Output Model.

Consider the input–output model (7.2):

$$y(z) = G(z)r(z) + \vartheta(z) \tag{7.41}$$

where the output error $\vartheta(z) = H_e(z)e(z)$ is *assumed* to be a colored noise generated as an output of a LTI system $H_e(z)$ driven by some zero-mean white noise process, instead of two white noise processes $w(k)$ and $v(k)$. The filter $H_e(z)$ is assumed to be a minimum phase whose numerator and the denominator are both monic polynomials (that is, the leading coefficient is unity). The error term is whitened by multiplying both sides by the inverse of the model that generates the error term $H_e^{-1}(z)$. Rewriting the resulting whitened equation so that with $y(z)$ is on the left-hand side and the rest on the right hand, we get

$$y(z) = \left(1 - H_e^{-1}(z)\right) y(z) + H_e^{-1}(z)G(z)r(z) + e(z) \tag{7.42}$$

The expression for the predictor is:

$$\hat{y}(z) = \left(1 - H_e^{-1}(z)\right) y(z) + H_e^{-1}(z)G(z)r(z) \tag{7.43}$$

After substituting $G(z) = \dfrac{N(z)}{D(z)}$, Eq. (7.42) will be exactly equal to that derived using the Kalman filter approach given by Eq. (7.30) if we assume $H_e(z) = \dfrac{F(z)}{D(z)}$. It also justifies the assumption that $H_e(z)$ is minimum phase and is a ratio of polynomials whose leading coefficients are unity. The colored noise approach is widely used as it gives directly an expression for the predictor without using the Kalman filter formulation. The justification for the colored form stems from the predictor or the innovation form.

Comment *The Kalman filter-based identification model may be interpreted to be an inverse system generating the innovation sequence $e(k)$ or a whitening filter realization of a state-space model driven disturbance and the measurement noise corrupting the measurement given by Eq. (7.1). Figure 7.4 shows the system and the Kalman filter connected in a cascade to generate the residual (which is a zero-mean white noise process with minimal variance) and the predictor. The input to the Kalman filter includes the measurement corrupted by colored noise generated by the disturbance and the measurement noise. Predictor Eq. (7.26) is asymptotically stable and its transient behavior is "good" even if (i) the system model $G(z)$ is unstable or (ii) system is stable but its transient response is poor.*

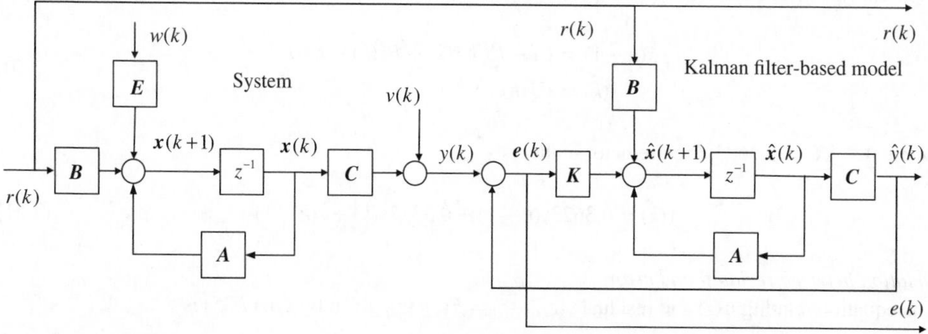

Figure 7.4 The system and the Kalman filter-based identification model

7.3.3 Illustrative Examples: Process, Predictor, and Innovation Forms

Examples of input–output models derived from the process form (7.9), and the innovation form (7.33) and the predictor form (7.34) are compared, and further the relationship between the equation error $\upsilon(k)$ for the process form (7.9) and the innovation form given by Eq. (7.35) is verified.

Example 7.6 First-order process form
Process form: The state-space model of the system is:

$$\begin{aligned} x(k+1) &= Ax(k) + Bu(k) + E_w w(k) \\ y(k) &= Cx(k) + v(k) \end{aligned} \tag{7.44}$$

where $A = 0.9$, $B = 1$, $C = 1$, $E_w = 1$, $Q = R = 1$.
ARMAX model: The ARMAX model is

$$y(k) = 0.9y(k-1) + r(k-1) + \upsilon(k) \tag{7.45}$$

The equation error $\upsilon(k)$ is:

$$\upsilon(k) = v(k) - 0.9v(k-1) + w(k-1) \tag{7.46}$$

Substituting for $\upsilon(k)$ the difference equation model is

$$y(k) = 0.9y(k-1) + r(k-1) + v(k) - 0.9v(k-1) + w(k-1) \tag{7.47}$$

Innovation form: The Kalman filter expressed in innovation form is:

$$\begin{aligned} \hat{x}(k+1) &= A\hat{x}(k) + Bu(k) + Ke(k) \\ y(k) &= C\hat{x}(k) + e(k) \end{aligned} \tag{7.48}$$

where the steady-state Kalman gain $K = 0.5377$.
ARMAX model: The ARMAX model is:

$$y(k) = 0.9y(k-1) + r(k-1) + e(k) - 0.3623e(k-1) \tag{7.49}$$

Predictor form: The Kalman filter in predictor form is:

$$\begin{aligned} \hat{x}(k+1) &= (A - KC)\hat{x}(k) + Br(k) + Ky(k) \\ \hat{y}(k) &= C\hat{x}(k) \end{aligned} \tag{7.50}$$

where $A - KC = 0.3623$.The predictor model is:

$$\hat{y}(k) = 0.3623\hat{y}(k-1) + 0.5377y(k) + r(k-1) \tag{7.51}$$

Relation between residual and error:
The equation relating $\upsilon(z)$ and residual $e(z)$, $\upsilon(z) = F(z)e(z)$, given by Eq. (7.35) is

$$v(k) - 0.9v(k-1) + w(k-1) = e(k) - 0.3623e(k-1) \tag{7.52}$$

Example 7.7 Second-order process form
Process form: the state-space model is given by:

$$\begin{aligned} \boldsymbol{x}(k+1) &= \begin{bmatrix} 1.7913 & 1 \\ -0.81 & 0 \end{bmatrix} \boldsymbol{x}(k) + \begin{bmatrix} 1 \\ 0 \end{bmatrix} r(k) + \begin{bmatrix} 1 \\ 0 \end{bmatrix} w(k) \\ y(k) &= [\,1 \quad 0\,]\boldsymbol{x}(k) + v(k) \end{aligned} \tag{7.53}$$

where $w(k)$ and $v(k)$ are zero-mean white noise processes with $Q = R = 1$.
ARMAX model: The ARMAX model is

$$\begin{aligned} y(k) &= 1.7913y(k-1) - 0.81y(k-2) + r(k-1) \\ &\quad + w(k-1) + v(k) - 1.7913v(k-1) + 0.81v(k-2) \end{aligned} \tag{7.54}$$

Predictor form: The predictor form (7.26) is given by

$$\begin{aligned} \hat{\boldsymbol{x}}(k+1) &= \begin{bmatrix} 0.7003 & 1 \\ -0.2121 & 0 \end{bmatrix} \hat{\boldsymbol{x}}(k) + \begin{bmatrix} 1 \\ 0 \end{bmatrix} r(k) + \begin{bmatrix} 1.0910 \\ -0.5979 \end{bmatrix} y(k) \\ \hat{y}(k) &= [\,1 \quad 0\,]\hat{\boldsymbol{x}}(k) \end{aligned} \tag{7.55}$$

ARMAX model: The ARMAX model is

$$\begin{aligned} \hat{y}(k) &= 0.7003\hat{y}(k-1) - 0.2121\hat{y}(k-2) \\ &\quad + 1.091y(k-1) - 0.5979y(k-2) + r(k-1) \end{aligned} \tag{7.56}$$

Innovation form: The innovation form is given by

$$\begin{aligned} \hat{\boldsymbol{x}}(k+1) &= \begin{bmatrix} 1.7913 & 1 \\ -0.81 & 0 \end{bmatrix} \hat{\boldsymbol{x}}(k) + \begin{bmatrix} 1 \\ 0 \end{bmatrix} r(k) + \begin{bmatrix} 1.0910 \\ -0.5979 \end{bmatrix} e(k) \\ y(k) &= [\,1 \quad 0\,]\boldsymbol{x}(k) + e(k) \end{aligned} \tag{7.57}$$

ARMAX model: The ARMAX model is:

$$\begin{aligned} y(k) &= 1.7913y(k-1) - 0.81y(k-2) + r(k-1) \\ &\quad + e(k) - 0.7003e(k-1) + 0.2121e(k-2) \end{aligned} \tag{7.58}$$

Relation between residual and error:
The equation relating $\upsilon(z)$ and residual $e(z)$, $\upsilon(z) = F(z)e(z)$, given by Eq. (7.35) is

$$\begin{aligned} w(k-1) + v(k) - 1.7913v(k-1) + 0.81v(k-2) \\ = e(k) - 0.7003e(k-1) + 0.2121e(k-2) \end{aligned} \tag{7.59}$$

Figure 7.5 compares the process form and the predictor forms of the second-order system. Subfigure (a) compares the output $y(k)$ generated by the process form and that generated by the predictor or the innovation form. Subfigure (c) compares the residual of the Kalman filter $e(z)$ and the filtered equation error $\dfrac{\upsilon(z)}{F(z)} = \dfrac{D(z)\vartheta(z)}{F(z)}$ given in Eq. (7.35). Subfigure (b) shows the auto-correlations of the residual and the filtered equation error and subfigure (d) shows the output and its prediction.

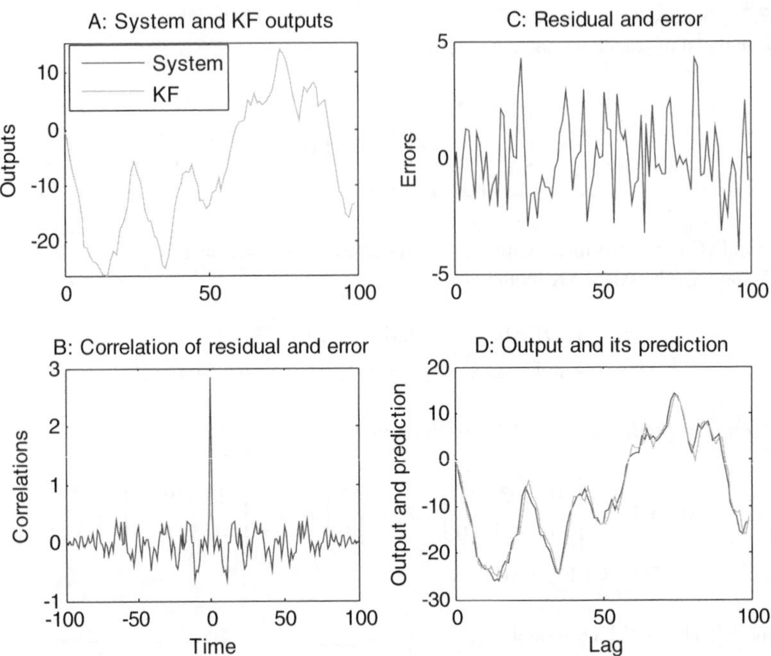

Figure 7.5 Comparison of the outputs of the process and the predictor forms

From the figure it can be concluded that the output and the equation error term generated by the process are identically equal to those generated by the innovation or the predictor form generated by the Kalman filter.

Remark It has been shown [5, 14] that the input–output models derived from the state-space model of the system, namely the transfer function $G(z)$ and the impulse response $h(k) = CA^{k-1}B$, and those derived from the Kalman filter-based identification model are identical if the assumptions of controllability and observability stated in Section 72.2 hold, and the input is persistently exciting as stated in Section 7.2.4.

7.3.4 Models for System Identification

The Kalman filter model is formed of the internal model of the system driven by the residual (the error between the system output and its prediction by the Kalman filter) [15].

A key requirement of an identification process is that the error between the actual system output and that generated by the assumed model is a zero-mean white noise process when both the system and its model are driven by the same input. The Kalman filter-based model meets this requirement while the process form (the "raw" system model structure) may not.

The state-space models of the Kalman filter, namely the innovation form (7.25) and the predictor form (7.26), and the residual model (7.29) derived from the state-space models are well suited for system identification compared to the "raw" process form. The error term in the Kalman filter-based model (7.31), or equivalently (7.33), is a colored noise (filtered white noise process), whereas the error term derived from the process form (7.7) is generated by two zero-mean white noise processes, namely the disturbance $w(k)$ and the measurement noise $v(k)$. The colored noise of the Kalman filter model may be easily whitened by an inverse filtering operation, and hence the innovation or the predictor form is preferable for system identification schemes. The predictor form is generally employed in applications where the system is poorly damped.

7.3.5 Identification Methods

Widely used and popular consistent identification methods, namely the prediction error and the high-order least squares, are developed using the residual model derived from the Kalman filter structure. An important requirement of an identification method is that it must give a consistent estimate. That is, the estimate must converge in probability to the true value when the number of data samples tends to infinity. Widely used and popular consistent identification methods include the following:

- Least-squares method.
- High-order least-squares method.
- Prediction error methods.
- Subspace identification methods.

Most of the identification methods, including the high-order least-squares method and the prediction error method, use the input–output residual model, while the subspace method uses the state-space Kalman filter model to derive the structure of the model set used in identification [5].

It is shown that the performance of the prediction error and that of the high-order least-squares method is very close. The quality of the performance of the identification methods is derived analytically for the high-order least-squares method as the prediction error methods and the subspace methods do not yield a closed-form solution for the estimates. The performance of the classical least-squares method is degraded when the equation error is a colored noise instead of a white noise process.

The choice of identification scheme depends upon the intended applications. In applications such as condition monitoring and fault diagnosis, the accuracy of the estimated parameters of the model should be high so as to meet the requirements of a low probability of false alarm and a high probability of correct decisions. For control design application, the quality of the identified model must be "good" in the sense that the closed-loop system formed of the controller, which is designed based on the identified model, meets the performance objectives. In a simulation-based scheme, the identified model must be accurate enough to predict the system behavior so that corrective action may be taken.

All of the identification methods are based on selecting a model, which is termed herein the *identification model*, so that its input–output dynamic behavior matches closely that of the given input–output data. The identification model is a two input, and one output, system where the inputs to the identification model are the system input and the system output (the output of the system is an input of the identification model), and its output is the predicted value of the system output. The parameters of the selected model structure are adjusted all-at-once (batch processing) or recursively updated (recursive processing) so that the output of the identification model is close to that of the system in some pre-specified measure such as the sum of the squares of the residuals. Figure 7.6 shows the system input–output data generated by the system to be identified, and the identification model. The parameters of the identification model are updated using an algorithm that is driven by the residual (error between the system output and that estimated by the identification model). The arrow indicates that the parameters of identification model are varied until $e(k)$ is "small."

In order to ensure that the identification model captures the behavior of the system from the given input–output data, which is generally corrupted by the measurement noise and the disturbances affecting the system, the structure of the identification model must be selected such that it includes the "true" model of the system. Generally, the initial structure of the identification model is derived from the physical laws governing the system, and the final model structure is determined using an iterative scheme consisting of identification, model validation, and selection of model order process.

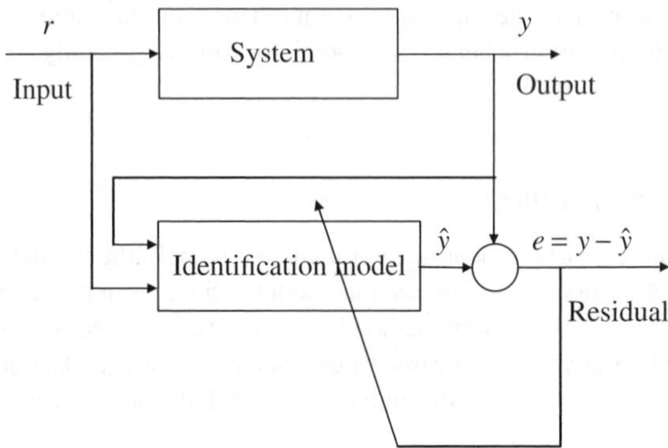

Figure 7.6 The identification model and its parameter update

The structure of the identification model is derived from that of the Kalman filter structure. Depending upon the intended application, either a state-space model or an input–output model is chosen. Further, the model may be based on the innovation or the predictor form:

- The least-squares, the high-order, and the prediction error methods employ an input–output identification model.
- The subspace method uses either the innovation or the predictor form of the state-space model.

7.3.5.1 Model Order Selection

Model order selection is a crucial step in identifying an accurate model of a system. An order of less than the true one will not capture the system behavior, while a model higher than the true value will capture both the system's dynamics as well as the effects of the noise and other artifacts.

The identification of a physical system is a challenging task as its model is generally stochastic, complex, and nonlinear. For practical purposes, the identified model needs to be simple and linear in a variety of practical applications such as controller design, condition monitoring, or fault diagnosis. In identifying a model of a physical system, the structure of the model may not be identical to that of the mathematical model derived from the physical laws due to various factors, including the presence of noise and fast dynamics.

The well-known Maximum Likelihood (*ML*) approach, although deemed best for parameter estimation for a selected model order, is unsatisfactory for model order selection as explained next. The optimal order $\hat{n}$, provided by the ML approach will be the maximum possible order since the quantity $\min_{\theta}\{\ell(y,\theta)\}$, where $\ell(y,\theta)$ is the log-likelihood function, is a monotonically decreasing function of the order n. In other words, the choice of an ML-optimal model order will result in over-fitting the data by also fitting the noisy data (instead of the noise-free data only) that is arbitrarily close to the estimated data as the order is increased. In general, the bias decreases and the variance of the estimate increases as the selected model order increases.

The approaches to model order selection are based on obtaining a proper tradeoff between a smaller bias and the larger variance with a higher model order, and the higher bias and smaller variance with a lower model order. The *Principle of Parsimony* is generally employed to obtain the tradeoff by using a criterion that includes a penalty term. The penalty term is a monotonic function of the number of estimated parameters to be estimated (which is function of the model order) so as to discourage over-fitting. A number of information-theoretic approaches have been proposed which address the model order selection problem in a general setup by, for example, penalizing not only the loss function but also the number of parameters. They include the well-known Akaike information criterion (AIC), Bayes Information criterion (BIC), and the minimum description length (MDL) criterion. All the above criteria are some function $g(.)$ of the log-likelihood function in either an additive or a multiplicative form, as shown below:

- *Additive form*

$$g\left(\log f(y\,|\,\boldsymbol{\theta})\right) = -\log f(y\,|\,\boldsymbol{\theta}) + \alpha \tag{7.60}$$

- *Multiplicative form*

$$g\left(\log f(y\,|\,\boldsymbol{\theta})\right) = \beta\left(-\log f(y\,|\,\boldsymbol{\theta})\right) \tag{7.61}$$

where $\log f(y\,|\,\theta)$ is the log-likelihood function of y given the parameter $\boldsymbol{\theta}$, M is the number of parameters, α and β are penalty terms which are monotonic functions of the number of parameters to be estimated.

Given an input–output data, a set of models is identified for a selected model order. For each model, the tradeoff measure $g\left(\log f(y\,|\,\boldsymbol{\theta})\right)$ is computed. The preferred model is the one with the minimum value of the measure $g\left(\log f(y\,|\,\boldsymbol{\theta})\right)$. The optimal model order $\hat{n}$ is:

$$\hat{n} = \arg\min\left\{g\left(\log f(y\,|\,\boldsymbol{\theta})\right)\right\} \tag{7.62}$$

7.4 Least-Squares Method

The Least-Squares (LS) method is widely used as it is (i) computationally efficient and (ii) there is a closed-form solution for the estimate and the quality of the estimate such as the bias and the covariance of the parameter estimation error. The LS estimate may be computed in batch or recursive processing mode. In the batch mode all the input–output data is stored and processed together at one time to compute the estimate, whereas in the recursive mode only a few data are processed at a time, and the estimate is available after several iterations. In current data computing devices, including microcomputers and digital signal processors, thanks to the higher speed of computation and the larger memory, the batch processing mode may be used even for online applications.

We will focus on the innovation form as the coefficients of the denominator of the system transfer function enter explicitly in the ARMAX model as given by Eq. (7.33):

$$\begin{aligned} y(k) &= -\sum_{i=1}^{n} a_i y(k-i) + \sum_{i=1}^{n} b_i r(k-i) + \upsilon(k) \\ \upsilon(k) &= e(k) + \sum_{i=1}^{n} c_i e(k-i) \end{aligned} \tag{7.63}$$

Linear Regression Model

Rewriting Eq. (7.63) as an inner product of a data vector formed of the inputs and vector formed of the unknown parameters we get:

$$y(k) = \boldsymbol{\psi}^T(k)\,\boldsymbol{\theta} + \upsilon(k) \tag{7.64}$$

where $\psi(k)$ is a $Mx1$ with $M = 2n$ data vector, which is formed of the past inputs, $r(k-i) : i - 1, 2, \dots, n$ and past outputs $y(k-i), i := 1, 2, \dots, n$, n is the order of the system:

$$\psi^T(k) = [\,-y(k-1) \quad -y(k-2) \quad . \quad -y(k-n) \quad r(k-1) \quad . \quad r(k-n)\,] \tag{7.65}$$

θ is a $Mx1$ vector of unknown model parameter given by

$$\theta = [\,a_1 \quad a_2 \quad . \quad a_n \quad b_1 \quad b_2 \quad . \quad b_n\,]^T \tag{7.66}$$

7.4.1 Linear Matrix Model: Batch Processing

A linear matrix model is given by

$$\boldsymbol{y}(k) = \boldsymbol{H}(k)\boldsymbol{\theta} + \boldsymbol{\upsilon}(k) \tag{7.67}$$

where $\boldsymbol{y}(k)$ is a $Nx1$ output vector, and $\boldsymbol{\upsilon}(k)$ is a $Nx1$ error vector given by $\boldsymbol{\upsilon}(k) = [\,\upsilon(k)\ \upsilon(k-1)\ .\ \upsilon(k-N+1)\,]^T$, $\boldsymbol{y}(k) = [\,y(k)\ y(k-1)\ .\ y(k-N+1)\,]^T$, and $\boldsymbol{H}(k)$ a NxM matrix termed data matrix, and $N \geq M$. The data matrix $\boldsymbol{H}(k)$ is a matrix formed by stacking the data vector, $\psi^T(k-i) : i = 0, 1, 2, \dots, N-1$, is given by:

$$\boldsymbol{H}(k) = \begin{bmatrix} -y(k-1) & -y(k-2) & . & -y(k-n) & r(k-1) & . & r(k-n) \\ -y(k-2) & -y(k-3) & . & -y(k-n-1) & r(k-2) & . & r(k-n-1) \\ . & . & . & . & . & . & . \\ -y(k-N) & -y(k-N-1) & . & -y(k-n-N+1) & r(k-N) & . & r(k-n-N+1) \end{bmatrix} \tag{7.68}$$

Comment *This generic linear model (7.67) finds application in many branches of science and engineering, including signal and speech processing, de-convolution and geophysical modeling, and covers a wide class of estimation problems in control, communication, and signal processing involving state-space mode, input–output models including the transfer function, difference equations, and linear regression models.*

7.4.2 The Least-Squares Estimate

The data matrix $\boldsymbol{H}(k)$ of the linear model (7.67) is not a constant matrix. It is time varying, as it is a function of the input–output data. Many of the results presented in the chapter on least-squares estimation need to be modified to handle the non-constant data matrix.

7.4.2.1 Objective Function

The objective function is to minimize the mean value of the squared residuals, $\frac{1}{N}\sum_{k=0}^{N-1} \upsilon^2(k-i)$, where the residual is the error between the measurements y and its estimate obtained using the assumed linear model $\boldsymbol{H\theta}$. Using the linear model (7.67) the objective function may be stated as follows:

$$\min_\theta \left\{ \frac{1}{N} \left(\boldsymbol{y}(k) - \boldsymbol{H}(k)\hat{\boldsymbol{\theta}}\right)^T \left(\boldsymbol{y}(k) - \boldsymbol{H}(k)\hat{\boldsymbol{\theta}}\right) \right\} \tag{7.69}$$

The estimate $\hat{\theta}$ of θ is given:

$$\frac{1}{N}\left(\boldsymbol{H}(k)^T\boldsymbol{H}(k)\right)\hat{\boldsymbol{\theta}} = \frac{1}{N}\boldsymbol{H}^T(k)\boldsymbol{y}(k) \tag{7.70}$$

Using the expression (7.68), $\boldsymbol{H}(k)^T\boldsymbol{H}(k)$ may be expressed in terms of estimates of the auto-correlation functions of $\boldsymbol{y}(k)$ and $\boldsymbol{r}(k)$, and cross-correlation functions of $\boldsymbol{y}(k)$ and $\boldsymbol{r}(k)$ denoted respectively as $\hat{\boldsymbol{r}}_{yy}$, $\hat{\boldsymbol{r}}_{rr}$, and $\hat{\boldsymbol{r}}_{yr}$, the term $\boldsymbol{H}(k)^T\boldsymbol{H}(k)$ on the left-hand side in Eq. (7.71) becomes:

$$\frac{1}{N}\boldsymbol{H}(k)^T\boldsymbol{H}(k) = \begin{bmatrix} \hat{\boldsymbol{r}}_{yy} & \hat{\boldsymbol{r}}_{yr} \\ \hat{\boldsymbol{r}}_{yr}^T & \hat{\boldsymbol{r}}_{rr} \end{bmatrix} \tag{7.71}$$

To illustrate the structure of the MxM matrix given by Eq. (7.71) clearly and compactly, an expression for the second-order system, $n = 2$, is given below:

$$\begin{bmatrix} \hat{\boldsymbol{r}}_{yy} & \hat{\boldsymbol{r}}_{yr} \\ \hat{\boldsymbol{r}}_{yr}^T & \hat{\boldsymbol{r}}_{rr} \end{bmatrix} = \frac{1}{N}$$

$$\begin{bmatrix} \sum_{i=1}^{N} y^2(k-i) & \sum_{i=1}^{N} y(k-i)y(k-i-1) & -\sum_{i=1}^{N} y(k-i)r(k-i) & -\sum_{i=1}^{N} y(k-i)y(k-i-1) \\ \sum_{k=1}^{N} y(k-i-1)y(k-i) & \sum_{k=1}^{N} y(k-i-1)^2 & -\sum_{k=1}^{N} y(k-i-1)r(k-i) & -\sum_{k=1}^{N} y(k-i-1)r(k-i-1) \\ -\sum_{i=1}^{N} y(k-i)r(k-i) & -\sum_{k=1}^{N} y(k-i-1)r(k-i) & \sum_{i=1}^{N} r^2(k-i) & \sum_{i=1}^{N} r(k-i)r(k-i-1) \\ -\sum_{i=1}^{N} y(k-i)r(k-i-1) & -\sum_{k=1}^{N} y(k-i-1)r(k-i-1) & \sum_{i=1}^{N} r(k-i)r(k-i-1) & \sum_{k=1}^{N} r(k-i-1)^2 \end{bmatrix} \tag{7.72}$$

Similarly $\frac{1}{N}\boldsymbol{H}^T(k)\boldsymbol{y}(k)$ becomes

$$\frac{1}{N}\boldsymbol{H}^T(k)\boldsymbol{y}(k) = \begin{bmatrix} -\frac{1}{N}\sum_{i=1}^{N} y(k-i)y(k-i+1) \\ -\frac{1}{N}\sum_{i=1}^{N} y(k-i-1)y(k-i+1) \\ \frac{1}{N}\sum_{i=1}^{N} r(k-i)y(k-i+1) \\ \frac{1}{N}\sum_{i=1}^{N} r(k-i-1)y(k-i+1) \end{bmatrix} \tag{7.73}$$

The solution $\hat{\theta}$ to the normal equation (7.70) exists and is unique if the following assumptions hold.

Assumptions *The input $r(k)$ is quasi-stationary and persistently exciting of order $M = 2n$, and the number of data points N tends to infinity.*

7.4.2.2 The Least-Squares Estimate of the Parameter and the Output

In view of the above assumptions, (i) the estimates of the auto-correlations $\hat{\boldsymbol{r}}_{yy}$, $\hat{\boldsymbol{r}}_{rr}$, and $\hat{\boldsymbol{r}}_{yr}$ tend to their corresponding correlation matrices, and (ii) $\boldsymbol{H}(k)^T\boldsymbol{H}(k)$ is non-singular. The least-squares estimate $\hat{\theta}$ exists and is given by

$$\hat{\theta} = \left(\frac{1}{N}\left(\boldsymbol{H}(k)^T\boldsymbol{H}(k)\right)\right)^{-1}\left(\frac{1}{N}\boldsymbol{H}^T(k)\boldsymbol{y}(k)\right) \tag{7.74}$$

Taking the limit as $N \to \infty$ the terms on the right-hand side become:

$$\lim_{N\to\infty}\frac{1}{N}\boldsymbol{H}(k)^T\boldsymbol{H}(k) = \begin{bmatrix} \boldsymbol{r}_{yy} & \boldsymbol{r}_{yr} \\ \boldsymbol{r}_{yr}^T & \boldsymbol{r}_{rr} \end{bmatrix} \tag{7.75}$$

$$\lim_{N\to\infty}\frac{1}{N}\boldsymbol{H}^T(k)\boldsymbol{y}(k) = \frac{1}{N}\begin{bmatrix} -r_{yy}(1) \\ . \\ -r_{yy}(n) \\ r_{ry}(1) \\ . \\ r_{rr}(n) \end{bmatrix} \tag{7.76}$$

where $\hat{\boldsymbol{r}}_{rr} \to \boldsymbol{r}_{rr}$, $\hat{\boldsymbol{r}}_{yy} \to \boldsymbol{r}_{yy}$, and $\hat{\boldsymbol{r}}_{yr} \to \boldsymbol{r}_{yr}$. The correlation matrices $\boldsymbol{r}_{yy}$, $\boldsymbol{r}_{rr}$, and $\boldsymbol{r}_{yr}$ are defined similarly to that given by Eq. (7.10):

$$r_{rr}(m) = \lim_{N\to\infty}\frac{1}{N}\sum_{k=0}^{N-1} r(k)r(k-m) = E\left[r(k)r(k-m)\right] \tag{7.77}$$

$$r_{yy}(m) = \lim_{N\to\infty}\frac{1}{N}\sum_{k=0}^{N-1} y(k)y(k-m) = E\left[y(k)y(k-m)\right] \tag{7.78}$$

$$r_{yr}(m) = \lim_{N\to\infty}\frac{1}{N}\sum_{k=0}^{N-1} y(k)r(k-m) = E\left[y(k)r(k-m)\right] \tag{7.79}$$

For clarity and compactness, a second-order example is used to illustrate the structure of Eq. (7.74):

$$\begin{bmatrix} \boldsymbol{r}_{yy} & \boldsymbol{r}_{yr} \\ \boldsymbol{r}_{yr}^T & \boldsymbol{r}_{rr} \end{bmatrix} = \begin{bmatrix} r_{yy}(0) & r_{yy}(1) & -r_{yr}(0) & -r_{yr}(1) \\ r_{yy}(1) & r_{yy}(0) & -r_{yr}(1) & -r_{yr}(0) \\ -r_{yr}(0) & -r_{yr}(1) & r_{rr}(0) & r_{rr}(1) \\ -r_{yr}(1) & -r_{yr}(0) & r_{rr}(1) & r_{rr}(0) \end{bmatrix}, \text{ and } \lim_{N\to\infty}\frac{1}{N}\boldsymbol{H}^T(k)\boldsymbol{y}(k) = \begin{bmatrix} -r_{yy}(1) \\ -r_{yy}(2) \\ r_{ry}(1) \\ r_{ry}(2) \end{bmatrix}$$

Note: In the sequel we will assume that the number of data samples N is finite, but sufficiently large so that the matrix $\frac{1}{N}\boldsymbol{H}(k)^T\boldsymbol{H}(k)$ approximates a deterministic and non-singular matrix. As the number N appears in both the numerator and denominator on the right-hand side terms of Eq. (7.74), namely $(\frac{1}{N}(\boldsymbol{H}(k)^T\boldsymbol{H}(k)))^{-1}$ and $(\frac{1}{N}\boldsymbol{H}^T(k)\boldsymbol{y}(k))$, it may be canceled and hence the explicit dependence on the number of data samples N is suppressed and the expression of the estimate is given by:

$$\hat{\boldsymbol{\theta}} = \left(\boldsymbol{H}(k)^T\boldsymbol{H}(k)\right)^{-1}\boldsymbol{H}^T(k)\boldsymbol{y}(k) \tag{7.80}$$

$$\hat{\boldsymbol{\theta}} = \boldsymbol{\theta} + \left(\boldsymbol{H}(k)^T\boldsymbol{H}(k)\right)^{-1}\boldsymbol{H}^T(k)\boldsymbol{\upsilon}(k) \tag{7.81}$$

Using the expression for the estimate (7.80) and (7.81), we will analyze the performance for a sufficiently large but finite number of data samples when (i) the equation error is correlated with a $\boldsymbol{H}(k)$ zero-mean process, and (ii) when it is a zero-mean white noise process so that it is uncorrelated with $\boldsymbol{H}(k)$.

The least-squares estimate, denoted $\hat{\boldsymbol{y}}_{LS}(k)$, of the system output $\boldsymbol{y}(k)$ using the expression for parameter estimate (7.80) and the model (7.67) is given by:

$$\hat{\boldsymbol{y}}_{LS}(k) = \boldsymbol{H}(k)\hat{\boldsymbol{\theta}} = \boldsymbol{P}_r(k)\boldsymbol{y}(k) \tag{7.82}$$

The residual, denotes $\boldsymbol{e}_{LS}(k)$ defined as $\boldsymbol{e}_{LS}(k) = \boldsymbol{y}(k) - \boldsymbol{y}_{LS}(k)$ using Eqs. (7.67) and (7.82) becomes:

$$\boldsymbol{e}_{LS}(k) = (\boldsymbol{I} - \boldsymbol{P}_r(k))\boldsymbol{y}(k) \tag{7.83}$$

where $\boldsymbol{P}_r(k) = \boldsymbol{H}(k)(\boldsymbol{H}^T(k)\boldsymbol{H}(k))^{-1}\boldsymbol{H}^T(k)$ is the projection operator, which satisfies: $\boldsymbol{P}_r^m = \boldsymbol{P}_r\ for\ m = 1, 2, 3, \ldots,$ and the eigenvalues of $\boldsymbol{I} - \boldsymbol{P}_r$ are only ones and zeros.

Comment *The estimate of the output $\hat{\boldsymbol{y}}_{LS}(k)$ captures all the information contained in the input–output data. The ability to capture all the information depends upon (i) the richness of the identification input as measured by the order of the persistence of excitation, and the "richness" of the assumed identification model structure, as determined by whether the model of the actual system to be identified is a member of the set of models generated by the assumed identification model structure. What is "left over" after identification is the residual $\boldsymbol{e}_{LS}(k)$. If the residual is a zero-mean white noise process, it implies that the identified model has captured completely the system model. The "leftover" is a non-information bearing zero-mean white noise residual. On the other hand, if the residual is correlated noise, then the identified model has not captured the true system model. In this case, the "leftover" is an information bearing colored noise residual.*

7.4.2.3 Quality of Estimate: Correlated Error

The colored noise process is generated as an output of a "coloring filter" driven by a zero-mean white noise process. The coloring filter, denoted $H_e(z)$, is a FIR filter given by:

$$\upsilon(z) = H_e(z)e(z) \tag{7.84}$$

where $H_e(z) = c_0 + \sum_{i=1}^{n} c_i z^{-i}$ is the coloring filter.

7.4.3 Quality of the Least-Squares Estimate

The performance of the least squares approaches to system identification is analyzed in terms of the bias and the covariance of the parameter estimates and the variance of the residual. The quality of the least squares estimate is derived when the equation is uncorrelated and correlated with the data matrix. Main results are established in the Appendix in a series of lemmas and corollaries.

The performance of the least-squares method is given for the case when the data samples are large but finite and the data samples are infinite. When the data sample is finite and large it is assumed that terms $\frac{1}{N}\boldsymbol{H}(k)^T\boldsymbol{H}(k)$ and $\frac{1}{N}\boldsymbol{H}^T(k)\boldsymbol{y}(k)$ are deterministic matrices formed of the correlation functions $r_{yy}(m)$ and $r_{yr}(m)$. Further, the performance for the case when the equation error $\boldsymbol{\upsilon}(k)$ is uncorrelated, and when it is correlated with the data matrix $\boldsymbol{H}^T(k)$, are considered separately.

Case 1: *The equation error is uncorrelated:* $E[\boldsymbol{H}^T(k)\boldsymbol{\upsilon}(k)] = 0$
Finite sample properties

- The parameter estimate is unbiased:

$$E\left[\hat{\theta}\right] = \theta \tag{7.85}$$

- The estimate is efficient, that is the covariance of the parameter estimation error the Cramér–Rao lower bound:

$$cov\left(\hat{\theta}\right) = \sigma_\upsilon^2\left(\boldsymbol{H}(k)^T\boldsymbol{H}(k)\right)^{-1} \tag{7.86}$$

- The mean-squared residual $\sigma_{LSres}^2 = E\left[\boldsymbol{e}_{LS}(k)\boldsymbol{e}_{LS}^T(k)\right]$ is:

$$\sigma_{LSres}^2 = \sigma_\upsilon^2\left(1 - \frac{M}{N}\right) \tag{7.87}$$

where $M = rank(\boldsymbol{H})$ is the rank of the data matrix, which is assumed to be of full rank.

Asymptotic properties:
The asymptotic properties of the estimates as $N \to \infty$ are listed below:

- $\hat{\theta}$ is a consistent estimator of θ:

$$\hat{\theta} \to \theta \text{ as } N \to \infty \tag{7.88}$$

- As N tends to infinity, the residual tends to the equation error: $\boldsymbol{e}_{LS}(k) \to \boldsymbol{\upsilon}(k)$

$$\boldsymbol{e}_{LS}(k) \to \boldsymbol{\upsilon}(k) \text{ as } N \to \infty \tag{7.89}$$

Comments Residual: *The residual is a good indicator of the quality of the identification method. If the equation is a zero-mean white noise process, then the residual is also a zero-mean white noise process indicating that all the information contained in the model has been extracted in obtaining the estimates. The residual contains what is left in the measurement after extracting its estimate. A better visual indicator of the performance of the identification is the auto-correlation of the residual, which will be a delta function if the residual is white noise.*

The expression of the variance of the residual σ_{LSres}^2 *given by Eq. (7.87) may be used to estimate the variance of the equation error* σ_υ^2 *when the equation error is a zero-mean white noise process.*

Parameter estimate: *The parameter estimate is consistent, unbiased, and efficient. The covariance of parameter estimate achieves the Cramér–Rao inequality. The covariance of the estimate depends upon the "signal to noise ratio" which is the ratio of the variance of equation error and the square of the minimum singular value of the data matrix.*
Case 2: The equation error is uncorrelated: $E[\boldsymbol{H}^T(k)\boldsymbol{v}(k)] \neq 0$

- *The estimate $\hat{\theta}$ is biased:*

$$E[\hat{\boldsymbol{\theta}}] = \boldsymbol{\theta} + (\boldsymbol{H}(k)^T\boldsymbol{H}(k))^{-1}E[\boldsymbol{H}^T(k)\boldsymbol{v}(k)] \tag{7.90}$$

- *A bound on the parameter estimation error is:*

$$\|E[\hat{\boldsymbol{\theta}}] - \boldsymbol{\theta}\| \leq \|(\boldsymbol{H}(k)^T\boldsymbol{H}(k))^{-1}\|\|E[\boldsymbol{H}^T(k)\boldsymbol{v}(k)]\| \tag{7.91}$$

- *The covariance of the parameter estimation error is*

$$\text{cov}(\hat{\boldsymbol{\theta}}) = (\boldsymbol{H}(k)^T\boldsymbol{H}(k))^{-1}E[\boldsymbol{H}^T(k)\boldsymbol{v}(k)\boldsymbol{v}^T(k)\boldsymbol{H}(k)](\boldsymbol{H}(k)^T\boldsymbol{H}(k))^{-1} \tag{7.92}$$

- *A bound on the covariance of the parameter estimation error is:*

$$\|\text{cov}(\hat{\boldsymbol{\theta}})\| \leq \frac{\sigma_{\max}(E[\boldsymbol{H}^T(k)\boldsymbol{v}(k)\boldsymbol{v}^T(k)\boldsymbol{H}(k)])}{\sigma_{\min}^4(\boldsymbol{H}(k))} \tag{7.93}$$

- *An expression for the residual is:*

$$\boldsymbol{e}_{LS}(k) = (\boldsymbol{I} - \boldsymbol{H}(k)(\boldsymbol{H}(k)^T\boldsymbol{H}(k))^{-1}\boldsymbol{H}^T(k))\boldsymbol{v}(k) \tag{7.94}$$

Comments *The performance of the least squares method is degraded in the presence of the colored noise equation error. The parameter estimate is not consistent, biased, and is not efficient (the covariance of the estimated does not achieve Cramér–Rao lower bound).*

One way to overcome this poor performance of the least-squares method, a two-stage least squares method, termed a high-order least-squares method, *is introduced in a later section. In the first stage, a high-order model with a sufficient number of parameters to capture both the system and the noise model is first estimated using the least-squares method. In the second stage the higher-order estimated model is reduced to a lower order using a frequency weighted least-squares method.*

7.4.4 Illustrative Example of the Least-Squares Identification

The quality of identification of the least-squares identification is illustrated using a simulated example with white and colored noise.

Example 7.8 Identification with white and colored noise
Model of the system

$$y(k) = -a_1y(k-1) - a_2y(k-2) + b_1r(k-1) + c_0v(k) + c_1v(k-1) + c_2v(k) \tag{7.95}$$

Table 7.1 Covariance of the parameter estimation error

Covariance of parameter estimation error cov $(\hat{\theta})$	
Equation error is a white noise	Equation error is a colored noise
$10^{-3}\begin{bmatrix} 0.0950 & -0.2805 & 0.2851 \\ -0.2805 & 0.8284 & -0.8420 \\ 0.2851 & -0.8420 & 0.8559 \end{bmatrix}$	$\begin{bmatrix} 0.0301 & -0.0278 & -0.0023 \\ -0.0278 & 0.0257 & 0.0021 \\ -0.0023 & 0.0021 & 0.0002 \end{bmatrix}$

where $\theta = [\, a_1 \;\; a_2 \;\; b_1 \,]^T$, $a_1 = -1.6$, $a_2 = 0.8$ and $b_1 = 1$, $M = 3$. The estimate $\hat{\theta} = [\, \hat{a}_1 \;\; \hat{a}_2 \;\; \hat{b}_1 \,]^T$ is obtained for the two cases:

a) $\upsilon(k) = v(k)$, $c_0 = 1$, $c_1 = 0$, and $c_2 = 0$, and
b) $\upsilon(k) = v(k) - v(k-1) + 0.2v(k)$

where $v(k)$ is a zero-mean white noise process with unit variance, $c_0 = 1$, $c_1 = -1$, and $c_2 = 0.2$. The case (i) corresponds to the white noise equation error whereas case (ii) corresponds to colored noise. The number of data samples and the variance of the equation error were $N = 1000$ and $\sigma_\upsilon^2 = 1$. The input for system identification was a PRBS waveform.

Results of the evaluation:
The covariance of the parameter estimation error is given in Table 7.1 when the equation error is white and colored noise.

The true feature vector θ and the estimates of the feature vector $\hat{\theta}$, when the equation error is white and colored noise are shown in Table 7.2.

Figure 7.7 shows (i) the step responses obtained with the true model parameters, their estimates with white noise and their estimates with colored noise, (ii) the auto-correlations of the white and colored noise, and (iii) the auto-correlations of the residuals when the equation error is white and colored noise. Subfigure (a) shows the step responses of the system, and estimated step responses when the equation error is white and colored noise. Subfigures (b) and (c) show the auto-correlations of the white and colored noise respectively. Subfigures (d) and (e) show the auto-correlations of the residuals with white and colored noise respectively.

Comments *In the case of a white noise equation error, the estimate is unbiased (7.85), the estimate of the step response is close to that of the system, and the residual is close to the equation error as established in Eq. (7.89). The parameter estimation error and its covariance are both negligibly small, as shown in Tables 7.1 and 7.2.*

In the case of a colored noise equation error, however, the estimate is biased as given in Eq. (7.90). Neither the estimate of the step response is close to the true one, nor is the residual close to the colored

Table 7.2 Feature vector and the estimates

θ	True	Estimate with white noise	Estimate with colored noise
a_1	−1.6	−1.5948	−1.3546
a_2	0.8	0.7840	0.5733
b_1	1.0	1.0296	0.9813

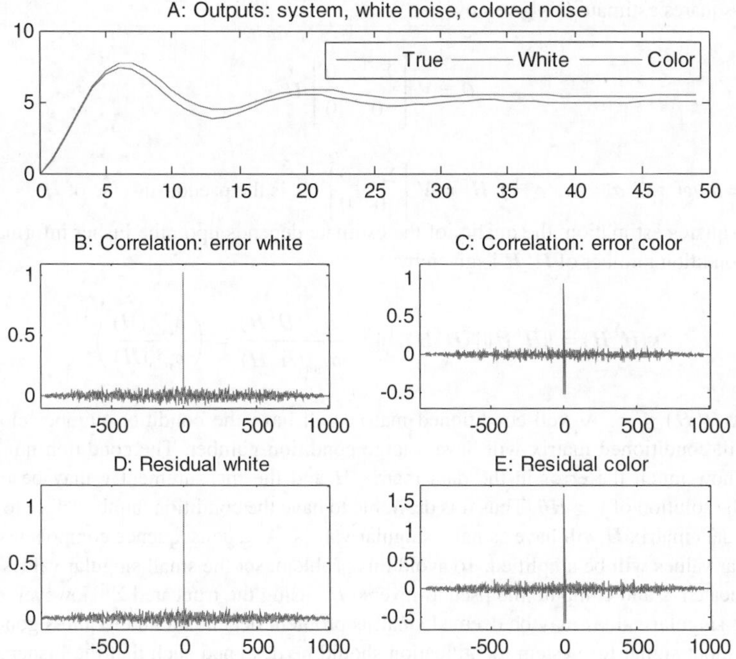

Figure 7.7 The step response and the correlation of the residuals

noise in view of correlation of the colored noise with the data matrix. The parameter estimation error and its covariance are both large, as shown in Tables 7.1 and 7.2.

The auto-correlation of the residual is a good indicator of the performance of the least squares estimator if the equation error is a zero-mean white noise process. If the least squares estimate is optimal, then the auto-correlation function is a delta function. The variance of the equation error is the maximum value of the auto-correlation function, which occurs at zero lag. In the presence of the colored noise, the auto-correlation of the residual is not a delta function.

7.4.5 Computation of the Estimates Using Singular Value Decomposition

Singular Value Decomposition (SVD) is one of the most robust algorithms for computing the inverse of square or rectangular matrices including those which are ill-conditioned and rank deficient (are not of full rank). The maximum and the minimum singular values provide quantitative information about the structure of the set of linear equations, such as the condition number, the norm of the matrix, and the norm of its pseudo-inverse.

The *NxM* matrix $\boldsymbol{H}$ can be decomposed using SVD as:

$$\boldsymbol{H} = \boldsymbol{U}\boldsymbol{S}\boldsymbol{V}^T \tag{7.96}$$

where $\boldsymbol{U}$ is *NxN* unitary matrix of rank $r \leq \min(M, N)$, $\boldsymbol{U}^T\boldsymbol{U} = \boldsymbol{U}\boldsymbol{U}^T = \boldsymbol{I}$, $\boldsymbol{V}$ is a *MxM* unitary matrix $\boldsymbol{V}\boldsymbol{V}^T = \boldsymbol{V}^T\boldsymbol{V} = \boldsymbol{I}$, $\boldsymbol{S}$ is a *NxM* rectangular matrix given by, $\boldsymbol{S} = \begin{bmatrix} \Sigma & 0 \\ 0 & 0 \end{bmatrix}$, $\Sigma = diag[\sigma_1 \quad \sigma_2 \ \sigma_3 \ . \ \sigma_r]$, $\{\sigma_i\}$ are the singular values of $\boldsymbol{H}$ which are positive $\sigma_1 \geq \sigma_2 \geq \sigma_3 \geq .. \geq \sigma_r > 0$ and $\sigma_{r+1} = \sigma_{r+2} = \ldots = \sigma_M = 0$.

The least-squares estimate $\hat{\theta}$ is given by

$$\hat{\theta} = V \begin{bmatrix} \Sigma^{-1} & 0 \\ 0 & 0 \end{bmatrix} U^T y \tag{7.97}$$

where $\Sigma^{-1} = diag(\sigma_1^{-1} \ \sigma_2^{-1} \ \ldots \ \sigma_r^{-1})$, $H^{\dagger} = V \begin{bmatrix} \Sigma^{-1} & 0 \\ 0 & 0 \end{bmatrix} U^T$ is the pseudo-inverse of H.

In the least-squares estimation, the quality of the estimate depends upon the Fisher information matrix $H^T H$. The condition number of $H^T H$ is given by:

$$\kappa(H^T H) = \|H^T H\| \|(H^T H)^{-1}\| = \frac{\sigma_{\max}(H^T H)}{\sigma_{\min}(H^T H)} = \left(\frac{\sigma_{\max}(H)}{\sigma_{\min}(H)}\right)^2 \tag{7.98}$$

where $1 \leq \kappa(H^T H) \leq \infty$. A well-conditioned matrix will have the condition number close to unity, whereas an ill-conditioned matrix will have a large condition number. The condition number gives a measure of how much the error in the data matrix H and the measurement y may be magnified in computing the solution of $y = H\theta$. Thus it is desirable to have the condition number close to 1. A poorly conditioned data matrix H will have "small" singular values. As a consequence components of $H^{\dagger}$ with small singular values will be amplified. To avoid this problem, set the small singular values to zero and obtain a truncated Σ and compute the pseudo-inverse $H^{\dagger}$ using the truncated Σ. However, determining which of the singular values may be deemed small is problem dependant and requires good judgment. Further, the input signal for system identification should be designed such that the Fisher information matrix $H^T H$ is well conditioned.

7.4.6 Recursive Least-Squares Identification

Computation of the least-squares estimate may be performed using batch or recursive processing. With the ever increasing memory size and the speed of the present day computing devices, batch processing may be employed in online applications using computationally efficient and robust singular value decomposition. However, in many online applications, recursive processing is preferred. The recursive scheme is commonly derived from the batch processing or independently using the Kalman filter approach.

The Kalman filter is designed for estimating the states of a linear time varying system recursively by fusing the *a priori* knowledge of the model and the *a posteriori* information provided by the noisy measurement data so that the covariance of the estimation error is minimized. The Kalman filter is reformulated to estimate the unknown constant parameters. The state-space model of the unknown parameter and the (time-varying) algebraic equation relating the input–output data and the unknown parameter is a linear regression model. Since θ is unknown constant vector $\theta(1) = \theta(2) = \cdots = \theta(k)$, its dynamic model is merely $\theta(k+1) = \theta(k)$.

The state-space model of the constant unknown feature vector θ using Eqs. (7.63), (7.64), and (7.65) has the following form:

$$\begin{aligned} \theta(k+1) &= \theta(k) + w(k) \\ y(k) &= \psi^T(k)\,\theta(k) + \upsilon(k) \end{aligned} \tag{7.99}$$

where $\upsilon(k)$ is the measurement noise and $w(k)$ is a $Mx1$ fictitious disturbance which is assumed to be zero-mean white noise process. Since the performance of the Kalman filter depends upon the ratio of the covariance $Q = E[w(k)w^T(k)]$ of the disturbance and the variance R of the measurement noise, for convenience the variance of the noise is assumed to be unity,$R = 1$, and the covariance of the disturbance Q is chosen by a tradeoff between a fast convergence of the estimate and the large covariance of the estimation error. The smaller the $\boldsymbol{Q}$, the smaller will be the estimation error covariance, but the slower will be the convergence of the estimate. An important advantage of including the fictitious disturbance is that the covariance of the estimation error $\boldsymbol{P} = \text{cov}\left(\hat{\boldsymbol{\theta}}\right)$, and hence the Kalman gain $\boldsymbol{K}(k)$, does not converge to zero. Otherwise the Kalman filter has to be restarted whenever the gain and the covariance converge to zero, which is inconvenient in practice. The covariance matrix $\boldsymbol{Q}$ may be chosen to be diagonal with identical or non-identical elements. If the performance requirements, namely the speed of convergence and the covariance of the estimation error, are different for different elements of the feature vector $\boldsymbol{\theta}$, then the diagonal elements $\boldsymbol{Q}$ may not be identical. The estimate $\hat{\boldsymbol{\theta}}(k)$ of $\boldsymbol{\theta}$ is obtained recursively from the following set of equations:

- Residual

$$e(k) = y(k) - \boldsymbol{\psi}^T(k)\hat{\boldsymbol{\theta}}(k-1)$$

- Recursive estimator

$$\hat{\boldsymbol{\theta}}(k) = \hat{\boldsymbol{\theta}}(k-1) + \boldsymbol{K}(k)(y(k) - \boldsymbol{\psi}^T(k)\hat{\boldsymbol{\theta}}(k-1))$$

- Kalman gain

$$\boldsymbol{K}(k) = \boldsymbol{P}(k-1)\boldsymbol{\psi}(k)(\boldsymbol{\psi}^T(k)\boldsymbol{P}(k-1)\boldsymbol{\psi}(k)+1)^{-1}$$

- Riccati equation

$$\boldsymbol{P}(k) = \boldsymbol{P}(k-1) - \boldsymbol{P}(k-1)\boldsymbol{\psi}(k)(\boldsymbol{\psi}^T(k)\boldsymbol{P}(k-1)\boldsymbol{\psi}(k)+1)^{-1}\boldsymbol{\psi}^T(k)\boldsymbol{P}(k-1) + \boldsymbol{Q}$$

Figure 7.8 shows a block diagram of a typical recursive identification algorithm for recursive least-squares, the high-order recursive least squares, and the recursive prediction error algorithm. The model of the system and the parameter update algorithm will differ for different methods. The arrow indicates that the parameters of identification model are varied until $e(k)$ is "small."

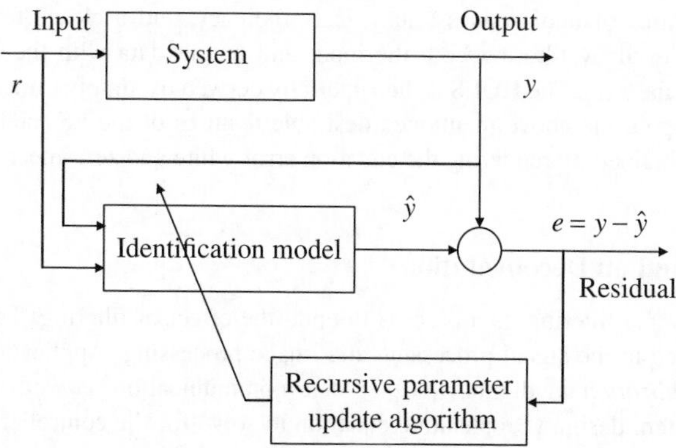

Figure 7.8 Recursive algorithm

7.5 High-Order Least-Squares Method

The least-squares method is computationally simple and the estimate may be expressed in closed form. The use of a high-order model for system identification was inspired by the seminal paper by [12] where it was shown that using a higher than the true order in estimating a signal model improves the accuracy. It is efficient, that is the covariance of the parameter estimation meets the Cramér–Rao lower bound, when the equation error is a zero-mean white noise process. However, the performance of the least-squares method degrades when the equation error is a colored noise process. Taking a cue from [12], to overcome this problem, the "high-order least-squares method," is used. It is a two-stage least squares method, is introduced. The high-order least-squares approach consists of the following steps:

Stage I: Identification of a high-order model
The least squares approach is used identify a high-order model from the input–output data

Stage II: Model order reduction
A number of reduced order models are estimated: The high-order model is reduced using a frequency weighted least-squares approach. A number of reduced order models are derived for different choices of the model orders.

Model order selection: A model order selection criterion such as Akaike information criterion is used to select one from the number of reduced order models.

7.5.1 *Justification for a High-Order Model*

The performance of HOLS approaches that of PEM as the number of terms in the approximation approaches infinity, while preserving the attractive features of the LS approach, namely its simplicity and provision of a closed-form (or analytic) solution to the parameter estimation problem. Unlike the PEM, the HOLS method involves an equation error that is linear in the unknown parameters, is computationally efficient, and yields a closed-form expression of the parameter estimates and their errors.

7.5.1.1 Whitening Operation

The objective here is to exploit the fact the Kalman filter-based system model allows for the lumping of both the disturbance and measurement noise into a single noisy signal, that is, the equation error. Direct application of the LS method for the estimation of the parameters of such a model would thus lead to unbiased and inefficient parameter estimates. The aim here is therefore to de-correlate the colored equation error $\upsilon(z)$ from the data matrix prior to applying the popular LS method so as to recover the latter's attractive features of unbiasedness, statistical efficiency, and a closed-form of the parameter estimate. This is achieved by filtering both the input and output data with the inverse of the stable Kalman filter polynomial $F(z)$. The HOLS is then naturally derived by simply truncating $F(z)$ up to some high order, while keeping the above-mentioned desirable features of the LS method. Consequently, a high-order model is obtained by rendering the equation error white and zero-mean.

7.5.1.2 Background on Deconvolution

Deconvolution, or inverse filtering, is a process to undo the effect of filtering. The concept of deconvolution is widely used in the signal processing and image processing. Applications of deconvolution include *channel equalization* in digital voice or data communication, *equalization* of room or car acoustics, control system design (Diophantine equation to solve for the controller parameters), system identification, oil exploration, geophysics, image enhancement and restoration, echo cancellation, noise cancellation, and undoing the distortion of speech waveform caused by filtering action of transmission medium. The noise cancellation includes cancellation of 60 Hz power signal, acoustic noise cancellation in cars, planes, and so on.

The deconvolution problem is formulated as follows: given two polynomials $\alpha(z)$ and $\beta(z)$ find a polynomial $\gamma(z)$ such that $\alpha(z)\gamma(z) = \beta(z)$. It essentially involves undoing the filtering operation of $\alpha(z)$ to recover $\gamma(z)$. It is also called *inverse filtering* since $\gamma(z) = \dfrac{\beta(z)}{\alpha(z)}$: $\gamma(z)$ is obtained by performing inverse filtering. The important consideration in inverse filtering is the stability of the inverse filter $\dfrac{1}{\alpha(z)}$. The inverse filter must be stable, otherwise the resulting operation would be meaningless.

7.5.1.3 Least-Squares Approach: Frequency Domain Deconvolution

Commonly used approaches to deconvolution include polynomial division and the least-squares approach to the deconvolution formulated in the frequency domain. Deconvolution, or polynomial division, is the inverse operation of convolution. This method is very sensitive to variations in the coefficients, and hence the frequency-domain approach is preferred as it is practical.

The solution of the equation is obtained by finding the best fit in the frequency domain between $\alpha(z)\gamma(z)$ and $\beta(z)$ is obtained at a specified set of frequencies $z_i = e^{j2\pi f_i}$ where the frequencies $\{f_i\}$ are chosen to cover the frequency region of interest, such as bandwidth of the system, which is usually a low-frequency region. The problem may be posed as follows:

Find the coefficients $\{\gamma_i\}$ of the polynomial $\gamma(z) = \sum_{i=0}^{M} \gamma_i z^{-i}$ such that the error $\beta(z)\gamma(z) - \alpha(z)$ is minimized:

$$\min_{\gamma} \sum_{i=1}^{N} \left| \sum_{j=0}^{M} \gamma_j z_i^{-j} \alpha(z_i) - \beta(z_i) \right|^2 \tag{7.100}$$

The solution is the frequency-domain version of the least-squares solution (normal equation):

$$\begin{bmatrix} \beta(z_1) \\ \beta(z_2) \\ . \\ \beta(z_N) \end{bmatrix} = \begin{bmatrix} \alpha(z_1) & z_1^{-1}\alpha(z_1) & z_1^{-2}\alpha(z_1) & . & z_1^{-M}\alpha(z_1) \\ \alpha(z_2) & z_2^{-1}\alpha(z_2) & z_2^{-2}\alpha(z_2) & . & z_2^{-N}\alpha(z_2) \\ . & . & . & . & . \\ \alpha(z_N) & z_N^{-1}\alpha(z_N) & z_N^{-2}\alpha(z_N) & . & z_N^{-M}\alpha(z_N) \end{bmatrix} \begin{bmatrix} \gamma_0 \\ \gamma_1 \\ . \\ \gamma_M \end{bmatrix} \tag{7.101}$$

Thus an IIR filter $\dfrac{\beta(z)}{\alpha(z)}$ is approximated by a FIR filter $\gamma(z)$ given by:

$$\frac{\beta(z)}{\alpha(z)} \approx \gamma(z) = \sum_{i=1}^{M} \gamma_i z^{-i} \tag{7.102}$$

7.5.1.4 High-Order Linear Regression Model

The innovation form (7.29) is expressed in the form of a high-order polynomial by performing deconvolution. Deconvolving the polynomials $D(z)$ by $F(z)$ and the polynomials $N(z)$ by $F(z)$ yields:

$$\begin{aligned} D(z) &= \left(1 + \sum_{i=1}^{n_h} a_{hi} z^{-i}\right) F(z) + \Delta_D(z) \\ N(z) &= \left(\sum_{i=0}^{n_h} b_{hi} z^{-i}\right) F(z) + \Delta_N(z) \end{aligned} \tag{7.103}$$

where n_h is the selected high order, $\{a_{hi}\}$ and $\{b_{hi}\}$ are coefficients in the high-order approximations, and $\Delta_D(z)$ and $\Delta_N(z)$ are error terms resulting from truncation of the coefficients of FIR approximation to a finite number of terms n_h. Since $F(z)$ is a stable polynomial (roots of the polynomial are strictly inside the unit circle), we can select an order n_h high enough so that the approximation errors $\Delta_D(z)$ and $\Delta_D(z)$ become negligibly small. Substituting for $N(z)$ and $D(z)$ using Eq. (7.103), the innovation form (7.30) is converted into a high-order polynomial model:

$$\begin{aligned} y(z) &= \left(1 - \frac{D(z)}{F(z)}\right) y(z) + \frac{N(z)}{F(z)} r(z) + e(z) \\ &\approx -\sum_{i=1}^{n_h} a_{hi} z^{-i} y(z) + \sum_{i=1}^{n_h} b_{hi} z^{-i} r(z) + e(z) \end{aligned} \tag{7.104}$$

The high-order model of the system transfer function becomes

$$G_h(z) = \frac{N_h(z)}{D_h(z)} = \frac{\sum_{i=0}^{n_h} b_{hi} z^{-i}}{1 + \sum_{i=1}^{n_h} a_{hi} z^{-i}} \tag{7.105}$$

where $N_h(z) = \sum_{i=0}^{n_h} b_{hi} z^{-i}$ and $D_h(z) = 1 + \sum_{i=1}^{n_h} a_{hi} z^{-i}$ are the numerator and denominator polynomials of $G_h(z)$.

The high-order linear regression model becomes:

$$y(k) \approx -\sum_{i=1}^{n_h} a_{hi} y(k-i) + \sum_{i=1}^{n_h} b_{hi} r(k-i) + e(k) \tag{7.106}$$

The $2n_h x1$ feature vector $\boldsymbol{\theta}_h$ is:

$$\boldsymbol{\theta}_h = [\,a_{h1} \quad a_{h2} \quad . \quad a_{hn_h} \quad b_{h1} \quad b_{h1} \quad . \quad b_{hn_h}\,]^T \tag{7.107}$$

Using the least-squares approach the estimate $\hat{\boldsymbol{\theta}}_h$ of the high dimensional feature vector $\boldsymbol{\theta}_h$, similar to that given in Eq. (7.80), becomes:

$$\hat{\boldsymbol{\theta}}_h = \left(\boldsymbol{H}_h(k)^T \boldsymbol{H}_h(k)\right)^{-1} \boldsymbol{H}_h^T(k) \boldsymbol{y}(k) \tag{7.108}$$

where $\hat{\boldsymbol{\theta}}_h = [\,\hat{a}_{h1} \ \hat{a}_{h2} \ . \ \hat{a}_{hn_h} \ \hat{b}_{h1} \ \hat{b}_{h1} \ . \ \hat{b}_{hn_h}\,]^T$, and $\boldsymbol{H}_h(k)$ is $Nx2n_h$ data matrix defined similarly to Eq. (7.68). The estimate $\hat{\boldsymbol{y}}(k)$ of the output $\boldsymbol{y}(k)$ for the least squares method given by Eq. (7.82) may be extended to this case:

$$\hat{\boldsymbol{y}}(k) = \boldsymbol{H}_h(k) \hat{\boldsymbol{\theta}}_h = \boldsymbol{P}_{hr}(k) \boldsymbol{y}(k) \tag{7.109}$$

where $\boldsymbol{P}_{hr}(k) = \boldsymbol{H}_h(k)(\boldsymbol{H}_h^T(k)\boldsymbol{H}_h(k))^{-1}\boldsymbol{H}_h^T(k)$ and the expression for the residual $\boldsymbol{e}_{res}(k) = \boldsymbol{y}(k) - \hat{\boldsymbol{y}}(k)$ similarly to that of Eq. (7.83) becomes

$$\boldsymbol{e}_{res}(k) = (\boldsymbol{I} - \boldsymbol{P}_{hr}(k))\boldsymbol{y}(k) \tag{7.110}$$

7.5.1.5 Performance of the High-Order Least-Squares Method

Thanks to the high-order model approximation, the equation error $e(k)$ approaches the zero-mean white noise process order n_h is sufficiently large. Hence the quality of the estimate of the high least squares method will be the same as that of the least squares when the equation error is a zero-mean white noise process.

Finite Sample Properties

- The parameter estimated is unbiased and efficient:

$$\begin{aligned} E\left[\hat{\boldsymbol{\theta}}_h\right] &= \hat{\boldsymbol{\theta}}_h \\ \text{cov}\left(\hat{\boldsymbol{\theta}}_h\right) &= (\boldsymbol{H}_h(k)^T\boldsymbol{H}_h(k))^{-1}\sigma_e^2 \end{aligned} \tag{7.111}$$

- The variance of the residual for the least-squares method given by Eq. (7.86) also holds for the high-order least-squares method:

$$\sigma_{res}^2 = \sigma_e^2\left(1 - \frac{M_h}{N}\right) \tag{7.112}$$

where σ_{res}^2 is the variance of the residual $\boldsymbol{e}_{res}(k)$ and $M_h = 2n_h$ assuming that $\boldsymbol{H}_h$ has a full rank.

Asymptotic Properties

The asymptotic properties as $N \to \infty$ are listed below:

- $\hat{\boldsymbol{\theta}}_h$ is a consistent estimator of $\boldsymbol{\theta}_h$:

$$\hat{\boldsymbol{\theta}}_h \to \boldsymbol{\theta}_h \text{ as } N \to \infty \tag{7.113}$$

- As N tends to infinity, the residual tends to the equation error: $\boldsymbol{e}_{res}(k) \to \boldsymbol{e}(k)$

$$\boldsymbol{e}_{res}(k) \to \boldsymbol{e}(k) \text{ as } N \to \infty$$

Comment *The expression for the variance of the residual given by Eq. (7.112) has an important practical application. It may be used to estimate the variance σ_e^2 of the equation error **e**(k) which in this case is a zero-mean white noise process.*

7.5.1.6 Estimate of the High-Order Transfer Function

The estimate $\hat{G}_h(z)$ of the high-order transfer function $G_h(z)$ is given by:

$$\hat{G}_h(z) = \frac{\hat{N}_h(z)}{\hat{D}_h(z)} = \frac{\sum_{i=0}^{n_h} \hat{b}_{hi}z^{-i}}{1 + \sum_{i=1}^{n_h} \hat{a}_{hi}z^{-i}} \tag{7.114}$$

Comments *If the order n_h is chosen to be high enough (i) $\Delta_D(z)$ and $\Delta_N(z)$ will be negligibly small, (ii) $N_h(z)F(z)$ will contain $N(z)$, and $\hat{D}_h(z)F(z)$ will contain $D(z)$, and (iii) $\hat{G}_h(z)$ will be arbitrarily close to $G(z)$.*

There is no need to know a priori *information such as (i) the covariances of the disturbance and measurement noise, Kalman gain, (ii) Kalman polynomial $F(z)$, and (iii) the number of terms in the truncation of the high-order model. The solution to this problem of the unknowns is to choose a high order and iterate the selected high order till the residual from the least-squares identification is a zero-mean white noise process. Since the residual is a white noise process Eq. (7.106) will be satisfied.*

7.5.1.7 Illustrative Example: High-Order Model Using Deconvolution

The derivation of a high-order model by whitening the equation error is illustrated using examples of a first- and a second-order system with colored noise.

Example 7.9 First-order system

$$y(k) = -a_1 y(k-1) + b_1 r(k-1) + e(k) + c_1 e(k-1) \tag{7.115}$$

where $a_1 = -0.9$, $b_1 = 1$ and $c_1 = 0.5$, $D(z) = 1 - 0.9z^{-1}$, $N(z) = z^{-1}$, $F(z) = 1 - 0.5z^{-1}$. Using deconvolution and selecting the order $n_h = 4$ so that the truncation error is negligible yields:

$$\frac{D(z)}{F(z)} = \frac{1-0.9z^{-1}}{1-0.5z^{-1}} = 1 - 0.4\mathrm{z}^{-1} - 0.2\mathrm{z}^{-2} - 0.1\mathrm{z}^{-3} - 0.05\mathrm{z}^{-4}$$

$$\frac{N(z)}{F(z)} = \frac{z^{-1}}{1-0.5z^{-1}} = z^{-1} + 0.5\mathrm{z}^{-2} + 0.25\mathrm{z}^{-3} + 0.125\mathrm{z}^{-4}$$

The high-order linear regression model becomes

$$\begin{aligned} y(k) &= 0.4y(k-1) + 0.2y(k-2) + 0.1y(k-3) + 0.05y(k-4) \\ &\quad r(k-1) + 0.5r(k-2) + 0.25r(k-3) + 0.125r(k-4) + e(k) \end{aligned} \tag{7.116}$$

Example 7.10 Second-order system
The linear regression model is given by;

$$y(k) = -a_1 y(k-1) - a_2 y(k-2) + b_1 r(k-1) + b_1 r(k-1) + e(k) + c_1 e(k-1) + c_2 e(k-2) \tag{7.117}$$

where $a_1 = -1.6$, $a_2 = 0.8$, $b_1 = 1$, $b_2 = 0$ and $c_1 = 1$, $c_2 = -1$, $c_3 = 0.2$, $D(z) = 1 - 1.6z^{-1} + 0.8z^{-2}$, $N(z) = z^{-1}$, $F(z) = 1 - z^{-1} + 0.2z^{-2}$.

Using deconvolution and selecting the order $n_h = 7$ so that the truncation error is negligible yields:

$$\frac{D(z)}{F(z)} = \frac{1-1.6z^{-1}+0.81z^{-1}}{1-z^{-1}+0.2z^{-2}} = \sum_{i=1}^{n_h} a_{hi} z^{-i}$$

$$\frac{N(z)}{F(z)} = \frac{z^{-1}}{1-z^{-1}+0.2z^{-2}} = \sum_{i=0}^{n_h} b_{hi} z^{-1}$$

The high-order linear regression model becomes

$$y(k) = -\sum_{i=1}^{n_h} a_{hi} y(k-i) + \sum_{i=1}^{n_h} b_{hi} r(k-i) + e(k) \tag{7.118}$$

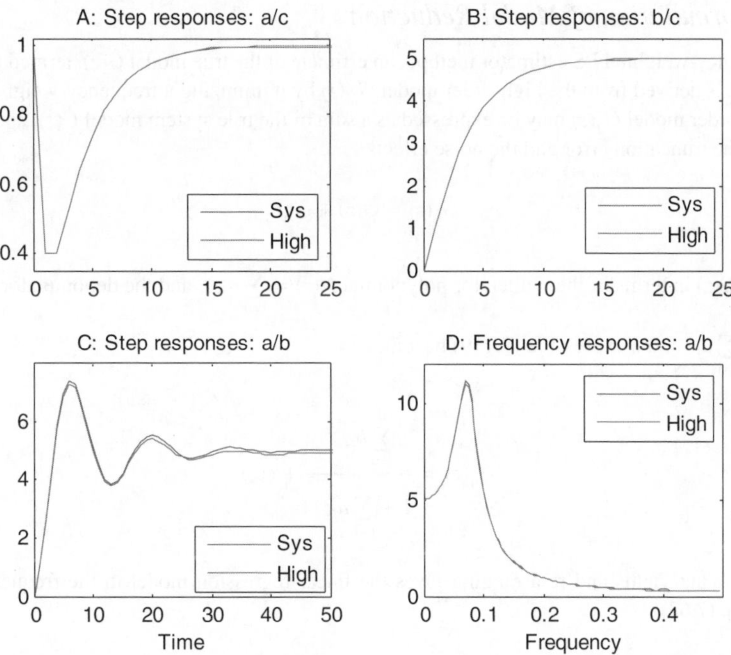

Figure 7.9 The step and frequency responses of the high-order least squares method

Figure 7.9 shows the performance of the high-order system obtained using the frequency-domain deconvolution approach. Subfigure (a) compares the step response of the FIR approximation model obtained from deconvolution $D(z)$ by $F(z)$ with the IIR model $\frac{D(z)}{F(z)}$ while subfigure (b) compares the step response of the FIR approximation model obtained from deconvolution $B(z)$ by $F(z)$ with the IIR model $\frac{B(z)}{F(z)}$. Subfigures (c) and (d) compare respectively the step and frequencies of the high-order model $G_h(z)$ and the original system model $G(z)$.

Comment *Both approaches, namely the frequency-domain deconvolution and the polynomial division, match the actual system responses if the elected order is high. However, thanks to the appropriate selection of the frequencies, the performance of the frequency-domain deconvolution approach is superior when the order is not high enough. The step and the frequency responses of the high order model derived from the deconvolution match the actual system responses.*

7.5.2 Derivation of a Reduced-Order Model

As mentioned earlier, the commonly-used approach for obtaining a reduced-order model includes the balanced realization approach and the frequency-weighted LS estimator approach. The balanced realization hinges upon the equality of the controllability and observability Grammians. The diagonal entries are the singular values of the Hankel matrix. The ratio of the maximum to the minimum singular values is the minimum.

In the frequency-weighted LS estimator method, an estimate of the true model $G(z)$, termed the reduced-order model, is derived from the high-order model $G_h(z)$ by minimizing a frequency-weighted residual [6, 8].

7.5.3 Formulation of Model Reduction

In the frequency-weighted LS estimator method, an estimate of the true model $G(z)$, termed the reduced-order model, is derived from the high-order model $G_h(z)$ by minimizing a frequency-weighted residual. The higher-order model $\hat{G}_h(z)$ may be expressed as a sum of the true system model $G(z)$, and the artifact $g(z)$ due to the truncation error and the noise effects:

$$\hat{G}_h(z) = G(z) + g(z) \tag{7.119}$$

Expressing $G(z)$ in terms of the numerator polynomial $N(z) = \sum_{i=0}^{n} b_i z^{-i}$ and the denominator polynomial $D(z) = 1 + \sum_{i=1}^{n} a_i z^{-i}$, the expression (7.119) becomes:

$$\hat{G}_h(z) = \frac{\sum_{i=0}^{n} b_i z^{-i}}{1 + \sum_{i=1}^{n} a_i z^{-i}} + g(z) \tag{7.120}$$

Cross-multiplying yields and re-arranging gives the linear regression model in the frequency domain similar to Eq. (7.63):

$$\hat{G}_h(z) = -\sum_{i=1}^{n_r} a_{ri} z^{-i} \hat{G}_h(z) + \sum_{i=0}^{n_r} b_{ri} z^{-i} + \varepsilon(z) \tag{7.121}$$

where $\varepsilon(z) = D(z)g(z)$.

The problem of identifying the system model $G(z)$ from the higher-order model $\hat{G}_h(z)$ is posed as a frequency-weighted least-squares estimation problem. The frequency weights are determined *a posteriori* from the frequency response of the high-order model assuming that (i) $G(z)$ has a smooth rational spectrum, (ii) $G(z)$ is dominant in $\hat{G}_h(z)$ over the bandwidth of the system, and (ii) the dominant peaks in the high-order model $\hat{G}_h(z)$ come from $G(z)$. In the Appendix, details of the model reduction scheme [6, 8] are given for completeness.

7.5.4 Model Order Selection

A number of reduced-order models for the selected model orders are identified from the high-order model $\hat{G}_h(z)$. The correct model order is selected using the AIC. The loss function denoted $LF(\boldsymbol{\theta})$ of AIC is:

$$LF(\boldsymbol{\theta}) = -2l(\boldsymbol{\theta} \mid y) + 2M_r \tag{7.122}$$

where $M_r = 2n_r$ is the number of parameters to be estimated, namely, the number of elements in $M_r x1$ vector $\boldsymbol{\theta}$, n_r is the selected model order. The loss function using AIC is given by:

$$LF(\boldsymbol{\theta}) = \log\left(\hat{\sigma}_{res}^2\right) + \frac{2M_r}{N} \tag{7.123}$$

where $\hat{\sigma}_{res}^2 = \frac{1}{N}\sum_{k=0}^{N-1} \varepsilon^2(k)$ is the estimate of the variance of the residual.

7.5.5 *Illustrative Example of High-Order Least-Squares Method*

The high-order least-squares approach of first identifying a high-order model, and then obtaining a reduced-order model of appropriate model order, which is selected using AIC criterion, is illustrated using the correlated noise model.

Example 7.11 Correlated equation error
Consider the model given in Example 7.10.

$$y(k) = 1.6y(k-1) - 0.8y(k-2) + r(k-1) + e(k) - e(k-1) + 0.2e(k-2) \tag{7.124}$$

The variance of the zero-mean white noise process was unity. The number of data samples was $N = 1024$.

Results of the high-order least-squares method:

- *Identified high-order model*: The high-order model was chosen to be a 7th order linear regression model. The number of data samples used was $N = 1024$.
- *Reduced-order model*: Reduced-order models were derived from the high-order model for the selected orders of 2, 3, 4 and 5 using the frequency-weighted approach
- *Model order selection*: The four reduced-order models were analyzed using the AIC measure. The second-order model was found to have the minimal AIC measure among the four derived reduced-order models.

The identified second-order model $\hat{G}(z)$ is given by:

$$\hat{G}(z) = \frac{-0.0578 + 1.070z^{-1} + 0.0055z^{-2}}{1 - 1.603z^{-1} + 0.8050z^{-2}} \tag{7.125}$$

The identified model was very close to the true model $G(z) = \dfrac{1.070z^{-1}}{1 - 1.6z^{-1} + 0.81z^{-2}}$.

The performance of the two-stage identification of selecting a high order model of 7th order and reducing it to selected orders of 2, 3, 4, and 5 is shown in Figure 7.10. Subfigure (a) shows the step responses of the system, high-order model, and all four reduced-order models, while subfigure (b) gives the AIC measure for each of the four reduced-order models. Subfigures (c), (d), (e), and (f) show the pole-zero maps of the four reduced-order models.

Key points on the performance of the proposed HOLS method:
The performance of the two-stage identification of selecting a high-order model of 7th order and reducing it to selected orders of 2, 3, 4, and 5 is shown in Figure 7.10. Subfigure (a) shows the step responses of the system, high-order model, and all four reduced-order models, while subfigure (b) gives the AIC measure for each of the four reduced-order models. Subfigures (c), (d), (e), and (f) show the pole-zero maps of the four reduced-order models.

- The step responses of the system, high-order model and all four reduced-order models are almost identical.
- It is interesting to note that all poles of the reduced-order models of the selected orders 2, 3, 4, and 5 contain poles which are close to the true poles of the system.
- The extraneous poles are due to the artifact $g(z)$ and the noise effect. The extraneous poles located in the right-half of the z-plane are approximately canceled by the zeros for the selected orders greater than the correct identified second-order model. However for the 5th model shown in subfigure (f) of Figure 7.10, there are two complex conjugate poles in the left-half plane which are not canceled.

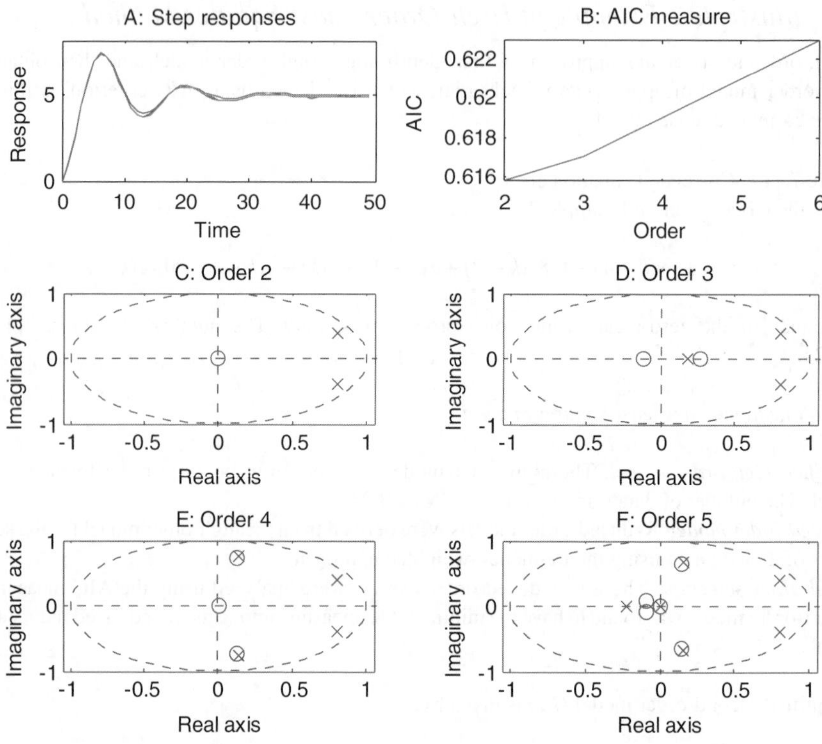

Figure 7.10 The performance of the high-order least squares method

- The AIC criterion identifies the correct reduced-order model.
- The order of the high-order model should be low enough to ensure that the data matrix is well conditioned, yet large enough to ensure that the equation error is whitened, showing that the high-order model has captured completely the dynamic behavior of the system.
- The order of the high-order model should be low enough to ensure that the data matrix is well conditioned, yet large enough to ensure that the equation error is whitened, that is the high-order model has captured completely the dynamic behavior of the system. The condition number for the 7th order high-order model was 21.05. The condition number $k(\boldsymbol{H})$ given by Eq. (7.97) increases with the increase in the model order n_h. The condition numbers for selected orders $n_h = 2, 3, 4, \ldots, 12$ were respectively, $1.60, 3.19, 5.64, 18.31, 21.05, 23.39, 27.64, 31.16, 34.16, 36.32$.

7.5.6 Performance of the High-Order Least-Squares Scheme

The high-order least squares method performs well in the presence of the correlated equation error in the sense that (i) the estimates tend to the true value as data samples tendsto infinity, (ii) estimates are unbiased, (iii) the estimates have closed-form solution, and (iv) the computational algorithm is numerically robust.

The order of the high-order model should be selected such that it is low enough to ensure that the data matrix is well conditioned yet large enough to ensure the high-order model has captured completely the dynamic behavior of the system. The higher the selected order, the larger the condition number.

The performance of the high-order method based on extensive simulation is comparable to that of the prediction error method. The estimates are unbiased, and efficient (achieve the Cramér–Rao lower bound). The performance of the identification of the reduced-order model from the high-order model is simple and gives an accurate estimate of the system model. It also provides a clear understanding of the model order selection process. There are approximate pole-zero cancellations if the selected order is higher than the correct order, as shown in Figure 7.10. If the location of the poles of the system is in the right-half plane, indicating that the system is a low pass, one may select the model order such that all the poles are in the right-half plane [5].

7.6 The Prediction Error Method

7.6.1 Residual Model

Using the expression for the Kalman filter residual given by Eq. (7.29) we get:

$$e(z) = y(z) - \hat{y}(z) = \frac{D(z)}{F(z)}y(z) - \frac{N(z)}{F(z)}r(z) \tag{7.126}$$

Expressing in term of the coefficients of the polynomials $N(z)$, $D(z)$, and $F(z)$ yields:

$$e(z) = \frac{\left(1 + \sum_{i=1}^{n} a_i z^{-i}\right) y(z) - \sum_{i=1}^{n} b_i z^{-i} r(z)}{1 + \sum_{i=1}^{n} c_i z^{-i}} \tag{7.127}$$

From the above expression, it can be deduced that (i) residual $e(z)$ is stable as $F(z)$ is stable and (ii) the residual is a nonlinear function of the unknown parameters, namely the coefficients of $N(z)$, $D(z)$, and $F(z)$. Expressing (7.127) as the linear regression model (ARMAX model) we get:

$$e(k) = y(k) + \sum_{i=1}^{n} a_i y(k-i) - \sum_{i=1}^{n} b_i r(k-i) - \sum_{i=1}^{n} c_i e(k-i) \tag{7.128}$$

Let θ_a be an augmented feature vector which includes the coefficients $\{b_i\}$ of numerator polynomial $N(z)$, and the coefficients $\{a_i\}$ of the denominator polynomial $D(z)$ of the system transfer function $G(z)$ as well as the coefficients $\{c_i\}$ of the polynomial $F(z)$. The augmented $M_a x1$ vector θ_a, where $M_a = 3n$, is given by:

$$\theta_a = [\, a_1 \quad a_2 \quad . \quad a_n \quad b_1 \quad b_2 \quad . \quad b_n \quad c_1 \quad c_2 \quad . \quad c_n \,]^T \tag{7.129}$$

7.6.2 Objective Function

The objective of prediction error method (PEM) is to minimize the following quadratic function of the prediction error $V_N = \frac{1}{2}\sum_{k=1}^{N} e^2(k)$:

$$\min_{\theta_a}\{V_N\} \tag{7.130}$$

where $e(z) = F^{-1}(z)\,(D(z)y(z) - N(z)r(z))$. Since V_N depends nonlinearly on unknown parameter θ_a, the PEM therefore involves a nonlinear optimization problem and the estimates are obtained iteratively using the popular Newton–Raphson method or some gradient-based method.

Assumptions

- *The true model is a member of the set of identification models generated by Eq. (7.127).*
- *The input and the output are quasi-stationary stochastic processes.*
- *The error term $e(k)$ is a zero-mean white noise process.*
- *The input is persistently exciting of sufficient order.*
- *$F(z)$ is a stable polynomial.*

Lemma 7.1 *If the above assumptions hold, then the prediction error estimate $\hat{\theta}_a$ of θ_a is asymptotically unbiased and is asymptotically efficient (achieves the Cramér-Rao lower bound).*

Proof: The proof is given in [7].

7.6.3 Iterative Prediction Algorithm

The Newton–Raphson method is generally employed to estimate the prediction error estimates iteratively by minimizing the loss function at the time instant k:

$$\min_{\theta_a}\left\{ V_N(k) = \frac{1}{2}\sum_{i=1}^{k} e^2(i) \right\} \tag{7.131}$$

The optimal estimate $\hat{\theta}_a(k)$ of θ_a at the kth iteration is obtained by setting the gradient of $V_N(k)$ to zero:

$$\boldsymbol{p}(k) = \frac{\delta V_N(k)}{\delta \theta_a} = 0 \tag{7.132}$$

The minimizing (7.131) at each iteration yields the following iterative equation:

$$\hat{\theta}_a(k) = \hat{\theta}_a(k-1) - \boldsymbol{P}(k-1)\boldsymbol{p}(k-1) \tag{7.133}$$

$\boldsymbol{p}(k) = \dfrac{\delta V_N(k)}{\delta \theta_a}$ is a $M_a x1$ vector of the gradient of $V_N(k)$, $\boldsymbol{P}(k) = \left(\dfrac{\partial^2 V_N(k)}{\partial \theta_a^2}\right)^{-1}$ is the $M_a x M_a$ inverse of the Hessian matrix $\dfrac{\partial^2 V_N(k)}{\partial \theta_a^2}$. Let us now compute the expressions for the gradient and the Hessian matrix using the expressions for the filtered input $r_F(z) = \dfrac{r(z)}{F(z)}$, the filtered output $y_F(z) = \dfrac{y(z)}{F(z)}$, and the filtered residual $e_F(z) = \dfrac{e(z)}{F(z)}$ given by

$$\begin{aligned} y_F(k) &= y(k) - \sum_{i=1}^{n} c_i y_F(k-i) \\ r_F(k) &= r(k) - \sum_{i=1}^{n} c_i r_F(k-i) \\ e_F(k) &= e(k) - \sum_{i=1}^{n} c_i e_F(k-i) \end{aligned} \tag{7.134}$$

Using the expression for the residual given by Eqs. (7.126) and (7.127), we get:

$$e(k) = y_F(k) + \sum_{i=1}^{n} a_i y_F(k-i) - \sum_{i=1}^{n} b_i r_F(k-i) \tag{7.135}$$

The gradient $\boldsymbol{p}(k) = \frac{\delta V_N(k)}{\delta \theta_a}$ is obtained by differentiating $V(k)$ with respect to a_i, b_i, and c_i:

$$\begin{aligned} \frac{\delta V(k)}{\delta a_i} &= \sum_{i=1}^{k} e(i) y_F(k-i) \\ \frac{\delta V(k)}{\delta b_i} &= -\sum_{i=1}^{k} e(i) r_F(k-i) \\ \frac{\delta V(k)}{\delta c_i} &= -\sum_{i=1}^{k} e(i) e_F(k-i) \end{aligned} \tag{7.136}$$

The last equation governing the filtered residual in Eq. (7.136) is derived from the last expression in Eq. (7.134). Let us define an augmented $1xM_a$ data vector $\boldsymbol{\psi}_a^T(k)$ as:

$$\boldsymbol{\psi}_a^T(k) = \begin{bmatrix} -y_F(k-1) & . & -y_F(k-n) & r_F(k-1) & . & r_F(k-n) & e_F(k-1) & . & e_F(k-n) \end{bmatrix} \tag{7.137}$$

Using the expressions (7.136) and (7.137), partial derivative $\boldsymbol{p}(k) = \dfrac{\delta V_N(k)}{\delta \theta_a}$ becomes:

$$\boldsymbol{p}(k) = -\sum_{i=1}^{k} e(i) \boldsymbol{\psi}_a(i) \tag{7.138}$$

The Hessian matrix $\boldsymbol{P}^{-1}(k) = \frac{\partial^2 V_N(k)}{\partial \theta_a^2}$ is obtained by partial differentiation of $\boldsymbol{p}(k)$ given by:

$$\boldsymbol{P}^{-1}(k) = \sum_{i=1}^{k} \boldsymbol{\psi}_a(i) \boldsymbol{\psi}_a^T(i) \tag{7.139}$$

The gradient and the Hessian matrix given by Eqs. (7.138) and (7.139) involve summation terms obtained from the initial iteration until the present $i = 1, 2, \ldots, k$. Expressing Eq. (7.138) in an iterative form yields:

$$\boldsymbol{p}(k) = \boldsymbol{p}(k-1) - e(k) \boldsymbol{\psi}_a(k) \tag{7.140}$$

Since the optimal $\hat{\boldsymbol{\theta}}_a(k-1)$ is a solution of Eq. (7.132) at the time instant $k-1$, $\boldsymbol{p}(k-1) = \boldsymbol{0}$ and hence

$$\boldsymbol{p}(k) = -e(k) \boldsymbol{\psi}_a(k) \tag{7.141}$$

The iterative form for the Hessian matrix (7.139) is:

$$\boldsymbol{P}^{-1}(k) = \boldsymbol{P}^{-1}(k-1) + \boldsymbol{\psi}_a(k) \boldsymbol{\psi}_a^T(k) \tag{7.142}$$

Using the matrix inversion lemma yields:

$$\boldsymbol{P}(k) = \boldsymbol{P}(k-1) - \boldsymbol{P}(k-1) \boldsymbol{\psi}_a(k) \left[1 + \boldsymbol{\psi}_a^T(k) \boldsymbol{P}(k-1) \boldsymbol{\psi}_a(k)\right]^{-1} \boldsymbol{\psi}_a^T(k) \boldsymbol{P}(k-1) \tag{7.143}$$

Summarizing, the iterative prediction error algorithm is

$$\begin{aligned}
\hat{\theta}_a(k) &= \hat{\theta}_a(k-1) + \boldsymbol{K}(k)e(k) \\
\boldsymbol{K}(k) &= \boldsymbol{P}(k)\boldsymbol{\psi}_a(k) \\
\boldsymbol{P}(k) &= \boldsymbol{P}(k-1) - \boldsymbol{P}(k-1)\boldsymbol{\psi}_a(k)\left[1 + \boldsymbol{\psi}_a^T(k)\boldsymbol{P}(k-1)\boldsymbol{\psi}_a(k)\right]^{-1}\boldsymbol{\psi}_a^T(k)\boldsymbol{P}(k-1)
\end{aligned} \tag{7.144}$$

7.6.4 Family of Prediction Error Algorithms

There are a family of algorithms which are simpler but approximate versions of the Newton–Raphson algorithm (7.144) including the Least Mean Squares (LMS) and pseudo-linear regression (PLR) method. The LMS algorithm is a gradient-based method and is a special case of the Newton–Raphson algorithm when the Hessian matrix is an identity matrix given by

$$\hat{\theta}_a(k) = \hat{\theta}_a(k-1) + \boldsymbol{K}(k)e(k) \tag{7.145}$$

where $\boldsymbol{K}(k) = \dfrac{1}{\|\boldsymbol{\psi}_a(k)\|^2}$.

The PLR algorithm is essentially a least squares method adapted to the ARMAX model (7.128) with the augmented parameter vector (7.129) and the augmented data vector (7.137). In the data vector the residual $e(k)$ is replaced by its previous estimate $\hat{e}(k) = y(k) - \hat{y}(k-1)$.

$$\begin{aligned}
\hat{\theta}_a(k) &= \hat{\theta}_a(k-1) + \boldsymbol{K}(k)e(k) \\
e(k) &= y(k) - \boldsymbol{\psi}_a^T(k)\hat{\theta}_a(k-1) \\
\boldsymbol{K}(k) &= \boldsymbol{P}(k)\boldsymbol{\psi}_a(k)\left[1 + \boldsymbol{\psi}_a^T(k)\boldsymbol{P}(k-1)\boldsymbol{\psi}_a(k)\right]^{-1} \\
\boldsymbol{P}(k) &= \boldsymbol{P}(k-1) - \boldsymbol{P}(k-1)\boldsymbol{\psi}_a(k)\left[1 + \boldsymbol{\psi}_a^T(k)\boldsymbol{P}(k-1)\boldsymbol{\psi}_a(k)\right]^{-1}\boldsymbol{\psi}_a^T(k)\boldsymbol{P}(k-1)
\end{aligned} \tag{7.146}$$

7.7 Comparison of High-Order Least-Squares and the Prediction Error Methods

The performance of the least squares, the high-order least squares, and the prediction error methods are compared.

The least squares method is simple, computationally efficient, and gives a closed-form solution of the estimate. One can determine analytically the statistics of the errors including the variance of the residual and the covariance of the parameter estimation error. However, its performance degrades in the presence of colored noise equation error as a result of correlation between the data vector and the equation error. The estimates will have bias. The degradation may be acceptable for some applications.

One of the main difficulties in system identification is that the equation error may be correlated with the data vector formed of the past inputs and the past outputs. The prediction error and the high-order least-squares approaches overcome the performance degradation resulting from this cross-correlation.

The prediction error method and the high-order least-squares method are both designed to whiten the equation error. The high-order least-squares method uses FIR approximation of the IIR filters, $\frac{D(z)}{F(z)}$ and $\frac{N(z)}{F(z)}$, at the cost of estimating larger number coefficients of the high-order model, so that the computationally efficient least-squares method may be employed. The prediction error method, on the other hand, estimates the coefficients of the numerator $N(z)$ and the coefficients of denominator $D(z)$

of the system, as well as the coefficients of the coloring filter $F(z)$. Unlike the case of the high-order least-squares method, due to the presence of an inverse of the coloring filter $F(z)$, the residual is a nonlinear function of the unknown parameters of the system as well as those of the coloring filter. As a result, the prediction error method is computational burdensome.

The order of the high-order model should be low enough to ensure that the data matrix is well conditioned, yet large enough to ensure that the equation error is whitened, that is the high-order model has captured completely the dynamic behavior of the system.

Summarizing, the high-order least-squares method estimates a large number of parameters using the computationally simple least-squares algorithm, whereas the prediction error method estimates the unknown parameters using a computationally burdensome iterative algorithm. The prediction error method may be used to estimate the unstable system as the predictor is asymptotically stable. The data matrix of high-order least-squares must be well conditioned.

7.7.1 *Illustrative Example: LS, High Order LS, and PEM*

The performances of the identification using the least-squares, the high-order least-squares, and the prediction error methods were compared using a simulated example given by Eq. (7.124). The estimates of the feature vector and the step response of the identified models were compared with those of the (true) system. The number of data samples was $N = 5000$.

Example 7.12 *Model for the prediction error:*
The state-space model (7.1) is given by

$$\begin{aligned} \boldsymbol{x}(k+1) &= \begin{bmatrix} 1.6 & 1 \\ -0.8 & 0 \end{bmatrix} \boldsymbol{x}(k) + \begin{bmatrix} 1 \\ 0 \end{bmatrix} r(k) + \begin{bmatrix} 1 \\ 0 \end{bmatrix} w(k) \\ y(k) &= [\,1 \quad 0\,]\boldsymbol{x}(k) + v(k) \end{aligned} \tag{7.147}$$

where $v(k)$ and $w(k)$ are both zero-mean white noise processes with unit variance, $D(z) = 1 - 1.6z^{-1} + 0.8z^{-2}$ and $N(z) = z^{-1}$. The poles are $\lambda\,(A) = 0.8 \pm j0.4$
The Kalman filter:
The Kalman gain $\boldsymbol{K} = \begin{bmatrix} 0.9175 \\ -0.5661 \end{bmatrix}$, $F\,(z) = 1 + c_1 z^{-1} + c_2 z^{-2}$, $c_1 = -0.6825$, $c_2 = 0.2339$. The poles are $\lambda\,(\boldsymbol{A} - \boldsymbol{KC}) = 0.3412 \pm j0.3428$

The innovation form (7.25) is given by

$$\begin{aligned} \hat{\boldsymbol{x}}(k+1) &= \begin{bmatrix} 1.6 & 1 \\ -0.8 & 0 \end{bmatrix} \hat{\boldsymbol{x}}(k) + \begin{bmatrix} 1 \\ 0 \end{bmatrix} r(k) + \begin{bmatrix} 0.9175 \\ -0.5661 \end{bmatrix} e(k) \\ e\,(k) &= y(k) - [\,1 \quad 0\,]\hat{\boldsymbol{x}}(k) \end{aligned} \tag{7.148}$$

The predictor form (7.26) is:

$$\begin{aligned} \hat{\boldsymbol{x}}(k+1) &= \begin{bmatrix} 0.6825 & 1 \\ -0.2339 & 0 \end{bmatrix} \hat{\boldsymbol{x}}(k) + \begin{bmatrix} 1 \\ 0 \end{bmatrix} r(k) + \begin{bmatrix} 0.9175 \\ -0.5661 \end{bmatrix} y(k) \\ \hat{y}(k) &= [\,1 \quad 0\,]\hat{\boldsymbol{x}}(k) \end{aligned} \tag{7.149}$$

The predictor model and the regression model are given by

$$\hat{y}(k) = 0.6825\hat{y}(k-1) - 0.2239\hat{y}(k-2) + 0.9175y(k-1) - 0.5661y(k-2) + r(k-1) \quad (7.150)$$

$$y(k) = 1.6y(k-1) - 0.8y(k-2) + r(k-1) + e(k) - 0.6825e(k-1) + 0.2239e(k-2) \quad (7.151)$$

where the variance σ_e^2 of the zero-mean equation error $e(k)$ is $\sigma_e^2 = 1.5205$.

The model for the least-squares method:

The linear regression model is given by Eq. (7.151). The feature vector and the data vector are

$$\boldsymbol{\theta} = [\,-1.6 \quad 0.8 \quad 1\,]^T \quad (7.152)$$

$$\boldsymbol{\psi}^T(k) = [\,-y(k-1) \quad -y(k-2) \quad r(k-1)\,] \quad (7.153)$$

Model for the prediction error method:

Expressing the model (7.124) in the form of (7.127) yields:

$$e(z) = \frac{(1-1.6z^{-1}+0.8z^{-2})y(z)}{1-0.6825z^{-1}+0.2339z^{-2}} - \frac{z^{-1}r(z)}{1-0.6825z^{-1}+0.2339z^{-2}} \quad (7.154)$$

Expressing in terms of the filtered input, output, and the residual using Eq. (7.135) we get

$$e(k) = y_F(k) - 1.6y_F(k-1) + 0.8y_F(k-2) - r_F(k-1) \quad (7.155)$$

Where filtered data is

$$\begin{aligned} y_F(k) &= y(k) + 0.6825y_F(k-1) - 0.2339y_F(k-2) \\ r_F(k) &= r(k) + 0.6825r_F(k-1) - 0.2339r_F(k-2) \\ e_F(k) &= e(k) + 0.6825e_F(k-1) - 0.2339e_F(k-2) \end{aligned} \quad (7.156)$$

The augmented 5x1 feature vector (7.129) and the augmented 1x5 data vector (7.137) are:

$$\boldsymbol{\theta}_a = [\,-1.6 \quad 0.8 \quad 1 \quad -0.6825 \quad 0.2339\,]^T \quad (7.157)$$

$$\boldsymbol{\psi}_a^T(k) = [\,-y_F(k-1) \quad -y_F(k-2) \quad r_F(k-1) \quad e_F(k-1) \quad e_F(k-2)\,] \quad (7.158)$$

The model for the high-order least-squares method:

The high-order model given by Eq. (7.118) is

$$y(k) = -\sum_{i=1}^{7} a_{hi}y(k-i) + \sum_{i=1}^{7} b_{hi}r(k-i) + e(k) \quad (7.159)$$

The high order was $n_h = 7$.

Table 7.3 shows the quality of the estimates using the least-squares, the high-order least-squares, and the prediction error methods. The estimates $\{\hat{a}_i\}$ and $\{\hat{b}_i\}$ of the plant parameter coefficients $\{a_i\}$ and $\{b_i\}$ are given when the true values were $a_1 = -1.6$, $a_2 = 0.8$, and $b_1 = 1$.

In the PEM, the estimate of the coloring filter $F(z) = 1 - 0.6825z^{-1} + 0.2339z^{-2}$ was $\hat{F}(z) = 1 - 0.6779z^{-1} + 0.2443z^{-2}$.

Figure 7.11 shows the performance of the LS, HOLS, and PEM methods. Subfigures (a), (b), and (c) show respectively the step responses of the system computed using the LS, the HOLS, and the PEM. The true step response of the system is also shown for comparison. This figure clearly shows the superior

Table 7.3 Least squares, high-order least squares, and prediction error estimates

	$[1 \;\; \hat{a}_1 \;\; \hat{a}_2]$	$[\hat{b}_0 \;\; \hat{b}_1 \;\; \hat{b}_2]$
Least squares	[1 −1.2886 0.5029]	[0 1.0069 0]
High-order least squares	[1 −1.5999 0.7929]	[−0.0342 1.0383 −0.0317]
Prediction error	[1 −1.6015 0.7984]	[01.0060 − 0.0313]

performance of the HOLS over the LS and its closeness in performance to the most popular PEM, when tackling practical identification problems where the noise and disturbance signals involved are colored, rather than white as they are usually assumed.

Comment *It is relatively simple to use the least-squares identification method for the model based on the Kalman filter structure as the equation error is colored noise generated by the innovation sequence. As the equation error for the HOLS method is a zero-mean white noise process, the least-squares estimates of the parameters of the high-order model are unbiased and satisfy the Cramér–Rao lower band. The performance of the reduced-order model derived from the high-order model is dependent upon the selection of appropriate frequencies. With the use of software to select the appropriate frequencies automatically, the estimated reduced-order model was close to the true model. The high-order least-squares method is computationally efficient and yields a closed-form solution, and its performance is very close to that of the widely used prediction error method.*

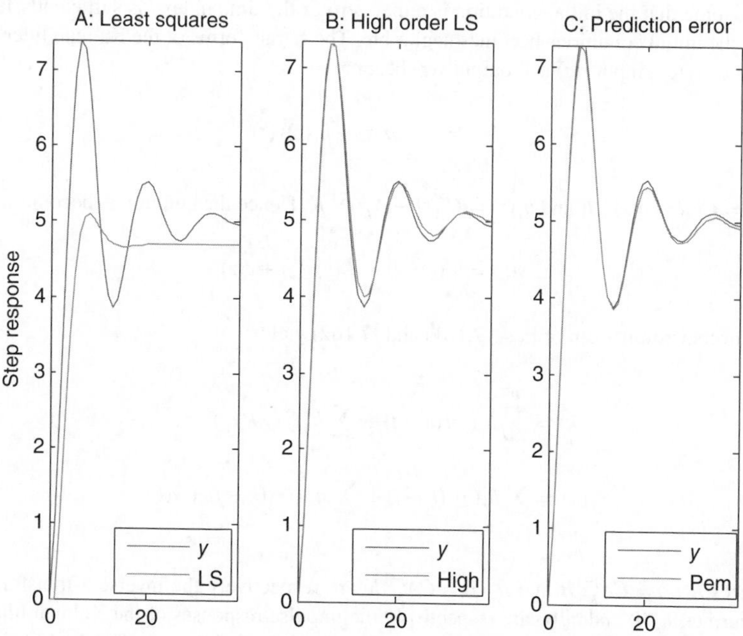

Figure 7.11 Comparison of the step responses with different identification methods

7.8 Subspace Identification Method

The subspace identification problem consists of estimating the system matrices $(\boldsymbol{A}, \boldsymbol{B}, \boldsymbol{C})$ of the state-space model from the given set of input and output measurement data. Either the innovation form or the predictor form of the Kalman filter can be employed as the identification model since either of them can represent the input and output measurement data $\{r(k)\}$ and $\{y(k)\}$ exactly. Therefore, one has the option to use any of these forms for convenience. From the given input–output measurements, the state-space model $(\boldsymbol{A}, \boldsymbol{B}, \boldsymbol{C}, \boldsymbol{K})$ is estimated if the innovation form is chosen, whereas the state-space model $(\boldsymbol{A} - \boldsymbol{KC}, \boldsymbol{B}, \boldsymbol{C}, \boldsymbol{K})$ is estimated if the predictor form is selected. The structure of the identified state-space model will, in general, be equivalent to the actual state-space model. The identified state-space model may be transformed using a similarity transformation so that the true and identified models may be compared.

We will focus here on the predictor form. Consider the predictor form relating the estimate of the state to the input and output of the system given in Eq. (7.26).

7.8.1 *Identification Model: Predictor Form of the Kalman Filter*

Consider the predictor form relating the estimate of the state and the input and the output of the system given in Eq. (7.26). Rewriting for an arbitrary time instant i, yields:

$$\begin{aligned} \hat{\boldsymbol{x}}(i+1) &= \boldsymbol{A}_K \hat{\boldsymbol{x}}(i) + \boldsymbol{B} r(i) + \boldsymbol{K} y(i) \\ \hat{y}(i) &= \boldsymbol{C} \hat{x}(i) \\ y(i) &= \hat{y}(i) + e(i) \end{aligned} \tag{7.160}$$

where $\boldsymbol{A}_K = \boldsymbol{A} - \boldsymbol{KC}$.

We will assume that the initial condition is either zero or the time index i is sufficiently large so that the effect of the initial conditions becomes negligible. The z-transforms of the Kalman filter output $\hat{y}(z)$ in terms of the system input $r(z)$ and output $y(z)$ become:

$$\hat{y}(z) = h_r(z) r(z) + h_y(z) y(z) \tag{7.161}$$

where $h_r(z) = \boldsymbol{C}(z\boldsymbol{I} - \boldsymbol{A}_K)^{-1}\boldsymbol{B}$ and $h_y(z) = \boldsymbol{C}\left(z\boldsymbol{I} - \boldsymbol{A}_K\right)^{-1}\boldsymbol{K}$. Hence the output $y(z)$ becomes:

$$y(z) = h_r(z) r(z) + h_y(z) y(z) + e(z) \tag{7.162}$$

Taking the inverse z-transform of Eqs. (7.161) and (7.162) yields:

$$\begin{aligned} \hat{y}(k) &= \sum_{j=1}^{k} h_r(j) r(k-j) + \sum_{j=1}^{k} h_y(j) y(k-j) \\ y(k) &= \sum_{j=1}^{k} h_r(j) r(k-j) + \sum_{j=1}^{k} h_y(j) y(k-j) + e(k) \end{aligned} \tag{7.163}$$

The sequences $h_r(j) = \boldsymbol{C}\boldsymbol{A}_K^{j-1}\boldsymbol{B}$ and $h_y(j) = \boldsymbol{C}\boldsymbol{A}_K^{j-1}\boldsymbol{K}$ are respectively the inverse z-transforms of $h_r(z)$ and $h_y(z)$. Further, $h_r(j)$ and $h_y(j)$ are respectively the impulse responses of the Kalman filter at the jth time instant when the reference input $r(j)$ and the system output $y(j)$ are both delta functions, that is, $r(j) = \delta(j)$ and $y(j) = \delta(j)$. The sequences $\{\boldsymbol{C}\boldsymbol{A}_K^{j-1}\boldsymbol{B}\}$ and $\{\boldsymbol{C}\boldsymbol{A}_K^{j-1}\boldsymbol{K}\}$ are termed the Markov parameters of the Kalman filter or observer.

Assumption *Since the predictor form (7.160) is asymptotically stable, there exists a finite time instant M such that this impulse response sequence may be assumed to be negligible for all time instants equal to or greater than or equal to M, that is $h_r(j) = 0$ and $h_y(j) = 0$ for all $j \geq M$.*

Assuming that $k \geq M$, and truncating the impulse response sequences $\{h_r(j)\}$ and $\{h_y(j)\}$, Eq. (7.163) becomes:

$$\begin{aligned} \hat{y}(k) &\approx \sum_{j=1}^{M} h_r(j)r(k-j) + \sum_{j=1}^{M} h_y(j)y(k-j) \\ y(k) &\approx \sum_{j=1}^{M} h_r(j)r(k-j) + \sum_{j=1}^{M} h_y(j)y(k-j) + e(k) \end{aligned} \tag{7.164}$$

The assumption of having both impulse response sequences vanish for $k \geq M$ will ensure that Eq. (7.164) have a finite number of impulse response coefficients to be estimated.

Comment *It is interesting to note that the linear regression model for the subspace method given by Eq. (7.164) is similar to that of the high-order LS method. A key point worth pointing out here is that in the case of the HOLS method, the rational polynomials $\frac{D(z)}{F(z)}$ and $\frac{N(z)}{F(z)}$ are approximated by high-order polynomials, whereas in the SM method, it is the inverse z-transforms of $h_r(z) = \mathbf{C}(z\mathbf{I} - \mathbf{A}_K)^{-1}\mathbf{B}$ and $h_y(z) = \mathbf{C}(z\mathbf{I} - \mathbf{A}_K)^{-1}\mathbf{K}$, or equivalently, their corresponding impulse responses $\{h_r(j)\}$ and $\{h_y(j)\}$, that are truncated. The required state-space model is finally determined from the estimates of impulse responses $\{h_r(j)\}$ and $\{h_y(j)\}$.*

7.8.1.1 Estimation of the State-Space Model

The objective of the subspace method is to estimate the state-space model $(\mathbf{A}_K, \mathbf{B}, \mathbf{C}, \mathbf{K})$. The estimation is to be obtained in two steps. The impulse response sequences (Markov parameters) are estimated, and then the state-space model parameters $\mathbf{A}_K$, $\mathbf{B}$, $\mathbf{K}$, and $\mathbf{C}$ are estimated from the extended observability and extended controllability matrices, which are derived from the impulse response sequences.

7.8.1.2 Estimation of the Impulse Response Sequences

Thanks to the truncation of the impulse response sequences, that is setting to zero the impulse response sequences which are negligible, there are $2M$ number of the unknown impulse response sequences to be estimated. Hence there should be at least $2N$ equations. Let us collect $N \geq 2M$ data $y(k-i)$, and $\hat{y}(k-i)$, and $e(k-i) = y(k-i) - \hat{y}(k-i)$ for $i = 0, 2, 3, \ldots, N-1$. The $Nx1$ vectors formed the outputs $\hat{\mathbf{y}}(k)$, $\mathbf{y}(k)$, and the residuals $\mathbf{e}(k) = \mathbf{y}(k) - \hat{\mathbf{y}}(k)$ are defined as follows:

$$\mathbf{y}(k) = \begin{bmatrix} y(k) \\ y(k-1) \\ y(k-2) \\ . \\ y(k-N)1) \end{bmatrix}, \quad \hat{\mathbf{y}}(k) = \begin{bmatrix} \hat{y}(k) \\ \hat{y}(k-1) \\ \hat{y}(k-2) \\ . \\ \hat{y}(k-N+1) \end{bmatrix}, \quad \mathbf{e}(k) = \begin{bmatrix} e(k) \\ e(k-1) \\ e(k-2) \\ . \\ e(k-N+1) \end{bmatrix}$$

7.8.1.3 Batch Processing

We will assume that the present and the past the time instants $k-i$, $i = 0, 2, 3, \ldots, N-1$ are sufficiently large so that (i) the effect of initial condition $\{\mathbf{A}_K^i \hat{\mathbf{x}}(0)\}$ may be neglected and (ii) the impulse response sequences $\{h_r(j)\}$ and $\{h_y(j)\}$ may be truncated to some finite number N. Let us collect the present and the past N predicted outputs $\hat{y}(k-i) : i = 0, 1, 2, \dots, N-1$. Using Eq. (7.164) we get the following vector-matrix model:

$$\begin{aligned} \hat{\boldsymbol{y}}(k) &= \boldsymbol{R}(k)\boldsymbol{h}_r + \boldsymbol{Y}(k)\boldsymbol{h}_y \\ \boldsymbol{y}(k) &= \boldsymbol{R}(k)\boldsymbol{h}_r + \boldsymbol{Y}(k)\boldsymbol{h}_y + \boldsymbol{e}(k) \end{aligned} \tag{7.165}$$

where $\boldsymbol{R}(k)$ and $\boldsymbol{Y}(k)$ NxM matrices are formed of the past values of $\{r(i)\}$ and $\{y(i)\}$, $\boldsymbol{h}_r$ and $\boldsymbol{h}_y$ are $Nx1$ unknown Markov parameters to be estimated. They are defined as follows:

$$\boldsymbol{R}(k) = \begin{bmatrix} r(k-1) & r(k-2) & r(k-3) & . & r(k-M) \\ r(k-2) & r(k-3) & r(k-4) & . & r(k-1-M) \\ r(k-3) & r(k-4) & r(k-5) & . & r(k-2-M) \\ . & . & . & . & . \\ r(k-N) & r(k-N+1) & r(k-N+2) & . & r(k-N+M-1) \end{bmatrix},$$

$$\boldsymbol{h}_r(k) = \begin{bmatrix} h_r(1) \\ h_r(2) \\ h_r(3) \\ . \\ h_r(M) \end{bmatrix} = \begin{bmatrix} \boldsymbol{CB} \\ \boldsymbol{CA}_K\boldsymbol{B} \\ \boldsymbol{CA}_K^2\boldsymbol{B} \\ . \\ \boldsymbol{CA}_K^{M-1}\boldsymbol{B} \end{bmatrix}$$

$$\boldsymbol{Y}(k) = \begin{bmatrix} y(k-1) & y(k-2) & y(k-3) & . & y(k-M) \\ y(k-2) & y(k-3) & y(k-4) & . & y(k-1-M) \\ y(k-3) & y(k-4) & y(k-5) & . & y(k-2-M) \\ . & . & . & . & . \\ y(k-N) & y(k-N-1) & y(k-N-2) & . & y(k-N+M-1) \end{bmatrix},$$

$$\boldsymbol{h}_y(k) = \begin{bmatrix} h_y(1) \\ h_y(2) \\ h_y(3) \\ . \\ h_y(M) \end{bmatrix} = \begin{bmatrix} \boldsymbol{CK} \\ \boldsymbol{CA}_K\boldsymbol{K} \\ \boldsymbol{CA}_K^2\boldsymbol{K} \\ . \\ \boldsymbol{CA}_K^{M-1}\boldsymbol{K} \end{bmatrix}$$

Using Eq. (7.165) yields the batch processing model similar to that given by Eqs. (7.67) and (7.68):

$$\boldsymbol{y} = \boldsymbol{H}_p\theta_p + \boldsymbol{e}_p \tag{7.166}$$

where $\boldsymbol{H}_p$ is a $Nx2M$ matrix given by:

$$\boldsymbol{H}_p = \begin{bmatrix} y(k-1) & y(k-2) & . & y(k-M) & r(k-1) & . & r(k-M) \\ y(k-2) & y(k-3) & . & y(k-M-1) & r(k-2) & . & r(k-M-1) \\ . & . & . & . & . & . & . \\ y(k-N) & y(k-N-1) & . & y(k-N-M+1) & r(k-N) & . & r(k-N-M+1) \end{bmatrix}, \theta_p = \begin{bmatrix} \boldsymbol{h}_y \\ \boldsymbol{h}_r \end{bmatrix}$$

It is assumed that $\boldsymbol{H}_p$ is of full rank, and $N \geq M$. The high-order least-squares approach similar to Eqs. (7.69) and (7.74) is employed to obtain a closed-form solution for the estimate of the impulse response sequence $\boldsymbol{\theta}_p$ as shown below:

$$\hat{\boldsymbol{\theta}}_p = (\boldsymbol{H}_p)^{\dagger}\boldsymbol{y} \tag{7.167}$$

The estimate $\hat{\boldsymbol{\theta}}_p$ of $\boldsymbol{\theta}_p$ is unbiased. The pseudo-inverse $(\boldsymbol{H}_p)^{\dagger}$ is computed using SVD given by Eq. (7.97).

7.8.1.4 Estimation of the State-Space Model

The state-space model of the Kalman filter $\boldsymbol{A}_K$, $\boldsymbol{B}$, $\boldsymbol{K}$, and $\boldsymbol{C}$ are estimated from the estimate of the impulse response sequence

$$\hat{\boldsymbol{\theta}}_p = \begin{bmatrix} \hat{\boldsymbol{h}}_y \\ \hat{\boldsymbol{h}}_r \end{bmatrix} \tag{7.168}$$

$$\hat{\boldsymbol{h}}_r(k) = \begin{bmatrix} \hat{h}_r(1) \\ \hat{h}_r(2) \\ \hat{h}_r(3) \\ . \\ \hat{h}_r(M) \end{bmatrix} \quad \text{and} \quad \hat{\boldsymbol{h}}_y(k) = \begin{bmatrix} \hat{h}_y(1) \\ \hat{h}_y(2) \\ \hat{h}_y(3) \\ . \\ \hat{h}_y(M) \end{bmatrix}.$$

An interesting approach using SVD is used to estimate the state-space model using the NxN Hankel matrices, denoted $\boldsymbol{H}_y$ and $\boldsymbol{H}_r$, obtained respectively from the $Nx1$ Markov parameter vectors $\boldsymbol{h}_r$ and $\boldsymbol{h}_y$. The Hankel matrices are given by:

$$\boldsymbol{H}_r(0) = \begin{bmatrix} \boldsymbol{CB} & \boldsymbol{CA}_K\boldsymbol{B} & . & \boldsymbol{CA}_K^{M-1}\boldsymbol{B} \\ \boldsymbol{CA}_k\boldsymbol{B} & \boldsymbol{CA}_K^2\boldsymbol{B} & . & \boldsymbol{CA}_K^{M}\boldsymbol{B} \\ . & . & . & . \\ \boldsymbol{CA}_K^{M-1}\boldsymbol{B} & \boldsymbol{CA}_K^{M}\boldsymbol{B} & . & \boldsymbol{CA}_K^{2M-2}\boldsymbol{B} \end{bmatrix}; \boldsymbol{H}_y(0) = \begin{bmatrix} \boldsymbol{CK} & \boldsymbol{CA}_K\boldsymbol{K} & . & \boldsymbol{CA}_K^{M-1}\boldsymbol{K} \\ \boldsymbol{CA}_k\boldsymbol{K} & \boldsymbol{CA}_K^2\boldsymbol{K} & . & \boldsymbol{CA}_K^{M}\boldsymbol{K} \\ . & . & . & . \\ \boldsymbol{CA}_K^{M-1}\boldsymbol{K} & \boldsymbol{CA}_K^{M}\boldsymbol{K} & . & \boldsymbol{CA}_K^{2M-2}\boldsymbol{K} \end{bmatrix}$$

$$\boldsymbol{H}_r(1) = \begin{bmatrix} \boldsymbol{CA}_K\boldsymbol{B} & \boldsymbol{CA}_K^2\boldsymbol{B} & . & \boldsymbol{CA}_K^{M}\boldsymbol{B} \\ \boldsymbol{CA}_K^2\boldsymbol{B} & \boldsymbol{CA}_K^2\boldsymbol{B} & . & \boldsymbol{CA}_K^{M+1}\boldsymbol{B} \\ . & . & . & . \\ \boldsymbol{CA}_K^{M}\boldsymbol{B} & \boldsymbol{CA}_K^{M+1}\boldsymbol{B} & . & \boldsymbol{CA}_K^{2M-1}\boldsymbol{B} \end{bmatrix}; \boldsymbol{H}_y(1) = \begin{bmatrix} \boldsymbol{CA}_K\boldsymbol{K} & \boldsymbol{CA}_K^2\boldsymbol{K} & . & \boldsymbol{CA}_K^{M}\boldsymbol{K} \\ \boldsymbol{CA}_K^2\boldsymbol{K} & \boldsymbol{CA}_K^2\boldsymbol{K} & . & \boldsymbol{CA}_K^{M+1}\boldsymbol{K} \\ . & . & . & . \\ \boldsymbol{CA}_K^{M}\boldsymbol{K} & \boldsymbol{CA}_K^{M+1}\boldsymbol{K} & . & \boldsymbol{CA}_K^{2M-1}\boldsymbol{K} \end{bmatrix}$$

7.8.1.5 SVD of the Hankel Matrix

The Hankel matrix $\boldsymbol{H}_r(0)$ can be expressed as a product of two matrices, namely the extended observability matrix $\boldsymbol{Ob} = \begin{bmatrix} \boldsymbol{C} & \boldsymbol{CA}_K & \boldsymbol{CA}_K^2 & . & \boldsymbol{CA}_K^{M-1} \end{bmatrix}^T$ and the extended controllability matrix

$\boldsymbol{Cb} = \begin{bmatrix} \boldsymbol{B} & \boldsymbol{A}_K\boldsymbol{B} & \boldsymbol{A}_K^2\boldsymbol{B} & . & \boldsymbol{A}_K^{M-1}\boldsymbol{B} \end{bmatrix}$:

$$\begin{bmatrix} \boldsymbol{CB} & \boldsymbol{CA}_K\boldsymbol{B} & . & \boldsymbol{CA}_K^{M-1}\boldsymbol{B} \\ \boldsymbol{CA}_k\boldsymbol{B} & \boldsymbol{CA}_K^2\boldsymbol{B} & . & \boldsymbol{CA}_K^{M}\boldsymbol{B} \\ . & . & . & . \\ \boldsymbol{CA}_K^{M-1}\boldsymbol{B} & \boldsymbol{CA}_K^{M}\boldsymbol{B} & . & \boldsymbol{CA}_K^{2M-2}\boldsymbol{B} \end{bmatrix} = \begin{bmatrix} \boldsymbol{C} \\ \boldsymbol{CA}_K \\ . \\ \boldsymbol{CA}_K^{M-1} \end{bmatrix} \begin{bmatrix} \boldsymbol{B} & \boldsymbol{A}_K\boldsymbol{B} & \boldsymbol{A}_K^2\boldsymbol{B} & . & \boldsymbol{A}_K^{M-1}\boldsymbol{B} \end{bmatrix} \tag{7.169}$$

$$\boldsymbol{H}_r(0) = \boldsymbol{U}_r\boldsymbol{S}_r\boldsymbol{V}_r^T \tag{7.170}$$

where $\boldsymbol{U}_r$ and $\boldsymbol{V}_r$ are MxM unitary matrices, $\boldsymbol{U}_r^T\boldsymbol{U}_r = \boldsymbol{I}$ and $\boldsymbol{V}_r^T\boldsymbol{V}_r = \boldsymbol{I}$, $\boldsymbol{S}_r = \begin{bmatrix} \boldsymbol{\Sigma}_r & \boldsymbol{0} \\ \boldsymbol{0} & \boldsymbol{0} \end{bmatrix}$, $\Sigma_r = diag[\, \sigma_{r1} \;\; \sigma_{r2} \; . \; \sigma_{rn} \,], \sigma_{r1} \geq \sigma_{r2} \geq \sigma_{r3} \geq \ldots \geq \sigma_{rn}$.

Expressions for the extended observability and the extended controllability matrices in terms of SVD matrices are given by:

$$\begin{aligned} \boldsymbol{Ob} &= \boldsymbol{U}_r\boldsymbol{\Sigma}^{1/2} \\ \boldsymbol{Cb} &= \Sigma^{1/2}\boldsymbol{V}_r^T \end{aligned} \tag{7.171}$$

7.8.1.6 Expressions for State-Space Model in Terms of the SVD

The $\boldsymbol{B}$ matrix is the first column of $\boldsymbol{Cb}$ while the $\boldsymbol{C}$ matrix is the first row $\boldsymbol{Ob}$.

The Hankel matrix $\boldsymbol{H}_r(1)$ can be expressed in terms of the SVD of $\boldsymbol{H}_r(0)$ and $\boldsymbol{A}_K$ as:

$$\boldsymbol{H}_r(1) = \boldsymbol{U}_r\boldsymbol{\Sigma}^{1/2}\boldsymbol{A}_K\Sigma^{1/2}\boldsymbol{V}_r^T \tag{7.172}$$

Hence the matrix $\boldsymbol{A}_K$ may be derived from $\boldsymbol{H}_r(1)$ as follows:

$$\boldsymbol{A}_K = \boldsymbol{\Sigma}^{-1/2}\boldsymbol{U}_r^T\boldsymbol{H}_r(1)\,\boldsymbol{V}_r\boldsymbol{\Sigma}^{-1/2} \tag{7.173}$$

7.8.1.7 Estimates of the State-Space Model

Let $\hat{\boldsymbol{H}}_r, \hat{\boldsymbol{U}}_r, \hat{\boldsymbol{V}}_r, \hat{\boldsymbol{\Sigma}}_r$ be the estimates respectively of $\boldsymbol{H}_r, \boldsymbol{U}_r, \boldsymbol{V}_r, \boldsymbol{\Sigma}_r$ derived from the least squares estimates $\hat{\boldsymbol{h}}_r$ and $\hat{\boldsymbol{h}}_y$ from Eq. (7.168). Then the estimates of the Kalman filter state-space model $(\hat{\boldsymbol{A}}_K, \hat{\boldsymbol{B}}, \hat{\boldsymbol{C}}, \hat{\boldsymbol{K}})$, are given by:

$$\hat{\boldsymbol{A}}_K = \hat{\boldsymbol{\Sigma}}^{-1/2}\hat{\boldsymbol{U}}_r^T\hat{\boldsymbol{H}}_r(1)\,\hat{\boldsymbol{V}}_r\hat{\boldsymbol{\Sigma}}^{-1/2} \tag{7.174}$$

The estimate $\hat{\boldsymbol{B}}$ and $\hat{\boldsymbol{C}}$ are respectively the first column of the estimate of the observability matrix $\boldsymbol{Cb}$ and the first row of the estimate of $\boldsymbol{Ob}$. Similarly $\hat{\boldsymbol{K}}$ is estimated from the SVD of Section 7.8.2.

The predictor form is numerically efficient when the system model is poorly damped. Further, as the eigenvalues of $\boldsymbol{A}_K$ of the predictor model $(\boldsymbol{A}_K, \boldsymbol{B}, \boldsymbol{K}, \boldsymbol{C})$ are located closer to the origin of the z-plane compared to those of the matrix $\boldsymbol{A}$ in the innovation model $(\boldsymbol{A}, \boldsymbol{B}, \boldsymbol{K}, \boldsymbol{C})$, the truncation error in the approximation of the impulse response is lower.

The innovation form may be derived from the prediction by analyzing the expression of the Kalman filter residual (7.126) using the definition of the system transfer function $G(z)$. Substituting $\frac{N(z)}{F(z)} = \frac{D(z)}{F(z)}\frac{N(z)}{F(z)}$ yields:

$$e(z) = \frac{D(z)}{F(z)} y(z) - \frac{D(z)}{F(z)} G(z) r(z) \tag{7.175}$$

From Eqs. (7.164) and (7.175), we can express the impulse responses $h_r(k)$ and $h_y(k)$ as:

$$\begin{aligned} h_y(k) &= \mathbb{Z}^{-1}\left(\frac{D(z)}{F(z)}\right) \\ h_r(k) &= \mathbb{Z}^{-1}\left(\frac{N(z)}{F(z)}\right) = \mathbb{Z}^{-1}\left(\frac{D(z)}{F(z)} G(z)\right) \end{aligned} \tag{7.176}$$

The impulse response $h_r(k)$ is a convolution of the impulse response $h_y(k)$ and $h_I(k)$:

$$h_r(k) = conv(h_y(k), h(k)) \tag{7.177}$$

where $h_I(k) = \mathbb{Z}^{-1}\left(\frac{N(z)}{D(z)}\right)$ is the impulse response of the system, that is $h_I(k) = \boldsymbol{C}\boldsymbol{A}^{k-1}\boldsymbol{B}$. The impulse response $h(k)$ may be derived from the estimate of the impulse response estimate $\hat{\theta}_p = (\boldsymbol{H}_p)^{\dagger}\boldsymbol{y}$ of the prediction error model given by Eq. (7.168). Deconvolving $\hat{\boldsymbol{h}}_r$ by $\hat{\boldsymbol{h}}_y$ using:

$$\hat{h}_I(z) = \frac{\hat{h}_r(z)}{\hat{h}_y(z)} \tag{7.178}$$

where $\hat{h}_r(z)$, $\hat{h}_y(z)$, and $\hat{h}_I(z)$ are the z-transforms of the respective estimates of the impulse responses. Let the Hankel matrices associated with $\hat{h}_I(z)$ be $\boldsymbol{H}_I(0)$ and $\boldsymbol{H}_I(1)$. The estimates of state-space model of the innovation form $(\hat{\boldsymbol{A}}, \hat{\boldsymbol{B}}, \hat{\boldsymbol{C}})$ may be used if we replace $\boldsymbol{H}_r(0)$, $\hat{\boldsymbol{H}}_r(0)$, $\boldsymbol{H}_r(1)$, and $\hat{\boldsymbol{H}}_r(1)$, respectively by $\boldsymbol{H}_I(0)$, $\hat{\boldsymbol{H}}_I(0)$, $\boldsymbol{H}_I(1)$ and $\hat{\boldsymbol{H}}_I(1)$ in Eqs. (7.170), (7.172), (7.173), and (7.174).

Comment *There are mainly three design parameters: namely horizon M, the number of data samples N, and the threshold value for the truncation number of singular values (by determining which of the values may be considered low). One common way in practice is to evaluate a set of combinations and then use model validation to determine the best values. Subspace system identification was used early on in industry since it fits very well with model predictive control.*

7.8.1.8 Illustrated Example Subspace Method

Example 7.13 *Model for the prediction error:*
The state-space model given in Example 7.12 is employed with a view to compare the performance of the subspace method with other approaches. The predictor form (7.26) is:

$$\begin{aligned} \hat{\boldsymbol{x}}(k+1) &= \boldsymbol{A}_K\hat{\boldsymbol{x}}(k) + \boldsymbol{B}r(k) + \boldsymbol{K}y(k) \\ \hat{y}(k) &= \boldsymbol{C}\hat{\boldsymbol{x}}(k) \\ y(k) &= \hat{y}(k) + e(k) \end{aligned} \tag{7.179}$$

where $A_K = A - KC$. The objective of the subspace method is to estimate the state-space predictor model A_K, B, C, and K. The disturbances $v(k)$ and $w(k)$ are both zero-mean white noise processes with unit variance.

Results of the evaluation:
The prediction error model is given by

$$\begin{aligned}\hat{x}(k+1) &= \begin{bmatrix} 0.6825 & 1 \\ -0.2339 & 0 \end{bmatrix}\hat{x}(k) + \begin{bmatrix} 1 \\ 0 \end{bmatrix} r(k) + \begin{bmatrix} 0.9175 \\ -0.5661 \end{bmatrix} y(k) \\ y(k) &= [\,1 \quad 0\,]\hat{x}(k) + e(k)\end{aligned} \tag{7.180}$$

where the variance of the residual $e(k)$ was $\sigma_e^2 = 3.5329$, $N = 7000$, and $M = 10$. The identified state-space model is given by:

$$\hat{A}_K = \begin{bmatrix} 0.4829 & 0.3673 \\ -0.3673 & 0.1442 \end{bmatrix}, \hat{B} = \begin{bmatrix} 1.0289 \\ 0.2363 \end{bmatrix}, \hat{K} = \begin{bmatrix} 0.8726 \\ 0.2248 \end{bmatrix}, \hat{C} = [\,1.0289 \quad -0.2363\,]$$

It can be verified that (A_K, B, C, K), and $(\hat{A}_K, \hat{B}, \hat{C}, \hat{K})$ are *similar*. The poles of the actual and the estimated models were respectively $\lambda(A_K) = 0.3412 \pm j0.3428$ and $\lambda(\hat{A}_K) = 0.3136 \pm j0.3260$.

Model for the innovation form:
The innovation form is given by

$$\begin{aligned}\hat{x}(k+1) &= \begin{bmatrix} 1.6 & 1 \\ -0.8 & 0 \end{bmatrix}\hat{x}(k) + \begin{bmatrix} 1 \\ 0 \end{bmatrix} r(k) + \begin{bmatrix} 0.9175 \\ -0.5661 \end{bmatrix} e(k) \\ y(k) &= [\,1 \quad 0\,]\hat{x}(k) + e(k)\end{aligned} \tag{7.181}$$

where the variance of the residual $e(k)$ was $\sigma_e^2 = 3.5329$, $N = 5000$, and $M = 10$. The identified state-space model is given by:

$$\hat{A} = \begin{bmatrix} 0.8552 & 0.4565 \\ -0.3599 & 0.7476 \end{bmatrix}, \hat{B} = \begin{bmatrix} 0.0018 \\ 0.0041 \end{bmatrix}, \hat{K} = \begin{bmatrix} 0.0020 \\ 0.0008 \end{bmatrix}, \hat{C} = \begin{bmatrix} 430.9706 & 56.5989 \end{bmatrix}.$$

It can be verified that (A_K, B, C, K), and $(\hat{A}_K, \hat{B}, \hat{C}, \hat{K})$ are *similar*. The poles of the actual and the estimated models were respectively $\lambda\,(A) = 0.8 \pm j0.4$ and $\lambda(\hat{A}) = 0.8013 \pm j0.4018$.

Figure 7.12 shows the actual and estimated impulse responses and step responses of the prediction model and the step responses of the innovation model. Subfigures (a) and (b) show results of the prediction form while subfigure (c) shows the result of the innovation form. Subfigure (a) shows respectively the true and the estimated impulse response sequences θ_p and $\hat{\theta}_p$ given by Eq. (7.168), and subfigure (b) shows respectively the step responses of the true and the identified models (A_K, B, C) and $(\hat{A}_K, \hat{B}, \hat{C})$.

7.9 Summary

System Model

$$\begin{aligned}x(k+1) &= Ax(k) + Br(k) + E_w w(k) \\ y(k) &= Cx(k) + v(k)\end{aligned}$$

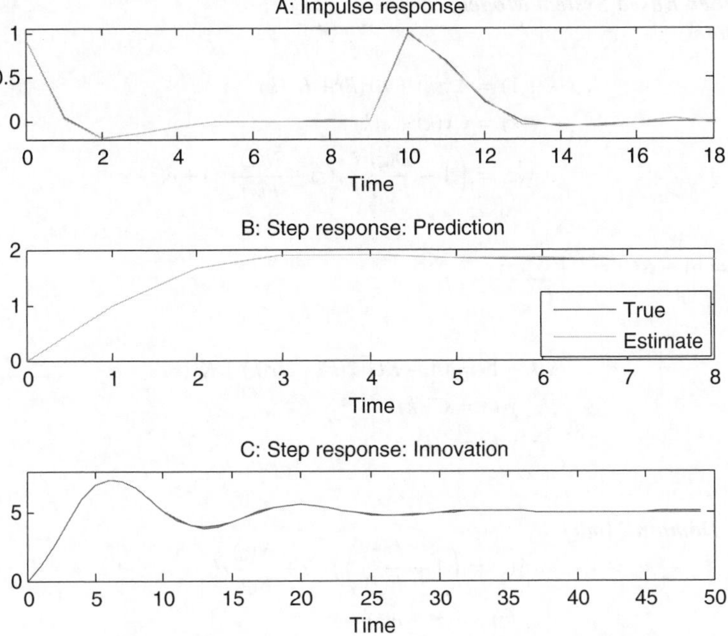

Figure 7.12 Responses of the prediction and innovations forms

Difference equation model

$$y(k) = -\sum_{i=1}^{n} a_i y(k-i) + \sum_{i=1}^{n} b_i u(k-i) + v(k)$$

Persistence of Excitation

Definition *A signal $\{r(k)\}$ with power spectral density $P_{rr}(f)$ is said to be persistently exciting of order M_r if*

- *The $M_r x M_r$ correlation matrix, $\mathbf{r}_{rr}$ is positive definite*

$$\mathbf{r}_{rr} = \begin{bmatrix} r_{rr}(0) & r_{rr}(1) & r_{rr}(2) & . & r_{rr}(M_r-1) \\ r_{rr}(1) & r_{rr}(0) & r_{rr}(1) & . & r_{rr}(M_r-2) \\ r_{rr}(2) & r_{rr}(1) & r_{rr}(0) & . & r_{rr}(M_r-3) \\ . & . & . & . & . \\ r_{rr}(M_r-1) & r_{rr}(M_r-2) & r_{rr}(M_r-3) & . & r_{rr}(0) \end{bmatrix} > 0$$

- $P_{rr}(f) \neq 0$ *for* $\{f_i : i = 1, 2, \ldots, M_r\}$

Kalman Filter-Based System Models

Innovation form

$$\begin{aligned}\hat{\boldsymbol{x}}(k+1) &= \boldsymbol{A}\hat{\boldsymbol{x}}(k) + \boldsymbol{B}r(k) + \boldsymbol{K}e(k)\\ y(k) &= \boldsymbol{C}\hat{\boldsymbol{x}}(k) + e(k)\\ y(z) &= \left(1 - \frac{D(z)}{F(z)}\right) y(z) + \frac{N(z)}{F(z)} r(z) + e(z)\end{aligned}$$

where $F(z) = |\boldsymbol{A} - \boldsymbol{KC}| = 1 + \sum_{i=1}^{n} c_i z^{-i}$.

Predictor form

$$\begin{aligned}\hat{\boldsymbol{x}}(k+1) &= (\boldsymbol{A} - \boldsymbol{KC})\hat{\boldsymbol{x}}(k) + \boldsymbol{B}r(k) + \boldsymbol{K}y(k)\\ \hat{y}(k) &= \boldsymbol{C}\hat{x}(k)\\ y(k) &= \hat{y}(k) + e(k)\end{aligned}$$

Frequency Domain Model

$$\begin{aligned}\hat{y}(z) &= \left(1 - \frac{D(z)}{F(z)}\right) y(z) + \frac{N(z)}{F(z)} r(z)\\ y(z) &= \hat{y}(z) + e(z)\end{aligned}$$

Kalman filter residual and the equation error

$$\upsilon(z) = N_w(z)w(z) + D(z)v(z) = F(z)e(z)$$

Identification methods

System identification schemes include the following:

- Least squares method.
- High-order least-squares method.
- Prediction error methods.
- Subspace identification methods.

All of the identification methods are based on selecting a model, which is termed herein the *identification model*, so that its input–output dynamic behavior matches closely of that of the given input–output data. Depending upon the intended application, either a state-space model or an input–output model is chosen. Further the model may be based on the innovation or the predictor form. The least-squares, the high-order, and the prediction error methods employ the input–output identification model, whereas the subspace method uses either the innovation of the predictor form of the state space model. An overview of the various identification methods under a one umbrella of Kalman filter based model is shown in Figure 7.13.

Least-Squares Method

$$y(k) = \boldsymbol{H}(k)\boldsymbol{\theta} + \upsilon(k)$$

$$\boldsymbol{H}(k) = \begin{bmatrix} -y(k-1) & -y(k-2) & . & -y(k-n) & r(k-1) & . & r(k-n)\\ -y(k-2) & -y(k-3) & . & -y(k-n-1) & r(k-2) & . & r(k-n-1)\\ . & . & . & . & . & . & .\\ -y(k-N) & -y(k-N-1) & . & -y(k-n-N+1) & r(k-N) & . & r(k-n-N+1)\end{bmatrix}$$

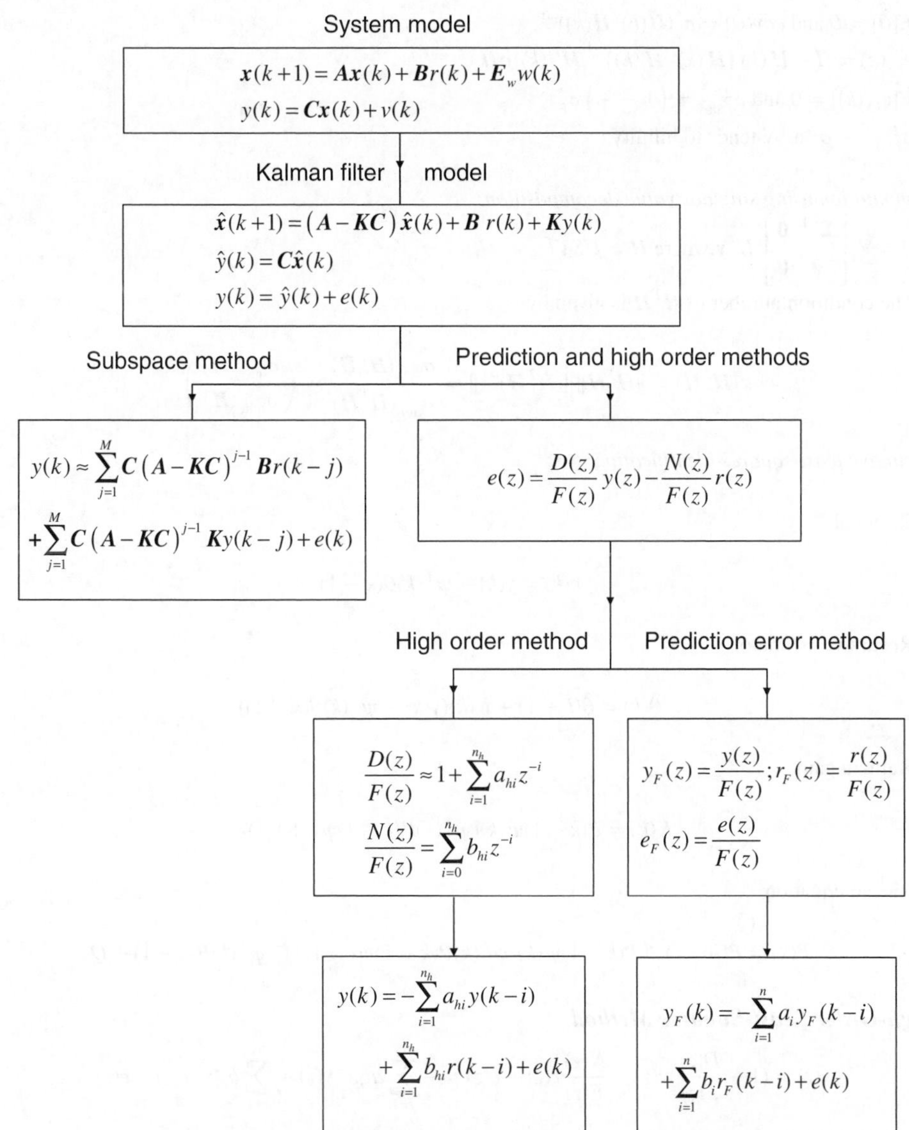

Figure 7.13 Overview of various identification methods

The least-squares estimate

$$\min_{\theta} \left\{ \frac{1}{N} \left(\boldsymbol{y}(k) - \boldsymbol{H}(k)\hat{\theta} \right)^T \left(\boldsymbol{y}(k) - \boldsymbol{H}(k)\hat{\theta} \right) \right\}$$

$$\hat{\theta} = \left(\boldsymbol{H}(k)^T \boldsymbol{H}(k) \right)^{-1} \boldsymbol{H}^T(k) \boldsymbol{y}(k)$$

- $\hat{\theta} \rightarrow \theta$ as N tends to infinity

- $E[\hat{\boldsymbol{\theta}}] = \boldsymbol{\theta}$ and $cov(\hat{\boldsymbol{\theta}}) = \sigma_v^2(\boldsymbol{H}(k)^T\boldsymbol{H}(k))^{-1}$.
- $\boldsymbol{e}_{LS}(k) = (\boldsymbol{I} - \boldsymbol{H}(k)\left(\boldsymbol{H}(k)^T\boldsymbol{H}(k)\right)^{-1}\boldsymbol{H}^T(k))\boldsymbol{v}(k)$
- $E[\boldsymbol{e}_{LS}(k)] = 0$ and $\sigma_{LSres}^2 = \left(1 - \frac{M}{N}\right)\sigma_v^2$
- $\sigma_{LSres}^2 \to \sigma_v^2$ as N tends to infinity.

Computation using singular value decomposition:

$\hat{\theta} = \boldsymbol{V}\begin{bmatrix}\boldsymbol{\Sigma}^{-1} & \boldsymbol{0}\\ \boldsymbol{0} & \boldsymbol{0}\end{bmatrix}\boldsymbol{U}^T\boldsymbol{y}$ where $\boldsymbol{H} = \boldsymbol{USV}^T$

The condition number of $\boldsymbol{H}^T\boldsymbol{H}$ is given by:

$$\boldsymbol{\kappa}(\boldsymbol{H}^T\boldsymbol{H}) = \|\boldsymbol{H}^T\boldsymbol{H}\|\|(\boldsymbol{H}^T\boldsymbol{H})^{-1}\| = \frac{\sigma_{\max}(\boldsymbol{H}^T\boldsymbol{H})}{\sigma_{\min}(\boldsymbol{H}^T\boldsymbol{H})} = \left(\frac{\sigma_{\max}(\boldsymbol{H})}{\sigma_{\min}(\boldsymbol{H})}\right)^2$$

Recursive least-squares identification:

- Residual

$$e(k) = y(k) - \boldsymbol{\psi}^T(k)\hat{\boldsymbol{\theta}}(k-1)$$

- Recursive estimator

$$\hat{\boldsymbol{\theta}}(k) = \hat{\boldsymbol{\theta}}(k-1) + \boldsymbol{K}(k)(y(k) - \boldsymbol{\psi}^T(k)\hat{\boldsymbol{\theta}}(k-1))$$

- Kalman gain

$$\boldsymbol{K}(k) = \boldsymbol{P}(k-1)\boldsymbol{\psi}(k)(\boldsymbol{\psi}^T(k)\boldsymbol{P}(k-1)\boldsymbol{\psi}(k) + 1)^{-1}$$

- Riccati equation

$$\boldsymbol{P}(k) = \boldsymbol{P}(k-1) - \boldsymbol{P}(k-1)\boldsymbol{\psi}(k)(\boldsymbol{\psi}^T(k)\boldsymbol{P}(k-1)\boldsymbol{\psi}(k) + 1)^{-1}\boldsymbol{\psi}^T(k)\boldsymbol{P}(k-1) + \boldsymbol{Q}$$

High-Order Least-Squares Method

$$y(z) = (1 - \frac{D(z)}{F(z)})y(z) + \frac{N(z)}{F(z)}r(z) + e(z) = -\sum_{i=1}^{n_h} a_{hi}z^{-i}y(z) + \sum_{i=1}^{n_h} b_{hi}z^{-i}r(z) + e(z)$$

The least-squares estimate $\hat{\boldsymbol{\theta}}_h$ of $\boldsymbol{\theta}_h = [\,a_{h1}\ \ a_{h2}\ \ .\ \ a_{hn_h}\ \ b_{h1}\ \ b_{h1}\ \ .\ \ b_{hn_h}\,]^T$ is given by

$$\hat{\boldsymbol{\theta}}_h = (\boldsymbol{H}_h(k)^T\boldsymbol{H}_h(k))^{-1}\boldsymbol{H}_h^T(k)\boldsymbol{y}(k)$$

where $\boldsymbol{H}_h(k)$ is a $Nx2n_h$ data matrix formed of the past inputs and the past outputs. The high-order model of the system transfer function becomes

$$G_h(z) = -\sum_{i=1}^{n} a_i z^{-i}G_h(z) + \sum_{i=0}^{n} b_i z^{-i} + \varepsilon(z)$$

where $\varepsilon(z)$ is an error term.

Reduced-order model

Expressing in the frequency domain version of the batch least squares model yields:

$$\begin{aligned} G_h(z) &= \boldsymbol{H}_G(z)\boldsymbol{\theta} + \boldsymbol{\varepsilon}(z) \\ \hat{\theta} &= \left(\boldsymbol{H}_G^T(z)\boldsymbol{H}_G(z)\right)^{-1}\boldsymbol{H}_G^T(z)\hat{\boldsymbol{G}}_h(z) \end{aligned}$$

The prediction error method

$$e(z) = y(z) - \hat{y}(z) = \frac{D(z)}{F(z)}y(z) - \frac{N(z)}{F(z)}r(z)$$

Objective function

The objective of prediction error method (PEM) is to minimize the following quadratic function of the prediction error $V_N = \frac{1}{2}\sum_{k=1}^{N} e^2(k)$

$\min_{\theta_a}\{V_N\}$ where $\boldsymbol{\theta}_a = [a_1\ a_2\ .\ a_n\ b_1\ b_2\ .\ b_n\ c_1\ c_2\ .\ c_n]^T$.

$$\begin{aligned} y^F(k) &= y(k) - \sum_{i=1}^{n} c_i y^F(k-i) \\ r^F(k) &= r(k) - \sum_{i=1}^{n} c_i r^F(k-i) \\ e^F(k) &= e(k) - \sum_{i=1}^{n} c_i e^F(k-ai) \\ e(k) &= y^F(k) + \sum_{i=1}^{n} a_i y^F(k-i) - \sum_{i=1}^{n} b_i r^F(k-i) \end{aligned}$$

The data vector $\boldsymbol{\psi}_a^T(k)$ becomes:

$$\boldsymbol{\psi}_a^T(k) = \left[-y^F(k-1)\quad .\quad -y^F(k-n)\quad r^F(k-1)\quad .\quad r(k-n)\quad e^F(k-1)\quad .\quad e^F(k-n)\right]$$

The prediction error algorithm is iterative as the objective function is a nonlinear function of the unknown parameter $\boldsymbol{\theta}_a$. The estimate $\hat{\boldsymbol{\theta}}_a$ is computed iteratively using the Newton–Raphson method:

$$\begin{aligned} &\hat{\boldsymbol{\theta}}_a(k) = \hat{\boldsymbol{\theta}}_a(k-1) + \boldsymbol{K}(k)e(k) \\ &\boldsymbol{K}(k) = \boldsymbol{P}(k)\boldsymbol{\psi}_a(k) \\ &\boldsymbol{P}(k) = \boldsymbol{P}(k-1) - \boldsymbol{P}(k-1)\boldsymbol{\psi}_a(k)\left[1 + \boldsymbol{\psi}_a^T(k)\boldsymbol{P}(k-1)\boldsymbol{\psi}_a(k)\right]^{-1}\boldsymbol{\psi}_a^T(k)\boldsymbol{P}(k-1) \end{aligned}$$

Family of prediction error algorithms

$$\begin{aligned} &\hat{\boldsymbol{\theta}}_a(k) = \hat{\boldsymbol{\theta}}_a(k-1) + \boldsymbol{K}(k)e(k) \\ &e(k) = y(k) - \boldsymbol{\psi}_a^T(k)\hat{\boldsymbol{\theta}}_a(k-1) \\ &\boldsymbol{K}(k) = \boldsymbol{P}(k)\boldsymbol{\psi}_a(k)\left[1 + \boldsymbol{\psi}_a^T(k)\boldsymbol{P}(k-1)\boldsymbol{\psi}_a(k)\right]^{-1} \\ &\boldsymbol{P}(k) = \boldsymbol{P}(k-1) - \boldsymbol{P}(k-1)\boldsymbol{\psi}_a(k)\left[1 + \boldsymbol{\psi}_a^T(k)\boldsymbol{P}(k-1)\boldsymbol{\psi}_a(k)\right]^{-1}\boldsymbol{\psi}_a^T(k)\boldsymbol{P}(k-1) \end{aligned}$$

Subspace Identification Method

$$\hat{\boldsymbol{x}}(i+1) = \boldsymbol{A}_K\hat{\boldsymbol{x}}(i) + \boldsymbol{B}r(i) + \boldsymbol{K}y(i)$$
$$\hat{y}(i) = \boldsymbol{C}\hat{x}(i)$$
$$\hat{y}(k) = \sum_{j=1}^{M} h_r(j)r(k-j) + \sum_{j=1}^{M} h_y(j)y(k-j)$$
$$y(k) = \sum_{j=1}^{M} h_r(j)r(k-j) + \sum_{j=1}^{M} h_y(j)y(k-j) + e(k)$$

The sequence $h_r(j) = \boldsymbol{C}\boldsymbol{A}_K^{j-1}\boldsymbol{B}$ and $h_y(j) = \boldsymbol{C}\boldsymbol{A}_K^{j-1}\boldsymbol{K}$ which are termed Kalman filter or observer Markov parameters.

Batch processing

$$\boldsymbol{y} = \boldsymbol{H}_p\boldsymbol{\theta}_p + \boldsymbol{e}_p$$

where $\boldsymbol{H}_p$ is a $Nx2M$ matrix given by:

$$\boldsymbol{H}_p = \begin{bmatrix} y(k-1) & y(k-2) & . & y(k-M) & r(k-1) & . & r(k-M) \\ y(k-2) & y(k-3) & . & y(k-M-1) & r(k-2) & . & r(k-M-1) \\ . & . & . & . & . & . & . \\ y(k-N) & y(k-N-1) & . & y(k-N-M+1) & r(k-N) & . & r(k-N-M+1) \end{bmatrix}, \boldsymbol{\theta}_p = \begin{bmatrix} \boldsymbol{h}_y \\ \boldsymbol{h}_r \end{bmatrix}$$

$$\boldsymbol{h}_r(k) = \begin{bmatrix} h_r(1) \\ h_r(2) \\ h_r(3) \\ . \\ h_r(M) \end{bmatrix} = \begin{bmatrix} \boldsymbol{CB} \\ \boldsymbol{CA}_K\boldsymbol{B} \\ \boldsymbol{CA}_K^2\boldsymbol{B} \\ . \\ \boldsymbol{CA}_K^{M-1}\boldsymbol{B} \end{bmatrix}, \boldsymbol{h}_y(k) = \begin{bmatrix} h_y(1) \\ h_y(2) \\ h_y(3) \\ . \\ h_y(M) \end{bmatrix} = \begin{bmatrix} \boldsymbol{CK} \\ \boldsymbol{CA}_K\boldsymbol{K} \\ \boldsymbol{CA}_K^2\boldsymbol{K} \\ . \\ \boldsymbol{CA}_K^{M-1}\boldsymbol{K} \end{bmatrix}, \boldsymbol{y}(k) = \begin{bmatrix} y(k) \\ y(k-1) \\ y(k-2) \\ . \\ y(k-N)1) \end{bmatrix}$$

It is assumed that $\boldsymbol{H}_p$ is of full rank, and $N \geq M$. A closed-form solution for the estimate the impulse response sequence $\boldsymbol{\theta}_p$ as shown below:

$$\hat{\boldsymbol{\theta}}_p = (\boldsymbol{H}_p)^{\dagger}\boldsymbol{y}$$

The estimate $\hat{\boldsymbol{\theta}}_p$ of $\boldsymbol{\theta}_p$ is unbiased. The pseudo-inverse $(\boldsymbol{H}_p)^{\dagger}$ is computed using SVD.

Estimation of the state-space model

$$\boldsymbol{H}_r(0) = \begin{bmatrix} \boldsymbol{CB} & \boldsymbol{CA}_K\boldsymbol{B} & . & \boldsymbol{CA}_K^{M-1}\boldsymbol{B} \\ \boldsymbol{CA}_K\boldsymbol{B} & \boldsymbol{CA}_K^2\boldsymbol{B} & . & \boldsymbol{CA}_K^M\boldsymbol{B} \\ . & . & . & . \\ \boldsymbol{CA}_K^{M-1}\boldsymbol{B} & \boldsymbol{CA}_K^M\boldsymbol{B} & . & \boldsymbol{CA}_K^{2M-2}\boldsymbol{B} \end{bmatrix}; \boldsymbol{H}_r(1) = \begin{bmatrix} \boldsymbol{CA}_K\boldsymbol{B} & \boldsymbol{CA}_K^2\boldsymbol{B} & . & \boldsymbol{CA}_K^M\boldsymbol{B} \\ \boldsymbol{CA}_K^2\boldsymbol{B} & \boldsymbol{CA}_K^2\boldsymbol{B} & . & \boldsymbol{CA}_K^{M+1}\boldsymbol{B} \\ . & . & . & . \\ \boldsymbol{CA}_K^M\boldsymbol{B} & \boldsymbol{CA}_K^{M+1}\boldsymbol{B} & . & \boldsymbol{CA}_K^{2M-1}\boldsymbol{B} \end{bmatrix};$$

$$\boldsymbol{H}_r(0) = \boldsymbol{U}_r\boldsymbol{S}_r\boldsymbol{V}_r^T$$
$$\boldsymbol{Ob} = \boldsymbol{U}_r\boldsymbol{\Sigma}^{1/2}$$
$$\boldsymbol{Cb} = \boldsymbol{\Sigma}^{1/2}\boldsymbol{V}_r^T$$
$$\hat{\boldsymbol{A}}_K = \hat{\boldsymbol{\Sigma}}^{-1/2}\hat{\boldsymbol{U}}_r^T\hat{\boldsymbol{H}}_r(1)\hat{\boldsymbol{V}}_r\hat{\boldsymbol{\Sigma}}^{-1/2}$$

The estimate $\hat{B}$ and $\hat{C}$ are respectively the first column of estimate of the observability matrix $\boldsymbol{Cb}$ and the first row of the estimate of $\boldsymbol{Ob}$.

Innovation form

The innovation form is derived from the predictor form by exploiting that the impulse response $h_r(j) = \boldsymbol{C}\boldsymbol{A}_K^{j-1}\boldsymbol{B}$ is a convolution of the impulse response $h_y(j) = \boldsymbol{C}\boldsymbol{A}_K^{j-1}\boldsymbol{K}$ and the impulse response of the system $h_I(j) = \boldsymbol{C}\boldsymbol{A}^{j-1}\boldsymbol{B}$.

7.10 Appendix: Performance of the Least-Squares Approach

The performance of the least-squares approaches to system identification analyzed in terms the bias, and the covariance of the parameter estimates and the variance of the residual. The quality of the least-squares estimate is derived when the equation is uncorrelated and correlated with the data matrix. Main results are established in a series of lemmas and corollaries that are listed below:

Case I: Equation error is a zero-mean white noise process. It shown in a series of lemmas:

- *Lemma 7.2*: The estimate is consistent if the equation error is uncorrelated with the data matrix.
- *Lemma 7.3*: The estimate is unbiased $E[\hat{\boldsymbol{\theta}}] = \boldsymbol{\theta}$.
- *Lemma 7.4*: The covariance of the estimation error is $cov(\hat{\boldsymbol{\theta}}) = \sigma_v^2(\boldsymbol{H}(k)^T\boldsymbol{H}(k))^{-1}$.
- *Corollary 7.4:* the bound on the covariance is:

$$\|cov(\hat{\boldsymbol{\theta}})\| \leq \frac{\sigma_v^2}{\sigma_{\min}^2(\boldsymbol{H}(k))}.$$

- *Lemma 7.5:* An expression of the residual is: $\boldsymbol{e}_{LS}(k) = (\boldsymbol{I} - \boldsymbol{H}(k)(\boldsymbol{H}(k)^T\boldsymbol{H}(k))^{-1}\boldsymbol{H}^T(k))\boldsymbol{v}(k)$.
- *Corollary 7.5:* The residual is a zero-mean process: $E[\boldsymbol{e}_{LS}(k)] = 0$.
- *Lemma 7.6*: The covariance of the residual is $\sigma_{LSe}^2 = (1 - \frac{M}{N})\sigma_v^2$ and $\sigma_{LSe}^2 \to \sigma_v^2$ as $N \to \infty$.
- *Lemma 7.7:* The residual converges to the equation error: $\boldsymbol{e}_{LS}(k) \to \boldsymbol{v}(k)$.

Case II: Equation error is a correlated noise. It shown in a series of lemmas:

- *Lemma 7.8:* The parameter estimate $\hat{\boldsymbol{\theta}}$ is biased:

$$E[\hat{\boldsymbol{\theta}}] = \boldsymbol{\theta} + (\boldsymbol{H}(k)^T\boldsymbol{H}(k))^{-1}E[\boldsymbol{H}^T(k)\boldsymbol{v}(k)].$$

- *Corollary 7.8:* A bound in the estimation error is

$$\|E[\hat{\boldsymbol{\theta}}] - \boldsymbol{\theta}\| \leq \frac{\|E[\boldsymbol{H}^T(k)\boldsymbol{v}(k)]\|}{\sigma_{\min}^2(\boldsymbol{H}(k))}.$$

- *Lemma 7.9:* $\text{cov}(\hat{\boldsymbol{\theta}}) = (\boldsymbol{H}(k)^T\boldsymbol{H}(k))^{-1}E[\boldsymbol{H}^T(k)\boldsymbol{v}(k)\boldsymbol{v}^T(k)\boldsymbol{H}(k)](\boldsymbol{H}(k)^T\boldsymbol{H}(k))^{-1}$.
- *Corollary 7.9:* A bound on the covariance is:

$$\|\text{cov}(\hat{\boldsymbol{\theta}})\| \leq \frac{\sigma_{\max}(E[\boldsymbol{H}^T(k)\boldsymbol{v}(k)\boldsymbol{v}^T(k)\boldsymbol{H}(k)])}{\sigma_{\min}^4(\boldsymbol{H}(k))}.$$

7.10.1 Correlated Error

In Lemma 7.3 and its Corollary, it is shown that if the equation error is correlated then the estimate will be biased, and bound on the bias as given ine Corollary 7.3. Expressions for the covariance of the parameter estimation error and its bound are respectively given in Lemma 7.4 and its Corollary.

7.10.2 Uncorrelated Error

The case of an uncorrelated white noise equation error is considered. In Lemma 7.5 it shown that the estimate is unbiased, and in *Lemma 7.6* an expression for covariance of the estimation error is given. A bound on the covariance is given in Lemma 7.5. For the case of a zero-mean white noise equation error, an expression relating the residual and the equation error is derived in Lemma 7.7. In its corollary it is shown that the residual is a zero mean process. In Lemma 7.8 it is shown that the variance of the residual is related to the variance of the white noise by a constant term which depends upon the ratio of the rank of the data matrix and the number of data samples. Further, when the data samples tend to infinity, then the variance of the residual tends to that of the equation error. In Lemma 7.9 it is shown that the residual tends to the equation error when the data samples tend to infinity if the equation error is uncorrelated with the data matrix.

Relevant equations are repeated here for the sake of convenience in following the various steps in the derivations. A linear matrix model is given by

$$\boldsymbol{y}(k) = \boldsymbol{H}(k)\boldsymbol{\theta} + \boldsymbol{\upsilon}(k) \tag{7.182}$$

The least-squares estimate $\hat{\boldsymbol{\theta}}$ exists and is given by

$$\hat{\boldsymbol{\theta}} = \left(\frac{1}{N}\left(\boldsymbol{H}(k)^T \boldsymbol{H}(k)\right)\right)^{-1}\left(\frac{1}{N}\boldsymbol{H}^T(k)\boldsymbol{y}(k)\right) \tag{7.183}$$

If the number of data samples is finite, as N appears in both the numerator and denominator terms on the right-hand side, it may be canceled, and hence the explicit dependence on the number of data samples N may be suppressed in the above equations:

$$\hat{\boldsymbol{\theta}} = \left(\boldsymbol{H}(k)^T \boldsymbol{H}(k)\right)^{-1}\boldsymbol{H}^T(k)\boldsymbol{y}(k) \tag{7.184}$$

$$\hat{\boldsymbol{\theta}} = \boldsymbol{\theta} + \left(\boldsymbol{H}(k)^T \boldsymbol{H}(k)\right)^{-1}\boldsymbol{H}^T(k)\boldsymbol{\upsilon}(k) \tag{7.185}$$

Quality of estimate: $\upsilon(k)$ is a white noise process

Lemma 7.2 *If $\lim_{N\to\infty}\frac{1}{N}\boldsymbol{H}^T(k)\boldsymbol{\upsilon}(k) = 0$, then the estimate $\hat{\boldsymbol{\theta}}$ converges to the true value $\boldsymbol{\theta}$:*

$$\hat{\boldsymbol{\theta}} \to \boldsymbol{\theta} \text{ as } N \to \infty \tag{7.186}$$

Proof: Follows from Eq. (7.185).

Remark In order to ensure the estimate is *consistent*, that is, the estimate approaches asymptotically the true parameter, the equation error must be uncorrelated with the data matrix. Implications of this condition are analyzed in the next section.

7.10.3 Correlation of the Error and the Data Matrix

The condition of the Lemma 7.2 *implies* that the equation error $\upsilon(k)$ must be uncorrelated with the past outputs $y(k)$ as well as the past inputs $r(k)$ as follows:

$$r_{y\upsilon}(m) = \lim_{N\to\infty}\frac{1}{N}\sum_{i=1}^{N} y(k)\upsilon(k-m) = 0 \tag{7.187}$$

$$r_{r\upsilon}(m) = \lim_{N\to\infty}\frac{1}{N}\sum_{i=1}^{N} r(k)\upsilon(k-m) = 0 \tag{7.188}$$

for $m = 1, 2, 3 \ldots, n$. These conditions are satisfied if $\upsilon(k)$ is a zero-mean white noise process.

Lemma 7.3 *The least-squares estimate is unbiased:*

$$E\left[\hat{\theta}\right] = \theta \tag{7.189}$$

Proof: Taking expectation of Eq. (7.81) yields:

$$E\left[\hat{\theta}\right] = \theta + E[\left(\boldsymbol{H}(k)^T \boldsymbol{H}(k)\right)^{-1} \boldsymbol{H}^T(k)\,\boldsymbol{\upsilon}(k)] \tag{7.190}$$

As $\boldsymbol{H}(k)^T \boldsymbol{H}(k)$ is deterministic, and $\boldsymbol{\upsilon}(k)$ and $\boldsymbol{H}(k)$ are independent random variables, we get:

$$E\left[\hat{\theta}\right] = \theta + \left(\boldsymbol{H}(k)^T \boldsymbol{H}(k)\right)^{-1} E\left[\boldsymbol{H}^T(k)\right] E\left[\boldsymbol{\upsilon}(k)\right] \tag{7.191}$$

Since $E\left[\boldsymbol{\upsilon}(k)\right] = 0$, the result follows.

Lemma 7.4 *The covariance of the estimation error is given by:*

$$cov\left(\hat{\theta}\right) = \sigma_{\upsilon}^2 \left(\boldsymbol{H}(k)^T \boldsymbol{H}(k)\right)^{-1} \tag{7.192}$$

Proof: Using expression (7.81) we get:

$$\left(\hat{\theta} - \theta\right)\left(\hat{\theta} - \theta\right)^T = \left(\boldsymbol{H}(k)^T \boldsymbol{H}(k)\right)^{-1} \boldsymbol{H}^T(k)\,\boldsymbol{\upsilon}(k)\,\boldsymbol{\upsilon}^T(k)\,\boldsymbol{H}(k)\left(\boldsymbol{H}(k)^T \boldsymbol{H}(k)\right)^{-1} \tag{7.193}$$

Taking expectation yields:

$$\mathrm{cov}\left(\hat{\theta}\right) = E[\left(\hat{\theta} - \theta\right)\left(\hat{\theta} - \theta\right)^T] = (\boldsymbol{H}(k)^T \boldsymbol{H}(k))^{-1} \Phi(k)\,(\boldsymbol{H}(k)^T \boldsymbol{H}(k))^{-1} \tag{7.194}$$

where $\Phi(k) = E[\boldsymbol{H}^T(k)\,E[\boldsymbol{\upsilon}(k)\,\boldsymbol{\upsilon}^T(k)]\boldsymbol{H}(k)]$. Since $\boldsymbol{\upsilon}(k)$ and $\boldsymbol{H}(k)$ are independent random variables, the inner expectation is with respect to $\boldsymbol{\upsilon}(k)$ while the outer expectation is with respect to $\boldsymbol{H}(k)$. Since $E\left[\boldsymbol{\upsilon}(k)\,\boldsymbol{\upsilon}^T(k)\right] = \boldsymbol{I}\sigma_{\upsilon}^2$ and $\boldsymbol{H}^T(k)\,\boldsymbol{H}(k)$ is deterministic, we get

$$\Phi(k) = \sigma_{\upsilon}^2 E[\boldsymbol{H}^T(k)\,\boldsymbol{H}(k)] = \sigma_{\upsilon}^2 \boldsymbol{H}^T(k)\,\boldsymbol{H}(k) \tag{7.195}$$

Substituting for $\Phi(k)$ we get:

$$cov\left(\hat{\theta}\right) = \sigma_{\upsilon}^2 \left(\boldsymbol{H}(k)^T \boldsymbol{H}(k)\right)^{-1} \boldsymbol{H}(k)^T \boldsymbol{H}(k)\left(\boldsymbol{H}(k)^T \boldsymbol{H}(k)\right)^{-1} \tag{7.196}$$

Simplifying, we get Eq. (7.192)

Corollary 7.4

$$\left\| cov\left(\hat{\theta}\right) \right\| \leq \frac{\sigma_{\upsilon}^2}{\sigma_{\min}^2\left(\boldsymbol{H}(k)\right)} \tag{7.197}$$

Remark The covariance of the estimation error satisfies the Cramér–Rao inequality, and hence the least squares estimate is efficient in the case when the equation error is a white noise process.

7.10.4 Residual Analysis

The residual, $\boldsymbol{e}_{LS}(k) = \boldsymbol{y}(k) - \hat{\boldsymbol{y}}_{LS}(k)$ where $\hat{\boldsymbol{y}}_{LS}(k) = \boldsymbol{H}(k)\hat{\boldsymbol{\theta}}$, is compared with the equation error $\boldsymbol{\upsilon}(k)$, which is a zero-mean white noise process, and the mean and the variances are determined.

Lemma 7.5

$$\boldsymbol{e}_{LS}(k) = (\boldsymbol{I} - \boldsymbol{H}(k)(\boldsymbol{H}(k)^T\boldsymbol{H}(k))^{-1}\boldsymbol{H}^T(k))\boldsymbol{\upsilon}(k) \tag{7.198}$$

Proof: Using Eq. (7.80) we get:

$$\boldsymbol{e}_{LS}(k) = (\boldsymbol{I} - \boldsymbol{H}(k)\left(\boldsymbol{H}(k)^T\boldsymbol{H}(k)\right)^{-1}\boldsymbol{H}^T(k))\boldsymbol{y}(k) \tag{7.199}$$

Substituting $\boldsymbol{y}(k) = \boldsymbol{H}(k)\boldsymbol{\theta} + \boldsymbol{\upsilon}(k)$ and simplifying yields:

$$\boldsymbol{e}_{LS}(k) = (\boldsymbol{I} - \boldsymbol{H}(k)(\boldsymbol{H}(k)^T\boldsymbol{H}(k))^{-1}\boldsymbol{H}^T(k))\boldsymbol{\upsilon}(k) \tag{7.200}$$

Corollary 7.5

$$E[\boldsymbol{e}_{LS}(k)] = 0 \tag{7.201}$$

Proof:

$$E[\boldsymbol{e}_{LS}(k)] = E[\boldsymbol{\upsilon}(k)] - E[\boldsymbol{H}(k)(\boldsymbol{H}(k)^T\boldsymbol{H}(k))^{-1}\boldsymbol{H}^T\boldsymbol{\upsilon}(k)] \tag{7.202}$$

Since $\boldsymbol{\upsilon}(k)$ and $\boldsymbol{H}(k)$ are independent random variables and $E[\boldsymbol{\upsilon}(k)]$, $E[\boldsymbol{e}(k)] = 0$.

Lemma 7.6 *The variance of the residual* $\sigma^2_{LSe} = \frac{1}{N}E\left[\boldsymbol{e}^T_{LS}(k)\boldsymbol{e}_{LS}(k)\right]$ *is:*

$$\sigma^2_{LSe} = \left(1 - \frac{M}{N}\right)\sigma^2_{\upsilon} \tag{7.203}$$

Proof: Using the expression of the residual (7.198), and expressing the inner product $\boldsymbol{e}^T_{LS}(k)\boldsymbol{e}_{LS}(k)$ as a trace of the outer product: $\boldsymbol{e}^T_{LS}(k)\boldsymbol{e}_{LS}(k) = trace\{\boldsymbol{e}_{LS}(k)\boldsymbol{e}^T_{LS}(k)\}$ yields:

$$trace\left\{\frac{1}{N}\boldsymbol{e}_{LS}(k)\boldsymbol{e}^T_{LS}(k)\right\} = trace\left\{\frac{1}{N}\left(\boldsymbol{I} - \boldsymbol{P}_r(k)\right)\boldsymbol{\upsilon}(k)\boldsymbol{\upsilon}^T(k)\left(\boldsymbol{I} - \boldsymbol{P}_r(k)\right)\right\} \tag{7.204}$$

Since $trace\{\boldsymbol{ABC}\} = trace\{\boldsymbol{BCA}\} = trace\{\boldsymbol{CAB}\}$ we get

$$trace\left\{\frac{1}{N}\boldsymbol{e}_{LS}(k)\boldsymbol{e}^T_{LS}(k)\right\} = trace\left\{\frac{1}{N}\left(\boldsymbol{I} - \boldsymbol{P}_r(k)\right)^2\boldsymbol{\upsilon}(k)\boldsymbol{\upsilon}^T(k)\right\} \tag{7.205}$$

where $\boldsymbol{P}_r(k) = \boldsymbol{H}(k)\left(\boldsymbol{H}(k)^T\boldsymbol{H}(k)\right)^{-1}\boldsymbol{H}^T(k)$ is the projection operator. Since $\boldsymbol{\upsilon}(k)$ and $\boldsymbol{H}(k)$ are independent, taking expectation with respect to $\boldsymbol{\upsilon}(k)$ only, in other words taking conditional expectation, $E[(.)\,|\,\boldsymbol{H}(k)]$ yields:

$$\sigma^2_{LSe} = trace\left\{E\left[\frac{1}{N}\boldsymbol{e}_{LS}(k)\boldsymbol{e}^T_{LS}(k)\,|\,\boldsymbol{H}(k)\right]\right\} = \sigma^2_{\upsilon}\,trace\left\{\frac{1}{N}\left(\boldsymbol{I} - \boldsymbol{P}_r(k)\right)^2\right\} \tag{7.206}$$

Using the properties of the projection operator, namely $\boldsymbol{P}^2_r(k) = \boldsymbol{P}_r(k)$ and $trace\left\{\boldsymbol{I} - \boldsymbol{P}_r(k)\right\} = N - M$, and the variance of the residual σ^2_{LSe} becomes Eq. (7.203).

When N tends to infinity, the variance of the residual approaches that of the equation error.

$$\sigma_{LSe}^2 \to \sigma_\upsilon^2 \tag{7.207}$$

Lemma 7.7 *The residual converges to the equation error as* N *tends to infinity*

$$\boldsymbol{e}_{LS}(k) \to \boldsymbol{\upsilon}(k) \tag{7.208}$$

Proof: Consider the expression (7.198) relating $\boldsymbol{e}_{LS}(k)$ and $\boldsymbol{\upsilon}(k)$. Multiplying and dividing by N, the terms $\left(\boldsymbol{H}(k)^T \boldsymbol{H}(k)\right)^{-1}$ and $\boldsymbol{H}^T(k)\,\boldsymbol{\upsilon}(k)$, and re-arranging yields:

$$\boldsymbol{e}_{LS}(k) = \boldsymbol{\upsilon}(k) - \boldsymbol{H}(k)\left(\frac{\boldsymbol{H}(k)^T \boldsymbol{H}(k)}{N}\right)^{-1} \frac{\boldsymbol{H}^T(k)\,\boldsymbol{\upsilon}(k)}{N} \tag{7.209}$$

Since $\boldsymbol{\upsilon}(k)$ is a zero-mean white noise process, $\lim_{N\to\infty} it \left(\frac{\boldsymbol{H}^T(k)\boldsymbol{\upsilon}(k)}{N}\right) \to \boldsymbol{0}$ we get Eq. (7.208)

Lemma 7.8 *The least-squares estimate is biased:*

$$E\left[\hat{\boldsymbol{\theta}}\right] = \boldsymbol{\theta} + \left(\boldsymbol{H}(k)^T \boldsymbol{H}(k)\right)^{-1} E\left[\boldsymbol{H}^T(k)\,\boldsymbol{\upsilon}(k)\right] \tag{7.210}$$

Proof: Since $\boldsymbol{H}(k)^T \boldsymbol{H}(k)$ is deterministic, and $\boldsymbol{\upsilon}(k)$ and $\boldsymbol{H}(k)$ are correlated, we get:

$$E\left[\hat{\boldsymbol{\theta}}\right] = \boldsymbol{\theta} + \left(\boldsymbol{H}(k)^T \boldsymbol{H}(k)\right)^{-1} E\left[\boldsymbol{H}^T(k)\,\boldsymbol{\upsilon}(k)\right] \tag{7.211}$$

Corollary 7.8 *Let the norm of $\boldsymbol{H}(k)$ be $\|\boldsymbol{H}\| = \sigma_{\max}(\boldsymbol{H})$.*

$$\|E[\hat{\boldsymbol{\theta}}] - \boldsymbol{\theta}\| \le \frac{\|E[\boldsymbol{H}^T(k)\boldsymbol{\upsilon}(k)]\|}{\sigma_{\min}^2(\boldsymbol{H}(k))} \tag{7.212}$$

Proof Using $\|\boldsymbol{AB}\| \le \|\boldsymbol{A}\|\|\boldsymbol{B}\|$, we get

$$\|E[\hat{\boldsymbol{\theta}}] - \boldsymbol{\theta}\| \le \|(\boldsymbol{H}(k)^T\boldsymbol{H}(k))^{-1}\|\|E[\boldsymbol{H}^T(k)\boldsymbol{\upsilon}(k)]\| \tag{7.213}$$

Expressing the 2-norm in terms of the singular values, the corollary is proved.

Lemma 7.9 *The covariance of the estimation error is given by:*

$$\operatorname{cov}\left(\hat{\boldsymbol{\theta}}\right) = \left(\boldsymbol{H}(k)^T \boldsymbol{H}(k)\right)^{-1} E\left[\boldsymbol{H}^T(k)\,\boldsymbol{\upsilon}(k)\,\boldsymbol{\upsilon}^T(k)\,\boldsymbol{H}(k)\right]\left(\boldsymbol{H}(k)^T \boldsymbol{H}(k)\right)^{-1} \tag{7.214}$$

Proof: Using expression (7.185) we get:

$$\left(\hat{\boldsymbol{\theta}} - \boldsymbol{\theta}\right)\left(\hat{\boldsymbol{\theta}} - \boldsymbol{\theta}\right)^T = \left(\boldsymbol{H}(k)^T \boldsymbol{H}(k)\right)^{-1} \boldsymbol{H}^T(k)\,\boldsymbol{\upsilon}(k)\,\boldsymbol{\upsilon}^T(k)\,\boldsymbol{H}(k)\left(\boldsymbol{H}(k)^T \boldsymbol{H}(k)\right)^{-1} \tag{7.215}$$

Taking expectation $\operatorname{cov}(\hat{\boldsymbol{\theta}}) = E[(\hat{\boldsymbol{\theta}} - \boldsymbol{\theta})(\hat{\boldsymbol{\theta}} - \boldsymbol{\theta})^T]$ becomes Eq. (7.214):

Corollary 7.9

$$\|\operatorname{cov}(\hat{\boldsymbol{\theta}})\| \le \frac{\sigma_{\max}(E[\boldsymbol{H}^T(k)\boldsymbol{\upsilon}(k)\boldsymbol{\upsilon}^T(k)\boldsymbol{H}(k)])}{\sigma_{\min}^4(\boldsymbol{H}(k))} \tag{7.216}$$

Proof: Using $\|ABC\| \le \|A\|\|B\|\|C\|$, we get

$$\|\mathrm{cov}(\hat{\theta})\| \le \|(H(k)^T H(k))^{-1}\|\|E[H^T(k)v(k)v^T(k)H(k)]\|\|(H(k)^T H(k))^{-1}\| \tag{7.217}$$

Simplifying yields

$$\|\mathrm{cov}(\hat{\theta})\| \le \|(H(k)^T H(k))^{-1}\|^2\|E[H^T(k)v(k)v^T(k)H(k)]\| \tag{7.218}$$

Expressing the 2-norm in terms of the singular values and noting that $\|(A^TA)^{-1}\| = \sigma_{\max}((A^TA)^{-1}) = \sigma_{\min}(A^TA)$ yields:

$$\|\mathrm{cov}(\hat{\theta})\| \le \frac{\sigma_{\max}(E[H^T(k)v(k)v^T(k)H(k)])}{\sigma^4_{\min}(H(k))} \tag{7.219}$$

We have used the relation between the singular values of $H(k)^TH(k)$ and that of $H(k)$, namely $\sigma_{\min}(H(k)^TH(k)) = \sigma^2_{\min}(H(k))$, and $\|(H(k)^TH(k))^{-1}\|^2 = \sigma^4_{\min}(H(k))$.

7.11 Appendix: Frequency-Weighted Model Order Reduction

In the frequency-weighted LS estimator method, an estimate of the true model $G(z)$, termed the reduced-order model, is derived from the high-order model $G_h(z)$ by minimizing a frequency-weighted residual. The higher-order model $\hat{G}_h(z)$ may be expressed as a sum of the true system model $G(z)$, and the artifact $g(z)$ due to the truncation error and the noise effects:

$$\hat{G}_h(z) = G(z) + g(z) \tag{7.220}$$

Expressing $G(z)$ in terms of the numerator polynomial $N(z) = \sum_{i=0}^{n} b_i z^{-i}$ and the denominator polynomial $D(z) = 1 + \sum_{i=1}^{n} a_i z^{-i}$, we get:

$$\hat{G}_h(z) = \frac{\sum_{i=0}^{n} b_i z^{-i}}{1 + \sum_{i=1}^{n} a_i z^{-i}} + g(z) \tag{7.221}$$

Cross-multiplying yields and rearranging gives the linear regression model in the frequency domain similar to Eq. (7.63):

$$\hat{G}_h(z) = -\sum_{i=1}^{n_r} a_{ri} z^{-i}\hat{G}_h(z) + \sum_{i=0}^{n_r} b_{ri} z^{-i} + \varepsilon(z) \tag{7.222}$$

where $\varepsilon(z) = D(z)g(z)$. The problem of identifying the system model $G(z)$ from the higher-order model $\hat{G}_h(z)$ is posed as a frequency-weighted least squares estimation problem. Let $z_i = e^{j2\pi f_i} : i = 1, 2, \ldots, N$, and $\{f_i\}$ being the frequencies. These frequencies are determined *a posteriori* from the frequency response of the high-order model assuming that (i) $G(z)$ has a smooth rational spectrum (relatively to the jagged spectrum of the error term $\varepsilon(z_i)$), (ii) $G(z)$ is dominant in $\hat{G}_h(z)$ over the bandwidth of the system, and (iii) the dominant peaks in the high-order model $\hat{G}_h(z)$ come from $G(z)$. The number of samples must be greater than or equal to the number of unknowns, that is, $N \ge 2n_h$. The frequencies are selected using resonance frequencies and the smoothness of the system transfer function. Collecting N data samples, the frequency-domain model (7.222) is expressed in a vector-matrix form similar to the standard least squares model in the time domain (7.67):

$$\hat{\boldsymbol{G}}_h(z) = \boldsymbol{H}_G(z)\boldsymbol{\theta} + \boldsymbol{\varepsilon}(z) \tag{7.223}$$

where

$$\hat{\boldsymbol{G}}_h(z) = \begin{bmatrix} \hat{G}_h(z_1) \\ \hat{G}_h(z_2) \\ . \\ \hat{G}_h(z_N) \end{bmatrix}; \boldsymbol{H}_G(z) = \begin{bmatrix} -z_1^{-1}\hat{G}_h(z_1) & -z_1^{-2}\hat{G}_h(z_1) & . & -z_1^{-n}\hat{G}_h(z_1) & z_1^{-1} & z_1^{-2} & . & z_1^{-n} \\ -z_2^{-1}\hat{G}_h(z_2) & -z_2^{-2}\hat{G}_h(z_2) & . & -z_2^{-n}\hat{G}_h(z_2) & z_2^{-1} & z_2^{-2} & . & z_2^{-n} \\ . & . & . & . & . & . & . & . \\ -z_N^{-1}\hat{G}_h(z_N) & -z_N^{-2}\hat{G}_h(z_N) & . & -z_N^{-1}\hat{G}_h(z_N) & z_N^{-1} & z_N^{-2} & . & z_N^{-n} \end{bmatrix},$$

$$\boldsymbol{\varepsilon}(z) = \begin{bmatrix} \varepsilon_1(z_1) \\ \varepsilon_2(z_2) \\ . \\ \varepsilon_N(z_N) \end{bmatrix}$$

The $Mx1$ feature vector θ is estimated using the following objective function similar to Eq. (7.69):

$$\min_{\theta}(\hat{\boldsymbol{G}}_h(z) - \boldsymbol{H}_G(z)\boldsymbol{\theta})^H(\hat{\boldsymbol{G}}_h(z) - \boldsymbol{H}_G(z)\boldsymbol{\theta}) \tag{7.224}$$

where $\boldsymbol{A}^H = (\boldsymbol{A}^*)^T$ denotes Hermitian of $\boldsymbol{A}$, which is the conjugate-transpose of $\boldsymbol{A}$. Assuming $\boldsymbol{H}_G(z)$ has a full rank, the estimate $\hat{\boldsymbol{\theta}}$ of the feature vector of the system model $\boldsymbol{\theta}$ similar to Eq. (7.74) is given by:

$$\hat{\boldsymbol{\theta}} = (\boldsymbol{H}_G^T(z)\boldsymbol{H}_G(z))^{-1}\boldsymbol{H}_G^T(z)\hat{\boldsymbol{G}}_h(z) \tag{7.225}$$

Lemma 7.10 *Given the data* $G_h(z_i)\ :\ i = 1, 2, 3, \ldots, N$, *if*

$$g(z_i) = 0 \text{ and } G(z_i) \neq 0 \text{ for } i = 1, 2, 3, \ldots, N \tag{7.226}$$

where $N \geq 2n$. *Then the parameter* $\boldsymbol{\theta}$ *may be estimated accurately, that is* $\hat{\boldsymbol{\theta}} = \boldsymbol{\theta}$.

Proof: Using Eqs. (7.226), (7.223) become artifact-free, $\hat{\boldsymbol{G}}_h(z) = \boldsymbol{H}_G(z)\boldsymbol{\theta}$. From (7.225), we get $\hat{\boldsymbol{\theta}} = \boldsymbol{\theta}$.

Comment *The interpretation of the condition (7.226) is that spectrum* $G(z)$ *and that of* $g(z)$ *must not be coincident at some* N *frequency locations. For example if* $G(z)$ *and* $g(z)$ *are both non-zero at some frequency* f_i *then it is impossible to differentiate between the two when the only information available is their sum* $\hat{G}_h(z) = G(z) + g(z)$.

In practice the requirement of a non-overlapping condition may not be met, and assumptions have to be relaxed. It is assumed that and the samples of $g(z_i)$ and $G(z_i)$ are uncorrelated, which implies that they have no common frequencies at least at the selected N frequencies. In other words their spectra are orthogonal to each other. This will guarantee that the least squares estimate will be unbiased, that is, $E[\hat{\boldsymbol{\theta}}] = \boldsymbol{\theta}$. Further, the frequencies (over the bandwidth of the system) where the amplitudes of $\{\hat{G}_h(z_i)\} : i = 1, 2, 3, \ldots, N$ are maximum come from $\{G(z_i)\}$ and not $\{g(z_i)\}$, that is $\{G(z_i)\}$ dominates $g(z_i)$. For example the absolute value of the ratio of the amplitude $|G(z_i)|/|g(z_i)|$ is much greater than the one to be assessed experimentally or empirically.

7.11.1 Implementation of the Frequency-Weighted Estimator

In order to implement the estimator we need to select the frequencies $z_i = e^{j2\pi f_i} : i = 1, 2, \ldots, N$ from the spectrum of the measurements $\{\hat{G}_h(z_i)\}$ so that that the conditions of Lemma 7.10 are met. For the selection of these frequencies we will assume the following:

Assumptions

- *The frequency response of the system $G(z)$ dominates the frequency response of the artifact $g(z)$ in frequency regions $z_i = e^{j2\pi f_i} : i = 1, 2, \ldots, N$. The dominant peaks of the higher-order frequency response $G_h(z)$ come from $G(z)$.*
- *The frequency response of the system $G(z)$ has a rational spectrum. This implies that the frequency response of $G(z)$ is smoothly varying, as opposed to the noise spectrum which is jagged.*

7.11.2 Selection of the Frequencies

From system and signal theory, we know that the peaks and the valleys in the spectrum are associated with the locations of the poles and the zeros of $G(z)$, and further that the spectrum of $G(z)$ is smoother than that of the spectrum of the artifact $g(z)$ (which will be usually jagged). If the order of the system $G(z)$ is n, then there must be at least $n + 1$ frequency points where $G(z)$ has a greater relative magnitude. However, as $G(z)$ is corrupted by the noise artifact $g(z)$ there may be erroneous peaks in the spectrum of the high-order model that could be confused with the peaks $G(z)$. The dominant peaks in the high order model spectrum are associated with the $G(z)$ and not $g(z)$. The dominant peaks are distinguished from those of the noise artifact using an *energy measure*, derived from the area under some neighborhood of the peak value of the high-order model. The energy measure of the peaks is used to select a set of frequencies where the SNR is high and to determine the associated weights. The maximum number of peaks (local maxima) in the magnitude response of $G(z)$ is $(n + 1)/2$ and frequencies associated with the dominant peaks are selected. If the spectrum of $G(z)$ has no peaks, then the relative smoothness of $G(z)$ is exploited for locating the frequencies. A frequency-weighted least-squares fit is used to estimate the model of $G(z)$ such that the best fit to the measurements $\{G_h(z_i)\}$ at the selected frequencies is obtained. The increased relative weightings of the frequency intervals of the peaks of $G(z)$ have more influence on the estimated model in modeling dynamics of $G(z)$ at the expense of artifact $g(z)$. The algorithm is such that the peaks of the response and the frequency intervals where the measurement spectrum are relatively more smooth, are weighted more heavily and have more influence on the estimated signal model, and therefore this algorithm is termed the *weighted-frequency algorithm*. Application software has been developed to automatically locate these frequencies [8].

References

[1] Doraiswami, R. and Cheded, L. (2013) A unified approach to detection and isolation of parametric faults using a Kalman filter residuals. *Journal of Franklin Institute*, **350**(5), 938–965.

[2] Doraiswami, R. and Cheded, L. (2013) Fault diagnosis of a sensor network: a distributed filtering approach. *Journal of Dynamic Systems, Measurement and Control*, **135**(5), 1–10.

[3] Doraiswami, R., Diduch, C., and Tang, T. (2010) A new diagnostic model for identifying parametric faults. *IEEE Transactions on Control System Technology*, **18**(3), 533–544.

[4] Wahlberg, B., Jansson, M., Matsko, T., and Molander, M. (2007) Experiences from subspace identification - comments from process industry users and researches, 315–327, A. Chiuso et al. (Eds), *Estimation and Control, LNCIS 364*, Springer-Verlag, Berlin

[5] Qin, J. S. (2006) Overview of subspace identification. *Computer and Chemical Engineering*, **30**, 1502–1513.

[6] Doraiswami, R. (2005) A two-stage identification with application to control, feature extraction and spectral estimation. *IEEE Proceedings: Control Theory and Applications*, **152**(4), 379–386.

[7] Ljung, L. (1999) *System Identification: Theory for the User*, Prentice-Hall, New Jersey.

[8] Mallory, G. and Doraiswami, R. (1999) A filter for on-line estimation of spectral content. *IEEE Transactions on Instrumentation and Measurement*, **48**(6), 1047–1055.

[9] Mendel, J. (1995) *Lessons in Estimation Theory in Signal Processing, Communications and Control*, Prentice Hall, New Jersey.

[10] Kay, S. M. (1993) *Fundamentals of Signal Processing: Estimation Theory*, Prentice Hall, New Jersey.

[11] Soderstrom, T. and Stoica, P. (1989) *System Identification*, Prentice Hall, New Jersey.

[12] Kumaresan, R. and Tufts, D.W. (1982) Estimating exponentially damped sinusoids and pole-zero mapping. *IEEE Transactions on Acoustics, Speech and Signal Processing*, **30**(6), 833–840

[13] Porat, B. and Friedlander, B. (1987) On the accuracy of the Kumerasan-Tufts method for estimating complex damped exponentials. *IEEE Transactions on Acoustics, Speech, and Signal Processing*, **35**(2), 231–235.

[14] Astrom, K and Eykhoff, P. (1971) System identification – a survey. *Automatica*, **7**(2), 123–162.

[15] Doraiswami, R. and Cheded, L. (2012) Kalman filter for fault detection: an internal model approach. *IET Control Theory and Applications*, **6**(5), 1–11.

8

Closed Loop Identification

8.1 Overview

Identification of a system operating in a closed loop is considered. The objective of closed loop identification is to use routine operating data with dither signal (a low level noise signal) excitation to develop a dynamic model of the process. In practice, and for a variety of reasons (e.g., analysis, design, and control), it is often necessary to identify a system that must operate in a closed-loop fashion under some type of feedback control. These reasons could also include the design of a high performance controller, safety issues, the need to stabilize an unstable plant and/or improve its performance while avoiding the cost incurred through downtime if the plant were to be taken offline for test. In these cases, it is therefore necessary to perform closed-loop identification. Applications include aerospace, magnetic levitation, levitated micro-robotics, magnetically-levitated automotive engine valves, magnetic bearings, mechatronics, adaptive control of processes, satellite-launching vehicles or unstable aircraft operating in closed-loop and process control systems [1–4]. It is important to note that the two-stage approach has a wide applications, including closed-loop control systems formed of plants, actuators, sensors, and controllers, and closed-loop sensor networks where individual subsystems need to be identified accurately.

Closed-loop identification has attracted much attention due to the emerging area of joint identification and control. The key idea of joint identification and control strategy, as opposed to disjoint or separate identification and control strategy, is to identify and control with the objective of minimizing a global control performance criterion. This topic is often referred to as control relevant identification, iterative identification and control, and so on.

There are three basic approaches to closed-loop identification, namely a direct, an indirect, and a two-stage one [5,6]. Using a direct approach, there may be a bias in the identified subsystem models due mainly to the correlation between the input and the noise. Further, the open-loop plant may be unstable. The indirect approach is based on identifying the closed-loop system using the reference input and the subsystem output. The desired subsystem transfer function can then be deduced from the estimated closed-loop transfer function obtained from the algebraic relationship between the system's open loop and closed-loop transfer functions. However, the derivation of the desired subsystem transfer function from the closed-loop transfer function may be prone to errors if algebraic relationship governing the open and the closed loop transfer functions are not accurate. This may stem from the inaccuracies and the nonlinearities of the systems connected in cascade with the desired subsystem. Hence, the indirect approach may not be preferable. The two-stage approach gives a consistent estimate of the open-loop subsystem regardless of the assumed noise model structure as long as the feedback has linear structure.

Identification of Physical Systems: Applications to Condition Monitoring, Fault Diagnosis, Soft Sensor and Controller Design, First Edition. Rajamani Doraiswami, Chris Diduch and Maryhelen Stevenson.

It has been shown [6] that as the model order and the data record length tend to infinity, performance of all the approaches are identical. In this chapter, we address an important problem of evaluating the performance of the closed-loop identification scheme when the data record is finite. In practice, the input–output data record is finite, and the performance of closed-loop scheme may differ from performance when the data record is finite.

The state-space approach was adopted here as it is an elegant way to model a SIMO system. In such a system, transfer functions relating all combinations of the inputs and the outputs share a common denominator that characterizes the dynamics of the entire system. In order to ensure that the estimated transfer functions meet this denominator-wise consistency requirement, all the transfer functions are identified simultaneously using a MIMO identification scheme such as the Prediction Error Method (PEM), High-Order Least Squares (HOLS) method, or the Subspace Method (SM). Identifying the individual transfer functions separately using SISO schemes may not meet this consistency requirement governing the transfer functions due to the presence of noise, nonlinearity, and other artifacts. For example, the sensitivity and complementary sensitivity functions may have different denominators if identified separately.

A general case of a closed-loop system is formed of a cascade combination of subsystems in the forward path. The input and the output of each subsystem are measured, and the objective is to identify the subsystems. The subsystems are subject to noise (disturbances and the measurement noise).

8.1.1 Kalman Filter-Based Identification Model

Most of the popular and widely used identification methods, including the High-Order Least-Squares (HOLS) method, the prediction method (PEM), and Subspace Method (SM) use the structure of the identification model set based on the Kalman filter [5, 7]. The residual of the Kalman filter takes over the combined role of the disturbance and the measurement of noise affecting the system. As the Kalman filter residual, termed also an innovation process, is a zero-mean white process, one can develop an appropriate identification scheme using the vast literature on statistical estimation theory for parameter estimation in a filtered white noise process (or colored noise) [8].

8.1.2 Closed-Loop Identification Approaches

There are three approaches to the identification of the subsystem operating in a closed-loop configuration, namely the direct, the indirect, and the two-stage approach (also termed joint input–output approach).

- *Direct approach*: In the direct approach, the subsystem model is estimated directly from its input and its output. This approach to identifying a subsystem from its input and output data suffers from the problem of correlation between the subsystem input, and disturbance and the noise affecting the subsystem output. Because of this correlation, there may be bias errors in the identified subsystem models, especially if a simple scheme such as the least squares method is employed. A consistent and efficient identification scheme, such as the prediction error method, may be used to overcome the problem of correlations between the input and the output.
- *Indirect approach*: The indirect approach identifies some closed-loop transfer function and determines the open loop subsystem using the perfect knowledge of the (linear) system in the feedback path (e.g., linear controller). The problem with the physical system is that the controller may contain various delimiters, anti-windup functions, and other nonlinearities even if the controller parameters (e.g., Proportional Integral and Derivative (PID) parameters) are known.
- *The two-stage approach* : The two-stage approach identifies the subsystem by first identifying the transfer functions relating the reference input and the system outputs. Then the estimated outputs of the identified transfer functions are used in the second stage to identify the subsystem (unlike the case of a direct approach where the subsystem is identified from the actual input and the actual output of the subsystem). The two-stage approach gives consistent estimate of the open-loop subsystem regardless

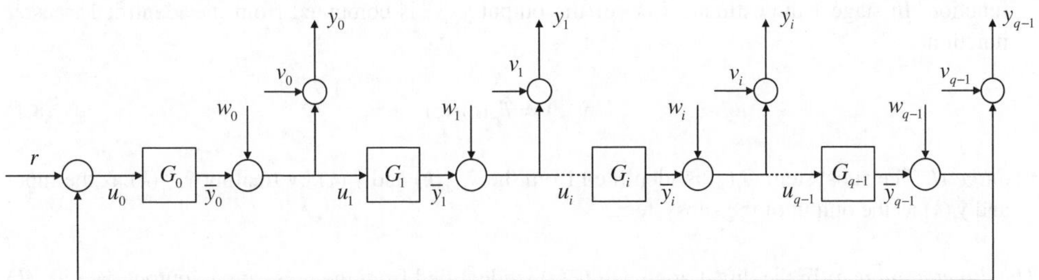

Figure 8.1 Typical closed-loop system

of the assumed noise model structure as long as the feedback has linear structure. The quality of the identification in the first stage is assumed accurate enough so that the estimated input to each subsystem is uncorrelated with the noise and the disturbance.

First, the Single Input and Multiple Output (SIMO) system is considered, and the results are extended to Multiple Input and Multiple Output (MIMO) system. A brief outline of the prediction error and the subspace methods are given. Detailed derivations may be found in excellent books and research articles [5–9]. The high-order least-squares method, however, is given a detailed treatment.

8.2 Closed-Loop System

We will consider a class of unity feedback SIMO closed-loop systems formed of cascade-connected subsystems. A subsystem may be a controller, an amplifier, an actuator, a sensor, or other device. This class of closed loop-systems includes process control systems, position control systems, mechatronics systems, and power generating systems.

A typical closed loop is formed of cascade-connected subsystems $G_i : i = 0, 1, 2, \ldots q$ in the forward path shown in Figure 8.1, where $u_i(k)$ is the input and $y_i(k)$ is the measured output of the subsystem G_i, $w_i(k)$ is the disturbance affecting the subsystem, and $v_i(k)$ is the measurement noise.

In a typical closed-loop control system there will be four subsystems, namely the controller, amplifier, the actuator, and the plant represented respectively by $G_0(z)$, $G_1(z)$, $G_2(z)$, and $G_3(z)$, $u_0(k)$ is the tracking error, $u_1(k)$ is the input to the amplifier, $u_2(k)$ is the input to actuator, and $u_4(k)$ is the input to the plant, $y_4(k)$ is the output.

Objective

The objective is to identify the subsystems $\{G_i(z)\}$ operating in closed loop. The measurement data includes reference input $r(k)$, and the subsystem output measurements. $\{y_i(k)\} : i = 0, 1, 2, \ldots q - 1$.

Approaches

There are three approaches to closed-loop identification, namely the direct approach, an indirect and a two-stage one. We will first focus on the two-stage identification approach.

8.2.1 Two-Stage and Direct Approaches

The two-stage approach: The subsystems $\{G_i(z)\}$ are identified in two stages:

- *Stage I*: The closed-loop transfer functions relating the reference input $r(k)$ and the outputs $y_i(k)$ are identified. Let $T_{ri}(z)$ be the transfer function relating $r(z)$ and $y_i(z)$, and $\hat{T}_{ri}(z)$ be the estimated transfer function. In stage I the estimate $\hat{y}_i(k)$ of the output $y_i(k)$ is computed from the identified transfer function

$$\hat{y}_i(z) = \hat{T}_{ri}(z)r(z) \tag{8.1}$$

- *Stage II:* The subsystem $G_i(z)$ is identified from the $\hat{y}_{i-1}(k)$ and $\hat{y}_i(k)$ by treating $\hat{y}_{i-1}(k)$ as the input and $\hat{y}_i(k)$ as the output of the subsystem.

The direct approach: In the direct approach $G_i(z)$ is identified from the noisy input–output data $y_{i-1}(k)$ and $y_i(k)$. Any of the high performance identifications, such as the prediction error, the high-order least squares, or the subspace method may be used.

8.3 Model of the Single Input Multi-Output System

Consider the SIMO model shown in Figure 8.1. The linear regression model of subsystems $G_i(z)$: $i = 1, 2, 3, \ldots q-1$ is given by:

$$\begin{aligned} \bar{y}_i(k) &= \sum_{\ell=1}^{n_i} \alpha_{i\ell}\bar{y}_i(k-\ell) + \sum_{\ell=1}^{n_i} \beta_{i\ell}u_i(k-\ell) \\ y_i(k) &= \bar{y}_i(k) + w_i(k) + v_i(k) \\ u_i(k) &= y_{i-1}(k) + w_{i-1}(k) \end{aligned} \tag{8.2}$$

The model of the subsystem $G_0(z)$ is:

$$\begin{aligned} \bar{y}_0(k) &= \sum_{\ell=1}^{n_0} \alpha_{0\ell}\bar{y}_0(k-\ell) + \sum_{\ell=1}^{n_0} \beta_{0\ell}u_0(k-\ell) \\ y_0(k) &= \bar{y}_0(k) + w_0(k) + v_0(k) \\ u_0(k) &= r(k) - \bar{y}_{q-1}(k) - w_{q-1}(k) \end{aligned} \tag{8.3}$$

8.3.1 State-Space Model of the Subsystem

The state-space model of the MIMO system is obtained from the state-space model of the subsystems. The state-space model $(\boldsymbol{A}_i, \boldsymbol{B}_i, \boldsymbol{C}_i)$ of the subsystem $G_i(z)$, $i = 1, 2, 3, \ldots, q-1$

$$\begin{aligned} \boldsymbol{x}_i(k+1) &= \boldsymbol{A}_i\boldsymbol{x}_i(k) + \boldsymbol{B}_iu_i(k) \\ y_i(k) &= \boldsymbol{C}_i\boldsymbol{x}_i(k) + w_i(k) + v_i(k) \\ u_i(k) &= \boldsymbol{C}_{i-1}\boldsymbol{x}_{i-1}(k) + w_{i-1}(k \end{aligned} \tag{8.4}$$

Substituting for $u_i(k)$ in terms of the states $\boldsymbol{x}_{i-1}(k)$ of the subsystem G_{i-1}, and the disturbance $w_{i-1}(k)$ we get:

$$\begin{aligned} \boldsymbol{x}_i(k+1) &= \boldsymbol{A}_i\boldsymbol{x}_i(k) + \boldsymbol{B}_i\boldsymbol{C}_{i-1}\boldsymbol{x}_{i-1}(k) + \boldsymbol{B}_iw_{i-1}(k) \\ y_i(k) &= \boldsymbol{C}_i\boldsymbol{x}_i(k) + w_i(k) + v_i(k) \end{aligned} \tag{8.5}$$

where $\boldsymbol{x}_i(k)$ is a $n_i x1$ vector, $u_i(k)$, $y_i(k)$, $w_i(k)$, and $v_i(k)$ are all scalars denoting respectively the input, the output measurement, the disturbance, and the measurement noise.

The state-space model $(\boldsymbol{A}_0, \boldsymbol{B}_0, \boldsymbol{C}_0)$ of the subsystem $G_0(z)$ with unity feedback is:

$$\begin{aligned}\boldsymbol{x}_0(k+1) &= \boldsymbol{A}_0\boldsymbol{x}_0(k) + \boldsymbol{B}_0 r(k) - \boldsymbol{B}_0\boldsymbol{C}_{q-1}x_{q-1}(k) - \boldsymbol{B}_0 w_{q-1}(k)\\ y_0(k) &= \boldsymbol{C}_0\boldsymbol{x}_0(k) + w_0(k) + v_0(k)\end{aligned} \tag{8.6}$$

8.3.2 State-Space Model of the Overall System

Using Eqs. (8.5) and (8.6), the state-space model of the overall closed-loop system $(\boldsymbol{A}, \boldsymbol{B}, \boldsymbol{C})$ relating the reference input $r(k)$ and the output measurements $\{y_i(k)\}$ is given below.

$$\begin{aligned}\boldsymbol{x}(k+1) &= \boldsymbol{A}\boldsymbol{x}(k) + \boldsymbol{B}r(k) + \boldsymbol{E}_w\boldsymbol{w}(k)\\ \boldsymbol{y}(k) &= \boldsymbol{C}\boldsymbol{x}(k) + \boldsymbol{F}_w\boldsymbol{w}(k) + \boldsymbol{F}_v\boldsymbol{v}(k)\end{aligned} \tag{8.7}$$

where $\boldsymbol{x}(k)$ is $nx1$, $n = \sum_{i=0}^{q-1} n_i$ state vector, $\boldsymbol{y}(k)$, $\boldsymbol{w}(k)$, and $\boldsymbol{v}(k)$ are respectively $qx1$ measurement, disturbance, and measurement noise vectors given by:

$$\boldsymbol{x}(k) = \begin{bmatrix} \boldsymbol{x}_0(k) \\ \boldsymbol{x}_1(k) \\ . \\ \boldsymbol{x}_{q-1}(k) \end{bmatrix},\ \boldsymbol{y}(k) = \begin{bmatrix} y_0(k) \\ y_1(k) \\ . \\ y_{q-1}(k) \end{bmatrix},\ \boldsymbol{w}(k) = \begin{bmatrix} w_{q-1}(k) \\ w_0(k) \\ . \\ w_{q-2}(k) \end{bmatrix},\ \boldsymbol{v}(k) = \begin{bmatrix} v_0(k) \\ v_1(k) \\ . \\ v_{q-1}(k) \end{bmatrix};\ \boldsymbol{C} = \begin{bmatrix} C_0 \\ C_1 \\ . \\ C_{q-1} \end{bmatrix}$$

$$\boldsymbol{A} = \begin{bmatrix} \boldsymbol{A}_0 & 0 & 0 & . & -\boldsymbol{B}_0\boldsymbol{C}_{q-1} \\ \boldsymbol{B}_1\boldsymbol{C}_0 & \boldsymbol{A}_1 & 0 & . & 0 \\ 0 & \boldsymbol{B}_2\boldsymbol{C}_1 & \boldsymbol{A}_2 & . & 0 \\ . & . & . & . & . \\ 0 & 0 & 0 & \boldsymbol{B}_{q-1}\boldsymbol{C}_{q-2} & \boldsymbol{A}_{q-1} \end{bmatrix};\ \boldsymbol{B} = \begin{bmatrix} \boldsymbol{B}_0 \\ 0 \\ 0 \\ . \\ 0 \end{bmatrix},\ \boldsymbol{E}_w = \begin{bmatrix} -\boldsymbol{B}_0 \\ \boldsymbol{B}_1 \\ \boldsymbol{B}_2 \\ . \\ \boldsymbol{B}_{q-1} \end{bmatrix},\ \boldsymbol{F}_w = \boldsymbol{F}_v = \begin{bmatrix} 1 \\ 1 \\ 1 \\ . \\ 1 \end{bmatrix}$$

Assumptions

- *$(\boldsymbol{A}, \boldsymbol{B})$ and $(\boldsymbol{A}, \boldsymbol{E}_w)$ are controllable and $(\boldsymbol{A}, \boldsymbol{C})$ is observable,*
- *It is assumed that the disturbance $\{w_i(k)\}$and the measurement noise $\{v_i(k)\}$ are zero-mean uncorrelated white noise processes.*
- *The reference input $r(k)$ is uncorrelated with $\{w_i(k)\}$ and $\{v_i(k)\}$.*
- *The closed-loop system is asymptotically stable.*

8.3.3 Transfer Function Model

The transfer function model of the system relating the reference input $r(z)$, the disturbance $\boldsymbol{w}(z)$, and the measurement noise $\boldsymbol{v}(z)$ to the output $\boldsymbol{y}(z)$ is given by:

$$y(z) = \frac{\boldsymbol{N}(z)}{D(z)} r(z) + \left(\frac{\boldsymbol{N}_w(z)}{D(z)} + \boldsymbol{F}_w\right)\boldsymbol{w}(z) + \boldsymbol{F}_v\boldsymbol{v}(z) \tag{8.8}$$

where $\dfrac{\boldsymbol{N}(z)}{D(z)} = \boldsymbol{C}(zI - \boldsymbol{A})^{-1}\boldsymbol{B}$, and $\dfrac{\boldsymbol{N}_w(z)}{D(z)} = \boldsymbol{C}(zI - \boldsymbol{A})^{-1}\boldsymbol{E}_w$. Rewriting by cross-multiplying by $D(z)$, we get:

$$D(z)\boldsymbol{y}(z) = \boldsymbol{N}(z)r(z) + \boldsymbol{\upsilon}(z) \tag{8.9}$$

where $\upsilon(z) = (N_w(z) + D(z)F_w)w(z) + D(z)F_v v(z)$ is the equation error formed of two colored noise processes generated by the disturbance $w(z)$ and the measurement noise $v(z)$.

8.3.4 Illustrative Example: Closed-Loop Sensor Network

Consider an example of a closed loop sensor network formed of three cascade-connected subsystems:

Subsystem $G_0(z) = \dfrac{0.1z^{-1}}{1 - 0.5z^{-1} + 0.8z^{-2}}$

$$\begin{aligned} x_0(k+1) &= \begin{bmatrix} 0.5 & -0.8 \\ 1 & 0 \end{bmatrix} x_0(k) + \begin{bmatrix} 1 \\ 0 \end{bmatrix} u_0(k) \\ y_0(k) &= \begin{bmatrix} 0.1 & 0 \end{bmatrix} x_1(k) \end{aligned} \tag{8.10}$$

Subsystem $G_1(z) = \dfrac{0.1z^{-1}}{1 - 0.7z^{-1} + 0.9z^{-2}}$

$$\begin{aligned} x_1(k+1) &= \begin{bmatrix} 0.7 & -0.9 \\ 1 & 0 \end{bmatrix} x_1(k) + \begin{bmatrix} 1 \\ 0 \end{bmatrix} u_1(k) + \begin{bmatrix} 1 \\ 0 \end{bmatrix} w_0(k) \\ y_1(k) &= \begin{bmatrix} 0.1 & 0 \end{bmatrix} x_1(k) + v_0(k) \end{aligned} \tag{8.11}$$

Subsystem $G_2(z) = \dfrac{0.1z^{-1}}{1 - 0.9z^{-1} + 0.81z^{-2}}$

$$\begin{aligned} x_2(k+1) &= \begin{bmatrix} 0.9 & -0.81 \\ 1 & 0 \end{bmatrix} x_2(k) + \begin{bmatrix} 1 \\ 0 \end{bmatrix} u_2(k) + \begin{bmatrix} 1 \\ 0 \end{bmatrix} w_1(k) \\ y_2(k) &= \begin{bmatrix} 0.1 & 0 \end{bmatrix} x_2(k) + v_1(k) \end{aligned} \tag{8.12}$$

Unity feedback:

$$u_0(k) = r(k) - y_2(k) \tag{8.13}$$

The closed-loop system:

$$\begin{aligned} x(k) &= \begin{bmatrix} 0.5 & -0.8 & 0 & 0 & -0.1 & 0 \\ 1 & 0 & 0 & 0 & 0 & 0 \\ 0.1 & 0 & 0.7 & -0.9 & 0 & 0 \\ 0 & 0 & 1 & 0 & 0 & 0 \\ 0 & 0 & 0.1 & 0 & 0.9 & -0.81 \\ 0 & 0 & 0 & 0 & 1 & 0 \end{bmatrix} x(k) + \begin{bmatrix} 1 \\ 0 \\ 0 \\ 0 \\ 0 \\ 0 \end{bmatrix} r(k) + \begin{bmatrix} 0 & 0 \\ 0 & 0 \\ 1 & 0 \\ 0 & 0 \\ 0 & 1 \\ 0 & 0 \end{bmatrix} \begin{bmatrix} w_0(k) \\ w_1(k) \end{bmatrix} \\ y(k) &= \begin{bmatrix} 1 & 0 & 0 & 0 & 0 & 0 \\ 0 & 0 & 1 & 0 & 0 & 0 \\ 0 & 0 & 0 & 0 & 1 & 0 \end{bmatrix} x(k) + \begin{bmatrix} v_0(k) \\ v_1(k) \\ v_2(k) \end{bmatrix} \end{aligned} \tag{8.14}$$

where $x(k) = \begin{bmatrix} x_0(k) \\ x_1(k) \\ x_2(k) \end{bmatrix}$; $y(k) = \begin{bmatrix} y_0(k) \\ y_1(k) \\ y_2(k) \end{bmatrix}$

Frequency-domain model:

$$D(z)\mathbf{y}(z) = \mathbf{N}(z)r(z) + \boldsymbol{\upsilon}(z) \tag{8.15}$$

where $\boldsymbol{\upsilon}(z)$ is the equation error which is colored noise generated by the zero-mean white noise processes, namely, the disturbances $w_0(z)$ and $w_1(z)$and the measurement noise $v_0(z)$, $v_1(z)$ and $v_2(z)$ given by

$$\boldsymbol{\upsilon}(z) = \begin{bmatrix} \upsilon_0(z) \\ \upsilon_1(z) \\ \upsilon_2(z) \end{bmatrix} = \mathbf{N}_w(z)\begin{bmatrix} w_0(z) \\ w_1(z) \end{bmatrix} + D(z)\begin{bmatrix} 1 & 0 & 0 \\ 0 & 1 & 0 \\ 0 & 0 & 1 \end{bmatrix}\begin{bmatrix} v_0(z) \\ v_1(z) \\ v_2(z) \end{bmatrix} \tag{8.16}$$

$$D(z) = 1 - 2.1z^{-1} + 3.94z^{-2} - 3.826z^{-3} + 3.2895z^{-4} - 1.4661z^{-5} + 0.5832z^{-6},$$

$$\mathbf{N}(z) = \begin{bmatrix} N_0(z) \\ N_1(z) \\ N_2(z) \end{bmatrix} = \begin{bmatrix} z^{-1} & -1.6z^{-2} & 2.34z^{-3} & -1.377z^{-4} & 0.7229z^{-5} & 0 \\ 0 & 0.1z^{-2} & -0.09z^{-3} & 0.081z^{-4} & 0 & 0 \\ 0 & 0 & 0.01z^{-3} & 0 & 0 & 0 \end{bmatrix},$$

$$\mathbf{N}_w(z) = \begin{bmatrix} \mathbf{N}_{w1}(z) & \mathbf{N}_{w2}(z) \end{bmatrix};\ \mathbf{N}_{w1}(z) = \begin{bmatrix} 0 & 0 & -0.01z^{-3} & 0 & 0 \\ z^{-1} & -1.4z^{-2} & 2.06z^{-3} & -1.125z^{-4} & 0.648z^{-5} \\ 0 & 0.1z^{-2} & -0.05z^{-3} & 0.08z^{-4} & 0 \end{bmatrix},$$

$$\mathbf{N}_{w2}(z) = \begin{bmatrix} 0 & -0.1z^{-2} & 0.07z^{-3} & -0.09z^{-4} & 0 \\ 0 & 0 & -0.01z^{-3} & 0 & 0 \\ 1z^{-1} & -1.2z^{-2} & 2.05z^{-3} & -1.01z^{-4} & 0.72z^{-5} \end{bmatrix}.$$

Note that all the closed-loop subsystems relating the reference input $r(z)$ to the outputs $y_i(z)$, $i = 0, 1, 2$ have the same denominator $D(z)$.

Figure 8.2 shows the outputs of the system $y_i(k)$, the auto-correlation of the equation error $\upsilon_i(k)$, and the cross-covariances of the system outputs. The reference input is a pseudo-random binary signal. The output $y_i(k)$ and the noise-free system output are shown.

Subfigures (a), (b), and (c) on the left show respectively the outputs $y_0(k)$, $y_1(k)$, and $y_2(k)$. The noise-free outputs are shown for comparison.

Subfigures (d), (e), and (f) on the middle show the auto-correlation function of the equation errors $\upsilon_0(k)$, $\upsilon_1(k)$, and $\upsilon_2(k)$. The auto-correlation functions are normalized so that their maximum values are all unity.

Subfigures (g), (h), and (i) on the right show the cross-correlation of $y_0(k)$, $y_1(k)$, and $y_2(k)$ with the disturbance $\mathbf{w}(k)$.

Comments *The auto-correlation functions of the equation errors are colored noise, as can be deduced from the Figure 8.2. Hence, the structure of the system model (8.15) may not be used for identification, as the equation error is not a zero-mean white noise process. In the subsequent sections, the Kalman filter-based identification model is derived such that the equation error is a zero-mean white noise process.*

The system outputs $y_0(k)$, $y_1(k)$, and $y_2(k)$ are correlated with the disturbance $\mathbf{w}(k)$ due to the closed-loop action as can be deduced from the right most subfigures. Hence a scheme to identify the subsystems $G_i(z)$: $i = 0, 1, 2$ directly from the system outputs must take into account the correlations that exist between the input and the output of the subsystems. The two- stage identification scheme is widely used to overcome this problem of correlation between the subsystem input and its output.

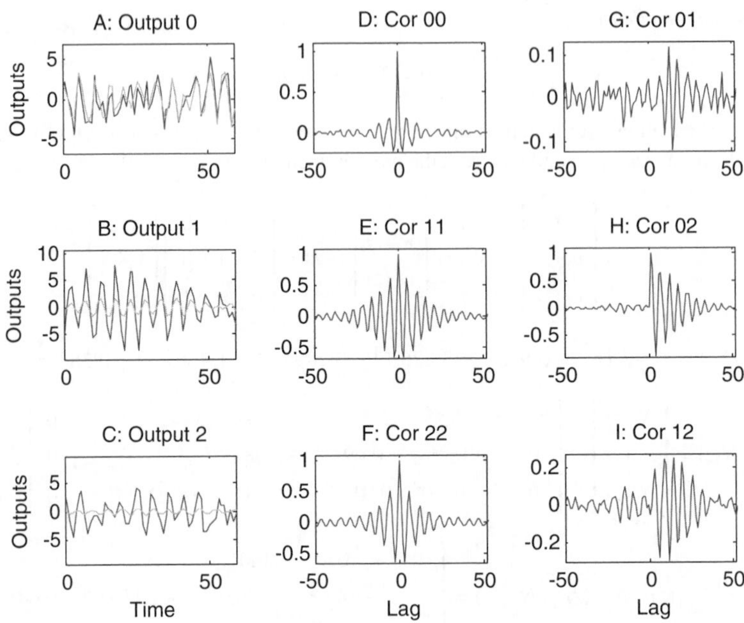

Figure 8.2 The system outputs and the auto-correlations of the equation error

8.4 Kalman Filter-Based Identification Model

In our case, the equation error $\boldsymbol{\upsilon}(z)$ is the sum of two-colored noise process as shown in Eq. (8.9). If the objective function is chosen as a function of $\boldsymbol{\upsilon}(z)$, the performance of the identification will be degraded. To overcome this problem, we will use a Kalman filter-based model structure to generate a zero-mean white noise equation error.

The residual of the Kalman filter takes over the combined role of the disturbance and the measurement noise affecting the system. The principle role of the Kalman filter-based model is to help in whitening the equation error in the linear regression model for identification, including the high-order least-squares and prediction error methods.

It may be worthwhile reviewing the Kalman filter model, which will help derive the structure of the identification model set.

8.4.1 State-Space Model of the Kalman Filter

The Kalman filter for the system (8.7) is given by:

$$\begin{aligned}\hat{\boldsymbol{x}}(k+1) &= (\boldsymbol{A}-\boldsymbol{KC})\hat{\boldsymbol{x}}(k)+\boldsymbol{B}r(k)+\boldsymbol{Ky}(k)\\ \hat{\boldsymbol{y}}(k) &= \boldsymbol{C}\hat{\boldsymbol{x}}(k)\\ \boldsymbol{e}(k) &= \boldsymbol{y}(k)-\hat{\boldsymbol{y}}(k)\end{aligned} \tag{8.17}$$

where the $\boldsymbol{e}(k) = [\,e_0(k)\;\; e_1(k)\;\; .\;\; e_{q-1}(k)\,]^T$ is a $qx1$ residual, $\boldsymbol{K} = [\,\boldsymbol{K}_0\;\; \boldsymbol{K}_1\;\; .\;\; \boldsymbol{K}_{q-1}\,]$ is nxq Kalman gain matrix, $\hat{\boldsymbol{y}}(k)$ is the $qx1$ is predictor of the output $\boldsymbol{y}(k)$ and the residuals $e_i(k) : i = 0, 1, 2, \ldots, q-1$ uncorrelated zero-mean white noise process (or innovation processes).

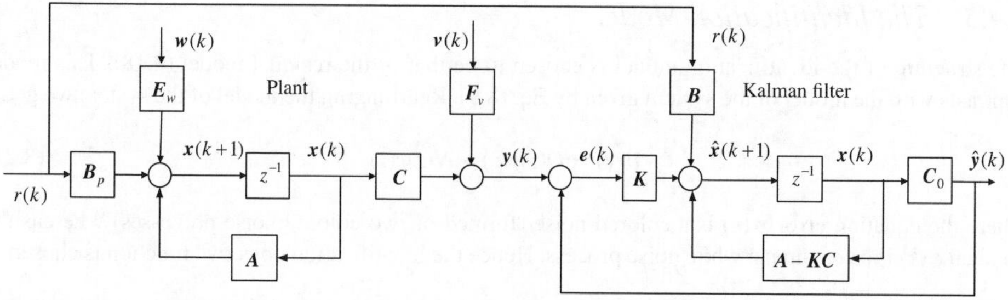

Figure 8.3 Role of the Kalman filter: whitening the equation error

The inputs to the Kalman filter are reference input $r(k)$ and the $qx1$ output $\mathbf{y}(k)$, and the output is the $qx1$ residual $\mathbf{e}(k)$, whereas the input to the system is $r(k)$ and its output is $qx1$ output $\mathbf{y}(k)$, which is corrupted by a colored noise process $\upsilon(k)$. The Kalman filter may be viewed as a *whitening filter*: the input to the Kalman filter is $y(k)$ and its output is a zero-mean white noise process $e(k)$. Figure 8.3 shows the role of the Kalman filter in whitening the equation error. The equation error $\boldsymbol{\upsilon}(k)$ given by Eq. (8.9) is a colored noise as shown in subfigures (d), (e), and (f) of Figure 8.2 whereas the residual output of the Kalman filter $\mathbf{e}(k)$ is a zero-mean white noise process.

8.4.2 Residual Model

The z-transfer expression relating the Kalman filter inputs, namely the reference input $r(z)$, system output measurements $\mathbf{y}(z)$ to the Kalman filter residual output $\mathbf{e}(z)$ is:

$$\mathbf{e}(z) = \left(\mathbf{I} - \frac{\mathbf{N}_y(z)}{F(z)}\right)\mathbf{y}(z) - \frac{\mathbf{N}_r(z)}{F(z)}r(z) \tag{8.18}$$

where $\frac{\mathbf{N}_r(z)}{F(z)}$ is a$qx1$ vector transfer function relating $r(z)$ to $\mathbf{e}(z)$, and $\frac{\mathbf{N}_y(z)}{F(z)}$ is qxq a matrix transfer function relating $\mathbf{y}(z)$ to $\mathbf{e}(z)$ defined by:

$$\begin{aligned}\frac{\mathbf{N}_r(z)}{F(z)} &= \mathbf{C}(z\mathbf{I} - \mathbf{A} + \mathbf{KC})^{-1}\mathbf{B} \\ \frac{\mathbf{N}_y(z)}{F(z)} &= \mathbf{C}(z\mathbf{I} - \mathbf{A} + \mathbf{KC})^{-1}\mathbf{K}\end{aligned} \tag{8.19}$$

where $F(z) = |z\mathbf{I} - \mathbf{A} + \mathbf{KC}|$ is the characteristic polynomial of the Kalman filter, termed the *Kalman polynomial*, $\mathbf{N}_r(z)$ is $qx1$ vector polynomial and $\mathbf{N}_y(z)$ is qxq matrix polynomials, and $F(z) = 1 + \sum_{i=1}^{n} c_i z^{-i}$, and $\{c_i\}$ are the coefficients.

The model (8.18) relating the residual to the system input and the output is termed the *residual model*.

Scalar system:

In the case when the overall system is a scalar: $y(k)$ is a scalar instead of a $qx1$ vector, the residual model reduces to the following scalar model:

$$e(z) = \frac{D(z)}{F(z)}y(z) - \frac{N(z)}{F(z)}r(z) \tag{8.20}$$

where $G(z) = \frac{N(z)}{D(z)}$ is the transfer function of the system, $\mathbf{I} - \mathbf{N}_{ye}(z) = \frac{D(z)}{F(z)}$; $\mathbf{N}_{re}(z) = N(z)$.

8.4.3 The Identification Model

The structure of the identification model is chosen to be that of the residual model (8.18). This model contrasts with the model of the system given by Eq. (8.9). Rearranging the model of the system we get:

$$\boldsymbol{\upsilon}(z) = D(z)\boldsymbol{y}(z) - \boldsymbol{N}(z)r(z) \tag{8.21}$$

where the equation error $\boldsymbol{\upsilon}(k)$ is a colored noise (formed of two colored noise processes) whereas the residual $\boldsymbol{e}(k)$ is a zero-mean white noise process. Hence the identification objective function is chosen to be the sum of the squares of the residual $\sum_{i=0}^{N-1} e^2(k-i)$ instead of the sum of the squares of the equation error $\sum_{i=0}^{N-1} \upsilon^2(k-i)$. Expressing in the vector the objective function, denoted J, becomes:

$$J = \boldsymbol{e}^T(k)\boldsymbol{e}(k) \tag{8.22}$$

where $\boldsymbol{e}(k)$ is given by the residual model. Let unknown parameters in the residual model be a vector, denoted $\boldsymbol{\theta}_{res}$, which is formed of the coefficients of (i) the matrix polynomial $\boldsymbol{N}_{ye}(z)$, (ii) the matrix polynomial $\boldsymbol{N}_{re}(z)$, and (iii) the Kalman polynomial $F(z)$. The identification problem may be formulated as follows:

$$\min_{\theta_{res}}\{J = \boldsymbol{e}^T(k)\boldsymbol{e}(k)\}\text{such that } \boldsymbol{e}(k) \text{ satisfies (8.18)} \tag{8.23}$$

8.5 Closed-Loop Identification Schemes

We will consider three important identification approaches, namely the Prediction Error Method (PEM), the High-Order Least-Squares (HOLS), and the subspace method (SM) methods. The prediction error and the subspace methods were developed in earlier chapters, and hence only the high-order least-squares method is developed here for the single-input and multi-output case.

8.5.1 The High-Order Least-Squares Method

The HOLS method is based on deriving a high-order linear regression model from the residual model so that the equation error is a zero-mean white noise process. The parameters of the high-order model, namely the coefficients of the numerator and the denominator polynomials are estimated using the simple and computationally efficient least squares method. The estimates of the parameters are obtained in a closed form.

Rearranging the residual model (8.18) yields:

$$\boldsymbol{y}(z) = \frac{\boldsymbol{N}_y(z)}{F(z)}\boldsymbol{y}(z) + \frac{\boldsymbol{N}_r(z)}{F(z)}r(z) + \boldsymbol{e}(z) \tag{8.24}$$

Dividing each of the dividend matrix polynomials $\boldsymbol{N}_y(z)$ and $\boldsymbol{N}_r(z)$ by the devisor polynomial $F(z)$, and truncating the quotient to some finite but large number of terms n_h so that the remainder term is negligible yields:

$$\begin{aligned}\frac{\boldsymbol{N}_y(z)}{F(z)} &\approx \sum_{\ell=1}^{n_h} \boldsymbol{A}_{h\ell} z^{-\ell}\\ \frac{\boldsymbol{N}_r(z)}{F(z)} &\approx \sum_{\ell=0}^{n_h} \boldsymbol{b}_{h\ell} z^{-\ell}\end{aligned} \tag{8.25}$$

where n_h is the selected high order, $\boldsymbol{A}_{h\ell}$ is a qxq matrix, and $\boldsymbol{b}_{h\ell}$ is $qx1$ vector coefficients in the high-order FIR approximations of the rational polynomials $\frac{\boldsymbol{N}_y(z)}{F(z)}$ and $\frac{\boldsymbol{N}_r(z)}{F(z)}$ respectively,

$$\boldsymbol{A}_{h\ell} = \begin{bmatrix} a_{h11\ell} & a_{h12\ell} & . & a_{h1q\ell} \\ a_{h21\ell} & a_{h22\ell} & . & a_{h2q\ell} \\ . & . & . & . \\ a_{hq1\ell} & a_{hq2\ell} & . & a_{hqq\ell} \end{bmatrix}; \boldsymbol{b}_{h\ell} = \begin{bmatrix} b_{h1\ell} \\ b_{h2\ell} \\ . \\ b_{hq\ell} \end{bmatrix}, \ell = 1, 2, 3, \ldots, n_h.$$

Since $F(z)$ is a stable (or Hurwitz) polynomial (roots of the polynomial are strictly inside the unit circle), we can select an order n_h high enough so that the approximation errors become negligible and the residual model (8.24) becomes:

$$\boldsymbol{y}(z) \approx \sum_{\ell=1}^{n_h} \boldsymbol{A}_{h\ell} z^{-\ell} \boldsymbol{y}(z) + \sum_{\ell=1}^{n_h} \boldsymbol{b}_{h\ell} z^{-\ell} r(z) + \boldsymbol{e}(z) \tag{8.26}$$

Assumption *In the HOLS method, we assume the following:*
A1: *In the modeling phase, n_h is large enough to ensure that the errors in the approximation of the rational polynomials by high-order polynomials are negligible. Hence, we may assume that the approximation sign in Eq. (8.26) may be replaced by equality sign, and we get:*

$$\boldsymbol{y}(z) = \sum_{\ell=1}^{n_h} \boldsymbol{A}_{h\ell} z^{-\ell} \boldsymbol{y}(z) + \sum_{\ell=1}^{n_h} \boldsymbol{b}_{h\ell} z^{-\ell} r(z) + \boldsymbol{e}(z) \tag{8.27}$$

Taking the inverse z-transform yields:

$$\boldsymbol{y}(k) = \sum_{\ell=1}^{n_h} \boldsymbol{A}_{h\ell} \boldsymbol{y}(k-\ell) + \sum_{\ell=1}^{n_h} \boldsymbol{b}_{h\ell} r(k-\ell) + \boldsymbol{e}(k) \tag{8.28}$$

Expressing the high-order regression model (8.28) relating the elements $y_i(k)$ of the $qx1$ output vector $\boldsymbol{y}(k)$, *the input $r(k)$, and the element $e_i(k)$ of the $qx1$ residual vector* $\boldsymbol{e}(k)$ *we get*

$$y_i(k) = \sum_{j=1}^{q} \sum_{\ell=1}^{n_h} a_{hij\ell} y_j(k-\ell) + \sum_{\ell=1}^{n_h} b_{hi\ell} r(k-\ell) + e_i(k) \tag{8.29}$$

Because of this assumption, the equation error $\boldsymbol{e}(k)$ *is a zero-mean white noise sequence, although it may not be identically distributed. That is, each element of the $qx1$ equation error $e(k)$ is a zero-mean independent and identically distributed white noise process. The variance of each may be different.*

A2: *In the identification phase, both the number of data samples N and selected order n_h are large, but $\frac{n_h}{N}$ is finite and small so that the quality of the estimates is superior.*

Comment *Equation (8.28) is termed the* polynomial model . *The polynomial model is similar to that used widely for the identification of multi-input–multi-output, and single-input–multi-output ARMA models [2].*

8.5.1.1 Polynomial Model: Batch Processing

Given values $k-i$, $i=1,2,3,\ldots,N$ we can express in a linear matrix model given by:

$$\mathbf{y}_i(k) = \mathbf{H}_{pi}(k)\boldsymbol{\theta}_{pi} + \mathbf{e}_i(k) \tag{8.30}$$

where $\mathbf{y}_i(k)$ is a $Nx1$ output vector, and $\mathbf{e}_i(k)$ is an $Nx1$ error vector, $\mathbf{H}_{pi}$ is $Nx(q+1)n_h$ matrix formed of output s $\{\mathbf{y}_i(k-\ell)\}$ and the inputs $r(k-\ell)$, $\boldsymbol{\theta}_{pi}$ is $(q+1)n_h x1$ vector formed of the elements $\{a_{hij\ell}\}$ of qxq matrices $\{\mathbf{A}_{h\ell}\}$ elements of vectors $\{b_{hi\ell}\}$ of the $qx1$ of the vector $\mathbf{b}_{h\ell}$,

$$\mathbf{y}_i(k) = \begin{bmatrix} y_i(k) \\ y_i(k-1) \\ . \\ y_i(k-N+1) \end{bmatrix}; \mathbf{e}_i(k) = \begin{bmatrix} e_i(k) \\ e_i(k-1) \\ . \\ e_i(k-N+1) \end{bmatrix}; \mathbf{H}_{pi} = \begin{bmatrix} \mathbf{H}_{y1} & \mathbf{H}_{y2} & . & \mathbf{H}_{yq} & \mathbf{H}_r \end{bmatrix}; \boldsymbol{\theta}_{pi} = \begin{bmatrix} \boldsymbol{\theta}_{ahi} \\ \boldsymbol{\theta}_{bhi} \end{bmatrix};$$

$$\mathbf{H}_{y_i} = \begin{bmatrix} y_i(k-1) & y_i(k-2) & . & y_i(k-n_h) \\ y_i(k-2) & y_i(k-3) & . & y_i(k-n_h-1) \\ . & . & . & . \\ y_i(k-N) & y_i(k-N-1) & . & y_i(k-N-n_h+1) \end{bmatrix}; \boldsymbol{\theta}_{ahij} = \begin{bmatrix} a_{hij1} \\ a_{hij2} \\ . \\ a_{hijn_h} \end{bmatrix}; \boldsymbol{\theta}_{ahi} = \begin{bmatrix} \boldsymbol{\theta}_{ahi1} \\ \boldsymbol{\theta}_{ahi2} \\ . \\ \boldsymbol{\theta}_{ahiq} \end{bmatrix};$$

$$\mathbf{H}_r = \begin{bmatrix} r(k-1) & r(k-2) & . & r(k-n_h) \\ r(k-2) & r(k-3) & . & r(k-n_h-1) \\ . & . & . & . \\ r(k-N) & r(k-N-1) & . & r(k-N-n_h+1) \end{bmatrix}; \boldsymbol{\theta}_{bhi} = \begin{bmatrix} b_{hi1} \\ b_{hi2} \\ . \\ b_{hin_h} \end{bmatrix};$$

The $Nx1$ equation error $\mathbf{e}_i(k)$ is a zero-mean white noise process with variance $E[e_i^2(k)] = \sigma_i^2$.

8.5.1.2 Estimate Using the Least-Squares

Since the equation error is a zero-mean white noise process thanks to the high-order approximation, the estimate $\hat{\boldsymbol{\theta}}_p$ of the unknown polynomial model parameter $\boldsymbol{\theta}_p$ may be computed using the least squares approach. The estimate of the parameter is given by:

$$\hat{\boldsymbol{\theta}}_{pi} = \left(\mathbf{H}_{pi}^T(k)\mathbf{H}_{pi}(k)\right)^{-1} \mathbf{H}_{pi}^T(k)\mathbf{y}_i(k) \tag{8.31}$$

Using the polynomial model (8.30) and the estimated model parameter given by Eq. (8.31), the estimate of the system output, denoted $\hat{\mathbf{y}}_{pi}(k)$, is:

$$\hat{\mathbf{y}}_{pi}(k) = \mathbf{H}_{pi}(k)\hat{\boldsymbol{\theta}}_{pi} \tag{8.32}$$

Comments *The structure of the polynomial model is derived from the structure of the matrix transfer function of the predictor form of the Kalman filter (8.17). The matrix transfer function relates the predicted output $\hat{\mathbf{y}}(k)$ (of $\mathbf{y}(k)$) to the reference input $r(k)$ and the system output $\mathbf{y}(k)$ so that the predicted output $\hat{\mathbf{y}}_i(k)$ of the Kalman filter matches the estimate $\hat{\mathbf{y}}_{pi}(k)$ obtained from the polynomial model, that is $\hat{\mathbf{y}}_i(k) = \hat{\mathbf{y}}_{pi}(k)$.*

This generic linear model (8.30) finds application in many branches of science and engineering, including signal and speech processing, de-convolution and geophysical modeling, and covers a wide class of estimation problems in control, communication, and signal processing involving the state-space mode, input–output models including the transfer function, difference equation, and the linear regression models.

8.5.1.3 Performance of Stage I Identification

The performance of the *stage I* identification may be analyzed considering finite samples and infinite data samples.

Finite sample behavior: Assuming that the order n_h is large, the number of data samples N is very large so that $\frac{n_h}{N}$ is very small, the quality of the parameter estimates are as follows:

- The parameter estimate is unbiased:

$$E[\hat{\theta}_p] = \theta_p \tag{8.33}$$

- The estimate of the covariance of the parameter estimate $\hat{\theta}_h$ satisfy the inequality:

$$E[(\hat{\theta}_{pi} - \theta_{pi})(\hat{\theta}_{pi} - \theta_{pi})^T] = \sigma_i^2 \left(\boldsymbol{H}_p^T(k)\boldsymbol{H}_p(k)\right)^{-1} \tag{8.34}$$

  where $\sigma_i^2 = E\left[e_i^2(k)\right]$. The covariance of the estimation error satisfies Cramér–Rao inequality and hence is efficient.

Asymptotic behavior:
The asymptotic properties as $N \to \infty$ are listed below:

- The parameter estimate converges to the true value as:

$$\lim_{N\to\infty} it \left\{\hat{\theta}_p\right\} = \theta_p \tag{8.35}$$

- The output estimation error tends to a zero-mean white noise process $\boldsymbol{e}_p(k)$:.

$$\boldsymbol{y}(k) - \hat{\boldsymbol{y}}_p(k) \to \boldsymbol{e}_p(k) \tag{8.36}$$

Comment *When the number of data samples is sufficiently large, the quality of the estimates approaches the ideal case in the sense that the estimates capture all the information from the measurement, and what is left over from identification is the information-less zero-mean white noise process.*

8.5.1.4 Selection of the High Order

The high order n_h that will ensure the equation error $\boldsymbol{e}(k)$ given by Eq. (8.28) is a zero-mean white noise process depends upon the ratio Q/R where Q is the covariance of the disturbance, and R is the variance of the measurement noise. The larger the ratio Q/R, the larger will be the Kalman gain and the roots of the Kalman polynomial $F(z)$ will be closer to the origin. In this case, the order n_h will be close to the true order of the system. In the extreme case when the roots are all at the origin the order will be equal to that of the system $n_n = n$. On the other hand, the smaller the ratio Q/R, the smaller will be the Kalman gain and the roots of the Kalman polynomial $F(z)$ will be closer to that of the plant. In this case, the order n_h will be larger.

A practical guideline for selecting the high order n_h is that the selected order should be large enough to ensure that the identification error is a zero-mean white noise process (or equivalently the auto-correlation of the error is a delta function). The output identification error approaches the Kalman filter residual if the selected model order is sufficiently large and the number of data samples N tends to infinity:

The identification error denoted $\mathbf{e}_{pi}(k)$is given by

$$\mathbf{e}_{pi}(k) = \mathbf{y}_i(k) - \hat{\mathbf{y}}_{pi}(k) \tag{8.37}$$

where $\mathbf{e}_{pi}(k) = [\, e_{pi}(k) \;\; e_{pi}(k-1) \;\; . \;\; e_{pi}(k-N+1)\,]^T$ is a $Nx1$ vector. The identification error $\mathbf{e}_{pi}(k)$ is an indicator of the performance of the identification. If the identification model, which in this case is polynomial model (8.27), has completely captured the static and the dynamic behavior of the residual model (8.24), then the identification error will be a zero-mean white noise process. Equivalently, the auto-correlation function of $e_{pi}(k)$ will be a delta function. Assuming that the auto-correlation function is ergodic the auto-correlation satisfies:

$$r_{epi}(m) = \lim_{N\to\infty} it \frac{1}{N} \sum_{k=0}^{N-1} e_{pi}(k) e_{pi}(k-m) = \delta(m) \tag{8.38}$$

It can be deduced from Eq. (8.25) that the identification error approaches the Kalman filter residual if the selected model order is sufficiently large and the number of data samples N tends to infinity:

$$e_{pi}(k) \to e_i(k) \tag{8.39}$$

8.5.1.5 Illustrated Example: Identification of Sensor Net

The closed-loop sensor net given in previous section is identified. The identification model is chosen to be the polynomial model given by Eq. (8.29) as its structure is based on that of the Kalman filter. The least squares estimate of the parameters $\hat{\theta}_{pi}$ is obtained from Eq. (8.31).

The system is identified for different choice of high order n_h. The input $r(k)$ is a pseudo-random binary signal, the covariance of the disturbances is $Q = 1$, and the variance of the measurement noise $R = 1$.

Figure 8.4 shows the estimates $\hat{\mathbf{y}}_{pi}(k)$ and $\hat{\mathbf{y}}_i(k)$: $i = 0, 1, 2$ of the system outputs obtained respectively from (i) the polynomial model given by Eq. (8.32) and (ii) the Kalman filter given by Eq. (8.17). The subfigures (a), (b), and (c) show respectively $\hat{\mathbf{y}}_{pi}(k)$ and $\hat{\mathbf{y}}_i(k)$ for $i = 0, 1, 2$. The three subfigures on the left show the estimates when the selected high-order model equals the true order $n_h = n = 6$ and those on the right correspond to the case when the elected model order is higher, $n_h = 2n = 12$.

Figure 8.5 shows the auto-correlation of (i) the Kalman filter residual $\mathbf{e}_i(k)$ and (ii) the equation error $\mathbf{e}_{pi}$ of the system outputs obtained respectively from (i) the polynomial model given by Eq. (8.37) and (ii) the Kalman filter given by Eq. (8.18). Subfigures (a), (b), and (c), and subfigures (d), (e), and (f) show respectively the auto-correlation function of the residual $\mathbf{e}_i(k)$ and that of the identification error $\mathbf{e}_{pi}(k)$ for $i = 0, 1, 2$. The subfigures on the left show the auto-correlations when the selected high-order model equals the true order $n_h = n = 6$ and those on the right correspond to the case when the selected model order is higher, $n_h = 2n = 12$.

Comments *The polynomial model captures completely the predictor form of the Kalman filter, as can be deduced from Figure 8.4. The predicted outputs of the Kalman filter match the outputs of the polynomial model whose structure is based on the transfer function model of the Kalman filter relating predicted output to the reference input and the plant outputs to the polynomial model output.*

The auto-correlation of the identification error $\mathbf{e}_{pi}(k)$ *is a delta function when the selected model order is high,* $n_h \geq 2n$. *The identification error* $\mathbf{e}_{pi}(k)$ *is "close" to residual of the Kalman filter residual* $\mathbf{e}(k)$ *as*

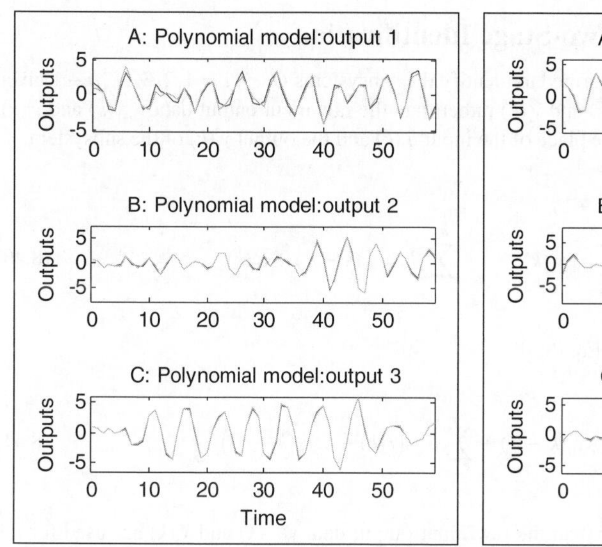

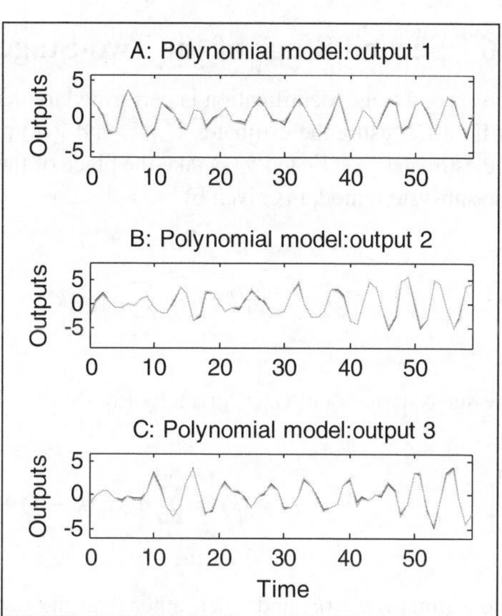

Figure 8.4 The estimates using polynomial and the Kalman filter models for $n_h = n$ and $n_h = 2n$

shown in Eq. (8.28). The auto-correlation and the variance (auto-correlation evaluated at the zero lag) of $\mathbf{e}_{pi}(k)$ *are close to those of* $\mathbf{e}(k)$ *as shown in Eq. (8.39). It may be worth comparing the correlation of the equation error when (i) the structure of the identification model is based on the system model shown in Figure 8.1 and (ii) when it is based on the Kaman filter shown in Figure 8.3. The equation error is a zero-mean white noise process when the identification model is based on the Kalman filter structure and the model order is high.*

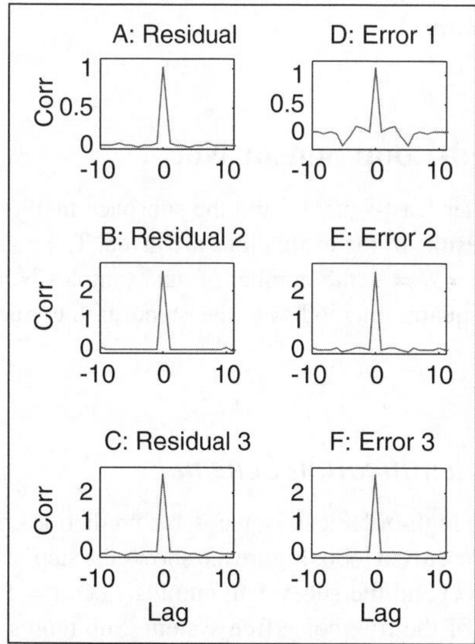

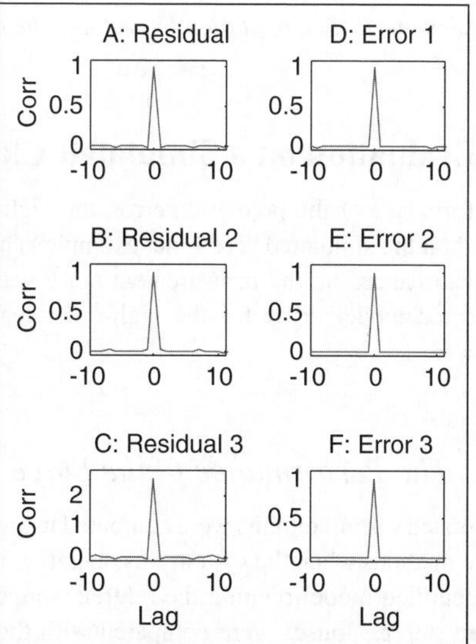

Figure 8.5 Auto-correlation of the residual and the equation error for $n_h = n$ and $n_h = 2n$

8.6 Second Stage of the Two-Stage Identification

The second stage identification is performed to identify the subsystems $G_i(z)$: $i = 1, 2, 3, \ldots, q-1$ given by Eq. (8.2) using the estimates $\hat{y}_{i-1}(k)$ and $\hat{y}_i(k)$ rather than the raw input output data $y_{i-1}(k)$ and $y_i(k)$. The estimates $\hat{y}_{i-1}(k)$ and $\hat{y}_i(k)$ take the place of the input $u_i(k)$ and the output $\bar{y}_i(k)$ of the subsystem. The subsystem model is given by:

$$\hat{y}_i(k) = -\sum_{\ell=1}^{n_i} \alpha_{i\ell}\hat{y}_i(k-\ell) + \sum_{\ell=1}^{n_i} \beta_{i\ell}\hat{y}_{i-1}(k-\ell) + e_i(k) \tag{8.40}$$

The subsystem model $G_0(z)$ given by Eq. (8.3)

$$\hat{y}_0(k) = \sum_{\ell=1}^{n_0} \alpha_{0\ell}\hat{y}_0(k-\ell) + \sum_{\ell=1}^{n_0} \beta_{0\ell}\left(r(k) - \hat{y}_{q-1}(k-\ell)\right) \tag{8.41}$$

The estimates $\hat{y}_{i-1}(k)$ and $\hat{y}_i(k)$ rather than the raw input output data $y_{i-1}(k)$ and $y_i(k)$ are used for the identification of the subsystems. The estimates $\hat{y}_{i-1}(k)$ and $\hat{y}_i(k)$ take the place of the input $u_i(k)$ and the output $\bar{y}_i(k)$ of the subsystems $G_i(z)$: $i = 1, 2, 3, \ldots, q-1$. For the subsystem $G_0(z)$ the input $r(k) - \hat{y}_{q-1}(k)$ and output $\hat{y}_0(k)$ replace the actual $u_0(k)$ and the output $\bar{y}_0(k)$ respectively.

A simple and computationally efficient least squares method is generally used, as the equation error of the linear regression model of the subsystem is a zero-mean white noise process.

Comment *In general, the prediction error gives superior performance as it uses an* accurate *Kalman filter, whereas both the high-order least-squares and the subspace methods use approximate versions of the residual and the Kalman filter models respectively. In the high-order least-squares method, the rational polynomials are approximated by high-order polynomial, and in the subspace method the impulse sequences are truncated. However, the prediction error method does not give a closed-form solution and it is difficult to evaluate its performance, as analytical expressions are not available. One has to use numerical computation to evaluate the performance.*

8.7 Evaluation on a Simulated Closed-Loop Sensor Net

The performance of the prediction error, the high-order least-squares, and the subspace methods were evaluated on the simulated sensor net example. The results of the evaluations are given. The covariance of the disturbance and the measurement noise was $Q = R = 1$, the number of data samples N = 7000. The selected model order for the high-order least-squares method was the same as the true order, $n_h = n = 6$.

8.7.1 The Performance of the Stage I Identification Scheme

The estimated system outputs were computed using the high-order least-squares, the prediction error, and subspace methods when the system was operating in the closed loop. Figure 8.6 shows the step responses of the identified model relating the reference input $r(k)$ and the subsystem outputs $y_i(k)$, $i = 1, 2$. The estimated step responses were compared with those of the true noise-free system. Subfigures (a) and (b), subfigures (c) and (d), and subfigures (e) and (f) compare respectively the true step responses with those estimated using (i) the high-order least squares, (ii) the prediction error, and (iii) the subspace methods.

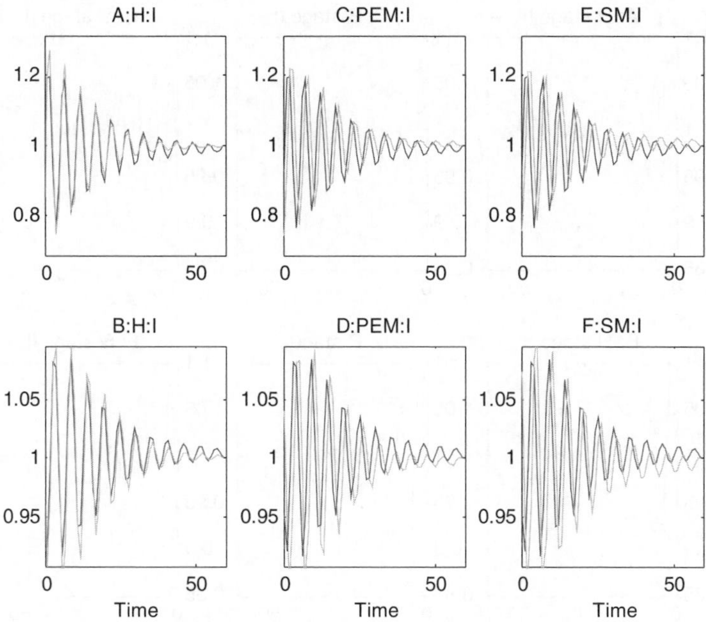

Figure 8.6 Stage I: the true and the estimated closed-loop step responses

8.7.2 The Performance of the Stage II Identification Scheme

The subsystems $G_2(z)$ and $G_3(z)$ were identified using the estimated outputs $\hat{y}_1(k)$, $\hat{y}_2(k)$ and $\hat{y}_3(k)$ from the stage I identification. In the direct approach, the actual noisy outputs $y_1(k)$, $y_2(k)$, and $y_3(k)$ are used. The subsystem $G_2(z)$ was identified by treating $\hat{y}_1(k)$ as the input and $\hat{y}_2(k)$ as the output, while the subsystem $G_3(z)$ was identified by treating $\hat{y}_2(k)$ as the input and $\hat{y}_3(k)$ as the output. In order to evaluate the performance of the second stage identification, the true step responses of the subsystems were compared with those of the identified ones. Figure 8.7 shows true and estimated step responses of the identified subsystem models. Subfigures (a)and (b), subfigures (c) and (d), and subfigures (e) and (f) compare respectively the true step responses with those estimated using (i) the high-order least squares, (ii) the prediction error, and (iii) the subspace methods.

Table 8.1 below gives the estimated transfer functions of the subsystem.

Table 8.1 The true and the estimated subsystem

True	$G_1(z) = \dfrac{0.1z^{-1}}{1-0.7z^{-1}+0.9z^{-2}}$	$G_2(z) = \dfrac{0.1z^{-1}}{1-0.9z^{-1}+0.81z^{-2}}$
HOLS	$\hat{G}_1(z) = \dfrac{0.0889z^{-1}+0.0181z^{-2}}{1-0.6681z^{-1}+0.8846z^{-2}}$	$\hat{G}_2(z) = \dfrac{0.0819z^{-1}-0.0074z^{-2}}{1-0.8649z^{-1}+0.8215z^{-2}}$
PEM	$\hat{G}_1(z) = \dfrac{0.09z^{-1}-0.0055z^{-2}}{1-0.6895z^{-1}+0.9157z^{-2}}$	$\hat{G}_2(z) = \dfrac{0.071z^{-1}-0.0156z^{-2}}{1-0.8559z^{-1}+0.8390z^{-2}}$
SM	$\hat{G}_1(z) = \dfrac{0.0896z^{-1}-0.0006z^{-2}}{1-0.6932z^{-1}+0.9150z^{-2}}$	$\hat{G}_2(z) = \dfrac{0.0678z^{-1}-0.0170z^{-2}}{1-0.8531z^{-1}+0.8375z^{-2}}$
Direct	$\hat{G}_1(z) = \dfrac{0.0888z^{-1}-0.0076z^{-2}}{1-0.7002z^{-1}+0.8964z^{-2}}$	$\hat{G}_2(z) = \dfrac{0.0858z^{-1}-0.0007z^{-2}}{1-0.9z^{-1}+0.8161z^{-2}}$

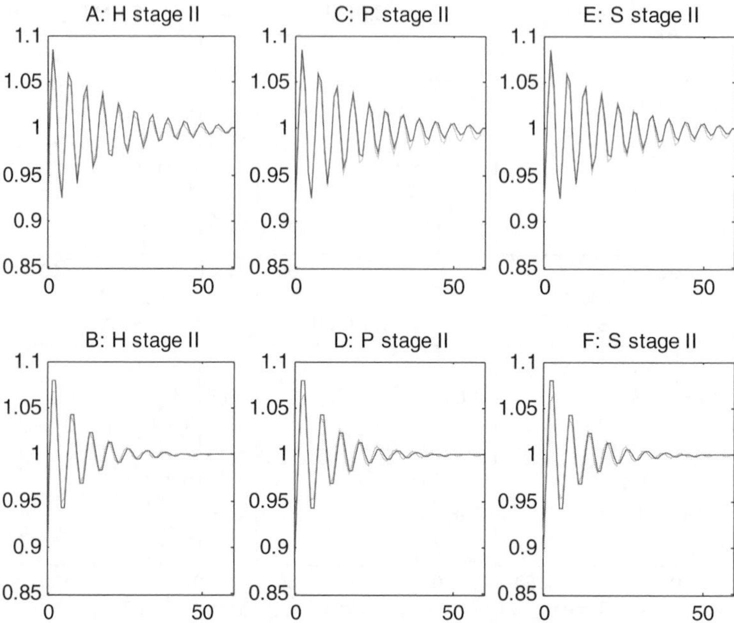

Figure 8.7 Stage II: the true and the estimated subsystem step responses

Comments *The performance of the high-order least-squares and the subspace method were close to that of the prediction error method. The performance of all the methods degrades when the ratio of the variances of the disturbance and the measurement noise is low.*

It should be emphasized that, for ensuring high performance of the identification method, the structure of the identification model should be based on that of the Kalman filter predictor form.

The high-order least-squares method is more practical in view of the following:

- *Its performance is close to that of the prediction error method.*
- *The closed-form expression for performance metrics such as the output estimation error and the covariance of the parameter estimation error may be derived.*
- *The high order selected to ensure that the equation error is a zero-mean white noise is only moderately higher than the true order.*
- *The high-order least squares method retains the simplicity and the efficiency of the least squares method.*

8.8 Summary

Approaches to Closed-loop System Identification

There are three approaches to closed loop identification, namely the direct approach, an indirect and a two-stage one. We will first focus on the two-stage identification approach.

The two-stage approach: The subsystems $\{G_i(z)\}$ are identified in two stages:

- *Stage I*: The closed-loop transfer functions relating the reference input $r(k)$ and the outputs $y_i(k)$ are identified. Let $T_{ri}(z)$ be the transfer function relating $r(z)$ and $y_i(z)$, and $\hat{T}_{ri}(z)$ be the estimated transfer function. In stage I the estimate $\hat{y}_i(k)$ of the output $y_i(k)$ is computed from the identified transfer function

$$\hat{y}_i(z) = \hat{T}_{ri}(z)\, r(z) \tag{8.42}$$

- *Stage II:* The subsystem $G_i(z)$ is identified from the $\hat{y}_{i-1}(k)$ and $\hat{y}_i(k)$ by treating $\hat{y}_{i-1}(k)$ as the input and $\hat{y}_i(k)$ as the output of the subsystem.

  The direct approach: In the direct approach $G_i(z)$ is identified from the noisy input–output data $y_{i-1}(k)$ and $y_i(k)$. Any of the high performance identifications, such as the prediction error, the high order least squares, or the subspace method may be used.

Model of the Single-input–multi-output System

- The state-space model $(\boldsymbol{A}_i, \boldsymbol{B}_i, \boldsymbol{C}_i)$ of the subsystem $G_i(z)$, $i = 1, 2, 3, \ldots, q-1$

$$\begin{aligned} \boldsymbol{x}_i(k+1) &= \boldsymbol{A}_i\boldsymbol{x}_i(k) + \boldsymbol{B}_i\boldsymbol{C}_{i-1}\boldsymbol{x}_{i-1}(k) + \boldsymbol{B}_i w_{i-1}(k) \\ y_i(k) &= \boldsymbol{C}_i\boldsymbol{x}_i(k) + w_i(k) + v_i(k) \end{aligned}$$

- The state-space model $(\boldsymbol{A}_0, \boldsymbol{B}_0, \boldsymbol{C}_0)$ of the subsystem $G_0(z)$ with unity feedback is:

$$\begin{aligned} \boldsymbol{x}_0(k+1) &= \boldsymbol{A}_0\boldsymbol{x}_0(k) + \boldsymbol{B}_0 r(k) - \boldsymbol{B}_0\boldsymbol{C}_{q-1}x_{q-1}(k) - \boldsymbol{B}_0 w_{q-1}(k) \\ y_0(k) &= \boldsymbol{C}_0\boldsymbol{x}_0(k) + w_0(k) + v_0(k) \end{aligned}$$

State-space Model of the Overall System

$$\begin{aligned} \boldsymbol{x}(k+1) &= \boldsymbol{A}\boldsymbol{x}(k) + \boldsymbol{B}r(k) + \boldsymbol{E}_w\boldsymbol{w}(k) \\ \boldsymbol{y}(k) &= \boldsymbol{C}\boldsymbol{x}(k) + \boldsymbol{F}_w\boldsymbol{w}(k) + \boldsymbol{F}_v\boldsymbol{v}(k) \end{aligned}$$

Transfer Function Model

The transfer function model of the system relating the reference input $r(z)$, the disturbance $\boldsymbol{w}(z)$, and the measurement noise $\boldsymbol{v}(z)$ to the output $\boldsymbol{y}(z)$ is given by:

$$\boldsymbol{y}(z) = \frac{\boldsymbol{N}(z)}{D(z)} r(z) + \left(\frac{\boldsymbol{N}_w(z)}{D(z)} + \boldsymbol{F}_w\right)\boldsymbol{w}(z) + \boldsymbol{F}_v\boldsymbol{v}(z)$$

Rewriting we get:

$$D(z)\boldsymbol{y}(z) = \boldsymbol{N}(z)r(z) + \boldsymbol{\upsilon}(z) \quad \text{where} \quad \boldsymbol{\upsilon}(z) = \left(\boldsymbol{N}_w(z) + D(z)\boldsymbol{F}_w\right)\boldsymbol{w}(z) + D(z)\boldsymbol{F}_v\boldsymbol{v}(z)$$

Kalman Filter Model: Prediction Error form

$$\begin{aligned} \hat{\boldsymbol{x}}(k+1) &= (\boldsymbol{A} - \boldsymbol{K}\boldsymbol{C})\hat{\boldsymbol{x}}(k) + \boldsymbol{B}r(k) + \boldsymbol{K}\boldsymbol{y}(k) \\ \hat{\boldsymbol{y}}(k) &= \boldsymbol{C}\hat{\boldsymbol{x}}(k) \\ \boldsymbol{e}(k) &= \boldsymbol{y}(k) - \hat{\boldsymbol{y}}(k) \end{aligned}$$

Residual Model

$$\boldsymbol{e}(z) = \left(\boldsymbol{I} - \frac{\boldsymbol{N}_y(z)}{F(z)}\right)\boldsymbol{y}(z) - \frac{\boldsymbol{N}_r(z)}{F(z)} r(z)$$

$$\frac{\boldsymbol{N}_r(z)}{F(z)} = \boldsymbol{C}(z\boldsymbol{I} - \boldsymbol{A} + \boldsymbol{K}\boldsymbol{C})^{-1}\boldsymbol{B}$$
$$\frac{\boldsymbol{N}_y(z)}{F(z)} = \boldsymbol{C}(z\boldsymbol{I} - \boldsymbol{A} + \boldsymbol{K}\boldsymbol{C})^{-1}\boldsymbol{K}$$

where $F(z) = 1 + \sum_{i=1}^{n} c_i z^{-i}$

The Identification Model

The structure of the identification model is chosen to be that of the residual model.

The High-order Least-Squares Method

Rearranging the residual model yields:

$$\boldsymbol{y}(z) = \frac{\boldsymbol{N}_y(z)}{F(z)}\boldsymbol{y}(z) + \frac{\boldsymbol{N}_r(z)}{F(z)}r(z) + \boldsymbol{e}(z)$$

Dividing each of the dividend matrix polynomials $\boldsymbol{N}_y(z)$ and $\boldsymbol{N}_r(z)$ by the devisor polynomial $F(z)$, and truncating the quotient to some finite but large number of terms n_h so that the remainder term is negligible yields:

$$\frac{\boldsymbol{N}_y(z)}{F(z)} \approx \sum_{\ell=1}^{n_h} \boldsymbol{A}_{h\ell} z^{-\ell}$$
$$\frac{\boldsymbol{N}_r(z)}{F(z)} \approx \sum_{\ell=0}^{n_h} \boldsymbol{b}_{h\ell} z^{-\ell}$$

Assumption

In the HOLS method, we assume the following:

A1: In the modeling phase, n_h is large enough to ensure that the errors in the approximation of the rational polynomials by high-order polynomials are negligible

$$\boldsymbol{y}(k) = \sum_{\ell=1}^{n_h} \boldsymbol{A}_{h\ell}\boldsymbol{y}(k-\ell) + \sum_{\ell=1}^{n_h} \boldsymbol{b}_{h\ell} r(k-\ell) + \boldsymbol{e}(k)$$

In the identification phase, both the number of data samples N and selected order n_h are large, and $\frac{n_h}{N}$ is finite and small so that the quality of the estimates is superior.

Polynomial Model: Batch Processing

$$\boldsymbol{y}_i(k) = \boldsymbol{H}_{pi}(k)\boldsymbol{\theta}_{pi} + \boldsymbol{e}_i(k)$$

Estimate Using the Least-Squares

$$\hat{\boldsymbol{\theta}}_{pi} = \left(\boldsymbol{H}_{pi}^T(k)\boldsymbol{H}_{pi}(k)\right)^{-1} \boldsymbol{H}_{pi}^T(k)\boldsymbol{y}_i(k)$$

$$\hat{\boldsymbol{y}}_{pi}(k) = \boldsymbol{H}_{pi}(k)\hat{\boldsymbol{\theta}}_{pi}$$

Performance of Stage I Identification

The parameter estimate is unbiased:

$$E[\hat{\theta}_p] = \theta_p$$

- The estimate of the covariance of the parameter estimate $\hat{\theta}_h$ satisfy the inequality:

$$E[(\hat{\theta}_{pi} - \theta_{pi})(\hat{\theta}_{pi} - \theta_{pi})^T] = \sigma_i^2 \left(\boldsymbol{H}_p^T(k)\boldsymbol{H}_p(k)\right)^{-1}$$

where $\sigma_i^2 = E\left[e_i^2(k)\right]$. The covariance of the estimation error satisfies Cramér–Rao inequality and hence is efficient.

Asymptotic Behavior

- The parameter estimate converges to the true value as:

$$\lim_{N\to\infty} it \left\{\hat{\theta}_p\right\} = \theta_p$$

- The output estimation error tends to a zero mean white noise process $\boldsymbol{e}_p(k)$:

$$\boldsymbol{y}(k) - \hat{\boldsymbol{y}}_p(k) \to \boldsymbol{e}_p(k)$$

Selection of the High Order

A practical guideline for the selection of the high order n_h is that the selected order should be large enough to ensure that the identification error is a zero-mean white noise process. The identification error approaches the Kalman filter residual if the selected model order is sufficiently large and the number of data samples N tends to infinity:

$$\boldsymbol{e}_p(k) \to \boldsymbol{e}(k)$$

Second Stage of the Two-stage Identification

$$\hat{y}_i(k) = -\sum_{\ell=1}^{n_i} \alpha_{i\ell}\hat{y}_i(k-\ell) + \sum_{\ell=1}^{n_i} \beta_{i\ell}\hat{y}_{i-1}(k-\ell) + e_i(k): i = 1, 2, \ldots, q-1$$

$$\hat{y}_0(k) = \sum_{\ell=1}^{n_0} \alpha_{0\ell}\hat{y}_0(k-\ell) + \sum_{\ell=1}^{n_0} \beta_{0\ell}\left(r(k) - \hat{y}_{q-1}(k-\ell)\right).$$

References

[1] Shahab, M. and Doraiswami, R. (2009) A novel two-stage identification of unstable systems. Seventh International Conference on Control and Automation (ICCA 2010), Christ Church, New Zealand.

[2] Ljung, L. (1999) *System Identification: Theory for the User*, Prentice-Hall, New Jersey.

[3] Huang, B. and Shah, S.L. (1997) Closed-loop identification: a two step approach. *Journal of Process Control*, **7**, 425–438.

[4] Qin, J. S. (2006) Overview of subspace identification. *Computer and Chemical Engineering*, **30**, 1502–1513.

[5] Ljung, L. (1999) *System Identification: Theory for the User*, Prentice-Hall, New Jersey.

[6] Forssell, U. and Ljung, L. (1999) Closed loop identification revisited. *Automatica*, **35**, 1215–1241.

[7] Soderstrom, T. and Stoica, P. (1989) *System Identification*, Prentice Hall, New Jersey.

[8] Mendel, J. (1995) *Lessons in Estimation Theory in Signal Processing, Communications and Control*, Prentice-Hall, New Jersey.

[9] Qin, J.S. (2006) Overview of subspace identification. *Computer and Chemical Engineering*, **30**, 1502–1513.

[10] Wahlberg, B., Jansson, M., Matsko, T., and Molander, M. (2007) Experiences from subspace identification-comments from process industry users and researches., 315–327. A. Chiuso et al. (Eds), *Estimation and Control, LNCIS 364*, Springer-Verlag Berlin.

[11] Astrom, K. and Eykhoff, P. (1971) System identification-a survey. *Automatica*, **7**, 123–162.

[12] Doraiswami, R. and Cheded, L. (2012) Kalman filter for fault detection: an internal model approach. *IET Control Theory and Applications*, **6**(5), 1–11.

9

Fault Diagnosis

9.1 Overview

Fault diagnosis of physical systems remains a challenging problem and continues to be a subject of intense research both in industry and in academia in view of the stringent and conflicting requirements in practice for a high probability of correct detection and isolation, low false alarm probability, and timely decision on the fault status. The basic idea behind the model-based approach for fault diagnosis is to generate a signal termed a residual which, in the ideal case, is zero when there is no fault and non-zero otherwise. The ideal case refers to the situation where the model of the system is precisely known and there are no disturbances or measurement noise affecting the system. In practice, however, the system is hardly free from such disturbances or measurement noise and these are either partially or totally unknown, thus making the derived model at best an approximation of the real system. Fault diagnosis of a system refers to two tasks, namely the detection and isolation of a fault. First the presence or an absence of a fault is asserted, and if there is a fault the faulty device or a subsystem is determined. As the complexity of engineering systems increases, fault diagnosis, condition-based maintenance, and health monitoring become of vital practical importance to ensure key advantages, such as the system's reliability and performance sustainability, reduction of down time, low operational costs, and personnel safety.

A unified approach to both detection and isolation of a fault is presented based on the Kalman filter residual. The residual of the Kalman filter is a zero-mean white noise process if and only if there is no fault. If there is a fault, the residual will have an additive fault-indicating term. The fault-indicating term is function of the variation in the diagnostic parameter.

The fault detection problem is posed as a binary hypothesis testing problem and the threshold value is chosen as an acceptable tradeoff between the correct detection and false alarm probabilities. The fault isolation problem is similarly posed as a multiple hypothesis testing problem. The hypotheses include a single fault in a subsystem, simultaneous faults in two subsystems, and so on until simultaneous faults in all subsystems. For a single fault, a closed-form solution is presented.

The problem of fault diagnosis consists of the following subproblems:

- identification for fault diagnosis
- residual generation
- fault detection
- fault isolation.

Identification of Physical Systems: Applications to Condition Monitoring, Fault Diagnosis, Soft Sensor and Controller Design, First Edition. Rajamani Doraiswami, Chris Diduch and Maryhelen Stevenson.

9.1.1 Identification for Fault Diagnosis

The accuracy and the reliability of the identified model is crucial to the superior performance of the model-based applications (such as fault diagnosis, soft sensor and performance monitoring). A model termed a diagnostic model is identified, which completely captures the static and the dynamic behavior of the nominal fault-free and fault-bearing systems. The system is identified by performing a number of experiments to cover all likely operating scenarios rather than merely identifying the system at a given operating point. Emulators, which are transfer blocks, are included at the accessible points in the system, such as inputs, the outputs, or both, in order to mimic the likely operating scenarios. This is similar in spirit to the artificial neural network approachm where a training set comprising data obtained from a number of representative operating scenarios is presented so as to capture completely the behavior of the system.

9.1.2 Residual Generation

There are various approaches to the generation of residuals, including the Kalman filter or observer-based approaches, parameter estimation methods, and parity vector methods [1–10]. There are two types of fault models employed, namely additive and parametric (or multiplicative) types. In the additive type, a fault is modeled as an additive exogenous input to the system, whereas in the parametric type, a fault is modeled as a change in the parameters which completely characterizes the fault behavior of the subsystems. These parameters are termed diagnostic parameters. The Kalman filter is most widely and successfully used for additive fault detection, while for parametric (multiplicative) faults, model identification-based schemes are employed [4–10]. Herein, the Kalman filter residual is employed for both fault detection and isolation since (i) the Kalman filter residual is zero in the statistical sense if and only if there is no fault, and (ii) its performance is robust to plant and measurement noise affecting the system output [1–3].

9.1.3 Fault Detection

A number of approaches to fault detection have been proposed in the literature, including the Bayes decision theoretic approach, the classical Neyman–Pearson approach, the Maximum *A Posteriori* (MAP) approach and the Maximum Likelihood (ML) approach. Usually in radar and sonar applications, the Neyman–Pearson approach is employed [11]. The Bayesian approach has been widely used in many applications in recent times due mainly to the availability of large amount of data and high computing power.

The Bayesian approach is employed herein as it encompasses all other approaches. The decision theoretic problem is posed as a problem of hypotheses testing where one of the M hypotheses where $M \geq 2$ has to be selected. If $M = 2$, it is called a binary hypotheses testing problem. A probabilistic model is used to characterize hypotheses using conditional probabilities and *a priori* probabilities. The conditional probability is the probability that the measurements have been generated under one of the hypotheses. The *a priori* probability is the probability that the given hypothesis is true. The decision strategy to choose between the hypotheses is based on minimizing the Bayes risk, which quantifies the costs associated with correct and incorrect decisions. The minimization of the Bayes risk yields the Bayes decision strategy. In the case of the binary hypotheses, it reduces to the likelihood ratio test. The decision between the two hypotheses is based on comparing the likelihood ratio, which is the ratio of the conditional probabilities under the two hypotheses, to a threshold value.

The Bayes criterion cannot be applied directly if the parameters such as the mean or the variance of the PDFs are unknown. The unknown parameters are estimated *a posteriori* from data. The resulting problem is termed the *composite hypotheses testing* problem. In the case of binary hypotheses, the likelihood ratio test becomes the Generalized Likelihood Ratio Test (GLRT). Assuming that the PDF of the residuals is Gaussian, a function of the residual and the reference input termed *test statistics* is derived to help discriminate between the faulty and no-fault operating regimes [12].

9.1.4 Fault Isolation

Since a system consists of an interconnection of subsystems, then, for ease of analysis, each subsystem is modeled as a transfer function that may represent a physical entity of the system, such as a sensor, actuator, controller, or other system component, that is subject to faults. Parameters, termed herein as *diagnostic parameters*, are selected so that they are capable of monitoring the health of the subsystems, and may be varied either directly or indirectly (using a fault emulator) during the offline identification phase [1,4]. An emulator is a transfer function block which is connected at the input with a view to inducing faults in a subsystem which may arise as a result of variations in the phase and the magnitude of the transfer function of a subsystem. An emulator may take the form of a gain or a filter to induce gain or phase variations. A fault occurs within a subsystem when one or more of its diagnostic parameters vary. A variation in the diagnostic parameter does not necessarily imply that the subsystem has failed, but it may lead to a potential failure resulting in poor product quality, shut down, or damage to subsystem components. Hence a proactive action, such as condition-based preventive maintenance, must be taken prior to the occurrence of a fault. The fault detection capability of the Kalman filter residual is extended to the task of fault isolation. It is shown that the residual is a linear function in each of the diagnostic parameters when other parameters are kept constant, that is it is multi-linear function of the diagnostic parameters. A vector, termed *influence vector*, plays a crucial role in the fault isolation process. The influence vector is made of elements that are partial derivatives of the feature vector with respect to each diagnostic parameter. The influence vectors are estimated offline by performing a number of experiments. Each experiment consists of perturbing the diagnostic parameters one at a time, and the influence vector is estimated from the best least-squares fit between the residual and the diagnostic parameter variations. When diagnostic parameters are not accessible, as in the case of plant model parameters for example, the fault emulator parameters are perturbed instead. It is shown that the residual is a zero-mean white noise process if and only if there is no fault. If there is a fault, the residual is a non-zero-mean process, and is expressed as a sum of a deterministic component, temed fault indicator, and a zero-mean stochastic component. The spectral content of a fault parameter is identical to that of the reference input and this property is used in the composite hypothesis testing strategy [12].

The fault detection and isolation scheme is evaluated on (i) a simulated position control system to isolate the sensor, actuator, and plant faults, and (ii) a physical laboratory-scale two-tank process control system to isolate leakage, actuator, and sensor faults.

9.2 Mathematical Model of the System

The model of the system is expressed in a state-space form, as it is convenient in that it handles multivariate, time-varying, and nonlinear systems and lends itself to powerful analysis, design, and realization (or implementation) of filters and controllers. Let the state-space model of a nominal fault-free system be given by:

$$\begin{aligned} \boldsymbol{x}(k+1) &= \boldsymbol{A}_0\boldsymbol{x}(k) + \boldsymbol{B}_0 r(k) + \boldsymbol{E}_w w(k) \\ y(k) &= \boldsymbol{C}_0\boldsymbol{x}(k) + v(k) \end{aligned} \tag{9.1}$$

where $\boldsymbol{x}(k)$ is the $nx1$ state vector, $y(k)$ the measured output, $r(k)$ the input to the system, $w(k)$ the disturbance or plant noise, $v(k)$ the measurement noise, $\boldsymbol{A}_0$ a $n \times n$ matrix, $\boldsymbol{B}_0$, $\boldsymbol{C}_0$, and $\boldsymbol{E}_w$ matrices of compatible dimensions. It is assumed that $r(k)$, $y(k)$, $w(k)$, and $v(k)$ used here are scalars, as this will allow us to focus mainly on fault diagnosis without being side-tracked by the technicalities involved in the algebraic manipulations of multivariable system matrices.

We shall assume that the system $(\boldsymbol{A}_0, \boldsymbol{C}_0)$ is observable. If the system is observable, then all its states may be estimated from the input–output data. The plant noise $w(k)$ and the measurement noise $v(k)$ are assumed to be zero-mean white noise processes. If, however, $w(k)$ and $v(k)$ are colored noise processes, then they must have rational spectra. This would then require that both $w(k)$ and $v(k)$ have to be modeled as outputs of a *known* linear system driven by a zero-mean white noise process. In this case, it is always possible to rewrite the given model with its colored noise inputs, as an augmented system containing the dynamics of the zero-mean white noise-driven schemes that generate these colored noise inputs. The covariances of $w(k)$ and $v(k)$ are defined by:

$$E[w^T w] = Q \quad \text{and} \quad E[v^T v] = R \tag{9.2}$$

where $Q > 0$ and $R \geq 0$.

9.2.1 Linear Regression Model: Nominal System

The linear regression model of the nominal fault-free system derived from the state-space model (9.1) is given by:

$$y(k) = \boldsymbol{\psi}^T(k)\boldsymbol{\theta}^0 + \upsilon_0(k) \tag{9.3}$$

where $\upsilon_0(k)$ is the effect of $w(k)$ and $v(k)$ on the output $y(k)$; $\upsilon_0(z) = |z\boldsymbol{I} - \boldsymbol{A}_0|(\boldsymbol{C}_0(z\boldsymbol{I} - \boldsymbol{A}_0)^{-1}\boldsymbol{E}_w w(z) + v(z))$, $\boldsymbol{\psi}^T(k)$ is data vector, $\boldsymbol{\theta}^0$ is a nominal $Mx1$ feature vector formed of the coefficients of the denominator coefficients $\{a_i^0\}$ and the numerator coefficients $\{b_i^0\}$ of the nominal system transfer function.

$$\boldsymbol{\psi}^T(k) = [-y(k-1) \quad -y(k-2) \quad . \quad -y(k-n) \quad r(k-1) \quad r(k-2) \quad . \quad r(k-n)] \tag{9.4}$$

$$\boldsymbol{\theta}^0 = \begin{bmatrix} a_1^0 & a_2^0 & . & a_{n_a}^0 & b_1^0 & b_2^0 & . & b_{n_b}^0 \end{bmatrix}^T \tag{9.5}$$

9.3 Model of the Kalman Filter

The state-space model of the Kalman filter is configured as a closed-loop system which is (i) an exact copy of the nominal (fault-free) model of the system (9.1), denoted by $(\boldsymbol{A}_0, \boldsymbol{B}_0, \boldsymbol{C}_0)$, (ii) driven by the output estimation error $e(k)$, and (iii) is stabilized by the Kalman gain $\boldsymbol{K}_0$ [3]:

$$\begin{aligned} \hat{\boldsymbol{x}}(k+1) &= \boldsymbol{A}_0\hat{\boldsymbol{x}}(k) + \boldsymbol{B}_0 r(k) + \boldsymbol{K}_0(y(k) - \boldsymbol{C}_0\hat{\boldsymbol{x}}(k)) \\ \hat{y}(k) &= \boldsymbol{C}_0\hat{\boldsymbol{x}}(k) \\ e(k) &= y(k) - \hat{y}(k) \end{aligned} \tag{9.6}$$

where $\hat{\boldsymbol{x}}(k)$ is the estimate of the state $\boldsymbol{x}(k)$, $\hat{y}(k)$ the estimate of the output $y(k)$, and $e(k)$ the residual (or an innovation) process.

Comments *The Kalman filter estimates the system states by fusing the information provided by the residual $e(k)$ and the* a priori *information contained in the model $(\boldsymbol{A}_0, \boldsymbol{B}_0)$. If the ratio Q/R is large, the Kalman filter will rely on the measurement to generate the residual. The Kalman filter in this case will be robust (or less sensitive) to model uncertainties. If, on the other hand, the ratio Q/R is small, the*

Kalman filter will rely on the model to generate the residual. The Kalman filter in this case will be robust (or less sensitive) to the measurement noise.

In applications such as fault diagnosis, it is desirable that the Kalman filter be highly sensitive (or least robust) to model uncertainties. To achieve this, the ratio Q/R must be very small, so that the dynamic model $(\boldsymbol{A}_0, \boldsymbol{B}_0)$ is given the maximum importance compared to the measurement. In practice this requirement is met if the nominal model of the system is identified accurately. The requirement of high sensitivity (or the lack of robustness) of the Kalman filter to model uncertainty in the above stated applications is in striking contrast to that of a controller, which has to be highly insensitive (highly robust) to model uncertainties.

9.4 Modeling of Faults

A fault in a system may be classified as additive of a multiplicative type fault [5].

The state-space model when subject to both additive and multiplicative faults becomes:

$$\begin{aligned} \boldsymbol{x}(k+1) &= \boldsymbol{A}\boldsymbol{x}(k) + \boldsymbol{B}r(k) + \boldsymbol{E}_w w(k) + \boldsymbol{E}_f \boldsymbol{f}(k) \\ y(k) &= \boldsymbol{C}\boldsymbol{x}(k) + v(k) + \boldsymbol{F}_f \boldsymbol{f}(k) \end{aligned} \tag{9.7}$$

where $\boldsymbol{A} = \boldsymbol{A}_0 + \delta\boldsymbol{A}$; $\boldsymbol{B} = \boldsymbol{B}_0 + \delta\boldsymbol{B}$; $\boldsymbol{C} = \boldsymbol{C}_0 + \delta\boldsymbol{C}$, $\delta\boldsymbol{A}$, $\delta\boldsymbol{B}$, and $\delta\boldsymbol{C}$ corresponds to deviations in the parameters of the process, the actuator, and the sensor respectively. A fault resulting from these variations $\delta\boldsymbol{A}$, $\delta\boldsymbol{B}$, and $\delta\boldsymbol{C}$ is termed a multiplicative or parametric fault while the additive exogenous input $\boldsymbol{f}(k)$ is termed an additive fault. Thus $\boldsymbol{A}$, $\boldsymbol{B}$, $\boldsymbol{C}$ matrices and $\boldsymbol{f}(k)$ capture, respectively, the effects of a process, actuator, and sensor faults.

An additive faults for a typical control system results in an 3×1 unknown exogenous input vector $\boldsymbol{f}(k)$ to the system with scalar components $f_s(k)$, $f_a(k)$, and $f_p(k)$ representing sensor fault, actuator fault, and the process fault respectively, $\boldsymbol{f}(k) = \begin{bmatrix} f_s(k) & f_a(k) & f_p(k) \end{bmatrix}^T$. The entry vectors of these inputs in the dynamical and algebraic equations are respectively denoted by a $n \times 3$ matrix $\boldsymbol{E}_f$ and 1×3 vector F_f. The fault input $\boldsymbol{f}(k)$ is may be constant or a function of time representing a fault such as a constant bias or an offset or a ramp type drift in sensors or actuators. The additive fault does not cause instability in the (linear) closed-loop system whereas a multiplicative fault may cause instability.

9.4.1 Linear Regression Model

The state-space model (9.7) may be expressed in the linear regression form given by:

$$y(k) = \boldsymbol{\psi}^T(k)\boldsymbol{\theta} + y_f(k) + \upsilon(k) \tag{9.8}$$

where $y_f(k)$ and $\upsilon(k)$ are respectively the inverse z-transforms of $N_f(z)\boldsymbol{f}(z) + D(z)\,\boldsymbol{E}_f\boldsymbol{f}(z)$ and $N_w(z)w(z) + D(z)v(z)$ respectively, $\boldsymbol{\theta}$ is a $Mx1$ feature vector formed of the coefficients of the denominator coefficients $\{a_i\}$ and the numerator coefficients $\{b_i\}$ of the transfer function of the faulty system:

$$\boldsymbol{\theta} = [a_1 \quad a_2 \quad . \quad a_n \quad b_1 \quad b_2 \quad . \quad b_n]^T \tag{9.9}$$

The deviation in the feature vector $\Delta\boldsymbol{\theta}$ is given by:

$$\Delta\boldsymbol{\theta} = \boldsymbol{\theta} - \boldsymbol{\theta}^0 \tag{9.10}$$

Expressing in terms of the deviations in the denominator and numerator coefficients we get:

$$\Delta\theta = \theta - \theta^0 = \begin{bmatrix}\Delta a_1 & \Delta a_2 & . & \Delta a_{n_a} & \Delta b_1 & \Delta b_2 & . & \Delta b_{n_b}\end{bmatrix}^T \tag{9.11}$$

where $\Delta a_i = a_i - a_i^0$ and $\Delta b_i = b_i - b_i^0$.

9.5 Diagnostic Parameters and the Feature Vector

The system is modeled as an interconnection of subsystems that are each subject to faults as shown in Figure 9.1, where $G_i(z)$: $i = 0, 1, 2, 3, \dots, s$ is a subsystem, $w_i(k)$ is the disturbance, and $v(k)$ is the measurement noise.

A subsystem may be a process (or a plant), a controller, an actuator, or another device, and is characterized by parameters termed *diagnostic parameters* $\{\gamma^i\}$. A subsystem is said to be faulty if a diagnostic parameter deviates from its nominal fault-free value. A variation in the diagnostic parameter does not necessarily imply that the subsystem has failed, but may provide an early warning that a potential failure may be in the offing and thus may cause adverse effects such as poor product quality, shutdown, and damage to subsystem components or possibly danger to operating personnel. Hence a proactive action, based for example on condition-based preventive maintenance, must be taken to avert the occurrence of potential faults. The diagnostic parameters $\{\gamma^i\}$ are selected so that they are capable of monitoring the health of the subsystem, and may be varied either directly or indirectly using an emulator. The diagnostic parameter γ^i is the parameter vector of the subsystem $G_i(z)$ if its parameters are accessible. Otherwise γ^i is parameter vector of an emulator $E_i(z)$ associated with the subsystem $G_i(z)$. A model, termed a *diagnostic model*, is developed that governs the mapping between the system input, the system output, and the diagnostic parameters.

The fault diagnosis scheme uses a two-stage process. During the first stage, a series of experiments are conducted on the overall system, in which measurement data is collected and used to identify the diagnostic model. The identification experiments are conducted by perturbing the diagnostic parameters of each $G_i(z)$ and measuring the overall response of the system. The purpose of the identification stage is to estimate the diagnostic model of the overall system when the models of the subsystems are unknown. In the second stage, a Kalman filter is designed for the nominal fault-free identified model. The Kalman filter residual is shown to capture the deviation of the diagnostic parameters. When a fault in one or more subsystems occurs during online operation it will inject artifacts into the residual that are then used for fault detection and isolation.

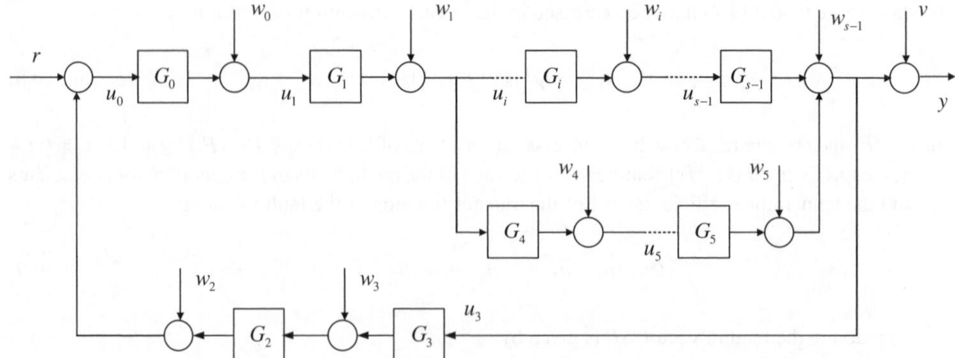

Figure 9.1 Cascade, parallel, and feedback interconnected subsystems

Consider a discrete-time system formed of a cascade, parallel, and feedback interconnection of subsystems characterized by q_1x1 diagnostic parameter vector γ^i. The *diagnostic parameter vector* γ for the entire interconnected system is a $(q \times 1)$ vector that augments $\{\gamma^i, i = 1, 2, \ldots, s\}$ for all subsystems, $G_i(z), i = 1, 2, \ldots, s$, that are subject to failure. The overall $q \times 1$ diagnostic parameter vector γ formed of all the subsystem diagnostic parameters $\{\gamma^i\}$ is given by:

$$\gamma = [(\gamma^1)^T \quad (\gamma^2)^T \quad (\gamma^3)^T \quad \ldots \quad (\gamma^s)^T]^T = [\gamma_1 \quad \gamma_2 \quad \ldots \quad \gamma_q]^T \tag{9.12}$$

where $s_i = \{\gamma_\ell \in \gamma^i\}$ is a subset of the elements of γ which are associated with γ^i. If there is a variation in any of the elements of the set $\gamma_\ell \in s_i$, then the subsystem $G_i(z)$ is faulty.

Let γ^0 be the nominal fault-free parameter value and its variation $\Delta\gamma$ as given below:

$$\Delta\gamma = \gamma - \gamma^0 \tag{9.13}$$

The deviation in the feature vector $\Delta\boldsymbol{\theta}$ will be some nonlinear function of the deviation diagnostic parameter $\Delta\gamma$, $\Delta\boldsymbol{\theta} = \boldsymbol{\varphi}(\Delta\gamma)$ where $\boldsymbol{\varphi}(.)$ is some Mx1 nonlinear function. The nonlinear function $\boldsymbol{\varphi}(\Delta\gamma)$ relating $\Delta\boldsymbol{\theta}$ and $\Delta\gamma$ can be expressed as a power series in $\{\Delta\gamma_i\}$ with a finite number of terms as given below

$$\Delta\boldsymbol{\theta} = \boldsymbol{\varphi}(\Delta\gamma) = \sum_i \boldsymbol{\Omega}_i \Delta\gamma_i + \sum_{i,j} \boldsymbol{\Omega}_{ij} \Delta\gamma_{ij} + \sum_{i,j,k} \boldsymbol{\Omega}_{ijk} \Delta\gamma_{ijk} \cdots + \sum_{1,2,3,\ldots q} \boldsymbol{\Omega}_{123..q} \Delta\gamma_{123..q} \tag{9.14}$$

where $\Delta\gamma_{ij} = \Delta\gamma_i\Delta\gamma_j$, $\Delta\gamma_{ijk} = \Delta\gamma_i\Delta\gamma_j\Delta\gamma_k$, $\Delta\gamma_{123\ldots q} = \Delta\gamma_1\Delta\gamma_2\Delta\gamma_3 \ldots \Delta\gamma_q$, $\boldsymbol{\Omega}_i$, $\boldsymbol{\Omega}_{ij}$, $\boldsymbol{\Omega}_{ijk}$, $\ldots$ $\boldsymbol{\Omega}_{123..q}$ denote the first, second, third and up to the qth partial derivative of $\boldsymbol{\theta}$ with respect to $\{\gamma_i\}$; $\boldsymbol{\Omega}_i = \dfrac{\partial\boldsymbol{\theta}}{\partial\gamma_i}$, $\boldsymbol{\Omega}_{ij} = \dfrac{\partial^2\boldsymbol{\theta}}{\partial\gamma_i\partial\gamma_j}$, $\boldsymbol{\Omega}_{123..q} = \dfrac{\partial^q\boldsymbol{\theta}}{\partial\gamma_1\partial\gamma_2\partial\gamma_3 \ldots \partial\gamma_q}$. The power series expansion terms include products of the deviations $\{\Delta\gamma_i\}$ taken one-at-a-time, two-at-a-time, and so on up to q-at-a-time and the number of terms in the series is finite. The feature vector deviation $\Delta\boldsymbol{\theta}$ is a multilinear in $\Delta\gamma$ (linear in each of its elements γ_i when the rest of the elements $\gamma_j : j \neq i$ are kept constant). The partial derivative terms $\boldsymbol{\Omega}_i$, $\boldsymbol{\Omega}_{ij}$, $\boldsymbol{\Omega}_{ijk}$, $\ldots$ and $\boldsymbol{\Omega}_{123..q}$ which are the Jacobians of the feature vector $\boldsymbol{\theta}$ with respect to the diagnostic parameter γ, are termed *influence vectors*. Named as such, these vectors influence the direct effect of parameter variations on the feature vector. The influence vectors completely characterize the diagnostic model. In general, if all combinations of γ_i are to be included, then the number m of influence vectors in the set $\{\boldsymbol{\Omega}_i, \boldsymbol{\Omega}_{ij}, \boldsymbol{\Omega}_{ijk}, \ldots \boldsymbol{\Omega}_{123\ldots q}\}$ is given by

$$m = 2^q = 1 + \sum_{i=1}^{q} \binom{q}{i} \tag{9.15}$$

In practice, however, the number of relevant influence vectors is generally much lower.

The influence vectors, which are the Jacobians of the feature vector $\boldsymbol{\theta}$ with respect to the elements of the diagnostic parameters $\boldsymbol{\gamma}$, influence the effect of parameter variation on the feature vector. The power series expansion terms include products of the deviations $\{\Delta\gamma_i\}$ taken one-at-a-time, two-at-a-time, and so on up to q-at-a-time and the number of terms in the series is finite.

Comments *The expression (9.14) shows that $\Delta\boldsymbol{\theta} = \boldsymbol{\varphi}(\Delta\gamma)$ is nonlinear in the diagnostic parameters $\{\Delta\gamma_i\}$. This type of nonlinearity is termed* multi-linear. *It is separately linear in each $\Delta\gamma_i$ when the rest of the deviations in diagnostic parameters are zero, $\Delta\gamma_j = 0 : j \neq i$. It is not strictly a Taylor series expansion in $\{\Delta\gamma_i\}$, as only a finite number of terms in the expansion is required to accurately represent the nonlinear function $\boldsymbol{\varphi}(\Delta\gamma)$.*

If a discrete-time equivalent of a continuous time system is used, then the diagnostic parameters may not enter the feature vector multi-linearly in the discrete-time equivalent model, although they are multi-linear in the continuous time model. In this case the mapping is governed by the exponential, $A_d(\gamma) = \exp(A_c(\gamma)T_s) = I + \sum_i \frac{A_c^i T_s^i}{i!}$, *where* A_c *and* A_d *are respectively the continuous time and the discrete-time equivalent state-space matrices, and* T_s *is the sample period. To explicitly account for these effects, as well as system nonlinearity, one may include quadratic and higher order terms in the power series expansion. If the sample period is small compared to the dominant time constant then the effect of higher order terms becomes insignificant. To explicitly account for these effects, as well as system nonlinearity, one may include quadratic and higher order terms in the power series expansion.*

The diagnostic model is nonlinear in the diagnostic parameter variations, and is completely characterized by the influence vectors $\mathbf{\Omega}_i$, $\mathbf{\Omega}_{ij}$, $\mathbf{\Omega}_{ijk}$, $\ldots$ *and* $\mathbf{\Omega}_{123..q}$*:*

- $\mathbf{\Omega}_i$ *models the effect of the parameter variation* $\Delta\gamma_i$ *on the output* γ,
- $\mathbf{\Omega}_{ij}$ *models the effect of the parameter variation* $\Delta\gamma_{ij}$ *on the output y,*
- $\mathbf{\Omega}_{ijk}$ *models the effect of the parameter variation* $\Delta\gamma_{ijk}$ *on the output y,*
- *and so on.*

In those cases where the structure of a mathematical model agrees with that of the physical system, then one knows beforehand the dimension of $\boldsymbol{\theta}$ *and the number of terms in the set* $\mathbf{\Omega} = \{\mathbf{\Omega}_i, \mathbf{\Omega}_{ij}, \mathbf{\Omega}_{ijk}, \ldots \mathbf{\Omega}_{123\ldots q}\}$. *In this case, only the relevant partial derivative terms need to be estimated.*

9.6 Illustrative Example

Figure 9.2 below shows a position control system formed of subsystems (i) a PID controller with gains k_p, k_I, and k_d, (ii) an actuator which is an amplifier of gain k_A, (ii) position sensor of gain k_θ, (iv) velocity sensor of gain k_ω and a plant which is dc motor with gain k_1 and time constant α. The nominal values of k_A, k_ω, and k_θ are unity, $\alpha = 0.8$, $k_p = 0.05$, $k_I = 0.0023$, $k_d = 0.5$. Subsystems which are subject to fault include the actuator, the plant, the velocity sensor and the position sensor, and are characterized respectively by diagnostic parameters $\gamma_1 = k_A$, $\gamma_2 = \alpha$, $\gamma_3 = k_\omega$ and $\gamma_4 = k_\theta$. The nominal values of the diagnostic parameter $\gamma_i^0 = 1, i = 1, 2, 3, 4$. Variations in the diagnostic parameters $\gamma = [\gamma_1 \quad \gamma_2 \quad \gamma_3 \quad \gamma_4]^T$ induce faults in the system. The system is subject to load disturbance $w(k)$ and the measurement noise $v(k)$.

9.6.1 Mathematical Model

$$x(k+1) = \begin{bmatrix} 1 & 1 & 0 \\ -k_1 k_p k_A k_\theta & \alpha - k_1 k_A (k_d k_\omega + k_p k_\theta) & k_A k_I k_1 \\ -k_\theta & -k_\theta & 1 \end{bmatrix} x(k) + \begin{bmatrix} 0 \\ k_p \\ 1 \end{bmatrix} r(k) + \begin{bmatrix} 0 \\ k_1 \\ 0 \end{bmatrix} w(k) \tag{9.16}$$

$$y(k) = \begin{bmatrix} k_\theta & k_\theta & 0 \end{bmatrix} x(k) + v(k)$$

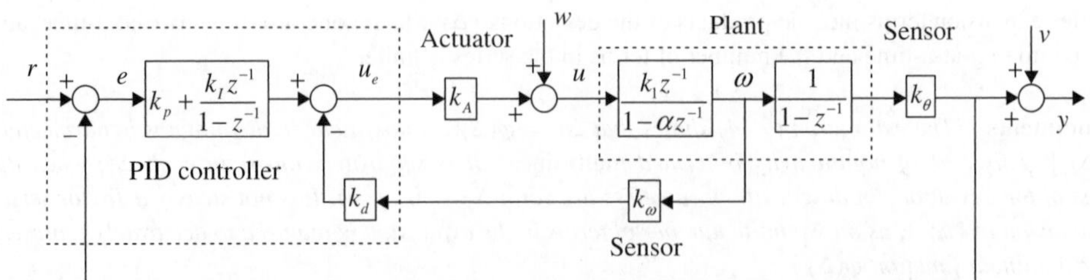

Figure 9.2 Position control system

The overall transfer function of the system $G(z)$ derived from physical law is of third order and is given by:

$$G(z) = \frac{b_1 z^{-1} + b_2 z^{-2}}{1 + a_1 z^{-1} + a_2 z^{-2} + a_3 z^{-3}} \tag{9.17}$$

where

$$b_1 = k_1 k_p k_A k_\theta;\ b_2 = (k_I - k_p)\, k_1 k_A k_\theta;\ a_1 = k_d k_1 k_\omega k_A - \alpha - 2 + k_1 k_p k_\theta k_A;$$
$$a_2 = 2(\alpha - k_d k_1 k_\omega k_A) + 1 + (k_I - k_p)\, k_1 k_A k_\theta;\ a_3 = -\alpha + k_d k_\omega k_A k_1$$

9.6.2 *Feature Vector and the Influence Vectors*

In the ideal case where the structure of a mathematical model and that of the physical system are the same, then one knows beforehand the expressions for the feature vector and the influence vectors in terms of the parameters of the subsystems and the emulators. The feature vector θ is a 5×1 vector derived from the expressions of the numerator and the denominator polynomials of the transfer function $G(z)$ given in Eq. (9.17). The map $\theta = \varphi(\gamma)$ relating θ and γ is as follows:

$$\theta = \begin{bmatrix} a_1 \\ a_2 \\ a_3 \\ b_1 \\ b_2 \end{bmatrix} = \begin{bmatrix} k_d k_1 \gamma_1 \gamma_3 - \gamma_2 - 2 + k_1 k_p \gamma_1 \gamma_4 \\ 2(\gamma_2 - k_d k_1 \gamma_1 \gamma_3) + 1 + (k_I - k_p)\, k_1 \gamma_1 \gamma_4 \\ -\gamma_2 + k_d k_1 \gamma_1 \gamma_3 \\ k_1 k_p \gamma_1 \gamma_4 \\ (k_I - k_p) k_1 \gamma_1 \gamma_4 \end{bmatrix} \tag{9.18}$$

The nonlinear function $\theta = \boldsymbol{\varphi}(\boldsymbol{\gamma})$ is multi-linear in $\{\gamma_i\}$, that is, it is a function of monomials of the elements of γ, namely γ_2, $\gamma_1\gamma_3$, $\gamma_1\gamma_4$. The influence vectors are restricted to the set $\boldsymbol{\Omega} = [\boldsymbol{\Omega}_1 \quad \boldsymbol{\Omega}_2 \quad \boldsymbol{\Omega}_3 \quad \boldsymbol{\Omega}_4 \quad \boldsymbol{\Omega}_{13} \quad \boldsymbol{\Omega}_{14}]$. The influence vectors $\boldsymbol{\Omega}_2$, $\boldsymbol{\Omega}_{13}$, and $\boldsymbol{\Omega}_{14}$ will not be functions of the diagnostic parameter γ. However, $\boldsymbol{\Omega}_2, \boldsymbol{\Omega}_2, \ldots$, and $\boldsymbol{\Omega}_2$ will be functions of the nominal diagnostic parameters $\boldsymbol{\gamma}^0$. Expressions for the influence vectors $\boldsymbol{\Omega}_1 = \dfrac{\delta\theta}{\delta\gamma_1}$, $\boldsymbol{\Omega}_2 = \dfrac{\delta\theta}{\delta\gamma_2}$, $\boldsymbol{\Omega}_3 = \dfrac{\delta\theta}{\delta\gamma_3}$, and $\boldsymbol{\Omega}_4 = \dfrac{\delta\theta}{\delta\gamma_4}$ are given by:

$$\Omega_1 = \begin{bmatrix} k_d k_1 \gamma_3^0 + k_1 k_p \gamma_4^0 \\ -2k_d k_1 \gamma_3^0 + (k_I - k_p) k_1 \gamma_4^0 \\ k_d k_1 \gamma_3^0 \\ k_1 k_p \gamma_4^0 \\ (k_I - k_p) k_1 \gamma_4^0 \end{bmatrix};\quad \Omega_2 = \begin{bmatrix} -1 \\ 2 \\ -1 \\ 0 \\ 0 \end{bmatrix};\quad \Omega_3 = \begin{bmatrix} k_d k_1 \gamma_1^0 \\ -2k_d k_1 \gamma_1^0 \\ k_d k_1 \gamma_1^0 \\ 0 \\ 0 \end{bmatrix};\quad \Omega_4 = \begin{bmatrix} k_1 k_p \gamma_1^0 \\ (k_I - k_p) k_1 \gamma_1^0 \\ 0 \\ k_1 k_p \gamma_1^0 \\ (k_I - k_p) k_1 \gamma_1^0 \end{bmatrix}$$

As $\boldsymbol{\Omega}_{12} = \dfrac{\delta^2\theta}{\delta\gamma_1\delta\gamma_2} = \mathbf{0}$, $\boldsymbol{\Omega}_{23} = \dfrac{\delta^2\theta}{\delta\gamma_2\delta\gamma_3} = \mathbf{0}$, $\boldsymbol{\Omega}_{24} = \dfrac{\delta^2\theta}{\delta\gamma_2\delta\gamma_3} = \mathbf{0}$, and $\boldsymbol{\Omega}_{34} = \dfrac{\delta^2\theta}{\delta\gamma_3\delta\gamma_4} = \mathbf{0}$, only the expressions for the influence vectors ; $\boldsymbol{\Omega}_{13} = \dfrac{\delta^2\theta}{\delta\gamma_1\delta\gamma_3}$, $\boldsymbol{\Omega}_{14} = \dfrac{\delta^2\theta}{\delta\gamma_1\delta\gamma_4}$ are given below:

$$\boldsymbol{\Omega}_{13} = \begin{bmatrix} k_d k_1 \\ -2k_d k_1 \\ k_d k_1 \\ 0 \\ 0 \end{bmatrix} \quad \boldsymbol{\Omega}_{14} = \begin{bmatrix} k_1 k_p \\ (k_I - k_p) k_1 \\ 0 \\ k_1 k_p \\ (k_I - k_p) k_1 \end{bmatrix}.$$

Note: the maximum number of influence vectors in the set $\{\boldsymbol{\Omega}_i, \boldsymbol{\Omega}_{ij}, \boldsymbol{\Omega}_{ijk}, \ldots \boldsymbol{\Omega}_{123\ldots q}\}$ when the number of subsystem $q = 4$ is $2^4 = 16$ as shown in Eq. (9.15). However in this example the relevant non-zero influence vectors are merely 6.

9.7 Residual of the Kalman Filter

The background on the Kalman filter is given in an earlier chapter. The residual of the Kalman filter $e(k) = y(k) - \hat{y}(k)$ in the presence of additive and multiplicative type faults can be expressed as:

$$e(z) = \frac{\boldsymbol{\psi}^T(z)\Delta\boldsymbol{\theta} + y_f(z)}{F_0(z)} + \frac{\upsilon(z)}{F_0(z)} \tag{9.19}$$

where $F_0(z) = |z\boldsymbol{I} - \boldsymbol{A}_0 + \boldsymbol{K}_0\boldsymbol{C}_0|$ is termed the Kalman polynomial. Variations $\delta\boldsymbol{A}$, $\delta\boldsymbol{B}$, and $\delta\boldsymbol{C}$ in the state-space model (9.7) map in to the feature vector variation $\Delta\boldsymbol{\theta}$ given by Eq. (9.11).

If there is no model mismatch, that is $\boldsymbol{A}_0 = \boldsymbol{A}, \boldsymbol{B}_0 = \boldsymbol{B}$, and $\boldsymbol{C}_0 = \boldsymbol{C}$, and the Kalman gain $\boldsymbol{K}_0$ is optimal, then the residual is a zero-mean white noise process. We will assume that the noisy component $e_0(k)$ of the residual $e(k)$ is a zero-mean white noise process under both fault and fault-free conditions.

In the presence of fault, that is when $\Delta\boldsymbol{\theta} \neq \boldsymbol{0}$ and $\boldsymbol{f} \neq \boldsymbol{0}$, from Eq. (9.19) it can be deduced that the residual $e(z)$ is a sum of two components, namely a bias term resulting from a fault $e_f(z)$ and the zero-mean white noise component $e_0(z)$, given by:

$$e(z) = e_f(z) + e_0(z) \tag{9.20}$$

where

$$e_0(z) = \frac{\upsilon(z)}{F_0(z)} \tag{9.21}$$

$$e_f(z) = \frac{\boldsymbol{\psi}^T(z)\Delta\boldsymbol{\theta} + y_f(z)}{F_0(z)} \tag{9.22}$$

where $e_f(z)$ is a fault-indicating term.

Substituting for $\Delta\boldsymbol{\theta} = \boldsymbol{\theta} - \boldsymbol{\theta}^0$ from Eq. (9.14), the fault indicator component $e_f(z)$ may be expressed explicitly in terms of the variations in the diagnostic parameter, $\Delta\gamma$ becomes:

$$e_f(z) = \frac{\boldsymbol{\psi}^T(z)\boldsymbol{\varphi}(\Delta\gamma) + y_f(z)}{F_0(z)} \tag{9.23}$$

Expressing $e_f(k)$ as a sum of two components, one induced by the multiplicative fault and the other by the additive fault, we get:

$$e_f(z) = e_F(z) + y_F(z) \tag{9.24}$$

where $e_F(z) = \boldsymbol{\psi}_F^T(z)\boldsymbol{\varphi}(\Delta\gamma)$ is the filtered multiplicative fault term, $y_F(z) = \dfrac{y_f(z)}{F_0(z)}$ is the filtered additive fault term, $\boldsymbol{\psi}_F^T(z) = \dfrac{\boldsymbol{\psi}^T(z)}{F_0(z)}$ is the filtered data vector, and $\dfrac{1}{F_0(z)}$ is the filter.

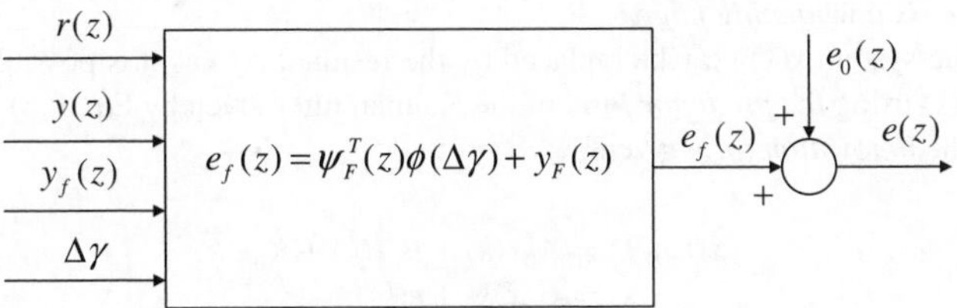

Figure 9.3 Diagnostic model

9.7.1 *Diagnostic Model*

A model that relates the Kalman filter residual $e(z)$ and the diagnostic parameter deviation $\Delta\gamma$ is termed herein the *diagnostic model*. From Eqs. (9.23) and (9.20) the diagnostic model becomes:

$$e(z) = \boldsymbol{\psi}_F^T(z)\boldsymbol{\varphi}(\Delta\gamma) + y_F(z) + e_0(z) \tag{9.25}$$

Figure 9.3 shows the diagnostic model relating the residual $e(z)$ to (i) the input $r(z)$ and the output $y(z)$, (ii) the additive fault input $y_f(z)$, and (iii) multiplicative fault $\Delta\gamma$.

9.7.2 *Key Properties of the Residual*

Recall the results from the chapter on the Kalman filter:

$$\lim_{k\to\infty} it\, E\left[e(k)\right] = 0 \text{ if and only if } \boldsymbol{\Delta\gamma} = \mathbf{0} \text{ and } \boldsymbol{f}(k) = \mathbf{0} \tag{9.26}$$

The auto-correlation function of the residual $r_{ee}(m)$ is a delta function:

$$r_{ee}(m) = E\left[e(k)e(k-m)\right] = \sigma_e^2\delta(m) = \begin{cases} \sigma_e^2 & m = 0 \\ 0 & else \end{cases} \text{ if and only if } \Delta\gamma = \mathbf{0} \text{ and } \boldsymbol{f}(k) = \mathbf{0} \tag{9.27}$$

where $\sigma_e^2 = E\left[e_0^2(k)\right]$ and m is the time-lag.

9.7.3 *The Role of the Kalman Filter in Fault Diagnosis*

- *The residual is the indicator of a fault:*
  a) *System is fault-free and the Kalman filter optimal*
     If the Kalman gain is optimal, then there is no fault if and only if the residual is a zero-mean white noise process, $e(k) = e_0(k)$ or equivalently the auto-correlation of the residual is not a function $r_{ee}(m) = \sigma_e^2\delta(m)$.
     The variance of the residual $E[e^2(k)]$ will be minimum.
  b) *System is faulty and the Kalman filter is optimal*:
     The residual is not a zero-mean white noise process, $e(k) \neq e_0(k)$, if and only if there is a fault. There will be a bias term $e_f(z)$, which is a function of the deviations in the diagnostic parameter.

- *The Kalman filter is a whitening filter*
  The output of the system $y(k)$ may be replaced by the residual $e(k)$ as it is possible to compute $e(k)$ from $y(k)$ and $r(k)$ using the *predictor form* of the Kalman filter given by Eq. (9.6) and $y(k)$ from $e(k)$ and $r(k)$ using the *innovation form* given by:

$$\begin{aligned}\hat{\boldsymbol{x}}(k+1) &= \boldsymbol{A}_0\hat{\boldsymbol{x}}(k) + \boldsymbol{B}_0 r(k) + \boldsymbol{K}_0 e(k)\\ y(k) &= \boldsymbol{C}_0\hat{\boldsymbol{x}}(k) + e(k)\end{aligned} \tag{9.28}$$

  In other words, $y(k)$ and $e(k)$ are *causally invertible* [7]. This property is exploited in many applications including fault diagnosis, as the residual $e(k)$ is a zero-mean white noise process. The Kalman filter may be viewed as a *whitening filter*: the input to the Kalman filter is $y(k)$ and its output is a zero-mean white noise process $e(k)$.
- *The Kalman filter is robust to noise and disturbances.*
  The filter's structure is error-driven, and because of its closed-loop configuration has an inherent ability to also attenuate disturbances.

In view of the above key properties, the diagnostic model (9.25) forms the backbone of the fault detection and fault isolation scheme.

9.8 Fault Diagnosis

Fault Diagnosis (FD) is based on the probabilistic analysis of the Kalman filter residual. Fault diagnosis comprises the following tasks:

- *Identification of the nominal model*: The nominal (fault-free) model of the system is identified using an efficient method such as the high-order least-squares method (a high-order model is identified, and a reduced-order model is a derived from the high-order identified model using a model reduction technique), the prediction error method, or the subspace method.
- *Generation of residual using Kalman filter*: A Kalman filter is designed for the identified nominal model to generate a residual.
- *Fault detection*
- *Estimation of the influence vectors and the additive fault*
- *Fault isolation*

A batch processing scheme is adopted here where residuals are collected in a sliding time window of length N and processed at each time instant. At each time instant k, the N residuals formed of the present and past $N-1$ residuals $\{e(k-i) : i = 0, 1, 2, \dots N-1\}$, are collected and the $Nx1$ residual vector $\boldsymbol{e}(k) = [\, e(k) \quad e(k-1) \quad e(k-2) \quad . \quad e(k-N+1)\,]^T$ is used. We will focus on fault detection and fault isolation assuming that the nominal fault-free model has be identified and the Kalman filter has been designed as given in Eq. (9.6).

9.9 Fault Detection: Bayes Decision Strategy

We assume the following:

- The noisy component $e_0(k)$ of the residual signal $e(k)$ is the same under both under both fault and fault-free conditions.
- $e_0(k)$ is a zero-mean identically distributed Gaussian white noise process.

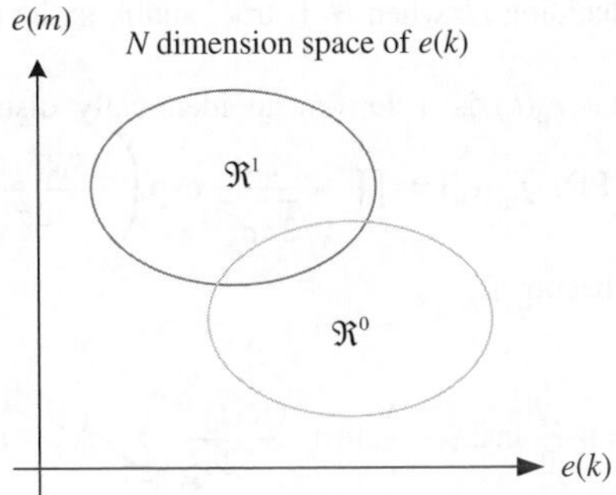

Figure 9.4 Binary pattern classifications

Let H_0 and H_1 be the two hypotheses indicating the absence and presence of a fault, respectively. Under the hypothesis H_0 the residual is merely a zero-mean white noise process denoted by a $N \times 1$ vector $\boldsymbol{e}_0(k) = [\, e_0(k) \quad e_0(k-1) \quad e_0(k-2) \quad . \quad e_0(k-N+1)\,]^T$, while under the hypothesis H_1 the residual will have an additive fault indicator term, namely a $N \times 1$ vector $\boldsymbol{e}_f(k) = [\, e_f(k) \quad e_f(k-1) \quad e_f(k-2) \quad . \quad e_f(k-N+1)\,]^T$.

Invoking the lemma (9.26), the diagnostic model (9.25) under these hypotheses becomes:

$$\boldsymbol{e}(k) = \begin{cases} \boldsymbol{e}_0(k) & : H_0 \\ \boldsymbol{e}_f(k) + \boldsymbol{e}_0(k) & : H_1 \end{cases} \tag{9.29}$$

9.9.1 *Pattern Classification Problem: Fault Detection*

The fault detection problem is a binary pattern classification problem. The Nx1 residual $\boldsymbol{e}(k)$ is located in a region $\mathfrak{R}^0$ if there is no fault (hypothesis H_0 holds), and in the region $\mathfrak{R}^1$ if there is a fault (hypothesis H_1 holds). In general there will be overlap between the two regions. The objective is to derive a strategy to assert whether the given residual $\boldsymbol{e}(k)$ lies in the region $\mathfrak{R}^0$ or in the region $\mathfrak{R}^1$. As there is a class overlap, the decision to locate the residual to one of the two regions will be prone to errors. The visual representation of the pattern classification problem of deciding between the two hypotheses is shown in Figure 9.4. The regions $\mathfrak{R}^0$ and $\mathfrak{R}^1$ are indicated by ellipses. We will formulate the fault detection problem as a binary hypothesis testing problem. The binary decision strategy using Bayes decision strategy states that:

$$\begin{cases} decide \quad H_0 \quad if \quad lr(\boldsymbol{e}) \le \gamma_{th} \\ decide \quad H_1 \quad if \quad lr(\boldsymbol{e}) > \gamma_{th} \end{cases} \tag{9.30}$$

where $lr(\boldsymbol{e})$ is the log-likelihood ratio defined by:

$$lr(\boldsymbol{e}) = \ln f_{e|H_1}(\boldsymbol{e}|H_1) - \ln f_{e|H_0}(\boldsymbol{e}|H_0), \tag{9.31}$$

$f_{e|H_1}(.)$ and $f_{e|H_0}(.)$ are respectively the conditional PDF of $\boldsymbol{e}(k)$ under the hypothesis H_i and H_0 given in Eq. (9.29), $\ln f_{e|H_1}(\boldsymbol{e}|H_1)$ and $\ln f_{e|H_0}(\boldsymbol{e}|H_0)$ are respectively the log-likelihood function under H_1 and H_0, and γ_{th} is the Bayesian threshold value given by

$$\gamma_{th} = \frac{(c_{10} - c_{00})p_0}{(c_{01} - c_{11})p_1} \tag{9.32}$$

c_{ij} is the cost associated with deciding H_i when H_j is true, and p_i is the prior probability, $p_0 = \Pr\{H_0\}$ and $p_1 = P\{H_1\}$.

Invoking the assumption that $e_0(k)$ is a zero-mean identically distributed Gaussian white noise process with variance $\sigma_{e_0}^2$ and PDF $f_{e_0}(\boldsymbol{e}_0) = \prod_{i=0}^{N-1} \frac{1}{\sqrt{2\pi\sigma_{e_0}^2}} \exp\left(-\frac{e_0^2(k-i)}{2\sigma_{e_0}^2} \right)$ log-likelihood functions $\ln f_{e|H_0}(\boldsymbol{e}|H_0)$ and $\ln f_{e|H_1}(\boldsymbol{e}|H_1)$ becomes:

$$\begin{aligned} \ln f_{e|H_0}(\boldsymbol{e}|H_0) &= -\frac{N}{2}\ln 2\pi - \frac{N}{2}\ln \sigma_{e_0}^2 - \frac{1}{2\sigma_{e_0}^2}\sum_{i=0}^{N-1} e^2(k-i) \\ \ln f_{e|H_1}(\boldsymbol{e}|H_1) &= -\frac{N}{2}\ln 2\pi - \frac{N}{2}\ln \sigma_{e_0}^2 - \frac{1}{2\sigma_{e_0}^2}\sum_{i=0}^{N-1} (e(k-i) - e_f(k-i))^2 \end{aligned} \tag{9.33}$$

The log-likelihood ratio $lr(\boldsymbol{e})$ given by Eq. (9.31) becomes after simplification:

$$lr(\boldsymbol{e}) = -\frac{1}{2\sigma_{e_0}^2}\sum_{i=0}^{N-1}\left[e_f^2(k-i) - 2e(k-i)e_f(k-i)\right] \tag{9.34}$$

9.9.2 *Generalized Likelihood Ratio Test*

The fault-indicating component $e_f(k)$ is generally unknown and hence a composite hypothesis testing scheme is employed where $e_f(k)$ is replaced by its maximum likelihood estimate in the log-likelihood ratio (9.34). In deriving this estimate, we use the fact that spectral contents of $e_f(k)$ will include those of the reference input $r(k)$ and the additive fault $\boldsymbol{y}_f(k)$. For the expression for the fault-indicating component $e_f(k)$ given by Eq. (9.23), assuming that (i) system is linear and is asymptotically stable and (ii) the fault is diagnosed when the system is in the steady-state (that is after all the transients have died), we can deduce the following:

- $e_f(k)$ will be a constant if $r(k)$ and $y_f(k)$ are both constants.
- If $r(k)$ is a sinusoid of frequency f_r and $y_f(k)$ is a constant, then $e_f(k)$ will be also be a sum of sinusoid of the same frequency f_r plus some constant.
- If $r(k)$ is a random process and $y_f(k)$ is a constant then $e_f(k)$ will be some colored random process with a constant mean.

Similarly we can deduce the spectral contents of $e_f(k)$ for other types of $r(k)$ and $y_f(k)$.

9.9.3 *Maximum Likelihood Estimate*

The ML estimate $\hat{e}_f(k)$ of $e_f(k)$ is obtained from the maximization of the log-likelihood function under the hypothesis H_1:

$$\hat{e}_f(k) = \arg\left(\max_{e_f}\{\ln f_{e|H_1}(\boldsymbol{e}|H_1)\}\right) \tag{9.35}$$

Since constant additive and multiplicative terms in the cost function do not affect the optimal estimate, we may set the additive term to be zero and the multiplicative term to be unity. Further, the maximization problem becomes a minimization problem in view of the negative cost function. Using Eq. (9.24), the maximization problem (9.35) becomes:

$$\hat{e}_f(k) = \arg\left(\min_{e_f} \left\{ \sum_{i=0}^{N-1} (e(k-i) - e_f(k-i))^2 \right\} \right) \tag{9.36}$$

The optimal cost function becomes (see Appendix)

$$lr(\boldsymbol{e}) = -\frac{1}{2\sigma_{e_0}^2} \sum_{i=0}^{N-1} \hat{e}_f^2(k-i) \tag{9.37}$$

Case 1 $e_f(k) = e_f$ is a constant ($r(k)$ and $y_f(k)$ are both constants)
Substituting $e_f(k) = e_f$, the optimization problem (9.36) becomes:

$$\hat{e}_f(k) = \arg\left(\min_{e_f} \left\{ \sum_{i=0}^{N-1} (e(k-i) - e_f))^2 \right\} \right) \tag{9.38}$$

Differentiating with respect to e_f and setting to zero we get the necessary condition

$$\sum_{i=0}^{N-1} e(k-i) - Ne_f = 0 \tag{9.39}$$

The optimal estimate is the mean value of the residual given by:

$$\hat{e}_f = \frac{1}{N} \sum_{i=0}^{N-1} e(k-1) \tag{9.40}$$

Replacing $e_f(k)$ by its ML estimate using Eq. (9.40), the log-likelihood ratio (9.34) becomes:

$$lr(\boldsymbol{e}) = \frac{N}{2\sigma_{e_0}^2} (\frac{1}{N} \sum_{i=0}^{N-1} e(k-1))^2 \tag{9.41}$$

Case 2 $e_f(k)$ is some time-varying function
In this case, the reference input $r(k)$ and additive fault $y_f(k)$ are both arbitrary functions of time. The solution of the minimization problem (9.38) simply becomes:

$$\hat{e}_f(k) = e(k) \tag{9.42}$$

Replacing $e_f(k)$ by its ML estimate $\hat{e}_f(k)$ using Eq. (9.42), the log-likelihood ratio (9.34) becomes:

$$lr(\boldsymbol{e}) = \frac{1}{2\sigma_{e_0}^2} \sum_{i=0}^{N-1} e^2(k-i) \tag{9.43}$$

Case 3 $r(k) = a_r \sin(2\pi f_r k)$ is sinusoid and $y_f = c_f$ is constant and f_r is known
In this case $\hat{e}_f(k) = a_f \sin(2\pi f_r k) + b_f \cos(2\pi f_r k) + c_f$. See **Appendix**.

The ML estimate is obtained from the minimization problem (9.36):

$$\hat{e}_f(k) = \arg\left(\min_{a_f,b_f,c_c}\left\{\sum_{i=0}^{N-1}\left[e(k-i) - a_f\sin(2\pi f_r(k-i)) - b_f\cos(2\pi f_r(k-i)) - c_f\right]^2\right\}\right) \tag{9.44}$$

In the Appendix it shown that

$$\begin{aligned} a_f &= \frac{2}{N}\sum_{i=0}^{N-1} e(k-i)\sin(2\pi f_r(k-i)) \\ b_f &= \frac{2}{N}\sum_{i=0}^{N-1} e(k-i)\cos(2\pi f_r(k-i)) \\ c_f &= \frac{1}{N}\sum_{i=0}^{N-1} e(k-i) \end{aligned} \tag{9.45}$$

Substituting for a_f, b_f,and c_f using Eq. (9.45) and replacing $e_f(k)$ by its ML estimate $\hat{e}_f(k)$, the log-likelihood ratio (9.34) becomes:

$$lr(\mathbf{e}) = \frac{N}{2\sigma_{e_0}^2}\left(\frac{a_f^2}{2} + \frac{b_f^2}{2} + c_f^2\right) \tag{9.46}$$

From the expression for a_f, b_f, and c_f given in Eq. (9.45), and using the definition of the periodogram which is an estimate of the power spectral density we get:

$$lr(\mathbf{e}) = \frac{1}{\sigma_{e_0}^2}\left(\hat{P}_{ee}(f_r) + \frac{1}{2}\hat{P}_{ee}(0)\right) \tag{9.47}$$

9.9.4 *Decision Strategy*

We may generalize the binary decision strategy (9.30) with a view to unify commonly used approaches, including the Bayes decision theoretic approach, the classical Neyman–Pearson approach, the Maximum *A Posteriori* (MAP) approach, and the Maximum Likelihood (ML) approach. The Bayes test statistics $lr(\mathbf{e})$ is replaced by an intuitive and practical test statistics, denoted $t_s(\mathbf{e})$, and the Bayes threshold γ_{th} is replaced by a more general threshold η_{th}. The unified decision strategy becomes:

$$t_s(\mathbf{e})\begin{cases} \leq \eta_{th} & \textit{no fault} \\ > \eta_{th} & \textit{fault} \end{cases} \tag{9.48}$$

Test statistics for a constant, sinusoid, and an arbitrary reference input $r(k)$, assuming the additive fault input $y_f(k)$ is a constant, may be deduced from the expressions for $lr(\mathbf{e})$ given by Eqs. (9.41), (9.43), and

(9.47), and are given by:

$$t_s(e) = \begin{cases} \left| \frac{1}{N} \sum_{i=0}^{N-1} e(k-i) \right| & r(k) = \text{constant} \\ \hat{P}_{ee}(f_r) + \frac{1}{2}\hat{P}_{ee}(0) & r(k) \text{ is a sinusoid} \\ \frac{1}{N} \sum_{i=0}^{N-1} e^2(k-i) & r(k) \text{ is an arbitrary signal} \end{cases} \tag{9.49}$$

The corresponding threshold value η_{th} is:

$$\eta_{th} = \begin{cases} \sqrt{\frac{2\gamma_{th}\sigma_{e0}^2}{N}} & r(k) = \text{constant} \\ 2\gamma_{th}\sigma_{e_0}^2 & r(k) \text{ is a sinusoid} \\ \gamma_{th}\sigma_{e_0}^2 & r(k) \text{ is an arbitrary signal} \end{cases} \tag{9.50}$$

9.9.5 Other Test Statistics

Other methods, such as the Neyman–Pearson method, are preferred when the *a priori* information, such as the *a priori* probabilities $p_0 = \Pr\{H_0\}$ and $p_1 = \Pr\{H_1\}$ and the cost associated with the decision $\{c_{ij}\}$, are not available.

Due the presence of noise and disturbances affecting the residual, there will be two types of errors, termed false alarm error or type I error and miss error or type II error. The false alarm error occurs when a "fault" is asserted when there is no fault, and the miss error occurs when "no fault" is asserted when there is a fault. These errors are quantified in terms of their probability of occurrence. The false alarm probability P_F is the probability of deciding the hypothesis H_1 when H_0 is true, and the probability of miss P_M, which is the result of asserting H_0 when H_1 is true. These probabilities are given by:

$$\begin{aligned} P_F(\eta) &= \Pr\{t_s(e) > \eta_{th}|H_0\} \\ P_M(\eta) &= \Pr\{t_s(e) \le \eta_{th}|H_1\} \end{aligned} \tag{9.51}$$

The probability of a correct decision, that is asserting H_0 when H_0 true and asserting H_1 when H_1 true, is given by:

$$P_D(\eta) = \begin{cases} \Pr\{t_s(e) \le \eta_{th}|H_0\} \\ \Pr\{t_s(e) > \eta_{th}|H_1\} \end{cases} \tag{9.52}$$

Since $P_M = 1 - P_D$, P_D and P_F suffice to specify all of the probabilities of interest. One type of error may dominate the other or both may be equal depending upon the choice of the threshold. It is challenging problem to select the threshold value. Depending upon the application, the available *a priori* information and the amount of a data, the following methods are widely used the compute an appropriate threshold value:

Bayesian method: The Bayesian threshold value γ_{th} is used if we have sufficient past data to estimate the associated costs $\{c_{ij}\}$ and prior probabilities p_0 and p_1. If not enough data is available, then choose $c_{00} = c_{11} = 0$ and $c_{01} = c_{10} = 1$ thereby penalizing (or rewarding) equally the incorrect (or correct) decisions and assigning equal prior probabilities to both decisions. In this case $\gamma = 1$.

Neyman–Pearson method: The threshold η_{th} is chosen so that the probability of correct decision is maximized under the constraint that the false alarm probability P_F is less than some specified value. In applications such as fault detection, restricting the false alarm probability is important in preventing unnecessary shutdowns of the system operation whenever a fault is asserted.

Min–max approach: The min–max criterion is to choose the prior probabilities with $c_{00} = c_{11} = 0$ and $c_{01} = c_{10} = 1$. The min–max criterion is to choose the threshold such that the false alarm probability is equal to the probability of a miss, that is, $P_M = P_F$.

Comment *The smallest failure that can be detected depends on the effect of model uncertainty and measurement noise. The essence of the failure detection problem is to generate a residual and choose a threshold value such that the widest possible failure detection zone is achieved. The threshold needs to be chosen large enough to eliminate false alarms, yet small enough to ensure sufficiently small failures are detected.*

9.10 Evaluation of Detection Strategy on Simulated System

The Kalman filter-based detection strategy is employed here to detect faults in the presence of noise and disturbance when the reference signal is (i) a constant, (ii) a sinusoid, and (iii) some arbitrary signal. The state-space model of the system (A, B, C) is given by:

$$\begin{aligned} x(k+1) &= \begin{bmatrix} 1.7913 & 1 \\ -0.81 & 0 \end{bmatrix} x(k) + \begin{bmatrix} 1 \\ 0 \end{bmatrix} r(k) + \begin{bmatrix} 1 \\ 0 \end{bmatrix} w(k) \\ y(k) &= \begin{bmatrix} 1 & 0 \end{bmatrix} x(k) + v(k) \end{aligned} \tag{9.53}$$

A process, an actuator, and a sensor fault were simulated by perturbing the $\boldsymbol{A}$, $\boldsymbol{B}$, and $\boldsymbol{C}$ matrices. The reference input $r(k)$ was chosen to be a sinusoid for the process fault, a constant for the sensor fault, and a colored noise for the actuator fault. Figure 9.5 shows the auto-correlation of the Kalman filter residual for different operating scenarios. Subfigures (a), (b), (c), and (d) show respectively the auto-correlation functions under no fault, process fault with $\Delta\boldsymbol{A} = 0.05$, sensor fault $\Delta\boldsymbol{C} = 0.1$, and the actuator fault $\Delta\boldsymbol{B} = 0.05$. The variance of the disturbance and the measurement noise were $Q = 0.1$ and $R = 100$. The auto-correlation function is a good indicator of the presence or absence of a fault. It is a delta function if and only if there is no fault, as established in Eq. (9.27). The fault-indicating component $e_f(k)$ has a spectral content identical to that of the reference input. For example, if the reference input is constant, sinusoid, or colored noise then $e_f(k)$ will also be constant or sinusoid or colored noise. The auto-correlation function will be the corresponding auto-correlation function of the residual, as shown in Figure 9.5.

9.11 Formulation of Fault Isolation Problem

The residual $e(k) = e_f(k) + e_0(k)$ given by Eqs. (9.20) and (9.23) contains complete information for fault isolation. A unified approach to both fault detection and fault isolation is employed. The fault isolation task is performed if a fault is asserted during the fault detection phase.

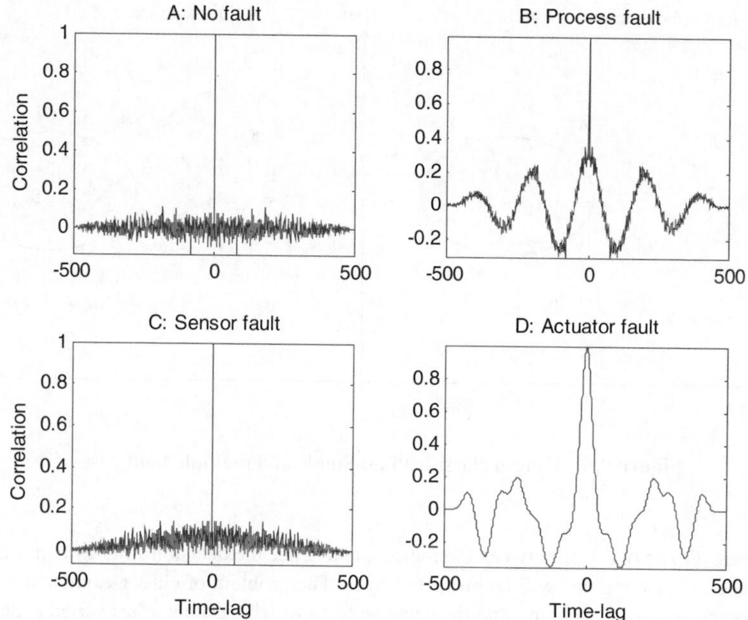

Figure 9.5 Auto-correlations for the fault-free and faulty cases

9.11.1 *Pattern Classification Problem: Fault Isolation*

The fault isolation is the problem of deciding which of the diagnostic parameters $\{\gamma_i\}$ has varied given the residual data $\boldsymbol{e}(k)$. We may classify the faults as a single or multiple faults by defining different hypotheses such as $H_i, H_{ij}, H_{ijk}, \ldots, H_{123.q}$:

- *Single fault*: H_i:$\Delta\gamma_i \neq 0,\ \Delta\gamma_j = 0\, j \neq i$
- *Two simultaneous faults*: H_{ij}:$\Delta\gamma_i \neq 0$ *and* $\Delta\gamma_j \neq 0$ *but* $\Delta\gamma_k = 0 :\ k \neq i, j$
- *Three simultaneous faults*: H_{ijk}: $\Delta\gamma_i \neq 0,\ \Delta\gamma_j \neq 0$ *and* $\Delta\gamma_k \neq 0$ *but* $\Delta\gamma_\ell = 0$: $\ell \neq i, j, k$

And so on until

- *Simultaneous faults in all*: $H_{123.q}$: $\Delta\gamma_i \neq 0$: $i = 1, 2, 3, \ldots, q$.

Let us define regions in the N-dimensional Euclidian space generated by the $Nx1$ residual $\boldsymbol{e}(k)$ under each of the above listed hypotheses as follows:

$$
\begin{aligned}
\Re_i &= \{\boldsymbol{e}(k) : H_i\ holds\} \\
\Re_{ij} &= \{\boldsymbol{e}(k) : H_{ij}\ holds\} \\
\Re_{ijk} &= \{\boldsymbol{e}(k) : H_{ijk}\ holds\} \\
&\ \ \vdots \\
\Re_{123.q} &= \{\boldsymbol{e}(k) : H_{123.q}\ holds\}
\end{aligned}
\tag{9.54}
$$

Figure 9.6 shows the regions $\{\Re_i\}$ for single faults, and the rest of the regions $\Re_{ij}$, $\Re_{ijk} \ldots \Re_{123.q}$ for multiple faults. The case of a single fault is shown on the left, while that of multiple faults is shown in the right.

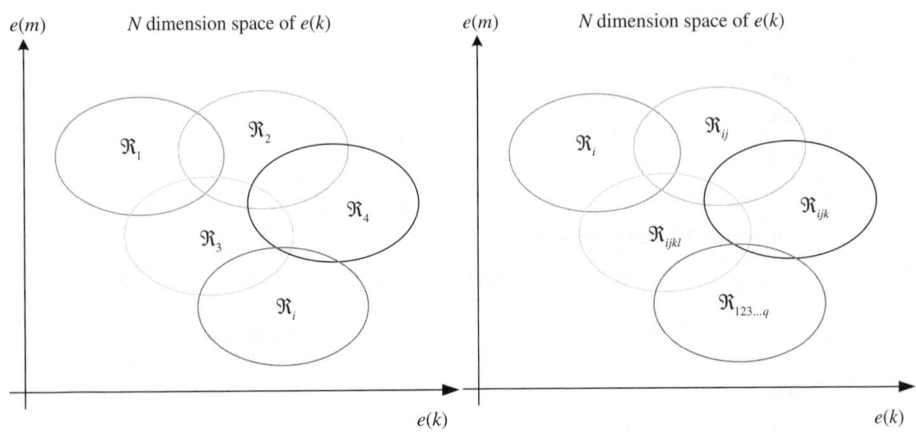

Figure 9.6 Pattern classification: Single and multiple fault cases

Generally there will be overlaps between the regions, and hence the decision strategy to assign the given residual to one of these regions will be prone to errors. The problem of class assignment is posed as a multiple hypothesis testing problem, and the decision as to which parameter has varied is derived using Bayes decision theory. It is similar to the approach used in the fault detection problem where a binary (instead of multiple) Bayes decision strategy is employed to decide between the presence and an absence of a fault.

The maximum likelihood approach based on the log-likelihood function $\ln f_{e|H_1}(\boldsymbol{e}|H_1)$ (for the hypothesis that there is a fault), is used for fault isolation. Note that for fault detection both the conditional PDFs, namely $\ln f_{e|H_1}(\boldsymbol{e}|H_1)$ and $\ln f_{e|H_0}(\boldsymbol{e}|H_0)$ are used, and the log-likelihood ratio $lr(\boldsymbol{e}) = \ln f_{e|H_1}(\boldsymbol{e}|H_1) - \ln f_{e|H_0}(\boldsymbol{e}|H_0)$ serves as the test statistics to assert the presence or an absence of a fault.

The fault isolation strategy is based on estimating the elements of the diagnostic parameter. If $|\Delta\gamma_i| \geq \varepsilon > 0$ is non-zero and greater than some specified threshold value ε, then subsystem associated with γ_i is asserted as faulty (single fault). On the other hand, if more than one diagnostic parameter is non-zero and greater than the threshold value, then the corresponding subsystems are asserted to be faulty (multiple faults).

9.11.2 *Formulation of the Fault Isolation Scheme*

The diagnostic parameter γ is estimated using the maximum likelihood approach when a fault is detected during the fault detection phase. The ML estimate of the elements of γ is obtained by maximizing the log-likelihood function $\ln f_{e|H_1}(\boldsymbol{e}|H_1)$ given by Eq. (9.33).

Consider the diagnostic model given by Eq. (9.25), where the expression for $\boldsymbol{\varphi}(\Delta\gamma)$ in terms of the influence vector and the deviations in the diagnostic parameters is given by Eq. (9.14). Substituting for $e(k)$, the log-likelihood function $\ln f_{e|H_1}(\boldsymbol{e}|H_1)$ given by Eq. (9.33) becomes:

$$\ln f_{e|H_1}(\boldsymbol{e}|H_1) = -\frac{N}{2}\ln 2\pi - \frac{N}{2}\ln\sigma_{e_0}^2 - \frac{1}{2\sigma_{e_0}^2}\sum_{i=0}^{N-1}\left(e(k-i) - \boldsymbol{\psi}_F^T(k-i)\boldsymbol{\varphi}(\Delta\gamma) - y_F(k-i)\right)^2 \quad (9.55)$$

In an optimization problem, additive and multiplicative constants in the cost function do not affect the optimal estimates, and the maximization problem becomes a minimization problem if the cost function is negative. Hence the relevant term in the cost function $\ln f_{e|H_1}(e|H_1)$ is essentially the last term in Eq. (9.55). This relevant term, termed the log-likelihood function of $\boldsymbol{\gamma}$ given $\boldsymbol{e}$, denoted $l(\gamma|e)$, is given by:

$$l(\gamma|\boldsymbol{e}) = \sum_{i=0}^{N-1} \left(e(k-i) - \boldsymbol{\psi}_F^T(k-i)\boldsymbol{\varphi}(\Delta\gamma) - y_F(k-i)\right)^2 \tag{9.56}$$

Expressing in a vector form yields:

$$l(\gamma|\boldsymbol{e}) = \left\|\boldsymbol{e}(k) - \boldsymbol{\Phi}(k)\boldsymbol{\varphi}(\Delta\gamma) - \boldsymbol{y}_F(k)\right\|^2 \tag{9.57}$$

where $\boldsymbol{e}(k)$, $\boldsymbol{e}_F(k) = \boldsymbol{\Phi}(k)\boldsymbol{\varphi}(\Delta\gamma)$, and $\boldsymbol{y}_F(k)$ are respectively $Nx1$ vectors formed from the residuals $\{e(k-i)\}$, the fault indicator functions $\{e_F(k-i)\}$, and additive faults $\{y_F(k-i)\}$, $\boldsymbol{\Phi}(k)$ is a NxM data matrix formed of the filtered data vectors $\left\{\boldsymbol{\psi}_F^T(k-i)\right\}$, $i = 0, 1, 2, \ldots, N-1$, and $\|\boldsymbol{x}(k)\| = \sqrt{\boldsymbol{x}^T(k)\boldsymbol{x}(k)}$ is the Euclidean or the 2-norm,

$$\boldsymbol{\Phi}(k) = \begin{bmatrix} \boldsymbol{\psi}_F^T(k) \\ \boldsymbol{\psi}_F^T(k-1) \\ \boldsymbol{\psi}_F^T(k-2) \\ . \\ \boldsymbol{\psi}_F^T(k-N+1) \end{bmatrix}, \boldsymbol{e}(k) = \begin{bmatrix} e(k) \\ e(k-1) \\ e(k-2) \\ . \\ e(k-N+1 \end{bmatrix}, \boldsymbol{e}_F(k) = \begin{bmatrix} e_F(k) \\ e_F(k-1) \\ e_F(k-2) \\ . \\ e_F(k-N+1 \end{bmatrix}, \boldsymbol{y}_F(k) = \begin{bmatrix} y_F(k) \\ y_F(k-1) \\ y_F(k-2) \\ . \\ y_F(k-N+1 \end{bmatrix}$$

Comment *The log-likelihood function $l(\gamma|\boldsymbol{e})$ is a function of the influence vectors and the additive fault, which are not known. Hence these unknowns influence vectors and additive faults have to be estimated first in the offline identification phase. Then in the operational phase, if a fault is detected, it is isolated by obtaining the maximum likelihood estimate of the diagnostic parameter by minimalizing $l(\gamma|\boldsymbol{e})$ given by Eq. (9.57) with respect to γ.*

9.11.3 *Fault Isolation Tasks*

Fault isolation is based on the analysis of the Kalman filter residual and is performed only if a fault is detected. Fault isolation comprises the following tasks:

- *Estimation of the influence vectors*: Using the diagnostic model (9.25), the set of influence vectors $\boldsymbol{\Omega} = \{\boldsymbol{\Omega}_i, \boldsymbol{\Omega}_{ij}, \boldsymbol{\Omega}_{ijk}, \ldots \boldsymbol{\Omega}_{123\ldots q}\}$ are estimated offline using parameter perturbed experiments.
- *Fault isolation using a sequential estimation scheme*: The residual is subject to a detailed analysis to estimate which of the diagnostic parameters $\gamma_i : i = 1, 2, 3, \ldots, q$ has varied by hypothesizing whether there is a variation in any single diagnostic parameter, or simultaneous variations on two or more diagnostics.

9.12 Estimation of the Influence Vectors and Additive Fault

The performance of the fault diagnosis depends upon the accuracy of the identified nominal model and the influence vectors. A reliable identification scheme is employed. The motivation for the proposed identification scheme stems from the intuition that if a model is identified under various operating scenarios, then the identified model is likely to capture the system's behavior in these operating regimes. This intuition is found to be true based on our evaluation of the proposed identification scheme on many simulated and diagnostic systems. The proposed scheme was inspired by the artificial neural network approach where a training set comprising data obtained from a number of representative scenarios is presented to the network to estimate the weights and, once trained, the network can then be used as a model for similar unseen scenarios. The approach may be considered as a kind of a "gray box model" in that it represents a compromise between the opaque black box model of the artificial neural network and the classical transparent identification approach based on an *a priori* known structure of the model.

The diagnostic model for fault diagnosis is completely characterized by a feature vector and influence vectors. The influence vectors are identified by performing a number of offline experiments. Each experiment consists of perturbing one or more elements of the diagnostic parameter. In order to reduce the computational burden, which one encounters in identification, an efficient recursive identification of the influence vector is proposed.

Consider the expression for the log-likelihood function of $\boldsymbol{\gamma}$ given $\boldsymbol{e}$ in Eq. (9.57). As the error term $\boldsymbol{e}_0(k)$ is a zero-mean white noise process (as can be deduced from the diagnostic model (9.25)), a simple and efficient least-squares approach is used to estimate the set of influence vectors $\boldsymbol{\Omega} = \{\boldsymbol{\Omega}_i, \boldsymbol{\Omega}_{ij}, \boldsymbol{\Omega}_{ijk}, \dots, \boldsymbol{\Omega}_{123\dots q}\}$.The set of influence vector $\boldsymbol{\Omega}$ and additive fault $y_F(k)$ is estimated by minimizing the log-likelihood function $l(\boldsymbol{\gamma}|\boldsymbol{e})$ given by:

$$[\hat{\boldsymbol{\Omega}}, \hat{y}_F] = \arg\left\{\min_{\boldsymbol{\Omega}, y_F}\{l(\gamma|\boldsymbol{e}) = \|\boldsymbol{e}(k) - \boldsymbol{\Phi}(k)\boldsymbol{\varphi}(\Delta\gamma) - \boldsymbol{y}_F(k)\|^2\}\right\} \tag{9.58}$$

where $\hat{\boldsymbol{\Omega}}$ and $\hat{\boldsymbol{y}}_F$ are the estimates of the influence vector $\boldsymbol{\Omega}$ and the additive fault $\boldsymbol{y}_F(k)$. A computationally simple recursive approach is used to estimate the influence vectors by performing a number of parameter perturbed experiments instead of computing them all at once by solving a complex optimization problem (9.58).

9.12.1 Parameter-Perturbed Experiment

To identify the set of influence vectors $\boldsymbol{\Omega}$, and additive fault inputs $\boldsymbol{y}_F$, a series of experiments is conducted that involves perturbing the diagnostic parameters of each subsystem and additive fault with an input excitation that takes on different amplitudes and spectral content. If there are no product terms then the parameters need to be perturbed one at a time. If there is a double product $\gamma_i\gamma_j$, then two parameters γ_i and γ_j need to be perturbed. For a triple product, $\gamma_i\gamma_j\gamma_k$, three parameters γ_i, γ_j, and γ_k need to be perturbed, and so on. For each experiment, a data record, $\{\, e(k-i) \quad r(k-i)\,\}, i = 0, 1, 2, \dots\ N-1$, is collected:

- *Single fault*: $H_i : \Delta\gamma_i \neq 0,\ \Delta\gamma_j = 0\, j \neq i$

$$[\hat{\boldsymbol{\Omega}}_i, \hat{\boldsymbol{y}}_F] = \arg\left\{\min_{\boldsymbol{\Omega}_i y_F}\left\{\left\|\boldsymbol{e}(k) - \boldsymbol{\Phi}(k)\boldsymbol{\Omega}_i\Delta\gamma_i - \boldsymbol{y}_F(k)\right\|^2\right\}\right\} \tag{9.59}$$

- *Two simultaneous faults*: $H_{ij} : \Delta\gamma_i \neq 0$ *and* $\Delta\gamma_j \neq 0$ but $\Delta\gamma_k = 0 \;:\; k \neq i,j$

$$H_{ij} : [\hat{\boldsymbol{\Omega}}_{ij}, \hat{\boldsymbol{y}}_F] = \arg\left\{\min_{\boldsymbol{\Omega}_{ij} y_F}\left\{\left\|\boldsymbol{e}(k) - \boldsymbol{\Phi}(k)(\hat{\boldsymbol{\Omega}}_i\Delta\gamma_i + \hat{\boldsymbol{\Omega}}_j\Delta\gamma_j + \boldsymbol{\Omega}_{ij}\Delta\gamma_{ij} - \boldsymbol{y}_F(k)\right\|^2\right\}\right\} \tag{9.60}$$

- *Three simultaneous faults*: $H_{ijk} : \Delta\gamma_i \neq 0\,,\ \Delta\gamma_j \neq 0$ and $\Delta\gamma_k \neq 0$ but $\Delta\gamma_\ell = 0 : \ell \neq i, j, k$.

$$[\hat{\boldsymbol{\Omega}}_{ijk}, \hat{\boldsymbol{y}}_F] = \arg\left\{ \min_{\boldsymbol{\Omega}_{ijk}, \boldsymbol{y}_F} \left\{ \left\| \boldsymbol{e}(k) - \boldsymbol{\Phi}(k)\left(\sum_i \hat{\boldsymbol{\Omega}}_i \Delta\gamma_i + \sum_{i,j} \hat{\boldsymbol{\Omega}}_{ij} \Delta\gamma_{ij} + \boldsymbol{\Omega}_{ijk} \Delta\gamma_{ijk} \right) - \boldsymbol{y}_F(k) \right\|^2 \right\} \right\} \tag{9.61}$$

And so on until simultaneous faults in all: $H_{123.q} : \Delta\gamma_i \neq 0$ for all i are introduced:

$$\left[\hat{\boldsymbol{\Omega}}_{ijk}, \hat{\boldsymbol{y}}_F\right] =$$

$$\arg\left\{ \min_{\boldsymbol{\Omega}_{123.q} \boldsymbol{y}_F} \left\{ \left\| \boldsymbol{e}(k) - \boldsymbol{\Phi}(k)\left(\sum_i \hat{\boldsymbol{\Omega}}_i \Delta\gamma_i + \sum_i \hat{\boldsymbol{\Omega}}_{ij} \Delta\gamma_{ij} + \sum_i \hat{\boldsymbol{\Omega}}_{ijk} \Delta\gamma_{ijk} \ldots + \boldsymbol{\Omega}_{123.q} \Delta\gamma_{1.q} \right) - \boldsymbol{y}_F(k) \right\|^2 \right\} \right\} \tag{9.62}$$

9.12.2 *Least-Squares Estimates*

We will focus mainly on multiplicative fault assuming that $\boldsymbol{y}_F(k) = \boldsymbol{0}$. Closed form solutions for the estimates are given in the Appendix. The estimates of the influence vectors using Eqs. (9.59)–(9.62) are given below:

$$\hat{\boldsymbol{\Omega}}_i = (\boldsymbol{\Phi}^i(k)\Delta\gamma_i)^\dagger \boldsymbol{e}^i(k) \tag{9.63}$$

$$\hat{\boldsymbol{\Omega}}_{ij} = (\boldsymbol{\Phi}^{ij}(k)\Delta\gamma_{ij})^\dagger (\boldsymbol{e}^{ij}(k) - \boldsymbol{\Phi}^{ij}(k)(\hat{\boldsymbol{\Omega}}_i \Delta\gamma_i + \hat{\boldsymbol{\Omega}}_j \Delta\gamma_j)) \tag{9.64}$$

$$\hat{\boldsymbol{\Omega}}_{ijk} = (\boldsymbol{\Phi}^{ijk}(k)\Delta\gamma_{ijk})^\dagger \left(\boldsymbol{e}^{ijk}(k) - \boldsymbol{\Phi}^{ijk} \left(\sum_i \hat{\boldsymbol{\Omega}}_i \Delta\gamma_i + \sum_{i,j} \hat{\boldsymbol{\Omega}}_{ij} \Delta\gamma_{ij} \right) \right) \tag{9.65}$$

And so on until:

$$\hat{\boldsymbol{\Omega}}_{123.q} = (\boldsymbol{\Phi}^{123.q}(k)\Delta\gamma_{123.q})^\dagger \left(\boldsymbol{e}^{123.q}(k) - \boldsymbol{\Phi}^{123.q} \left(\sum_i \hat{\boldsymbol{\Omega}}_i \Delta\gamma_i \sum_{i,j} \hat{\boldsymbol{\Omega}}_{ij} \Delta\gamma_{ij} + \cdots + \hat{\boldsymbol{\Omega}}_{123..q} \Delta\gamma_{123...q} \right) \right) \tag{9.66}$$

Comments *Batch recursive least squares estimates of the set of influence vectors* $\boldsymbol{\Omega} = \left\{ \boldsymbol{\Omega}_i, \boldsymbol{\Omega}_{ij}, \boldsymbol{\Omega}_{ijk}, \ldots, \boldsymbol{\Omega}_{123\ldots q} \right\}$ *exist and are unique under the following conditions:*

- *The data matrices* $\boldsymbol{\Phi}^i(k)$, $\boldsymbol{\Phi}^{ij}(k)$, $\boldsymbol{\Phi}^{ijk}(k), \ldots,$ $\boldsymbol{\Phi}^{123.q}(k)$ *must all have full rank. This implies that the inputs are persistently exciting for each of the perturbation experiments.*
- *The experiments include all perturbations in the diagnostic parameters taken* one at a time, two at a time, three at a time *and so on until all parameter combinations are perturbed.*

9.13 Fault Isolation Scheme

A Bayesian multiple composite hypotheses testing scheme is employed. It is based on determining which of the diagnostic parameters $\gamma_i : i = 1, 2, 3, \ldots, q$ has varied. The widely used maximum likelihood method, which is efficient and unbiased, is employed herein to estimate the variation $\Delta\boldsymbol{\gamma}$. The ML estimate $\Delta\hat{\boldsymbol{\gamma}}$ of $\Delta\boldsymbol{\gamma}$ is obtained by minimizing the log-likelihood function $l(\boldsymbol{\gamma}|\boldsymbol{e})$ given by:

$$\Delta\hat{\gamma} = \arg\left\{\min_{\Delta\gamma} \|\boldsymbol{e}(k) - \boldsymbol{\Phi}(k)\boldsymbol{\varphi}(\Delta\gamma)\|^2\right\} \tag{9.67}$$

The fault indicator function $\boldsymbol{e}_f(k) = \boldsymbol{\Phi}(k)\boldsymbol{\varphi}(\Delta\gamma)$ is a nonlinear function of $\{\Delta\gamma_i\}$ as shown in Eqs. (9.14) and (9.24), and as result there is no closed-form solution to the minimization problem.

9.13.1 Sequential Fault Isolation Scheme

One approach is to solve the nonlinear equation sequentially. A criterion for the "best fit" between the residual $\boldsymbol{e}(k)$ and its estimate $\boldsymbol{\Phi}(k)\boldsymbol{\varphi}(\Delta\hat{\boldsymbol{\gamma}})$ is defined as follows:

$$\|(\boldsymbol{e}(k) - \boldsymbol{\Phi}(k)\boldsymbol{\varphi}(\Delta\hat{\gamma}))\|^2 \leq \varepsilon \tag{9.68}$$

where $\varepsilon > 0$ is a specified measure of the acceptable estimation error. Using the criterion of the best fit, the sequential approach is as follows:

- First, hypothesize single faults by assuming that only one of the diagnostic parameters has varied, $\Delta\gamma_i \neq 0$ and $\Delta\gamma_j = 0$ for all $j \neq i$, and the optimal cost (sum of the squares of the residual) for a single fault is obtained.
- If none of the estimates $\Delta\hat{\gamma}_i$ meets the best fit criterion, then hypothesize a double fault. Two diagnostic parameters have varied simultaneously, that is assume any one pair $\Delta\gamma_i\Delta\gamma_j \neq 0$ with the rest of the products $\Delta\gamma_k\Delta\gamma_\ell = 0$ for all k and ℓ, and the optimal cost for a double fault is obtained.
- If none of the estimated pairs $\Delta\hat{\gamma}_i\Delta\hat{\gamma}_j$ satisfy best fit criterion, hypothesize simultaneous variations in any triplet $\gamma_i\gamma_j\gamma_k$, and the optimal cost for a triple fault is obtained.
- If none of the triplets $\Delta\hat{\gamma}_i\Delta\hat{\gamma}_j\Delta\hat{\gamma}_k$ give the best fit, repeat the process until we reach a stage where the estimates give the best fit.

The smallest number of faults for which the optimal cost is acceptably small, as given in Eq. (9.68), is then chosen. The optimal costs under various hypotheses are listed below:

$\boldsymbol{H}_i$: The diagnostic parameter γ_i, has changed,

$$l_i^0 = \|\boldsymbol{e}(k) - \boldsymbol{\Phi}(k)\boldsymbol{\Omega}_i\Delta\hat{\gamma}_i\|^2 \tag{9.69}$$

$\boldsymbol{H}_{ij}$: The diagnostic parameters γ_i and γ_j have changed,

$$l_{ij}^0 = \|\boldsymbol{e}(k) - \boldsymbol{\Phi}(k)(\boldsymbol{\Omega}_i\Delta\hat{\gamma}_i + \boldsymbol{\Omega}_j\Delta\hat{\gamma}_j + \boldsymbol{\Omega}_{ij}\Delta\hat{\gamma}_{ij})\|^2 \tag{9.70}$$

$\boldsymbol{H}_{ijk}$: The diagnostic parameters γ_i, γ_j and γ_k have changed,

$$l_{ijk}^0 = \|\boldsymbol{e} - \boldsymbol{\Phi}(k)[\boldsymbol{\Omega}_i\Delta\hat{\gamma}_i + \boldsymbol{\Omega}_j\Delta\hat{\gamma}_j + \boldsymbol{\Omega}_k\Delta\hat{\gamma}_k + \boldsymbol{\Omega}_{ij}\Delta\hat{\gamma}_{ij} + \boldsymbol{\Omega}_{ik}\Delta\hat{\gamma}_{ik} + \boldsymbol{\Omega}_{jk}\Delta\hat{\gamma}_{jk} + \boldsymbol{\Omega}_{ijk}\Delta\hat{\gamma}_{ijk}]\|^2 \tag{9.71}$$

$\boldsymbol{H}_{123\ldots q}$: All the diagnostic parameters $\gamma_1, \gamma_2, \gamma_3 \ldots \gamma_q$ have changed:

$$l^0_{123..q} = \left\| \boldsymbol{e} - \boldsymbol{\Phi}(k) \left[\sum_i \boldsymbol{\Omega}_i \Delta\hat{\gamma}_i + \sum_{i,j} \boldsymbol{\Omega}_{ij} \Delta\hat{\gamma}_{ij} + \sum_{i,j,k} \boldsymbol{\Omega}_{ijk} \Delta\hat{\gamma}_{ijk} + \cdots + \boldsymbol{\Omega}_{123..q} \Delta\hat{\gamma}_{123..q} \right] \right\|^2 \quad (9.72)$$

9.13.2 *Isolation of the Fault*

The optimal costs under each of the above hypotheses satisfy the following inequality:

$$l^0_{1223.q} \leq \ldots \leq l^0_{ijk\ell} \leq l^0_{ijk} \leq l^0_{ij} \leq l^0_i \quad (9.73)$$

- If $l^0_i \leq \varepsilon$, then there is single fault in γ_i asserted.
- If $l^0_{ij} \leq \varepsilon$, then there are simultaneous faults in γ_i and γ_j asserted.
- If $l^0_{ijk} \leq \varepsilon$, then there are simultaneous faults in γ_i, γ_j and γ_k asserted.
- And so on.
- If $l^0_{123.q} \leq \varepsilon$, then there are simultaneous faults all $\gamma_i : i = 1, 2, 3, \ldots .q$ asserted.

We will focus herein on a single fault resulting from the variation of single diagnostic parameter with the rest of the diagnostic parameters unchanged.

9.14 Isolation of a Single Fault

There is a closed-form solution when there is a single fault and for multiple faults there are no closed-form solutions and the optimal solutions are computed iteratively.

Consider the single fault case where only one of the q parameters of $\{\gamma_i, i = 1, 2, \ldots q\}$ changes at a time, and the value remains constant during the execution of the detection and isolation logic. We assume that the poles and zeros of $G_i(z)$ are simple. If the poles are not simple, then the feature vector is no longer affine in the diagnostic parameters.

9.14.1 *Fault Discriminant Function*

The fault isolation problem is posed as a pattern classification problem in Figure 9.6. There are q classes generated by the variations in the diagnostic parameters $\gamma_i : i = 1, 2, 3, \ldots, q$. For the case of single fault, the expression for the residual model given by Eq. (9.25) after setting $y_F(k) = 0$ becomes:

$$\boldsymbol{e}(k) = \boldsymbol{\Phi}(k)\boldsymbol{\Omega}_i \Delta\gamma_i + \boldsymbol{e}_0(k) \quad (9.74)$$

The objective is to design a fault discrimination function, which is a function of the residual $\boldsymbol{e}(k)$, to assert to which regions $\mathfrak{R}_i : i = 1, 2, 3, \ldots, q$ the residual is most likely to belong. A discriminant function is derived from the optimal cost function l^0_i for the hypothesis of a single fault $\boldsymbol{H}_i$ and the associated residual given by Eq. (9.69). As the residual vector is affine in $\Delta\gamma_i$, the optimal solution says that the distance between the vector $\boldsymbol{e}(k)$ and the hyper plane generated by the columns of the residual influence matrix $\boldsymbol{\Phi}(k)\boldsymbol{\Omega}_i \Delta\gamma_i$ must be a minimum. Since the *size* of the fault $\Delta\gamma_i(k)$, is unknown, a composite hypothesis testing scheme is used in which we substitute the unknown $\Delta\gamma_i(k)$ by its least-squares estimate. The estimate $\Delta\hat{\gamma}_i$ is:

$$\Delta\hat{\gamma}_i(k) = (\boldsymbol{\Phi}(k)\boldsymbol{\Omega}_i\Delta\gamma_i))^{\dagger}\boldsymbol{e}(k) \tag{9.75}$$

Substituting for $\Delta\hat{\gamma}_i$ gives

$$l_i^0 = \left\|\boldsymbol{e}(k) - \boldsymbol{\Phi}(k)\boldsymbol{\Omega}_i\Delta\gamma_i\right\|^2 = \|\boldsymbol{e}(k)\|^2\left(1 - \cos^2\varphi_i(k)\right) \tag{9.76}$$

where

$$\cos^2\varphi_i(k) = \left(\frac{\langle\boldsymbol{e}(k), \boldsymbol{\Phi}(k)\boldsymbol{\Omega}_i\rangle}{\|\boldsymbol{e}(k)\|\ \|\boldsymbol{\Phi}(k)\boldsymbol{\Omega}_i\|}\right)^2 \tag{9.77}$$

And $\langle\boldsymbol{x},\boldsymbol{y}\rangle$ denotes inner-product of $\boldsymbol{x}$ and $\boldsymbol{y}$.

If a fault has been detected, then hypothesis, H_i signifying a change in $\gamma_i(k)$, is asserted true if i is the value of j where $\{\cos^2\varphi_j(k), j = 1, 2, 3, \dots, q\}$ is maximum:

$$i = \arg\left(\max_j\left\{\cos^2\varphi_j(k)\right\}\right) \tag{9.78}$$

This says that the hypothesis of a fault in γ_i is correct when the angle between the residual vector $\boldsymbol{e}(k)$ and the vector, $\boldsymbol{\Phi}(k)\boldsymbol{\Omega}_i$ is minimum. Alternatively we may state that hypothesis H_i is asserted (that is, only one diagnostic parameter γ_i has varied), if the residual and its estimate are maximally aligned. The function $\cos^2\varphi_j(k)$, which is derived by maximizing the log-likelihood function $l(\boldsymbol{\gamma}|\boldsymbol{e})$ for the case when there is single fault, is the *discriminant function* for fault isolation.

9.14.2 Performance of Fault Isolation Scheme

In the Appendix, the performance of $\cos^2\varphi_j(k)$ is analyzed. Substituting for $\boldsymbol{e}(k)$ using Eq. (9.77) and exploiting the fact that $\boldsymbol{\Phi}(k)\boldsymbol{\Omega}_i$ is uncorrelated with $\boldsymbol{e}_0(k)$, the expression for $\cos^2\varphi_j(k)$ becomes:

$$\cos^2\varphi_j(k) = \left(\frac{\langle\boldsymbol{\Phi}(k)\boldsymbol{\Omega}_i, \boldsymbol{\Phi}(k)\boldsymbol{\Omega}_j\rangle}{\|\boldsymbol{\Phi}(k)\boldsymbol{\Omega}_i\|\|\boldsymbol{\Phi}(k)\boldsymbol{\Omega}_j\|}\right)^2 \frac{1}{\left(1 + \sigma_{0i}^2\right)} \tag{9.79}$$

where $\sigma_{0i}^2 = \dfrac{\sigma_0^2}{\left\|\boldsymbol{\Phi}(k)\boldsymbol{\Omega}_i\right\|(\Delta\gamma_i)^2}$.

Using the Cauchy–Schwarz inequality and assuming that the data matrix $\boldsymbol{\Phi}(k)$ has maximal rank, it is shown in the Appendix that the cosine-squared of the angle between the vectors $\boldsymbol{\Phi}(k)\boldsymbol{\Omega}_i$ and $\boldsymbol{\Phi}(k)\boldsymbol{\Omega}_j$, denoted $\cos^2\varphi_{ij}(k)$, satisfies:

$$\cos^2\varphi_{ij}(k) = \left(\frac{|\langle\boldsymbol{\Phi}(k)\boldsymbol{\Omega}_i, \boldsymbol{\Phi}(k)\boldsymbol{\Omega}_j\rangle|}{\|\boldsymbol{\Phi}(k)\boldsymbol{\Omega}_i\|\|\boldsymbol{\Phi}(k)\boldsymbol{\Omega}_j\|}\right)^2 \le 1 \text{ for all } i \text{ and } j \tag{9.80}$$

And with equality if and only if the influence vectors are aligned, that is,

$$\boldsymbol{\Omega}_i = c\,\boldsymbol{\Omega}_j \tag{9.81}$$

where c is a constant.

We deduce from Eqs. (9.79) and (9.80) that $\cos^2 \varphi_j(k)$ satisfies:

$$\cos^2 \varphi_j(k) \begin{cases} \leq \dfrac{1}{\left(1+\sigma_{0i}^2\right)} & \text{if } j \neq i \\ = \dfrac{1}{\left(1+\sigma_{0i}^2\right)} & \text{if and only if } i = j \end{cases} \tag{9.82}$$

9.14.3 Performance Issues and Guidelines

9.14.3.1 Issues

Ideal No Noise Case

In the ideal no noise case when γ_i varies, and $|\Delta\gamma_i| = \varepsilon > 0$ where ε is arbitrarily small, then $\cos^2 \varphi_i(k) = 1$ will be maximum and the estimate $\Delta\hat{\gamma}_i$ of the fault size $\Delta\gamma_i$ will be accurate, $\Delta\hat{\gamma}_i = \Delta\gamma_i$.

Noisy Case

Due to the presence of noise and disturbances, the performance of the fault isolation scheme may be degraded, especially in the low signal to noise ratio case. The scenarios include the following:

Inaccurate fault size estimation: Due to the errors in estimating the influence vectors, and the noise and disturbances affecting the residual data, $\max_j\{\cos^2 \varphi_j(k)\} < 1$ and the fault size estimate may not be accurate $\Delta\hat{\gamma}_i \neq \Delta\gamma_i$.

Non-unique solution: Although only one device is faulty, more than one device may be asserted to be faulty including the correct one as the fault isolation criterion may pinpoint more than device, that is, the solution of the maximization problem $\arg\left\{\max_j \left\{\cos^2 \varphi_j(k)\right\}\right\}$ is not unique.

Incorrect fault isolation: A healthy subsystem or a device may be asserted as faulty. That is, if γ_i has varied, the isolation criteria may assert that $\gamma_\ell : \ell \neq i$ is faulty: $\ell = \arg\left(\max_j\{\cos^2 \varphi_j(k)\}\right) : \ell \neq i$.

9.14.3.2 Guidelines

From Eq. (9.79), one may interpret σ_{0i}^2 as the *noise to signal ratio*. In the presence of model uncertainty and measurement noise, we deduce the following.

The change in the diagnostic parameter must exceed the noise to signal ratio,

$$(\Delta\gamma_i)^2 \gg \frac{\sigma_0^2}{\|\boldsymbol{\Phi}(k)\Omega_i\|^2} \tag{9.83}$$

The data matrix $\boldsymbol{\Phi}(k)$ must have full rank and its norm must be large enough to ensure a small noise to signal ratio, that is, $\|\boldsymbol{\Phi}(k)\Omega_i\|^2 \gg \sigma_0^2$. It is well known [8] that $\sigma_{\min}(A) \leq \|A\| \leq \sigma_{\max}(A)$, where $\sigma_{\min}$ and $\sigma_{\max}$ denote the minimum and maximum singular values of A. Therefore,

$$\sigma_{\min}(\boldsymbol{\Phi}(k))\,\|\boldsymbol{\Omega}_i\| \leq \|\boldsymbol{\Phi}(k)\boldsymbol{\Omega}_i\| \leq \sigma_{\max}(\boldsymbol{\Phi}(k))\,\|\boldsymbol{\Omega}_i\| \tag{9.84}$$

Hence, to ensure that Eq. (9.83) holds, the data matrix must have a smaller range of singular values and a large minimum singular value, that is,

$$\frac{\sigma_{\min}\{\mathbf{\Phi}(k)\}}{\sigma_{\max}\{\mathbf{\Phi}(k)\}} \approx 1 \text{ and } \sigma_{\min}\{\mathbf{\Phi}(k)\} \gg 0 \tag{9.85}$$

9.14.3.3 Isolability

Isolability is a measure of the ability to distinguish and isolate a fault in γ_j, from a fault in $\gamma_i, i \neq j$. Isolability is improved for a larger difference between φ_i and $\left\{\varphi_j, j \neq i\right\}$. It can deduce form (9.80) and (9.81) that the diagnostic parameters, $\{\gamma_i, i = 1, 2, ...q\}$, are isolable if and only if $\mathbf{\Phi}(k)$ has maximal rank and the influence vectors $\{\mathbf{\Omega}_i, i = 1, 2, \ldots, q\}$ are linearly independent in the following sense:

$$\cos^2 \varphi_{ij}(k) = \left(\frac{|\langle \mathbf{\Phi}(k)\mathbf{\Omega}_i, \mathbf{\Phi}(k)\mathbf{\Omega}_j\rangle|}{\|\mathbf{\Phi}(k)\mathbf{\Omega}_i\|\|\mathbf{\Phi}(k)\mathbf{\Omega}_j\|}\right)^2 = 1 \text{ if and only if } i = j \tag{9.86}$$

The fault isolation is essentially a pattern classification problem. There are q classes generated by the variations in the diagnostic parameters γ_i. The function $\cos^2 \varphi_j(k)$, which is derived by maximizing the log-likelihood function $l(\boldsymbol{\gamma}|\boldsymbol{e})$ for the case when there is single fault, is chosen as the discriminant function.

9.15 Emulators for Offline Identification

The estimation of the influence vector is based on performing a number of parameter perturbed experiments. In practice, the diagnostic parameters of the subsystems may not be accessible except, for example, subsystems such as controllers. To meet the requirement of the accessibility of the subsystem parameters, an emulator is connected in cascade with the system model at either the input or output or both during the identification stage to mimic the fault behavior of the associated subsystem. The emulator parameter plays the role of the diagnostic parameters. During the identification, the diagnostic parameters of the subsystem are set equal to their nominal (fault-free) values and the emulator parameters are varied instead. The behavior of the nominal fault-free system when the emulator parameters are perturbed (during the identification phase) will emulate the behavior of the system when its diagnostic parameters vary (during the operational phase).

A fault may arise either as a result of variations in the phase and magnitude of the transfer function of a subsystem, in which case it termed a non-parametric fault, or as a result of variations in the parameters characterizing the subsystem, in which case it is then termed a parametric fault.

Consider an interconnected system formed of a number of subsystems denoted by their transfer functions $\left\{G_i{:}i = 1, 2, \ldots\ m\right\}$. Each subsystem may represent a physical entity such as a sensor, actuator, controller, or any other system component that is subject to parametric faults, and may be affected by noise or disturbance inputs as illustrated in Figure 9.1. In an ideal case, the parameters which characterize the behavior of a subsystem may be identified online, and the identified parameters may then be used to monitor its faults. We will consider the case of a physical system where it is not possible to identify online each of the subsystem models from the available input–output data. In other words, this case will exemplify those where it is difficult to determine how the parameters of the subsystems enter the identified model.

Consider an ith subsystem whose transfer function may change due to some degradation or the occurrence of a fault or may vary from one operating regime to another. Further, the model of the subsystem may differ from the true model due to un-modeled dynamics or a host of other unspecified effects [13]. One may represent such a variation from the nominal transfer function by the following multiplicative perturbation model:

$$G_i(z) = G_{0i}(z)(1 + \Delta G_i(z)) \tag{9.87}$$

where $G_{i0}(z)$ is the nominal transfer function and $\Delta G_i(z)$ is the perturbation. Except for a physical system such as a controller, the parameters of $G_i(z)$ are assumed to be not accessible. In the case of a controller, the controller also plays the role of an emulator as its parameters are accessible. A fault in a subsystem may be diagnosed by monitoring either (i) a variation in the phase and magnitude of the transfer function or (ii) variations in the parameters characterizing the subsystem. The former gives a macroscopic picture of a fault, while the latter gives microscopic details of the faults. For example, a fault in an electric motor may be detected from the macroscopic picture of its frequency response or from the microscopic details of the variations in its diagnostic parameters, such as its inductance, resistance, and inertia. An emulator can emulate these two types of fault manifestation in a subsystem (the gain and the phase variations, or variations in its parameters) to which it is connected in cascade. Consider the ith perturbed subsystem $G_i(z)$ given by Eq. (9.87). The multiplicative type perturbation $(1 + \Delta G_i(z))$ in the subsystem is simulated by an emulator $E_i(z)$ connected in cascade with the subsystem at an input or at an output:

$$G_i(z) = G_{0i}(z)(1 + \Delta G_i(z)) = G_{0i}(z)E_i(z)$$

9.15.1 Examples of Emulators

During the identification phase, the emulator parameters play the role of the diagnostic parameters of the subsystem. For convenience, the emulator parameter is denoted by the same symbol as the diagnostic parameter of the associated subsystem system.

The first and high-order filters cover special cases including the all-pass and FIR filters. The general form of the emulators is given below;

$$E_i(z) = \begin{cases} \gamma_1 & \text{gain} \\ \dfrac{\gamma_{0i} + \gamma_{1i}z^{-1}}{1 + \gamma_{2i}z^{-1}} & \text{first-order filter} \\ \prod_j \dfrac{\gamma_{0ij} + \gamma_{1ij}z^{-1}}{1 + \gamma_{2ij}z^{-1}} & \text{high-order filter} \end{cases} \tag{9.88}$$

9.15.2 Emulators for Multiple Input-Multiple-Output System

The emulators for a MIMO system are shown in Figure 9.7 below, where u_i and y_i are respectively the measured input and the output, G_i, k_{Ai}, and k_{si} are respectively the subsystem, the actuator, and the sensor. The emulator $E_i(z) = \dfrac{\gamma_{0i} + \gamma_{1i}z^{-1}}{1 + \gamma_{2i}z^{-1}}$ for the subsystem $G_i(z)$ is driven by the input u_i. Alternatively, one may also include the emulator at the measured output y_i. The constant gain emulators γ_i and γ_j are connected in cascade with the sensor and actuators. The emulators and their associated devices (subsystem, sensor, or actuator), which are connected in cascade, are shaded in the same color to easily locate the emulator-device pairs.

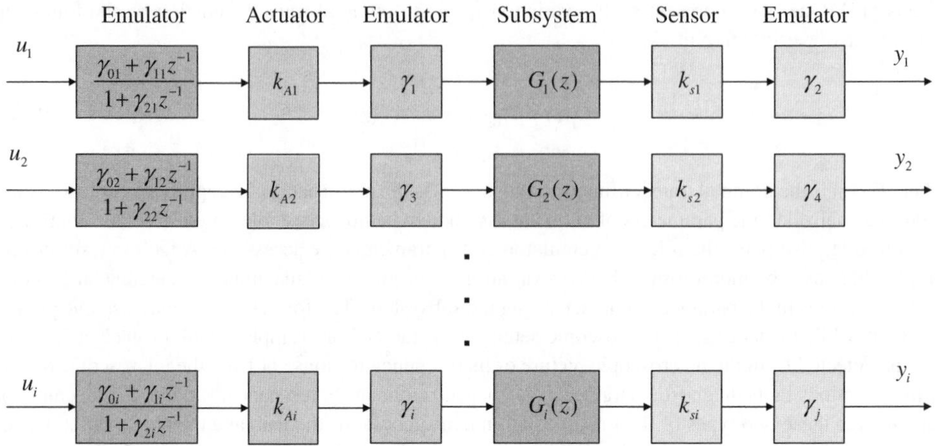

Figure 9.7 Emulators for a MIMO system

9.15.3 *Role of an Emulator*

The subsystem model during the operational phase (9.87) may be expressed in terms of the emulator during the operational phase as:

$$G_i(z) = G_{0i}(z)(1 + \Delta G_i(z)) = G_{0i}(z)E_i(z) \tag{9.89}$$

The emulators are connected only during the system identification phase to mimic variations in the parameters of the system including the subsystems, sensors, and actuators. The nominal values of the emulators are set such that the emulator transfer function is preferably unity (or "close" to unity), by choosing, for example, the nominal $E_i(z) = 1$. During the identification phase the diagnostic parameters are varied one at a time by setting the rest of the diagnostic parameters at their nominal values. The nominal values of emulators during the offline identification and online monitoring stages are chosen as follows:

- For inducing a fault in the subsystem G_i, its diagnostic parameters γ_{0i}, γ_{1i}, or γ_{2i} of the associated emulator $E_i(z)$ are varied with the rest of the emulators set at unity, $\gamma_i = \gamma_i^0 = 1$ for all i and $E_j(z) = 1$ for all $j \neq i$.
- For inducing a fault in the sensor k_{si}, the diagnostic parameter γ_i of the associated emulator is varied with rest of the emulators set at unity: $\gamma_j = \gamma_j^0 = 1{:}j \neq i$, and $E_i(z) = 1$ for all i.
- For inducing a fault in the actuator k_{Ai}, the diagnostic parameter γ_j of the associated emulator is varied with rest of the emulators set at unity: $\gamma_i = \gamma_i^0 = 1$ for $i \neq j$, and the $E_i(z) = 1$ for all i.

In practice, however, one may choose the nominal values of the diagnostic parameters of the emulators such that their effect on the static and the dynamical behaviors of the system is negligible, and the structure (order of the system) of the system is unchanged during both the identification and the operational stages. For example, choose the nominal dynamical transfer function of the emulator $E_i(z) \approx 1$. With this choice, the structure of the system (order of the system) will be unchanged.

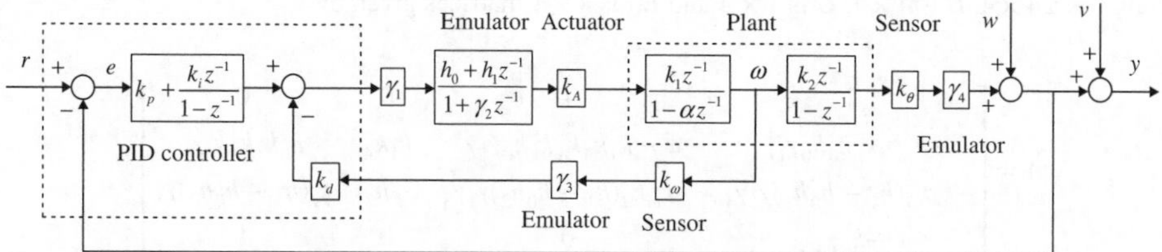

Figure 9.8 Position control system with fault emulators

9.15.4 *Criteria for Selection*

The criteria for a selection of an emulator transfer function $E_i(z)$, which is designed to emulate a fault in the subsystem $G_i(z)$, includes the following:

- The diagnostic parameters of the emulator must be capable of simulating faults in the that may likely occur in the subsystem during the operational phase of the system.
- The influence vectors of the emulator are maximally aligned with those of the subsystem so that a fault in a subsystem is isolated correctly.

The structure of the emulator may be selected from the general form given by Eq. (9.88). The diagnostic parameters of the emulator induce a fault in the subsystem, and its influence vector is used to isolate which of the subsystems has varied. For example, the influence vector $\mathbf{\Omega}_i = \dfrac{\delta\theta}{\delta\gamma_i}$ associated with a diagnostic parameter γ_i of the emulator $E_i(z)$ should be maximally aligned with the influence vector $\mathbf{\Omega}_{\alpha_i} = \dfrac{\delta\theta}{\delta\alpha_i}$ of the parameter α_i of the subsystem $G_i(z)$. The diagnostic parameter γ_i emulates variations of the parameter α_i of the subsystem, and $\mathbf{\Omega}_i$ is used to indicate that the subsystem $G_i(z)$ is faulty. In general the diagnostic parameters of the emulator $E_i(z)$ induce variations in the phase and gain of the subsystem $G_i(z)$ and these variations in $G_i(z)$ are captured only by the influence vectors associated to the emulator $E_i(z)$ (and not by those of the emulators $E_j(z) : j \neq i$).

The role of the emulators in inducing a fault during the offline identification stage, and the influence vectors in fault isolation during the operating regime, is explained using an illustrated example in the next section.

9.16 Illustrative Example

Consider the position control example (9.16). In order to develop identification of the nominal feature vector $\boldsymbol{\theta}^0$ and the influence vector $\mathbf{\Omega}$, and subsequently a fault diagnosis scheme, emulators $E_i, i = 1,2,3,4$ are included to simulate and isolate faults in the actuator, the plant, the velocity sensor, and the position sensor, as shown Figure 9.8.

9.16.1 *Mathematical Model*

The state-space model is:

$$\begin{aligned} \boldsymbol{x}(k+1) &= \boldsymbol{A}\boldsymbol{x}(k) + \boldsymbol{B}r(k) + \boldsymbol{E}_w w(k) \\ y(k) &= \boldsymbol{C}\boldsymbol{x}(k) + v(k) \end{aligned} \tag{9.90}$$

where $\boldsymbol{A}$ is a 4×4, $\boldsymbol{B}$ is 4×1, $\boldsymbol{C}$ is 1×4 and $\boldsymbol{E}_w$ is 4×1 matrices given by

$$\boldsymbol{A} = \begin{bmatrix} 1 & 1 & 0 & 0 \\ -k_1 k_A k_p k_\theta h_0 \gamma_1 \gamma_4 & \alpha - k_1 k_A k_\omega k_d h_0 \gamma_1 \gamma_3 & k_1 k_A & h_0 k_1 k_A k_I \gamma_1 \\ -k_p k_\theta (h_1 - h_0 h_2)\gamma_1 \gamma_4 & -k_\omega k_d (h_1 - h_0 h_2)\gamma_1 \gamma_3 & -h_2 & k_I (h_1 - h_0 h_2)\gamma_1 \\ -k_\theta \gamma_4 & 0 & 0 & 1 \end{bmatrix};$$

$$\boldsymbol{B} = \begin{bmatrix} 0 \\ \gamma_1 h_0 k_1 k_A k_p \\ (h_1 - h_0 \gamma_2)\gamma_1 k_p \\ 1 \end{bmatrix} \boldsymbol{E}_w = \begin{bmatrix} 0 \\ 0 \\ 0 \\ 1 \end{bmatrix}; \ \boldsymbol{C} = \begin{bmatrix} k_\theta \gamma_4 & 0 & 0 & 0 \end{bmatrix}$$

9.16.2 *Selection of Emulators*

The actuator k_A, the velocity sensor k_ω, and the position sensor k_θ were modeled as constant gain transfer functions. Since the subsystem of the plant $\dfrac{k_1 z^{-1}}{1 - \alpha z^{-1}}$ is subject to a fault as result of variation of the parameter α, and since the plant parameter α is not accessible, an emulator, denoted $E_2(z)$, is connected at the output of the PID controller to mimic variations in the plant parameter α. The emulator is chosen to be a first-order filter, and the structure of the filter is determined such that the influence vector associated with the emulator parameter is maximally aligned with that of the plant parameter α. The emulator for the plant is given by:

$$E_2(z) = \frac{1 + \gamma_2^0 z^{-1}}{1 + \gamma_2 z^{-1}} \tag{9.91}$$

The emulators E_1, E_3, and E_4 to induce faults respectively in the actuator k_A, the velocity sensor k_ω, and the position sensor k_θ are chosen to be static constant gain transfer functions given by:

$$E_1 = \gamma_1 \tag{9.92}$$

$$E_2 = \gamma_2 \tag{9.93}$$

$$E_3 = \gamma_3 \tag{9.94}$$

All the emulators are connected in cascade with respective devices. The nominal diagnostic parameters are chosen such that during the operating regime they have negligible effect on the static and the dynamic behavior. The nominal parameters of the emulators E_1, E_3, and E_4 were chosen to be unity, $\gamma_i^0 = 1 {:} i = 1, 3, 4$. In the case of the plant emulator, the nominal diagnostic parameters of $E_2(z)$ were chosen to be approximately unity. They were not set equal to unity merely to simplify the development of the fault simulation and isolation software, as the order of the overall system remains unchanged (order is four) during the identification and the operational phase.

9.16.3 *Transfer Function Model*

$$G(z) = \frac{b_2 z^{-2} + b_3 z^{-3} + b_4 z^{-4}}{1 + a_1 z^{-1} + a_2 z^{-2} + a_3 z^{-3} + a_4 z^{-4}} \tag{9.95}$$

Let $p_1 = k_1 k_d k_\omega k_A$; $p_2 = k_1 k_A k_p k_\theta$; $p_3 = k_1 k_A k_\theta (k_I - k_p)$

$$\theta = \begin{bmatrix} a_1 \\ a_2 \\ a_3 \\ a_4 \\ b_2 \\ b_3 \\ b_4 \end{bmatrix} = \begin{bmatrix} p_1 h_0 \gamma_1 \gamma_3 + h_2 - \alpha - 2 \\ p_1 h_1 \gamma_1 \gamma_3 - \alpha h_2 - 2(p_1 h_0 \gamma_1 \gamma_3 + h_2 - \alpha) + 1 + p_2 h_0 \gamma_1 \gamma_4 \\ -2(p_1 h_1 \gamma_1 \gamma_3 - h_2 \alpha) + p_1 h_0 \gamma_1 \gamma_3 + h_2 - \alpha + p_2 h_1 \gamma_1 \gamma_4 + p_3 h_0 \gamma_1 \gamma_4 \\ p_1 h_1 \gamma_1 \gamma_3 - \alpha h_2 + p_3 h_1 \gamma_1 \gamma_4 \\ p_2 h_0 \gamma_1 \gamma_4 \\ p_2 h_1 \gamma_1 \gamma_4 + p_3 h_0 \gamma_1 \gamma_4 \\ p_3 h_1 \gamma_1 \gamma_4 \end{bmatrix} \tag{9.96}$$

The influence vectors are given by

$$\mathbf{\Omega}_1 = \begin{bmatrix} p_1 h_0 \gamma_3 \\ p_1 h_1 \gamma_3 - 2 p_1 h_0 \gamma_3 + p_2 h_1 \gamma_4 \\ -2 p_1 h_1 \gamma_3 + p_1 h_0 \gamma_3 + p_2 h_1 \gamma_4 + p_3 h_0 \gamma_4 \\ p_1 h_1 \gamma_3 + p_3 h_1 \gamma_4 \\ p_2 h_0 \gamma_4 \\ p_2 h_1 \gamma_4 + p_3 h_0 \gamma_4 \\ p_3 h_1 \gamma_4 \end{bmatrix}; \quad \mathbf{\Omega}_2 = \begin{bmatrix} 1 \\ -\alpha - 2 \\ 2\alpha + 1 \\ -\alpha \\ 0 \\ 0 \\ 0 \end{bmatrix}; \quad \mathbf{\Omega}_3 = \begin{bmatrix} p_1 h_0 \gamma_1 \\ \gamma_1 (p_1 h_1 \gamma_1 - 2 p_1 h_0) \\ -\gamma_1 (2 p_1 h_1 \gamma_1 - p_1 h_0) \\ p_1 h_1 \gamma_1 \\ 0 \\ 0 \\ 0 \end{bmatrix};$$

$$\mathbf{\Omega}_4 = \begin{bmatrix} 0 \\ p_2 h_0 \gamma_1 \\ \gamma_1 (p_2 h_1 \gamma_1 + p_3 h_0) \\ p_3 h_1 \gamma_1 \\ p_2 h_0 \gamma_1 \\ \gamma_1 (p_2 h_1 \gamma_1 + p_3 h_0) \\ p_3 h_1 \gamma_1 \end{bmatrix}, \quad \mathbf{\Omega}_\alpha = \frac{\delta \theta}{\delta \alpha} = \begin{bmatrix} -1 \\ -\gamma_2 + 2 \\ 2\gamma_2 - 1 \\ -\gamma_2 \\ 0 \\ 0 \\ 0 \end{bmatrix} \quad \mathbf{\Omega}_{13} = \begin{bmatrix} p_1 h_0 \\ p_1 h_1 - 2 p_1 h_0 \\ -2 p_1 h_1 + p_1 h_0 \\ p_1 h_1 \\ 0 \\ 0 \\ 0 \end{bmatrix}; \quad \mathbf{\Omega}_{14} = \begin{bmatrix} 0 \\ p_2 h_0 \\ p_2 h_1 + p_3 h_0 \\ p_3 h_1 \\ p_2 h_0 \\ p_2 h_1 + p_3 h_0 \\ p_3 h_1 \end{bmatrix};$$

$\mathbf{\Omega}_i$ is a function of $\gamma_j : j \neq i$, and $\mathbf{\Omega}_\alpha$ is a function of γ_2 whereas $\mathbf{\Omega}_{ij}$ is not a function of the diagnostic parameters.

9.16.4 Role of the Static Emulators

The role of the static emulators E_1, E_3, and E_4 in inducing and isolating a fault in the actuator, the velocity sensor, and the position sensor respectively can be deduced from the expression of the feature vector θ and $\Delta\theta$ given respectively by Eqs. (9.96) and (9.14), the influence vector $\mathbf{\Omega}_i$ derived from θ.

9.16.4.1 Identification Phase

In the identification phase, the actuator, the velocity sensor and the position sensor gains are at their nominal value (no-fault regime). In view of Eq. (9.89), the feature vector θ will be a function of $k_A^0 \gamma_1$, $k_\omega^0 \gamma_3$, and $k_\theta^0 \gamma_4$ during the identification phase, while during the operational phase θ will be function of

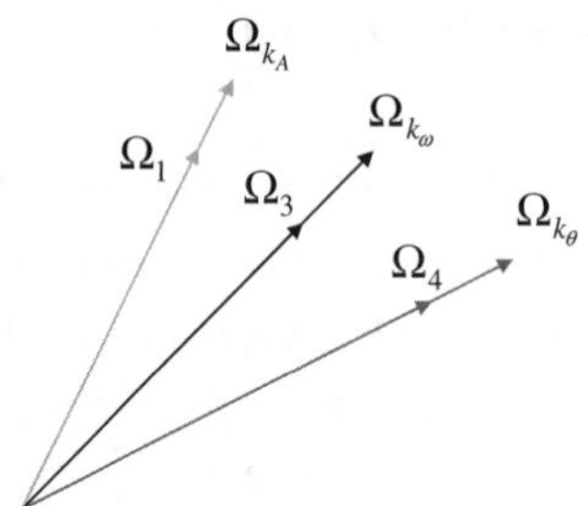

Figure 9.9 Perfect alignments of influence vectors

$k_A\gamma_1^0$, $k_\omega\gamma_3^0$, and $k_\theta\gamma_4^0$. Using Eq. (9.89), expressions relating the diagnostic parameters γ_1, γ_3, and γ_4, and associated static gains k_A, k_ω, and k_θ become:

$$\begin{aligned} k_A &= k_A^0(1+\Delta k_A) = k_A^0\gamma_1 = k_A^0(\gamma_1^0+\Delta\gamma_1) \\ k_\omega &= k_\omega^0(1+\Delta k_\omega) = k_\omega^0\gamma_3 = k_\omega^0(\gamma_3^0+\Delta\gamma_3) \\ k_\theta &= k_\theta^0(1+\Delta k_\theta) = k_\theta^0\gamma_4 = k_\theta^0(\gamma_4^0+\Delta\gamma_4) \end{aligned} \tag{9.97}$$

where k_A^0, k_ω^0, and k_θ^0 are the respectively the nominal fault-free values of the gains k_A, k_ω, and k_θ. Assuming $\gamma_1^0 = \gamma_2^0 = \gamma_3^0 = 1$, the variations Δk_A, Δk_ω, and Δk_θ that are likely to occur during the operational phase are simulated during the offline identification phase by inducing similar variations in the associated diagnostic parameters, $\Delta\gamma_1 = \Delta k_A$, $\Delta\gamma_3 = \Delta k_\omega$, and $\Delta\gamma_4 = \Delta k_\theta$. The variation in the feature vector $\Delta\boldsymbol{\theta}$ will be function of $\Delta\gamma_i$.

9.16.4.2 Fault Isolation Phase

In the operational phase the diagnostic parameters are set equal to one, $\gamma_1^0 = \gamma_2^0 = \gamma_3^0 = 1$ and the feature vector $\boldsymbol{\theta}$ will be a function of k_A, k_ω, and k_θ. Variations in k_A, k_ω, and k_θ will be captured during the fault isolation phase by the respective influence vectors $\boldsymbol{\Omega}_1$, $\boldsymbol{\Omega}_3$, and $\boldsymbol{\Omega}_4$.

Let $\boldsymbol{\Omega}_{k_A} = \dfrac{\delta\boldsymbol{\theta}}{\delta k_A}$, $\boldsymbol{\Omega}_{k_\omega} = \dfrac{\delta\boldsymbol{\theta}}{\delta k_\omega}$ and $\boldsymbol{\Omega}_{k_\theta} = \dfrac{\delta\boldsymbol{\theta}}{\delta k_\theta}$ be influence vectors associated with k_A, k_ω, and k_θ respectively. From the expression of the feature vector during the operation and the identification, it can be deduced that $\boldsymbol{\Omega}_1$, $\boldsymbol{\Omega}_3$, and $\boldsymbol{\Omega}_4$ will be collinear (perfectly aligned) with the influence vectors $\boldsymbol{\Omega}_{k_A}$, $\boldsymbol{\Omega}_{k_\omega}$, and $\boldsymbol{\Omega}_{k_\theta}$ respectively. That is, the angle between the influence vectors will be zero, or equivalently the cosine-squared of the angle will be unity:

$$\cos^2\phi_{1A} = \frac{\langle\boldsymbol{\Omega}_1,\boldsymbol{\Omega}_{k_A}\rangle}{\|\boldsymbol{\Omega}_1\|\|\boldsymbol{\Omega}_{k_A}\|} = 1; \cos^2\phi_{3\omega} = \frac{\langle\boldsymbol{\Omega}_2,\boldsymbol{\Omega}_{k_\omega}\rangle}{\|\boldsymbol{\Omega}_1\|\|\boldsymbol{\Omega}_{k_\omega}\|} = 1; \cos^2\phi_{1\theta} = \frac{\langle\boldsymbol{\Omega}_1,\boldsymbol{\Omega}_{k_\theta}\rangle}{\|\boldsymbol{\Omega}_1\|\|\boldsymbol{\Omega}_{k_\theta}\|} = 1 \tag{9.98}$$

where $\cos^2\phi_{1A}$, $\cos^2\phi_{1\omega}$, and $\cos^2\phi_{1\theta}$ are respectively the cosine-squared of the angle between the influence vectors $\boldsymbol{\Omega}_1$ and $\boldsymbol{\Omega}_{k_A}$, $\boldsymbol{\Omega}_3$ and $\boldsymbol{\Omega}_{k_\omega}$, and $\boldsymbol{\Omega}_4$ and $\boldsymbol{\Omega}_{k_\theta}$. Figure 9.9 shows the perfect alignments of the respective influence vectors.

9.16.5 Role of the Dynamic Emulator

The role of the dynamic emulator is not as straightforward as those of the static emulators as the model of the plant may not represent the true model, and the chosen emulator may not be able to mimic all the

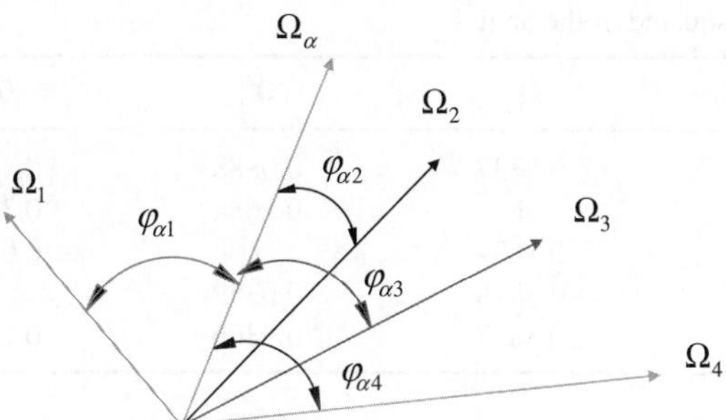

Figure 9.10 Maximal alignment

perturbations that occur in the true plant model. As a result we may be able to ensure that the influence victors of the dynamic emulators are collinear with those of the associated subsystem, as is the case with static emulators (9.98). We may, however, obtain maximal alignment in the sense that the angle between the influence vector of the emulator and the associated subsystem is minimum or the cosine-squared of the angle is maximum.

9.16.5.1 Identification Phase

During identification, variations in the diagnostic parameter γ_2 capture variations in the gain and the phase of the plant, which may occur as a result of variations in the plant parameter α during the operational phase of the system.

9.16.5.2 Fault Isolation Phase

The influence vector $\mathbf{\Omega}_2$ associated with diagnostic parameter γ_2 is capable of isolating only the plant when its parameters vary if the structure of the emulator is chosen such that the influence vector $\mathbf{\Omega}_2$ is maximally aligned with the influence vector $\mathbf{\Omega}_\alpha$ of the plant. That is, the angle between the two vectors $\mathbf{\Omega}_2$ and $\mathbf{\Omega}_\alpha$ is minimum compared to any other pair of influence vectors, $\mathbf{\Omega}_j$ and $\mathbf{\Omega}_\alpha$. Equivalently, this maximal alignment requirement may be expressed in terms of the cosine-squared of the angles as given below

$$\cos^2 \phi_{\alpha 2} = \max_j \left\{ \cos^2 \phi_{\alpha j} \right\} \tag{9.99}$$

where $\cos \phi_{\alpha j} = \dfrac{\langle \mathbf{\Omega}_\alpha, \mathbf{\Omega}_j \rangle}{\|\mathbf{\Omega}_\alpha\| \, \|\mathbf{\Omega}_\alpha\|}$ is cosine of the angle between $\mathbf{\Omega}_\alpha$ and $\mathbf{\Omega}_j$. Figure 9.10 shows the maximal alignment of $\mathbf{\Omega}_2$ with $\mathbf{\Omega}_\alpha$.

9.16.5.3 Simulation Results

The cosine-squared of the angle between the pairs of influences vectors $\mathbf{\Omega}_1$, $\mathbf{\Omega}_2$, $\mathbf{\Omega}_3$, $\mathbf{\Omega}_4$, $\mathbf{\Omega}_\alpha$ are given in the Table 9.1 below:

From the Table 9.1, it can be deduced that the influence vectors $\mathbf{\Omega}_1$, $\mathbf{\Omega}_2$, $\mathbf{\Omega}_3$, $\mathbf{\Omega}_4$, $\mathbf{\Omega}_\alpha$ are linearly independent, as $\cos^2 \phi_{ij} \neq 0$ and $\cos^2 \phi_{\alpha j} \neq 0$ for all i and j. From column 5 and row 2 or (column 2 and row 5) the cosine-squared of the angle between the influence vectors $\mathbf{\Omega}_\alpha$ and $\mathbf{\Omega}_2$ is maximum compared to the rest of the pairs, $\cos^2 \phi_{\alpha 2} = 0.8817 > \cos^2 \phi_{\alpha j} : j \neq 2$.

Table 9.1 The cosine-squared of the angles

	Ω_1	Ω_2	Ω_3	Ω_4	Ω_α
Ω_1	1	0.5337	0.0888	0.4072	0.6313
Ω_2	0.5337	1	0.1634	0.3818	0.8817
Ω_3	0.0888	0.1634	1	0.0359	0.3366
Ω_4	0.4072	0.3818	0.0359	1	0.2353
Ω_α	0.6313	0.8817	0.3366	0.2353	1

9.17 Overview of Fault Diagnosis Scheme

An overview of the fault diagnosis is given so as tie in all the issues involved in this complex and challenging task. Many of the equations are repeated here to avoid flipping pages to refer them. The steps involved in fault diagnosis are shown in Figure 9.11.

Step 1: System model: Model of the system subject to fault, and the output is affected disturbance and measurement noise (9.7) and (9.8).

Step 2: Generation of residual using Kalman filter: The Kalman filter is designed for the identified model $(\boldsymbol{A}_0, \boldsymbol{B}_0, \boldsymbol{C}_0)$ given by Eq. (9.6).

Step 3: *Diagnostic model* given by Eq. (9.25).

Step 4: *Fault detection*: Bayes decision strategy given by Eq. (9.48).

Step 5: *Maximum likelihood Estimation of influence vectors and the additive fault* by solving the optimization problem (9.58).

Step 6: Fault isolation scheme: Which of the diagnostic parameters have varied is obtained by solving Eq. (9.67).

9.18 Evaluation on a Simulated Example

The fault diagnosis scheme is evaluated on the position control example given by Eq. (9.90) in the section above. Faults in the actuator, the plant, the velocity sensor, and the position sensor are simulated by varying respectively the diagnostic parameters γ_1, γ_2, γ_3, and γ_4 of the associated emulators. The emulator (or diagnostic) parameters are varied only during the offline identification to estimate the influence vectors, and during the normal operating phase their parameters are chosen such that they have a negligible effect on the dynamic behavior of the system. During the normal operating phase, the emulators are set equal to one; $E_i = 1 : i = 1, 2, 3, 4$.

9.18.1 The Kalman Filter

The nominal fault-free model $(\boldsymbol{A}_0, \boldsymbol{B}_0, \boldsymbol{C}_0)$ given by Eq. (9.1) was identified. The Kalman filter (9.6) was designed using the identified nominal fault-free model. The Kalman gain $\boldsymbol{K}_0$ was computed for the disturbance and the measurement noise statistics Q and R.

9.18.2 The Kalman Filter Residual and Its Auto-correlation

Figure 9.12 shows the Kalman filter residual $\boldsymbol{e}(k)$, and its auto-correlation $r_{ee}(m) = \sum_{k-N+1}^{k} e(k)e(k-m)$ for (i) nominal (i.e., fault-free), (ii) actuator fault, (iii) plant fault, (iv) velocity sensor fault, and (v) the

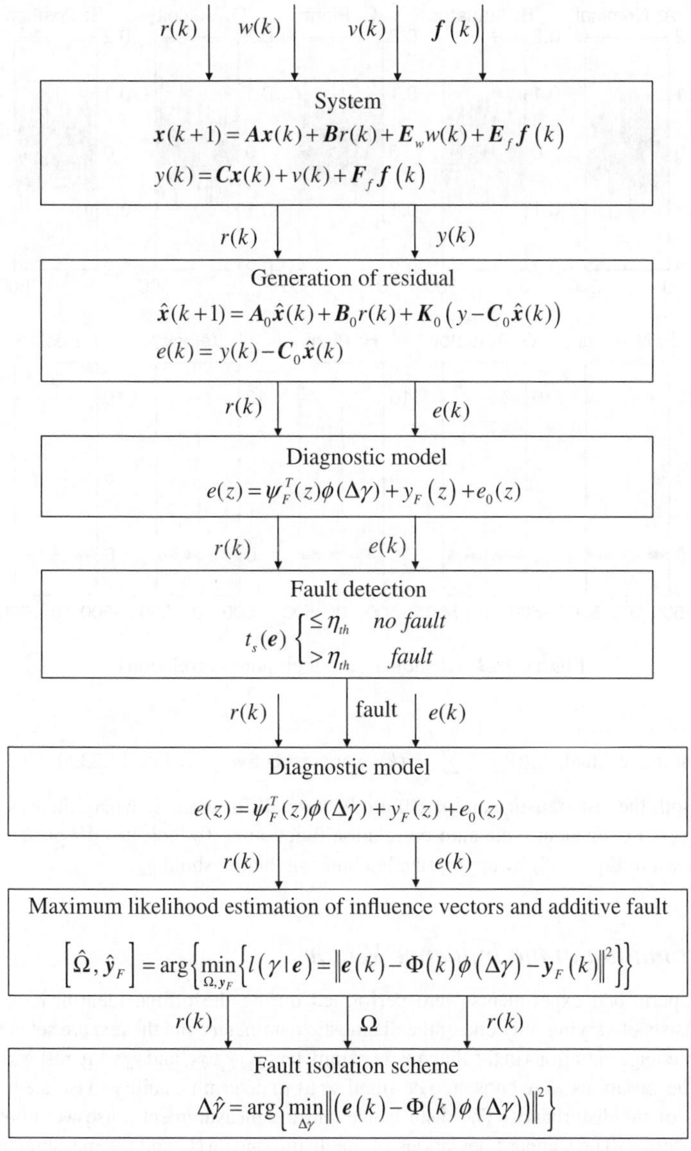

Figure 9.11 Overview of fault diagnosis scheme

position sensor fault. Subfigures (a), (b), (c), (d), and (e) show the residual and its norm in horizontal lines. The norm of the residual is the square root of its statistics $\sqrt{t_s(\boldsymbol{e})} = \|\boldsymbol{e}\|$ when the reference input $r(k)$ is an arbitrary signal as shown in Eq. (9.49). The subfigures (f), (g), (h), (i), and (j) show the corresponding auto-correlations of the residuals. The system and measurement noise variances have been set to Q = 0.5 and R = 0.001.

The norm $\|\boldsymbol{e}\|$ for the nominal, actuator, plant, velocity sensor, and the position sensors were respectively 0.0409, 0.0480, 0.0462, 0.0453, and 0.0490. The auto-correlation at zero lag, which is the sum

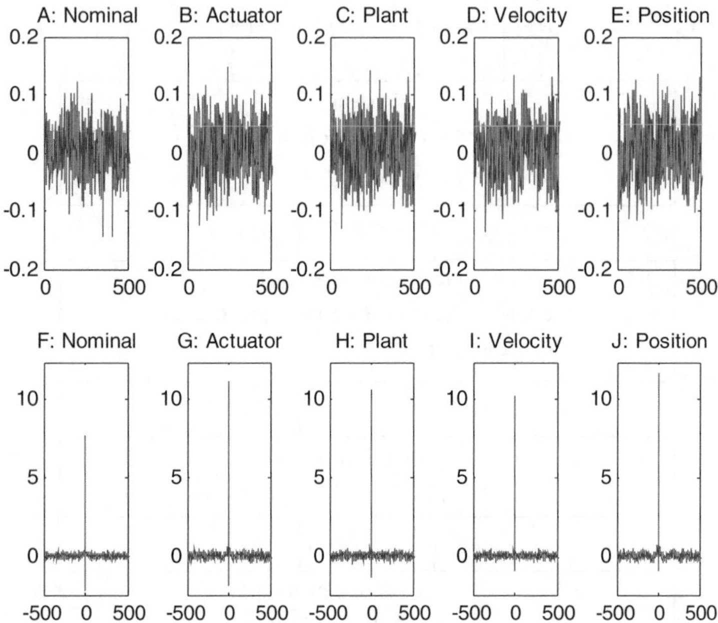

Figure 9.12 Residuals and their auto-correlations

of the squares of the residual, $r_{ee}(0) = \sum_{k-N+1}^{k} e^2(k)$, were respectively 7.7458, 11.2390, 10.7187, 10.1972, and 11.7176. Both the test statistics and $r_{ee}(0)$ are both small for the case where there is no fault. The test statistics $t_s(\boldsymbol{e})$ or equivalently the auto-correlation function $r_{ee}(0)$ indicates the presence or absence of a fault as shown in Eq. (9.48) by appropriately choosing the threshold η_{th}.

9.18.3 Estimation of the Influence Vectors

Ten parameter perturbed experiments were performed during the offline identification phase. Each experiment consists of varying only one of the diagnostic parameters and the rest are set at their nominal values. The percentage variations in the diagnostic parameters $\gamma_1, \gamma_2, \gamma_3$, and γ_4 were respectively 5%, 4%, 6%, and 5%. The variations were chosen to be small so as to determine ability to isolate incipient faults in the presence of the disturbance. The disturbance and the measurement noise were both zero-mean white noise processes. The standard deviations of the disturbance $w(k)$ and the measurement noise $v(k)$ were respectively $\sigma_w = 0.707$ (variance Q = 0.5) and $\sigma_v = 0.0316$ (variance R = 0.001). The number of data samples $N = 5000$.

The influence vectors $\{\boldsymbol{\Omega}_i\}$ for the actuator, plant, velocity sensor, and position sensor were estimated from the perturbed parameter experiments using Eq. (9.63).

9.18.4 Fault Size Estimation

Figure 9.13 shows the estimates of the fault size (estimate $\Delta\hat{\gamma}_i$ of the variation of the diagnostic parameter $\Delta\gamma_i$) versus the diagnostic parameter variations $\Delta\gamma_i$. Subfigures (a), (b), (c), and (d) of Figure 9.13 respectively show the variation of the true fault size $\{\Delta\gamma_j(k)\}$ and the estimated fault size $\Delta\hat{\gamma}_i(k)$ given by Eq. (9.75) as the diagnostic parameters of the actuator (γ_1), the plant parameter (α), the velocity

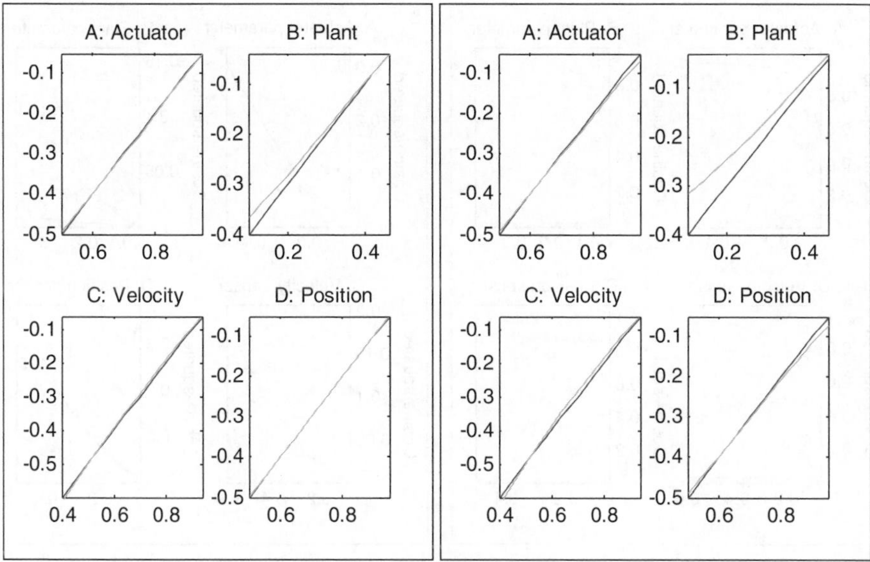

Figure 9.13 Fault size estimation

sensor (γ_3), and position sensor (γ_4) change. The estimated and true fault sizes are shown on the left of Figure 9.12, for the ideal no-noise case $Q = 0$ and $R = 0$, and, on the right, for $Q = 0.5$ and $R = 0.001$, respectively. The true fault size is shown by a black straight line for all cases. The estimate and the true variations were identical for the noise-free case and were close for the noisy case for the case of the actuator, the velocity sensor, and the position sensor. However, for the case of the inaccessible plant parameter, estimate $\Delta\hat{\alpha}$ is not accurate especially when the variations are small.

9.18.5 *Fault Isolation*

Using the estimates of the influence vectors, during the operating phase, the faults were isolated using the Bayes decision strategy (9.78). The subfigures (a), (b), (c), and (d) of Figure 9.14 on the left show the plot of discriminant function $\cos^2 \varphi_i(k)$ vs. γ_i for faults in the actuator, the plant, the velocity sensor, and the position sensor, respectively, when these faults were induced through the emulators. These figures correspond to the noise-free case. The four subfigures (a), (b), (c), and (d) on the right correspond to the noisy case generated by setting $Q = 0.5$ and $R = 0.001$. The plots for the actuator, the plant, the velocity sensor, and the position sensor faults are shown.

Any of the four subsystems (actuator, plant, velocity sensor, and position sensor) is asserted as faulty if its associated $\cos^2 \varphi_i(k)$ is maximum at γ_i as shown in Eq. (9.78). The faults were difficult to isolate in the low signal to noise ratio case as the plots are close to each other when the deviations $|\Delta\gamma_i|$ are small.

Comments *The fault isolation scheme using the discriminant function* $\cos^2 \varphi_i(k)$ *based on Bayes strategy was very effective in isolating an incipient fault in the presence of noise and disturbance. The performance of the isolation scheme depends upon the accuracy of the diagnostic model (9.25). The parameter perturbed experiments to identify the nominal model and the influence vectors were instrumental in obtaining an accurate diagnostic model. The emulators were able to mimic the variations of the parameters of the system during the identification stage, and the influence vectors associated with the emulator parameters were effective in isolating the fault during the operational phase.*

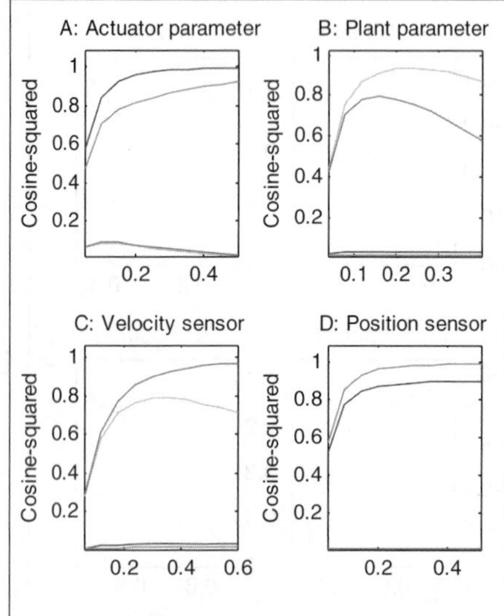

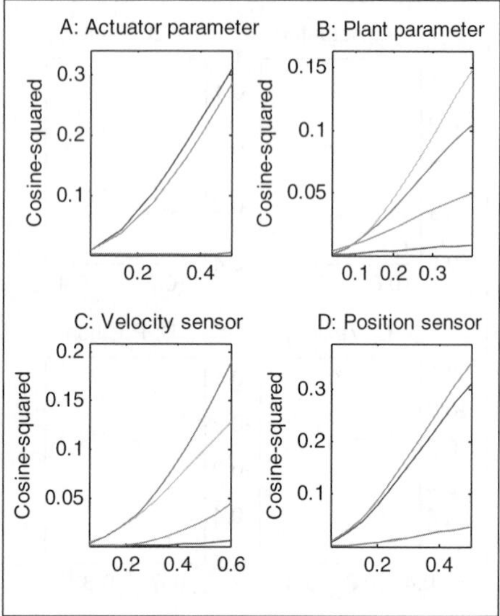

Figure 9.14 Plots of discriminant function: noise free and noisy cases

The Kalman filter played a key role in attenuating the effect of the noise and the disturbance affecting the data and unifying the important dual aim of detecting and isolating faults. The residual has a minimal variance if the Kalman gain is optimal. The filter's structure is error-driven, intuitive, and, because of its closed-loop configuration, has an inherent ability to also attenuate disturbances. The whiteness property of the residual is explored here in ensuring that a simple and computational efficient least squares method may be used, and further the filter's performance can be tuned online.

The estimation of the inaccessible plant parameter variation using the dynamic plant emulator is not accurate as compared to the case of the actuator and sensors whose emulators are static gains.

The performance of the isolation scheme degrades in the presence of low SNR and small fault size. The performance may be improved by increasing the number of data samples and performing a large number of parameter perturbed experiments to estimate the diagnostic model during the identification.

9.19 Summary

Mathematical model of the system: nominal fault-free system

State-space model

$$\begin{aligned} \boldsymbol{x}(k+1) &= \boldsymbol{A}_0\boldsymbol{x}(k) + \boldsymbol{B}_0 r(k) + \boldsymbol{E}_w w(k) \\ y(k) &= \boldsymbol{C}_0\boldsymbol{x}(k) + v(k) \end{aligned}$$

Linear regression model

$$y(k) = \boldsymbol{\psi}^T(k)\boldsymbol{\theta}^0 + \upsilon(k)$$

where $\boldsymbol{\theta}^0$ is the $M \times 1$ nominal feature vector, $\boldsymbol{\psi}^T(k) = [-y(k-1) \quad -y(k-2) \quad . \quad -y(k-n) \quad r(k-1) \quad r(k-2) \quad . \quad r(k-n)]$

Model of the Kalman Filter

Predictor form: residual generation

$$\begin{aligned}\hat{\mathbf{x}}(k+1) &= \mathbf{A}_0\hat{\mathbf{x}}(k) + \mathbf{B}_0 r(k) + \mathbf{K}_0(y(k) - \mathbf{C}_0\hat{\mathbf{x}}(k))\\ \hat{y}(k) &= \mathbf{C}_0\hat{\mathbf{x}}(k)\\ e(k) &= y(k) - \hat{y}(k)\end{aligned}$$

Innovation form: $e(k)$ to $y(k)$

$$\begin{aligned}\hat{\mathbf{x}}(k+1) &= \mathbf{A}_0\hat{\mathbf{x}}(k) + \mathbf{B}_0 r(k) + \mathbf{K}_0 e(k)\\ \hat{y}(k) &= \mathbf{C}_0\hat{\mathbf{x}}(k)\\ y(k) &= \hat{y}(k) + e(k)\end{aligned}$$

Modeling of Faults

The system model when subject to both additive and multiplicative fault becomes:

State-space model

$$\begin{aligned}\mathbf{x}(k+1) &= \mathbf{A}\mathbf{x}(k) + \mathbf{B}r(k) + \mathbf{E}_w w(k) + \mathbf{E}_f \mathbf{f}(k)\\ y(k) &= \mathbf{C}\mathbf{x}(k) + v(k) + \mathbf{F}_f \mathbf{f}(k)\end{aligned}$$

Linear regression model

$$y(k) = \boldsymbol{\psi}^T(k)\boldsymbol{\theta} + \upsilon(k) + y_f(k)$$

where $\boldsymbol{\theta}$ is the $M \times 1$ feature vector as a result of a multiplicative fault

Diagnostic Parameters and the Feature Vector

$$\Delta\boldsymbol{\theta} = \varphi(\Delta\gamma) = \sum_i \boldsymbol{\Omega}_i \Delta\gamma_i + \sum_{i,j} \boldsymbol{\Omega}_{ij}\Delta\gamma_{ij} + \sum_{i,j,k} \boldsymbol{\Omega}_{ijk}\Delta\gamma_{ijk} \ldots + \sum_{1,2,3,\ldots q} \boldsymbol{\Omega}_{123..q}\Delta\gamma_{123..q}$$

Diagnostic Model

$$e(z) = \boldsymbol{\psi}_F^T(z)\boldsymbol{\varphi}(\Delta\gamma) + y_F(z) + e_0(z)$$

Kalman Filter Residual

$\lim_{k\to\infty} it\, E\left[e(k)\right] = 0$ or equivalently $E\left[e(k)e(k-m)\right] = \sigma_e^2\delta(m) = \begin{cases}\sigma_e^2 & m=0\\ 0 & else\end{cases}$ if and only if there is no model mismatch, that is, $\mathbf{A}_0 = \mathbf{A}$, $\mathbf{B}_0 = \mathbf{B}$, and $\mathbf{C}_0 = \mathbf{C}$, or equivalently $\Delta\boldsymbol{\theta} \neq 0$ and the Kalman gain $\mathbf{K}_0$ is optimal.

The Role of the Kalman Filter in Fault Diagnosis

- The residual is the indicator of a fault:

$$\begin{aligned}e(k) &= e_0(k) \text{ if and only if the system is fault-free}\\ e(k) &= e_f(k) + e_0(k) \text{ if and only if the system is a fault}\end{aligned}$$

The variance of the residual $E\left[e_0^2(k)\right]$ will be minimum if

- *The Kalman filter is a whitening filter.*
- *The Kalman filter is robust to noise and disturbances.*

Fault Detection: Bayes Decision Strategy

$$\boldsymbol{e}(k) = \begin{cases} \boldsymbol{e}_0(k) & : H_0 \\ \boldsymbol{e}_f(k) + \boldsymbol{e}_0(k) & : H_1 \end{cases}$$

$$\begin{cases} decide \quad H_0 \ if \quad lr(\boldsymbol{e}) \le \gamma_{th} \\ decide \quad H_1 \ if \quad lr(\boldsymbol{e}) > \gamma_{th} \end{cases}$$

Composite Hypothesis Testing

$$lr(\boldsymbol{e}) = -\frac{1}{2\sigma_{e_0}^2} \sum_{i=0}^{N-1} \hat{e}_f^2(k-i) \text{ where} \hat{e}_f(k) = \arg\left(\min_{e_f} \left\{ \sum_{i=0}^{N-1} (e(k-i) - e_f(k-i))^2 \right\} \right).$$

Decision Strategy

$$t_s(\boldsymbol{e}) = \begin{cases} \left| \dfrac{1}{N} \displaystyle\sum_{i=0}^{N-1} e(k-i) \right| & r(k) = \text{constant} \\ \hat{P}_{ee}(f_r) + \dfrac{1}{2}\hat{P}_{ee}(0) & r(k) \text{ is a sinusoid} \\ \dfrac{1}{N} \displaystyle\sum_{i=0}^{N-1} e^2(k-i) & r(k) \text{ is an arbitrary signal} \end{cases}$$

The corresponding threshold value η_{th} is:

$$\eta_{th} = \begin{cases} \sqrt{\dfrac{2\gamma_{th}\sigma_{e_0}^2}{N}} & r(k) = \text{constant} \\ 2\gamma_{th}\sigma_{e_0}^2 & r(k) \text{ is a sinusoid} \\ \gamma_{th}\sigma_{e_0}^2 & r(k) \text{ is an arbitrary signal} \end{cases}$$

Formulation of Fault Isolation Problem

The estimates are obtained from:

$$[\hat{\boldsymbol{\Omega}}, \hat{\boldsymbol{y}}_F] = \arg\left\{ \min_{\boldsymbol{\Omega}, y_F} \left\{ l(\gamma|\boldsymbol{e}) = \left\| \boldsymbol{e}(k) - \boldsymbol{\Phi}(k)\boldsymbol{\varphi}(\Delta\gamma) - \boldsymbol{y}_F(k) \right\|^2 \right\} \right\}$$

where $\hat{\boldsymbol{\Omega}}$ and $\hat{\boldsymbol{y}}_F$ are the estimates of the influence vector $\boldsymbol{\Omega}$ and the additive fault $\boldsymbol{y}_F(k)$

Parameter-Perturbed Experiment

- *Single fault*: $H_i : \Delta\gamma_i \ne 0, \Delta\gamma_j = 0 j \ne i$

$$[\hat{\boldsymbol{\Omega}}_i, \hat{\boldsymbol{y}}_F] = \arg\left\{ \min_{\boldsymbol{\Omega}_i, y_F} \left\{ \left\| \boldsymbol{e}(k) - \boldsymbol{\Phi}(k)\boldsymbol{\Omega}_i \Delta\gamma_i - \boldsymbol{y}_F(k) \right\|^2 \right\} \right\}$$

- *Two simultaneous faults*: $H_{ij} : \Delta\gamma_i \ne 0$ and $\Delta\gamma_j \ne 0$ but $\Delta\gamma_k = 0 : k \ne i, j$

$$H_{ij} : [\hat{\boldsymbol{\Omega}}_{ij}, \hat{\boldsymbol{y}}_F] = \arg\left\{ \min_{\boldsymbol{\Omega}, y_F} \left\{ \left\| \boldsymbol{e}(k) - \boldsymbol{\Phi}(k)(\hat{\boldsymbol{\Omega}}_i \Delta\gamma_i + \hat{\boldsymbol{\Omega}}_j \Delta\gamma_j + \boldsymbol{\Omega}_{ij}\Delta\gamma_{ij}) - \boldsymbol{y}_F(k) \right\|^2 \right\} \right\}$$

- *Three simultaneous faults*: H_{ijk} : $\Delta\gamma_i \neq 0$, $\Delta\gamma_j \neq 0$ and $\Delta\gamma_k \neq 0$ but $\Delta\gamma_\ell = 0 : \ell \neq i,j,k.$

$$[\hat{\boldsymbol{\Omega}}_{ijk}, \hat{\boldsymbol{y}}_F] = \arg\left\{\min_{\boldsymbol{\Omega},\boldsymbol{y}_F}\left\{\left\|\boldsymbol{e}(k) - \boldsymbol{\Phi}(k)\left(\sum_i \hat{\boldsymbol{\Omega}}_i\Delta\gamma_i + \sum_{i,j}\hat{\boldsymbol{\Omega}}_{ij}\Delta\gamma_{ij} + .\boldsymbol{\Omega}_{ijk}\Delta\gamma_{ijk}\right) - \boldsymbol{y}_F(k)\right\|^2\right\}\right\}$$

And so on until

$$\left[\hat{\boldsymbol{\Omega}}_{ijk}, \hat{\boldsymbol{y}}_F\right] =$$
$$\arg\left\{\min_{\boldsymbol{\Omega},y_F}\left\{\left\|\boldsymbol{e}(k) - \boldsymbol{\Phi}(k)\left(\sum_i \hat{\boldsymbol{\Omega}}_i\Delta\gamma_i + \sum_i \hat{\boldsymbol{\Omega}}_{ij}\Delta\gamma_{ij} + \sum_i \hat{\boldsymbol{\Omega}}_{ijk}\Delta\gamma_{ijk}\ldots + \boldsymbol{\Omega}_{123.q}\Delta\gamma_{1.q}\right) - \boldsymbol{y}_F(k)\right\|^2\right\}\right\}$$

In the rest of the fault isolation tasks we will assume that the additive fault is absent and concentrate only on the multiplicative faults.

Least-Squares Estimates:

$$\begin{aligned}
\hat{\boldsymbol{\Omega}}_i &= (\boldsymbol{\Phi}^i(k)\Delta\gamma_i)^\dagger \boldsymbol{e}^i(k) \\
\hat{\boldsymbol{\Omega}}_{ij} &= (\boldsymbol{\Phi}^{ij}(k)\Delta\gamma_{ij})^\dagger(\boldsymbol{e}^{ij}(k) - \boldsymbol{\Phi}^{ij}(k)(\hat{\boldsymbol{\Omega}}_i\Delta\gamma_i + \hat{\boldsymbol{\Omega}}_j\Delta\gamma_j)) \\
\hat{\boldsymbol{\Omega}}_{ijk} &= (\boldsymbol{\Phi}^{ijk}(k)\Delta\gamma_{ijk})^\dagger\left(\boldsymbol{e}^{ijk}(k) - \boldsymbol{\Phi}^{ijk}\left(\sum_i \hat{\boldsymbol{\Omega}}_i\Delta\gamma_i - \sum_{i,j}\hat{\boldsymbol{\Omega}}_{ij}\Delta\gamma_{ij}\right)\right)
\end{aligned}$$

And so on.

Fault Isolation Scheme

$$\Delta\hat{\gamma} = \arg\{\min_{\Delta\gamma}\|\boldsymbol{e}(k) - \boldsymbol{\Phi}(k)\boldsymbol{\varphi}(\Delta\gamma)\|^2\}$$
$$\|(\boldsymbol{e}(k) - \boldsymbol{\Phi}(k)\boldsymbol{\varphi}(\Delta\hat{\gamma}))\|^2 \leq \varepsilon$$

Isolation of a Single Fault

$$\begin{aligned}
\boldsymbol{e}(k) &= \boldsymbol{\Phi}(k)\boldsymbol{\Omega}_i\Delta\gamma_i + \boldsymbol{e}_0(k) \\
\Delta\hat{\gamma}_i(k) &= (\boldsymbol{\Phi}(k)\boldsymbol{\Omega}_i\Delta\gamma_i))^\dagger \boldsymbol{e}(k) \\
l_i^0 &= \left\|\boldsymbol{e}(k) - \boldsymbol{\Phi}(k)\boldsymbol{\Omega}_i\Delta\gamma_i\right\|^2 = \|\boldsymbol{e}(k)\|^2(1 - \cos^2\varphi_i(k))
\end{aligned}$$

where $\cos^2\boldsymbol{\varphi}_i(k) = \left(\dfrac{\langle\boldsymbol{e}(k), \boldsymbol{\Phi}(k)\boldsymbol{\Omega}_i\rangle}{\|\boldsymbol{e}(k)\|\;\|\boldsymbol{\Phi}(k)\boldsymbol{\Omega}_i\|}\right)^2$

$$i = \arg\left(\max_j\left\{\cos^2\boldsymbol{\varphi}_j(k)\right\}\right)$$

Performance of Fault Isolation Scheme

Exploiting the fact that $\boldsymbol{\Phi}(k)\boldsymbol{\Omega}_i$ is uncorrelated with $\boldsymbol{e}_0(k)$, the expression for the discriminant function becomes:

$$\cos^2 \varphi_j(k) \begin{cases} \leq \dfrac{1}{(1+\sigma_{0i}^2)} & \textit{if } j \neq i \\ = \dfrac{1}{(1+\sigma_{0i}^2)} & \textit{if and only if } i = j \end{cases}$$

Emulators for Offline Identification

$$G_i(z) = G_{0i}(z)(1 + \Delta G_i(z))$$

$$G_i(z) = G_{0i}(z)(1 + \Delta G_i(z)) = G_{0i}(z)E_i(z)$$

The general form of the emulators is given below:

$$E_i(z) = \begin{cases} \gamma_1 & \text{gain} \\ \dfrac{\gamma_{0i} + \gamma_{1i}z^{-1}}{1 + \gamma_{2i}z^{-1}} & \text{first-order filter} \\ \prod_j \dfrac{\gamma_{0ij} + \gamma_{1ij}z^{-1}}{1 + \gamma_{2ij}z^{-1}} & \text{high-order filter} \end{cases}$$

9.20 Appendix: Bayesian Multiple Composite Hypotheses Testing Problem

Consider the following linear algebraic model:

$$\boldsymbol{e}(k) = \boldsymbol{\Phi}(k)\boldsymbol{\Omega}_i\Delta\gamma_i + \boldsymbol{e}_0(k) \tag{9.100}$$

where $\boldsymbol{e}(k)$ is an $N \times 1$ vector, $\boldsymbol{e}_0(k)$ an $N \times 1$ zero-mean Gaussian random variable, $\boldsymbol{\Phi}(k)\boldsymbol{\Omega}_i$ an Nx1 vector, and $\Delta\gamma_i$ a scalar. The log-likelihood function l_j is:

$$l_j = (\boldsymbol{e}(k) - \boldsymbol{\Phi}(k)\,\boldsymbol{\Omega}_j\Delta\gamma_j(k))^T(\boldsymbol{e}(k) - \boldsymbol{\Phi}(k)\,\boldsymbol{\Omega}_j\Delta\gamma_j(k)) \tag{9.101}$$

The optimal solutions for (i) the index ℓ of the most likely hypothesis H_ℓ, (ii) the best least-squares estimate $\Delta\hat{\gamma}_\ell(k)$ of the unknown diagnostic parameter variation $\Delta\gamma_\ell(k)$ (a size of the fault), and (iii) the optimal error measure J_ℓ^* are respectively as follows:

$$\ell = \arg\left\{\min_j J_j\right\} \tag{9.102}$$

$$\Delta\hat{\gamma}_\ell(k) = (\boldsymbol{\Phi}(k)\,\boldsymbol{\Omega}_\ell)^\dagger \boldsymbol{e}(k) \tag{9.103}$$

$$l_\ell^0 = \min_j l_j = \|\boldsymbol{e}(k)\|^2\,(1 - \cos^2\varphi_j(k)) \tag{9.104}$$

where $(.)^\dagger$ stands for the pseudo-inverse of (.), $\cos^2\varphi_j(k) = \left[\dfrac{\langle e(k), \boldsymbol{\Phi}(k)\,\boldsymbol{\Omega}_j\rangle}{\|e(k)\|\,\|\boldsymbol{\Phi}(k)\,\boldsymbol{\Omega}_j\|}\right]^2$, and $\langle \boldsymbol{x}, \boldsymbol{y}\rangle$ denotes the inner-product of $\boldsymbol{x}$ and $\boldsymbol{y}$.

Proof: The first step is to obtain a linear least-squares estimate $\Delta\hat{\gamma}_j(k)$ of the unknown $\Delta\gamma_j(k)$ using the model (9.100). The estimate is given by:

$$\Delta\hat{\gamma}_j(k) = (\boldsymbol{\Phi}(k)\,\boldsymbol{\Omega}_j)^{\dagger}\boldsymbol{e}(k) = \left(\boldsymbol{\Omega}_j^T\boldsymbol{\Phi}^T(k)\,\boldsymbol{\Phi}(k)\,\boldsymbol{\Omega}_j\right)^{-1}\boldsymbol{\Omega}_j^T\boldsymbol{\Phi}^T(k)\boldsymbol{e}(k) \tag{9.105}$$

Substituting $\Delta\gamma_j(k)$ by $\Delta\hat{\gamma}_j(k)$ in the expression for l_j yields the following optimal measure l_j^0:

$$l_j^0 = (\boldsymbol{e}(k) - \boldsymbol{\Phi}(k)\,\boldsymbol{\Omega}_j(\boldsymbol{\Phi}(k)\,\boldsymbol{\Omega}_j)^{\dagger}\boldsymbol{e}(k))^T(\boldsymbol{e}(k) - \boldsymbol{\Phi}(k)\,\boldsymbol{\Omega}_j(\boldsymbol{\Phi}(k)\,\boldsymbol{\Omega}_j)^{\dagger}\boldsymbol{e}(k)) \tag{9.106}$$

Simplifying (59), we get

$$l_j^0 = \boldsymbol{e}^T(k)(I - \boldsymbol{\Phi}(k)\,\boldsymbol{\Omega}_j(\boldsymbol{\Phi}(k)\,\boldsymbol{\Omega}_j)^{\dagger})\boldsymbol{e}(k) \tag{9.107}$$

Using the expression for the pseudo-inverse in Eq. (9.105), we get

$$l_j^0 = \boldsymbol{e}^T(k)\boldsymbol{e}(k) - \boldsymbol{e}^T(k)\boldsymbol{\Phi}(k)\,\boldsymbol{\Omega}_j(\boldsymbol{\Omega}_j^T\boldsymbol{\Phi}^T(k)\boldsymbol{\Phi}(k)\,\boldsymbol{\Omega}_j)^{-1}\boldsymbol{\Omega}_j^T\boldsymbol{\Phi}^T(k)\boldsymbol{e}(k) \tag{9.108}$$

As $(\boldsymbol{\Omega}_j^T\boldsymbol{\Phi}(k)\boldsymbol{\Phi}(k)\,\boldsymbol{\Omega}_j)^{-1}$ is a scalar, expressing it in terms of an inner product and a norm leads to:

$$l_j^0 = \|\boldsymbol{e}(k)\|^2\,(1 - \frac{\left|\langle \boldsymbol{e}(k), \boldsymbol{\Phi}(k)\,\boldsymbol{\Omega}_j\rangle\right|^2}{\|\boldsymbol{\Phi}(k)\,\boldsymbol{\Omega}_j\|^2\|\boldsymbol{e}(k)\|^2}) = \|\boldsymbol{e}(k)\|^2\,(1 - \cos^2\boldsymbol{\varphi}_j(k)) \tag{9.109}$$

9.21 Appendix: Discriminant Function for Fault Isolation

Consider the expression for $\cos^2\boldsymbol{\varphi}_j(k) = \left[\dfrac{\langle \boldsymbol{e}(k), \boldsymbol{\Phi}(k)\,\boldsymbol{\Omega}_j\rangle}{\|\boldsymbol{e}(k)\|\,\|\boldsymbol{\Phi}(k)\,\boldsymbol{\Omega}_j\|}\right]^2$. Substituting $\boldsymbol{e}(k) = \boldsymbol{\psi}^T(k)\boldsymbol{\Omega}_i\Delta\gamma_i + \boldsymbol{e}_0(k)$, we get:

$$\cos^2\varphi_i(k) = \left[\frac{\langle \boldsymbol{\Phi}(k)\boldsymbol{\Omega}_i\Delta\gamma_i + \boldsymbol{e}_0(k), \boldsymbol{\Phi}(k)\boldsymbol{\Omega}_j\rangle}{\|\boldsymbol{\Phi}(k)\boldsymbol{\Omega}_i\Delta\gamma_i + \boldsymbol{e}_0(k)\|\,\|\boldsymbol{\Phi}(k)\boldsymbol{\Omega}_j\|}\right]^2 \tag{9.110}$$

Since $\boldsymbol{\Phi}(k)\boldsymbol{\Omega}_i$ is uncorrelated with $\boldsymbol{e}(k)$, that is $\langle \boldsymbol{e}_0(k), \boldsymbol{\Phi}(k)\boldsymbol{\Omega}_i\rangle = 0$, we get:

$$\cos^2\boldsymbol{\varphi}_i(k) = \frac{|\langle \boldsymbol{\Phi}(k)\boldsymbol{\Omega}_i\Delta\gamma_i, \boldsymbol{\Phi}(k)\boldsymbol{\Omega}_j\rangle|^2}{\|\boldsymbol{\Phi}(k)\boldsymbol{\Omega}_i\Delta\gamma_i + \boldsymbol{e}_0(k)\|^2\|\boldsymbol{\Phi}(k)\boldsymbol{\Omega}_j\|^2} \tag{9.111}$$

Simplifying yields:

$$\cos^2\boldsymbol{\varphi}_i(k) = \frac{|\langle \boldsymbol{\Phi}(k)\boldsymbol{\Omega}_i\Delta\gamma_i, \boldsymbol{\Phi}(k)\boldsymbol{\Omega}_j\rangle|^2}{(\|\boldsymbol{\Phi}(k)\boldsymbol{\Omega}_i\Delta\gamma_i\|^2 + \sigma_0^2)\|\boldsymbol{\Phi}(k)\boldsymbol{\Omega}_j\|^2} \tag{9.112}$$

Dividing and multiplying by $|\Delta\gamma_i|^2$ and simplifying yields:

$$\cos^2\varphi_j(k) = \frac{|\langle\mathbf{\Phi}(k)\mathbf{\Omega}_i, \mathbf{\Phi}(k)\mathbf{\Omega}_j\rangle|^2}{(\|\mathbf{\Phi}(k)\mathbf{\Omega}_i\|^2 + \left(\frac{\sigma_0}{\Delta\gamma_i}\right)^2)\|\mathbf{\Phi}(k)\mathbf{\Omega}_j\|^2} = \left(\frac{\langle\mathbf{\Phi}(k)\mathbf{\Omega}_i, \mathbf{\Phi}(k)\mathbf{\Omega}_j\rangle}{\|\mathbf{\Phi}(k)\mathbf{\Omega}_i\|\|\mathbf{\Phi}(k)\mathbf{\Omega}_j\|}\right)\frac{1}{\left(1+\sigma_{0i}^2\right)} \tag{9.113}$$

where $\sigma_{0i}^2 = \left(\dfrac{\sigma_0}{\mathbf{\Phi}(k)\mathbf{\Omega}_i\Delta\gamma_i}\right)^2$

9.22 Appendix: Log-likelihood Ratio for a Sinusoid and a Constant

Case 3: $r(k) = a_r\sin(2\pi f_r k)$ is sinusoid and $y_f = c_f$ is constant and f_r is known

In this case $\hat{e}_f(k) = a_f\sin(2\pi f_r k) + b_f\cos(2\pi f_r k) + c_f$.

9.22.1 *Determination of a_f, b_f, and c_f*

The ML estimate is obtained from:

$$\hat{e}_f(k) = \arg\left(\min_{a_f,b_f,c_c}\left\{\sum_{i=0}^{N-1}\left[e(k-i) - a_f\sin(2\pi f_r(k-i)) - b_f\cos(2\pi f_r(k-i)) - c_f\right]^2\right\}\right) \tag{9.114}$$

Differentiating with respect to a_f yields the following necessary condition:

$$\sum_{i=0}^{N-1}\left[e(k-i) - a_f\sin(2\pi f_r(k-i)) - b_f\cos(2\pi f_r(k-i)) - c_f\right]\sin(2\pi f_r(k-i)) = 0 \tag{9.115}$$

Simplifying we get:

$$\begin{aligned}&\sum_{i=0}^{N-1} e(k-i)\sin(2\pi f_r(k-i)) - a_f\sum_{i=0}^{N-1}\sin^2(2\pi f_r(k-i))\\ &\quad - b_f\sum_{i=0}^{N-1}\cos(2\pi f_r(k-i))\sin(2\pi f_r(k-i)) - c_f\sum_{i=0}^{N-1}\sin(2\pi f_r(k-i)) = 0\end{aligned} \tag{9.116}$$

Using the trigonometric identities $\sin A\cos A = \dfrac{1}{2}\sin 2A$, $\cos^2 A = \dfrac{1}{2}[1+\cos 2A]$, and $\sin^2 A = \dfrac{1}{2}[1-\cos 2A]$, and dividing by N we get:

$$\begin{aligned}&\frac{1}{N}\sum_{i=0}^{N-1} e(k-i)\sin(2\pi f_r(k-i)) - \frac{a_f}{2} + \frac{1}{2N}\sum_{i=0}^{N-1}\cos(4\pi f_r(k-i))\\ &\quad -\tfrac{b_f}{2N}\sum_{i=0}^{N-1}\sin(4\pi f_r(k-i)) - \frac{c_f}{N}\sum_{i=0}^{N-1}\sin(2\pi f_r(k-i)) = 0\end{aligned} \tag{9.117}$$

Differentiating with respect to b_f yields the following necessary condition:

$$\sum_{i=0}^{N-1}\left[e(k-i) - a_f\sin(2\pi f_r(k-i)) - b_f\cos(2\pi f_r(k-i)) - c_f\right]\cos(2\pi f_r(k-i)) = 0 \tag{9.118}$$

Using the trigonometric identities and dividing by N yields:

$$\begin{aligned}&\frac{1}{N}\sum_{i=0}^{N-1} e(k-i)\cos(2\pi f_r(k-i)) - \frac{a_f}{2N}\sum_{i=0}^{N-1}\sin(4\pi f_r(k-i))\\&-\frac{b_f}{2}-\frac{1}{2N}\sum_{i=0}^{N-1}\cos(4\pi f_r(k-i)) - \frac{c_f}{N}\sum_{i=0}^{N-1}\cos(2\pi f_r(k-i)) = 0\end{aligned} \tag{9.119}$$

Differentiating with respect to c_f yields the following necessary condition:

$$\sum_{i=0}^{N-1}\left[e(k-i) - a_f\sin(2\pi f_r(k-i)) - b_f\cos(2\pi f_r(k-i)) - c_f\right] = 0 \tag{9.120}$$

Dividing by N yields:

$$\frac{1}{N}\sum_{i=0}^{N-1} e(k-i) - \frac{a_f}{N}\sum_{i=0}^{N-1}\sin(2\pi f_r(k-i)) - \frac{b_f}{N}\sum_{i=0}^{N-1}\cos(2\pi f_r(k-i)) - c_f \tag{9.121}$$

Assume that N is large so that $\frac{1}{N}\sum_{i=0}^{N-1}\sin(2\pi f_r(k-i)) = 0$, $\frac{1}{N}\sum_{i=0}^{N-1}\cos(2\pi f_r(k-i)) = 0$, $\frac{1}{N}\sum_{i=0}^{N-1}\sin(4\pi f_r(k-i)) = 0$ and $\frac{1}{N}\sum_{i=0}^{N-1}\cos(4\pi f_r(k-i)) = 0$. Using the above approximations and simplifying, Eqs. (9.117), (9.119), and (9.121), the estimates a_f, b_f, and c_f become:

$$\begin{aligned}a_f &= \frac{2}{N}\sum_{i=0}^{N-1} e(k-i)\sin(2\pi f_r(k-i))\\ b_f &= \frac{2}{N}\sum_{i=0}^{N-1} e(k-i)\cos(2\pi f_r(k-i))\\ c_f &= \frac{1}{N}\sum_{i=0}^{N-1} e(k-i)\end{aligned} \tag{9.122}$$

9.22.2 *Determination of the Optimal Cost*

Consider the log-likelihood ratio $lr(\mathbf{e}) = -\frac{1}{2\sigma_{e_0}^2}\sum_{i=0}^{N-1}\left[\hat{e}_f^2(k-i) - 2e(k-i)\hat{e}_f(k-i)\right]$. Substituting

$\hat{e}_f(k) = a_f\sin(2\pi f_r k) + b_f\cos(2\pi f_r k) + c_f$ and simplifying yields

$$lr(\mathbf{e}) = \frac{N}{2\sigma_{e_0}^2}\left(\frac{a_f^2}{2} + \frac{b_f^2}{2} + c_f^2\right) \tag{9.123}$$

From the expression for a_f, b_f, and c_f given in Eq. (9.122) using the definition the periodogram which is estimate of the power spectral density we get

$$lr(\mathbf{e}) = \frac{1}{\sigma_{e_0}^2}\left(\hat{P}_{ee}(f_r) + \frac{1}{2}\hat{P}_{ee}(0)\right) \tag{9.124}$$

References

Fault Diagnosis

[1] Doraiswami, R. and Cheded, L. (2013) A unified approach to detection and isolation of parametric faults using a Kalman filter residuals. *Journal of Franklin Institute*, **350**(5), 938–965.
[2] Doraiswami, R. and Cheded, L. (2013) Fault diagnosis of a sensor network: a distributed filtering approach. *Journal of Dynamic Systems, Measurement and Control*, **135**(5), 1–10.
[3] Doraiswami, R. and Cheded, L. (2012) Kalman Filter for fault detection: an internal model approach, *IET Control Theory and Applications*, **6**(5), 1–11.
[4] Doraiswami, R., Diduch, C. and Tang, T. (2010) A new diagnostic model for identifying parametric faults, *IEEE Transactions on Control System Technology*, **18**(3), 533–544.
[5] Ding, S. (2008) *Model-based Fault Diagnosis Techniques: Design Schemes*, Springer-Verlag.
[6] Isermann, R. (2006) *Fault Diagnosis Systems: An Introduction from Fault Detection to Fault Tolerance*, Springer-Verlag.
[7] Silvio, S., Fantuzzi, C., and Patton, R.J. (2003) *Model-based Diagnosis Using Identification Techniques, Advances in Industrial Control*, Springer-Verlag.
[8] Patton, R.J., Frank, P.M., and Clark, R.N. (2000) *Issues in Fault Diagnosis for Dynamic Systems*, Springer-Verlag.
[9] Gertler, J.F. (1998) *Fault Detection and Diagnosis in Engineering Systems*, Marcel-Dekker Inc.
[10] Patton, R.J. and Chen, J. (1991) A review of parity space approaches to fault diagnosis, IFAC/IMACS Symposium on Fault Detection, Supervision and Safety for Technical Processes – Safe Process ' 91, Baden-Baden, Germany, 65–82.

Detection and Estimation Theory

[11] Kay, S.M. (1998) *Fundamentals of Signal Processing: Detection Theory*, Prentice Hall, New Jersey.
[12] Kay, S.M. (1993) *Fundamentals of Signal Processing: Estimation Theory*, Prentice Hall, New Jersey.

Model Perturbations

[13] Zhou, K. Doyle, J., and Glover, K. (1996) *Robust Optimal Control*, Prentice Hall, New Jersey.

10

Modeling and Identification of Physical Systems

10.1 Overview

The theoretical underpinnings of identification and its applications are thoroughly verified by extensive simulations and very well corroborated by practical implementation on laboratory-scale systems including (i) a two-tank process control system, (ii) a magnetically-levitated system, and (iii) a mechatronic control system. In this chapter, mathematical models of the physical system derived from physical laws are given. The input–output data obtained from experiments performed on these physical systems are used to evaluate the performance of the identification, fault diagnosis, and soft sensor schemes. The closed-loop identification scheme developed in the earlier chapters is employed.

10.2 Magnetic Levitation System

A magnetic levitation system is a nonlinear and unstable system. Identification and control of the magnetic levitation system has been a subject of research in recent times in view of its applications to transportation systems, magnetic bearings used to eliminate friction, magnetically levitated micro robot systems, and magnetic levitation-based automotive engine valves. It poses a challenge for both identification and controller design [1–5]. A schematic of the magnetic levitation is shown in Figure 10.1, where there is an electromagnetic coil and a steel ball. The steel ball is levitated in the air by an upward electromagnetic force and the downward force due to the gravity. The electromagnetic field is generated by the current in the coil with a soft iron core.

10.2.1 Mathematic Model of a Magnetic Levitation System

The magnetic levitation system is a nonlinear closed-loop system formed of the continuous time controller (cascade connected with an amplifier) and a plant as shown in Figure 10.2. The plant consists of a ball levitating in the air as a result of two balancing forces: the gravitational force pulls the ball down while an electromagnetic force balances the ball by pulling it up. The electromagnetic field is generated by an electromagnetic coil. The force balance is achieved by the closed-loop controller action. The controller is driven by the error between the reference input r and the position of the ball y and the output of the controller (control input) u drives the coil.

Identification of Physical Systems: Applications to Condition Monitoring, Fault Diagnosis, Soft Sensor and Controller Design, First Edition. Rajamani Doraiswami, Chris Diduch and Maryhelen Stevenson.

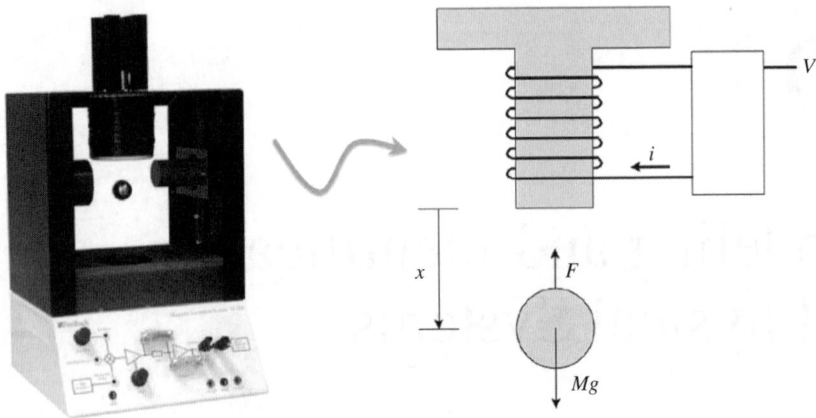

Figure 10.1 Magnetic levitation system

10.2.1.1 Nonlinear Model of the Plant

The electromagnetic force $f_e(t)$ acting on the steel ball is given by:

$$f_e(t) = k_e \frac{V^2(t)}{X^2(t)} \tag{10.1}$$

The gravitation force f_g on the steel ball is:

$$f_g = mg \tag{10.2}$$

where $V(t)$ is the voltage applied to the coil, $X(t)$ is vertical distance of the ball from some reference position X_0, and k_e is a constant, and m is mass of the ball. The electromagnetic force $f_e(t)$ and the gravitational force f_g are indicated in Figure 10.2 by F and Mg respectively.

Using Newton's law of motion, the acceleration $\ddot{X}$ of the ball is related to the net force $f_e - f_g$ acting on it as shown below:

$$\ddot{X}(t) = k\frac{V^2(t)}{X^2(t)} - g \tag{10.3}$$

where $k = \dfrac{k_e}{m}$.

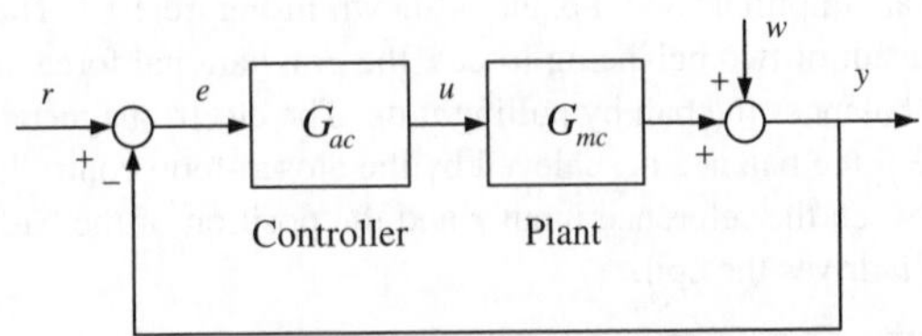

Figure 10.2 Closed-loop system

10.2.2 Linearized Model

Let $x(t) = X(t) - X_0$ and $v(t) = V(t) - V_0$ be respectively incremental values of the position and the voltage, and X_0 and V_0 are the respective nominal values. Linearizing Eq. (10.3) yields:

$$\ddot{x}(t) = a_{mc0}x(t) + b_{mc0}v(t) \tag{10.4}$$

where $a_{mc0} = 2k\dfrac{V_0^2(t)}{X_0^3(t)}$ and $b_{mc0} = 2k\dfrac{V_0(t)}{X_0^2(t)}$. Note that a_{mc0} is positive as the reference direction of X_0 is negative as indicated in Figure 10.2. The transfer function of the magnetic levitation system $G_{mc}(s)$ relating the current and the position is given by:

$$G_{mc}(s) = \frac{x(s)}{v(s)} = \frac{b_{mc0}}{s^2 - a_{mc0}} \tag{10.5}$$

where a_{mc0} and b_{mc0} are the coefficients of the transfer function, $x(s)$ and $v(s)$ the Laplace transforms of $x(t)$ and $v(t)$ respectively. The magnetic levitation system is unstable with poles symmetrically located on either side of the imaginary axis. The poles are real and are symmetrically located about the imaginary axis:

$$\begin{aligned} p_{c1} &= \sqrt{a_{m0}} \\ p_{c2} &= -p_{c1} = -\sqrt{a_{m0}} \end{aligned} \tag{10.6}$$

where p_{c1} and p_{c2} are the poles of the continuous-time model.

10.2.2.1 Controller

The controller is connected in cascade to an amplifier and the controller and the amplifier are treated as a single entity termed the controller. The controller is a phase lead circuit and the amplifier is modeled as a static gain. The controller is driven by the error between the reference input and the sensed position of the ball. The transfer function $G_{ac}(s)$ of the controller and amplifier combination takes the form:

$$G_{ac}(s) = \frac{v(s)}{e(s)} = \frac{b_{ac1}s + b_{ac0}}{s + a_{ac0}} \tag{10.7}$$

where a_{ac0}, b_{aco}, and b_{ac1} are the coefficients of the transfer function, $e(s)$ is the Laplace transform of the error $e(t)$.

10.2.2.2 Position Sensor Output

The position $x(t)$ of the ball is measured by an induction sensor given by:

$$y(t) = x(t) + w(t) \tag{10.8}$$

where $w(t)$ is the combination of the measurement noise and the disturbance.

10.2.2.3 Closed-Loop System

The closed-loop transfer function $T_{yc}(s)$ formed of the controller $G_{ac}(s)$ and the magnetic levitation system $G_{mc}(s)$, which relates the output $y(s)$ and the reference input $r(s)$, is given by:

$$y(s) = T_{yc}(s)(r(s) - w(s)) \tag{10.9}$$

where $T_{yc}(s) = \dfrac{G_{ac}(s)G_{ma}(s)}{1 + G_{ac}(s)G_{ma}(s)} = \dfrac{b_{mc0}(b_{ac1}s + b_{ac0})}{(s^2 - a_{mc0})(s + a_{ac0}) + b_{mc0}(b_{ac1}s + b_{ac0})}$, $y(s)$, $r(s)$ and $w(s)$ are respectively the Laplace transforms of $y(t)$, $r(t)$, and the disturbance $w(t)$. Simplifying using the expressions (10.5) and (10.9) yields:

$$T_{yc}(s) = \frac{b_{c1}s + b_{c0}}{s^3 + a_{c2}s^2 + a_{c1}s + a_{c0}} \tag{10.10}$$

where $b_{c0} = b_{mc0}b_{ac0}$, $b_{c1} = b_{mc0}b_{ac1}$, $a_{c2} = a_{ac0}$, $a_{c1} = -a_{mc0} + b_{c1}$, $a_{c0} = -a_{ac0}a_{mc0} + b_{c0}$.

10.2.3 *Discrete-Time Equivalent of Continuous-Time Models*

The continuous time system is expressed by an equivalent discrete-time model for the purpose of system identification using sampled input–output data.

10.2.3.1 Transfer Function of the Magnetic Levitation System

The discrete-equivalent of the continuous-time transfer function given by Eq. (10.5) is:

$$G_m(z) = \frac{x(z)}{v(z)} = \frac{b_{m1}z^{-1} + b_{m2}z^{-2}}{1 + a_{m1}z^{-1} + a_{m2}z^{-2}} \tag{10.11}$$

where $x(z)$ and $v(z)$ are respectively the z-transforms of the discrete-time signals $x(k)$ and $v(k)$. The poles of the discrete-time model are related those of the continuous-time model (10.6) as:

$$\begin{aligned} p_{di} &= e^{p_{c1}T_s} \\ p_{d2} &= e^{p_{c2}T_s} = e^{-p_{c1}T_s} = \frac{1}{p_{d1}} \end{aligned} \tag{10.12}$$

where p_{d1} and p_{d2} are the poles of the discrete-time model and T_s is the sampling period. Note that the poles of the discrete-time equivalent model are real and satisfy reciprocal symmetry with respect to the unit circle, $p_{d1}p_{d2} = 1$, whereas the poles of the continuous-time model are located symmetrically with respect to the imaginary axis, $p_{c2} = -p_{c1}$.

10.2.3.2 Transfer Function of the Controller and Amplifier

The *discrete equivalent* of the continuous-time transfer function given by Eq. (10.7) is:

$$G_a(z) = \frac{v(z)}{e(z)} = \frac{b_{a0} + b_{a1}z^{-1}}{1 + a_{a1}z^{-1}} \tag{10.13}$$

where $e(z)$ and $v(z)$ are respectively the z-transforms of the discrete-time signals $e(k)$ and $v(k)$.

The expression for the output given by Eq. (10.9) becomes:

$$y(z) = T_y(z)(r(z) - w(z)) \tag{10.14}$$

where $T_y(z) = \dfrac{G_c(z)G_m(z)}{1 + G_c(z)G_m(z)}$ is the discrete equivalent of the continuous-time transfer function denoted $T_c(z)$.

10.2.3.3 Closed-Loop Transfer Functions

Complementary Sensitivity Function

Using Eqs. (10.11) and (10.13), the expression for the closed-loop transfer function $T_y(z)$, termed the complementary sensitivity function is:

$$T_y(z) = \frac{\left(b_{m1}z^{-1} + b_{m2}z^{-2}\right)\left(b_{a0} + b_{a1}z^{-1}\right)}{\left(1 + a_{m1}z^{-1} + a_{m2}z^{-2}\right)\left(1 + a_{a1}z^{-1}\right) + \left(b_{m1}z^{-1} + b_{m2}z^{-2}\right)\left(b_{a0} + b_{a1}z^{-1}\right)} \tag{10.15}$$

Expressing in a compact form yields:

$$T_y(z) = \frac{b_1z^{-1} + b_2z^{-2} + b_3z^{-3}}{1 + a_1z^{-1} + a_2z^{-2} + a_3z^{-3}} \tag{10.16}$$

where $b_1 = b_{a0}b_{m1}$, $b_2 = b_{a0}b_{m2} + b_{a1}b_{m1}$; $b_3 = b_{a1}b_{m2}$, $a_1 = a_{m1} + a_{a1} + b_1$, $a_2 = a_{m2} + a_{m1}a_{a1} + b_2$, $a_3 = a_{m2}a_{a1} + b_3$.

Substituting the expression for $T_y(z)$ in Eq. (10.14), and taking the inverse z-transform, the linear regression model becomes:

$$y(k) + a_1y(k-1) + a_2y(k-2) + a_3y(k-3) = b_1r(k-1) + b_2r(k-2) + b_3r(k-3) + \upsilon(k) \tag{10.17}$$

where $\upsilon(k)$ is the equation error which is colored given by:

$$\upsilon(k) = -(b_1w(k-1) + b_2w(k-2) + b_3w(k-3)) \tag{10.18}$$

Sensitivity Function

The input sensitivity function $S_u(z)$ is given by

$$S_u(z) = \frac{u(z)}{r(z)} = \frac{G_c(z)}{1 + G_c(z)G_m(z)} \tag{10.19}$$

Using the expressions for $G_a(z)$ and $G_m(z)$ given by Eqs. (10.11) and (10.13) yields

$$S_u(z) = \frac{\left(b_{a0} + b_{a1}z^{-1}\right)\left(1 + a_{m1}z^{-1} + a_{m2}z^{-2}\right)}{\left(1 + a_{m1}z^{-1} + a_{m2}z^{-2}\right)\left(1 + a_{a1}z^{-1}\right) + \left(b_{m1}z^{-1} + b_{m2}z^{-2}\right)\left(b_{a0} + b_{a1}z^{-1}\right)} \tag{10.20}$$

Simplifying yields:

$$S_u(z) = \frac{b_{u1}z^{-1} + b_{u2}z^{-2} + b_{u3}z^{-3}}{1 + a_1z^{-1} + a_2z^{-2} + a_3z^{-3}} \tag{10.21}$$

where $b_{ui} : i = 1, 2, 3$ is the numerator coefficient obtained from simplifying the numerator of Eq. (10.20). The expression for the control input $u(z)$ becomes:

$$u(z) = S_u(z)(r(z) - w(z)) \tag{10.22}$$

Substituting the expression for $S_u(z)$ given by Eq. (10.21) and taking the inverse z-transform, the linear regression model becomes:

$$u(k) + a_1 u(k-1) + a_2 u(k-2) + a_3 u(k-3) = b_{u1} r(k-1) + b_{u2} r(k-2) + b_{u3} r(k-3) + \upsilon_u(k) \tag{10.23}$$

where $\upsilon_u(k)$ is the equation error which is colored given by:

$$\upsilon_u(k) = -(b_{u1} w(k-1) + b_{u2} w(k-2) + b_{u3} w(k-3)) \tag{10.24}$$

Comments *From the expressions of the complementary sensitivity function (10.15) and the input sensitivity function (10.20), we can deduce the following:*

- *The zeros of the complementary sensitivity function $T_y(z)$ are the zeros of the controller $G_c(z)$ and the levitation system $G_m(z)$.*
- *The zeros of the input sensitivity function $S_u(z)$ contain the poles of the levitation system $G_m(z)$.*

The above relationships governing the zeros of the closed-loop transfer functions, and the zeros and the poles of the open-loop subsystems, play an important role in cross-checking the accuracy of the identification method.

10.2.4 Identification Approach

10.2.4.1 Identification Objective

The magnetic levitation is a closed-loop system and the objective is to identify the plant $G_m(z)$. Since the plant is unstable, the identification has to be performed in closed-loop configuration. The application for the identification includes the design of a controller for the magnetic levitation.

There are two popular approaches, namely the direct approach and the two-stage approach [6–8].

10.2.4.2 Direct Approach

The input–output data of the plant $u(k)$ and $y(k)$ are employed to identify the plant $G_m(z)$ using an open-loop identification method. A direct approach using the classical least-squares approach generally yields poor estimates due to the presence of disturbance $w(k)$ circulating in the closed-loop system. It can be deduced from Eqs. (10.17) and (10.23) that the output $y(k)$ and the input $u(k)$ are correlated with the disturbance $w(k-i) : i = 1, 2, 3$. Hence an indirect approach, based on the two-stage identification scheme is employed instead.

10.2.4.3 Two-Stage Approach

- In Stage 1, the input sensitivity function $S_u(z)$ and the complementary sensitivity function $T_y(z)$ are identified using the linear regression models (10.23) and (10.17). Let $\hat{S}_u(z)$ and $\hat{T}_y(z)$ be the estimated models of $S_u(z)$ and $T_y(z)$ respectively. The estimate $\hat{u}(k)$ of $u(k)$ and the estimated $\hat{y}(k)$ of $y(k)$

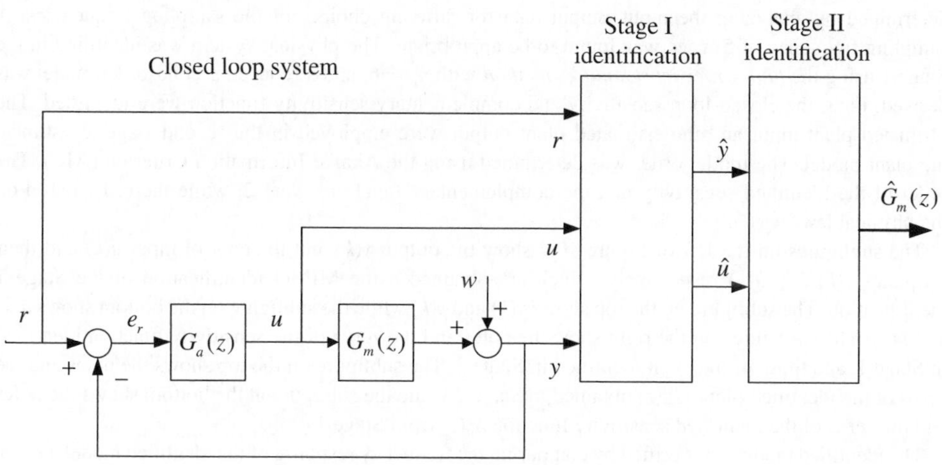

Figure 10.3 Two-stage identification method

are computed from the identified sensitivity function and the complementary sensitivity function respectively:

$$\begin{aligned}\hat{u}(z) &= \hat{S}_u(z)r(z)\\ \hat{y}(z) &= \hat{T}_y(z)r(z)\end{aligned} \tag{10.25}$$

- In Stage 2, the subsystem $G_m(z)$ is then identified using the estimated input $\hat{u}(k)$ and the output $\hat{y}(k)$ (rather than the actual ones) obtained from the first stage. The linear regression model relating $\hat{u}(k)$ and $\hat{y}(k)$:

$$\hat{y}(k) + \hat{a}_{m1}\hat{y}(k-1) + \hat{a}_{m2}\hat{y}(k-2) = b_{m1}\hat{u}(k-1) + b_{m2}\hat{u}(k-2) \tag{10.26}$$

where $\hat{a}_{m1}$, $\hat{a}_{m2}$, $\hat{b}_{m1}$, and $\hat{b}_{m2}$ are respectively the estimates of the coefficients a_{m1}, a_{m2}, b_{m1}, and b_{m2} of the transfer function $G_m(z)$ given by Eq. (10.11). The linear regression model (10.26) is used for identifying $G_m(z)$. Figure 10.3 shows the two-stage identification scheme.

Comments *The two-stage scheme ensures that the estimated sensitivity and the closed-loop sensitivity functions are consistent, that is they are associated with a closed-loop system formed of the plant to be identified and other subsystems including the controllers, actuators, and sensors. For example, the denominator polynomials of $\hat{S}_u(z)$ and $\hat{T}_y(z)$ are identical, the zeros of $\hat{S}(z)$ and $\hat{T}(z)$ are respectively equal to the poles and the zeros of the estimated plant $\hat{G}_m(z)$.*

In the case when the plant is unstable, unstable poles of the plant are captured by the zeros of the sensitivity function in Stage 2 under a very mild restriction on the the Stage 1 identification, namely the the estimated plant input and the output are bounded. Thanks to boundedness of the estimated output of Stage 1, the Stage 2 identification is robust.

10.2.5 Identification of the Magnetic Levitation System

A linearized model of the system was identified as closed-loop using a LABVIEW data A/D and D/A device. The reference input was a rich probing signal, random binary sequence. The plant model was identified from the closed-loop input–output data of the plant. An appropriate sampling frequency was determined by analyzing the input–output data for different choices of the sampling frequencies. A sampling frequency of 5 msec was found to be appropriate. The physical system was identified in the Stage 1 using the *high-order least squares method* with $n_h = 6$, and a reduced second-order model was derived. First, the closed-loop sensitivity and complementary sensitivity function were identified. The estimated plant input and the estimated plant output were employed in the second stage to estimate the plant model. The model order was determined using the Akaike Information Criterion (AIC). The order of the identified sensitivity and the complementary functions were 2, while the order based on the physical laws was 3.

The subfigures on the left of Figure 10.4 show the output $y(k)$ and the control input $u(k)$ and their estimates $\hat{y}(k)$ and $\hat{u}(k)$ respectively, which are obtained using MIMO identification of the Stage 1 identification. The subfigure on the top shows $y(k)$ and $\hat{y}(k)$ while the subfigure on the bottom shows $u(k)$ and $\hat{u}(k)$. The subfigures on the right show the poles and the zeros of the sensitivity function identified in Stage 1, and those of the plant identified in Stage 2. The subfigure on the top shows the poles and the zeros of the identified plant $\hat{G}_m(z)$ obtained in Stage 2 while the subfigure at the bottom shows the poles and the zeros of the identified sensitivity function $\hat{S}_u(z)$ from Stage 1.

The identified model was verified by comparing the frequency response of the identified model, $\hat{G}(j\omega)$, with the estimate of the transfer function, $\hat{G}_{freq}(j\omega)$, obtained using a non-parametric identification method using the Fourier transforms of the plant input and the plant output:

$$\hat{G}_{freq}(j\omega) = \frac{Y(j\omega)}{U(j\omega)} \tag{10.27}$$

Figure 10.5 shows the comparison of the frequency response of the identified model, $\hat{G}(j\omega)$, and estimate of the transfer function $\hat{G}_{freq}(j\omega)$.

The identified sensitivity function was:

$$\hat{S}_u(z) = \frac{-1.7124z^{-1} + 1.833z^{-2}}{1 - 1.7076z^{-1} + 0.7533z^{-2}} \tag{10.28}$$

The poles and the zeros of $\hat{S}_u(z)$ were $0.8538 \pm j0.1560$ and 1.0706 respectively. The identified plant model was:

$$\hat{G}_m(z) = \frac{-0.2472z^{-1} + 0.2400z^{-2}}{1 - 1.9139z^{-1} + 0.9028z^{-2}} \tag{10.29}$$

The poles and the zeros of $\hat{G}_m(z)$ were 1.0706 and 0.8433, and 0.97 respectively. Subfigures on the right of Figure 10.5 show the poles and the zeros of $\hat{G}_m(z)$ at the top and those of $\hat{S}_u(z)$ at the bottom.

10.2.5.1 Model Validation

Unlike the case of the simulated system, there is a need to validate the identified model. The following are the guidelines used in the verification process:

- A reliable model-order selection criterion, namely AIC, was employed to determine the appropriate structure of the magnetic levitation system.
- The structure of the model identified derived from the physical laws given by Eq. (10.11) was employed. The plant has two real poles, one stable and the other unstable. Further, the poles are reciprocals of each other.
- The zeros of the sensitivity function should contain the unstable pole(s) of the plant.
- The frequency responses of the plant computed using two entirely different approaches should be close to each other. In this case, a non-parametric approach was employed to compare the frequency response obtained using the proposed model-based scheme.

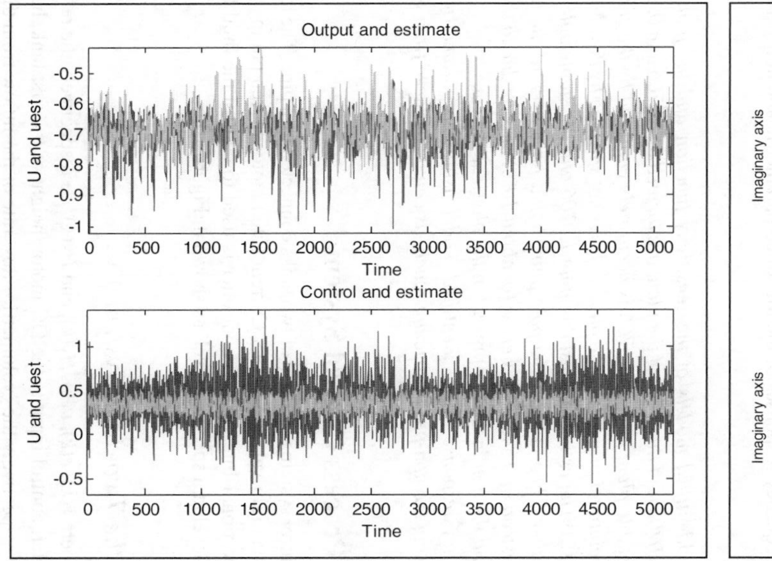

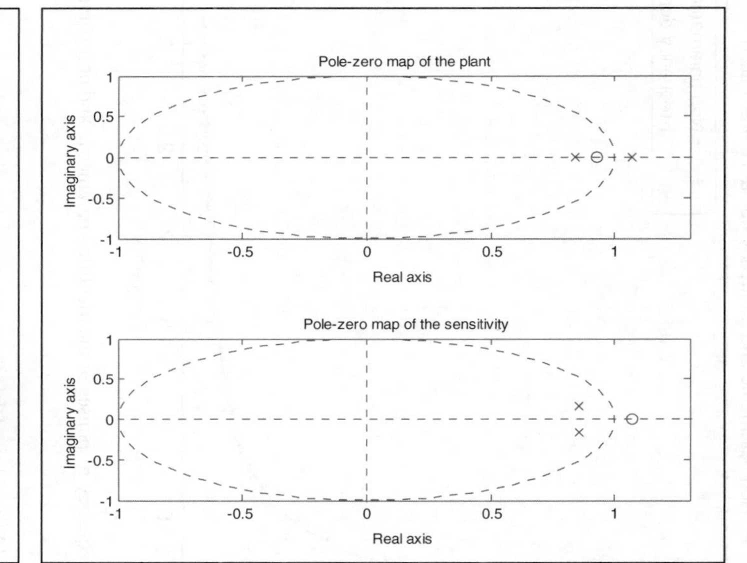

Figure 10.4 The output and the control input and their estimates and the pole-zero maps

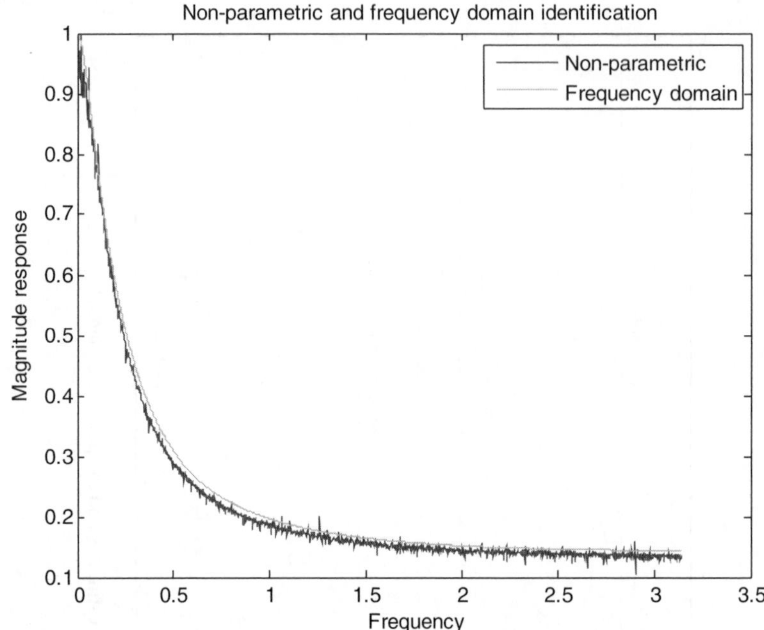

Figure 10.5 The frequency responses of the identified parametric and non-parametric models

Comments *From the identified models of the sensitivity function and the plant, the zeros of the sensitivity function are 1.0706 and the poles of the plant are located at 1.0706 and 0.8433. It is therefore clearly seen that the unstable plant pole at 1.0706 is accurately captured by the zero of the sensitivity function. However the reciprocal property of the plant poles is not satisfied by the identified plant model, as indicated by Eq. (10.12). The estimated unstable pole 1.0706 and the stable pole 0.8433 are clearly not reciprocal of each other. Theoretically, the pole pair should have been at the locations of 1.0706 and 0.9341 (1/1.0706) instead of 0.8433. This error (of about 10%) may be due to the complexity of the model, noise artifacts, and the nonlinearities present in the physical system.*

The estimated frequency response of the identified model using the parametric approach closely matches that obtained using the non-parametric approach, as can be deduced from Figure 10.6.

10.3 Two-Tank Process Control System

The two-tank process control system is formed of two tanks connected by a pipe. The leakage is simulated in the tank by opening the drain valve. A Direct Current (DC) motor-driven pump supplies the fluid to the first tank and a Proportional Integral (PI) controller is used to control the fluid level in the second tank by maintaining the level at a specified level, as shown in Figure 10.6.

10.3.1 Model of the Two-Tank System

Consider Figure 10.7, where $r(k)$, $e(k)$, $u(k)$, $H_1(k)$, and $H_2(k)$ are respectively the reference input, error input driving the controller, control input to the DC motor, height of the first tank, height of the second tank; Q_i, Q_o, and Q_ℓ are respectively the (volumetric) flow rate of the inflow from the pump to the first tank, outflow from the second tank, and leakage outflow from the pipe connecting the two tanks; A_1 and

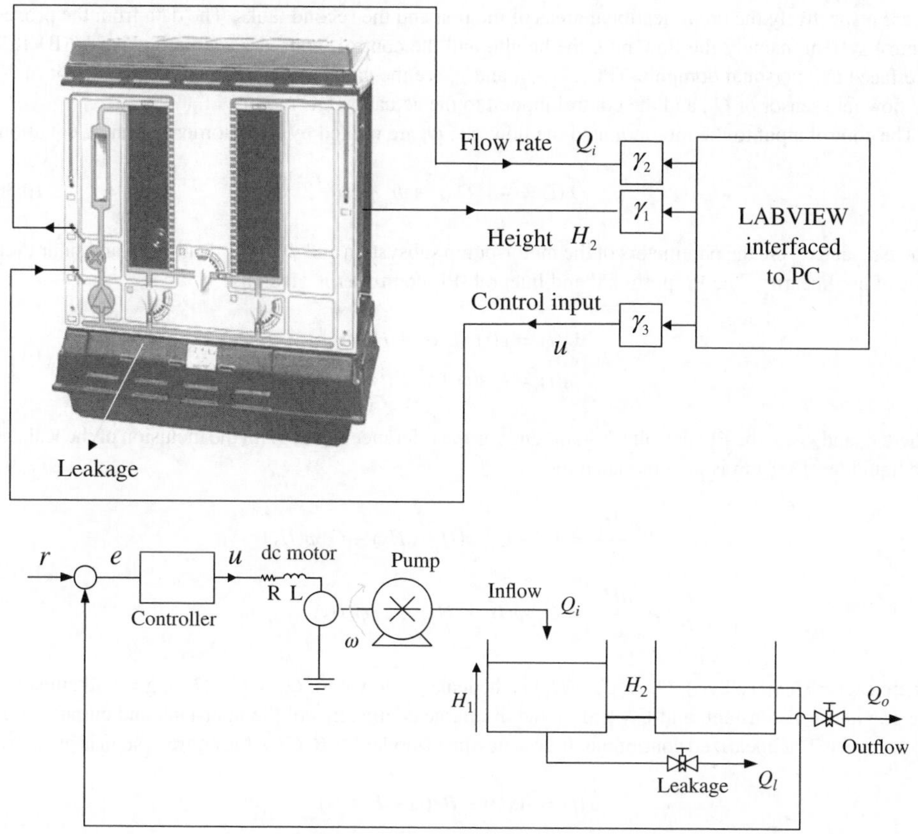

Figure 10.6 Two-tank process control system

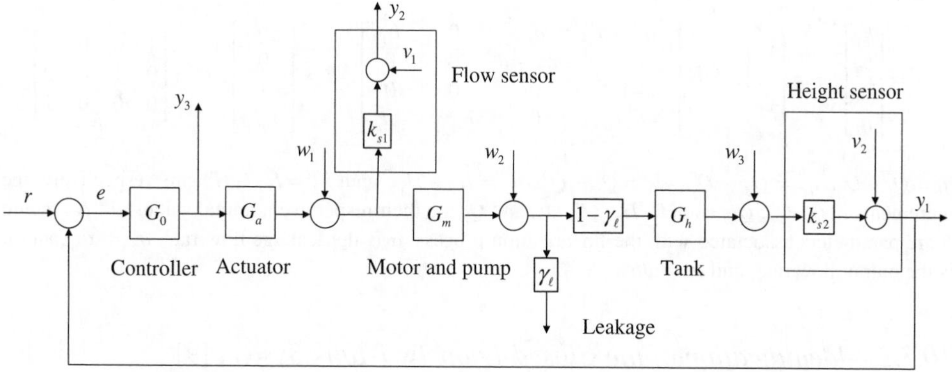

Figure 10.7 Sensor network: closed-loop two-tank process control system

A_2 are respectively the cross-sectional areas of the first and the second tanks. The data from the process control system, namely the flow rate, the height, and the control input, are acquired using LABVIEW interfaced to a personal computer (PC). γ_1, γ_2, and γ_3 are the gains associated with height sensor of H_2, the flow rate sensor of Q_i, and the control input u to the actuator (DC motor).

The control input to the motor, u, and the flow rate Q_i are related by a first-order nonlinear equation

$$\dot{Q}_i(t) = -a_m Q_i(t) + b_m \phi(u) \tag{10.30}$$

where a_m and b_m are the parameters of the motor-pump subsystem and $\phi(u)$ is a dead-band and saturation-type of nonlinearity. The Proportional and Integral (PI) controller is given by:

$$\begin{aligned} \dot{x}_3(t) &= e(t) = r(t) - H_2(t) \\ u(t) &= k_p e(t) + k_I x_3(t) \end{aligned} \tag{10.31}$$

where k_p and k_I are the PI controller's gains and r is the reference input. With the inclusion of the leakage, the liquid level system is now modeled by:

$$\begin{aligned} A_1 \frac{dH_1}{dt} &= Q_i - C_{12}\varphi(H_1 - H_2) - C_\ell \varphi(H_1) \\ A_2 \frac{dH_2}{dt} &= C_{12}\varphi(H_1 - H_2) - C_0 \varphi(H_2) \end{aligned} \tag{10.32}$$

where $\varphi(.) = sign(.)\sqrt{2g(.)}$, $Q_\ell = C_\ell \varphi(H_1)$ is the leakage flow rate, $Q_0 = C_0\varphi(H_2)$, g = 980 cm/sec^2 is the gravitational constant, and C_{12} and C_o the discharge coefficients of the inter-tank and output valves, respectively. The linearized continuous-time state-space model $(\boldsymbol{A}, \boldsymbol{B}, \boldsymbol{C})$ of the entire system is given by:

$$\begin{aligned} \dot{\boldsymbol{x}}(t) &= \boldsymbol{A}\boldsymbol{x}(t) + \boldsymbol{B}r(t) + \boldsymbol{E}_w w(t) \\ y_i(t) &= \boldsymbol{C}_i \boldsymbol{x}(t) + v_i(t) \end{aligned} \tag{10.33}$$

where $\boldsymbol{x}(t)$ is the 4×1 state, $y(t)$ is 3×1 the output measurement formed of (i) the height of the second tank $y_1(k)$, (ii), the input flow rate $y_2(k)$, and (iii) the control input $y_3(k)$, $w(k)$ is the disturbance affecting the system, $v_i(k)$ is the measurement noise; $\boldsymbol{A}$ is 4×4, $\boldsymbol{B}$ is 4×1, and $\boldsymbol{C}_i$ is 1×4 matrix:

$$\boldsymbol{x} = \begin{bmatrix} h_1 \\ h_2 \\ x_3 \\ q_i \end{bmatrix}, \ \boldsymbol{y} = \begin{bmatrix} y_1 \\ y_2 \\ y_3 \end{bmatrix}, \ \boldsymbol{A} = \begin{bmatrix} -a_1 - \alpha & a_1 & 0 & b_1 \\ a_2 & -a_2 - \beta & 0 & 0 \\ -1 & 0 & 0 & 0 \\ -b_m k_p & 0 & b_m k_I & -a_m \end{bmatrix}, \ \boldsymbol{B} = \begin{bmatrix} 0 \\ 0 \\ 1 \\ b_m k_p \end{bmatrix}, \ \boldsymbol{C} = \begin{bmatrix} 1 & 0 & 0 & 0 \\ 0 & 1 & 0 & 0 \\ 0 & 0 & 0 & 1 \end{bmatrix}$$

$q_i = Q_i - Q_i^0$, $q_\ell = Q_\ell - Q_\ell^0$, $q_0 = Q_o - Q_o^0$, $h_1 = H_1 - H_1^0$, and $h_2 = H_2 - H_2^0$ are respectively the increments in Q_i, Q_ℓ, Q_o, and H_1^0, H_2^0, Q_i^0, Q_ℓ^0, and Q_o^0 are their respective nominal values; a_1, a_2, α, and β are parameters associated with the linearization process, α is the leakage flow rate, $q_\ell = \alpha h_1$, and β is the output flow rate, and $q_o = \beta h_2$.

10.3.2 *Identification of the Closed-Loop Two-Tank System* [9]

The two-tank closed-loop process control system be considered as a simple sensor network as shown in Figure 10.7.

The sensor network is formed of the controller G_0, the actuator (amplifier) G_a, DC motor and pump G_m, and the combination of two tanks G_h.

The linearized discrete-time model of the system is identified from the sample reference input $r(k)$ and the sampled measurements $y_i(k)$: $i = 1, 2, 3$.

10.3.2.1 Objective

The objective is to identify the transfer functions of the three subsystems $G_i(z)$: $i = 0, 1, 2$, which are defined by $G_0(z) = G_{eu}(z)$ relating the error $e(k)$ and the control input $u(k)$, $G_1(z) = G_{uq}(z)$, relating the control input $u(k)$ and the flow rate $q_i(k)$, and $G_2(z) = G_{qh}(z)$, relating the flow rate $q_i(k)$ and the height $h_2(k)$.

The application for the identification includes the fault diagnosis of the sensor network using a bank of Kalman filters. The Kalman filters are designed using the identified fault-free subsystem models.

10.3.2.2 Identification Approach

There are two popular approaches, namely the direct approach and the two-stage approach.

Direct Approach

The input–output data $(r(k) - y_3(k), y_2(k))$, $(y_3(k), y_2(k))$, and $(y_2(k), y_1(k))$ are employed to identify respectively the subsystems $G_0(z)$, $G_1(z)$, and $G_2(z)$. Hence an indirect approach, based on our proposed two-stage identification scheme. is employed instead.

Two-Stage Approach

- In Stage 1, the MIMO closed-loop system is identified using data formed of the reference input r, and the subsystems' outputs measured by the three available sensors, $y_i(k)$: $i = 1, 2, 3$.
- In Stage 2, the subsystems $\{G_i\}$ are then identified using the ith subsystem's estimated input and output measurements (rather than the actual ones) obtained from the first stage.

10.3.2.3 Stage 1: Identification of the Closed-Loop System

The subspace method was employed as it can handle both the MIMO system and the model order selection seamlessly. The estimate of the 4×4 matrix transfer function of the MIMO closed-loop transfer function is given by:

$$[\hat{e}(z) \quad \hat{u}(z) \quad \hat{f}(z) \quad \hat{h}(z)]^T = D^{-1}(z)N(z)\,r(z) \tag{10.34}$$

$$N = \begin{bmatrix} 1.9927 & -191.5216z^{-1} & 380.4066z^{-2} & -190.8783z^{-3} \\ 0.0067 & -1.2751z^{-1} & 2.5526z^{-2} & -1.2842z^{-3} \\ -183.5624 & 472.5772z^{-1} & -394.4963z^{-2} & 105.4815z^{-3} \\ -0.9927 & 189.1386z^{-1} & -378.6386z^{-2} & 190.4933z^{-3} \end{bmatrix} \tag{10.35}$$

$$D = 1.0000 \quad -2.3830z^{-1} \quad +1.7680z^{-2} \quad -0.3850z^{-3}$$

The zeros of the sensitivity function, relating the reference input r to the error e are 1.02 and 1.0.

Figure 10.8 shows the estimation of the error, the control input, flow rate, and the height. Subfigures on the left, show the estimate of the error and its estimate on the top, and the control input and its estimate at the bottom. Subfigures on the right, show the estimate of the flow rate and its estimate on the top, and the height and its estimate at the bottom.

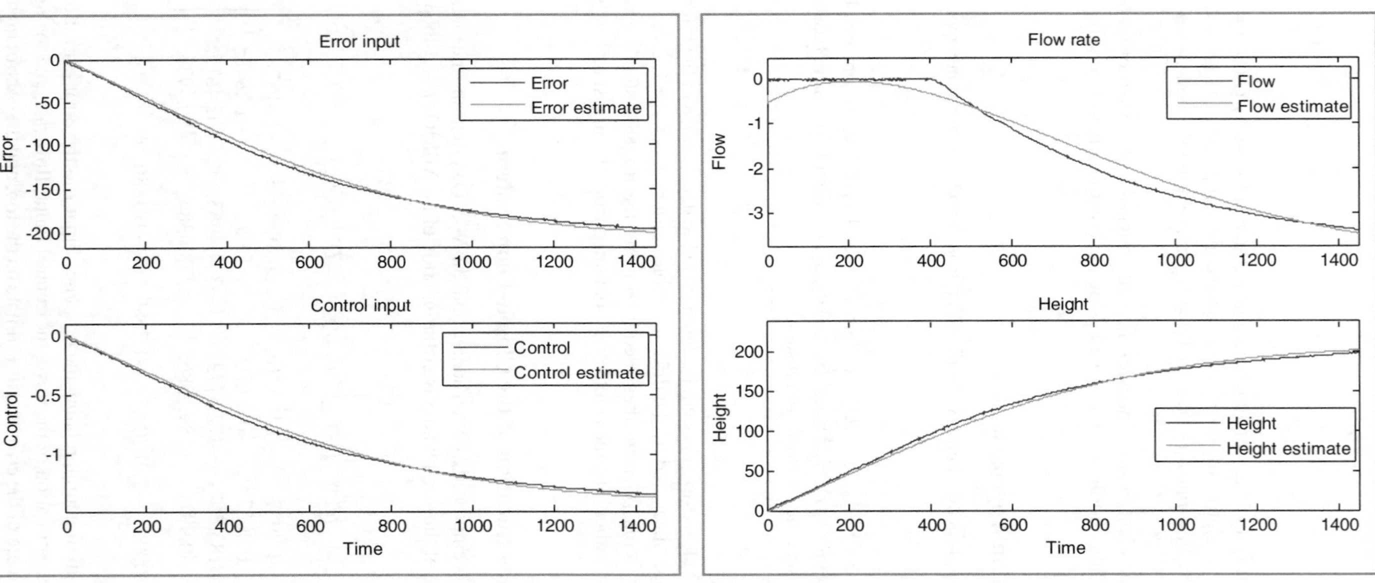

Figure 10.8 The error, control input, flow rate and the height, and their estimates

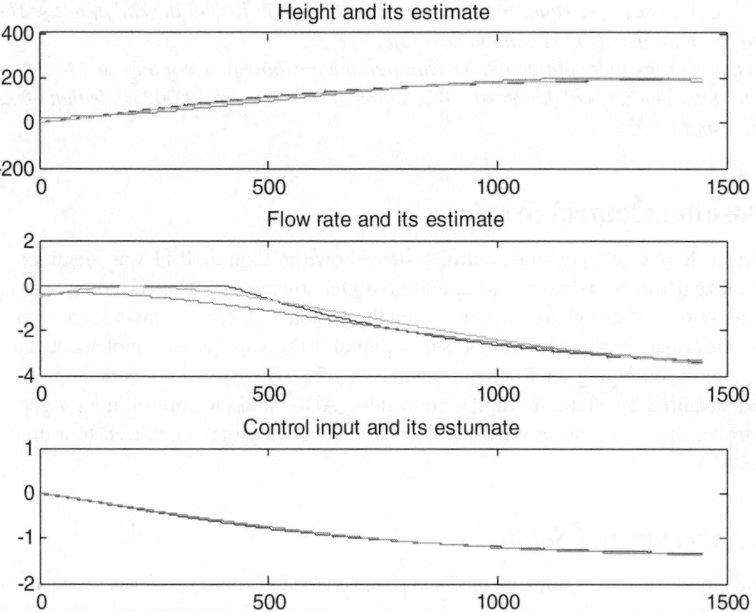

Figure 10.9 The height, flow rate, and control input and their estimates from Stages 1 and 2

10.3.2.4 Stage 2: Identification

This stage yields the following three open-loop transfer functions that are identified using their respective input–output estimates generated by the Stage-1 identification process:

$$\hat{G}_0(z) = \frac{\hat{u}(z)}{\hat{e}(z)} = 0.0067 + \frac{0.4576z^{-1}}{1 - z^{-1}} \tag{10.36}$$

$$G_1(z) = \frac{\hat{q}_i(z)}{\hat{u}(z)} = \frac{0.0104z^{-1}}{1 - 0.9968z^{-1}} \tag{10.37}$$

$$\hat{G}_3(z) = \frac{\hat{h}_2(z)}{\hat{q}_i(z)} = \frac{0.7856z^{-1}}{1 - 1.0039z^{-1}} \tag{10.38}$$

Figure 10.9 shows the combined plots of the actual values of the height, flow rate, and control input, and their estimates from both Stage 1 and 2.

From this figure, we can conclude that the results are on the whole excellent, especially for both the height and control input.

Comments

- *The two-tank level system is highly nonlinear, as can be clearly seen especially from the flow rate profile in Figure 10.9 located at the top right corner. This is saturation-type nonlinearity.*
- *The subsystems $G_0(z)$ and $G_1(z)$ representing respectively the PI controller and the transfer function relating the flow rate to the tank height are both unstable with a pole at unity representing an integral action. The estimated transfer functions $\hat{G}_0(z)$ and $\hat{G}_1(z)$ have captured these unstable poles. Although*

the pole of $G_1(z)$ is exactly equal to unity, the pole of $\hat{G}_1(z)$ is 1.0039 instead of unity. The error may be due to the nonlinearity effect on the flow rate.

- *The zeros of the sensitivity function have captured the unstable poles of the open-loop unstable plant with some error. The zeros of the sensitivity function are 1.0178 and 1.0002 while that of the subsystem poles are 1 and 1.0039.*

10.4 Position Control System

A laboratory-scale physical position control system shown in Figure 10.11 was identified. The system consists of an actuator, namely a PWM amplifier, a DC armature-controlled motor, a position sensor, and a velocity sensor. The angular position V_θ and the angular velocity V_T are the sensor measurements. The input to the armature of the DC motor is u. A digital PID controller was implemented on a PC using a real-time rapid prototyping environment, namely MATLAB® Real-time workshop. The outputs V_θ and V_T were acquired by analog-to-digital converters (ADC) and the control input u generated by the digital controller drives the pulse width modulator (PWM) amplifier interfaced to a digital-to-analog converter (DAC).

10.4.1 Experimental Setup

The identification scheme was implemented and tested on the DC servo system as shown in Figure 10.10. The motor was driven by a PWM amplifier. A tacho-generator and quadrature position encoder provided measurements of angular velocity and position respectively. The control input to the PWM amplifier, u, was generated by a digital to analog converter, DAC, on the target PC. The velocity sensor voltage was applied to the input of an analog to digital converter, ADC, and the position sensor was interfaced to an incremental position decoder on the target PC. A host PC and target PC were used as part of a rapid prototyping system that included MATLAB®/SIMULINK, Real Time Workshop, MS Visual C++, and xPC Target. The target PC boots a real-time kernel which permits feedback and signal processing algorithms to be downloaded from the host PC and executed in real time. The host PC and target PC communicate through a communication channel used for downloading compiled code from the host PC and exchanging commands and data.

10.4.2 Mathematical Model of the Position Control System

A simplified block diagram of the servo system based on a first-order continuous time model for the motor velocity dynamics appears in Figure 10.11.

10.4.2.1 State-Space Model

$$
\begin{aligned}
x(k+1) &= \begin{bmatrix} 1 & 1 & 0 \\ -k_\theta k_A k_p & \alpha - k_1 k_A k_\omega k_d & k_i k_A \\ -k_\theta & 0 & 1 \end{bmatrix} x(k) + \begin{bmatrix} 0 \\ k_p k_A k_1 \\ 1 \end{bmatrix} r(k) + \begin{bmatrix} 0 \\ 1 \\ 0 \end{bmatrix} w(k) \\
y(k) &= [k_\theta \quad 0 \quad 0] x(k) + v(k)
\end{aligned}
\tag{10.39}
$$

10.4.2.2 Identification Experiment

The physical position control system was identified off-line using a square wave reference input $r(k)$ of frequency 0.5 Hz, the sampling period $T_s = 0.001$ sec., and the number of data samples $N = 2000$. The high-order least squares method was used. Since the dynamics of the physical system contain uncertainty in the form of unmodeled dynamics, including nonlinear effects such as friction, backlash, and saturation,

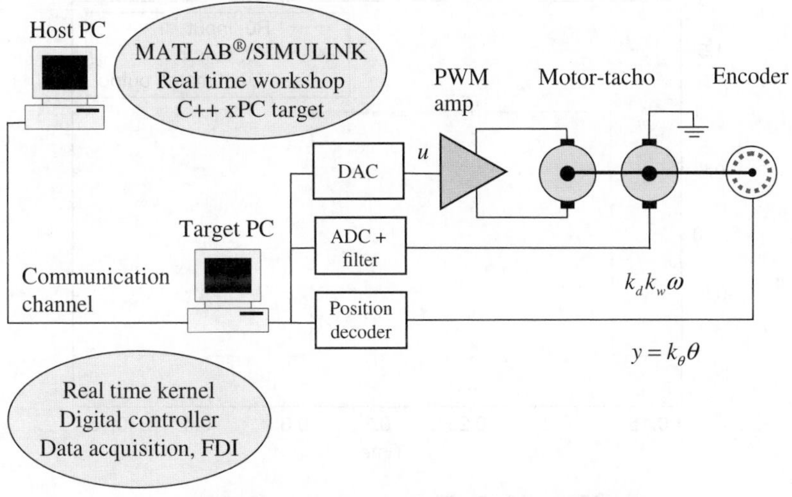

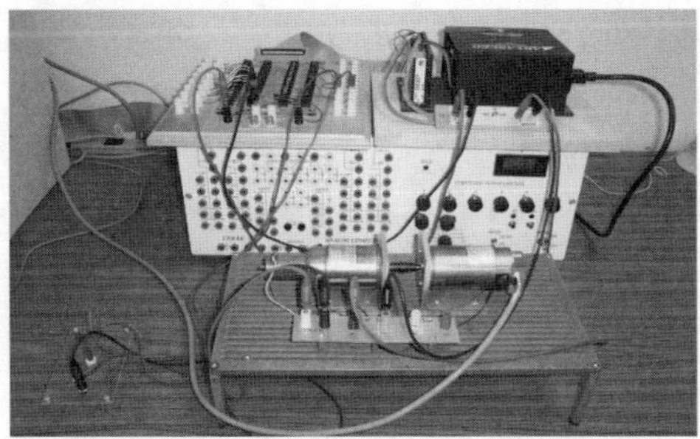

Figure 10.10 Experimental setup

the order of the identified model was selected by analyzing the model fit for a number of different orders that range from 3 to 25, and a 10th order model was selected.

Figure 10.12 shows the output of the system $y(k)$ (denoted Y0), the estimated output $\hat{y}(k)$ (denoted Y), and the reference input $r(k)$ (denoted R0). Since the dynamics of the physical system contain uncertainty in the form of unmodeled dynamics, including nonlinear effects such as friction, backlash, and saturation, the order of the identified model was selected by analyzing the model fit for a number of different orders that range from 3 to 25, and a tenth order model was selected. The probing input was a 0.5 Hz square wave, the sample frequency was 100 Hz and each data record contained 1000 samples.

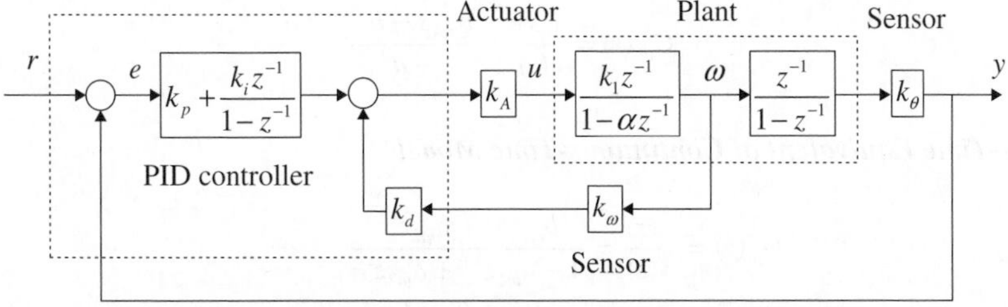

Figure 10.11 Block diagram of the position control system

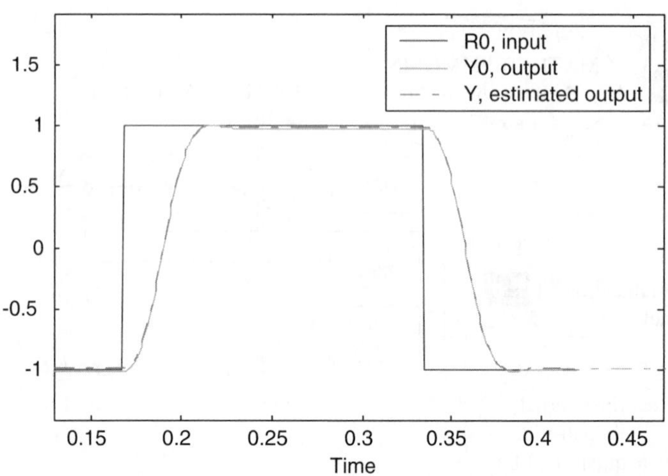

Figure 10.12 The output and its estimate

10.5 Summary

Magnetic Levitation System

The system is unstable:

$$\ddot{X}(t) = k\frac{V^2(t)}{X^2(t)} - g$$

Linearized Model

$$\begin{aligned}\ddot{x}(t) &= a_{mc0}x(t) + b_{mc0}v(t)\\ y(t) &= x(t) + w(t)\\ G_{mc}(s) &= \frac{x(s)}{v(s)} = \frac{b_{mc0}}{s^2 - a_{mc0}}\end{aligned}$$

Poles of the system are $p_{c1} = \sqrt{a_{m0}}$ *and* $p_{c2} = -p_{c1} = -\sqrt{a_{m0}}$

Controller

$$G_{ac}(s) = \frac{v(s)}{e(s)} = \frac{b_{ac1}s + b_{ac0}}{s + a_{ac0}}$$

Discrete-Time Equivalent of Continuous-Time Model

$$G_m(z) = \frac{x(z)}{v(z)} = \frac{b_{m1}z^{-1} + b_{m2}z^{-2}}{1 + a_{m1}z^{-1} + a_{m2}z^{-2}}$$

$$p_{di} = e^{p_{c1}T_s} \text{ and } p_{d2} = e^{p_{c2}T_s} = e^{-p_{c1}T_s} = \frac{1}{p_{d1}}$$

Note that the poles of the discrete-time equivalent model are real and satisfy reciprocal symmetry with respect to the unit circle, $p_{d1}p_{d2} = 1$, whereas the poles of the continuous-time model are located symmetrically with respect to the imaginary axis, $p_{c2} = -p_{c1}$.

Identification Approach

A closed-loop identification based on two-stage approach is employed.

Model Validation

Unlike the case of simulated system, there is a need to validate the identified model. The following are the guidelines used in the verification process:

- A reliable model-order selection criterion, namely AIC, was employed to determine the appropriate structure of the magnetic levitation system.
- The structure of the model identified derived from the physical laws given was employed. The plant has two real poles, one stable and the other unstable. Further, the poles are reciprocals of each other.
- The zeros of the sensitivity function should contain the unstable pole(s) of the plant.
- The frequency responses of the plant computed using two entirely different approaches should be close to each other. In this case, a non-parametric approach was employed to compare the frequency response obtained using the proposed model-based scheme.

Two-Tank Process Control System

Model of the two-tank system

$$\dot{Q}_i(t) = -a_m Q_i(t) + b_m \phi(u)$$

The Proportional and Integral (PI) controller is given by:

$$\begin{aligned}
\dot{x}_3(t) &= e(t) = r(t) - H_2(t) \\
u(t) &= k_p e(t) + k_I x_3(t) \\
A_1 \frac{dH_1}{dt} &= Q_i - C_{12}\varphi(H_1 - H_2) - C_\ell \varphi(H_1) \\
A_2 \frac{dH_2}{dt} &= C_{12}\varphi(H_1 - H_2) - C_0 \varphi(H_2)
\end{aligned}$$

The linearized continuous-time state-space model $(\boldsymbol{A}, \boldsymbol{B}, \boldsymbol{C})$ of the entire system is given by:

$$\begin{bmatrix} \dot{h}_1 \\ \dot{h}_2 \\ \dot{x}_3 \\ \dot{q}_i \end{bmatrix} = \begin{bmatrix} -a_1 - \alpha & a_1 & 0 & b_1 \\ a_2 & -a_2 - \beta & 0 & 0 \\ -1 & 0 & 0 & 0 \\ -b_m k_p & 0 & b_m k_I & -a_m \end{bmatrix} \begin{bmatrix} h_1 \\ h_2 \\ x_3 \\ q_i \end{bmatrix} + \begin{bmatrix} 0 \\ 0 \\ 1 \\ b_m k_p \end{bmatrix} r(t) + \begin{bmatrix} 0 \\ 1 \\ 0 \\ b_m k_p \end{bmatrix} w(t)$$

$$y(t) = Cx(t) + v(t)$$

Identification of the closed-loop two-tank system

- The two-tank level system is highly nonlinear, as can be clearly seen especially from the flow rate profile in Figure 10.9 located at the top right corner. This is saturation-type nonlinearity.
- The subsystems $G_0(z)$ and $G_1(z)$ representing respectively the PI controller and the transfer function relating the flow rate to the tank height are both unstable with a pole at unity representing an integral action. The estimated transfer functions $\hat{G}_0(z)$ and $\hat{G}_1(z)$ have captured these unstable poles. Although

the pole of $G_1(z)$ is exactly equal to unity, the pole of $\hat{G}_1(z)$ is 1.0039 instead of unity. The error may be due to the nonlinearity effect on the flow rate.

- The zeros of the sensitivity function have captured the unstable poles of the open-loop unstable plant with some error. The zeros of the sensitivity function are 1.0178 and 1.0002 while those of the subsystem poles are 1 and 1.0039.

Position Control System

State-space model

$$x(k+1) = \begin{bmatrix} 1 & 1 & 0 \\ -k_\theta k_A k_p & \alpha - k_1 k_A k_\omega k_d & k_i k_A \\ -k_\theta & 0 & 1 \end{bmatrix} x(k) + \begin{bmatrix} 0 \\ k_p k_A k_1 \\ 1 \end{bmatrix} r(k) + \begin{bmatrix} 0 \\ 1 \\ 0 \end{bmatrix} w(k)$$

$$y(k) = [k_\theta \quad 0 \quad 0]x(k) + v(k)$$

Identification experiment [10, 11]

The high-order least squares method was used. Since the dynamics of the physical system contain uncertainty in the form of unmodeled dynamics, including nonlinear effects such as friction, backlash, and saturation, the order of the identified model was selected by analyzing the model fit for a number of different orders that range from 3 to 25, and a 10th order model was selected.

References

[1] Valle, R., Neves, F., and Andrade, R.D. (2012) Electromagnetic levitation of a disc. *IEEE Transactions on Education*, **55**(2), 248–254.

[2] Shahab, M. and Doraiswami, R. (2009) A novel two-stage identification of unstable systems. Seventh International Conference on Control and Automation (ICCA 2010), Christ Church, New Zealand.

[3] Craig, D. and Khamesee, M.B. (2007) Black box model identification of a magnetically levitated microboyic system. *Smart Materials and Structures*, **16**, 739–747.

[4] Peterson, K., Grizzle, J., and Stefanpolou, A. (2006).Nonlinear Magnetic Levitatiobn of Automotive Engine Valves. *IEEE Transactions Control Systems Technology*, **14**(2) 346–354.

[5] Galvao, R.K., Yoneyama, T., Araujo, F., and Machado, R. (2003) A simple technique for identifying a linearized model for a didactic magnetic levitation system. *IEEE Transactions on Education*, **46**(1), 22–25.

[6] Forssell, U. and Ljung, L. (1999) Closed loop identification revisited. *Automatica*, **35**(7), 1215–1241.

[7] Ljung, L. (1999) *System Identification: Theory for the User*, Prentice Hall, New Jersey.

[8] Huang, B. and Shah, S.L. (1997) Closed-loop identification: a two step approach. *Journal of Process Control*, **7**(6), 425–438.

[9] Doraiswami, R.L., Cheded, L., and Khalid, M.H. (2010) *Sequential Integration Approach to Fault Diagnosis with Applications: Model-Free and Model-Based Approaches*, VDM Verlag Dr. Muller Aktiengesellschaft & Co. KG.

[10] Liu, Y., Diduch, C., and Doraiswmi, R. (2003) Modeling and identification for fault diagnosis. In 5th IFAC Symposium on Fault Detection, Supervision and Safety of RTechnical Process, Safeprocess 2003, Washington D.C.

[11] Doraiswami, R., Diduch, C., and Kuehner, J. (2001) Failure Detection and Isolation: a new paradigm. Proceedings of the American Control Conference, Arlington, USA.

11

Fault Diagnosis of Physical Systems

11.1 Overview

Model-based fault diagnosis of physical systems is presented in this chapter. A physical system is an interconnection of subsystems including the actuators, sensors, and plants. The fault diagnosis scheme consists of the following tasks:

Selection of diagnostic parameters: The diagnostic parameters are selected so that they are capable of monitoring the health of the subsystems, and may be varied either directly or indirectly (using a fault emulator) during the offline identification phase [1, 2].

Identification of the system model: A reliable identification scheme is used to obtain a model of the system that captures completely the static and dynamic behavior under various potential fault scenarios including sensor, actuator, and subsystem faults [1–7].

Design and implementation of the Kalman filter: The Kalman filter is designed using the identified nominal fault-free model. The Kalman filter residual is employed for both fault detection and isolation since (i) the Kalman filter residual is zero in the statistical sense if and only if there is no fault, and (ii) its performance is robust to plant and measurement noise affecting the system output. The Kalman gain $\boldsymbol{K}_0$ was tuned on line (as the covariances of the disturbance and measurement noise were unknown) so that Kalman filter residual is a zero-mean white noise process with a minimal variance [8].

Estimation of influence vectors: The influence vectors, which are partial derivatives of the residual with respect to each diagnostic parameter, are estimated offline by performing a number of experiments. Each experiment consists of perturbing the diagnostic parameters one at a time, and the influence vector is estimated from the best least-squares fit between the residual and the diagnostic parameter variations. When the diagnostic parameters are not accessible, as in the case of plant model parameters, for example, the fault emulator parameters are perturbed instead [1, 2].

Fault diagnosis: The decision to select between the hypothesis that the system has a fault and the alternative hypothesis that it does not is extremely difficult in practice as the statistics of both the noise corrupting the data and the model that generated these data are not known precisely. To effectively discriminate between these two important decisions (or hypotheses), the Bayes decision strategy is employed here as it allows for the inclusion of the information about the cost associated with the

---

Identification of Physical Systems: Applications to Condition Monitoring, Fault Diagnosis, Soft Sensor and Controller Design, First Edition. Rajamani Doraiswami, Chris Diduch and Maryhelen Stevenson.

decision taken, and the *a priori* probability of the occurrence of a fault. The fault detection and isolation problem is posed as a multiple hypothesis testing problem. The Bayes decision strategy is developed exploiting the fact that the residual is *a zero-mean wh*ite noise process if and only if the system and the Kalman filter models are identical, that is if there is no model mismatch and therefore no fault. A fault in the subsystem is isolated and asserted if the correlation between the measured residual and one of a number of hypothesized residual estimates is maximum [1–4].

The mathematical background on fault diagnosis is given in earlier chapters on the Kalman filter and fault diagnosis. Case studies in fault diagnosis of two-tank process control system and the position control are presented.

11.2 Two-Tank Physical Process Control System [9]

11.2.1 Objective

The objective is to detect and isolate leakages, actuator faults, and liquid-level sensor faults. In order to simulate faults in the physical system, static fault emulators are connected in cascade with the height sensor, the flow rate sensor, the actuator, and the leakage drain pipe as shown in Figure 11.1, which is the block diagram of the process control system. The emulator parameters γ_1, γ_2, and γ_3 are connected in cascade with the leakage drain pipe, the actuator, and the height sensor during the offline identification stage.

Figure 11.1 is a block diagram of the two-tank process control system where G_0 is the controller, G_1 is the actuator, G_2 tank; r, e, u, f, w, and y are respectively the reference input, error, flow rate, the measurement noise and the disturbance, and measured height y.

11.2.2 Identification of the Physical System

The physical two-tank fluid system is nonlinear with a dead-band nonlinearity and fast dynamics.

The process data was collected at sampling time $T_s = 0.5$ sec. The objective of the controller is to reach a reference height of 20 cm in the second tank. During this process, several faults were introduced, such as leakage faults, sensor faults, and actuator faults. Leakage faults were introduced through the pipe clogs of the system, knobs between the first and the second tank, and so on. Sensor faults were simulated by introducing a gain in the circuit as if there was a fault in the level sensor of the tank. Actuator faults were simulated by introducing a gain in the setup for the actuator that comprises the motor and pump. A Proportional Integral (PI) was employed in order to track the desired reference height. Due to the inclusion of faults, the controller was unable to track the desired level. The power of

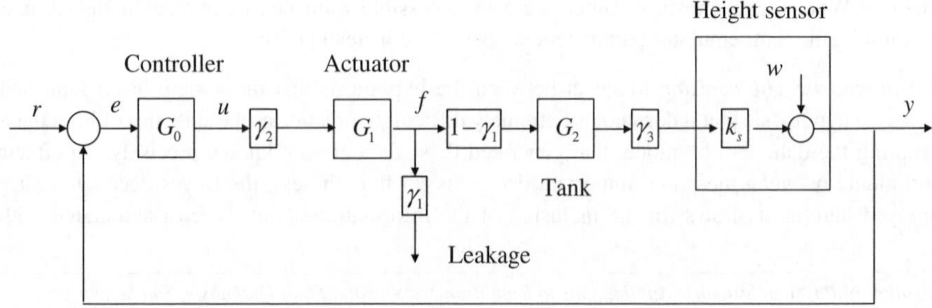

Figure 11.1 Two-tank process control system with emulators

the motor was increased from 5 volts to 18 volts in order to provide it with the maximum throttle to reach the desired level under various operating scenarios including faults. Due to the closed loop PI control action, the tracking performance under emulator parameter variations was good, but it compromised the performance of the fault diagnosis scheme.

The physical two-tank fluid system is nonlinear with a dead-band nonlinearity and fast dynamics. The identified model order is different from that of the model derived from the physical laws. The identified model was essentially a second-order system with a delay, even though the theoretical model is of a fourth order. Such a discrepancy is due to the inability of the identification scheme to capture the system's fast dynamics, especially in low-SNR scenarios. Using the fault-free model together with the covariance of the measurement noise R, and the plant noise covariance Q, the Kalman filter model was designed. To obtain an estimate, a number of experiments were performed under different operating scenarios to tune the Kalman gain K to obtain an optimal performance which ensures the generation of a white noise residual with a minimal variance.

Various types of faults, including the leakage, actuator fault, liquid-level sensor fault, and the flow sensor fault, were all introduced. These faults were emulated by varying γ_ℓ for the opening of the drainage valve that causes a leakage fault, varying the gain block γ_a connected to the actuator input, and varying γ_s connected to the liquid-level sensor output. The National Instruments LABVIEW package was employed to collect these data.

The reference input was chosen to be a step input. An offline perturbed-parameter experiment was performed to estimate the influence vector [1, 2].

The nominal fault-free model relating the reference input $r(k)$ and the output $y(k)$ was identified from the step response of the system. A high-order least-squares method with $n_h = 6$ was employed. A second reduced-order model was derived from the high-order identified model.

Figure 11.2 shows the plots of the step responses when the diagnostic parameters γ_ℓ, γ_a, and γ_s were varied one at a time to induce leakage, actuator, and sensor faults. Subfigures (a), (b), and (c) in Figure 11.2 show the leakage faults, actuator faults, and sensor faults induced by varying γ_1, γ_2, and γ_3, respectively. These three plots show the normal and faulty cases. Small, medium, and large faults were simulated by varying the diagnostic parameters by 25%, 50%, and 75% respectively and are shown. The normal case is also shown.

11.2.3 Fault Detection

At time instant k, the $N = 189$ residuals formed of the present and past $(N-1)$ residuals $e(k-i) : i = 0, 1, 2, \ldots, N-1$, are collected. Let H_0 and H_1 be the two hypotheses indicating the absence and presence of a fault, respectively. The fault detection strategy for the case when the reference input is constant was employed here. The fault detection strategy is:

$$t_s(\boldsymbol{e}) \begin{cases} \leq \eta & \textit{no fault} \\ > \eta & \textit{fault} \end{cases} \tag{11.1}$$

where $\boldsymbol{e}(k) = [\, e(k)\ e(k-1)\ e(k-2)\ .\ e(k-N+1)\,]^T$, and $t_s(\boldsymbol{e}) = \left| \frac{1}{N} \sum_{i=k-N+1}^{k} e(i) \right|$ is the statistics.

Figure 11.3 shows the Kalman filter residual and its auto-correlation for the following cases: (i) nominal (or fault-free), (ii) leakage fault, (iii) actuator fault, and (iv) sensor fault. The test statistic value is the lowest and the auto-correlation is that of a zero-mean white noise for the nominal (fault-free) case. The subfigures (a), (b), (c), and (d) of Figure 11.3 show the residuals and their test statistics shown as straight lines, whereas the subfigures (e), (f), (g), and (h) show the corresponding auto-correlations.

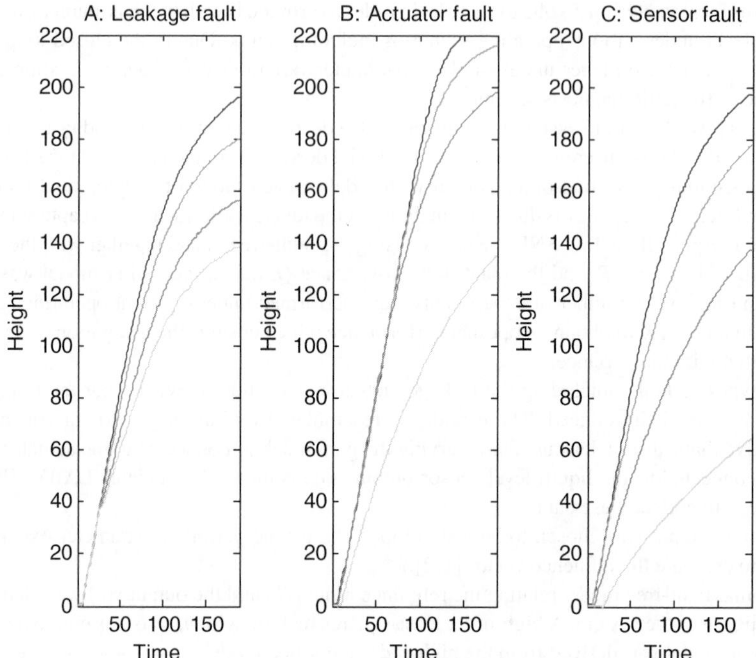

Figure 11.2 Liquid level for normal and fault scenarios

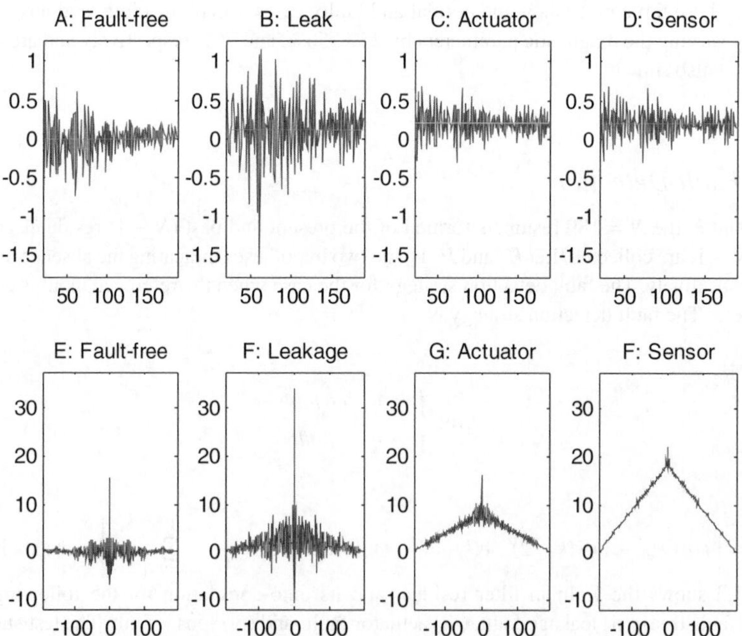

Figure 11.3 The residuals and their auto-correlations for the fault-free and faulty scenarios

11.2.4 Fault Isolation

The influence vectors $\{\Omega_i\}$ for the leakage, actuator, and sensor faults were estimated by perturbing the diagnostic parameters one at a time during the offline identification phase. During the operating phase, the faults were isolated using the Bayes decision strategy:

$$\ell = \arg\left(\max_j\{\cos^2\varphi_j(k)\}\right) \tag{11.2}$$

where $\cos^2\varphi_i(k)$ is the cosine of the angle between the residual $\boldsymbol{e}(k)$ and its estimate $\psi^T(k)\Omega_i$ given by:

$$\cos^2\varphi_J(k) = \left[\frac{\langle \boldsymbol{e}(k), \boldsymbol{\psi}^T(k)\Omega_j\rangle}{\|\boldsymbol{e}(k)\|\,\|\boldsymbol{\psi}^T(k)\Omega_j\|}\right]. \tag{11.3}$$

As was done before, the leakage, actuator, and sensor faults were all introduced by varying the diagnostic parameters γ_1, γ_2, and γ_3. Three values for $\gamma_i : i = 1, 2, 3$ were chosen, namely 0.25, 0.50, and 0.75, with the nominal value $\gamma_i^0 = 1$, to simulate "small," "medium," and "large" faults, respectively. The Bayes decision strategy was employed to assert the fault type, that is, leakage or actuator or sensor fault, by computing their three quantities $\cos^2\varphi_i(k)$, $i = 1, 2, 3$ associated with the leakage fault, actuator fault, and sensor fault, respectively. The decision on which fault is asserted to have occurred is made based on which $\cos^2\varphi_i(k)$ is maximum, that is, the leakage fault is asserted if $i = 1$, actuator fault if $i = 2$, and sensor fault if $i = 3$. Figure 11.4 shows the plot of $\cos^2\varphi_i(k)$ vs. γ_i for leakage, actuator, and sensor faults when $\gamma_i : i = 1, 2, 3$ were varied as described above. The maximum values of $\cos^2\varphi_i(k)$ were normalized to unity. The subfigures (a), (b), and (c) in Figure 11.4 show the plots of $\cos^2\varphi_i(k)$ with respect γ_i for the

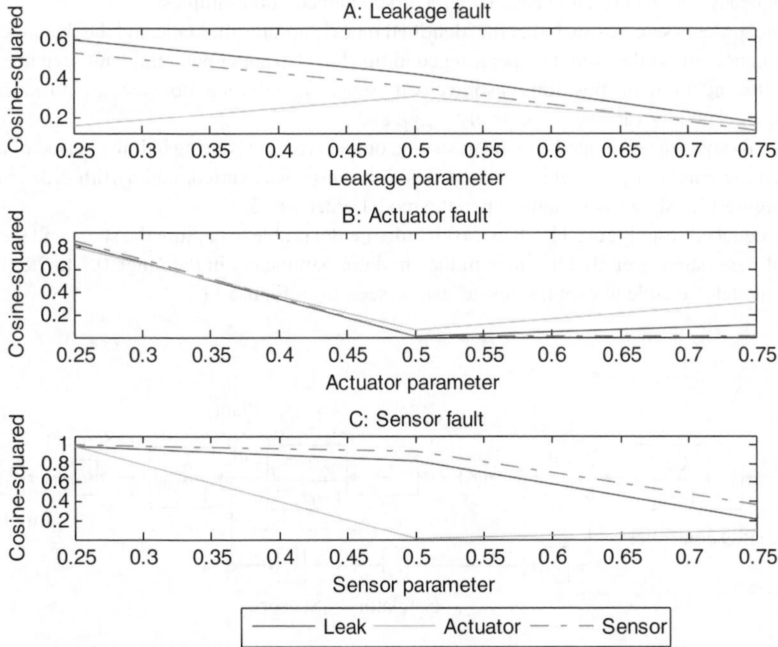

Figure 11.4 Plots of $\cos^2\varphi_i(k)$ vs. γ_i for leakage, actuator, and sensor faults

leakage, actuator, and sensor faults, respectively. The maximum values of the "cos" quantities are used to accurately isolate the faults.

Comment *The leakage fault may easily be isolated from the rest of the faults, while this is not so for both the actuator and sensor faults unless the fault size is large compared to noise floor, as their* $\cos^2 \varphi_i(k)$ *values are very close to each other.*

11.3 Position Control System

11.3.1 The Objective

The objective is to detect and isolate a position sensor, a velocity sensor, and an actuator fault.

In order to simulate faults in the physical system, static fault emulators are connected in the cascade position sensor, the velocity sensor, and the actuator as shown in Figure 11.5 which is a block diagram of the position control system. The emulator parameters γ_1, γ_2, and γ_3 are connected in cascade with the position sensor K_s, velocity sensor K_v, and the actuator K_a.

11.3.2 Identification of the Physical System [2, 10–12]

In general, a physical system is highly complex and nonlinear and, as such, defies mathematical modeling, and this physical system is no exception. The PWM amplifier exhibits saturation-type nonlinearity and further, the tachometer-based velocity sensor is very noisy. Since the dynamics of the physical system contain uncertainty in the form of unmodeled dynamics, including nonlinear effects such as friction, backlash, and saturation, the order of the identified model was selected by analyzing the model fit for a number of different orders that range from 3 to 25. The probing input was a 0.5 Hz square wave, the sample frequency was 100 Hz, and each data record contained 1000 samples.

The model order is selected such that the identified model captures the static and the dynamic behavior of the system not only at the nominal operating condition but also under potential fault scenarios. Various types of faults, including the position sensor, velocity sensor, and the actuator were all introduced. These faults were emulated by varying respectively γ_1, γ_2, and γ_3.

Figure 11.6 shows the estimated and the measured outputs (outputs during half the period of the square wave) when the emulator parameter γ_1 of the position sensor was varied, and a 10th order model was selected. Figure 11.7 shows the results when the model order was 5.

It can be deduced from Figure 11.6 that a 10th order model is able to capture the static and the dynamic behavior of the system over all variations in the emulator parameters in the range 0.7 to 2.8, whereas a fifth order model is unable to capture this, as can be seen from Figure 11.7.

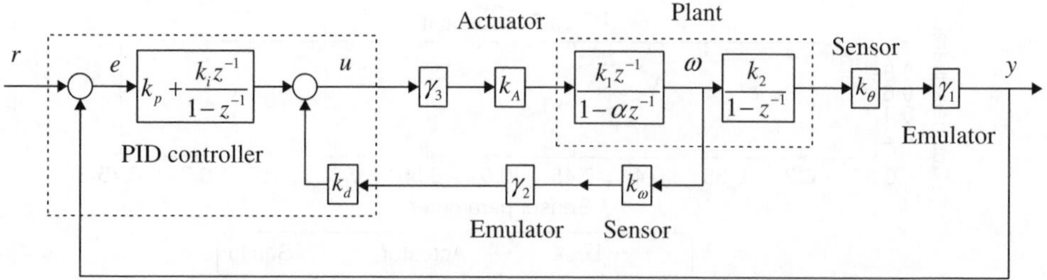

Figure 11.5 Block diagram of the position control system

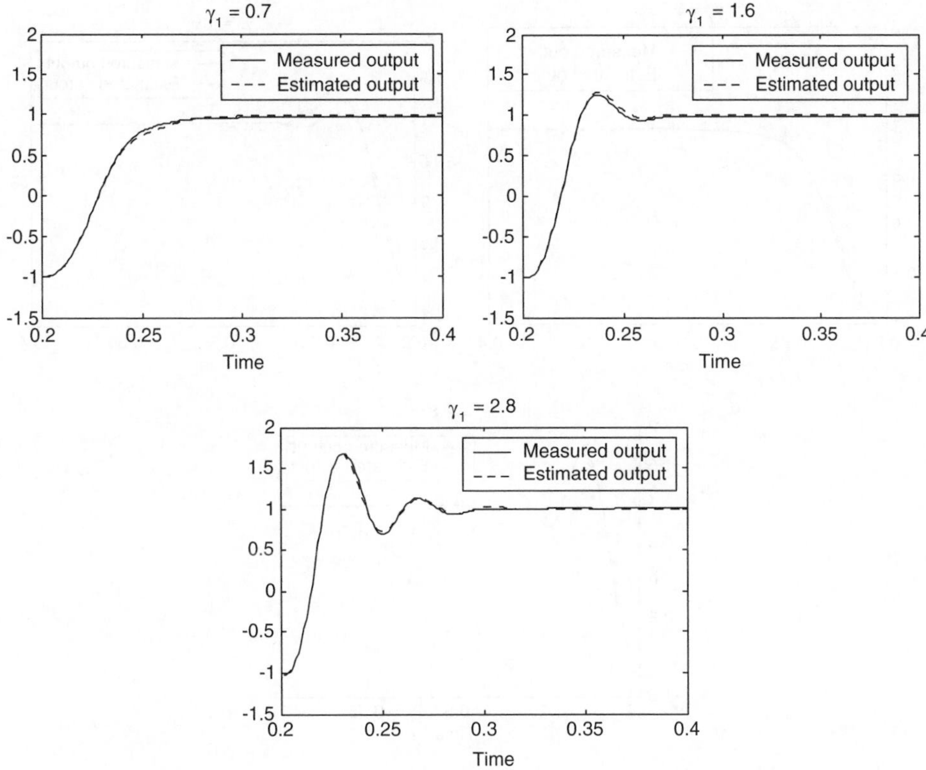

Figure 11.6 Output and its estimated when a 10th order model was selected

The performance of a 10th order model, identified using the scheme outlined herein, is evaluated considering the following three cases:

Case 1: The model is identified by performing a single experiment when the diagnostic parameters are at their nominal values and is validated when one of the diagnostic parameter varies and all others are fixed. This is the conventional scheme.

Case 2: The model is identified using many experiments, perturbing one of the diagnostic parameters while keeping the others fixed at their nominal values.

Case 3: The model is identified using many experiments, perturbing all of the diagnostic parameters. This is the proposed scheme. The mean-squared error is minimum over a wide range of the variations of γ_1.

Figure 11.8 below shows the mean-squared error for the Case 1, 2, and 3 when one of the diagnosed parameters changes. It was found that the conventional scheme of Case 1 is particularly prone to modeling errors and the proposed scheme of Case 3 is the best.

Kalman filter design

The nominal fault-free model relating the reference input $r(k)$ and the output $y(k)$ was identified from the step response of the system. Using the fault-free identified model together with the covariance of the measurement noise R, and the plant noise covariance Q, the Kalman filter model was designed.

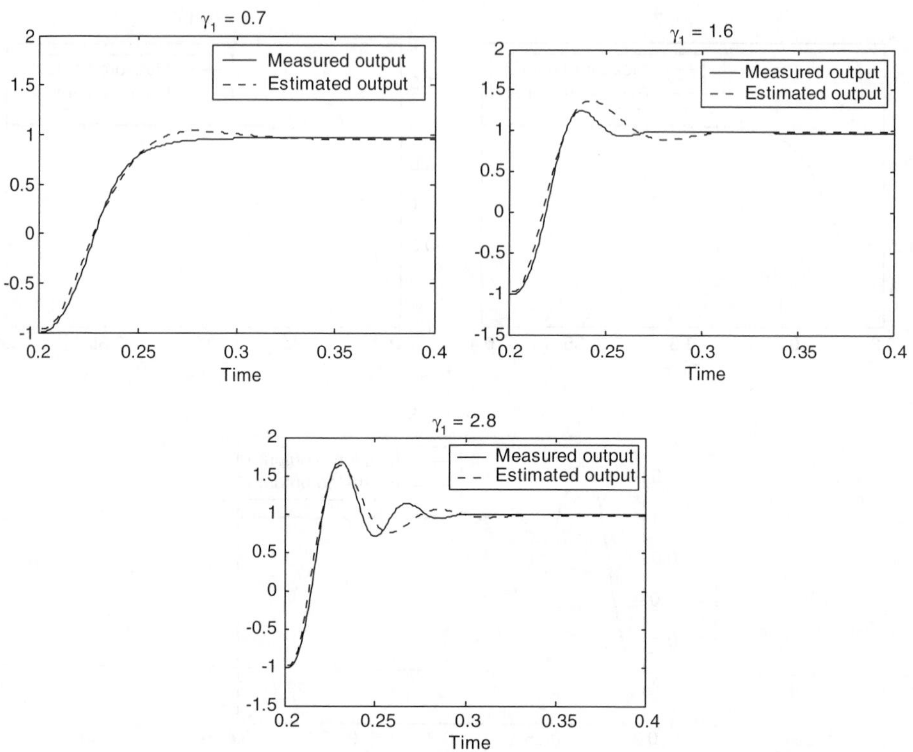

Figure 11.7 Output and its estimated when a fifth order model was selected

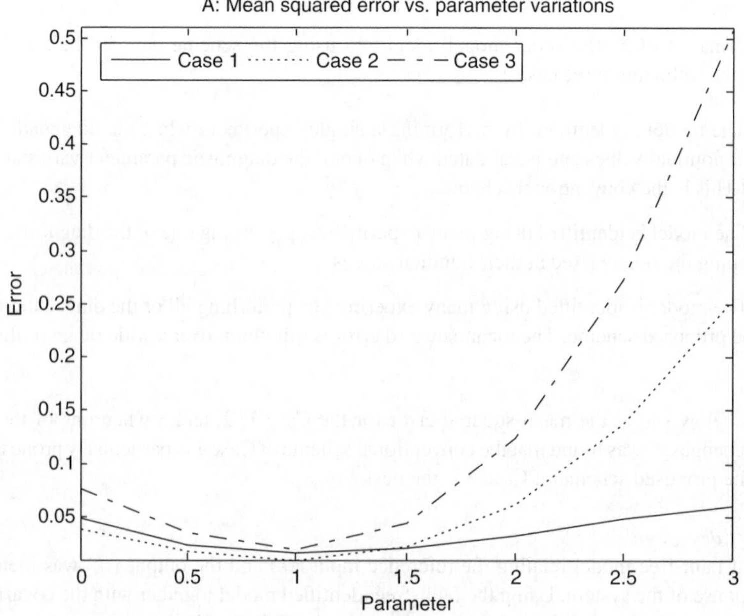

Figure 11.8 Model error: (i) Case 1, (ii) Case 2, (iii) Case 3

Table 11.1 Detectability and fault size estimation error

Emulator parameter	γ_1	γ_2	γ_3
Threshold η_{th}	0.45	0.55	0.45
Minimum variation $\Delta\gamma_i$	5%	6.3%	10%
Maximum fault size estimation error	0.04	0.04	0.06

11.3.3 Detection of Fault

The Bayes detection strategy is given by:

$$t_s(e) \begin{cases} \leq \eta_{th} & no\,fault \\ > \eta_{th} & fault \end{cases} \tag{11.4}$$

Where the statistics (for a square wave input) $t_s(e) = \frac{1}{N}\sum_{i=0}^{N-1} e^2(k-i)$. A fault is detected whenever the moving average of the residual energy exceeds a threshold value.

Faults were injected by changing the emulator parameters $\{\gamma_i\}$ stepwise in small increments. Table 11.1 gives the performance of the Bayes detection strategy. The threshold value, η_{th}, the minimum variation $\Delta\gamma_i$in the emulator parameter $\gamma_i$that can be detected accurately, and the maximum error in estimating the fault size. We can see that the worst-case minimal variation $\Delta\gamma_i$ that can be detected is 10% for the actuator fault. The maximum fault size estimation error is 0.06 for the actuator fault.

11.3.4 Fault Isolation

This says that the hypothesis of a fault in γ_i is correct when the angle between the residual vector $\boldsymbol{e}(k)$ and the vector, $\boldsymbol{\Phi}(k)\boldsymbol{\Omega}_i$ is minimum. Alternatively we may state that hypothesis H_i is asserted (that is only one diagnostic parameter γ_i has varied), if the residual and its estimate are maximally aligned in the sense defined by Eq. (11.2). The function $\cos^2 \boldsymbol{\varphi}_j(k)$, which is derived by maximizing the log-likelihood function $l(\boldsymbol{\gamma} \mid \boldsymbol{e})$ for case when there is single fault, is the *discriminant function* for fault isolation.

The actual and the estimated faults are shown in Figure 11.9. Each of three faults, in sequence, increase in a stepwise fashion to 1.5 times the nominal value and then decrease suddenly back to the nominal value of unity. The implementation indicates that the methodology is able to capture incipient faults and sudden faults. The subfigure on the top shows the output $y(k)$ when the reference input $r(k)$ is a square wave. During time interval [1 70], the emulator parameter γ_2 was varied (the value the actuator gain is denoted by k_v). During the next time interval [80 150], the emulator parameter γ_1 was varied (the value the sensor gain is denoted by k_s). During time interval [170 240], the emulator parameter γ_3 was varied (the value of the actuator gain is denoted by k_a). The actual emulator and its estimate are indicated in bold and dashed lines.

11.3.5 Fault Isolability

Isolability is a measure of the ability to distinguish and isolate a fault in γ_j, from a fault in $\gamma_i, i \neq j$. Isolability is improved for a larger difference between φ_i, and $\{\varphi_j, j \neq i\}$. The fault isolability measure is given by:

$$\cos^2 \varphi_{ij}(k) = \left(\frac{|\langle \boldsymbol{\Phi}(k)\boldsymbol{\Omega}_i, \boldsymbol{\Phi}(k)\boldsymbol{\Omega}_j \rangle|}{\|\boldsymbol{\Phi}(k)\boldsymbol{\Omega}_i\| \|\boldsymbol{\Phi}(k)\boldsymbol{\Omega}_j\|} \right)^2 \tag{11.5}$$

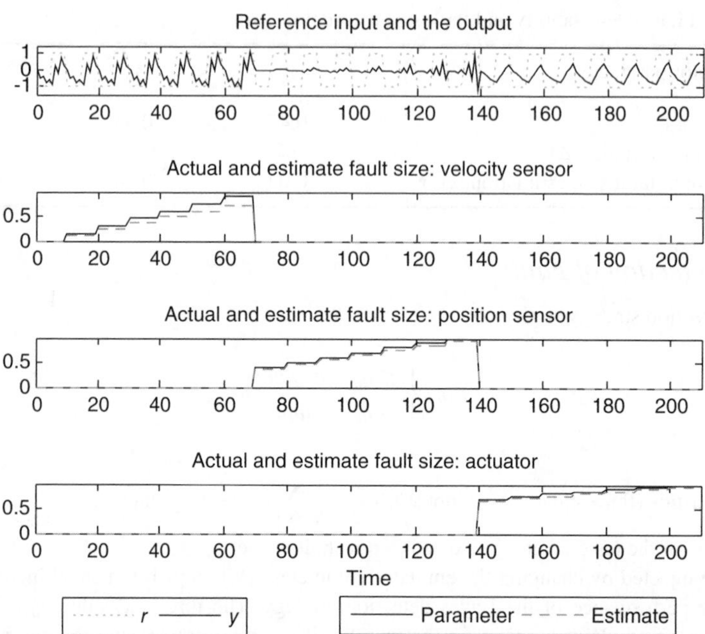

Figure 11.9 Fault isolation showing the faults, (solid) and the generated fault estimates (dashed)

The isolability measures are listed in Table 11.2. It can be seen that all faults can be isolated, however, γ_2 versus γ_3 has poorer isolability.

Comments *The results of the evaluation on the physical systems show acceptable performance even though the physical system is nonlinear, complex, and is subject to uncertainty in the form of unmodeled dynamics, including nonlinear effects such as friction, backlash, and saturation. The proposed identification scheme plays a key role in ensuring detection and isolation capability. The identified model is accurate over a wide range of variation in the diagnostic parameters. We have only considered a single input and a single output system where fault diagnosis capability is constrained by the availability of only one output measurement for a given input.*

A reliable offline identification of the physical system based on performing a number of experiments by perturbing the emulator parameters to imitate the fault operating regimes was crucial to the performance of the fault diagnosis.

It is interesting to note that although the physical system is nonlinear and complex the fault diagnosis scheme based on linear model appears to perform very well. This is, in part, due to the fact that incipient ("small" variations in the diagnostic parameters) were faults that were introduced and highly reliable identification based on performing a number of parameter perturbed experiments.

Table 11.2 Worst-case isolability measures

γ_i	γ_i	$\max(\lvert\cos\varphi_{ij}\rvert)$
γ_2	γ_1	0.6255
γ_2	γ_3	0.8005
γ_1	γ_3	0.7817

11.4 Summary

A fault is modeled as a variation in the diagnostic parameters. Emulator parameters play the role of diagnostic parameters to mimic normal and fault scenarios. A unified approach to fault detection and isolation was used based on Kalman filter residuals. Thanks to the identification scheme based on the emulator parameter-perturbed experiments, the identified model captures not only the input–output behavior of the system but also the diagnostic parameter input to the system output, and as a result the performance of the fault detection and isolation was acceptable.

References

[1] Doraiswami, R. and Cheded, L. (2013) A unified approach to detection and isolation of parametric faults using a Kalman filter residuals. *Journal of Franklin Institute*, **350**(5), 938–965.

[2] Doraiswami, R., Diduch, C. and Tang, T. (2010) A new diagnostic model for identifying parametric faults. *IEEE Transactions on Control System Technology*, **18**(3), 533–544.

[3] Doraiswami, R. and Cheded, L. (2012) Kalman filter for fault detection: an internal model approach. *IET Control Theory and Applications*, **6**(5), 1–11.

[4] Doraiswami, R. and Cheded, L. (2013) Fault diagnosis of a sensor network: a distributed filtering approach. *Journal of Dynamic Systems, Measurement and Control*, **135**(5).

[5] Ljung, L. (1999) *System Identification: Theory for the User*. Prentice Hall, New Jersey.

[6] Forssell, U. and Ljung, L. (1999) Closed loop identification revisited. *Automatica*, **35**, 1215–1241.

[7] Qin, J.S. (2006) Overview of subspace identification. *Computer and Chemical Engineering*, **30**, 1502–1513.

[8] Brown, R.G. and Hwang, P.Y. (1997) *Introduction to Random Signals and Applied Kalman Filtering*. John Wiley and Sons.

[9] Doraiswami, R.L., Cheded, L., and Khalid, M.H. (2010) *Sequential Integration Approach to Fault Diagnosis with Applications: Model-free and Model- Based Approaches*.VDM Verlag Dr. Muller Aktiengesellschaft & Co. KG.

[10] Mallory, G. and Doraiswami, R. (1997) A frequency domain identification scheme for control and fault diagnosis. *Transactions of the ASME: Journal of Dynamic Systems, Measurement and Control*, **119**(49), 48–56.

[11] Doraiswami, R., Diduch, C., and Kuehner, J. (2001) Failure Detection and Isolation: A New Paradigm. Proceedings of the American Control Conference, Arlington, USA.

[12] Liu, Y., Diduch, C., and Doraiswmi, R. (2003) Modeling and Identification for Fault Diagnosis. 5th IFAC Symposium on Fault Detection, Supervision and Safety of RTechnical Process, Safeprocess 2003, Washington D.C.

12

Fault Diagnosis of a Sensor Network

12.1 Overview

A model-based approach is developed for fault diagnosis of a sensor network formed of cascade, parallel, and feedback combination of subsystems and sensors. The objective is to detect and isolate a fault in any of the subsystems and measurement sensors which are subject to disturbances and/or measurement noise. The approach hinges on the use of a bank of Kalman filters (KF) to detect and isolate faults. Each KF is driven by either a pair (i) of consecutive sensor measurements or (ii) of a reference input and a measurement. It is shown that the KF residual is a reliable indicator of a fault in subsystems and sensors located in the path between the pair of the KF's input. A simple and efficient procedure is developed that analyzes each of the associated paths and leads to both the detection and isolation of any fault that occurs in the paths analyzed. The scheme is successfully evaluated on several simulated examples and a physical fluid system exemplified by a benchmarked laboratory-scale two-tank system to detect and isolate faults including sensor, actuator, and leakage ones. Further, its performance is compared with those of the Artificial Neural Network and fuzzy logic-based model-free schemes.

In general, there are two broad classes in fault diagnosis: model-free and model-based [1–21]. The former class includes tools based on Limit checking, Visual, and Plausibility (LVP) analysis, Artificial Neural Network (ANN), and Fuzzy Logic (FL). All these approaches have their own strengths and weaknesses, especially when applied to physical systems. A model-free approach is capable of detecting a possible fault quickly, unraveling its root cause(s), and isolating it. Its lack of dependence on a model imparts to it an equally attractive freedom from the usual model-related difficulties, such as identifying the required model, dealing with the presence of nonlinearities and structural complexities. However, these advantages are realized at a cost that could have various facets depending on the tool used. For neural networks, there is a lack of transparency, a need for a sufficient amount of training data covering most, if not all, operational scenarios, and a possibly lengthy training time. Fuzzy logic techniques, though less opaque than neural networks, suffer from the difficulty of deriving precise rules that distill an expert's knowledge of the application domain and which are necessary to drive the fuzzy inference engine. The Adaptive Neuro-Fuzzy Inference System (ANFIS) combines the advantages of the ANN and FL approaches so that the decision strategy is no longer a black box but a highly transparent set of if-then rules which are extracted from the set of representative and sufficient faulty and fault-free data. Both the ANN and ANFIS are universal approximators. On the other hand, given the availability of an appropriate model, the model-based method is transparent and provides a complete and accurate

Identification of Physical Systems: Applications to Condition Monitoring, Fault Diagnosis, Soft Sensor and Controller Design, First Edition. Rajamani Doraiswami, Chris Diduch and Maryhelen Stevenson.

diagnostic picture by exploiting a wealth of readily available and powerful analysis and design tools. The model-based approach is based on the use of Kalman filtering and system identification [1–13]. A bank of Kalman filter-based approaches is widely used for detecting and isolating sensor and subsystem faults, especially when the outputs of the subsystems are measurable [2, 4, 6]. The complex problem of detecting and isolating faults is divided into a number of simple problems. Each Kalman filter detects and isolates a subset of sensor and subsystem faults.

In recent years, Sensor Networks (SNs) are widely used for information collection and monitoring solution for a variety of applications. The sensor networks using wireless, wired, or their combination is being developed for a wide variety of practical applications, including military applications such as battlefield surveillance, many industrial and civilian application areas, including industrial process monitoring and control, machine health monitoring, environment and habitat monitoring, healthcare applications, and traffic control [22–25]. They are also used to monitor the integrity of a network of pipes carrying various fluids (water, oil, gas or other) and in target-tracking applications [25]. The advent and proliferation of both wireless sensors and embedded processors and the consequent deployment of intelligent sensors have undoubtedly been the key drivers behind the ever-increasing use of SNs.

Faults occurring to subsystems comprising the SN are common due to inherent malfunction of the subsystem itself and the harsh environment where the subsystems are located. Fault diagnosis of SN is still a challenging problem and continues to be a subject of intense research both in industry and in academia in view of (a) the stringent and conflicting requirements in practice for a high probability of correct diagnosis, (b) a low false alarm probability, (c) a timely decision on the fault status, and (d) devices and subsystems whose fault have to be diagnosed are distributed over a large geographical region.

There are a number of distributed fault detection schemes to detect faulty sensors in the SN. Its extension to the case where the sensor network formed of interconnected dynamic subsystems and sensors spread over a wide geographical region is considered herein. The extended class of SN finds applications to present day interconnected systems such as power utilities with a number of power generating stations connected to transmission networks, refineries with networks of pipes connecting the refining processes and tanks, desalination plants, process control, and chemical industries. The fault diagnosis of the extended class of SN is developed herein.

The main objective of this chapter is to design an effective strategy for monitoring the integrity and proper operation of a sensor network, such as the one depicted in Figure 12.1, in the face of faults

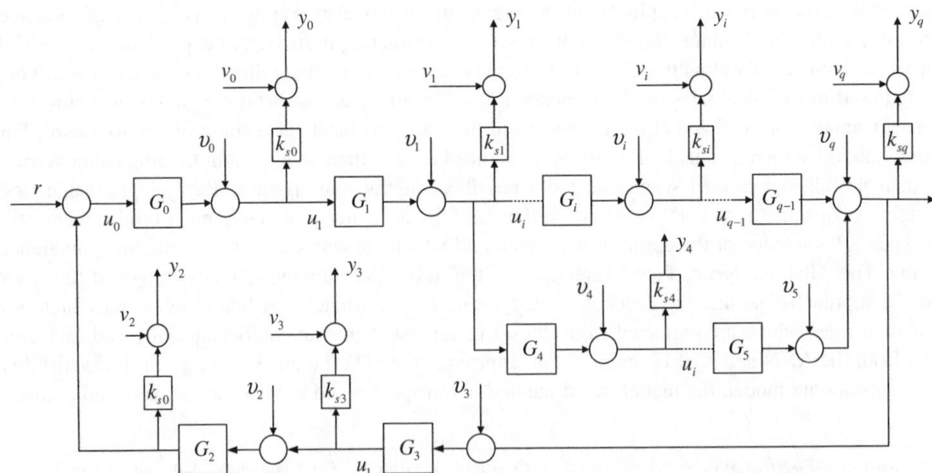

Figure 12.1 A sensor network formed of interconnections of subsystems $\{G_i\}$ and sensors $\{k_{si}\}$

occurring in the interconnected network formed of sensors and subsystems. In this scheme, a faulty sensor or a faulty subsystem is automatically detected, isolated, and located within the interconnected system, using both the measurement data collected, and the topography of the sensor network used.

The model-based Kalman filter-based as well model-free approaches, including the ANN and ANFIS using Takagi–Sugeno fuzzy systems [26, 27], were evaluated on simulated as well as physical systems.

This chapter is based primarily on the work presented in [2].

12.2 Problem Formulation

A typical sensor net of interconnected dynamic subsystems and sensors is shown in Figure 12.1, where $\{G_i\}$ are subsystems, $\{k_{si}\}$ are sensors, $\{y_i\}$ are output measurements, and r is the reference input to the closed-loop SN. A node "i" is comprised of a subsystem represented by a dynamic model G_i with disturbance w_i, and a sensor denoted by a static gain k_{si} with measurement noise v_i.

The subsystem model is:

$$y_i(z) = G_i(z)u_1(z) + v_i(z) \tag{12.1}$$

However, in a practical sensor network, the sensor is modeled as a gain and a delay given by

$$y_i(z) = k_{si}(z)u_i(z) + v_i(z) \tag{12.2}$$

where y_i, u_i, and v_i are respectively the output measurement, input to the subsystem G_i, and the measurement noise; $k_{si}(z) = k_{si}z^{-\tau_i}$, k_{si} is the gain and τ_i is the time-delay as shown in Figure 12.1.

Without any loss of generality, the proposed fault diagnosis scheme is developed for a typical closed-loop feedback sensor network formed of cascade-connected subsystems G_i: $i = 0, 1, 2, \ldots q$ in the forward path shown in Figure 12.2. The subsystem G_i and the sensor are in general dynamical systems.

12.3 Fault Diagnosis Using a Bank of Kalman Filters

A distributed fault diagnosis scheme is employed and it hinges on the use of a bank of Kalman Filters (KFs) to detect and isolate a simple fault and multiple faults. The Kalman filter is employed as it meets the key requirement of fault detection, namely its residual is a zero-mean white noise process if and only if there is no fault in both the subsystems and sensors located in the path between the pair of the

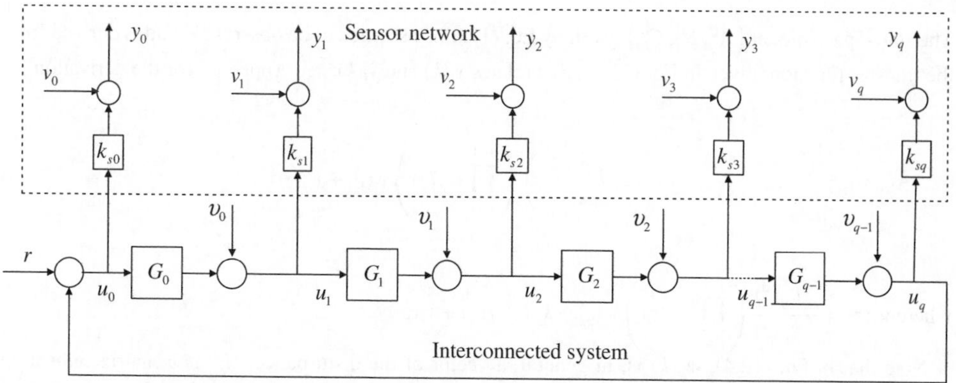

Figure 12.2 A typical closed-loop feedback sensor network

KF's inputs. If there is a fault, the residual will then an additive fault-indicating term. Each KF is driven by either a pair of (i) consecutive sensor measurements, y_{i-1} and y_i with y_{i-1} as an input and y_i as the desired output, as well as (ii) the reference input r and the measurement y_i, with r as an input and y_i as the desired output. The pairs in (i) and (ii) listed above are associated, respectively with the forward and feedback paths of the closed-loop sensor network.

An identified nominal (fault-free) model forms the core of the Kalman filter. As both the accuracy and reliability of the identified model are crucial for a superior fault diagnosis performance, a more accurate identification scheme is employed. The nominal system model is identified by performing a number of parameter perturbed experiments to mimic various likely operating scenarios (instead of the conventional scheme based on identifying the nominal mode from a single experiment). Emulators, which are transfer function blocks, are included at the accessible points in the system such as at the input, the output, or both to mimic the likely operating scenarios. The model thus identified is highly reliable and accurate as it captures the behavior of the system not only at the given operating point but also in its neighborhood.

12.4 Kalman Filter for Pairs of Measurements

The Kalman filter using $\{y_i(k), y_j(k)\}$ is governed by the following equations:

$$\begin{aligned} \hat{x}_{ij}(k+1) &= A_{ij}^0\hat{x}_{ij}(k) + B_{ij}^0 y_i(k) + K_{ij}^0\left(y_j(k) - C_{ij}^0\hat{x}_{ij}(k)\right) \\ e_{ij}(k) &= y_j(k) - C_{ij}^0\hat{x}_{ij}(k) \end{aligned} \tag{12.3}$$

where $e_{ij}(k)$ is the residual; K_{ij}^0 is the nominal Kalman gain, $\left(A_{ij}^0, B_{ij}^0, C_{ij}^0\right)$ is the state-space model of the system formed of subsystems and sensors located between the measurement nodes of $y_i(k)$ and $y_j(k)$, that is:

$$\begin{aligned} x_{ij}(k+1) &= A_{ij}^0 x_{ij}(k) + B_{ij}^0 y_i(k) + \begin{bmatrix} B_{ij}^0 & E_{ij}^0 \end{bmatrix}\begin{bmatrix} v_i(k) \\ w_{ij}(k) \end{bmatrix} \\ y_j(k) &= C_{ij}^0 x_{ij}(k) + v_j(k) \end{aligned} \tag{12.4}$$

The state-space model $\left(A_{ij}^0, B_{ij}^0, C_{ij}^0\right)$ given by Eq. (12.4) is assumed to be observable and is derived from the transfer function, given in Eq. (12.5), that relates $y_i(k)$ and $y_j(k)$: see Appendix for the derivation

$$y_j(z) = \left(\frac{k_{sj}(z)}{k_{si}(z)}\right)\left(\prod_{\ell=i}^{j-1} G_\ell(z)\right) y_i(z) + v_{ij}(z) \tag{12.5}$$

where $v_{ij}(z) = -\dfrac{k_{sj}(z)}{k_{si}(z)}\left(\prod_{\ell=i}^{j-1} G_\ell(z)\right) v_i(z) + k_{sj}(z)\upsilon_{ij}(z) + v_j(z)$.

Note that in Eq. (12.4), $w_{ij}(k)$ is, in general, a vector of the disturbances, E_{ij}^0 is a matrix indicating the entry points of the disturbances, v_{ij} depicts the effect of the disturbances $w_{ij}(k)$ and the measurement noises v_i and v_j that affect the output y_j.

12.5 Kalman Filter for the Reference Input-Measurement Pair

The Kalman filter using the reference input-measurement pair $\{r(k), y_i(k)\}$ is given by:

$$\begin{aligned}\hat{x}_{rj}(k+1) &= A^0_{rj}\hat{x}_{ri}(k) + B^0_{rj}r(k) + K^0_{rj}\left(y_j(k) - C^0_{rj}\hat{x}_{rj}(k)\right)\\ e_{rj}(k) &= y_j(k) - C^0_i\hat{x}_{rj}(k)\end{aligned} \tag{12.6}$$

where $e_{rj}(k)$ is the residual; K^0_{rj} the Kalman gain and the triplet $\left(A^0_{rj}, B^0_{rj}, C^0_{rj}\right)$ the state-space model of the system formed of subsystems and sensors located between the measurement nodes of $r(k)$ and $y_j(k)$, and is assumed to be observable. Similar to Eq. (12.4), the state-space model $\left(A^0_{ij}, B^0_{ij}, C^0_{ij}\right)$ given below in Eq. (12.7) is derived from the transfer function of Eq. (12.8) that relates $r(k)$ and $y_j(k)$

$$\begin{aligned}x_{rj}(k+1) &= A^0_{rj}x_{rj}(k) + B^0_{rj}r(k) + E^0_{rj}w_{rj}(k)\\ y_j(k) &= C^0_{rj}x_{rj}(k) + v_j(k)\end{aligned} \tag{12.7}$$

$$y_j(z) = k_{sj}(z)\left(\frac{\prod_{\ell=0}^{i} G_\ell(z)r(z)}{1 + \prod_{\ell=0}^{q-1} G_\ell(z)}\right) + v_{rj}(z) \tag{12.8}$$

where $v_{rj}(z) = v_j(z) + k_{sj}(z)\upsilon_{rj}(z)$; $w_{rj}(k)$ is, in general, a vector of the disturbances, E^0_{rj} is a matrix indicating the entry points of the disturbances, v_{rj} represents the effect of the disturbances $w_{rj}(k)$ and the measurement noise v_j affecting the output y_j. See Appendix for the derivation of Eq. (12.8).

12.6 Kalman Filter Residual: A Model Mismatch Indicator

Using the results in the chapter on the Kalman filter, any residual generated by the Kalman filter used for a particular pair of nodes may be expressed in terms of its associated model mismatch term $\Delta G(z) = G_0(z) - G(z)$ where $G_0(z)$ and $G(z)$ are respectively the actual and the nominal transfer functions of the system for that particular pair of nodes. We shall now give the expressions of the residuals, for the two pairs envisaged herein, namely a pair of measurements and a reference input-measurement pair.

12.6.1 Residual for a Pair of Measurements

Recall, that the residual for a pair of measurements $\{y_i(k), y_j(k)\}$ can be decomposed into two terms: A deterministic fault (or model mismatch)-indicator term (with the subscript "*f*") and a noise term (with the subscript "0"), as shown below.

$$e_{ij}(z) = e_{fij}(z) + e_{0ij}(z) \tag{12.9}$$

$$e_{fij}(z) = \Delta G_{ij}(z)\left(F^{-1}_{ij0}(z)D_{ij0}(z)y_i(z)\right) \tag{12.10}$$

$$e_{0ij}(z) = F^{-1}_{ij0}(z)D_{ij0}(z)\upsilon_{ij}(z) \tag{12.11}$$

where $\Delta G_{ij}(z) = \left(\frac{k^0_{sj}(z)}{k^0_{si}(z)}\right)\left(\prod_{\ell=1}^{j-1} G^0_\ell(z)\right) - \left(\frac{k_{sj}(z)}{k_{si}(z)}\right)\left(\prod_{\ell=1}^{j-1} G_\ell(z)\right)$, $G^0_\ell(z)$ and k^0_{sj} are the nominal subsystem and the sensor gain respectively.

12.6.1.1 Residual for Reference Input-measurement Pair

Similarly for this pair and with reference to the Appendix, we can write that:

$$e_{rj}(z) = e_{frj}(z) + e_{0rj}(z) \tag{12.12}$$

$$e_{frj}(z) = \Delta G_{rj}(z)\left(F_{rj0}^{-1}(z)D_{rj0}(z)r(z)\right) \tag{12.13}$$

$$e_{0rj}(z) = F_{rj0}^{-1}(z)D_{rj0}(z)v_{rj}(z) \tag{12.14}$$

where $\Delta G_{rj}(z) = k_{sj}^{0}(z)\left(\dfrac{\prod_{\ell=0}^{j-1} G_{\ell}^{0}(z)}{1+\prod_{\ell=0}^{q-1} G_{\ell}^{0}(z)}\right) - k_{sj}(z)\left(\dfrac{\prod_{\ell=0}^{j-1} G_{\ell}(z)}{1+\prod_{\ell=0}^{q-1} G_{\ell}^{0}(z)}\right)$;

Remarks $e_{fij}(z)$ ($e_{frj}(z)$) and $e_{0ij}(z)$ ($e_{0rj}(z)$) are the filtered versions of the input and the noise respectively, where the filter's transfer function is given by $F_{0}^{-1}(z)D_{0}(z)$. Thus the residual $e_{ij}(z)$ ($e_{rj}(z)$) is the sum of two components, one of which $e_{fij}(z)$ ($e_{fij}(z)$) is the *model-mismatch indicator* and the other $e_{0ij}(z)$ ($e_{0rj}(z)$) is assumed to be a zero-mean white noise process. The model-mismatch indicator is the deviation in the output of the system, caused by a fault-induced perturbation in the system model, and hence is employed herein to monitor the status of the system.

In the case where the sensor is modeled as a gain and delay given by Eq. (12.2), Eqs. (12.12)–(12.14) hold. Hence we may use the time-delay model for the sensor using the proposed bank of Kalman filters approach.

From the properties of the Kalman filter residuals, we deduce the following:

- The model-mismatch indicator $e_{fij}(z)$ ($e_{fij}(z)$) is zero if and only if there is no fault. If there is a fault, the mismatch indicator will belong to the same class as that of the reference input signal. For example, it will be constant, a sinusoid, and a colored noise respectively if the reference input is constant, a sinusoid, and a colored noise.
- The auto-correlation of the residual $e_{ij}(z)$ ($e_{rj}(z)$) is a better visual indicator than the residual as the correlation operation attenuates the noise component and shows clearly the characteristics of the auto-correlation of the mismatch indicator. If there is no fault the auto-correlation of the residual will be a delta function. In the faulty case, the auto-correlation of $e_{ij}(z)$ ($e_{rj}(z)$) will not be a delta function. For example, it will be triangular with a peak value at the zero lag for a constant, a sinusoid for a sinusoid, and an impulse response of the coloring filter for a colored noise reference input
- If there is a model mismatch, the residual signal $e(k)$ would be a normally-distributed stochastic process with a nonzero mean value $e_f(k)$.

12.7 Bayes Decision Strategy

The well-established binary Bayes decision strategy is employed as it is a gold standard in decision-making in the face of uncertainties. A test statistic, which is a function of the residual, its PDF, and a class of reference inputs, is derived and compared with a threshold value to assert the fault status of the subsystems and sensors located in the path. A subsystem or a sensor is assigned "binary 1" if the test statistic exceeds the threshold value, a "0" if it is less than the threshold value, and a "don't care" character x if the subsystem or a sensor is not located in the path under test. A fault in a subsystem or in a sensor may further be cross-checked by analyzing different paths between sensor nodes, thereby increasing the reliability of the proposed scheme. By analyzing the binary assignments from all the local nodes, the faulty subsystems and sensors in the SN may then be determined using simple binary arithmetic operations. Using the Bayes decision strategy, faults in subsystems and sensors are effectively detected and isolated as explained next. Let $H_0^{rj}(H_0^{ij})$ and $H_1^{rj}(H_1^{ij})$ be two hypotheses indicating respectively the absence and presence of a fault. The Bayes decision strategy then takes the general form:

$$t_s(\boldsymbol{e}) \begin{cases} \leq \eta & \textit{no fault} \\ > \eta & \textit{fault} \end{cases} \tag{12.15}$$

where $t_s(\boldsymbol{e})$ is the test statistics of the residual $\boldsymbol{e}$ and η is the threshold value computed, taking into account the variance of the noise term $e_0(k)$, prior probabilities of the two hypotheses, the cost associated with correct and wrong decisions, and the probability of a false alarm. The determination of the threshold η is given in [3]. In practice, however, the physical systems are nonlinear and the noise and disturbance statistics are unknown. We used online tuning of the threshold using all the data collected so as to cover various faults and determine an appropriate value of the threshold which would give a high probability of a correct decision with an acceptable false alarm.

Test statistics for a reference input, that is either a constant or a sinusoid of frequency f_0 or an arbitrary signal, are all listed below for completeness [28]:

$$t_s(\boldsymbol{e}) = \begin{cases} \left| \dfrac{1}{N} \sum\limits_{i=k-N+1}^{k} e(i) \right| & r(k) = \text{constant} \\ P_{ee}(f_0) & r(k) \text{ is a sinusoid} \\ \dfrac{1}{N} \sum\limits_{i=k-N+1}^{k} e^2(i) & r(k) \text{ is an arbitrary signal} \end{cases} \tag{12.16}$$

where $P_{ee}(f_0)$ is the power spectral density of the sinusoid. Bayes decision strategies for the reference input- measurement pair and the measurement-measurement pair take the following respective forms:

$$t_s(\boldsymbol{e}_{rj}) \begin{cases} \leq \eta_{rj} & \textit{decide } H_0^{rj} & : \textit{no fault} \\ > \eta_{rj} & \textit{decide } H_1^{rj} & : \textit{fault} \end{cases} \tag{12.17}$$

$$t_s(\boldsymbol{e}_{ij}) \begin{cases} \leq \eta_{ij} & \textit{decide } H_0^{ij} & : \textit{no fault} \\ > \eta_{ij} & \textit{decide } H_1^{ij} & : \textit{fault} \end{cases} \tag{12.18}$$

12.8 Truth Table of Binary Decisions

Our objective here is to detect simple or multiple faults and isolate the faulty subsystem(s) and sensor(s). We will first restrict ourselves to merely pinpointing the faulty components without estimating the fault sizes. To meet this requirement, a binary logic approach is employed:

- If hypothesis H_0^{rj} or H_0^{ij} is true from Eq. (12.17) or (12.18), respectively, then all the components located in the path between the pair of nodes are assigned a binary zero indicating none of the components is faulty. The hypotheses are then treated as false and denoted by a zero, that is,: $H_0^{rj} = 0$ or $H_0^{ij} = 0$.
- If hypothesis H_1^{rj} or H_1^{ij} is true, then at least one component is faulty. This is indicated by assigning a binary one to all the components located in the path. The hypotheses are then treated as true and denoted by a one, that is,: $H_1^{rj} = 1$ or $H_1^{ij} = 1$.
- Components, which are not located in the path between the pair of nodes or the hypothesis is not known, are assigned a "don't care" character x, which can assume either 0 or 1.

Based on the above convention, a truth table can then be derived whose columns indicate the components under test and whose rows indicate the status of the fault at these components. The derivation of such a table is illustrated below, but prior to that let us derive two key Propositions on the possibility of detecting and isolating single and multiple faults.

Proposition 12.1 Simple faults in both subsystems and sensor, and only one component may be faulty at any one time

- *If there are* q *subsystems,* G_i *,* $i = 0, 1, 2, \dots, q - 1$*, in a sensor network, then a single fault occurring in any one of the* q *subsystems at any one time can be detected and isolated.*
- *If there are* $(q + 1)$ *sensors,* k_{si} *,* $i = 0, 1, 2, \dots, q$*, in a sensor network, then a single fault occurring in any one of the* $(q + 1)$ *any one time can be detected and isolated.*

Proposition 12.2 Multiple faults in sensors (subsystems) with no fault in any subsystems (sensors)

- *If there are* $(q + 1)$ *sensors,* k_{si} *,* $i = 0, 1, 2, \dots, q$*, then provided there are no faults in the subsystems* G_i $i = 0, 1, 2, \dots q - 1$*, then multiple faults occurring simultaneously in any subset of sensors, of size* i *with* $1 \leq i \leq (q + 1)$ *can be detected and isolated.*
- *If there are* q *subsystems, then* G_i *,* $i = 0, 1, 2, \dots, (q - 1)$*, then, provided there are no faults the sensors used, that is,* k_{si} *,* $i = 0, 1, 2, \dots, q$*, then multiple faults occurring simultaneously in any subset of the subsystems, of size* i *with* $1 \leq i \leq q$ *can be detected and isolated.*

Proof: The proof of both propositions follows directly from a binary logic analysis of the various paths analyzed and is clearly illustrated by the following examples.

Assumptions

A1. *The reference input must be rich enough to excite all the modes of the subsystems so that all types of faults resulting from variations in the subsystem parameters may be detected. For example, a constant or sinusoidal reference input can only detect a fault resulting from variations in the steady-state gain or gain at the input frequency.*

A2. *A sensor and a subsystem cannot be faulty at the same time.*

A3. *Multiple faults: more than one sensor can be faulty if none of the subsystems is faulty and more than one subsystem can be faulty if none of the sensors is faulty.*

Note here that in the case of single or multiple sensor fault (in the absence of any subsystem fault) there is redundancy in the sensor fault isolation phase as faults are decided by analyzing separate residuals associated with (i) pairs of relevant measurements, and (ii) the pairs made of the reference input and relevant measurements. Hence these two separate decisions may be cross-checked for accuracy by verifying whether both analyzes yield the same results. In this work, only consecutive pairs of measurements are employed, as explained next.

Comments *During the operating regime, a constant reference input, which is normally employed, may be adequate. It is important, however, to note that the reference input was selected to be dynamically "rich" enough during the offline identification phase so as to excite different modes of the subsystems. The system was identified by performing a number of experiments in which various sensor and subsystem faults were emulated. The model thus identified captures a system model that is robust to variations in* *the subsystem and sensor parameter variations [1–3], [8]. The identified model, which is reliable and accurate, was then employed in the Kalman filter design.*

In the sensor networks, if the network is subject to delay and packet dropouts, these phenomena may be dealt with using the approach proposed in [24].

The Bayes binary decision strategy based scheme is computationally simple, efficient and intuitive, and importantly has a high probability of correct decision with a low false alarm, in view of the following:

a) *The complex fault diagnosis problem of detecting and isolating* m *subsystem and sensor faults from* m *output measurements is split into* m *simple FDI problems. Each Kalman filter generates a residual from two output measurements to detect and isolate faults in a subset of the subsystems and the sensors. This 'divide and conquer approach' contributes to the computational efficiency of the proposed scheme.*
b) *The results of fault diagnosis may be cross-checked for accuracy as there is redundancy in the number of paths that may be analyzed.*
c) *The proposed scheme can detect any 'small' deviation from the normal operating regime that may not yet be a fault at the present time but may potentially lead to fault if it is not detected and accommodated in time. The power of this scheme in detecting these incipient faults allows it to be a vital component in any preventive or predictive condition-based maintenance scheme that interested companies wish to develop for example, in the broad and important area of health monitoring.*

12.9 Illustrative Example

Consider a sensor network formed of subsystems G_0, G_1, and G_2, and sensors k_{s0}, k_{s1}, k_{s2}, and k_{s3}, as shown in Figure 12.3. We will assume that the sensor is static with a constant gain, and that at least one sensor is not faulty. The threshold η is determine assuming the measurement noise is zero-mean. Without any loss of generality, the sensor k_{s0} is assumed to be the fault-free one. We will consider two cases of fault scenarios given below, where some of the hypotheses are not given:

Case 1: $H_0^{01} = 1; H_1^{12} = 1; H_0^{r0} = 0; H_1^{r1} = 1; H_1^{r2} = 1; H_0^{r3} = 0$

Case 2: $H_0^{01} = 0; H_1^{12} = 1; H_0^{23} = 0; H_1^{r0} = 1; H_1^{r2} = 1; H_1^{r3} = 1$

Tables 12.1 and 12.2 show the truth tables where there are seven rows associated with fault statuses of all consecutive pairs of measurements $\{y_{i-1}, y_i\}: i = 1, 2, 3$ (denoted in the first column of both tables by $(i-1, i)$; $i = 1, 2, 3$) and all the reference input-to-measurements pairs $\{r, y_i\}: i = 0, 1, 2, 3$ (denoted in the first column of both tables by (r, i); $i = 0, 1, 2, 3$). In both tables, columns indicate both the subsystems

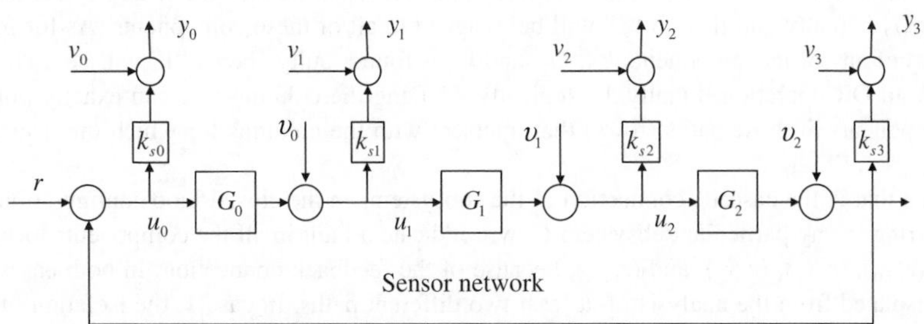

Figure 12.3 Sensor network example

Tables 12.1 and 12.2 Detection truth tables for Case 1 and Case 2 respectively

	Table 12.1								Table 12.2						
	G_0	G_1	G_2	k_{s0}	k_{s1}	k_{s2}	k_{s3}		G_0	G_1	G_2	k_{s0}	k_{s1}	k_{s2}	k_{s3}
(0,1)	1	x	x	0	1	x	x	(0,1)	0	x	x	0	0	x	x
(1,2)	x	1	x	x	1	1	x	(1,2)	x	1	x	x	1	1	x
(2,3)	x	x	x	x	x	x	x	(2,3)	x	x	0	x	x	0	0
(r,0)	0	0	0	0	x	x	x	(r,0)	1	1	1	0	x	x	x
(r,1)	1	1	1	x	1	x	x	(r,1)	x	x	x	x	x	x	x
(r,2)	1	1	1	x	x	1	x	(r,2)	1	1	1	x	x	1	x
(r,3)	0	0	0	x	x	x	0	(r,3)	1	1	1	x	x	x	1
d_f	0	0	0	0	1	1	0	d_f	0	1	1	0	0	0	0

and sensors used, that is, G_0, G_1, G_2, k_{s0}, $k_{s1}k_{s2}$; k_{s3}, and the rows indicate the two sets of pairs, that is, pairs of consecutive measurements and reference-measurement pairs. These two sets of pairs appear as part of the Bayes decisions $H_0^{rj}(H_1^{rj})$, $H_0^{(i-1)i}(H_1^{(i-1)i})$ that are associated with residuals $\{e_{(i-1)i}\}$ and $\{e_{rj}\}$. These Bayes decisions reflect the fault statuses of all of the seven components (three subsystems and four sensors) based on the Bayes binary decision strategy, with each row indicating "1" for a fault, a "0" for no fault and "x" for a "don't care condition," that is, either a "1" or a "0." Altogether, there are therefore seven components and seven pairs of nodes to be considered for fault isolation.

The intersection of each column with the elements of the seven rows indicates the fault status of the component in that particular column. By logically-ANDing the seven binary results in each column, a new binary value is obtained and stored in the decision vector d_f which occupies the last row in Tables 12.1 and 12.2 which is built on the assumption that subsystem and sensor faults cannot occur simultaneously.

Remarks The following explanations are provided here to further clarify how Tables 12.1 and 12.2 were generated. A KF is driven by the pair of measurements associated with the start and end of each analyzed path. For example, for path $(r, 3)$ in Table 12.1, the KF is driven by inputs r and y_3. The components within this path are G_0, G_1, G_2, and k_{s3}. If the KF residual test statistic is less than the threshold, thus indicating no fault, then a "0" is assigned to all the components within this path and an "x" is assigned to those outside the path. Similarly for path $(r, 3)$ in Table 12.2, the KF is driven by inputs r and y_3. In this case, the test statistic exceeds the threshold, thus indicating a fault. Here, we can only say that either G_0 or G_1 or G_2 or k_{s3} is faulty. Hence a "1" is assigned to G_0, G_1, G_2, and k_{s3}, indicating an OR operation, while components not in the path, that is, k_{s1} and k_{s2} are assigned an "x." Note that due to feedback, every path $(r, 1)$ $(r, 2)$ and $(r, 3)$ will be affected if any of the subsystems G_0, G_1, or G_2 is faulty and if so, a "1" will be assigned to all of these components. As for path $(2, 3)$, if the status of any of its components k_{s2}, G_2, and k_{s3} is found faulty, then a "1" will be assigned to all of them in an OR operation. Finally, by logically ANDing the columns, we can exactly pinpoint the faulty components in those paths (rows) that intersect with the columns for which the decision value d_f is "1."

Note also that in the cascade connection of the subsystems in the closed-loop configuration, a single fault occurring in any particular subsystem G_i will indicate a fault in all the components located along the paths (r, y_0), (r, y_1), (r, y_2), and (r, y_3), because of the feedback connection. In both cases 1 and 2, faults are isolated from the analysis of at least two different paths. In case 1, the isolation of the fault was reached from the analysis of the paths connecting the pairs of measurement nodes $(0, 1)$, $(1, 2)$, and $(2, 3)$; and the paths connecting the reference input and the measurement nodes $(r, 1)$ and $(r, 2)$.

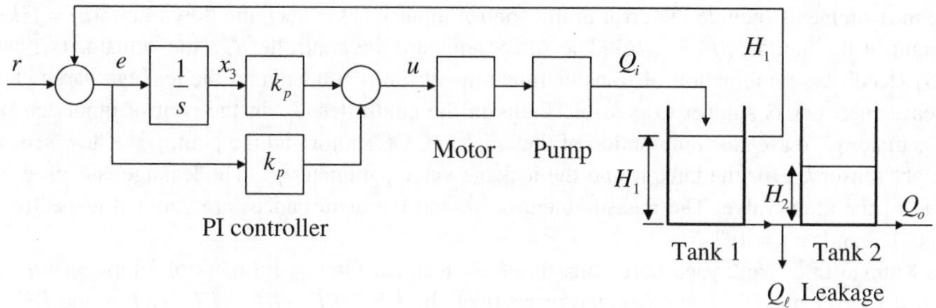

Figure 12.4 Schematics of the two-tank fluid system

In case 2, the isolation of the faults was confirmed from analyzing the pairs (1, 2), (2, 3); $(r, 0)$, $(r, 1)$; $(r, 2)$, $(r, 3)$.

The number of Kalman filters required is equal to the sum of the number of sensors and the number of subsystems whose faults have to be detected and isolated, where the subsystems refer to those whose outputs are measured. In this example, the number of measurement outputs is four, namely y_i: $i = 0, 1, 2, 3$. The number of sensors is also four, that is, k_{si}: $i = 0, 1, 2, 3$, and the number of subsystem is three, namely G_i: $i = 0, 1, 2$. Hence the total number of KFs is seven. There are seven subsystems and sensors and there are seven residuals, namely those associated with (0, 1), (1, 2), (2, 3), $(r, 0)$, $(r, 1)$, $(r, 2)$, $(r, 3)$. Hence we can determine the status of the seven subsystems and sensors from these seven residuals. In general, if there are m measurement outputs, there will be m sensors and $(m - 1)$ subsystems. The number of required Kalman filters would then be equal to $(2m - 1)$. The number of Kalman filters is linearly proportional to the number of measurement outputs.

12.10 Evaluation on a Physical Process Control System

The physical system under evaluation is a laboratory-scale process control system formed of two tanks connected by a pipe. The leakage is simulated in the tank by opening the drain valve. A DC motor-driven pump supplies the fluid to the first tank and Proportional Integral (PI) controller is used to control the flow rate in the second tank by maintaining the water height at the first tank at a specified level, as shown in Figure 12.4.

A sensor network representation of the two-tank system of Figure 12.4 is shown in Figure 12.5.

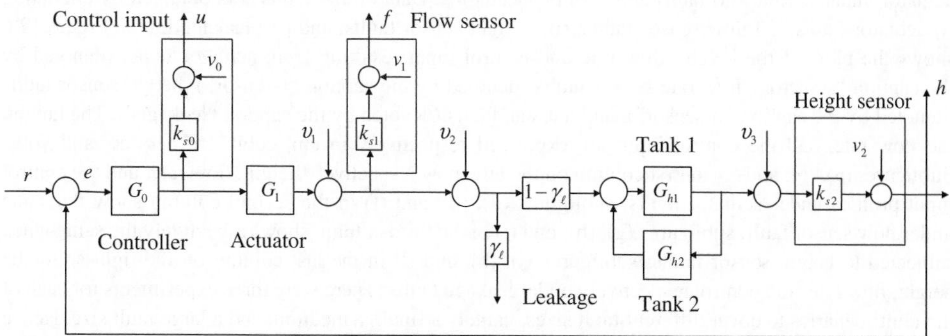

Figure 12.5 Two-tank fluid control system

The measurements include the error e, the control input $y_1(k) = u(k)$, the flow rate $y_2(k) = f(k)$, and the height of the liquid $y_3(k) = h_1(k)$. The subsystems are the controller G_0, the actuator G_1, and the tank G_2 (feedback combination of transfer functions of the two tanks and the leakage parameter γ_ℓ). The sensor network is subject to possible faults in the controller G_0, in the control input sensor k_{s0}, in the actuator G_1 (cascade combination of the amplifier, DC motor and the pump), the flow sensor k_{s1}, the height sensor k_{s2} for the tank 1, and the leakage valve parameter γ_ℓ. The leakage is caused by the opening of the drain valve. The measurement noise and the disturbances are denoted respectively by v_i: $i = 0, 1, 2$ and υ_i: $i = 1, 2, 3$.

The Kalman filter bank used here consists of six Kalman filters with pairs of I inputs (r, u), (r, f), (r, h), (e, u), (u, f), and (f, h), and denoted respectively by KF_{ru}, KF_{rf}, KF_{rh}, KF_{uf}, KF_{eu}, and KF_{uh}. The role of the Kalman filters is given below:

- KF_{ru} detects a fault in the controller G_0, or in the actuator G_1, or in the tank G_2, or in the controller sensor k_{s0}, or in the height sensor k_{s2}.
- KF_{rf} detects a fault in the controller G_0, or in the actuator G_1, or in the tank G_2, or in the height sensor k_{s2} or in the flow sensor k_{s1}.
- KF_{rh} detects a fault in the controller G_0, or in the actuator G_1, or in the tank G_2, or in the height sensor k_{s2}.
- KF_{eu} detects a fault in the controller G_0 or in the sensor k_{s0}.
- KF_{uf} detects a fault in the in the actuator G_1, in the controller sensor k_{s0}, or in the flow sensor k_{s1}.
- KF_{fh} detects a fault in the tank G_2, flow sensor k_{s1}, or height sensor k_{s2}.

12.11 Fault Detection and Isolation

The four simple faults considered are defined as follows: (i) an actuator fault in subsystem G_1, (ii) a flow sensor fault in k_{s1}, (iii) a height sensor fault in k_{s2}, and (iv) a leakage fault in subsystem G_2 (due to variations in γ_ℓ). As pointed out earlier, it is assumed that only a simple fault can occur at any given time instant. Each of the six Kalman filters used is designed using the identified nominal model relating its own pair of inputs.

The residuals of the Kalman filters are generated from all possible pairs of (i) sensor measurements $\{y_i(k), y_j(k)\}$ with $y_i(k)$ as input and $y_j(k)$ as the desired output as well as (ii) the reference input and measurements $\{r(k), y_j(k)\}$ with $r(k)$ as the input and $y_j(k)$ as the desired output. The pairs in (i) and (ii) listed above are associated respectively with the forward and feedback paths of the closed-loop sensor network.

A number of experiments were performed using National Instruments LABVIEW to acquire the required data on the height of the water level in the tank, the flow rate, and the control input to the actuator under normal (no fault) and "faulty" scenarios. Data on the faults was obtained by emulating (i) actuator faults, (ii) flow sensor faults, (iii) height sensor faults, and (iv) leakage faults. Figure 12.6 shows the plots of the height, flow rate and control input resulting from actuator faults (denoted by the caption "actuator," flow rate sensor faults (denoted by the caption "f-sensor"), height sensor faults (denoted by the caption "h-sensor"), and leakage faults (denoted by the caption "leakage"). The height, the flow rate, and the control input are expressed respectively in cm, cubic cm per sec, and volts. Subfigures (a), (b), and (c) in first column on the left show respectively height, flow rate, and the control input profiles under actuator faults; subfigures (d), (e), and (f) in the second column show the same under flow sensor fault; subfigures (g), (h), and (i) in the third column show respectively the same when subjected to height sensor faults; subfigures (j), (k), and (l) in the last column on the right show the height, flow rate, and control respectively under leakage faults. There were three experiments for each of the fault scenarios to cover different fault sizes, namely a small, a medium, and a large fault size. Hence there are three plots for each fault scenario in Figure 12.6. From the flow rate profiles, it can be deduced that the system is highly nonlinear, especially because of the saturation-type of nonlinearity it contains.

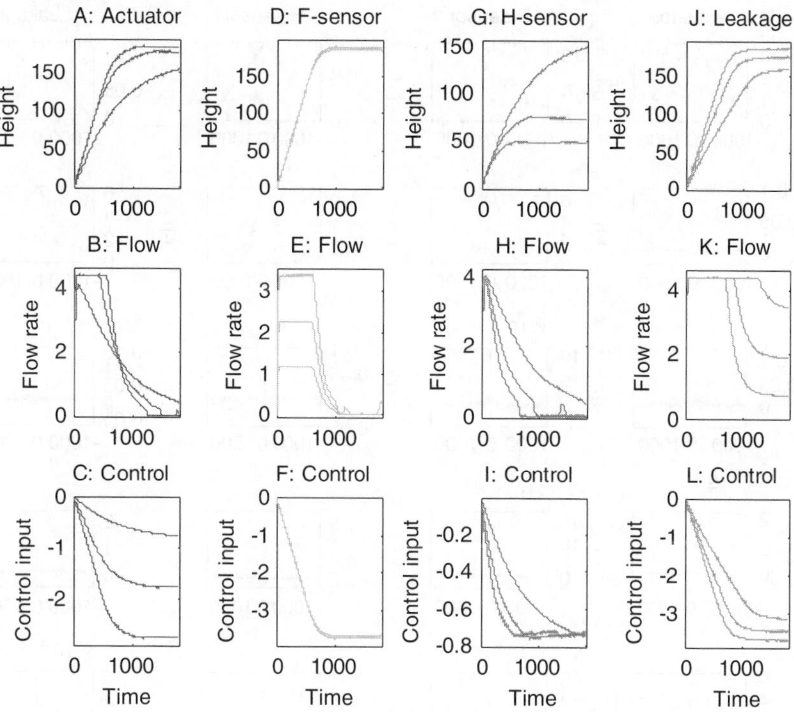

Figure 12.6 Plots of height, flow rate, control input

In our case the reference input is the constant and is the same for all fault scenarios and its value was chosen such that under fault-free (normal) operating scenario the saturation effect is absent: the height reaches the reference input value asymptotically thanks to the presence of PI controller. If however, there is fault, and as a consequence the flow rate hits the saturation limit, the height will be prevented from reaching asymptotically the reference input. It will reach a level less than the reference input. This can be clearly seen in the height profile.

For the data shown in Figure 12.6, and assuming a constant reference input, the test statistics have been computed based on Eq. (12.16), for the three experiments under actuator, flow sensor, height sensor, and leakage faults. As data is affected by nonlinearity effects, and the noise statistics were unknown, the threshold value η for the decision strategy Eq. (12.15) were determined from the fault data shown in Figure 12.6 such that 15% variations from nominal values of the sensor and subsystem parameters could be detected and isolated to ensure an acceptable false alarm probability and a high probability of correct fault diagnosis.

Figure 12.7 shows the result of the evaluation using the auto-correlation of residuals generated by the bank of Kalman filters as a result of an actuator, flow sensor, height sensor, and leakage faults. The auto-correlation of the residual instead of the residual is plotted as it gives a better visual picture of the presence or an absence of a fault. If there is no fault, the auto-correlation is ideally a delta function.

Figure 12.7 consists of four columns and five rows. The five subfigures in the first, the second, the third, and finally the fourth column show respectively the auto-correlation of the residuals under actuator, flow sensor, the height sensor, and the leakage faults. The first, the second, the third, the fourth, and the fifth row in each of the four columns show respectively the auto-correlation of the residuals generated by Kalman filters driven by the pairs of inputs (u, f), (f, h), (r, h), (r, f), and (r, u).

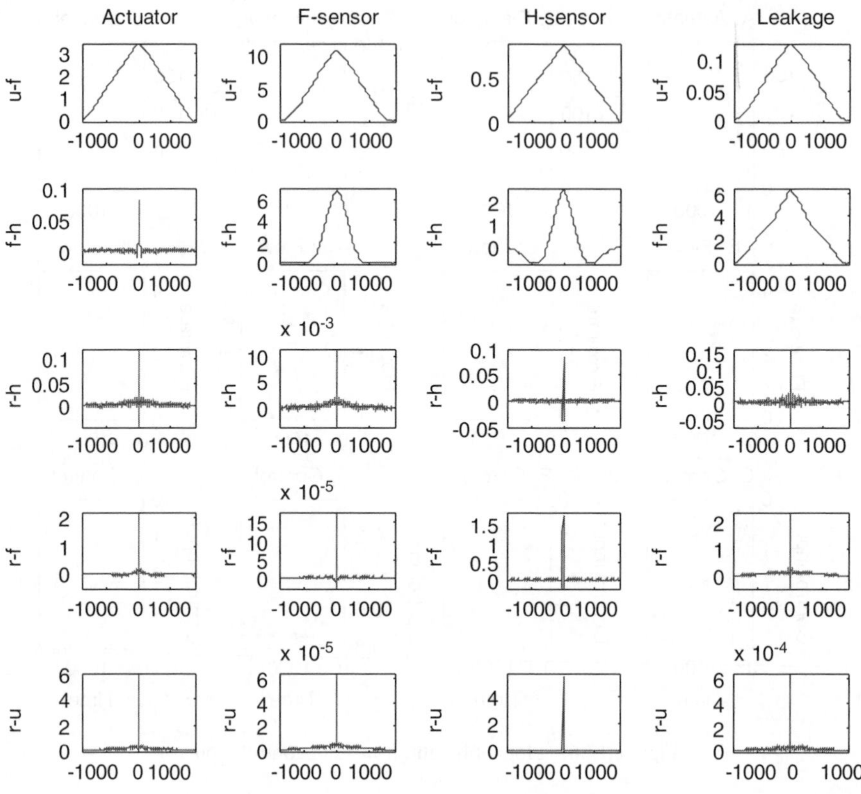

Figure 12.7 Auto-correlation of the residuals

The scaled-up (by a factor of 1000) versions of these test statistics are given in Table 12.3 below, where, for each fault scenario, the rows indicate the six paths analyzed, that is, $(e, u), (u, f)$, (f, h), (r, h), (r, f), and (r, u), and the columns indicate the three experiments carried out, that is: ex. 1, ex. 2, ex. 3.

Table 12.4 gives the truth tables for the four faults based on the test statistics of only experiment 3 (ex. 3) as shown in Table 12.3. As before, Bayes binary decisions for the paths (r, u), (r, f), (r, h), (e, u),

Table 12.3 Scaled-up test statistics of residuals for three experiments with four faults

	Actuator fault			Flow sensor fault			Height sensor fault			Leakage fault		
	ex.1	ex.2	ex.3	ex.1	ex.2	ex.3	ex.1	ex.2	ex.3	ex.1	ex.2	ex.3
(e,u)	0.01	0.01	0.01	0.01	0.01	0.01	0.01	0.01	0.01	0.01	0.01	0.01
(u,f)	0.96	1.84	2.66	3.10	3.18	3.33	0.93	0.85	0.82	3.48	3.41	3.15
(f,h)	0.04	0.21	0.29	1.76	1.27	0.83	0.51	1.32	1.63	0.23	1.24	2.33
(r,h)	0.59	56.28	70.87	77.68	75.71	78.61	6.94	100.7	140.2	71.04	45.0	3.27
(r,f)	0.04	0.12	0.11	1.52	1.06	0.47	0.15	1.11	1.37	0.73	2.10	3.65
(r,u)	1.91	3.61	5.26	6.83	6.77	6.84	1.88	2.04	2.0	6.69	6.18	5.3

Table 12.4 Bayes binary decisions

	Flow sensor fault						Height sensor fault						Leakage faults					
	G_0	k_{s0}	G_1	G_2	ks_1	ks_2	G_0	k_{s0}	G_1	G_2	ks_1	ks_2	G_0	k_{s0}	G_1	G_2	ks_1	ks_2
(r,u)	0	0	0	0	*x*	0	1	1	1	1	*x*	1	1	1	1	1	x	1
(r,f)	1	*x*	1	1	1	1	1	*x*	1	1	1	1	1	*x*	1	1	1	1
(r,h)	0	*x*	0	0	*x*	0	1	*x*	1	1	*x*	1	1	*x*	1	1	x	1
(e,u)	0	0	*x*	*x*	*x*	*x*	0	0	*x*	*x*	*x*	*x*	0	0	*x*	*x*	*x*	*x*
(u,f)	*x*	1	1	*x*	1	*x*	*x*	0	0	x	0	*x*	*x*	0	0	*x*	0	*x*
(f,h)	*x*	*x*	*x*	1	1	1	*x*	*x*	*x*	1	1	1	*x*	*x*	1	1	1	1
d_f	0	0	0	0	1	0	0	0	0	1	0	1	0	0	0	1	0	1

(u,f), and (f,h) are shown by a "1," "0," and "*x*" to indicate if there is a "fault," "no fault," and "not applicable (don't care status)", respectively:

- From the last row indicating the binary decision vector d_f, it can be seen that a fault in either the actuator or in the flow sensor are accurately isolated. However, in the "leakage or height sensor fault" scenario, Table 12.4 asserts that both the tank (G_2) and the height sensor (k_{s2}) are faulty, instead of pinpointing only the tank as being the sole faulty component. The reason for this is that both the leakage fault parameter (γ_ℓ), which affects the tank, and the height sensor appear together in cascade in all of the paths. Further, due to saturation type nonlinearity, when there is a leakage fault the PI controller is unable to maintain the reference height. In the case when the saturation limit is not reached under a fault condition, that is when the fault size is small (an incipient fault), there will be no problem in isolating the leakage from height sensor fault.
- In the cascade connection of the subsystems in the closed-loop configuration, a single fault occurring in any particular subsystem G_1, or G_2, or sensor k_{s2} will indicate a fault in all of the components located along the paths (r,u), (r,f), and (r,h).

Comment *The bank of the Kalman filters (KF) has successfully detected and isolated faults in subsystems and sensors thanks to the key property of the Kalman filter residual, namely the residual is a zero-mean white noise if and only if the model of the Kalman filter is identical to that of the system. This hinges on the internal model principle applied to KF [3]. It is interesting to note here that the FDI performance was found acceptable even though the linearized version of the two-tank system was used instead of its true nonlinear characteristics. However, this led to the estimation of the fault size being somewhat erroneous. It is worth pointing out that our main focus here was in the incipient fault detection and isolation for which the assumption of a linear model is acceptable. However, to handle larger excursions of the monitored outputs from their nominal values (i.e., to reliably estimate larger fault sizes), other excellent filtering strategies, such as the H-infinity and the energy-to-peak filtering [16] extended Kalman filter would be required. This will be focus of our future work.*

12.11.1 Comparison with Other Approaches

The performance of the proposed FDI scheme was compared with two model-free approaches, namely the ANN and fuzzy logic based scheme [14, 15], [26, 27] using the same fault data shown in Figure 12.6. The proposed bank of Kalman filters approach was found promising when applied to a benchmarked laboratory-scale two-tank system. The complex FDI problem of detecting and isolating m subsystem and sensor faults from m output measurements is split into m simple FDI problems. Each Kalman filter generates a residual from two output measurements. By analyzing the residual, a preliminary assertion is made as to which set of subsystems and sensors are likely to be faulty. By logical combination of all these preliminary assertions, a final assertion is made. This "divide and conquer approach" seems to perform well in the nonlinear process.

12.11.1.1 ANN Approach

Training data formed of the height and flow rate under leakage, actuator and sensor faults, as well as the classification of the fault types. The FDI performance of the ANN depends crucially upon the set of input–output data employed during the training phase. The training data should be sufficient in quantity and representative enough to cover all fault-free and faulty operating regimes. In practice, it is very difficult to cover all fault scenarios, especially the extreme cases involving disasters, for which data is either scarce or unavailable. The ANN approach suffers from the lack of transparency as the decision-making process is deeply embedded in the inner workings of the ANN, thus making the rationale behind the decisions taken rather unclear to the user. Nevertheless, the ANN is computationally fast and provides timely FDI.

12.11.1.2 ANFIS Approach

In the case of ANN and the model-based FDI scheme, the dynamic response (covering both transient and steady-state regions) of the system is presented. However, in the fuzzy-logic based approach, only the steady-state response under various operating regimes is presented. As a consequence of this, the fuzzy-logic approach is unable to detect a fault which manifests itself in the transient response but not in the steady-state response. ANFIS is a kind of neural network that is based on Takagi–Sugeno fuzzy inference system. Since it integrates both neural networks and fuzzy logic principles, it has the desirable potential to capture the benefits of both in a single framework. As such, it is regarded as a universal approximator, where the required set of fuzzy IF–THEN rules is developed automatically from the data presented to it.

The ANN and ANFIS perform well as long as they are trained with sufficient number of representative fault and fault-free data.

12.12 Summary

Subsystem and Sensor Models

$$y_i(z) = G_i(z)u_1(z) + v_i(z)$$
$$y_i(z) = k_{si}(z)u_i(z) + v_i(z)$$

Kalman Filter for Pairs of Measurements

$$\begin{aligned} x_{ij}(k+1) &= A_{ij}^0 x_{ij}(k) + B_{ij}^0 y_i(k) + \begin{bmatrix} B_{ij}^0 & E_{ij}^0 \end{bmatrix} \begin{bmatrix} v_i(k) \\ w_{ij}(k) \end{bmatrix} \\ y_j(k) &= C_{ij}^0 x_{ij}(k) + v_j(k) \end{aligned}$$

The Kalman filter model:

$$\begin{aligned}
\hat{x}_{ij}(k+1) &= A_{ij}^0 \hat{x}_{ij}(k) + B_{ij}^0 y_i(k) + K_{ij}^0 \left(y_j(k) - C_{ij}^0 \hat{x}_{ij}(k) \right) \\
e_{ij}(k) &= y_j(k) - C_{ij}^0 \hat{x}_{ij}(k) \\
x_{rj}(k+1) &= A_{rj}^0 x_{rj}(k) + B_{rj}^0 r(k) + E_{rj}^0 w_{rj}(k) \\
y_j(k) &= C_{rj}^0 x_{rj}(k) + v_j(k)
\end{aligned}$$

The Kalman filter model is:

$$\begin{aligned}
\hat{x}_{rj}(k+1) &= A_{rj}^0 \hat{x}_{ri}(k) + B_{rj}^0 r(k) + K_{rj}^0 \left(y_j(k) - C_{rj}^0 \hat{x}_{rj}(k) \right) \\
e_{rj}(k) &= y_j(k) - C_i^0 \hat{x}_{rj}(k)
\end{aligned}$$

Kalman Filter Residual: A Model-Mismatch Indicator

Expression of the residual

$$\begin{aligned}
e_{ij}(z) &= e_{fij}(z) + e_{0ij}(z) \\
e_{rj}(z) &= e_{frj}(z) + e_{0rj}(z)
\end{aligned}$$

where $e_{frj}(k)$ and $e_{fij}(k)$ are deterministic terms indicating model-mismatch if nonzero, and $e_{0rj}(k)$, $e_{0ij}(k)$ are zero-mean white noise processes.

Bayes Decision Strategy

$$t_s(\mathbf{e}) \begin{cases} \leq \eta & \textit{no fault} \\ > \eta & \textit{fault} \end{cases} \quad \text{where } t_s(\mathbf{e}) = \begin{cases} \left| \dfrac{1}{N} \displaystyle\sum_{i=k-N+1}^{k} e(i) \right| & r(k) = \text{constant} \\ P_{ee}(f_0) & r(k) \text{ is a sinusoid} \\ \dfrac{1}{N} \displaystyle\sum_{i=k-N+1}^{k} e^2(i) & r(k) \text{ is an arbitrary signal} \end{cases}$$

Truth Table of Binary Decisions

- If hypothesis H_0^{rj} or H_0^{ij} is true, then all the components located in the path between the pair of nodes are assigned a binary zero indicating none of the components is faulty.
- If hypothesis H_1^{rj} or H_1^{ij} is true, then at least one component is faulty.
- Components, which are not located in the path between the pair of nodes, are assigned a "don't care" character x, which can assume either 0 or 1.

12.13 Appendix

12.13.1 Map Relating $y_i(z)$ to $y_j(z)$

Consider the Figure 12.2. The map relating $u_j(z)$ and $u_i(z)$ is given by

$$u_j(z) = \left(\prod_{\ell=i}^{j-1} G_\ell(z) \right) u_i(z) + \upsilon_{ij}(z) \tag{12.19}$$

where $\upsilon_{ij}(z) = \left(\prod_{\ell=i+1}^{j-1} G_\ell(z) \right) \upsilon_i(z) + \left(\prod_{\ell=i+2}^{j-1} G_\ell(z) \right) \upsilon_{i+1}(z) + \ldots + G_{j-1}\upsilon_{j-2}(z) + \upsilon_{j-1}(z)$.

Since $u_i(z) = \dfrac{y_i(z) - v_i(z)}{k_{si}}$, substituting for $u_i(z)$ yields:

$$u_j(z) = \frac{1}{k_{si}(z)}\left(\prod_{\ell=i}^{j-1} G_\ell(z)\right) y_i(z) - \frac{1}{k_{si}(z)}\left(\prod_{\ell=i}^{j-1} G_\ell(z)\right) v_i(z) + \upsilon_{ij}(z) \tag{12.20}$$

From Figure 12.2, we infer that $y_j(z) = k_{sj}u_j(z) + v_j(z)$. Combining this expression with (12.20 then leads to the following expression relating $y_j(z)$ to $y_i(z)$ and the new noise term $v_{ij}(z)$ defined below:

$$y_j(z) = \left(\frac{k_{sj}(z)}{k_{si}(z)}\right)\left(\prod_{\ell=i}^{j-1} G_\ell(z)\right) y_i(z) + v_{ij}(z) \tag{12.21}$$

where $v_{ij}(z) = -\dfrac{k_{sj}(z)}{k_{si}(z)}\left(\displaystyle\prod_{\ell=i}^{j-1} G_\ell(z)\right) v_i(z) + k_{sj}(z)\upsilon_{ij}(z) + v_j(z)$.

12.13.2 Map Relating $r(z)$ to $y_j(z)$

Using Figure 12.2, and the superposition principle, the closed-loop system output u_j can be expressed as the sum of the outputs due to the reference input and the noise inputs υ_i , $i = 0, 1, 2, ..q - 1$, applied separately. Consider now the output $u_j(z)$ due only to the reference input r. Using the property of cascade and feedback combinations of the transfer functions $G_i(z)$, $i = 0.1, 2, ..q - 1$, the output due only to r and denoted by $u_{jr}(z)$ is:

$$u_{jr}(z) = \frac{\prod_{\ell=0}^{j-1} G_\ell(z)\, r(z)}{1 + \prod_{\ell=0}^{q-1} G_\ell(z)} \tag{12.22}$$

To derive the output, denoted by u_{ji}, which is due to the ith input noise component υ_i only, we need to first set all other input noise components υ_j, $j \neq i$ to zero. Next, by simply re-drawing the closed loop of Figure 12.2 in such a way that υ_i becomes the reference input (in place of r) and $u_{ji}(z)$ the system output (instead of u_q), then the expression of the output u_{ji} due to the ith input noise component υ_i only, can be readily derived in a similar way to that of (12.22), and is given below as:

$$u_{ji}(z) = \frac{\left(\prod_{\ell=i+1}^{j-1} G_\ell(z)\right) \upsilon_i(z)}{1 + \prod_{\ell=0}^{q-1} G_\ell(z)} \tag{12.23}$$

It is clear that the individual outputs ($u_{jk}(z)$, $k \neq i$) due to each of the remaining (q − 1) input noise components can be obtained in a similar way to that of (12.23).

Superposing the individual outputs due to the reference input r on the one hand, and to each of the q noise components, $\upsilon_i, i = 0, 1, 2, \ldots q-1$, on the other, as given by Eqs. (12.22) and (12.23) respectively, yields the following expression for the total input:

$$u_j(z) = u_{jr}(z) + \sum_{i=0}^{q-1} u_{ji}(z) = \frac{\prod_{\ell=0}^{j-1} G_\ell(z)\, r(z) + \sum_{i=0}^{q-1} \prod_{\ell=i+1}^{j-1} G_\ell(z)\upsilon_i(z)}{1 + \prod_{\ell=0}^{q-1} G_\ell(z)}$$
$$= \frac{\prod_{\ell=0}^{j-1} G_\ell(z)\, r(z)}{1 + \prod_{\ell=0}^{q-1} G_\ell(z)} + \upsilon_{rj}(z) \tag{12.24}$$

where $\upsilon_{rj}(z) = \dfrac{\sum_{i=0}^{q-1} \prod_{\ell=i+1}^{j-1} G_\ell(z)\, \upsilon_i(z)}{1 + \prod_{\ell=0}^{q-1} G_\ell(z)}$.

Using the relation between y_j and u_j, namely $y_j = k_{sj}(z)u_j + v_j$, the expression for y_j becomes:

$$y_j(z) = k_{sj}(z)\frac{\prod_{\ell=0}^{j-1} G_\ell(z) r(z)}{1 + \prod_{\ell=0}^{q-1} G_\ell(z)} + v_{rj}(z) \tag{12.25}$$

where $v_{rj}(z) = v_j(z) + k_{sj}(z)\upsilon_{rj}(z)$.

References

Fault Diagnosis: Model-Based

[1] Doraiswami, R. and Cheded, L. (2013) A unified approach to detection and isolation of parametric faults using a Kalman filter residuals. *Journal of Franklin Institute*, **350**(5), 938–965.

[2] Doraiswami, R. and Cheded, L. (2013) Fault diagnosis of a sensor network: a distributed filtering approach. *Journal of Dynamic Systems, Measurement and Control*, **135**(5), 1–10.

[3] Doraiswami, R. and Cheded, L. (2012) Kalman filter for fault detection: an internal model approach. *IET Control Theory and Applications*, **6**(5), 1–11.

[4] Xu, X. and Wang, X. (2010). Multiple sensors soft-failure diagnosis based on Kalman filter. *Signal and Image Processing: An International Journal (SIPIJ)*, **1**, 126–132.

[5] Ding, S. (2008) *Model-based Fault Diagnosis Techniques: Design Schemes*, Springer-Verlag.

[6] Kobayashi, T. and Simon, D. (2005). Evaluation of an enhanced bank of Kalman filters for in-flight aircraft engine sensor fault diagnosis. *Journal of Engineering for Gas and Power*, 497–504.

[7] Zhang, S., Toshiyuke, A., and Shoji, H. (2004). Gas Leakage Detection using Kalman Filter. International Conference on Signal Processing (ICSP), **3**, 2533–2536.

[8] Doraiswami, R., Diduch, C., and Tang, T. (2010) A new diagnostic model for identifying parametric faults. *IEEE Transactions on Control System Technology*, **8**(3), 533–544.

[9] Isermann, R. (2006) *Fault Diagnosis Systems An Introduction from Fault Detection to Fault Tolerance.* Springer-Verlag.

[10] Silvio, S., Fantuzzi, C., and Patton, R.J. (2003) *Model-based Diagnosis Using Identification Techniques, Advances in Industrial Control.* Springer-Verlag.

[11] Patton, R.J., Frank, P.M., and Clark, R.N. (2000) *Issues in Fault Diagnosis for Dynamic Systems.* Springer-Verlag.

[12] Patton, R.J. and Chen, J. (1991) A review of parity space approaches to fault diagnosis. IFAC/IMACS Symposium on Fault Detection, Supervision and Safety for Technical Processes - Safe Process ' 91, Baden-Baden, Germany, 65–82.

[13] Gertler, J. (1998) *Fault Detection and Diagnosis in Engineering Systems*. Marcel Dekker, Inc., New York.

Fault Diagnosis: Model-Free and Model-Based

[14] Doraiswami, R. Cheded, L. and Khalid, M.H. (2010) *Sequential Integration Approach to Fault Diagnosis with Applications: Model-free and Model- Based Approaches*. VDM Verlag Dr. Muller Aktiengesellschaft & Co. KG.

[15] Khouki, A., Khalid, H., Doraiswami, R., and Cheded, L. (2012). Fault detection and classification using kalman filter and hybrid neuro-fuzzy systems. *International Journal of Computer Applications*, **45**(12), 7–13.

[16] Zhang, H.S., Yang, H.S., and Mehr, A.S. (2012). On H_infinity filtering for discrete-time takagi-sugeno fuzzy systems. *IEEE Transactions on Fuzzy Systems*, **20**(2) 396–401.

[17] Brunone, B., Ferrante, M., and Meniconi, S. (2008). Portable pressure wavemaker for leak detection and pipe system characterization. *Journal of American Water Works Association*, **100** (4), 108–116.

[18] Campos-Delgadao, D., Palacios, E., and Espinoza-Trejo, D. (2008). Fault detection, isolation, accommod-tion for LTI systems based on GIMC structure. *Journal of Control Science and Engineering*, **2008**, 15. doi:10.1155/2008/853275.

[19] Witczak, M. (2006). Advances in model-based fault diagnosis with evolutionary algorithms and neural networks. *International Journal of Applied Mathematics and Computer Science*, **16**(1), 85–89.

[20] Shoukat, C., Jain, M., and Shah, S.L. (2006). Detection and quantification of valve stiction. Proceedings of 2006 American Control Conference, (pp. 2097–2106). Minneapolis.

[21] Awadalla, M. and Marcos, M. (2004). ANFIS-based diagnosis and location of stator insulation interturn faults in PM brushless DC motors. *IEEE Transactions on Energy Conversion*, **19**(4), 795–796.

Wireless Sensor Network

[22] Lewis, F. (2004) *Wireless Sensor Network, in Smart Environments: Technologies, Protocols and Applications*. John Wiley and Sons.

[23] Zhang, H., Yang, S. and Mehr, A. (2012) Robust H_infinity PID control for multivariable networked control systems with disturbance/noise attenuation. *International Journal of Robust and Nonlinear Control*, **22**(2), 183–204.

[24] Zhang, H., Yang, S.Y., and Mehr, A. (2011) Robust static output feedback and remote PID design for networked systems. *IEEE Transactions on Industrial Electronics*, **58**(12), 5396–5404.

[25] Process Leak Detection Diagnostics with Intelligent Differential Pressure Transmitter (2008). Emerson Process Management, Rosemount Inc, August 2008, pp. 1–9. www.rosemount.com.

Soft Computing

[26] Karry, F.O. and De Silva, C.W. (2004). *Soft Computing and Intelligent Design*. Addison Wesley.

[27] Haykin, S. (1999). *Neural Networks: A Comprehensive Foundation*. Prentice Hall, New Jersey.

Estimation Theory

[28] Kay, S.M. (1993) *Fundamentals of Signal Processing: Estimation Theory*. Prentice Hall PTR, New Jersey.

13

Soft Sensor

13.1 Review

A soft sensor and its application to a robust and fault tolerant control system is developed. Soft sensors are invaluable in industrial applications in which hardware sensors are either too costly to maintain or to physically access. Software-based sensors act as the virtual eyes and ears of operators and engineers looking to draw conclusions from processes that are difficult – or impossible – to measure with a physical sensor. With no moving parts, the soft sensor offers a maintenance-free method for fault diagnosis and process control. They are ideal for use in the aerospace, pharmaceutical, process control, mining, oil and gas, and healthcare industries [1, 2].

13.1.1 Benefits of a Soft Sensor

A soft sensor offers the following benefits:

- *Reduced cost and weight*: no physical equipment to purchase, repair or replace.
- *Reliability*: a hardware sensor with moving parts may be replaced by a soft sensor to avoid the problems of maintenance especially for system operating in hazardous environment or in inaccessible locations.
- *Product quality*: can estimate almost any desired variable, such as quality of a product (composition, texture, molecular weight etc.) indirectly using available measurements and the process model.
- Soft sensors are especially useful in data fusion, where measurements of different characteristics and dynamics are combined.
- It can be used for performance monitoring, fault diagnosis, as well as for implementing a controller estimating unmeasured plant outputs.

A soft sensor is a software algorithm based on an Artificial Neural Network, a Neuro-fuzzy system, Kernel methods (support vector machines), a multivariate statistical analysis, a Kalman filter, or other model-based or model-free approaches [3].

13.1.2 Kalman Filter

A model-based approach using the Kalman filter for the design of a soft sensor is proposed here. The Kalman filter is an optimal minimum variance estimator of the unknown variable from the noisy input and

---

Identification of Physical Systems: Applications to Condition Monitoring, Fault Diagnosis, Soft Sensor and Controller Design, First Edition. Rajamani Doraiswami, Chris Diduch and Maryhelen Stevenson.

the output of the system. The Kalman filter computes the estimate by fusing the *a posteriori* information provided by the measurement, and the *a priori* information contained in the model that generated the measurement. The estimate thus obtained is the best compromise between the estimates generated by the model and those obtained from the measurement, depending upon the plant noise and the measurement noise covariance.

The Kalman filter has wide applications, including performance monitoring, fault diagnosis. and controller implementation [4–8], and these additional abilities are exploited herein to develop a fault tolerant control system.

A Kaman filter is a copy of the mathematical model of the plant driven by the *residual*, which is the error between the measured output of the plant and its estimate generated by the Kalman filter. The Kalman gain is used as an effective design parameter to handle the uncertainty associated with the model of the physical system. Model uncertainty is effectively introduced in the determination of the gain by choosing a higher (lower) variance of the plant noise than the measurement noise variance if the dynamic part of the state-space model is less (more) reliable. In [6] an expression relating the residual and the deviation of the plant model from its nominal one is derived. This relationship is exploited herein to ensure high performance and stability by re-identifying the plant and re-designing the controller whenever the residual exceeds some threshold. Further tasks of performance monitoring and fault diagnosis are realized.

13.1.3 Reliable Identification of the System

The Kalman filter is designed using the identified nominal model of the system. Hence the performance of the soft sensor, and the soft sensor-based fault tolerant control system, critically depends upon the accuracy of the identified model. In general, a model of the physical system varies with operating conditions. A model identified at a given operating point may not be accurate when the operating condition changes. As result the performance of the soft sensor as well as the controller will be degraded. To overcome the performance degradation, a set of models in the neighborhood of a given operating point is generated by performing a number of experiments. A model termed *optimal nominal model* is identified, which is the best least-squares fit to the set of models thus obtained. To generate the set of models, emulators are connected at the input or at the output.

The nominal model is identified by considering the likely variations of the system around the normal operating point. The high-order least-squares identification scheme is used so that it minimizes the sum of the squares of the residuals not only at the given operating point but also around its neighboring points. The neighboring points are determined by varying the parameters of the emulators.

13.1.4 Robust Controller Design

Reliable identification of the system is also crucial to the performance of the controller. The controller is designed to ensure both stability and performance in the face of variation in the model of the system using the widely popular and effective robust control approach. Robust control originated in the 1980s and has gained prominence over classical control theory. Robust control theory considers the design of controller for a plant whose model is uncertain.

13.1.4.1 Emulator: Numerator-denominator Uncertainty Model

There are many approaches model uncertainty, with the most common cases taking a benchmark *nominal model* (generally obtained from the identification of the nominal plant) as a starting point and considering perturbations of this model. The model uncertainty associated identified model is modeled as perturbations in the numerator and the denominator polynomials to develop robust controllers using a mixed sensitivity H_∞ controller [9]. We use the emulator as the numerator-denominator uncertainty model, as it is a simple and effective. The mixed sensitivity H_∞ control design is sound, mature, and is geared to handle the problem of controller design when the plant model is uncertain, and has been successfully employed in practice in recent years [10–12].

A robust controller is determined by minimizing a performance measure for the worst-case model uncertainty. The performance measure is a tradeoff among the conflicting requirements for robust stability, performance, and control limitation. Although the robust controller is optimal for the chosen performance measure, the design is conservative as the uncertainty covers a wide range of all possible model deviations. High controller performance may not be achieved for a given operating regime, although stability is guaranteed for a wide operating range. To overcome this problem, adaptive robust control approach has emerged in recent years. The robust control scheme identifies the plant model and uses the identified model for designing a robust controller. There are two versions of the adaptive robust control scheme, namely the direct and the indirect. In the direct scheme, the identification and control tasks are be executed simultaneously, while in the indirect scheme identification and robust controller design are performed independently at different time intervals. The indirect control scheme may include health monitoring and prognostics [13]. The robust control theory has adopted the worst-case philosophy out of concerns for stability in the face of all model perturbations, as instability will have disastrous consequences. The robust adaptive control approach reduces the worst-case scenarios by reducing the model uncertainty by including the task of plant identification whenever the residual exceeds the threshold value.

13.1.5 Fault Tolerant System

In addition to the tasks of identification and robust controller design, an equally important task of performance and condition monitoring and prognostics is included. This will ensure high performance over a wide operating regime by excuting the identify-control design and implementation whenever the peformance degrades, and protect against unexpected variations in the parametemers resulting in failure of the system.

Performance monitoring and fault diagnosis of physical systems is a critical part of a reliable and high performance control system. The Kalman filter is used to monitor the performance and detect faults. Using the statistical hypotheses tests, the residual of the Kalman filter is analyzed to detect if the performance is acceptable or unacceptable.

The proposed soft sensor-based control system is evaluated on a simulated as well as laboratory-scale physical velocity control system where the angular velocity of the DC servo motor is estimated from the input to motor and the measure armature current. The performance of the control system is evaluated in the face of (i) measurement noise and load torque disturbances affecting the DC motor, and (ii) plant perturbations including the DC motor, current sensor, and the actuator (an amplifier).

13.2 Mathematical Formulation

The state-space model of a system is given by:

$$\begin{aligned} \boldsymbol{x}(k+1) &= \boldsymbol{A}\boldsymbol{x}(k) + \boldsymbol{B}u(k) + \boldsymbol{E}_w w(k) \\ \boldsymbol{y}(k) &= \boldsymbol{C}\boldsymbol{x}(k) + \boldsymbol{F}_v v(k) \\ y_r(k) &= \boldsymbol{C}_r\boldsymbol{x}(k) \end{aligned} \tag{13.1}$$

where $x(k)$ is a $n \times 1$ state, $u(k)$ is the scalar control input, $\boldsymbol{y}(k)$ is a $n_y \times 1$ vector formed of all measured (accessible) outputs, $\boldsymbol{v}$ is a measurement noise and w is a disturbance, $y_r(k)$ is the plant output that needs to be estimated as it is either inaccessible or not measured, $\boldsymbol{A}$, $\boldsymbol{B}$, $\boldsymbol{C}$, and $\boldsymbol{C}_r$ are respectively nxn, $nx1$,

$n_y \times n$, and $1 \times n$ matrices. v, w, and y_r are scalars; $\boldsymbol{E}_w$ and $\boldsymbol{F}_v$ are respectively $nx1$ disturbance and $n_y \times 1$ measurement noise entry vectors. The measurement noise v and disturbances are zero-mean white noise with variances Q and R respectively.

13.2.1 Transfer Function Model

The transfer function model of the system relating the reference input $r(z)$, the disturbance $\boldsymbol{w}(z)$, and the measurement noise $\boldsymbol{v}(z)$ to the output $\boldsymbol{y}(z)$ is given by:

$$\boldsymbol{y}(z) = \frac{\boldsymbol{N}(z)}{D(z)}u(z) + \frac{\boldsymbol{N}_w(z)}{D(z)}\boldsymbol{w}(z) + \boldsymbol{F}_v\boldsymbol{v}(z) \tag{13.2}$$

Where $\frac{\boldsymbol{N}(z)}{D(z)} = \boldsymbol{C}\,(zI - \boldsymbol{A})^{-1}\,\boldsymbol{B}$, and $\frac{\boldsymbol{N}_w(z)}{D(z)} = \boldsymbol{C}\,(zI - \boldsymbol{A})^{-1}\,\boldsymbol{E}_w$ are $n_y \times 1$ transfer matrices, $D(z) = |\boldsymbol{I}z - \boldsymbol{A}|$ is a scalar. Rewriting by cross-multiplying by $D(z)$, we get:

$$D(z)\boldsymbol{y}(z) = \boldsymbol{N}(z)u(z) + \boldsymbol{\upsilon}(z) \tag{13.3}$$

where $\boldsymbol{\upsilon}(z) = \boldsymbol{N}_w(z)\boldsymbol{w}(z) + D(z)\boldsymbol{F}_v\boldsymbol{v}(z)$ is the equation error formed of two colored noise processes generated by the disturbance $\boldsymbol{w}(z)$ and the measurement noise $\boldsymbol{v}(z)$. The $n_y \times 1$ matrix transfer function $\boldsymbol{G}(z)$ of the system derived from Eq. (13.3) is given by:

$$\boldsymbol{G}(z) = \boldsymbol{N}(z)D^{-1}(z) \tag{13.4}$$

where $\boldsymbol{N}(z)$ is the $n_y \times 1$ numerator (matrix) polynomial, and $D(z)$ is the denominator (scalar) polynomial of the matrix transfer function $\boldsymbol{G}(z)$.

13.2.2 Uncertainty Model

The structure and the parameters of a physical system may vary due to changes in the operating regime. The difference between the actual system and its model, termed *model uncertainty*, is considered in identification and subsequently in designing a controller based on the identified model.

Commonly, the transfer function model of the system is expressed as an additive or multiplicative combination of the assumed model and a perturbation term. The perturbation term represents the modeling error. A model, termed the numerator-denominator perturbation model, is employed herein, where the perturbation in the numerator and denominator polynomials are treated separately instead of clubbing together as a single perturbation of the overall transfer function [9]. This perturbation model is appropriate as both the design of the Kalman filter-based soft sensor and the controller for ensuring the closed-loop stability and performance hinge on the accuracy of the identified system model: the modeling error (error between the actual and the identified model) stems from the errors in the estimation of the numerator and the denominator coefficients.

13.2.2.1 Emulator: Numerator-denominator Perturbation Model

The numerator-denominator perturbation model [9] takes the following form:

$$\boldsymbol{G}(z) = \frac{\boldsymbol{N}(z)}{D(z)} = \frac{\left(\boldsymbol{I} + \Delta_N(z)\right)\boldsymbol{N}_0(z)}{\left(1 + \Delta_D(z)\right)D_0(z)} = \boldsymbol{G}_e(z)\boldsymbol{G}_0(z) \tag{13.5}$$

where $\boldsymbol{G}_0(z)$, is the nominal transfer function, $\boldsymbol{N}_0(z)$ is the nominal the numerator (matrix) polynomial, $D_0(z)$ is the nominal denominator (scalar) polynomial, and $\boldsymbol{G}_e(z)$ is the $n_y \times n_y$ multiplicative perturbation, termed the *emulator*;

$$\boldsymbol{G}_e(z) = \frac{\boldsymbol{I} + \Delta_N(z)}{1 + \Delta_D(z)} \tag{13.6}$$

$\Delta_N(z) \in RH_\infty$ and $\Delta_D(z) \in RH_\infty$ represent respectively pertutbations in the numerator and the denominator polynomials of the nominal model $\boldsymbol{G}_0(z)$, $\Delta_N(z)$ and $\Delta_D(z)$ are respectively a stable frequency dependent $n_y \times n_y$ matrix, and a scalar; $\boldsymbol{I}$ is a $n_y \times n_y$ identity matrix.

13.2.2.2 Selection of the Emulator Model

The emulator $\boldsymbol{G}_e(z)$ is chosen such that the perturbed model $\boldsymbol{G}(z)$ matches the actual model of the system. In many practical problems, for computational simplicity, the perturbation model is chosen to mimic the macroscopic behavior of the system characterized by gain and phase changes in the system transfer function. The $n_y \times n_y$ multiplicative perturbation $\boldsymbol{G}_e(z)$ is a diagonal matrix:

$$\boldsymbol{G}_e(z) = \begin{bmatrix} G_{e1}(z) & 0 & 0 & 0 \\ 0 & G_{e2}(z) & 0 & 0 \\ . & 0 & . & . \\ 0 & 0 & 0 & G_{en_y}(z) \end{bmatrix} \tag{13.7}$$

where $G_{ei}(z)$ is chosen to be a constant gain, a gain and a pure delay of d time instant, all-pass first-order filter or Blaschke product of all-pass first-order filters as shown in the list of choices given in Eq. (13.8). The choice depends upon the model order and the types of likely variations in the dynamic behavior of the system. The parameters of the emulator must be capable of simulating faults that may likely occur in the system during the operational phase.

$$G_{ei}(z) = \begin{cases} \gamma_i & \text{gain} \\ \gamma_i z^{-d} & \text{gain and pure delay} \\ \gamma_i \dfrac{\gamma_{i1} + z^{-1}}{1 + \gamma_{i1} z^{-1}} & \text{first order all pass} \\ \gamma_i \prod_j \dfrac{\gamma_{ij} + z^{-1}}{1 + \gamma_{ij} z^{-1}} & \text{Blaschke product} \end{cases} \tag{13.8}$$

The parameters γ_i, and γ_{ij} are termed herein as *emulator parameters*.

13.3 Identification of the System

The output $y_r(k)$ is inaccessible or not measured during the operational phase of the system. However, during the offline identification phase the output $y_r(k)$ is either measured (for example the angular velocity may be measured using a mechanical hardware such as a tacho generator) or computed from other outputs. The direct or indirect measurement of $y_r(k)$ will ensure that $y_r(k)$ during the identification phase ensures that the identified model captures accurately the map relating $y_r(k)$ to the input $u(k)$ and the measured output $\boldsymbol{y}(k)$.

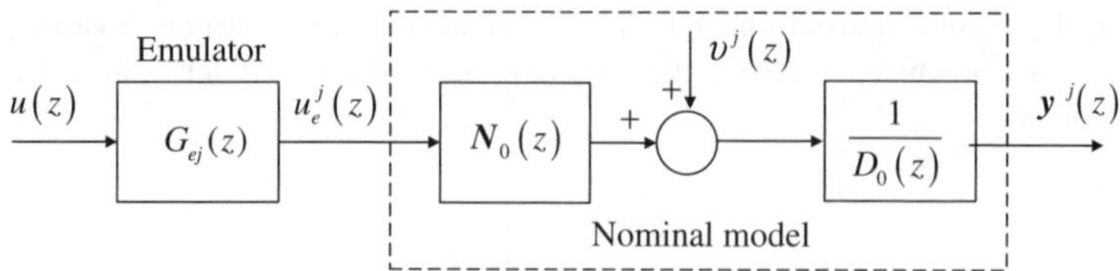

Figure 13.1 Emulation of operating scenarios: *j*th parameter perturbed experiment

In the other words, during offline identification, it is assumed that $y_r(k)$ is an element of the measured output vector $\mathbf{y}(k)$. For notational simplicity, whenever there is no confusion the augmented and the measured outputs are denoted by the same output variable $\mathbf{y}(k)$.

13.3.1 *Perturbed Parameter Experiment*

The performance of the soft sensor depends upon the accuracy of the identified nominal model, which is used to design the Kalman filter. A reliable identification scheme is employed. The system model is identified by performing a number of parameter-perturbed experiments. Each experiment consists of perturbing one or more emulator parameters. The input is chosen to be persistently exciting.

We can emulate operating scenarios by including the emulator $G_{ej}(z)$ at the input of the system $u(k)$, and varying the emulator parameters γ_j, and γ_{jk} as shown in the Figure 13.1. The experiment includes all the experiments which consist of perturbing one at a time all the parameters of the emulator $\mathbf{G}_e(z)$, and collecting the input data $u(k)$ (usually the input is chosen to be same for all experiments), and the output data $y(k)$.

Consider the *j*th experiment of perturbing the *j*th emulator parameter. The perturbed model of the system using Eq. (13.3) relating the *ith* output $y_i(z)$, the input $u(z)$, and the *ith* equation error $\upsilon_i(z)$ becomes:

$$D^j(z)y_i^j(z) = N_i^j(z)u(z) + \upsilon_i^j(z) \tag{13.9}$$

where $D^j(z)$ and $N_i^j(z)$ are the denominator and the numerator polynomials respectively resulting from the variation of the *j*th emulator parameter. The linear regression model for the *j*th experiment becomes:

$$y_i^j(k) = \left(\boldsymbol{\psi}_i^j(k)\right)^T \boldsymbol{\theta}_i^j + \upsilon_i^j(k) \quad i = 1, 2, 3, \dots, n_y \tag{13.10}$$

where $\boldsymbol{\psi}_i^j(k)$ is a $M \times 1$ data vector with $M = 2n$, which is formed of the past inputs, $u(k-\ell), \ell = 1, 2, \dots, n$ and past outputs $y_i^j(k-\ell), \ell = 1, 2, \dots, n$, n is the order of the system:

$$\left(\boldsymbol{\psi}_i^j(k)\right)^T = \left[-y_i^j(k-1) \quad -y_i^j(k-2) \quad . \quad -y_i^j(k-n) \quad u(k-1) \quad . \quad u(k-n)\right] \tag{13.11}$$

$\boldsymbol{\theta}_i^j$ is a $M \times 1$ vector of unknown model parameter given by:

$$\boldsymbol{\theta}_i^j = \left[a_1^j \quad a_2^j \quad . \quad a_n^j \quad b_{i1}^j \quad b_{i2}^j \quad . \quad b_{in}^j\right]^T \tag{13.12}$$

13.3.2 *Least-Squares Estimation*

Let $\hat{\boldsymbol{\theta}}_i^{opt}$ be an "optimal estimate" of the feature vector $\boldsymbol{\theta}_i$ of the "optimal nominal model," which is optimal in the sense that it minimizes the sum of the squares of the residuals from all the experiments

$j = 1, 2, 3, \ldots, N_{\exp}$, where $N_{\exp}$ is the number of experiments $N_{\exp}$:

$$\hat{\theta}_i^{opt} = \arg\left\{ \min_{\{\theta_i\}} \left\{ \sum_{j=1}^{N_{\exp}} \sum_{k=1}^{N} \left(y_i^j(k) - \left(\psi_i^j\right)^T (k)\theta_i \right)^T \left(y_i^j(k) - \left(\psi_i^j\right)^T (k)\theta_i \right) \right\} \right\} \tag{13.13}$$

The "optimal estimate" $\hat{y}_i^{jopt}(k)$ of the output $y_i^j(k)$ from the jth experiment is given by:

$$\hat{y}_i^{j\,opt}(k) = \left(\psi_i^j\right)^T (k)\hat{\theta}_i^{opt} \tag{13.14}$$

The sum of the squares of the error $e_i^j(k) = y_i^j(k) - \hat{y}_i^0(k)$ between the output and the nominal estimate is

$$\sum_{k=1}^{N} \left(y_i^j(k) - \hat{y}_i^{jopt}(k) \right)^2 \tag{13.15}$$

13.3.2.1 Conventional Identification Approach

In the conventional identification, the estimate $\hat{y}_i^{j0}(k)$ of the output $y_i^j(k)$ is obtained from the estimated of the nominal feature vector:

$$\hat{y}_i^{j0}(k) = \left(\psi_i^j\right)^T (k)\hat{\theta}_i^0 \tag{13.16}$$

where $\hat{\theta}_i^0$ is the least squares estimate of the feature vector of the system at the nominal operating point from performing a single experiment. The identification error (13.15) becomes:

$$\sum_{k=1}^{N} \left(y_i^j(k) - \hat{y}_i^{j0}(k) \right)^2 \tag{13.17}$$

13.3.3 Selection of the Model Order

The order of the model n must be high enough so that estimated output $\hat{y}_i^j(k)$ from the jth experiment matches the actual output denoted $\mathbf{y}^j(k)$ closely for all experiments: $j = 1, 2, 3, \ldots, N_{\exp}$.

The estimated nominal model is thus the best linear least-squares fit between the identified model and the actual model for all operating points in the neighborhood of the nominal point. The operating points in the neighborhood are simulated by varying the emulator parameters during the parameter perturbed experiments.

13.3.4 Identified Nominal Model

The "optimal nominal model" derived from the optimal estimate $\hat{\theta}_i^{opt} : i = 1, 2, 3, \ldots, n_y$ given by Eq. (13.13) becomes:

$$D^{opt}(z)\mathbf{y}(z) = \mathbf{N}^{opt}(z)u(z) + \mathbf{v}(z) \tag{13.18}$$

where $D^{opt}(z)$ and $N^{opt}(z)$ are derived from the elements of the estimated nominal feature vectors $\hat{\theta}_i^0 : i = 1, 2, \ldots, n_y$. The transfer function of optimal nominal model is:

$$G^{opt}(z) = \frac{N^{opt}(z)}{D^{opt}(z)} \tag{13.19}$$

From Eq. (13.19), the identified nominal state-space model, denoted (A_0, B_0, C_0), of the actual state-space model (A, B, C) given by Eq. (13.1) becomes:

$$\begin{aligned} x(k+1) &= A_0 x(k) + B_0 u(k) \\ y(k) &= C_0 x(k) \\ y_r(k) &= C_{r0} x(k) \end{aligned} \tag{13.20}$$

13.3.4.1 The Nominal Plant Model

Let the nominal plant model (A^0, B^0, C^0) be:

$$\begin{aligned} x(k+1) &= A^0 x(k) + B^0 u(k) + E_w w(k) \\ y(k) &= C^0 x(k) + F_v v(k) \\ y_r(k) &= C_r^0 x(k) \end{aligned} \tag{13.21}$$

Assumptions *It is assumed that (A^0, B^0) is controllable and (A^0, C^0) are both observable, so that a controller and a (steady-state) Kalman filter may be designed to meet the requirement of performance and stability.*

For notational simplicity the state-space of the actual, the nominal and the identified nominal models are indicated by same state $x(k)$.

13.3.5 Illustrative Example

A simple example of a second-order system is considered.

The nominal model of the system $G_0(z)$ is:

$$G_0(z) = \frac{N_0(z)}{D_0(z)} = \frac{b_{01} z^{-1}}{1 + a_{01} z^{-1}} \tag{13.22}$$

The emulator model $G_e(z)$ is:

$$G_e(z) = \frac{1 + \Delta_N(z)}{1 + \Delta_D(z)} = \frac{\gamma + z^{-1}}{1 + \gamma z^{-1}} \tag{13.23}$$

where $b_{01} = 1$, $a_{01} = 0.8$, γ is emulator parameter, $\Delta_N(z) = \gamma - 1 + z^{-1}$, and $\Delta_D(z) = \gamma z^{-1}$.

The actual model of the system $G(z)$ given by Eq. (13.5) becomes:

$$G(z) = \frac{N(z)}{D(z)} = G_e(z) G_0(z) = \frac{\gamma + z^{-1}}{1 + \gamma z^{-1}} \frac{b_{01} z^{-1}}{1 + a_{01} z^{-1}} = \frac{b_1 z^{-1} + b_2 z^{-2}}{1 + a_1 z^{-1} + a_2 z^{-2}} \tag{13.24}$$

where $b_1 = b_{01}\gamma$, $b_2 = b_{01} = 1$, $a_1 = a_{01} + \gamma = 0.8 + \gamma$, $a_2 = a_{01}\gamma = 0.8\gamma$.

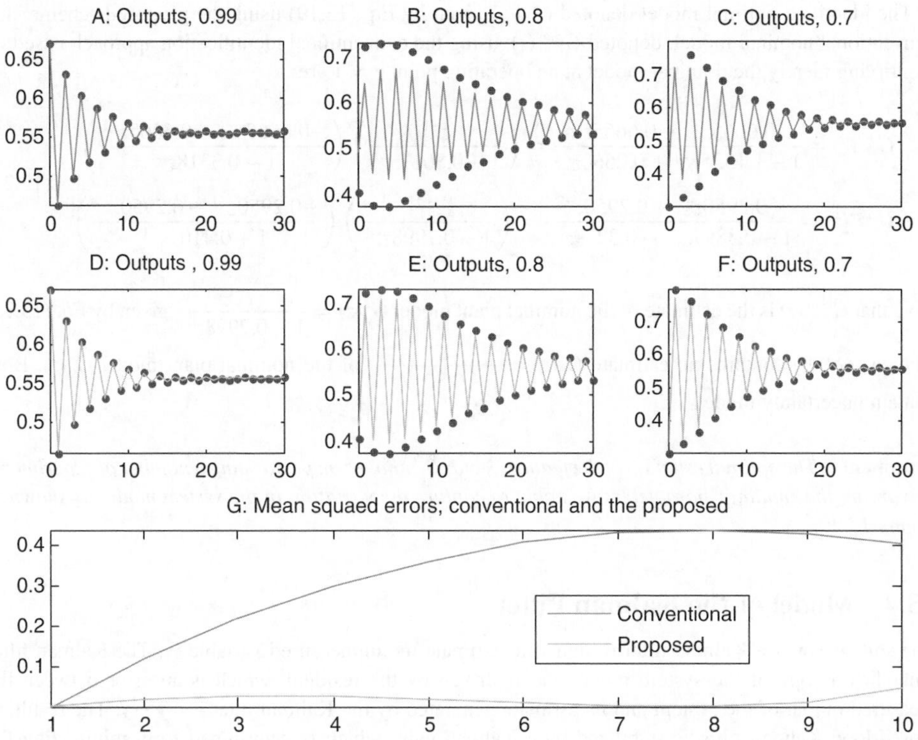

Figure 13.2 The output, the estimate, and identification error: conventional and proposed schemes

The output of the system $y(z)$ becomes:

$$y(z) = G(z)u(z) + \upsilon(z) \tag{13.25}$$

where $b_1 = 0.9$; $b_2 = 1$; $a_1 = 1.7$; $a_2 = 0.72$. The number of data samples $N = 100$; $\text{var}(\upsilon) = 0.01$, and the input was a square wave. Ten experiments were performed by varying the emulator parameter γ in the range 0.1 to 1 in steps of 0.1. The perturbed model (13.9) becomes:

$$\left(1 + \left(a_{10} + \gamma\right) z^{-1} + a_{10}\gamma z^{-2}\right) y^j(z) = \left(b_{01}\gamma z^{-1} + b_{10}z^{-2}\right) u(z) + \upsilon^j(z) \tag{13.26}$$

where $\gamma = 1 - 0.1(j - 1)$ for $j = 1, 2, 3, \ldots, 10$.

Figure 13.2 shows (i) the output $y^j(k)$, the optimal estimate $\hat{y}_i^{jopt}(k)$ using proposed scheme given by Eq. (13.14) and the estimate $\hat{y}_i^{j0}(k)$ using the conventional method given by Eq. (13.16) and (ii) the identification errors $\sum_{k=1}^{N} \left(y^j(k) - \hat{y}^{jopt}(k)\right)^2$ given by Eq. (13.15) for the proposed and $\sum_{k=1}^{N} \left(y^j(k) - \hat{y}^{j0}(k)\right)^2$ for the conventional given by Eq. (13.17).

Subfigures (a), (b), (c) on the top and the subfigures (d), (e), (f) in the middle show respectively the outputs (in dotted lines) and the optimal estimates (in solid lines) obtained when the emulator parameters were respectively $\gamma = 0.99$, $\gamma = 0.8$, $\gamma = 0.7$ for the conventional and the proposed identification schemes. The envelope in dots shows the true output. Subfigure (g) shows the identification errors using the proposed and the conventional schemes.

The identified optimal model denoted $G^{opt}(z)$ given by Eq. (13.19) using the proposed scheme, and conventional nominal model, denoted $G^{conv}(z)$, using the conventional identification approach based on identifying merely the nominal model at an operating point $\gamma = 1$ are:

$$\begin{aligned} G^{opt}(z) &= \frac{-0.6665z^{-1} + 0.6665z^{-2}}{1 - 1.1335z^{-1} + 0.2660z^{-2}} = \left(\frac{z^{-1}}{1 - 0.8017z^{-1}}\right)\left(\frac{-0.6665z^{-1} + 0.6665z^{-1}}{1 - 0.3318z^{-1}}\right) \\ G^{conv}(z) &= \frac{-0.2950z^{-1} + 0.2950z^{-2}}{1 - 0.3897z^{-1} - 0.3279z^{-2}} = \left(\frac{z^{-1}}{1 - 0.7998z^{-1}}\right)\left(\frac{-0.2950z^{-1} + 0.2950z^{-2}}{1 + 0.4101z^{-1}}\right) \end{aligned} \tag{13.27}$$

Note that $G^{conv}(z)$ is the estimate of the nominal plant model $\hat{G}_0(z) = \dfrac{z^{-1}}{1 - 0.7998z^{-1}}$ given by Eq. (13.22) whereas $G^{opt}(z)$ contains an estimate $\hat{G}_0(z) = \dfrac{z^{-1}}{1 - 0.8017z^{-1}}$ of the nominal plant model $G_0(z)$. Both contain uncertainty models.

Comment *The optimal estimate of the feature vector* $\hat{\theta}_i^{opt}$ *based on performing a number of experiments by varying the emulator parameters was able to capture the variation in the system model as shown in Figure 13.2.*

13.4 Model of the Kalman Filter

The soft sensor is a Kalman filter designed to estimate the unmeasured variable y_r. The Kalman filter embodies a copy of the system model and is driven by the residual, which is an error between the measured output of the system and its estimate generated by the Kalman filter $\boldsymbol{e} = \boldsymbol{y} - \hat{\boldsymbol{y}}$. The resulting closed-loop Kalman filter is stabilized by a Kalman gain, which is determined from minimizing the covariance of the state estimation error. The estimate $\hat{y}_r$ of the unmeasurable (inaccessible) variable y_r is an output of the Kalman filter. Figure 13.3 shows the plant and the Kalman filter. The inputs to the Kalman filter are the control input u and measured outputs of the plant $\boldsymbol{y}$. The Kalman filter contains a copy of the model of the plant, which is driven by the residual, an error between the plant output $\boldsymbol{y}$ and its estimate $\hat{\boldsymbol{y}}$. $\boldsymbol{K}_0$ is the Kalman gain which minimizes the covariance of the estimation error $\boldsymbol{e}$.

13.4.1 Role of the Kalman Filter

The Kalman filter not only estimates the unmeasured output but also plays the following important roles:

- *Soft sensor*: Estimates the unmeasured output to be regulated y_r.
- *Status monitor*: The states of the Kalman filter may be employed monitoring the status of the system.

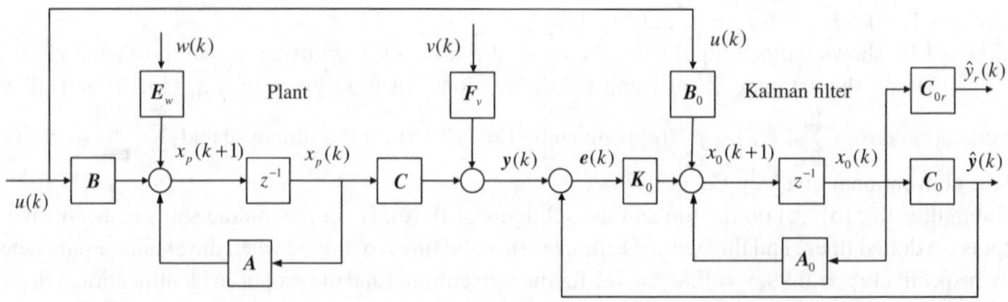

Figure 13.3 Kalman filter embodies the plant model driven by the residual

- *Model perturbation monitor*: The residual of the Kalman filer provides an estimate of the model perturbation, which is the deviation of the plant transfer function from that of the nominal one. This helps in achieving the following objectives;
  - o designing a robust H-infinity controller
  - o ensuring superior performance of the closed-loop system over a wide range of model perturbations. The plant is identified, and the controller redesigned whenever the residual exceeds some specified threshold. Frequent identification and the design of controller using the identified model will ensure superior controller performance
  - o the residual may be employed for fault diagnosis.

13.4.2 Model of the Kalman Filter

$$\begin{aligned} \boldsymbol{x}_0(k+1) &= \boldsymbol{A}_0\boldsymbol{x}_0(k) + \boldsymbol{B}_0 u(k) + \boldsymbol{K}_0\left(\boldsymbol{y}(k) - \hat{\boldsymbol{y}}(k)\right) \\ \hat{\boldsymbol{y}}(k) &= \boldsymbol{C}_0\boldsymbol{x}_0(k) \\ \hat{y}_r(k) &= \boldsymbol{C}_{0r}\boldsymbol{x}_0(k) \\ \boldsymbol{e}(k) &= \boldsymbol{y}(k) - \hat{\boldsymbol{y}}(k) \end{aligned} \tag{13.28}$$

where $\left(\boldsymbol{A}_0, \boldsymbol{B}_0, \boldsymbol{C}_0\right)$ is the identified nominal plant model $(\boldsymbol{A}, \boldsymbol{B}, \boldsymbol{C})$, $\boldsymbol{x}_0(k)$ is a $nx1$ state, $\hat{\boldsymbol{y}}(k)$ is an $n_y x1$ estimate of the plant output $\boldsymbol{y}(k)$. Rewriting the Kalman filter model after substituting for $\hat{\boldsymbol{y}}$ in the dynamical equation yields

$$\begin{aligned} \boldsymbol{x}_0(k+1) &= \left(\boldsymbol{A}_0 - \boldsymbol{K}_0\boldsymbol{C}_0\right)\boldsymbol{x}_0(k) + \boldsymbol{B}_0 u(k) + \boldsymbol{K}_0\boldsymbol{y}(k) \\ \hat{\boldsymbol{y}}(k) &= \boldsymbol{C}_0\boldsymbol{x}_0(k) \\ \hat{y}_r(k) &= \boldsymbol{C}_{0r}\boldsymbol{x}_0(k) \end{aligned} \tag{13.29}$$

13.4.3 Augmented Model of the Plant and the Kalman Filter

The augmented model $\left(\boldsymbol{A}_{pk}, \boldsymbol{B}_{pk}, \boldsymbol{C}_{pk}\right)$ formed of the plant and the Kalman filter relating the input $u(k)$ and the estimate of the inaccessible output $y_r(k)$ takes the following form:

$$\begin{aligned} \boldsymbol{x}_{pk}(k+1) &= \boldsymbol{A}_{pk}\boldsymbol{x}_{pk}(k) + \boldsymbol{B}_{pk}\boldsymbol{u}(k) + \boldsymbol{E}_{pk}w(k) + \boldsymbol{F}_{pk}v(k) \\ \hat{y}_r(k) &= \boldsymbol{C}_{pk}\boldsymbol{x}_{pk}(k) \end{aligned} \tag{13.30}$$

where $\boldsymbol{x}_{pk}(k) = \begin{bmatrix} \boldsymbol{x}(k) \\ \boldsymbol{x}_0(k) \end{bmatrix}$, $\boldsymbol{A}_{pk} = \begin{bmatrix} \boldsymbol{A} & \boldsymbol{0} \\ \boldsymbol{K}_0\boldsymbol{C}_p & \boldsymbol{A}_0 - \boldsymbol{K}_0\boldsymbol{C}_0 \end{bmatrix}$, $\boldsymbol{B}_{pk} = \begin{bmatrix} \boldsymbol{B} \\ \boldsymbol{B}_0 \end{bmatrix}$

$\boldsymbol{F}_{pk} = \begin{bmatrix} \boldsymbol{0} \\ \boldsymbol{K}_0\boldsymbol{F}_v \end{bmatrix}$, $\boldsymbol{E}_{pk} = \begin{bmatrix} \boldsymbol{E}_w \\ \boldsymbol{0} \end{bmatrix}$; $\boldsymbol{C}_{pk} = \begin{bmatrix} \boldsymbol{0} & \boldsymbol{C}_{0r} \end{bmatrix}$

13.5 Robust Controller Design

13.5.1 Objective

Design a robust controller such that the closed-loop control system meets the stability and performance requirements, and the output of the system, denoted $y(k)$, tracks a given reference input $r(k)$. In general, the output $y(k)$ may be an element of the measured output $\boldsymbol{y}(k)$ or $y_r(k)$, which the soft sensor measures. In this chapter, the output to be regulated is chosen to be $y(k) = y_r(k)$.

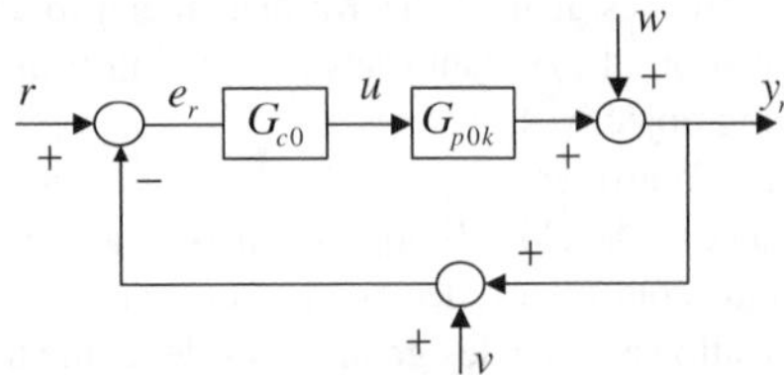

Figure 13.4 A closed-loop system formed of the augmented nominal plant and controller

13.5.2 Augmented Model

Let G_{pk} be the transfer function of the augmented state space model $\left(\boldsymbol{A}_{pk}, \boldsymbol{B}_{pk}, \boldsymbol{C}_{pk}\right)$.

13.5.2.1 Nominal Augmented Model

Let G_{p0k} be the transfer function of the augmented nominal plant $\left(\boldsymbol{A}_{pk0}, \boldsymbol{B}_{pk0}, \boldsymbol{C}_{pk0}\right)$ formed of the nominal plant $\left(\boldsymbol{A}^0, \boldsymbol{B}^0, \boldsymbol{C}_r^0\right)$ and the Kalman filter $\left(\boldsymbol{A}_0, \boldsymbol{B}_0, \boldsymbol{C}_{0r}\right)$.

13.5.3 Closed-Loop Performance and Stability

Let G_{c0} be the controller that stabilizes the nominal plant G_{p0k}; $r(k)$, $y(k)$, and $e_r(k)$ are the reference input, the output to be regulated, and the tracking error respectively; w is the disturbance at the plant output; v is the measurement or sensor noise. The closed-loop formed of the augmented nominal plant and the controller is shown in Figure 13.4.

Let $\boldsymbol{U}$ and $\boldsymbol{Z}$ be respectively a 3×1 input vector comprising r, w, and v, and a 2×1 output to be regulated formed of tracking error e_r and control input u given by

$$\boldsymbol{U} = [\, r \quad w \quad v\,]^T \tag{13.31}$$

$$\boldsymbol{Z} = [\, e_r \quad u\,]^T \tag{13.32}$$

Closed-loop transfer functions, which play significant role in the stability and performance of a control system, are three sensitivity functions of the closed-loop system formed of the nominal plant and nominal controller [8]. They are the sensitivity S_0, the (control) input sensitivity S_{u0}, and the complementary sensitivity T_0 given by:

$$\begin{aligned} S_0 &= \frac{1}{1 + G_{p0k} G_{c0}} \\ S_{u0} &= S_0 G_{c0} \\ T_0 &= 1 - S_0 \end{aligned} \tag{13.33}$$

Since $T_0 = 1 - S_0$, there are essentially two sensitivity functions, namely S_0 and S_{u0}, that determine the stability and the performance. The performance objective of a control system is to regulate the tracking error $e = r - y$ so that the steady-state tracking error is acceptable and its transient response meets the time and the frequency domain specifications respecting the technological constraints on the control input u so that, for example, the actuator is not saturated. A map relating the inputs r, w, and v and the outputs to be regulated namely e_r and u are:

$$e_r = S_0(r - w - v) \tag{13.34}$$

$$u = S_{u0}(r - v - w) \tag{13.35}$$

The transfer matrix relating $\boldsymbol{U}$ to $\boldsymbol{Z}$ is given by

$$\boldsymbol{Z} = \begin{bmatrix} S_0 & -S_0 & -S_0 \\ S_{u0} & -S_{u0} & -S_{u0} \end{bmatrix} \boldsymbol{U} \tag{13.36}$$

13.5.4 Uncertainty Model

The numerator-denominator perturbation model, which is similar to the emulator (13.5), considers the perturbation in the numerator and denominator polynomials separately instead of clubbing together as a single perturbation of the overall transfer function.

$$G_{pk} = N_{pk} D_{pk}^{-1} = \left(N_{p0k} + \Delta_N\right)\left(D_{p0k} + \Delta_D\right)^{-1} \tag{13.37}$$

where N_{p0k} and N_{pk} are the numerator polynomials; D_{p0k} and D_{pk} are the denominator polynomials respectively of G_{p0k} and G_{pk}; $\Delta_N \in RH_\infty$ and $\Delta_D \in RH_\infty$ are respectively frequency dependent relative perturbation in the numerator and the denominator polynomials [5]. The robust stability of the closed-loop system with plant model uncertainty is established using the small gain theorem.

Theorem 13.1 *Assume that G_{c0} internally stabilizes the nominal augmented plant G_{p0k}. Hence $S_0 \in RH_\infty$ and $S_{u0} \in RH_\infty$. Then closed-loop system is well posed and internally stable for all numerator and denominator perturbations*

$$\left\| \left[\Delta_N \quad \Delta_D \right] \right\|_\infty = \max_\omega \left\{ \sqrt{\Delta_N^2(j\omega) + \Delta_D^2(j\omega)} \right\} \leq 1/\gamma_0 \tag{13.38}$$

If and only if

$$\left\| \left[S_0 \quad S_{u0} \right] D_{p0}^{-1} \right\|_\infty < \gamma_0 \tag{13.39}$$

Proof: The SISO robust stability problem considered herein is a special case of MIMO case proved in [14].

Thus to ensure a robustly stable closed-loop system, the nominal sensitivity S_0 should be made small in frequency regions where the denominator uncertainty Δ_D is large, and the nominal complementary sensitivity S_{u0} should be made small in frequency regions where the numerator uncertainty Δ_N is large. Our objective is to design a controller G_{c0} such that robust performance and robust stability are achieved, that is, both the performance and stability hold for all plant model perturbations $\|[\,\Delta_N \quad \Delta_D\,]\| \leq 1/\gamma_0$ for some $\gamma_0 > 0$. Besides these requirements, we need to also consider technological constraints, especially the control input limitations. From Theorem 13.1 and Eq. (13.36) it is clear that the requirements for robust stability, performance and control limitation are inter-related:

- Robust performance for tracking with disturbance rejection. as well as robust stability in the face of denominator perturbations, requires small sensitivity function S_0 in the low-frequency region.
- Control input limitation and robust stability in the face of numerator perturbation require small control input sensitivity function S_{u0}.

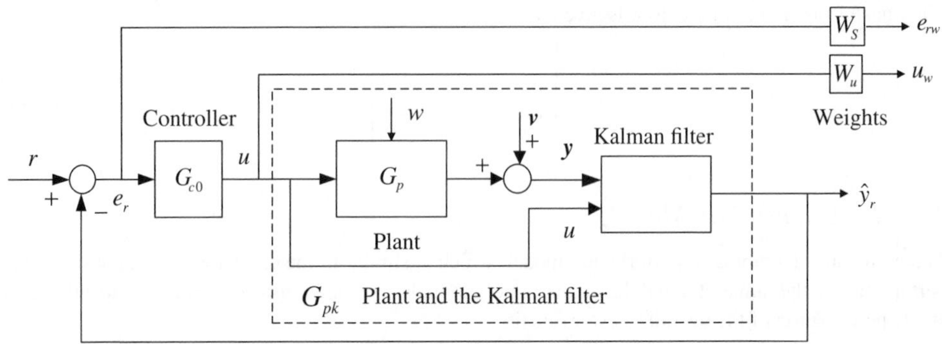

Figure 13.5 Mixed sensitivity weights for closed-loop system

13.5.5 Mixed-sensitivity Optimization Problem

With a view to address these requirements, let us select regulated outputs to be a frequency weighted tracking error e_{rw}, and a weighted control input u_w to meet respectively the requirements of performance and control input limitation.

$$\mathbf{Z}_w = [\, e_{rw} \quad u_w \,]^T \tag{13.40}$$

where $\mathbf{Z}_w$ is 2×1 vector output to be regulated, e_w and u_w are defined by their respective Fourier transforms: $e_{rw}(j\omega) = e_r(j\omega)W_S(j\omega)$ and $u_w(j\omega) = u(j\omega)W_u(j\omega)$. Figure 13.5 shows the weighted tracking error and the control input for the closed-loop system formed of the augmented plant G_{pk} comprising the plant and the Kalman filter. The frequency weights are chosen such that their inverses are the upper bound on the respective sensitive functions so that weighted sensitive functions are less than unity. The weighting functions $W_S(j\omega)$ and $W_u(j\omega)$ provide the tools to specify the tradeoff between robust performance and robust stability for a given application. For example, if robust performance (and robust stability to denominator perturbation Δ_D) is more important than the control input limitation, then the weighting function W_S is chosen to be large compared to W_u. To emphasize control limitation (and robust stability to numerator perturbation Δ_N) the weighting function W_u is chosen to be large compared to W_S. For steady-state tracking with disturbance rejection in the weighting function W_S one may include an approximate but "stable integrator" by choosing a pole close to zero or close to unity for the continuous-time and the discrete-time cases respectively [14]. Let T_{rz} be the nominal transfer matrix (when the plant perturbation $\Delta_0 = 0$) relating the reference input to the two frequency weighted outputs z_w, which is function of the G_{p0k} and G_{c0} given by

$$T_{rz}\left(G_{c0}, G_{p0k}\right) = D_{p0k}^{-1}\left[\, \bar{W}_S S_0 \quad \bar{W}_u S_{u0} \,\right]^T \tag{13.41}$$

where $\bar{W}_s = D_{p0k} W_s$ and $\bar{W}_u = D_{p0k} W_u$ so that the D_{p0k}^{-1} term appearing in the mixed sensitivity measure T_{rz} is canceled, thus yielding the following simplified measure $T_{rz} = \left[\, W_S S_0 \quad W_u S_{u0} \,\right]^T$. The mixed-sensitivity optimization problem for robust performance and stability in the H_∞ framework is then reduced to finding the controller G_{c0} such that:

$$\left\| T_{rz}\left(G_{c0}, G_{p0k}\right) \right\|_\infty = \left\| \left[\, W_S S_0 \quad W_u S_{u0} \,\right] \right\|_\infty = \max_{\omega} \left\{ \sqrt{\left(W_S S_0 (j\omega)\right)^2 + \left(W_u S_{u0} (j\omega)\right)^2} \right\} \leq \gamma < 1 \tag{13.42}$$

It is shown in [14] that the robustness condition (13.42) guarantees not only robust stability but also robust performance all $\|[\,\Delta_N \quad \Delta_D\,]\|_\infty \le 1/\gamma$: that is for all perturbations constrained by the inequality (13.38).

13.5.6 State-Space Model of the Robust Control System

Let the state-space model of the H-infinity robust controller G_{c0} be

$$\begin{aligned} \boldsymbol{x}_c(k+1) &= \boldsymbol{A}_c\boldsymbol{x}_c(k) + \boldsymbol{B}_c e_r(k) \\ u(k) &= \boldsymbol{C}_c\boldsymbol{x}_c(k) + D_c e_r(k) \end{aligned} \tag{13.43}$$

where $e_r(k) = r(k) - \hat{y}_r(k)$ is the tracking error. Substituting for $\hat{y}_r(k)$, the expression for the control input $u(k)$ becomes

$$u(k) = \boldsymbol{C}_c\boldsymbol{x}_c(k) - D_c\boldsymbol{C}_{0r}\boldsymbol{x}_0(k) + D_c r(k) \tag{13.44}$$

The closed-loop control system formed of the augmented plant $\left(\boldsymbol{A}_{pk}, \boldsymbol{B}_{pk}, \boldsymbol{C}_{pk}\right)$ given by Eq. (13.30) and the controller Eqs. (13.43) and (13.44) becomes

$$\begin{aligned} \boldsymbol{x}(k+1) &= \boldsymbol{A}\boldsymbol{x}(k) + \boldsymbol{B}r(k) + \boldsymbol{E}w(k) + \boldsymbol{F}v(k) \\ \hat{y}_r(k) &= \boldsymbol{C}_{0r}\boldsymbol{x}(k) \end{aligned} \tag{13.45}$$

where $\boldsymbol{A} = \begin{bmatrix} \boldsymbol{A}_p & -\boldsymbol{B}_0 D_c \boldsymbol{C}_{0r} & \boldsymbol{B}_p\boldsymbol{C}_c \\ \boldsymbol{K}_0\boldsymbol{C}_p & \boldsymbol{A}_0 - \boldsymbol{K}_0\boldsymbol{C}_0 - \boldsymbol{B}_0 D_c \boldsymbol{C}_{0r} & \boldsymbol{B}_0\boldsymbol{C}_c \\ 0 & -\boldsymbol{B}_c\boldsymbol{C}_{0r} & \boldsymbol{A}_c \end{bmatrix}$; $\boldsymbol{B} = \begin{bmatrix} \boldsymbol{B}_p D_c \\ \boldsymbol{B}_0 D_c \\ \boldsymbol{B}_c \end{bmatrix}$

$$\boldsymbol{E} = \begin{bmatrix} \boldsymbol{0} \\ \boldsymbol{K}_0\boldsymbol{F}_p \\ \boldsymbol{0} \end{bmatrix}, \boldsymbol{F} = \begin{bmatrix} \boldsymbol{F}_{pk} \\ \boldsymbol{0} \\ \boldsymbol{0} \end{bmatrix}; \boldsymbol{C} = \begin{bmatrix} \boldsymbol{0} & \boldsymbol{C}_{0r} & \boldsymbol{0} \end{bmatrix}$$

Figure 13.6 shows the closed-loop control system using a soft sensor. The unmeasured output of the plant y_r is substituted by its estimate $\hat{y}_r$ computed using the soft sensor (Kalman filter). The closed-loop system (13.45) is robustly stable as long as the perturbations of the nominal plant and the Kalman filter satisfy the robust stability condition (13.42) when perturbations in the numerator and the denominator polynomials of the augmented plant defined by Eq. (13.37) satisfy the inequality (13.38). The perturbation Δ_N in the numerator polynomial N_{pk} and perturbation Δ_D in the denominator polynomial D_{pk}

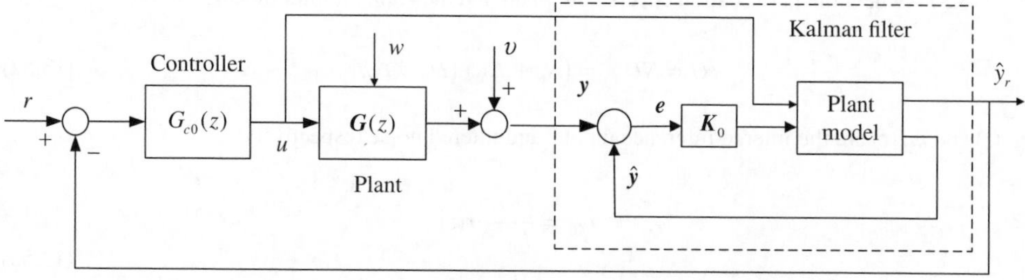

Figure 13.6 Closed-loop control with soft sensor

of the augmented nominal plant G_{p0k} result from the perturbations in the nominal state-space model $\left(\boldsymbol{A}_{p0k}, \boldsymbol{B}_{p0k}, \boldsymbol{C}_{p0k}\right)$.

13.6 High Performance and Fault Tolerant Control System

The Kalman filter residual is employed for achieving the additional tasks of high performance control system, performance monitoring, and fault diagnosis. The plant transfer function $G(z)$ relating the control input $u(z)$ and the regulated output $y(z)$ is:

$$G(z) = \frac{N(z)}{D(z)} \tag{13.46}$$

The plant output $y(z)$ is given by

$$y(z) = G(z)u(z) + \upsilon(z) \tag{13.47}$$

13.6.1 Residual and Model-mismatch

An expression relating the residual and the model perturbation takes the following form:

In [4], the relation between the residual $e = y - \hat{y}$ and the Kalman filter inputs u and y is shown to be

$$e(z) = \frac{D_0(z)}{F_0(z)}y(z) - \frac{N_0(z)}{F_0(z)}u(z) \tag{13.48}$$

where $F_0(z) = \det\left(z\boldsymbol{I} - \boldsymbol{A}_0 + \boldsymbol{K}_0\boldsymbol{C}_0\right)$, $N_0(z)$, $D_0(z)$ and $G_0(z)$ are the nominal numerator, nominal denominator, and nominal transfer functions respectively of $N(z)$, $D(z)$, and $G(z)$ given by Eq. (13.46). Further, an expression for the residual e of the Kalman filter for the closed-loop system formed of the plant, the Kalman filter, and the robust controller (13.45) is derived using the relation between the residual and model-mismatch for an open-loop system formed of the plant and the Kalman filter given in [4] and the expression (13.48). Expression for the residual is given by

$$e(z) = e_f(z) + e_0(z) \tag{13.49}$$

$$e_f(z) = S_{u0}(z)\Delta G r_{filt}(z) \tag{13.50}$$

$$e_0(z) = \upsilon_{filt}(z) \tag{13.51}$$

where the model-mismatch term $\Delta G = G - G_0$. Expressing plant model-mismatch ΔG in terms of perturbation in the numerator and the denominator polynomials of the plant transfer function similar to Eq. (13.37) (where the perturbation covers both plant and the Kalman filter models), we get

$$G = ND^{-1} = \left(N_0 + \Delta_N\right)\left(D_0 + \Delta_D\right)^{-1} \tag{13.52}$$

$r_{filt}(z)$ and $\upsilon_{filt}(z)$ are the filtered reference input r and filtered noise respectively:

$$\begin{aligned} r_{filt} &= \frac{D_0(z)}{F_0(z)}r(z) \\ \upsilon_{filt} &= \frac{D_0(z)}{F_0(z)}\upsilon(z) \end{aligned} \tag{13.53}$$

The residual expression (13.49) forms the basis for developing a high performance and fault tolerant control system. The model mismatch will be nonzero if the actual plant model deviates from the nominal model. The mean of the residual is an indicator of a model mismatch. We will assume that $e_0(k)$ continues to be a zero-mean Gaussian white noise process under both fault and fault-free conditions. It is shown in [4]

$$E[e] = 0 \text{ if and only if } \Delta G_p(z) \equiv 0 \tag{13.54}$$

13.6.2 Bayes Decision Strategy

We will formulate a binary hypothesis testing problem to decide between two hypotheses, H_0 and H_1, where H_0 denotes a normal (or an acceptable) operating regime and H_1 indicates an abnormal (or an unacceptable) operating regime. A batch processing scheme is adopted here where residuals are collected in a sliding time window of length N and processed at each time instant. At each time instant k, the N residuals formed of the present and past $N-1$ residuals, $e(k-i) : i = 0, 1, 2, \ldots N-1$, are collected:

$$\boldsymbol{e}(k) = \begin{bmatrix} e(k) & e(k-1) & e(k-2) & . & e(k-N+1) \end{bmatrix}^T \tag{13.55}$$

The Bayes decision strategy takes the general form

$$t_s(\boldsymbol{e}) \begin{cases} \leq \eta & \text{normal} \\ > \eta & \text{abnormal} \end{cases} \tag{13.56}$$

where $t_s(\boldsymbol{e})$ is the test statistics of the residual $\boldsymbol{e}$ and η is the threshold value computed taking into account the variance of the noise term $e_0(k)$, prior probabilities of the two hypotheses, the cost associated with correct and wrong decisions, and the probability of false alarm. Test statistics for a reference input $r(k)$ that is either a constant or a sinusoid of frequency f_0 or an arbitrary signal are all listed below [15]:

$$t_s(\boldsymbol{e}) = \begin{cases} \left| \dfrac{1}{N} \displaystyle\sum_{i=k-N+1}^{k} e(i) \right| & r(k) = \text{constant} \\ P_{ee}(f_0) & r(k) \text{ is a sinusoid} \\ \dfrac{1}{N} \displaystyle\sum_{i=k-N+1}^{k} e^2(i) & r(k) \text{ is an arbitrary signal} \end{cases} \tag{13.57}$$

13.6.3 High Performance Control System

The performance of the robust controller depends upon (i) the accuracy of the estimate $\hat{y}_r$ of the output to be regulated y_r and is generated by the Kalman filter-based soft sensor, (ii) noise and the disturbance affecting the output, and (iii) the plant model deviation ΔG. In order to meet the requirement of high performance and stability in the design stage, the control input sensitivity function $S_{u0}(z)$ is given an appropriate frequency weight $W_u(z)$ in the mixed sensitivity H-infinity design setting so that resulting controller will ensure that the residual is small in the face of plant deviation, as can be inferred from Eq. (13.50). When $S_{u0}(z)$ is small, then the residual will be small and hence the true and the estimated variable are close ensuring thereby acceptable controller performance. In the operational stage, the test statistics $t_s(\boldsymbol{e})$ is monitored using Eq. (13.56) as shown in Figure 13.7. When an abnormal operating regime is indicated, then the plant is re-identified and the robust controller is redesigned for the identified plant model. It is an *indirect adaptive control* scheme. The Kalman gain $\boldsymbol{K}_0$ is chosen optimally to ensure that covariance of the estimation error is minimum.

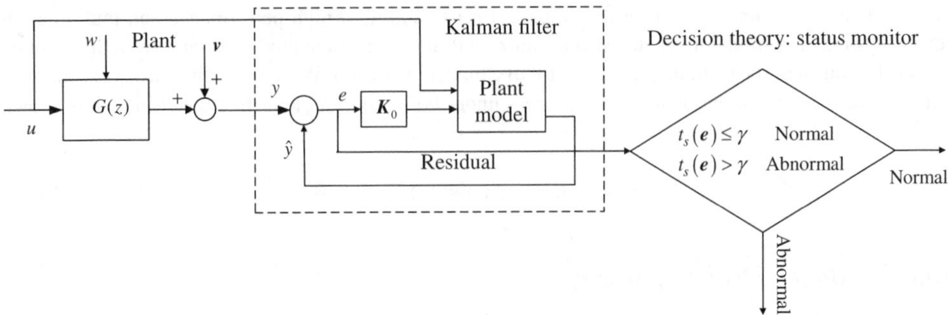

Figure 13.7 Monitoring the status of the system

13.6.4 Fault-Tolerant Control System

As before, the test statistics $t_s(\boldsymbol{e})$ may be used to monitor the status of the system. If there is a fault due to variations in the plant, sensor, or actuators, the decision strategy will indicate an abnormality. There are two options: one is to accommodate the model variations by re-identifying and controller redesign as explained in the previous section. If it cannot be accommodated, then the operation is halted until the faulty device is repaired.

Remark It is interesting to note that the design of the closed-loop system to be robust to plant perturbations hides incipient faults due to plant fault perturbations in the plant in view of Eq. (13.50).

13.7 Evaluation on a Simulated System: Soft Sensor

The proposed scheme is evaluated on a velocity control system using a DC servomotor. DC motors are versatile and extensively used in industry. Large DC motors are used in machine tools, printing presses, conveyors, fans, pumps, hoists, cranes, paper mills, textile mills, rolling mills, transit cars, locomotives, and so forth. Small DC motors are used primarily as control devices, such as servomotors for positioning and tracking. The DC motor system has two state variables, namely the angular velocity $y_r = \omega$ and armature current $y = i$. It is assumed that y_r is inaccessible and y is measured.

The objective is to estimate the angular velocity y_r using the Kalman filter. The Kalman filter contains a copy of the identified plant model, which is driven by the error between the measured plant output y and its estimate $\hat{y}$ namely the residual $y - \hat{y}$, and generates the estimate $\hat{y}_r$ of y_r. The soft sensor is thus a Kalman filter, which estimates the angular velocity y_r using the input u to the amplifier of the DC motor and armature current $y = i$. The armature current is measured using a static sensor. A static sensor is inexpensive and does not require maintenance. A software sensor (Kalman filter) replaces a hardware velocity sensor (e.g., tachometer).

The controller for the augmented system formed of the plant and the Kalman filter is designed using a robust H-infinity approach so that the closed-loop system is stable in the face of (i) an uncertainty in the mathematical model of the plant employed in designing the Kalman filter, (ii) the measurement noise v, and (iii) the load disturbance w affecting the system. Figure 13.8 shows the velocity control system formed of a DC servomotor, an amplifier, and a soft sensor to estimate the angular velocity of the motor ω. The steady-state Kalman gain $\boldsymbol{K}_0$ is chosen optimally to attenuate the noise v corrupting the current measurement and the load disturbance w acting on the DC motor. The parameters $L_a = 0.1$ and $R_a = 4.2$ are respectively the inductance and the resistance of the armature, $J_m = 0.00014$ and $b_m = 0.0000072$ are respectively the moment of inertia and damping coefficient, and $K_T = 0.1448$ is the torque constant, the sampling period $Ts = 0.01\text{sec}$.

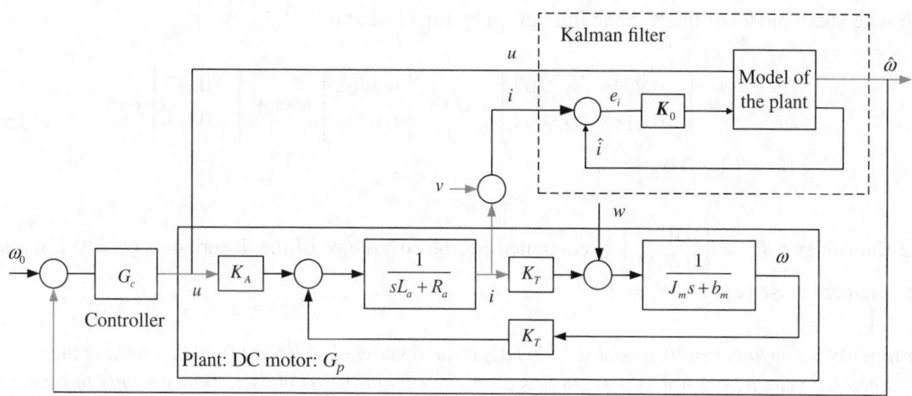

Figure 13.8 Soft sensor for a velocity control system

13.7.1 Offline Identification

During the identification phase, the angular velocity y_r is measured (in practice, during the identification phase the angular velocity may be measured using a hardware device, namely a tacho-generator). It is assumed that the plant model uncertainty is due to variations in the amplifier gain, the current sensor gain, and the tacho-generator gain.

The plant is identified by performing a number of experiments by varying the emulator parameters to mimic these variations. Emulators are chosen to be static gains $\gamma_i : i = 1, 2, 3$. The gains γ_1, γ_2, and γ_3 are connected in cascade with the amplifier, the current sensor, and the velocity sensor respectively as shown in Figure 13.9. The optimal feature vector $\boldsymbol{\theta}^{opt}$ is estimated using the least-squares method given by Eq. (13.13).

13.7.2 Identified Model of the Plant

The identified nominal state-space model $\left(\boldsymbol{A}^0, \boldsymbol{B}^0, \boldsymbol{C}^0\right)$ of the plant derived from the optimal feature vector $\boldsymbol{\theta}^{opt}$ given by Eq. (13.20) becomes:

$$
\begin{aligned}
\boldsymbol{x}(k+1) &= \begin{bmatrix} 0.9349 & 8.2346 \\ -0.0115 & 0.6009 \end{bmatrix} \boldsymbol{x}(k) + \begin{bmatrix} 0.4462 \\ 0.0796 \end{bmatrix} u(k) + \begin{bmatrix} 1 \\ 0 \end{bmatrix} w(k) \\
\boldsymbol{y}(k) &= [\,0 \quad 1\,]\boldsymbol{x}(k) + v(k) \\
y_r(k) &= [\,1 \quad 0\,]\boldsymbol{x}(k)
\end{aligned}
\tag{13.58}
$$

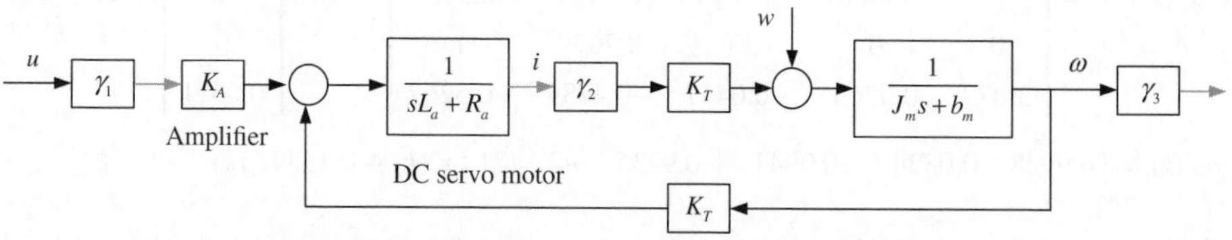

Figure 13.9 Identification by varying the emulator parameters

The state-space model of the Kalman filter given by Eq. (13.29) is

$$\boldsymbol{x}_0(k+1) = \begin{bmatrix} 0.9349 & 8.2209 \\ -0.0115 & 0.5998 \end{bmatrix} \boldsymbol{x}_0(k) + \begin{bmatrix} 0.4462 \\ 0.0796 \end{bmatrix} u(k) + \begin{bmatrix} 0.0137 \\ 0.0011 \end{bmatrix} \boldsymbol{y}(k) \tag{13.59}$$
$$\hat{y}_r(k) = [\,1 \quad 0\,]\boldsymbol{x}_0(k)$$

The Kalman gain $\boldsymbol{K}_0 = \begin{bmatrix} 0.0137 \\ 0.0011 \end{bmatrix}$ is computed for the covariance of the disturbance $Q_0 = 0.1$ and the measurement noise variance $R_0 = 100$.

Comments *The uncertainty model used in the physical system is different from that used in the case of the simulated velocity control system. In this case, the emulator was chosen to target only those model parameters that are likely to vary, while in the case of the simulated example an unstructured emulator model was employed. However, the design of the robust controller was similar to that of the simulated example.*

Recall that the Kalman filter computes the estimate by fusing the a posteriori *information provided by the measurement, and the* a priori *information contained in the model that generated the measurement. The covariance of the measurement noise and the covariance of the disturbance quantify the degree of belief associated with the measurement and model information, respectively. These covariances play a crucial role in the performance of the Kalman filter. The estimate of the state is obtained as the best compromise between the estimates generated by the model and those obtained from the measurement, depending upon the plant noise and the measurement noise covariances.*

As the covariance of the disturbance $Q_0 = 0.1$ is very small compared to that of the measurement noise variance $R_0 = 100$, that is Q/R is very small, the Kalman filter model $(\boldsymbol{A}^0, \boldsymbol{B}^0, \boldsymbol{C}^0)$ is assumed to be more accurate compared to the measurement y. This is reflected in the "size" of the Kalman gain $\boldsymbol{K}_0$ which is "small." As a result, the Kalman filter computes the estimate by giving more weight to the model than the noisy measurement.

13.7.3 Mixed-Sensitivity Optimization Problem

Frequency weights $W_S(z)$ associated with the sensitivity functions $S_0(z)$ and $W_u(z)$ associated with control input sensitivity $S_{u0}(z)$ were determined using a recursive design procedure, to obtain an acceptable tradeoff between the steady-state tracking requirement and control input limitation in the mixed sensitivity problem (13.42). The frequency weights were $W_s(z) = \dfrac{0.1}{1 - 0.99z^{-1}}$ emphasizing very low frequency for tracking a constant reference input and $W_u(z) = 0.1$. The H_∞ robust controller given by Eq. (13.43) is:

$$\boldsymbol{x}_c(k+1) = \begin{bmatrix} 0.99 & 0 & 0 & 0 & 0 \\ 0 & 0.7674 & 1 & 0 & 0 \\ 0 & -0.0667 & 0.7674 & 0.3354 & 0.205 \\ 0 & 0 & 0 & 0.7679 & 1 \\ 3.077 & 0.0363 & -0.0417 & -0.4734 & -0.5267 \end{bmatrix} \boldsymbol{x}_c(k) + \begin{bmatrix} 0.0395 \\ 0 \\ 0 \\ 0 \\ 0.0611 \end{bmatrix} e_r(k) \tag{13.60}$$
$$u(k) = \begin{bmatrix} 6.9928 & 0.0824 & -0.0947 & -0.9235 & -2.8421 \end{bmatrix} \boldsymbol{x}_c(k) + 0.1389 e_r(k)$$

As the controller has a pole at 0.99(which is close to the origin of the z-plane), a steady-state tracking of a constant reference input is ensured.

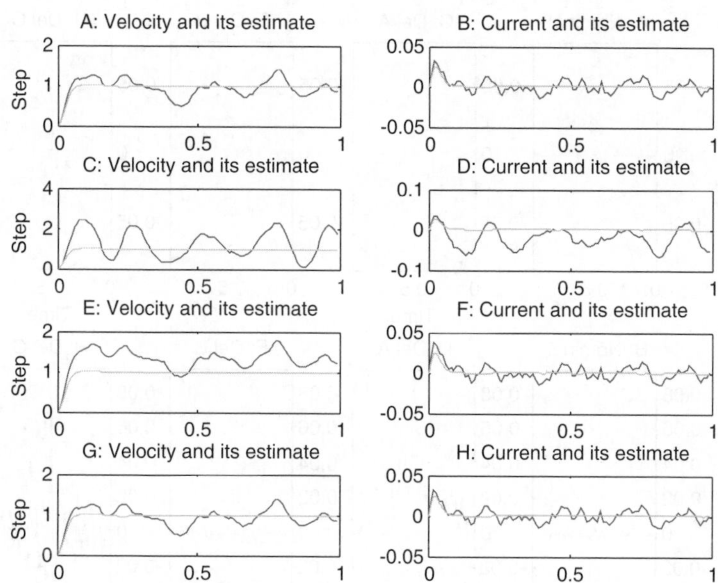

Figure 13.10 Velocity and current profiles and their estimates for different operating regimes

13.7.4 Performance and Robustness

Various operating regimes were simulated including:

- *Nominal operating regime*: The plant state model is equal to the nominal model $(\boldsymbol{A}_{p0}, \boldsymbol{B}_{p0}, \boldsymbol{C}_{p0})$.
- *Plant perturbations:* The plant state-space model $(\boldsymbol{A}_p, \boldsymbol{B}_p, \boldsymbol{C}_p)$ is subject to perturbations:
  - $\boldsymbol{A}_p = \boldsymbol{A}_{p0}(1 + \Delta_A)$
  - $\boldsymbol{B}_p = \boldsymbol{B}_{p0}(1 + \Delta_B)$
  - $\boldsymbol{C}_p = \boldsymbol{C}_{p0}(1 + \Delta_C)$

  where $\Delta_A = 0.2$, $\Delta B = 0.2$, and $\Delta_C = 0.2$ The results of normal and perturbed parameter experiments are shown in Figure 13.10. The current, the velocity, and their estimates from the Kalman filter are displayed under different operating regimes. The estimated velocity and the current generated by the Kalman filter and those from the plant are shown. Subfigures (a) and (b) show respectively the velocity and the current profiles under the normal operating regime. Similarly, subfigures (c) and (d), subfigures (e) and (f), and subfigures (g) and (h) show the plots of the estimated velocity and the estimated current respectively under perturbations $A_p = A_{p0}(1 + \Delta_A)$, $\boldsymbol{B}_p = \boldsymbol{B}_{p0}(1 + \Delta_B)$, and $\boldsymbol{C}_p = \boldsymbol{C}_{p0}(1 + \Delta_C)$. The tracking performance of the velocity control system using the estimated regulated variable $\hat{y}_r$ in the face of plant perturbations is very good. The steady-state tracking error is negligible and transient performance is good. Further, the Kalman filter estimate meets requirements of (i) accuracy and (i) low estimation error variance. Note that estimates of both current and velocity estimates are practically noise-free even though Kalman filter inputs, namely the current and control input, are both noisy.

13.7.5 Status Monitoring

The status of the system is monitored in the face of the plant perturbation, and in the presence of the noise and disturbances corrupting the plant output. The residual, which is an error between the current

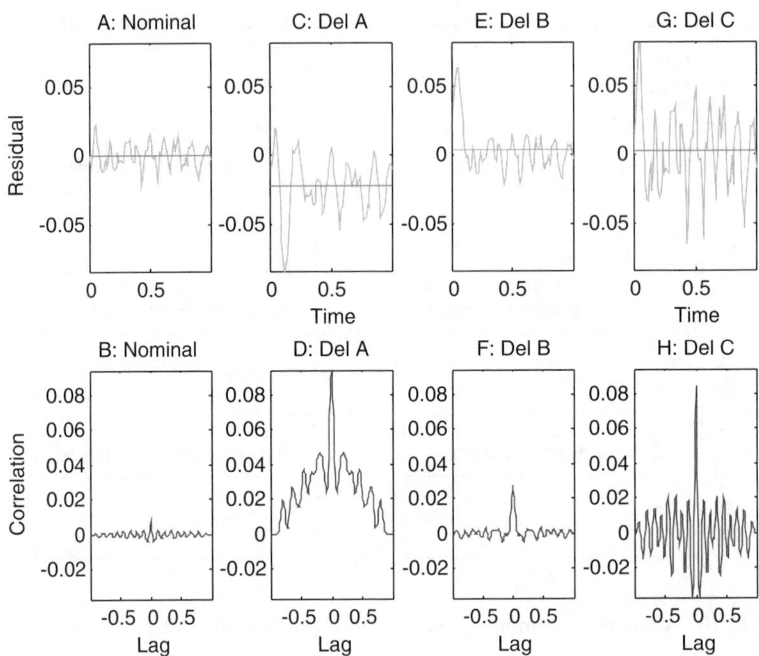

Figure 13.11 Residuals and their auto-correlations under plant perturbations

and its estimate using the Kalman filter, and the auto-correlations of the residual were also computed, as shown in Figure 13.11.

Subfigures (a) and (b) show respectively the residual and its auto-correlation under the normal operating regime. Similarly, subfigures (c) and (d), (e) and (f), and (g) and (h) show respectively the plots of the residual and its auto-correlation under perturbations $\boldsymbol{A}_p = \boldsymbol{A}_{p0}(1 + \Delta_A)$, $\boldsymbol{B}_p = \boldsymbol{B}_{p0}(1 + \Delta_B)$ and $\boldsymbol{C}_p = \boldsymbol{C}_{p0}(1 + \Delta_C)$. The mean of residual is nonzero under plant perturbations. The auto-correlation function of the residual visually enables us to distinguish the normal from abnormal operating conditions resulting from plant perturbations. The test statistics $t_s(\boldsymbol{e})$ under various plant perturbations are given in Table 13.1. It is important to emphasize that the performance of the Kalman filter operating in an open-loop configuration (when the Kalman filter estimate is not employed to drive the controller to close the loop) is poor in the face of plant perturbations. In other words, the performance of the Kalman filter is superior when operating in a closed-loop with its estimate driving the controller, as shown in Figure 13.8.

13.8 Evaluation on a Physical Velocity Control System

MATLAB® Real time workshop is used to implement and evaluate the performance of the proposed soft sensor in a real-time environment on a laboratory-scale physical velocity control system.

Table 13.1 The test statistics

Normal	$\boldsymbol{A}_p = \boldsymbol{A}_{p0}(1 + \Delta_A)$	$\boldsymbol{B}_p = \boldsymbol{B}_{p0}(1 + \Delta_B)$	$\boldsymbol{C}_p = \boldsymbol{C}_{p0}(1 + \Delta_C)$
0.0074	0.0244	0.0106	0.0235

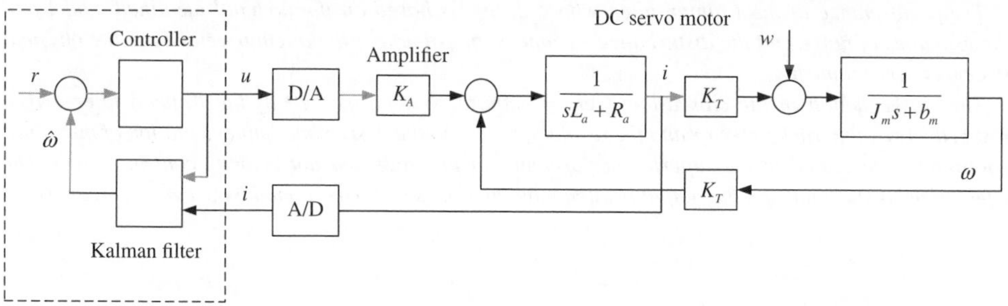

Figure 13.12 Physical DC motor with amplifier is interfaced to the real-time workshop

The physical velocity control system is similar to that of the simulated system given in the Section 13.7. A host PC is employed to design the controller using MATLAB®. The target PC is interfaced to an A/D to acquire the current from the DC motor, and the estimated angular velocity fed to the D/A converter I/O board, termed Data Translation DT 2821. The host PC downloads the executable code of the Simulink model in C++ to the target PC. The target PC executes the code in real time using the sampled input from A/D converter, and the soft sensor output is fed to the D/A converter. Figure 13.12 shows the laboratory-scale physical position control system interfaced to a personal computer using analog to digital and digital to analog converters, and its block diagram representation.

In order to identify the plant and to evaluate the performance, the tachometer was connected to the DC motor to measure the velocity merely for comparing the estimated velocity from the Kalman filter with the actual velocity. The plant was identified by performing a number of parameter-perturbed experiments. The static gain emulators were connected in cascade with the amplifier, the current sensor, and the velocity sensors, as shown in Figure 13.9.

Figure 13.13 shows the actual and the estimated current and the angular velocity. The figure on the left shows the velocity while that on the right shows the current profiles.

It can be deduced that the estimate of current and the angular velocity given by the Kalman filter closely matches that sensed by the current sensor and the tachometer respectively. The noise spikes in the current and the velocity are due to the commutator and the brush associated with the armature of the DC motor.

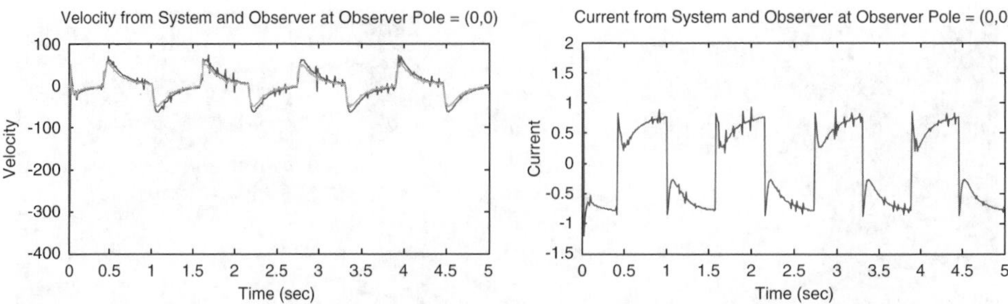

Figure 13.13 Velocity, current, and their estimates

Comments *The identified model of the plant using the proposed parameter perturbed identification scheme was robust in the face of model uncertainties associated with friction, nonlinearity, and variations in the operating points.*

The performance of the Kalman filter, whose design is based on the nominal identified model and the estimates of noise and the disturbance variances, in estimating the angular velocity of the physical system is very promising.

The closed action of the closed-loop velocity control system, as well as the Kalman filter driven respectively by the tracking error and the residual, thanks to the design and implementation of the robust controller, contributed to the superior performance of the fault tolerant velocity control system. The effect of the disturbance on the output is attenuated and the sensitivity to model mismatch is reduced.

13.9 Conclusions

The reliable identification scheme of the plant using a number of parameter-perturbed experiments by emulating likely fault scenarios is key to ensure high performance and robust stability in the face of model uncertainty and variation in the operating conditions

The robust controller based on minimizing frequency-weighted combinations of sensitivity and control input sensitivity functions in the mixed sensitivity H-infinity setting is effective in meeting the requirements of accuracy of the Kalman filter estimate, the performance and stability in the face of plant model perturbations. The tracking performance of the control system using the estimated regulated variable in the face of plant perturbations is highly promising. The steady-state tracking error is negligible and transient performance is good.

The Kalman filter residual is very effective in monitoring the status of the system. The auto-correlation function of the residual visually enables us to distinguish the normal from abnormal operating conditions resulting from plant perturbations.

The residual is employed in both the design and in operational stages. In the operational stage, whenever the test statistics indicate an abnormal operating regime, the plant is identified and the controller is redesigned to achieve a high performance control system. If the controller cannot accommodate to model the perturbations, a fault is indicated, and the system is shut down for repair.

The performance of a fault tolerant control system is highly promising in the face of noise and disturbance, model uncertainty, and variation in the operating conditions thanks to the reliable identification of the plant and the robust controller design. The Kalman filter plays a key role in providing a maintenance-free soft sensor to estimate unmeasured or inaccessible variables, and in monitoring the status of the system.

The proposed scheme was evaluated on simulated and physical laboratory-scale velocity control systems.

13.10 Summary

Mathematical Formulation

The state-space model of a system is given by:

$$\begin{aligned} \boldsymbol{x}(k+1) &= \boldsymbol{A}\boldsymbol{x}(k) + \boldsymbol{B}u(k) + \boldsymbol{E}_w w(k) \\ \boldsymbol{y}(k) &= \boldsymbol{C}\boldsymbol{x}(k) + \boldsymbol{F}_v v(k) \\ y_r(k) &= \boldsymbol{C}_r\boldsymbol{x}(k) \end{aligned}$$

Transfer Function Model

$$\boldsymbol{y}(z) = \frac{\boldsymbol{N}(z)}{D(z)}u(z) + \frac{\boldsymbol{N}_w(z)}{D(z)}\boldsymbol{w}(z) + \boldsymbol{F}_v\boldsymbol{v}(z)$$

Rewriting by cross-multiplying by $D(z)$,we get:

$$D(z)\boldsymbol{y}(z) = \boldsymbol{N}(z)u(z) + \boldsymbol{\upsilon}(z)$$

where $\boldsymbol{\upsilon}(z) = \boldsymbol{N}_w(z)\boldsymbol{w}(z) + D(z)\boldsymbol{F}_v\boldsymbol{v}(z)$

The $n_y \times 1$ matrix transfer function $\boldsymbol{G}(z)$ of the system is given by:

$$\boldsymbol{G}(z) = \boldsymbol{N}(z)D^{-1}(z)$$

Emulator: Numerator-denominator Perturbation Model

$$\boldsymbol{G}(z) = \frac{\boldsymbol{N}(z)}{D(z)} = \frac{\left(\boldsymbol{I} + \Delta_N(z)\right)}{\left(1 + \Delta_D(z)\right)}\frac{\boldsymbol{N}_0(z)}{D_0(z)} = \boldsymbol{G}_e(z)\boldsymbol{G}_0(z)$$

where $\boldsymbol{G}_0(z)$, is the nominal transfer function, $\boldsymbol{N}_0(z)$ is the nominal the numerator (matrix) polynomial, $D_0(z)$ is the nominal denominator (scalar) polynomial; $\boldsymbol{G}_e(z) = \left(\frac{\boldsymbol{I} + \Delta_N(z)}{1 + \Delta_D(z)}\right)$ is the $n_y x n_y$ *emulator*; $\Delta_N(z) \in RH_\infty$ and $\Delta_D(z) \in RH_\infty$

Selection of the Emulator Model

$$\boldsymbol{G}_e(z) = \begin{bmatrix} G_{e1}(z) & 0 & 0 & 0 \\ 0 & G_{e2}(z) & 0 & 0 \\ . & 0 & . & . \\ 0 & 0 & 0 & G_{en_y}(z) \end{bmatrix}$$

$$G_{ei}(z) = \begin{cases} \gamma_i & \text{gain} \\ \gamma_i z^{-d} & \text{gain and pure delay} \\ \gamma_i \dfrac{\gamma_{i1} + z^{-1}}{1 + \gamma_{i1} z^{-1}} & \text{first order all pass} \\ \gamma_i \prod\limits_j \dfrac{\gamma_{ij} + z^{-1}}{1 + \gamma_{ij} z^{-1}} & \text{Blaschke product} \end{cases}$$

The parameters γ_i and γ_{ij} are termed herein *emulator parameters*.

Identification of the System: Reliable Scheme

The perturbed model at the jth experiment relating the input $u(z)$ and the ith output $y_i^j(z)$ is:

$$D^j(z)y_i^j(z) = N_i^j(z)u(z) + \upsilon_i^j(z)$$

The linear regression model becomes:

$$y_i^j(k) = \left(\boldsymbol{\psi}_i^i(k)\right)^T \boldsymbol{\theta}_i^j + \upsilon_i^j(k) \quad i = 1, 2, 3, \dots, n_y$$

where $\left(\boldsymbol{\psi}_i^j(k)\right)^T = \left[-y_i^j(k-1) \quad -y_i^j(k-2) \quad . \quad -y_i^j(k-n) \quad u(k-1) \quad . \quad u(k-n)\right]\boldsymbol{\theta}_i^j$ is a $Mx1$ vector of unknown model parameter given by:

$$\boldsymbol{\theta}_i^j = \left[a_1^j \quad a_2^j \quad . \quad a_n^j \quad b_{i1}^j \quad b_{i2}^j \quad . \quad b_{in}^j\right]^T$$

Least-Squares Estimation

$$\hat{\boldsymbol{\theta}}_i^{opt} = \arg\left\{\min_{\{\theta_i\}}\left\{\sum_{j=1}^{N_{\exp}}\sum_{k=1}^{N}\left(y_i^j(k) - \left(\boldsymbol{\psi}_i^j\right)^T(k)\boldsymbol{\theta}_i\right)^T\left(y_i^j(k) - \left(\boldsymbol{\psi}_i^j\right)^T(k)\boldsymbol{\theta}_i\right)\right\}\right\}$$

$$\hat{y}_i^{j\,opt}(k) = \left(\boldsymbol{\psi}_i^j\right)^T(k)\hat{\boldsymbol{\theta}}_i^{opt}$$

Identified Nominal Model

The "optimal nominal model" derived from the optimal estimate $\hat{\boldsymbol{\theta}}_i^{opt}$ becomes:

$$D^{opt}(z)\mathbf{y}(z) = \mathbf{N}^{opt}(z)u(z) + \boldsymbol{\upsilon}(z)$$

where $D^{opt}(z)$ and $\mathbf{N}^{opt}(z)$ are derived from $\hat{\boldsymbol{\theta}}_i^0 : i = 1, 2, \dots, n_y$. The optimal nominal model is:

$$\mathbf{G}^{opt}(z) = \frac{\mathbf{N}^{opt}(z)}{D^{opt}(z)}$$

The identified nominal state-space model, denoted $\left(\mathbf{A}_0, \mathbf{B}_0, \mathbf{C}_0\right)$ of $(\mathbf{A}, \mathbf{B}, \mathbf{C})$ is:

$$\begin{aligned}
\mathbf{x}(k+1) &= \mathbf{A}_0\mathbf{x}(k) + \mathbf{B}_0 u(k) \\
\mathbf{y}(k) &= \mathbf{C}_0\mathbf{x}(k) \\
y_r(k) &= \mathbf{C}_{r0}\mathbf{x}(k)
\end{aligned}$$

The Nominal Plant Model

Let the nominal plant model $\left(\mathbf{A}^0, \mathbf{B}^0, \mathbf{C}^0\right)$ be:

$$\begin{aligned}
\mathbf{x}(k+1) &= \mathbf{A}^0\mathbf{x}(k) + \mathbf{B}^0 u(k) + \mathbf{E}_w w(k) \\
\mathbf{y}(k) &= \mathbf{C}^0\mathbf{x}(k) + \mathbf{F}_v v(k) \\
y_r(k) &= \mathbf{C}_r^0\mathbf{x}(k)
\end{aligned}$$

Model of the Kalman Filter

The soft sensor is a Kalman filter designed to estimate the unmeasured variable y_r.

$$
\begin{aligned}
\boldsymbol{x}_0(k+1) &= \boldsymbol{A}_0\boldsymbol{x}_0(k) + \boldsymbol{B}_0 u(k) + \boldsymbol{K}_0\left(\boldsymbol{y}(k) - \hat{\boldsymbol{y}}(k)\right)\\
\hat{\boldsymbol{y}}(k) &= \boldsymbol{C}_0\boldsymbol{x}_0(k)\\
\hat{y}_r(k) &= \boldsymbol{C}_{0r}\boldsymbol{x}_0(k)\\
\boldsymbol{e}(k) &= \boldsymbol{y}(k) - \hat{\boldsymbol{y}}(k)
\end{aligned}
$$

Augmented Model of the Plant and the Kalman Filter

The augmented model formed of the plant and the Kalman filter is:

$$
\begin{aligned}
\boldsymbol{x}_{pk}(k+1) &= \boldsymbol{A}_{pk}\boldsymbol{x}_{pk}(k) + \boldsymbol{B}_{pk}\boldsymbol{u}(k) + \boldsymbol{E}_{pk}w(k) + \boldsymbol{F}_{pk}v(k)\\
\hat{y}_r(k) &= \boldsymbol{C}_{pk}\boldsymbol{x}_{pk}(k)
\end{aligned}
$$

Robust Controller Design

Objective: Design a robust controller for the augmented plant and Kalman filter

Augmented plant:

Let $G_{pk}(z)$ and $G_{p0k}(z) = \dfrac{N_{p0k}(z)}{D_{p0k}(z)}$ be actual and the nominal augmented plant.

The closed-loop performance and stability:

$\begin{bmatrix} e_r \\ u \end{bmatrix} = \begin{bmatrix} S_0 & -S_0 & -S_0 \\ S_{u0} & -S_{u0} & -S_{u0} \end{bmatrix}\begin{bmatrix} r \\ w \\ v \end{bmatrix}$ where $S_0 = \dfrac{1}{1 + G_{p0k}G_{c0}}$; $S_{u0} = S_0 G_{c0}$ is the input sensitivity.

Uncertainty model:

$\boldsymbol{G_{pk} = N_{pk}D_{pk}^{-1} = \left(N_{p0k} + \Delta_N\right)\left(D_{p0k} + \Delta_D\right)^{-1}}$***Mixed-sensitivity Optimization Problem***

$\boldsymbol{Z}_w = \begin{bmatrix} e_{rw} & u_w \end{bmatrix}^T$ where $e_{rw}(j\omega) = e_r(j\omega)W_S(j\omega)$; $u_w(j\omega) = u(j\omega)W_u(j\omega)$

$$
T_{rz}\left(G_{c0},\ G_{p0k}\right) = \begin{bmatrix} W_S S_0 & W_u S_{u0} \end{bmatrix}^T
$$

Finding the controller G_{c0} such that:

$$
\|T_{rz}(G_{c0}, G_{p0k})\|_\infty = \left\|[\, W_S S_0 \quad W_u S_{u0}\,]\right\|_\infty = \max_{\omega}\left\{\sqrt{\left(W_S S_0\,(j\omega)\right)^2 + \left(W_u S_{u0}\,(j\omega)\right)^2}\right\} \le \gamma < 1
$$

State-Space Model of the Robust Control System

Let the state-space model of the *H*-infinity robust controller G_{c0} be

$$
\begin{aligned}
\boldsymbol{x}_c(k+1) &= \boldsymbol{A}_c\boldsymbol{x}_c(k) + \boldsymbol{B}_c(r(k) - \hat{y}_r(k))\\
u(k) &= \boldsymbol{C}_c\boldsymbol{x}_c(k) + D_c(r(k) - \hat{y}_r(k))
\end{aligned}
$$

The closed-loop control system

$$
\begin{aligned}
\boldsymbol{x}(k+1) &= \boldsymbol{A}\boldsymbol{x}(k) + \boldsymbol{B}r(k) + \boldsymbol{E}w(k) + \boldsymbol{F}v(k)\\
\hat{y}_r(k) &= \boldsymbol{C}_{0r}\boldsymbol{x}(k)
\end{aligned}
$$

$$\text{where } \boldsymbol{A} = \begin{bmatrix} \boldsymbol{A}_p & -\boldsymbol{B}_0\boldsymbol{D}_c\boldsymbol{C}_{0r} & \boldsymbol{B}_p\boldsymbol{C}_c \\ \boldsymbol{K}_0\boldsymbol{C}_p & \boldsymbol{A}_0 - \boldsymbol{K}_0\boldsymbol{C}_0 - \boldsymbol{B}_0\boldsymbol{D}_c\boldsymbol{C}_{0r} & \boldsymbol{B}_0\boldsymbol{C}_c \\ \boldsymbol{0} & -\boldsymbol{B}_c\boldsymbol{C}_{0r} & \boldsymbol{A}_c \end{bmatrix}; \ \boldsymbol{B} = \begin{bmatrix} \boldsymbol{B}_p\boldsymbol{D}_c \\ \boldsymbol{B}_0\boldsymbol{D}_c \\ \boldsymbol{B}_c \end{bmatrix}$$

$$\boldsymbol{E} = \begin{bmatrix} \boldsymbol{0} \\ \boldsymbol{K}_0\boldsymbol{F}_p \\ \boldsymbol{0} \end{bmatrix}, \ \boldsymbol{F} = \begin{bmatrix} \boldsymbol{F}_{pk} \\ \boldsymbol{0} \\ \boldsymbol{0} \end{bmatrix}; \ \boldsymbol{C} = \begin{bmatrix} \boldsymbol{0} & \boldsymbol{C}_{0r} & \boldsymbol{0} \end{bmatrix}$$

High Performance and Fault Tolerant Control System

$$y(z) = G(z)u(z) + \upsilon(z) \text{ where } F_0(z) = \det\left(z\boldsymbol{I} - \boldsymbol{A}_0 + \boldsymbol{K}_0\boldsymbol{C}_0\right).$$

The expression for the residual e is given by

$$\begin{aligned} e(z) &= e_f(z) + e_0(z) \\ e_f(z) &= S_{u0}(z)\Delta G_p r_{filt}(z) \\ e_0(z) &= \upsilon_{filt}(z) \end{aligned}$$

where the model mismatch term $\Delta G = G - G_0$.

$$G = ND^{-1} == (N_0 + \Delta_N)(D_0 + \Delta_D)^{-1}$$

$r_{filt}(z)$ and $\upsilon_{filt}(z)$ are the filtered reference input r and filtered noise respectively:

$$r_{filt} = \frac{D_0(z)}{F_0(z)}r(z); \quad \upsilon_{filt} = \frac{D_0(z)}{F_0(z)}\upsilon(z)$$

$$E[e] = 0 \text{ if and only if } \Delta G_p(z) \equiv 0$$

Bayes Decision Strategy

The Bayes decision strategy takes the general form

$$t_s(e) \begin{cases} \leq \eta & \text{normal} \\ > \eta & \text{abnormal} \end{cases}$$

where $e(k) = \begin{bmatrix} e(k) & e(k-1) & e(k-2) & . & e(k-N+1) \end{bmatrix}^T$, $t_s(\boldsymbol{e})$ is the test statistics of the residual $\boldsymbol{e}$ and η is the threshold value

$$t_s(\boldsymbol{e}) = \begin{cases} \left| \dfrac{1}{N} \sum\limits_{i=k-N+1}^{k} e(i) \right| & r(k) = cons\tan t \\ P_{ee}(f_0) & r(k) \ is \ a \ \sin usoid \\ \dfrac{1}{N} \sum\limits_{i=k-N+1}^{k} e^2(i) & r(k) \ is \ an \ arbitrary \ signal \end{cases}$$

References

[1] Kadlec, P., Gabrys, B., and Strandt, S. (2009) Data driven soft sensors in the process industry. *Computers and Chemical Engineering*, **33**, 795–814.

[2] Fortuna, L., Graziani, S., and Xibilia, G. (2007) *Soft Sensors for Monitoring and Control of Industrial Processes*. Springer-Verlag.

[3] Angelov, P. and Kordon, A. (2010) Adaptive inferential sensors based on evolving fuzzy models. *IEEE Transactions on Systems, Man and Cybernetics: Case Study*, **40**(2), 529–539.

[4] Doraiswami, R. and Cheded, L. (2013) A unified approach to detection and isolation of parametric faults using a Kalman filter residuals. *Journal of Franklin Institute*, **350**(5), 938–965.

[5] Doraiswami, R. and Cheded, L. (2013) Fault diagnosis of a sensor network: a distributed filtering approach. *Journal of Dynamic Systems, Measurement and Control*, **135**(5), 1–10.

[6] Doraiswami, R. and Cheded, L. (2012) Kalman filter for fault detection: an internal model approach. *IET Control Theory and Applications*, **6**(5), 1–11.

[7] Doraiswami, R., Diduch, C., and Tang, J. (2010) A new diagnostic model for identifying parametric faults. *IEEE Transactions on Control System Technology*, **18**(3), 533–544.

[8] Goodwin, G.C., Graeb, S.F, and Salgado, M.E. (2001) *Control System Design*. Prentice Hall, New Jersey.

[9] Kwakernaak, H. (1993) Robust control and H-inf optimization: tutorial paper. *Automatica*, **29**(2), 255–273.

[10] Cerone, V., Milanese, M. and Regruto, D. (2009) Yaw stability control design through mixed sensitivity approach. *IEEE Transactions on Control Systems Technology*, **17**(5), 1096–1104.

[11] Tan, W., Marquez, H.J., Chen, T., and Gooden, R. (2001) H infinity Control Design for Industrial Boiler, Proceedings of The American Control Conference, Virginia, USA.

[12] Doraiswami, R. and Cheded, L. (2012) Robust fault tolerant controller. IECON 2012 IEEE Industrial Electronics Society, Montreal.

[13] Yao, B. and Palmer, A. (2002) Indirect adaptive robust control of SISO nonlinear systems in semi-strict forms. 15th IFAC World Congress, Barcelona, Spain.

[14] Zhou, K., Doyle, J., and Glover, K. (1996) *Robust Optimal Control*. Prentice Hall, New Jersey.

[15] Kay, S.M. (1993) *Fundamentals of Signal Processing: Estimation Theory*. Prentice Hall PTR, New Jersey.

Index

Identification of Physical Systems: Applications to Condition Monitoring, Fault Diagnosis, Soft Sensor and Controller Design, First Edition. Rajamani Doraiswami, Chris Diduch and Maryhelen Stevenson.